# HANDBOOK OF
## CHEMICAL LASERS

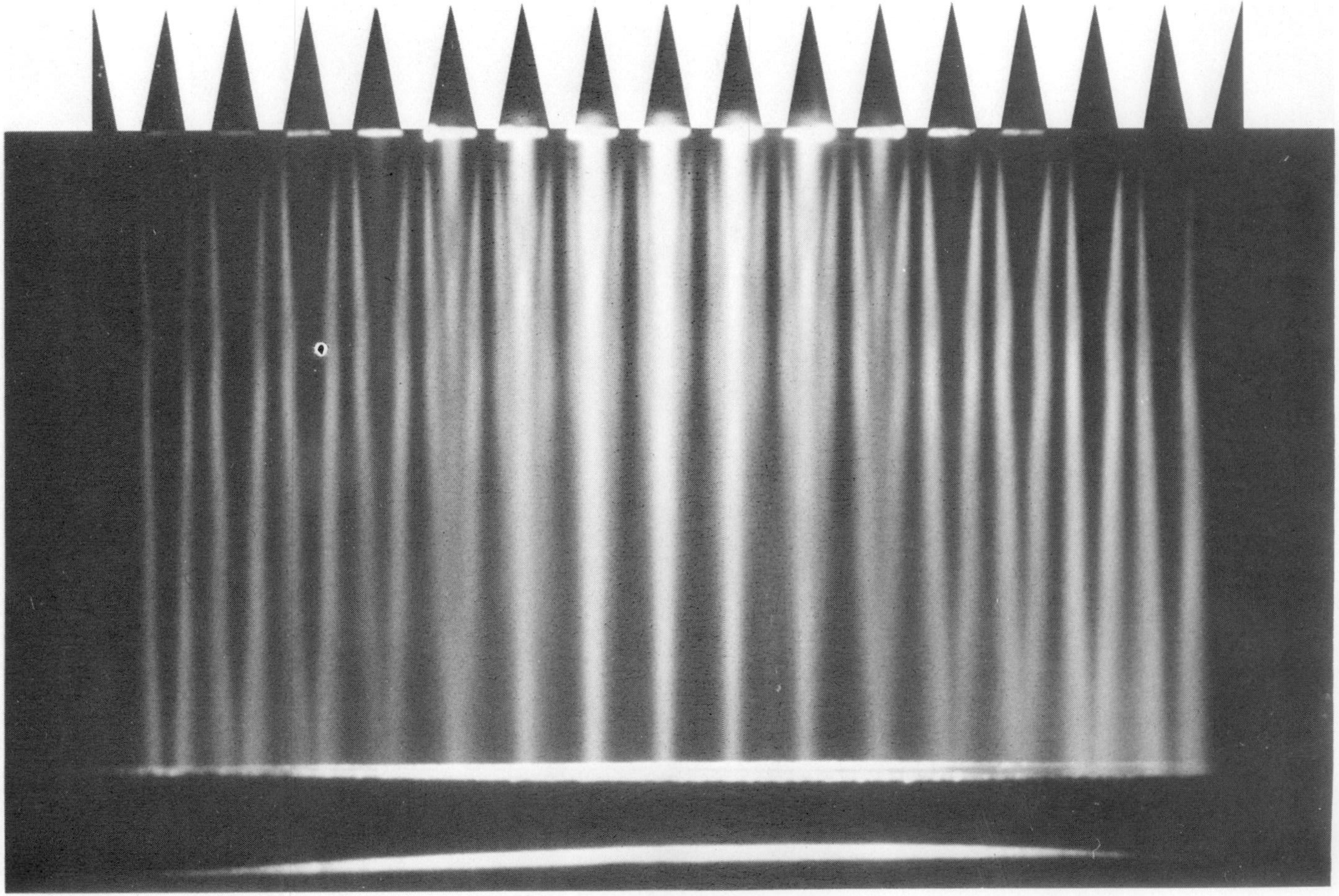

Plate 1

# HANDBOOK OF CHEMICAL LASERS

**Edited by**

**R. W. F. GROSS**

**J. F. BOTT**

**The Aerospace Corporation
Los Angeles, California**

**A WILEY-INTERSCIENCE PUBLICATION**

**JOHN WILEY & SONS, New York · London · Sydney · Toronto**

Copyright © 1976 by John Wiley & Sons, Inc.

*Library of Congress Cataloging in Publication Data*

Main entry under title:

Handbook of chemical lasers.

"A Wiley-Interscience publication."
Includes bibliographies and index.
1. Chemical lasers–handbooks, manuals, etc.
I. Gross, Rolf W. F., 1931-    II. Bott, Jerry F.
TA1690.H36 621.36'6 76-6865

ISBN 0-471-32804-9

Printed in the United States of America

10 9 8 7 6 5 4 3 2 1

# CONTRIBUTORS

J. R. AIREY, Naval Research Laboratory, Washington, D.C.

A. BEN-SHAUL, The Hebrew University, Jerusalem, Israel and Technische Universität München, München, Germany

J. F. BOTT, The Aerospace Corporation, El Segundo, California

B. R. BRONFIN, United Technologies Research Center, East Hartford, Connecticut

A. N. CHESTER, Hughes Research Laboratories, Malibu, California

R. A. CHODZKO, The Aerospace Corporation, Los Angeles, California

N. COHEN, The Aerospace Corporation, El Segundo, California

T. A. COOL, Cornell University, Ithaca, New York

G. EMANUEL, TRW Systems Group, Redondo Beach, California

G. GROHS, TRW Systems Group, Redondo Beach, California

R. W. F. GROSS, The Aerospace Corporation, El Segundo, California

G. L. HOFACKER, Technische Universität München, München, Germany

K. HOHLA, The Max-Planck-Institut für Plasmaphysik, Garching, Germany

W. Q. JEFFERS, McDonnell Douglas Research Laboratories, St. Louis, Missouri

R. J. JENSEN, Los Alamos Scientific Laboratory, Los Alamos, New Mexico

K. L. KOMPA, Max-Planck-Institut für Plasmaphysik, Garching, Germany

G. C. PIMENTEL, University of California, Berkeley, California

D. J. SPENCER, The Aerospace Corporation, El Segundo, California

S. N. SUCHARD, The Aerospace Corporation, El Segundo, California

R. L. WILKINS, The Aerospace Corporation, El Segundo, California

# PREFACE

In September 1964 an illustrious group of chemists, physicists, and laser specialists convened in San Diego, California to discuss the possibilities of nonequilibrium excitation and chemical pumping of lasers. The conference started out entirely speculative, since no chemical lasers existed at that time. At the end of the meeting, a young graduate student from Berkeley got up and claimed that he had observed the first laser pulses produced by a chemical reaction. The laser was the flash-photolysis iodine laser, and the student, Jerry Kasper from George Pimentel's group. With this dramatic event, chemical lasers were born. In the following ten years, the people who wrote this book have had the rare and exciting experience to be the participants in the development of an entirely new field from its basic inception to technical maturity.

This book collects the pieces of this work for the first time. We feel that we are at another turning point in the development of chemical lasers. Much of the basic scientific research, at least as far as the major chemical lasers HF, CO, and iodine are concerned, appears to be concluded. In the future, this work will pass on to the engineer, who will develop the large laser systems required for technical applications in laser chemistry, laser fusion, and materials processing.

To unlock the full possibilities of chemical lasers, three disciplines had to be combined: chemical kinetics of nonequilibrium reactions, gas dynamics of reactive flows, and laser physics of high-gain media. People working in any one of these fields are often unfamiliar with the "language" of the other two. It has, therefore, been our goal to make this book a kind of dictionary for the communication between these disciplines and a handbook for the future laser-systems engineer. It is not a basic textbook, but it attempts to equip the gas dynamicist with an understanding of kinetics and the physics of unstable cavities, the kineticist with a grasp of the all important gas dynamics of chemical lasers, and the laser expert with a feel for the parameters of their unusual media to enable them to accomplish their task together.

In keeping with this objective, the book starts with basic reviews of the three fundamental fields. The ability to control the kinetics and thermodynamics of highly exothermic reactions by fast supersonic flows places gas dynamics into a central role. This subject is thoroughly explored in the chapter by Grohs and Emanuel. The following three chapters review the work on pulsed chemical lasers, transfer lasers, and numerical modeling of chemical lasers. Chapters 9 and 10 are devoted to the theory of reactive collision mechanisms leading to the nonequilibrium vibrational excitation of the molecules that constitute the active medium of chemical lasers. A new, detailed review of the CO chemical laser and the first comprehensive collection of the work on the high power iodine laser are given in Chapters 11 and 12. Reed Jensen concludes the book with a description of the work on metal-atom oxidation lasers at Los Alamos.

There is a long list of colleagues, supporters, and friends who have contributed to this field and to this book and who do not appear by name. It is only just to acknowledge our debt to them in the beginning. Foremost are our present and former colleagues at Aerospace Corporation: D. Durran, W. Gaskill, R. R. Giedt, J. Herbelin, T. A. Jacobs, M. Kwok, H. Mirels, R. Varwig, and W. Warren. Without their encouragement, work, and patience this book could not have been written. We are also aware of a great debt to the U.S. Air Force and to DARPA, not only for financial support of much of the work on chemical lasers described in this book, but also for the personal encouragement, drive, and vision in the ups and downs of this research—in particular to M. Berta, P. Clark, E. Gerry, J. McCallum, R. Oglukian, J. Rich, M. Rogers, and L. Wilson. Finally the editors owe their gratitude to Kathleen Bregand for performing the tremendous task of editing the original manuscripts with charm and devotion.

R. W. F. GROSS

J. F. BOTT

*Los Angeles*
*March 1976*

# CONTENTS

ix

# HANDBOOK OF
## CHEMICAL LASERS

CHAPTER **1**

# WHAT IS A CHEMICAL LASER? AN INTRODUCTION

**GEORGE C. PIMENTEL**

**University of California**
**Berkeley, California**

**KARL. L. KOMPA**

**Max-Planck-Institut für Plasmaphysik**
**Garching, Germany**

## 1.  WHAT IS A CHEMICAL LASER?

Chemical lasers now have nearly a 10-year history. The first meeting on
the subject, in 1964, was organized by the American Optical Society and
was filled with high hopes for a rapid development of this field. Chemical-
laser emission was then experimentally realized for the first time in 1965
by Kasper and Pimentel[1] at Berkeley. The emission occurred in a
photolytically initiated hydrogen–chlorine explosion. Predating this discovery, the first photodissociation laser was described by the same
authors in 1964.[2] This laser is based on the formation of excited iodine
atoms in the photoinduced bond rupture of alkyl iodides.

   Such lasers have since provided a fruitful avenue for the study of
elementary reaction dynamics and have stimulated the investigation of
collisional energy-transfer processes. Most recently, chemical lasers have
also been put to use as specialized excitation sources for spectroscopic
and kinetic work.

   Interest in the technical development of chemical lasers only appeared
around 1970. This is, in part, due to their complicated theoretical
background. Not only is their multilevel pumping scheme complex, but
also the gain and spectral distribution change rapidly with time because of
efficient collisional deactivation processes. Thus before progress could be
made in either the theory or the technology of chemical lasers, concepts
from various disciplines had to be combined. However, beginning about
1969, increasing numbers of technically oriented papers on chemical
lasers have appeared in the literature. There have already appeared some
general reviews.[3-6] Continuous chemical lasers based on several reactions
are now in operation, and pulsed chemical lasers have been scaled up in
power to a technically interesting regime. Most of the information needed
to assess the potentials of chemical lasers is now available.

   This book, then, addresses itself to physicists, physical chemists, and laser engineers interested in chemical lasers. The general thrust is towards application-oriented physics and laser engineering with some consideration of the chemical background. For the purposes of this audience, we shall define a chemical laser as *a laser operating on a population inversion produced—directly or indirectly—in the course of an exothermic chemical reaction.* Furthermore, our definition shall extend to all types of chemiluminescent excitation, which include, for example, pumping in the course of photoinduced or even electron-impact-induced chemical bond rupture as well as by the radiative association of atoms or molecules. Finally, we shall also consider lasing action that occurs in an admixture to which the chemical energy is transferred prior to lasing.

In its simplest approach, the energy partitioning in exothermic reactions may be considered with reference to the usual picture of a reaction profile, as illustrated in Fig. 1.

It is apparent that a product molecule that has just emerged from the activated complex will not be at thermal equilibrium. It contains excess energy that can be given off either in *radiative* or in *collisional* processes, which are always in competition. The key question, though, is whether

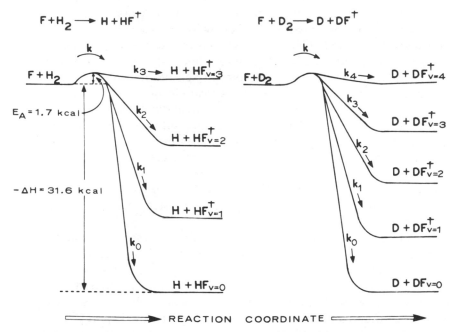

**Fig. 1.** Energy profiles for the reactions $F+H_2$ and $F+D_2$ ($k_v$ is the rate constant into $v$th vibrational state).

the reaction selectively channels part of the energy into one or more higher product states to create a population inversion. This can happen when some dynamical constraint in the course of the reaction prevents a purely statistical distribution of the energy.

A molecule can store energy in the electronic, vibrational, rotational, and (although of no concern here) translational degrees of freedom. However, the probability that energy will accumulate in these various degrees of freedom and then lead to chemical laser action differs considerably. Beside the need for a population inversion, the production of laser gain also depends upon the cross-section for stimulated emission, which is simply related to the Einstein coefficient of stimulated radiation $B_{12}$ and the line width $\Delta \nu$ of the transition

$$\sigma = \frac{B_{12}}{\Delta \nu c} = \frac{\lambda^2}{8 \pi \, \Delta \nu \tau}$$

Thus the chance that laser emission may dominate over collisional relaxation varies with the square of the emission wavelength and inversely with the spontaneous lifetime $\tau$ of the emitting state and the line width $\Delta \nu$.

These factors seem to be more favorable for chemical lasers based on electronic excitation than for those based upon vibrational excitation; electronic transitions involve smaller wavelengths and, in most cases, much shorter lifetimes [$I^*(^2P_{1/2} \rightarrow {}^2P_{3/2})$ is an exception]. Yet, chemically pumped vibrational lasers are common, and the electronic counterparts are rare. Of course, the energy required to excite transitions in the visible spectral region is high compared to typical chemical-reaction heats, that is, between 41 kcal/mol (700 nm) and 71 kcal/mol (400 nm). However, there are many chemical reactions that are sufficiently exothermic and, indeed, are known to produce chemiluminescence.

Plainly, loss mechanisms must be considered as well as gain factors. For vibrational excitation, collisional deactivation is the most significant; whereas, for electronic transitions, large $A$ factors, which raise the gain, can also limit the growth of population inversion through spontaneous emission.

Still, favorable reactions are known (for instance, some reactions where polyatomic molecules or small molecules containing metal atoms are involved), and success in the chemical pumping of electronic states can be expected. Nevertheless, only in the related group of photodissociation lasers has emission from electronic states been realized. Laser emission based on rotational transitions has also been observed,[7,8] but in no case has it been established definitively that chemical reactions provide the population inversion that caused the laser emission.

## 2. REACTION TYPES IN CHEMICAL LASERS

At this time, a relatively small number of chemical reactions are being intensively studied because they are the pumping reactions in efficient, high-energy lasers. The subsequent chapters will focus attention on these important systems. First, however, it is appropriate to review the range of reaction types that have successfully been used to pump chemical lasers, without regard to their possible applicability.

Table 1 lists the reaction types, following the categorization proposed by Pimentel.[9] For each of these types, prototype examples are discussed, a comprehensive listing is included, and generalizations are noted.

**Table 1.** Reaction types in chemical lasers

| Name | Prototype examples | Year | References |
|---|---|---|---|
| Three-atom Exchange | $H + Cl_2 \rightarrow HCl^\dagger + Cl$ | 1965 | 1 |
| | $F + H_2 \rightarrow HF^\dagger + H$ | 1967 | 11 |
| | $O + CS \rightarrow CO^\dagger + S$ (?) | 1966 | 12 |
| Abstraction | $F + CH_4 \rightarrow HF^\dagger + CH_3$ | 1968 | 13 |
| | $O + CS_2 \rightarrow CO^\dagger + S_2$ (?) | 1966 | 12 |
| Photodissociation | $CF_3I + h\nu \rightarrow CF_3 + I(^2P_{1/2})$ | 1964 | 2 |
| | $ClNO + h\nu \rightarrow NO^\dagger + Cl$ | 1966 | 14 |
| Elimination | | | |
| Radical combination | $CH_3 + CF_3 \rightarrow HF^\dagger + CH_2CF_2$ | 1968 | 15 |
| Insertion | $O(^1D) + CH_nF_{4-n} \rightarrow HF^\dagger + OCH_{n-1}F_{3-n}$ | 1971 | 16 |
| Addition | $NF + H_2C{=}CH_2 \rightarrow HF^\dagger + CH_3C{\equiv}N$ | 1972 | 17 |
| Photoelimination | $H_2C = CHCl + h\nu \rightarrow HCl^\dagger + HCCH$ | 1969 | 18 |

## 2.1. Three-Atom Exchange Reactions

The reaction between an atom $A$ and a diatomic molecule $BC$ will be exothermic if the $AB$ chemical bond is stronger than the $BC$ chemical bond. Such free radical reactions are generally extremely fast because the activation energies are small and there are no significant restrictive geometrical requirements for a collision to be reactive.

The first two chemical lasers based upon such three-atom exchange reactions used the $H + Cl_2$ [1,10] and the $F + H_2$ reactions:[11]

$$H + Cl_2 \rightarrow HCl^\dagger + Cl \quad -\Delta H = 45.1\,\text{kcal/mol} \quad E_A = 1.8\,\text{kcal/mol} \quad (1)$$

$$F + H_2 \rightarrow HF^\dagger + H \quad -\Delta H = 31.5\,\text{kcal/mol} \quad E_A = 1.7\,\text{kcal/mol} \quad (2)$$

The energy available to the reaction products $(-\Delta H + E_A)$ is enough to excite HCl to $v = 6$ in Reaction (1) and HF to $v = 3$ in Reaction (2). The

**Table 2.** Vibrational rate constants at room temperature

|  | $H+Cl_2{}^a$ | $F+H_2{}^b$ | $F+CH_4{}^c$ | $F+HI^c$ | $Cl+HI^d$ |
|---|---|---|---|---|---|
| $\Delta H$ | −45.1 | −31.5 | −32.6 | −62.6 | −32 |
| $E_A$ | 1.8 | 1.7 | 1.2 | 1.4 | 2 |
| $\%(-\Delta H+E_A)$ |  |  |  |  |  |
| In vibration | 39 | 66 | 68 | 56 | 65 |
|  | $h$ |  |  | $h$ |  |
| $k_6$ | $0.005^e$ | — | — | 1.07 | — |
| $k_5$ | $0.05^e$ | — | — | 1.69 | — |
|  |  |  |  |  | $h$ |
| $k_4$ | 0.1 | — | — | 1.27 | 2.29 |
|  |  | $h$ | $h$ |  |  |
| $k_3$ | 0.92 | 0.47 | 0.23 | 1.05 | 3.00 |
| $k_2$ | (1.00) | (1.00) | (1.00) | (1.00) | (1.00) |
| $k_1$ | 0.28 | 0.31 | 0.33 | 0.83 | 0.57 |
| $k_0$ | — | $0.0^f$ | $0.0^f$ | $0.4^f$ | — |
|  |  | $0.06^g$ |  |  |  |

[a] Ref. 19.
[b] Ref. 20.
[c] Ref. 21.
[d] Ref. 22.
[e] Ref. 23.
[f] Estimated, see Ref. 21.
[g] Private Communication, M. J. Berry; also, Ref. 24.
[h] Energy insufficient to excite higher levels.

actual distribution among the vibrational states is expressed in terms of relative rate constants $k_v$ in Table 2. These vibrational rate constants channel, on the average, 39% of the available energy into vibration in Reaction (1)[19] and 66% in Reaction (2).[20] In each case, the highest population is formed in the $v=2$ state, which provides highest laser gain in $v=2 \rightarrow v=1$ V–R transitions.

The observed laser emission from either of these reactions depends upon the populations initially determined by these $k_v$'s but as subsequently modified by rotational equilibration and vibrational deactivation processes. Usually, rotational equilibration is so rapid that the primitive rotational occupancies are not influential in the laser performance. In contrast, the role of vibrational-relaxation reactions varies greatly and depends upon operational conditions. In pulsed lasers vibrational relaxa-tion can be entirely negligible,[25] whereas, it is always important in cw lasers. This dependence derives from the high deactivation efficiencies of certain collision partners, which can be associated with attractive forces (e.g., hydrogen bonding) or with a close energy match between the vibrational levels of the two molecules involved, or both.[26,27]

Despite the vibrational-deactivation problems, quite a number of exo-thermic three-atom reactions have been used to initiate chemical-laser emission. The reason for this success is illustrated in Table 2, particularly by the $F+HI$ and $Cl+HI$ reactions. In each case, vibrational excitation is observed up to the highest level that is thermodynamically possible in more than 15% of the reactive collisions. Even the $H+Cl_2$ reaction, the least efficient of the well-studied three-atom reactions, channels 39% of its exothermicity this way. It is a useful empirical generalization that *in exothermic three-atom reactions involving free radicals, a large fraction of the energy appears in vibrational excitation of the new bond formed.*

Table 3 lists 13 three-atom exchange reactions that have been used to pump chemical lasers. The various initiation techniques—pulsed photo-lysis, pulsed discharge, and continuous flow—are referenced separately. In each case, the earliest references are included, but, for later references, no attempt at completeness is made. Particularly for the $F+H_2$ pumping reaction, a complete list of references would be very long.

**Table 3.** Three-atom reactions for which chemical-laser emission has been observed[a]

| Reaction Initiation Method: | Pulsed photolysis | Pulsed discharge | Continuous flow |
|---|---|---|---|
| $H+Cl_2 \rightarrow HCl^\dagger + Cl$ | 1, 28, 29 | 30, 31, 32 | 33 |
| $F+H_2 \rightarrow HF^\dagger + H$ | 11, 34, 35, 36, 37 | 30, 36, 38, 39, 40, 41, 42, 43 | 44, 45, 33, 46 |
| $Cl+HI \rightarrow HCl^\dagger + I$ | 47 | 48 | 33, 46 |
| $F+HX \rightarrow HF^\dagger + X$ <br> $X = I, Br, Cl$ | 49, 50 | 51 | 47, 33, 46 |
| $Cl+HBr \rightarrow HCl^\dagger + Br$ | 52 | — | 46 |
| $Br+HI \rightarrow HBr^\dagger + I$ | 49 | — | 33[b] |
| $H+F_2 \rightarrow HF^\dagger + F$ | — | 42 | 33 |
| $H+Br_2 \rightarrow HBr^\dagger + Br$ | — | 31, 53 | 33[b] |
| $O(^3P)+CH, CF, CN \rightarrow CO+H, F, N$ | 54 | — | — |
| $O+CS \rightarrow CO^\dagger + S$ (?) | 12 | 55, 56 | 57 |

[a] The numbers refer to the list of references.
[b] Experiment attempted, no laser emission observed.

## 2.2.  Abstraction Reactions

Alone among the halogens, fluorine atoms can abstract hydrogen from hydrocarbons (and other hydrides) in exothermic reactions. This is due to the especially high bond dissociation energy of HF, 135 kcal/mol, to be compared to those of the CH bond in methane, 101 kcal/mol; of HCl, 103 kcal/mol; and of HBr, 87 kcal/mol.

Furthermore, the activation energies for hydrogen abstraction by fluorine are quite low. In fact, energetically, the fluorine–methane reaction (3) is surprisingly similar to the fluorine–hydrogen reaction (2):

$$F + CH_4 \rightarrow HF^\dagger + CH_3 \qquad -\Delta H = 32.6 \text{ kcal/mol} \qquad E_A = 1.2 \text{ kcal/mol}$$
$$(3)$$

On the other hand, the relative masses of reactants and products, together with the several vibrational degrees of freedom of $CH_3$ and $CH_4$, seem to distinguish Reaction (3) from Reaction (2). Actually, the two reactions are startlingly alike in their $k_v$'s, as shown in the third column of Table 2, and in the fraction of available energy entering vibration.

Parker and Pimentel[13] demonstrated that every one of a dozen hydrocarbons studied provided HF laser emission because of the fluorine–RH abstraction reaction. Using pulsed discharges to produce fluorine atoms from $NF_3$, Pearson et al.[58] compared the laser power obtained from 12 additional hydrides, including five inorganic hydrides ($SiH_4$, $GeH_4$, $AsH_3$, $SbH_3$, and $B_2H_6$), some of which have quite low M–H bond energies (e.g., $AsH_3$, $D_0 = 61$ kcal/mol; $SiH_4$, $D_0 = 76$ kcal/mol). Surprisingly, the highest power was not obtained from these inorganic hydrides, but rather from $n$-$C_3H_8$, $n$-$C_4H_{10}$, and $(CH_3)_3CH$. What did appear with the inorganic hydrides were vibrational transitions from higher quantum numbers ($v = 4 \rightarrow 3$) made possible by the higher exothermicities.

Table 4 lists references to the variety of reactions $F + RH$ that have been used to initiate HF chemical-laser emissions. Of course, the significance of the emissions observed with pulse discharge initiation is obscured by the complex chemistry that occurs in a discharge.

Although clear-cut evidence is lacking, hydrogen-atom reactions may furnish another series of chemical-abstraction lasers. Thus Kompa et al.[64] suggest that the reaction $H + XeF_4 \rightarrow HF^\dagger + XeF_3$ contributes to the laser emission in the photolysis-initiated $XeF_4$–$H_2$ laser. In a similar way, Krogh and Pimentel[61] argue that the photolysis-initiated $ClF_3$–$H_2$ laser must involve H-atom abstraction reactions, including, perhaps, the chain-branching reaction, $H + ClF_3 \rightarrow HF^\dagger + Cl + 2F$. Rice and Jensen[65] propose that the reaction $H + ClN_3 \rightarrow HCl^\dagger + N_3$ contributes to the laser emission in the $ClN_3$–$H_2$ chemical laser. Finally, infrared chemiluminescence has

**Table 4.** HF laser emission from hydrogen abstraction by fluorine atoms[a] $F + RH \rightarrow HF^\dagger + R$

| RH[b] | Pulsed photolysis | Pulsed discharge |
|---|---|---|
| CH$_4$ | 13, 37, 60, 61 | 58, 62, 63 |
| C$_2$H$_6$ | 13, 60 | 58, 62, 64, 63 |
| $\begin{cases} n\text{-}C_nH_{2n+2} & n \geq 3; \text{ branched,} \\ \text{cyclic hydrocarbons} \end{cases}$ | 13 | 58, 63 |
| Olefins, aromatics | — | 58 |
| Halosubstituted hydrocarbons | 13, 59, 60 | 65 |
| Inorganic hydrides | — | 58 |

[a] The numbers refer to the list of references.
[b] Including SiH$_4$, GeH$_4$, AsH$_3$, SbH$_3$, and B$_2$H$_6$.

been observed for a number of H-atom abstraction reactions, identifying additional potential chemical-laser systems (including $H + O_3$,[66,67] ClNO,[68] SCl$_2$,[69,70] Cl$_2$O,[69,71] and F$_2$O[72]).

Oxygen atom reactions may also provide chemical reactions of the abstraction type that can activate chemical-laser emission. Thus Lin and Brus[73] proposed that O($^1D$) atoms react with C$_3$O$_2$ to produce CO$^\dagger$ in a highly exothermic reaction. More important, possibly, is the CS$_2$–O$_2$ laser, whose complex chemistry may involve pumping by the oxygen atom reactions (4a) and (4b)[74]:

$$O + CS_2 \rightarrow CO^\dagger + S_2 \qquad \Delta H = -70 \text{ kcal/mol} \qquad (4a)$$

$$O + OCS \rightarrow CO^\dagger + SO \qquad \Delta H = -47.5 \text{ kcal/mol} \qquad (4b)$$

$$O + CS \rightarrow CO^\dagger + S \qquad \Delta H = -75 \text{ kcal/mol} \qquad (4c)$$

This CS$_2$–O$_2$ laser was first discovered by Pollack[12] through the use of flash photolysis, but it can also be initiated with a pulsed discharge.[55,56] Intense emission can be obtained in all of the CO vibrational transitions from $v = 1 \rightarrow 0$ up to $v = 16 \rightarrow 15$.[74] There is a voluminous amount of literature on the CS$_2$–O$_2$ chemical laser, which is summarized by Wiswall et al.[57]

## 2.3. Elimination Reactions

Quite a number of exothermic bimolecular reactions produce a single molecular product that is sufficiently energy-rich to undergo subsequent fragmentation (unimolecular decomposition). When one of the fragments is an atom or a small molecule, the reaction is called an elimination reaction. Such reactions can be segregated according to the nature of the original bimolecular reaction, as shown in Table 1.

**Radical-Combination Reactions.** The first elimination chemical laser was based upon the $CH_3 + CF_3$ reaction,[15] which had been shown earlier by Giles and Whittle[75] to produce HF and difluoroethylene:

$$CH_3 + CF_3 \rightarrow CH_3CF_3^\dagger \rightarrow HF^\dagger + H_2C = CF_2 \tag{5}$$

$$\Delta H = -100 \text{ kcal/mol} \qquad \Delta H = +31 \text{ kcal/mol}$$

$$E_A \simeq 0 \qquad\qquad E_A \simeq 71 \text{ kcal/mol}$$

The intermediate, 1,1,1-trifluoroethane, is formed with vibrational energy equal to the carbon–carbon bond dissociation energy, of about 100 kcal/mol.[76] This is not sufficient energy for electronic excitation, so almost all of the energy must be lodged in vibrational excitation. This implies that there is more than enough internal energy to permit elimination of HF, which is endothermic by 31 kcal/mol, and to surmount the activation energy barrier to elimination, which is about 71 kcal/mol.[77] The overall exothermicity of Reaction (5) is sufficient to excite HF up to about $v = 6$. However, laser emission is detected only from $v = 1 \rightarrow 0$ and $2 \rightarrow 1$ transitions. Furthermore, the transition of highest gain at room temperature, $P_1(4)$, implies that the population ratio $N_1/N_0$ must be in the range of 0.58 to 0.90. This ratio below unity, which has proved to be characteristic of all of the elimination lasers now known, was substantiated by Clough et al.[78] through the use of ir chemiluminescence measurements. They find that Reaction (5) gives population ratios $N_2/N_1 = 0.43$ and $N_3/N_2 = 0.30$ and that only 13% of the available energy appears in HF vibrational excitation.

Probably the most thoroughly studied chemical-elimination laser is that involving $N$-difluoromethylamine[79–82]:

$$CH_3 + N_2F_4(NF_2) \rightarrow CH_3NF_2^\dagger + NF_2 \tag{6a}$$

$$CH_3NF_2^\dagger \rightarrow HF^\dagger + H_2C = NF^\dagger \tag{6b}$$

$$H_2C = NF^\dagger \rightarrow HF^\dagger + HCN \tag{6c}$$

At room temperature, methyl can react with both $N_2F_4$ and $NF_2$ to initiate HF elimination. However, Padrick and Pimentel[80] have presented evidence that only $N_2F_4$ is important at temperatures of 250°K or lower and that only the first elimination step, Reaction (6b), contributes to the laser emission. Although Reaction (6a) places 78 kcal/mol of internal energy into the product $CH_3NF_2^\dagger$, maximum gain for HF is observed in the $v = 1 \rightarrow 0$ transition. Equal-gain-temperature measurements show that $N_1/N_0 = 0.47$, which permits an estimate that, on the average,

5.5±0.5 kcal/mol is lodged in the HF vibrational mode. This is quite close to the average energy per vibrational degree of freedom in the parent molecule, of $78/15 - 5.2$ kcal/mol. This example suggests that the low vibrational excitation found in elimination reactions is due to a statistical distribution of the internal energy among all of the parent-molecule vibrations.

**Insertion Reactions.** Lin has pioneered the use of insertion–elimination reactions using $O(^1D)$ atoms.[83,84,85] For example, two of the reactions from which laser emission has been observed are:

$$O(^1D) + CHF_3 \rightarrow HOCF_3{}^\dagger \rightarrow HF^\dagger + F_2C{=}O \qquad \Delta H = -155 \text{ kcal/mol}$$
$$(7a)$$

$$O(^1D) + CHCl_3 \rightarrow HOCCl_3{}^\dagger \rightarrow HCl^\dagger + Cl_2C{=}O \quad \Delta H = -155 \text{ kcal/mol}$$
$$(7b)$$

Once again, the elimination reaction is quite exothermic, sufficient to excite HF to $v \approx 16$ in Reaction $(7a)$ and HCl to $v \approx 25$ in Reaction $(7b)$. Yet the highest gain is observed in HF transition $P_2(4)$ for Reaction $(7a)$ and in HCl transition $P_1(8)$ for Reaction $(7b)$. These gains indicate that $N_2/N_1 \approx 0.8 \pm 0.2$ for HF, and $N_1/N_0 \approx 0.50 \pm 0.04$ for HCl. Again, only a small fraction of the available energy appears in excitation of the eliminated HX molecule.

Other insertion–elimination reactions have been studied using $CH_2$ and NH, both of which are isoelectronic with oxygen atoms. Roebber and Pimentel[86] applied the zero-gain method to the reaction between $CH_2$ and $CHF_3$ and were able to show gain in the $v = 1 \rightarrow 0$ transition. Using the same techniques, Poole and Pimentel[87] have demonstrated HF gain in the lower transitions from the insertion reaction between NH and $CHF_3$.

**Addition Reactions.** Addition of a free radical to a double bond can result in significant excitation and, hence, elimination. The prototype example is that initiated by the photolysis of $NF_2$, producing NF and F in the presence of ethylene.[17] Both NF and F atoms can add and, subsequently, eliminate HF:

$$CH_2CH_2 + NF \rightarrow H_2C{-}CH_2^\dagger \rightarrow HF^\dagger + CH_3C{\equiv}N \qquad \Delta H = -116 \text{ kcal/mol}$$
$$\underset{\underset{F}{\overset{|}{N}}}{\diagdown\diagup}$$
$$(8a)$$

$$CH_2CH_2 + 2F \rightarrow H_2FC{-}CH_2F^\dagger \rightarrow HF^\dagger + H_2C{=}CHF$$
$$\Delta H = -135 \text{ kcal/mol} \quad (8b)$$

Padrick and Pimentel[17] favor Reaction ($8a$) as the more important process but both Reactions ($8a$) and ($8b$) involve addition. The system gives HF laser emission only in the $v = 1 \to 0$ transition. The highest gain transition, $P_1(6)$, places $N_1/N_0$ near 0.5.

**Summary.** In every one of the many chemically activated elimination reactions now known, only a small fraction of the available internal energy appears as HX vibrational excitation. Thus, these reactions cannot be expected to provide either high-gain or high-energy lasers. They do, however, provide unique information about energy distribution in unimolecular decomposition, since the internal energy can be accurately known and, by choice of reactants, can be varied in some cases.

## 2.4. Photoelimination Reactions

Vacuum ultraviolet photolysis of halo-substituted olefins causes HX elimination and provides another source of HX laser emission. In the first report of such lasers,[18] HCl and/or HF emission was demonstrated from every Cl- and F-substituted ethylene studied, including 10 cases. Both HCl and HF emissions are observed from $HClC{=}CHF$ photolysis. Laser emissions from $H_2C{=}CF_2$ and from $H_2C{=}CCl_2$ show that the H atom and the halogen atom can come from adjacent carbon atoms ($\alpha\beta$ elimination), but it is likely that both can come from the same carbon when possible ($\alpha\alpha$ elimination), as suggested by the observation of HF laser emission from $HFC{=}CCl_2$.[88]

The energetics of photoelimination from vinyl fluoride and vinyl chloride provide large amounts of energy to be distributed among the products:

$$H_2C{=}CHF + h\nu \to H_2C{=}CHF^* \to HF^\dagger + HC{\equiv}CH$$

$$\Delta H = -130 \text{ kcal/mol} \quad (9a)$$

$$H_2C{=}CHCl + h\nu \to H_2C{=}CHCl^* \to HCl^\dagger + HC{\equiv}CH$$

$$\Delta H \approx -120 \text{ kcal/mol} \quad (9b)$$

When it is assumed that no energy-degradation processes have occurred before elimination, these energies are sufficient to excite HF to $v \approx 14$ and HCl to $v \approx 18$. Yet, the predominant laser emission occurs in HF, $v = 2 \to 1$, and in HCl, $v = 3 \to 2$. In each case, the highest gain transition indicates $N_v/N_{v-1}$ ratios below unity, just as is obtained for the chemically-activated elimination lasers.

The HCl photoelimination lasers produced from the three geometrical isomers of the dichloroethylenes provide an especially interesting contrast because they all give the same final products[88-90]:

$$H_2C\!\!=\!\!CCl_2 + h\nu \rightarrow HCl^\dagger + HC\!\!\equiv\!\!CCl \qquad (10a)$$

$$cis\text{-}HClC\!\!=\!\!CClH + h\nu \rightarrow HCl^\dagger + HC\!\!\equiv\!\!CCl \qquad (10b)$$

$$trans\text{-}HClC\!\!=\!\!CHCl + h\nu \rightarrow HCl^\dagger + HC\!\!\equiv\!\!CCl \qquad (10c)$$

Despite the identical final products, the vibrational excitation of the HCl differs markedly among these three precursors. Consequently, the implication of the principle of microscopic reversibility is that, in the reverse reaction (addition of HCl to the monochloroacetylene), vibration excitation of the HCl will cause a selective preference for the 1,1-dichloroethylene product (Markovnikov addition). Since this represented the first evidence that vibrational excitation might have a stereospecific influence, the result was carefully verified.[89,90] There is no doubt that photoelimination of HCl from $H_2C\!\!=\!\!CCl_2$ gives $v = 3 \rightarrow 2$ emission with $N_3/N_2 = 0.85$ and $v = 2 \rightarrow 1$ emission with $N_2/N_1 = 1.08$; whereas, for $cis$-$HClC\!\!=\!\!CClH$, first emission is from $v = 2 \rightarrow 1$ with $N_2/N_1 = 0.76$, and the $v = 3 \rightarrow 2$ emission corresponds to $N_3/N_2 = 0.71$.[89] By microscopic reversibility, this implies that the ratio of the 1,1-product to the $cis$-1,2-product is a factor of 1.70 higher if HCl in $v = 3$ is added to HCCCl than if HCl in $v = 1$ is added.

No doubt many other photoelimination lasers can be developed through the use of vacuum ultraviolet photolysis (e.g., from halo-substituted ethanes and amines). It is also evident that low vibrational excitation will always be obtained, just as in the chemically activated elimination lasers. Furthermore, the detailed processes involved will be difficult to interpret clearly because of the uncertainty about energy-conversion processes that might occur prior to elimination (e.g., singlet–triplet crossover). Nevertheless, the photoelimination lasers furnish an important class of chemical lasers, particularly since they have revealed for the first time the possibility that stereospecific control of chemical reactions might be obtained through selective vibrational excitation of reactants.

## 2.5. Photodissociation Reactions

In this group of lasers, selective excitation (pumping) results from electronic excitation of a molecule, followed by a decomposition process. The simplest decomposition is chemical bond rupture. The initial excitation can be achieved through photolysis, but electron-impact excitation could also be effective. Thus there are a number of discharge-induced lasers, such as the HCN, CO, and $CO_2$ lasers, in which electronic excitation might be involved. However, these contributions are not well understood or even definitively verified because of the variety of processes that occur in the energy-rich environment of a plasma. Hence, we will not pursue these systems further here.

The first photodissociation laser resulted from the flash photolysis of trifluoromethyl iodide.[2] The iodine atoms are formed in the electronically excited $^2P_{1/2}$ state:

$$CF_3I + h\nu(2700 \text{ Å}) \rightarrow I^*(5p \, ^2P_{1/2}) + CF_3 \qquad (11a)$$

$$I^*(^2P_{1/2}) \rightarrow I(^2P_{3/2}) + h\nu(13{,}150 \text{ Å}, 7605 \text{ cm}^{-1}) \qquad (11b)$$

Kasper, Parker, and Pimentel[91] investigated nine alkyl iodides, and all except one, isopropyl iodide (2-$C_3H_7I$), gave the $^2P_{1/2} \rightarrow \, ^2P_{3/2}$ emission. The most intense emission is obtained from $CF_3I$, and Donovan and Husain[92] have subsequently shown why. Apparently, Reaction (11a) produces *only* $^2P_{1/2}$ iodine atoms; hence, a 100% population inversion results.

This iodine laser has been one of the most thoroughly studied chemical lasers, with particular emphasis on the possible scale-up to high powers.[93-100] This laser and its prospects are dealt with in detail in the chapter by Kompa and Hohla.

The bromine counterpart is the only other one that has been successfully operated. The $Br(4p \, ^2P_{1/2}) \rightarrow (4p \, ^2P_{3/2})$ transition at 3685 cm$^{-1}$ has been produced through photolysis of $IBr$[101] and of $CF_3Br$,[102] but with much lower power than achieved with the iodine systems. The chlorine and fluorine $^2P_{1/2} \rightarrow \, ^2P_{3/2}$ transitions, which lie respectively at 881 and 404 cm$^{-1}$, have not been observed.

Photodissociation of the alkali metal diatomics has been shown to give laser emission by the corresponding transitions ($np, \, ^2P_{3/2} \rightarrow \, ^2P_{1/2}$): for Rb at 4437 cm$^{-1}$, and for Cs at 3231 cm$^{-1}$.[103]

There are also photodissociation lasers that depend on the vibrational excitation of the dissociation products. Pollack[14] showed that photolytic-bond rupture in nitrosyl chloride, ClNO, produced emission in the NO vibrational transitions $v = 6 \rightarrow 5$ up to $v = 9 \rightarrow 8$. It is possible that a low-lying electronically excited state of NO is involved as an intermediate.[104] In a similar way, flash photolysis of cyanogen, $C_2N_2$, gives vibrationally excited CN, with laser emission in three $v = 4 \rightarrow 3$ transitions.[105] Both the ClNO and $C_2N_2$ photodissociation lasers have the interesting property of reversibility, so repetitive use is possible.

It can be predicted with confidence that there will be other photodissociation lasers, from both electronically and vibrationally excited states. Ultimately, the former will surely be the more important, as foreshadowed by the complete population inversion and high power of the photodissociation iodine laser.

## 2.6. Other Reaction Types

This section deals with new concepts as well as with hopes and prospects for future developments. The list of unsuccessful attempts to discover new chemical lasers is long, and we shall not attempt to review it. Instead, we will conclude by pointing to a few specially promising developments.

An obvious and important development to be expected is the extension of chemical lasers to shorter wavelengths. To a limited extent, this can be accomplished by operating on vibrational overtones $\Delta v = 2$ as has been already demonstrated for the molecules DF[106] and CO.[107] In the case of CO, the very high excitation obtained (up to $v - 16$) permits the speculation that visible emission might be reached. More significantly, however, the direct excitation of lasers in the optical region can be envisioned through the use of chemiluminescent processes with electronic excitation. Three promising types of pumping processes are:

1. *Atom transfer reactions*

$$A + BC \rightarrow AB + C^* \qquad (12a)$$

*or*

$$(AB)^* + C \qquad (12b)$$

2. *Radiative recombination of atoms or molecules*

$$A + B \rightarrow (AB)^* \rightarrow AB + h\nu \qquad (13)$$

3. *Dissociative recombination of ions with electrons*

$$(AB)^+ + e^- \rightarrow A + B^* \qquad (14)$$

Many chemiluminescent reactions of types $(12a)$ and $(12b)$ are known, including the reaction of lithium atoms with $I_2$,[108] of chlorine atoms with $Na_2$ and $K_2$,[109,110] of barium and strontium atoms with $Cl_2$,[111] of alkaline earth atoms—Mg, Ca, Sr, and Ba—with $NO_2$ and $N_2O$,[112] and, of sodium atoms with $F_2$ and $Cl_2$.[113]

Radiative recombination, reaction type (13), is not well understood, and the difficulties of using these reactions as pumping reactions have been discussed by Oraevskii.[114] If $A$ and $B$ are atoms (rather than molecules), in the absence of a third body, emission of a photon is a necessary condition to stabilize the complex. The excited molecule $(AB)^*$ is formed with a range of thermal energies, and it radiates in a very short time, so the emission spectrum is not sharp. This implies that the emission cross-section will tend to be low. Conditions are improved if $(AB)^*$ has a long lifetime, as may occur if $A$ and $B$ are molecules. Further advantage is provided by the effective decrease of radiative lifetime that is associated with stimulated emission:

$$\frac{1}{\tau_{rad,s}} = \frac{1}{\tau_{rad,o}}\left[1+\left(\frac{\lambda^3 I}{hc}\right)\right]$$

where $\tau_{rad,o}$ is the intrinsic radiative lifetime at wavelength $\lambda$, and $\tau_{rad,s}$ is the effective radiative lifetime under stimulated emission with light intensity $I$. The possibility of developing such a photorecombinative laser action has been considered in the formation of NO,[115,116] CN,[115] $N_2$,[115] and the halogens.[114]

The so-called "exciplex" association lasers can be regarded as a special case of reaction type (13). These lasers involve an electronically excited reactant $A^*$, which can react with $B$ to form a bound molecule $(AB)^*$, even though the unexcited reactants $A$ and $B$ do not form a comparably stable $AB$ molecule. The simplest example would be an inert gas atom, for example, Ar, which, on electronic excitation, assumes the valence orbital configuration and, hence, the chemistry of its adjacent alkali metal atom, K, or, with Rydberg excitation, that of the adjacent halogen atom, Cl. Such an excited $Ar^*$ can react to form full chemical bonds with some reactant partner $B$, even though the ground states of Ar and $B$ interact only with van der Waal's forces. More subtle changes might also be exploited, such as changes in acid or base strength on electronic excitation, or changes in the tendency to form charge transfer complexes.

This type of exciplex association laser may be involved in the operation of some dye lasers. More obvious examples are the ones just described involving the inert-gas atoms. Laser emission has been observed from $Xe_2^*$ (1680–1700 Å), $PbXe^*$, $NeXe^*$ (3547 Å), $CdAr^*$ (1773, 1520 Å) and $XeAr^*$ (1182 Å), where the asterisk refers to the initially excited atom.[117]

There are a number of other chemiluminescent phenomena under study by organic chemists that have potentialities for laser pumping. For example, oxalyl chloride reacts with hydrogen peroxide, probably involving a dioxetane intermediate, to produce electronically excited ketones. These and the other reaction types just mentioned indicate that there are promising new directions to explore.

## 3. INFLUENTIAL FACTORS IN CHEMICAL LASERS

In principle, a chemical-laser pumping reaction might be endothermic, but, in practice, exothermicity proves to be an important feature. Of course, the fundamental dynamics of the reaction must be intrinsically favorable to selective excitation. Then the reaction rate must be fast so that pumping can compete with a variety of deactivation and energy-loss processes. For cw lasers, mixing of the reactants is a significant limitation, again in competition with deactivation. Finally, it is helpful if reaction mechanisms and competing reactions are understood.

Each of these factors will be the subject of an entire chapter, so we will consider illustrative examples only.

## 3.1. Overall Reaction Rate

To reduce deactivation losses, rapid reactions have thus far been involved in all chemical lasers. This implies that the activation energy cannot be much more than a few times $kT$, the operating temperature of the system. For example, the rate constant at 298°K for the $F+H_2$ reaction (2) is $6.7 \times 10^{12}$ cm$^3$/mol-sec.[118] This rate constant can be assessed by calculating the implied number of collisions needed on the average for a reaction to occur. With the molecular dimensions characterized by the following van der Waal's radii: F, 1.3 Å: $H_2$, 1.4 Å; and $CH_4$, 2.0 Å, the $F+H_2$ rate constant requires that about one out of 40 $F-H_2$ collisions results in reaction. The even higher rate constant (298°K) for the $F+CH_4$ reaction,[119] $1.6 \times 10^{13}$ cm$^3$/mol-sec, requires on the average only about a dozen $F-CH_4$ collisions for reaction. This is typical of the free-radical reactions that have been successfully employed in chemical-laser pumping.

A contrast is provided by the reaction $Cl+H_2$, which has a rate constant of only $8.2 \times 10^9$ cm$^3$/mol-sec and a 5.5 kcal/mol activation energy.[120] This lowers the room-temperature rate constant by almost three orders of magnitude and accounts for the fact that the rather weak emission from the $H_2/Cl_2$ explosion laser is enhanced by raising the temperature.[28] In the explosion, the rapid pumping reaction $H+Cl_2$ is restrained by the rate of the slower $Cl+H_2$ reaction.

Finally, it is informative to compare radical–radical combination reactions, the initiation step for an elimination laser.[15,17] The $CF_3+CF_3$ reaction has a low activation energy[121] and that of $CH_3+CF_3$ is undoubtedly lower. Hence the formation of the vibrationally excited intermediate, $CH_3CF_3^\dagger$, is quite rapid. The laser pumping depends, however, upon the rate of elimination as well. This second essential step, a unimolecular decomposition, is also extremely rapid at the level of excitation achieved by radical combination. For example, HCl elimination from $C_2H_5Cl$ formed from $CH_3+CH_2Cl$ ($\sim$90 kcal/mol excitation) occurs with a half-time near $2 \times 10^{-10}$ sec,[122] HF elimination from $CH_2FCH_2F$

formed from $2CH_2F$ radicals ($\sim$90 kcal/mol excitation) with a half-time near $2 \times 10^{-9}$ sec, and HF elimination from $CH_3NF_2$ formed from $CH_3 +$ $NF_2$ ($\sim$64 kcal/mol excitation) with a half-time near $3 \times 10^{-11}$ sec.[123] These times are all relatively short compared to collisional times at modest pressures (say, below 100 Torr). Equally important, however, the elimination times are much longer than vibrational periods ($\sim 10^{-13}$). This implies that, prior to elimination, the internal energy may be redistributed, perhaps statistically, among some or all of the vibrational degrees of freedom. Such redistribution, which would have to occur without collisional help, might account for the low fraction of the available energy that is lodged in vibrational excitation of the HX elimination product.[15,17,78,80,83–85]

We can generalize that free-radical reactions provide favorable candidates for chemical-laser pumping because they tend to react with low activation energies. These low activation energies give fast reaction rates so that pumping can compete with the inevitable deactivation processes. In the elimination lasers, whether chemically or photolytically activated, the time for the unimolecular decomposition is important, as well. These times tend to be short compared to collision times, but long compared to vibrational periods.

## 3.2. Chemical Exothermicity

In principle, a thermoneutral reaction could give selective excitation in the products. If it did, however, there would necessarily be an activation energy at least as high as the excitation lodged in the products. As just indicated, an activation energy no more than a few times $kT$ can be tolerated because of deactivation processes. Hence, all of the known chemical lasers are pumped by exothermic reactions except those photo-induced reactions that are endothermic but initiated by a light quantum of energy far exceeding the endothermicity.

The energy to be distributed among the reaction products ($-\Delta H + E_A$) must exceed the excitation energy of the state being pumped. For the known diatomic V–R lasers, the $v = 1$ state lies 5 to 11 kcal/mol above the ground state (HF, 11.3 kcal/mol; DF, 8.3 kcal/mol; HCl, 8.2 kcal/mol; DCl, 6.0 kcal/mol; HBr, 7.3 kcal/mol; CO, 6.1 kcal/mol; and NO, 5.4 kcal/mol). This places a very modest lower limit on the exothermicity needed. In fact, the efficiency of vibrational pumping is so great in three-atom reactions that vibrational population inversions can be expected if ($-\Delta H + E_A$) exceeds 15 kcal/mol. For the more complex elimination reactions, only a small fraction of the exothermicity appears as vibration. Hence, much larger exothermicities have been involved in these chemical lasers.

One can speculate about the limits imposed on chemical-laser pumping by the highest exothermicities that can be hoped for. For example, for an $A + BC$ reaction, an unusually weak $BC$ bond is provided by that of $F_2$, $D_0 = 36$ kcal/mol, and the strongest MF bonds are those of HF and LiF, 135 and 137 kcal/mol, respectively. The difference, near 100 kcal/mol, is sufficient to excite HF vibrationally to $v = 11$. The same energy transferred into electronic excitation could excite, at most, a state 35,000 cm$^{-1}$ above the ground state, which corresponds to a near ultraviolet wavelength of 2860 Å. Even more exothermic reactions are known, however, as, for example, the $O + CH \rightarrow H + CO$ chemical-laser reaction,[54] with $\Delta H = -176$ kcal/mol. This corresponds to 61,500 cm$^{-1}$ ($\lambda = $ 1624 Å), an energy exceeding that needed for electronic excitation of most molecules (though, interestingly enough, not reaching the lowest known excited state of CO itself, the $^1\pi$ state at 65,075 cm$^{-1}$). Thus chemical pumping of electronically excited molecular states to give laser emission reaching into the ultraviolet is entirely possible.

## 3.3.  Reaction Mechanisms

The chemical lasers that are best understood are those in which a single reaction can be isolated for study, as has been the case for each of the five reactions listed in Table 2. However, in practical application, the pumping reaction is generally accompanied by other reactions in sequence. Understanding this sequence, the reaction mechanism, is a prerequisite to controlling and optimizing the chemical laser. Three examples will make this evident.

In the first chemical laser, the $H_2/Cl_2$ explosion system, chlorine atoms are produced photolytically, and the reaction proceeds via a chain mechanism[1,28]:

$$Cl_2 + h\nu \rightarrow 2Cl$$

$$Cl + H_2 \rightarrow HCl + H \qquad -\Delta H = -1 \text{ kcal/mol} \qquad E_A = 5.5 \text{ kcal/mol}$$
$$(15a)$$

$$H + Cl_2 \rightarrow HCl^\dagger + Cl \qquad -\Delta H = 45.1 \text{ kcal/mol} \qquad E_A = 1.8 \text{ kcal/mol}$$
$$(15b)$$

The mechanism has three important effects on the operation of the laser. First, the rate of the pumping reaction (15b) is limited by that of the slower, preceding reaction (15a) because of its activation energy of 5.5 kcal/mol. Second, the inversion due to vibrationally excited HCl produced in Reaction (15$b$) is necessarily diluted by "cold" HCl produced concurrently in Reaction (15$a$) [for which $(-\Delta H + E_A)$ is insufficient to excite HCl $(v = 1)$]. Third, the cyclic chain provided by Reactions (15$a$) and (15$b$) multiplies many-fold the number of pumping reaction steps (15$b$) that can be obtained for each chlorine atom produced. It is implied that only a small number of input photons in the form of initiating light might be sufficient to produce a far larger number of output photons, extracting most or all of the potential chemical energy from the $H_2/Cl_2$ mixture.

The second example to be considered is the analogous $H_2/F_2$ system. The pumping reaction (16), $F + H_2$, has been studied cleanly with a number of photolytic fluorine atom sources, the first being $UF_6$.[11,13,34] Since that time, other fluorine atom sources, including $F_2O$,[35] $XeF_4$,[64] $ClF_3$,[61] and $CF_3I$,[49] have been explored, and critical comparisons of them have been made.[124,125] By far the most important fluorine atom source, however, is fluorine itself, $F_2$. Just as for $H_2/Cl_2$, the reaction between $H_2$ and $F_2$ proceeds by a chain mechanism:

$$H + F_2 \rightarrow HF^\dagger + F \qquad -\Delta H = 98.0 \text{ kcal/mol} \qquad E_A = 2.4 \text{ kcal/mol} \tag{16a}$$

$$F + H_2 \rightarrow HF^\dagger + H \qquad -\Delta H = 31.5 \text{ kcal/mol} \qquad E_A = 1.7 \text{ kcal/mol} \tag{16b}$$

In contrast to Reaction (15$a$), Reaction (16$a$) is extremely exothermic and has almost as low an activation energy as Reaction (16$b$). Not only can Reaction (16$a$) contribute to the population inversion (for HF, up to $v = 10$), but it also introduces at least the possibility of chain branching:

$$HF^\dagger (v \geq 4) + F_2 \rightarrow HF (v = 0) + 2F \tag{16c}$$

There has been speculation about the participation of Reaction (16$c$),[126] but, as yet, there has been no definitive verification that it proceeds sufficiently rapidly to be important. Thus Tal'rose et al.[127] conclude "... the very fact of the presence of branching chains is not proved with that degree of certainty which is necessary ...." Of course, chain branching implies an accelerating pumping process, a desirable property both for initiation and for efficiency. Because of this potential importance, it is worthy of note that chain branching has also been postulated in the $H_2/ClF_3$ laser,[61] although, again, definitive proof is lacking.[128,129]

One more example provides sufficient evidence that complex reaction mechanisms can be involved in chemical-laser operation. In one of the earliest photolytic lasers, the $CS_2/O_2$ explosion system,[12] the kinetics are not yet perfectly clear, and, hence, the relative contributions of different possible pumping reactions are not known for sure. Suart et al.[130] postulate that Reaction (17c) accounts for pumping:

$$CS_2 + h\nu \rightarrow CS + S$$
$$S + O_2 \rightarrow SO + O$$
$$O + CS \rightarrow CO^{\dagger} + S \qquad -\Delta H = 75 \text{ kcal/mol} \qquad (17c)$$

Gregg and Thomas,[74] however, point out no fewer than four additional possible pumping reactions: (17a), (17b), (17d), and (17e), and they favor Reactions (17a) and (17e) as the most important:

$$O + CS_2 \rightarrow CO^{\dagger} + S_2 \qquad -\Delta H = 70 \text{ kcal/mol} \qquad (17a)$$
$$O + OCS \rightarrow CO^{\dagger} + SO \qquad -\Delta H = 47.5 \text{ kcal/mol} \qquad (17b)$$
$$SO + CS \rightarrow CO^{\dagger} + S_2 \qquad -\Delta H = 40 \text{ kcal/mol} \qquad (17d)$$
$$SO + O \rightarrow SO_2^* \qquad -\Delta H = 123 \text{ kcal/mol}$$
$$SO_2^* + CO \rightarrow CO^{\dagger}(v \leq 13) + SO_2 \qquad (17e)$$

The last reaction (17e) is of special interest since it involves energy transfer from an electronically excited molecule, $SO_2$. Certainly, a clearer understanding of the reaction mechanism would aid in the optimization of this chemical laser.

## 3.4. Deactivation Processes

In most practical applications of chemical lasers, deactivation (relaxation) processes have a strong influence on laser power and performance. The reason is, of course, that molecular collisions are intrinsically required to permit the pumping reaction to proceed, but then molecular collisions can deactivate the excited products as they are formed. The deactivation efficiency of different collision partners varies greatly; certain molecules are particularly effective. This effectiveness can be most readily visualized in terms of $Z$, the average number of collisions needed to cause deactivation, when conventional molecular sizes are assumed. Table 5 lists some experimental values of $Z$ for HCl and HF deactivation. These data are sufficient to reveal two factors that are at work: attractive forces between the colliding molecules and resonance transfer because of close energy match between their vibrational levels.

First, notice that argon and nitrogen are quite ineffective, each requiring on the average many tens of thousands of collisions to deactivate either $HF^{\dagger}$ or $HCl^{\dagger}$. At the other extreme, $H_2O$ seems to deactivate either molecule with only a few collisions. Similarly, HF deactivates $HF^{\dagger}$ in less than 60 collisions. These efficiencies indicate attractive forces, all

**Table 5.**   Deactivation of vibrationally excited HF and HCl by various collision partners

|  | HF | | HCl | |
|---|---|---|---|---|
|  | $Z^{a,c}$ | $\Delta E$ cm$^{-1}$ [b] | $Z^{a,d}$ | $\Delta E$ cm$^{-1}$ [b] |
| Ar | $>100 \times 10^3$ | — | $>3000 \times 10^3$ | — |
| $N_2$ | $56 \times 10^3$ | 1630 | $9.1 \times 10^3$ | 552 |
| $D_2$ | $2.7 \times 10^3$ | 971 | 143 | $-108$ |
| $H_2$ | 570 | $-198$ | $106 \times 10^3$ | $-1274$ |
| $CF_4$ | $19 \times 10^3$ | 2680 | — | — |
| $CH_4$ | 160 | 941 | 106 | $-30$ |
| DCl | — | — | $2 \times 10^3$ | 795 |
| HI | — | — | $1.3 \times 10^3$ | 656 |
| HBr | — | — | 167 | 327 |
| HF | 57 | 0 | — | — |
| $H_2O$ | 1.3 | 205 | 10 | $-871$ |
| $D_2O$ | 1.3 | 1173 | — | — |

[a] $Z$ is the average number of kinetic collisions needed.
[b] $\Delta E$ is the energy discrepancy (in cm$^{-1}$) of $\nu(HX)$ and nearest vibrational frequency of M.
[c] Ref. 26.
[d] Ref. 27.

resulting from hydrogen bonding, which cause long-lived complexes to form. As another example, $I_2$ is quite effective in deactivating $I^*(^2P_{1/2})$,[91,131] which is again explainable in terms of a long-lived $I_3$ complex.[132] Both for $I_3$ and the hydrogen-bonded interactions, the molecular complexes are bound by energies estimated to be in the range of 5 to 10 kcal/mol, which explains why deactivation efficiencies can decrease as temperature is raised.

The importance of energy matching is illustrated by the relative efficiencies of $H_2$ and $D_2$ in deactivating HF and HCl. For $D_2$, the vibrational energy discrepancy with respect to HF is 971 cm$^{-1}$; whereas, for $H_2$, the discrepancy is only 200 cm$^{-1}$ so that $D_2$ requires almost five times as many collisions as $H_2$. In contrast, the effects are just reversed (and much larger) for HCl deactivation; in this case, $D_2$ has the smaller energy discrepancy. The ineffectiveness of $H_2$ for HCl$^\dagger$ deactivation is accentuated even more because the energy mismatch places $H_2^\dagger$ at higher energy. This shows that the deactivating mechanisms for $H_2/HF^\dagger$ and $D_2/HCl^\dagger$ deactivation involve $V \rightarrow V$ energy transfer.

Because of these effects, there has been a heavy concentration of effort on the measurement of these deactivation rate constants, as will be discussed in detail in the chapter by Bott and Cohen.

## 4. EXPERIMENTAL TECHNIQUES

In this section, we consider briefly the experimental methods used to operate and study chemical lasers. For pulsed operation, three initiation methods are in use: pulsed photolysis, discharge, or electron beam bombardment. For continuous operation, glow discharge or thermal (pyrolytic) production of reactants can be used. Each of these will be considered briefly, and, thereafter, the use of chemiluminescence and chemical-laser gain to measure fundamental rate constants will be discussed.

### 4.1. Chemical-Laser Efficiency

When experimental techniques are compared, it is useful to have in mind the potential efficiency. For many practical applications, the efficiency of conversion of external-energy input into laser-energy output is of primary interest. To eliminate external-energy input, self-sustaining reactions, perhaps initiated by a small external trigger, are desirable. As just discussed, chain reactions will be most effective and chain-branching reactions even better. The condition needed is[133,134]

$$\eta \frac{Q_i}{Q_e} \frac{N}{N_0} > 1$$

where $\eta$ is the chemical efficiency of the laser; $Q_i$ and $Q_e$ are the chemical energy released and the electrical energy consumed, respectively; $N$ is the number of lasing molecules formed; and $N_0$ is the number of initiating molecules formed (so that $N/N_0$ is the average chain length). Values of $\eta$ over 50% are obtained for V–R excitation (see Table 2). Oraevskii[133] evaluates $N/N_0$ to be near five in pulsed operation of an $H_2/F_2$ laser, and it has been suggested that other chain carriers, such as $XeF_4$[64] or $ClF_3$,[61] might increase $N/N_0$, $Q_i/Q_e$, and/or $\eta$.

### 4.2. Flash-Photolysis Initiation

Since ultraviolet light is needed for flash-photolytic initiation, efficiency is poor. Another limiting factor (because of deactivation and radiative losses) is the potential pumping rate. For example, F atoms can be photolytically generated at a rate of about $10^{23}$ to $10^{24}$ atoms $cm^{-3}$ $sec^{-1}$.

Hence, the rate of formation of the population inversion in the $F + H_2$ reaction is limited by photolysis-light input rather than by the chemical-reaction rate.

Despite these practical limitations, most of the presently important chemical lasers were discovered by this technique,[1,2,11–18] and it is an important method for clarifying the operative reactions.[37,89,90] The reason for this success is that absorption cross-sections, details of the photodissociation processes, and competing deactivation processes can be well known and controlled.

### 4.3. Electrical-Discharge Initiation

Transverse discharge geometries have proved to be convenient for the excitation of chemical lasers. Although simple TEA discharges encounter problems with halogen-containing molecules that limit the rate of energy input, the overall efficiency of discharge-initiated chemical lasers can be high. Electrical efficiencies of >4% have been realized in discharge-induced HF lasers with 100-nsec pulses of 10-J energy.[135]

### 4.4. Electron-Beam Initiation

Since electron-beam controlled discharges can furnish pumping rates significantly in excess of $10^{24} \, cm^{-3} \, sec^{-1}$, the highest power outputs yet reported for chemical lasers have come from such methods. At the time of this writing, the highest HF energy output is $\sim 250 \, J$ in 60 nsec from electron-beam pumping of $SF_6/C_2H_6$ mixtures.[157]

### 4.5. Continuous Chemical Lasers

Initiation of a cw chemical laser is most readily accomplished with an electric discharge to produce a free-radical reactant.[33,44–46] Purely chemical initiation can be achieved, however, in $H_2/F_2$ chemical lasers by admixing NO to react with $F_2$, which produces FNO and F atoms to initiate the $H_2/F_2$ chain.[136] Pyrolysis of $F_2$ provides another purely chemical initiation, since the $H_2/F_2$ reaction itself can provide the thermal energy.

In performance, cw chemical lasers are limited by the rate of materials transport through the active region of the laser. For modest laser power, subsonic flow can be used, which implies rather simple technology. For the highest cw laser power, supersonic flows are needed, with significantly greater sophistication in design. The cw lasers will be discussed in detail in the chapter by Gross and Spencer.

## 4.6. Chemiluminescence

When a chemical reaction produces a non-Boltzmann energy distribution, radiation or chemiluminescence can result. Since this radiation depends upon the populations of upper and lower states, it can, in principle, be used to deduce rate constants into particular states. This is the fundamental information needed to predict and understand chemical-laser behavior.

In practice, the interpretation of chemiluminescence intensities are complicated by deactivation processes. There has been a long period of development,[66] principally in the laboratory of J. C. Polanyi,[10,19,20,22,23,68,137,138] which culminated in the "arrested relaxation" method.[139] Molecular flow of two reagent beams at pressures near $10^{-4}$ Torr is maintained by cryogenic pumping. Reaction occurs at the intersection of these two beams in the center of the vessel and the products are transported out of the active region with only a few secondary collisions. The success of the method is attested to by the observation of only partial equilibration of rotational excitation.[139] Now amply validated, the chemiluminescence method is actively contributing basic data relevant to chemical lasers from a number of laboratories, including those of Jonathan[21] and Setzer.[69]

## 4.7. Chemical-Laser Studies

Just as in chemiluminescence measurements, fundamental information from chemical-laser gain data can be extracted provided deactivation processes are under control. Here again, there has been a substantial period of development, this time in the laboratory of G. C. Pimentel.[1,2,11,37,89] Two methods were developed, each avoiding the necessity for absolute gain measurements. In the equal-gain-temperature method,[37,88,89] rotational equilibration is assured through addition of an inert gas, and then the temperature is varied until the time to threshold of two transitions is exactly equal. More recently, the zero-gain-temperature method has evolved,[25,86,89] which also requires rotational equilibration, but, when applicable, is simpler to interpret and more precise. At the present stage of development, these two methods are capable of revealing nuances of selective excitation that are not readily observed by alternative techniques, such as the temperature dependence of selective excitation[25] and the effect of $H_2$ rotation[140] in the important $F + H_2$ reaction.

Another approach to the interpretation of chemical-laser performance has been based upon computer simulations. The system of coupled rate equations describing the growth and decay of both the vibrational populations and the photon densities is solved analytically.[141–144] Since a large matrix of rate constants, both for pumping and for deactivation, is involved, the physical significance of the derived parameters must be examined critically, particularly if time to threshold or sequencing rates are used as input data and the gain-equals-loss approximation is used.[90]

## 4.8. Molecular Beams

The most primitive information about reaction dynamics is obtained from molecular-beam studies. Data are obtained that immediately reveal translational excitation of the products and, by subtraction, the total internal excitation in vibration plus rotation. For the $F + D_2$ reaction, it has even been possible to distinguish individual vibrational states of $DF$.[145] Thus extremely important and fundamental information about chemical-laser reactions can be expected from molecular-beam experiments.

## 4.9. Theory

A variety of theoretical models have been developed to help us understand and predict the energy distribution in the reactions that pump chemical lasers. One of these approaches can be described as an information-theoretical model that uses temperature-like distributions and entropy deficiencies to describe the inversion and gain characteristics in chemical lasers.[146–149] This technique, which will be presented in the chapter by Ben-Shaul and Hofacker, promises useful description of energy distributions among reaction products.

Fundamental calculations, mostly classical, have been carried out by many workers attempting to describe the reactive collision process. The reaction is considered to take place over a single potential-energy hypersurface, and the reaction probability and product-energy distribution are calculated by numerical integration of the classical equations of motion.

Unfortunately, an accurate knowledge of the potential-energy hypersurface is rarely available because of the difficulty of the required *ab initio* calculations. In fact, *ab initio* calculations that include configuration interaction are available only for two reaction surfaces presently important in lasers, $F + H_2$[150] and $H + F_2$.[151]

Without such a starting point, some semiempirical method must be used to ascertain the qualitative features of the hypersurface. Prominent among those in use is the London equation,[152,153] which led to the so-called LEPS equation. An interesting historic account is given by Polanyi[154] and reviews of the dependence of the calculated results upon the potential energy surface are presented by Polanyi,[154] Muckerman,[155] and Blais and Truhlar.[156] These theoretical methods will be discussed in detail in the chapter by Wilkins.

## 5. CONCLUSION

Chemical lasers have matured in their 10-year history. The literature references cited in this chapter represent only a fraction, possibly one-quarter, of the published work directly concerned with chemical lasers. Hence it is most significant that this chapter is the introduction to a book addressed to those concerned with applications. New pumping reactions are sure to be found and new potentialities can be expected; the future looks bright for chemical lasers.

## REFERENCES

1. J. V. V. Kasper and G. C. Pimentel, Phys. Rev. Lett. **14,** 352 (1965).
2. J. V. V. Kasper and G. C. Pimentel, Appl. Phys. Lett. **5,** 231 (1964).
3. M. S. Dzhidzhoev, V. T. Platonenko, and R. V. Khokhlov, Sov. Phys. Usp. **13,** 247 (1970).
4. N. G. Basov, V. I. Igoshin, J. I. Markin, and A. N. Oraevskii, Kvantovaja Elektronika **2,** 3 (1971).
5. P. H. Dawson and G. H. Kimball, "Chemical Lasers," in *Advances in Electronics and Electron Physics, Vol. 31* (Academic, New York, 1972).
6. K. L. Kompa, "Chemical Lasers," in *Fortschritte der Chemischen Forschung/Topics in Current Chemistry, Vol. 37* (Springer, Heidelberg, 1973).
7. T. F. Deutsch, Appl. Phys. Lett. **11,** 18 (1967).
8. D. P. Akitt and J. T. Yardley, IEEE J. Quant. Electron. **QE-6,** 113 (1970).
9. G. C. Pimentel, IEEE J. Quant. Electron. **QE-6,** 174 (1970).
10. J. C. Polanyi, J. Chem. Phys. **34,** 347 (1961).
11. K. L. Kompa and G. C. Pimentel, J. Chem. Phys. **47,** 857 (1967).
12. M. A. Pollack, Appl. Phys. Lett. **8,** 237 (1966).
13. J. H. Parker and G. C. Pimentel, J. Chem. Phys. **48,** 5273 (1968).
14. M. A. Pollack, Appl. Phys. Lett. **9,** 94 (1966).
15. M. J. Berry and G. C. Pimentel, J. Chem. Phys. **49,** 5190 (1968).
16. M. C. Lin, J. Phys. Chem. **75,** 3642 (1971).
17. T. D. Padrick and G. C. Pimentel, Appl. Phys. Lett. **20,** 167 (1972).
18. M. J. Berry and G. C. Pimentel, J. Chem. Phys. **51,** 2274 (1969).
19. K. G. Anlauf, D. S. Horne, R. G. Macdonald, J. C. Polanyi and K. B. Woodall, J. Chem. Phys. **57,** 1561 (1972).
20. J. C. Polanyi and K. B. Woodall, J. Chem. Phys. **57,** 1574 (1972).

21. N. Jonathan, C. M. Meliar-Smith, S. Okuda, D. H. Slater and D. Timlin, Mol. Phys. **22**, 561 (1971).

22. K. G. Anlauf, J. C. Polanyi, W. H. Wong, and K. B. Woodall, J. Chem. Phys. **49**, 5189 (1968).

23. P. D. Pacey and J. C. Polanyi, J. Appl. Opt. **10**, 1725 (1971).

24. M. J. Berry, J. Chem. Phys. **59**, 6229 (1973).

25. See, for example, R. D. Coombe and G. C. Pimentel, J. Chem. Phys. **59**, 251 (1973).

26. J. K. Hancock and W. H. Green, J. Chem. Phys. **57**, 4515 (1972).

27. H.-L. Chin and C. B. Moore, J. Chem. Phys. **54**, 4072 (1971); J. Chem. Phys. **54**, 4080 (1971).

28. P. H. Corneil and G. C. Pimentel, J. Chem. Phys. **49**, 1379 (1968).

29. N. G. Basov, V. V. Gromov, E. L. Koshelev, E. P. Markin, and A. N. Oraevskii, Zh. ETF Pis. Red. **9**, 250 (1969); JETP Lett. **9**, 147 (1969).

30. T. F. Deutsch, Appl. Phys. Lett. **10**, 234 (1969).

31. T. F. Deutsch, J. Quant. Electron. **3**, 419 (1967).

32. A. Henry, F. Bourcin, I. Arditi, R. Charneau, and J. Menard, C. R. Acad. Sci. **267**, 617 (1968).

33. T. A. Cool, R. R. Stephens, and J. A. Shirley, J. Appl. Phys. **41**, 4038 (1970).

34. K. L. Kompa, J. H. Parker, and G. C. Pimentel, J. Chem. Phys. **48**, 3821 (1968).

35. R. W. F. Gross, N. Cohen, and T. A. Jacobs, J. Chem. Phys. **48**, 3821 (1968).

36. O. M. Batovskii, G. K. Vasilijev, E. F. Makarov, and V. L. Tal'rose, Zh. ETF Pis. Red. **9**, 341 (1969); JETP Lett. **9**, 200 (1969).

37. J. H. Parker and G. C. Pimentel, J. Chem. Phys. **51**, 91 (1969).

38. N. G. Basov, L. V. Kulakov, E. P. Markin, A. I. Nikitin, and A. N. Oraevskii, Zh. ETF Pis. Red. **9**, 613 (1969); JETP Lett. **9**, 375 (1969).

39. T. F. Deutsch, IEEE J. Quant. Electron. **4**, 174 (1971).

40. W. H. Green and M. C. Lin, J. Chem. Phys. **54**, 3222 (1971).

41. M. C. Lin, J. Phys. Chem. **75**, 284 (1971).

42. S. N. Suchard, R. L. Kerber, G. Emanuel, and J. S. Whittier, J. Chem. Phys. **57**, 5065 (1972).

43. G. G. Dolgov-Savel'ev and A. A. Podminogin, Sov. J. Quant. Electron. **2**, 248 (1973).

44. D. J. Spencer, H. Mirels, T. A. Jacobs, and R. W. F. Gross, Int. J. Chem. Kinet. **1**, 493 (1969).

45. D. J. Spencer, H. Mirels, T. A. Jacobs, and R. W. F. Gross, Appl. Phys. Lett. **16**, 235 (1970).

46. D. I. Rosen, R. N. Sileo, and T. A. Cool, IEEE J. Quant. Electron. **QE-9**, 163 (1973).

47. J. R. Airey and S. F. McKay, Appl. Phys. Lett. **15**, 401 (1969).

48. C. B. Moore, IEEE J. Quant. Electron. **QE-4**, 52 (1968).

49. M. J. Berry, Ph.D. Dissertation, University of California, Berkeley, California (1970).

50. R. D. Coombe, G. C. Pimentel, and M. J. Berry, IEEE J. Quant. Electron. **QE-9**, 192 (1973).

51. W. H. Green and M. C. Lin, IEEE J. Quant. Electron. **QE-7**, 98 (1971).

52. J. R. Airey, J. Chem. Phys. **52,** 156 (1970).

53. I. Burak, Y. Noter, A. M. Ronn, and A. Szöke, Chem. Phys. Lett. **13,** 322 (1972).

54. L. E. Brus and M. C. Lin, J. Phys. Chem. **76,** 1429 (1972).

55. S. J. Arnold and G. H. Kimball, Appl. Phys. Lett. **15,** 351 (1969).

56. C. Wittig, J. C. Hassler, and P. D. Coleman, Appl. Phys. Lett. **16,** 117 (1970).

57. C. E. Wiswall, D. P. Ames, and T. J. Menne, IEEE J. Quant. Electron. **QE-9,** 181 (1973).

58. R. K. Pearson, J. O. Cowles, G. L. Hermann, D. W. Gregg, and J. R. Creighton, IEEE J. Quant. Electron. **QE-9,** 879 (1973).

59. J. H. Parker and G. C. Pimentel, J. Chem. Phys. **55,** 857 (1971).

60. L. E. Brus and M. C. Lin, J. Phys. Chem. **75,** 2546 (1971).

61. O. D. Krogh and G. C. Pimentel, J. Chem. Phys. **56,** 969 (1972).

62. W. H. Green and M. C. Lin, J. Chem. Phys. **54,** 3222 (1971); IEEE J. Quant. Electron. **QE-7,** 98 (1971)

63. T. V. Jacobson and G. H. Kimbell, IEEE J. Quant. Electron. **QE-9,** 173 (1973).

64. K. L. Kompa, P. Gensel, and J. Wanner, Chem. Phys. Lett. **3,** 210 (1969).

65. W. W. Rice and R. J. Jensen, J. Phys. Chem. **76,** 805 (1972).

66. J. D. McKinley, D. Garvin, and M. Boudart, J. Chem. Phys. **23,** 784 (1955).

67. P. E. Charters, R. G. Macdonald, and J. C. Polanyi, J. Appl. Opt. **10,** 1747 (1971).

68. J. K. Cashion and J. C. Polanyi, J. Chem. Phys. **35,** 600 (1961).

69. M. J. Perona, D. W. Setzer, and R. J. Johnson, J. Chem. Phys. **52,** 6384 (1970).

70. H. Heydtmann and J. C. Polanyi, J. Appl. Opt. **10,** 1738 (1971).

71. M. C. Lin, Chem. Phys. Lett. **7,** 209 (1970).

72. M. J. Perona, J. Chem. Phys. **54,** 4024 (1971).

73. M. C. Lin and L. E. Brus, J. Chem. Phys. **54,** 5423 (1971).

74. D. W. Gregg and S. J. Thomas, J. Appl. Phys. **39,** 4399 (1968).

75. R. D. Giles and E. W. Whittle, Trans. Faraday Soc. **61,** 1425 (1965).

76. J. W. Coomber and E. Whittle, Trans. Faraday Soc. **63,** 1394 (1967).

77. P. Cadman, M. Day, A. W. Kirk, and A. F. Trotman-Dickenson, Chem. Commun. **4,** 203 (1970).

78. P. N. Clough, J. C. Polanyi, and R. T. Taguchi, Can. J. Chem. **48,** 2919 (1970).

79. T. D. Padrick and G. C. Pimentel, J. Chem. Phys. **54,** 720 (1971).

80. T. D. Padrick and G. C. Pimentel, J. Phys. Chem. **76,** 3125 (1972).

81. L. E. Brus and M. C. Lin, J. Phys. Chem. **75,** 2546 (1971).

82. D. S. Ross and R. Shaw, J. Phys. Chem. **75,** 1170 (1971).

83. M. C. Lin, J. Phys. Chem. **75,** 3642 (1971).

84. M. C. Lin, J. Phys. Chem. **76,** 811 (1972).

85. M. C. Lin, J. Phys. Chem. **76,** 1425 (1972).

86. J. Roebber and G. C. Pimentel, IEEE J. Quant. Electron. **QE-9,** 201 (1973).

87. P. R. Poole and G. C. Pimentel, J. Chem. Phys. **63,** 1950 (1975).

88. M. J. Berry and G. C. Pimentel, J. Chem. Phys. **53,** 3453 (1970).

89. M. J. Molina and G. C. Pimentel, J. Chem. Phys. **56,** 3988 (1972).

90. M. J. Molina and G. C. Pimentel, IEEE J. Quant. Electron. **QE-9,** 64 (1973).

91. J. V. V. Kasper, J. H. Parker, and G. C. Pimentel, J. Chem. Phys. **43,** 1827 (1965).

92. R. J. Donovan and D. J. Husain, Trans. Faraday Soc. **62,** 11 (1966).

93. A. J. DeMaria and C. J. Ultee, Appl. Phys. Lett. **9,** 67 (1966).

94. M. A. Pollack, Appl. Phys. Lett. **8,** 36 (1966).

95. D. W. Gregg, R. E. Kidder, and C. V. Dobler, Appl. Phys. Lett. **13,** 297 (1968).

96. S. D. Velikanov, S. B. Korner, V. D. Nikolaev, M. V. Sinitsyn, Y. A. Solov'yev, and V. D. Urlin, A. N. SSSR Dokl **192,** 528 (1970); **15,** 478 (1970).

97. I. M. Beloutsova, O. B. Danilov, N. S. Kladivikova, and I. L. Yachnev, Zh. Tekh. Fiz. **40,** 1562 (1970); **15,** 1212 (1971).

98. D. E. O'Brien and J. R. Bowen, J. Appl. Phys. **42,** 1010 (1971).

99. P. Gensel, K. Hohla, and K. L. Kompa, Appl. Phys. Lett. **18,** 48 (1971).

100. V. A. Dudkin and V. I. Malyshev, K. Sp. F. **3,** 35 (1971).

101. C. R. Guiliano and L. D. Hess, J. Appl. Phys. **40,** 2428 (1969).

102. J. D. Campbell and J. V. V. Kasper, Chem. Phys. Lett. **10,** 436 (1971).

103. R. P. Sorokin and J. R. Lankard, J. Chem. Phys. **51,** 2929 (1969).

104. N. Basco and R. G. W. Norrish, Proc. Roy. Soc. **268,** 291 (1962).

105. M. A. Pollack, Appl. Phys. Lett. **9,** 230 (1966).

106. S. N. Suchard and G. C. Pimentel, Appl. Phys. Lett. **18,** 530 (1971).

107. F. G. Sadie, P. A. Buerger, and O. G. Malan, J. Appl. Phys. **43,** 2906 (1972).

108. K. Ljalikov and A. Terenin, Z. Phys. **40,** 107 (1926).

109. M. Polanyi and G. Schay, Z. Phys. Chem. **1,** 46 (1928); M. Polanyi, *Atomic Reactions* (Williams and Northgate, London, 1932).

110. W. S. Struve, T. Kitigawa, and D. R. Herschbach, J. Chem. Phys. **54,** 2759 (1971).

111. C. D. Jonah and R. N. Zare, Chem. Phys. Lett. **9,** 65 (1971).

112. C. D. Jonah, R. N. Zare, and Ch. Ottinger, J. Chem. Phys. **56,** 263 (1972).

113. D. O. Ham and H. W. Chang, Chem. Phys. Lett. **24,** 579 (1974).

114. A. N. Oraevskii, Sov. Phys. JETP **32,** 856 (1971).

115. R. A. Young, J. Chem. Phys. **40,** 1848 (1964).

116. See, for example, H. B. Palmer, J. Chem. Phys. **26,** 648 (1957); R. A. Young and G. A. St. John, J. Chem. Phys. **48,** 895 (1968).

117. N. G. Basov, V. A. Danilychev, Y. M. Popov, and D. D. Khodkevich, JETP Lett. **12,** 473 (1970). See also N. G. Basov, Laser Focus **7,** 30 (1971).

118. P. D. Mercer and H. O. Pritchard, J. Phys. Chem. **63,** 1468 (1959).

119. G. C. Fettis, J. H. Knox, and A. F. Trotman-Dickenson, J. Chem. Soc. 1064 (1960).

120. P. G. Ashmore and J. Chanmugam, Trans. Faraday Soc. **49,** 254 (1953).

121. T. Ogawa, G. A. Carlson, and G. C. Pimentel, J. Phys. Chem. **74,** 1090 (1970).

122. J. C. Hassler, D. W. Setzer, and R. L. Johnson, J. Chem. Phys. **45,** 3231 (1966).

123. D. S. Ross and R. Shaw, J. Phys. Chem. **75,** 1170 (1971).

124. M. J. Berry, Chem. Phys. Lett. **15,** 269 (1972).

125. G. G. Dolgov-Savel'ev, V. A. Polyakov, and G. M. Chumak, Sov. Phys. JETP **31,** 643 (1970).

126. N. G. Basov, V. I. Igoshin, E. P. Markin, and A. N. Oraevskii, Sov. J. Quant. Electron. **1,** 119 (1971).

127. V. L. Tal'rose, G. K. Vasil'ev, and O. M. Batovskii, Kinet. Katal. **11,** 277 (1970).

128. S. N. Suchard, J. Chem. Phys. **58,** 1269 (1973).

129. G. C. Pimentel, J. Chem. Phys. **58,** 1270 (1973).

130. R. D. Suart, P. H. Dawson, and G. H. Kimbell, J. Appl. Phys. **43,** 1022 (1972).

131. R. J. Donovan and D. Husain, Nature **206,** 171 (1965).

132. D. L. Bunker and N. Davidson, J. Amer. Chem. Soc. **80,** 5090 (1958).

133. A. N. Oraevskii, Trends in Physics 95 (June 20, 1973).

134. N. G. Basov, V. I. Igoshin, E. P. Markin, and A. N. Oraevskii, Sov. J. Quant. Electron, **1,** 119 (1971).

135. H. Pummer, W. Breitfeld, H. Wedler, G. Klement, and K. L. Kompa, Appl. Phys. Lett. **22,** 319 (1973).

136. T. A. Cool and R. R. Stephens, Appl. Phys. Lett. **16,** 55 (1970).

137. J. C. Polanyi, Quant. Spectrosc. Radiat. Transfer **3,** 471 (1963).

138. F. D. Findlay and J. C. Polanyi, Can. J. Chem. **42,** 2176 (1964); J. R. Airey, F. D. Findlay and J. C. Polanyi, Can. J. Chem. **42,** 2193 (1964).

139. K. G. Anlauf, P. J. Kuntz, D. H. Maylotte, P. D. Pacey, and J. C. Polanyi, Disc. Faraday Soc. 183 (1967).

140. R. D. Coombe and G. C. Pimentel, J. Chem. Phys. **59,** 1535 (1973).

141. J. R. Airey, J. Chem. Phys. **52,** 156 (1970).

142. R. L. Kerber, G. Emanuel, and J. S. Whittier, Appl. Opt. **11,** 112 (1972).

143. N. Cohen, T. A. Jacobs, G. Emanuel, and R. L. Wilkins, Int. J. Chem. Kinet. **1,** 551 (1969).

144. P. H. Corneil and J. V. V. Kasper, IEEE J. Quant. Electron. **QE-6,** 170 (1970).

145. T. P. Schafer, P. E. Siska, J. M. Parsons, F. P. Tully, Y. C. Wong, and Y. T. Lee, J. Chem. Phys. **53,** 3385 (1970).

146. R. B. Bernstein and R. D. Levine, J. Chem. Phys. **57,** 434 (1972).

147. A. Ben-Shaul, R. D. Levine, and R. B. Bernstein, J. Chem. Phys. **57,** 5427 (1972).

148. R. D. Levine, B. R. Johnson, and R. B. Bernstein, Chem. Phys. Lett. **19,** 1 (1973).

149. A. Ben-Shaul, G. L. Hofacker, and K. L. Kompa, J. Chem. Phys. **59,** 4664 (1973).

150. C. F. Bender, P. K. Pearson, S. V. O'Neil, and H. F. Schaefer III, J. Chem. Phys. **56,** 4626 (1972).

151. S. V. O'Neil, P. K. Pearson, H. F. Schaefer III, and C. F. Bender, J. Chem. Phys. **58,** 1126 (1973).

152. F. London, Z. Elektrochem. **35,** 552 (1929).

153. H. Eyring and M. Polanyi, Z. Phys. Chem. **B12,** 279 (1931).

154. J. C. Polanyi, Acct. Chem. Res. **5,** 161 (1972).

155. J. T. Muckerman, J. Chem. Phys. **56,** 2997 (1972).

156. N. C. Blais and D. G. Truhlar, J. Chem. Phys. **58,** 1090 (1972).

157. N. R. Greiner, L. S. Blair, E. L. Patterson, and R. A. Gerber, "100 Gigawatt $H_2$-$F_2$ Laser Initiated by an Electron Beam", Presented at the VIII International Conference on Quantum Electronics, San Francisco, California, June 1974.

CHAPTER **2**

# KINETICS OF HYDROGEN-HALIDE CHEMICAL LASERS

## N. COHEN

## J. F. BOTT

**The Aerospace Corporation**
**El Segundo, California**

To the kineticist, one of the most gratifying aspects of the burgeoning interest in chemical lasers in the past few years is the impetus it has given to the study of the detailed kinetics and dynamics of chemical and energy-transfer reactions. For example, 10 years ago there were no reliable experimental data on vibrational relaxation of molecules in vibrational levels higher than $v = 1$; today not only is it becoming almost routine to obtain such data, but also workers are giving a serious look at the problem of multiple quantum transitions as well, an aspect of the problem that was brushed aside a decade ago. Ten years ago, Polanyi and his co-workers were just smoothing out details in their pioneering work on the distribution of energy among the vibrational states of molecules formed in exothermic atom transfer reactions. Since then, several techniques, both theoretical and experimental, have been applied to numerous systems.

The system that has received the most attention in recent years is the HF laser system. In 1968 when the HF laser was starting to draw attention in our laboratory, modeling calculations were not feasible for want of sufficient kinetic data. Consequently, our first modeling efforts were devoted to the HCl laser system, for which considerably more data were available at the time.[1] Since then, studies of the kinetics of the HF system have been published with such frequency that it has been difficult to prepare an up-to-date review of the subject.[2] Work on the hydrogen halides still continues at such a pace that this review may be out of date by the time it is finally published. Hence we can only hope that it is a reliable indicator of the state of knowledge as of October 1973.

Although the chapter title might indicate a broader scope, we have concentrated on the kinetics of three systems: $H_2-F_2$, $H_2-Cl_2$, and $D_2-F_2$. Comments on kinetics of the analogous bromine and iodine systems are brief, limited to sufficient references to recent literature to enable the interested reader to pursue the subject conveniently. This is justified mainly by the current state of affairs in chemical lasers: As of yet there has been little interest in HBr and HI as laser systems. Second, we have limited our review to the kinetics of those reactions encountered in the simple halogen–hydrogen (deuterium) system, notwithstanding the fact that there has been much work on HF or HCl lasers generated in systems with alternate sources of H or F atoms.

For the large part, the contents of this review constitute an update of several earlier reviews[2,3] prepared in this laboratory. We have followed the same general practices here. In particular, we note that all rate coefficients are reported in units of cubic centimeters, moles, and seconds; energies are given in calories per mole. Rate coefficients are reported in the modified Arrhenius form, $k = AT^n \exp(-E/RT)$, although in some cases either $n$ or $E$ is equal to zero. This is done, even when there is no a priori justification, for ease of computer programming, as most programs have restrictions on the format of the input. Also, one of the most commonly used programs can handle several chaperones M in a single reaction provided that the $k_i^M$ all differ only by a constant temperature-independent factor. Therefore, we have occasionally taken minor liberties with data in order to arrive at sets of $k$'s for which this was true. Where our tamperings have altered the originally reported numbers by more than a few percent, we have so indicated in the text. We stress that our primary emphasis throughout is on presenting data for computational purposes, rather than on unraveling the theoretical significance of the findings. We have relied on JANAF data for thermochemical quantities, using the notation $K_{a,b} = k_a/k_b$.

We have listed the important reactions for HF, DF, and HCl laser modeling and our recommendations for their rate coefficients in Tables 20, 21, and 22. Many of the rates involving the higher vibrational levels have been simply extrapolated from $v = 1$ data, and some rates have been based upon theoretical estimates with no experimental data. The rate coefficients have been fitted with expressions that are accurate between 295 and 1000°K where most of the experimental data have been obtained. The important energy-transfer rates have weak-temperature dependences around 300°K, and extrapolations to 200°K should pose no great difficulties. Other energy-transfer processes are even less important at low temperatures. Pumping reactions have not been measured at lower than room temperature in most cases. Because of these uncertainties, rate coefficients should be used for chemical-laser modeling with caution. The importance of each rate should be established in any actual modeling, and the sensitivity to variations of the rate coefficient should be determined.

## 1. DISSOCIATION AND RECOMBINATION REACTIONS

Since almost all chemical-laser systems of interest operate at temperatures below 500°K, thermal dissociation of any of the molecules involved—with the possible exception of $F_2$—is negligible. On the other hand, the recombination processes are in most cases negligible not because of the temperature regime but because of the pressures. All atom–atom recombination rates are third-order and proceed with rate coefficients on the order of $10^{15\pm1}$ cm$^6$/mol$^2$-sec. Since recombination competes with fast bimolecular atom-transfer reactions that have rate coefficients on the order of $10^{11\pm2}$ cm$^3$/mol-sec, these reactions will in general not be important unless total pressures exceed 1 atm. Nevertheless, we include a cursory review here for the sake of completeness.

It should be noted, in preface, that there is a fundamental problem in relating experimentally measured recombination and dissociation rates to quantities appropriate in calculations involving chemical lasers. The difficulty stems from the fact that experimental studies are invariably conducted at conditions under which the internal energy level of the molecular species are populated according to a Boltzmann distribution; whereas, for a chemical-laser model, we are interested in rates into and out of individual V–R levels. In other words, although we know experimentally the recombination rate for $H + Cl + M \rightarrow HCl + M$, we do not know the state of the product HCl molecule. For computational purposes, we must make some assumption: Two extreme cases are (1) all vibrational levels are populated at the same rate; or (2) only the highest level (or levels) is populated directly, and lower levels are affected only by V–V and V–T(R) energy-transfer processes. If the individual levels are treated as separate chemical species and conventional thermochemical methods are used, assumption (1) implies that the dissociation rate coefficient increases as $v$ increases by the relation $k_{dis}^{v} = k_{dis}^{v-1} \times \exp[(E_v - E_{v-1})/RT]$, where $E_v$ represents the energy of the $v$th level above the zeroth vibrational level. Assumption (2) implies that no dissociation takes place out of lower levels. While the second is probably the more nearly correct picture, it imposes extra burdens on the calculation, since what used to be treated as a one-step process (dissociation of a ground-state molecule) now becomes a complex sequence—vibrational ladder climbing, followed by dissociation out of the top level—carefully chosen to give overall kinetic agreement with the simple phenomenological one-step process. This may require a major investment in terms of computation capability for handling a reaction of minor importance. Consequently, we suggest either adhering to alternative (1), or else suspending the normal thermodynamic relationship between forward and reverse rate coefficients and allowing combination to take place into all levels according to assumption (1), but dissociation only out of the $v = 0$ state. In our own laboratory, we follow (1), recognizing that physically it is the poorer model, but, practically speaking, that it is adequate for modeling purposes. In the comments that follow, we concern ourselves only with thermal dissociation and recombination under normal (i.e., nonlaser) circumstances, leaving to others to adapt the data according to whichever model of the cavity region best suits their needs.

## 1.1. Hydrogen and Deuterium Dissociation

Hydrogen dissociation and recombination rates have been widely studied since the first room-temperature measurements were made in 1929. The data have been summarized and reviewed critically by Baulch et al.[4] Shock-tube studies of the following three reactions have been made by Jacobs et al.[5] over a temperature range of 2900 to 4700°K:

$$H + H + Ar \rightleftharpoons H_2 + Ar \qquad (1a)$$

$$H + H + H_2 \rightleftharpoons H_2 + H_2 \qquad (1b)$$

$$H + H + H \rightleftharpoons H_2 + H \qquad (1c)$$

Their review of previous shock-tube data showed that the results of various experimenters disagreed by factors of about 3, 5, and 10, respectively, for Ar, H, and $H_2$ as chaperone gases. (The shock-tube experiments actually involved the measurement of hydrogen dissociation. However, it has been customary to report the results in terms of a recombination rate.) In all three cases, their results lie in the middle of the range of values. They reported $k_{1a} = 10^{18} T^{-1}$, $k_{1b} = 2.5 k_{1a}$, and $k_{1c} = 20 k_{1a}$. Their values for $k_{1a}$ and $k_{1b}$, extrapolated to room temperature, yield 3.3 and $8.2 \times 10^{15}$ cm$^6$/mol$^2$-sec, respectively. In four decades of measurements, the room-temperature value of $k_{1b}$ has ranged from $0.2 \times 10^{15}$ to $72 \times 10^{15}$. Most recently, Ham et al.[6] and Trainor et al.[7] have measured $k_{1a}$ and $k_{1b}$ in a flow system using an isothermal catalytic probe over the temperature range of 77 to 300°K and found temperature dependences of $T^{-0.81}$ and approximately $T^{-0.61}$, respectively, with values at 300°K of $3.3 \times 10^{15}$ and $2.9 \times 10^{15}$ cm$^6$/mol$^2$-sec. Their results, extrapolated to the temperature range of the experiments of Jacobs et al.,[5] agree with the results of the latter workers within 35%. Because the lower temperatures are of more interest in laser chemistry, we weight the results of Ham et al. heavily. The expressions $k_{1a} = 6.2 \times 10^{17} T^{-0.95}$ and $k_{1b} = 9.4 \times 10^{16} T^{-0.61}$ fit both high- and low-temperature data within about 10%. The numerous results are not tabulated here; further citations are given elsewhere.[4,5,7]

Of the several room-temperature measurements of $k_{1c}$, the most reliable seems to be that of Bennett and Blackmore.[8] They obtained an upper limit of $2.5 \times 10^{15}$ cm$^6$/mol$^2$-sec in a discharge flow system using ESR for atom detection. This suggests that $k_{1c}$ has a maximum value somewhere in the temperature range of 1000 to 3000°K, and is relatively unimportant at low temperatures. The expression $k_{1c} = 1.2 \times 10^{14} T^{1/2}$ is consistent with the upper limit of Bennett and Blackmore and agrees with the low-temperature end of the shock-tube experiments of Jacobs et al. Therefore, it is probably a reasonable expression in the temperature range of interest, although it is not useful above 3000°K. This expression is recommended in Table 20 and 22, subject to modification when more experimental data are available.

The recombination/dissociation of deuterium has been studied by three groups under shock-tube conditions. Jacobs et al.[9] who summarized the results of two earlier studies,[10,11] obtained for

$$D_2 + M \rightarrow 2D + M$$
$$k_2^{Ar} = 10^{18} \, T^{-1}, \quad k_2^D = 20 \, k_2^{Ar}, \quad \text{and} \quad k_2^{D_2} = 1.75 \, k_2^{Ar} \tag{2}$$

This value of $k_2^{Ar}$ is 1.4 times larger than that of Rink,[10] and approximately 1.3 times larger than that of Sutton.[11] Their $k_2^{D_2}$ is larger than the results of Rink and Sutton by factors of 1.75 and ~3, respectively; and $k_2^D$ is almost 2.9 times larger than Rink's value. Sutton measured a value for $k_2^D$ that agreed with that of Jacobs et al.[9] at about 3500°K but had a considerably larger temperature coefficient. Trainor et al.[7] obtained $k_2^{D_2} = 10^{17} \, T^{-0.67}$ from data at 77 and 298°K. This extrapolates to a value of $4.7 \times 10^{14}$ at 3000°K, which is 0.8 times the shock tube result of Jacobs et al.[9] and within their experimental error. Therefore, the value of $k_2^{D_2}$ of Trainor et al. seems reliable at both high and low temperatures. An early room temperature value of $k_2^D = 10^{15.88}$ was obtained by Amdur[12]; this is an order of magnitude smaller than the extrapolated results of Jacobs et al., if the $T^{-1}$ temperature dependence is assumed valid down to 300°K. However, Amdur's result for $H + H + H$ recombination is a factor of more than 2 smaller than the preferred recent result of Trainor et al.,[7] so possibly a remeasurement at room temperature would prove Amdur's result for the $D + D + D$ rate on the low side as well. Nevertheless, it does seem possible that $T^{-1}$ is too strong a temperature dependence for $k_2^D$ at lower temperatures than the shock-tube results, and since lower temperatures are of greater interest, the single room-temperature measurement should be taken into account. Therefore, we suggest a rate coefficient of $k_2^D = 3 \times 10^{17} \, T^{-1/2}$, which agrees with the shock-tube results of Jacobs et al. at the midpoint of their temperature range (4000°K) and is about three times larger than Amdur's room-temperature measurement. This evaluation is tentative, however, and clearly a better room-temperature measurement is needed. Other results for $k_2^{D_2}$ have been summarized by Baulch et al.[4]

## 1.2. Halogen Dissociation/Recombination

A thorough critical review of the dissociation/recombination of fluorine and chlorine was published by Lloyd[13] in 1971, and we have tried to minimize repetition of his work.

**Cl₂.**  Studies of the kinetics of the reaction

$$Cl_2 + M \rightleftarrows Cl + Cl + M \tag{3}$$

**Table 1.** $Cl_2 + Ar \underset{k_{-3}}{\overset{k_3}{\rightleftarrows}} 2Cl + Ar$

| Temperature, °K | % $Cl_2$ | $k_3 = AT^n \exp(-E/RT)$ | | | Reference |
|---|---|---|---|---|---|
| | | $\log A$ | $n$ | $E$ | |
| 1550–2650 | 4, 20, 25 | 15.13 | 0 | $49,060^a$ | 14 |
| 1710–3190 | 0.5 | 12.84 | 0 | $40,360^a$ | 15 |
| 1738–2583 | 2, 4 | 13.94 | 0 | 48,300 | 16 |
| | | 21.79 | −2.087 | 57,080 | |
| 1658–2370 | 5, 10 | 13.65 | 0 | 45,430 | 17 |
| 1738–2582 | 17 | $14.03^b$ | 0 | 48,300 | 18 |

$^a$ As computed by Lloyd.[13]
$^b$ Lloyd[13] computed $\log A = 14.01$; $E = 47,900$ from their data.

fall into two groups: high-temperature (1500–3200°K) $Cl_2$ dissociation and low-temperature (200–500°K) studies of Cl-atom recombination. Five shock-tube studies of $Cl_2$ dissociation in Ar have been published[14–18] (see Table 1); on the basis of his careful examination, Lloyd concluded that the work of Jacobs and Giedt[16] was the most reliable of the five and recommended their result of $k_3 = 10^{13.94} \exp(-48,500/RT)$ over the temperature range of 1500 to 3000°K for M = Ar. Because the apparent activation energy was considerably lower than the $Cl_2$ bond energy, Jacobs and Giedt proposed an alternative expression of $k_3 = 10^{21.79} T^{-2.087} \exp(-D/RT)$ where $D = 57,080$ cal/mol.

Five low-temperature recombination studies were reviewed by Lloyd (see Table 2): those of Chiltz et al.,[19] Linnett and Booth,[20] Bader and Ogryzol,[21] Hutton and Wright,[22,23] and Clyne and Stedman.[24] Since each of the first four studies involved procedures of questionable precision, Lloyd favored the results of Clyne and Stedman, suggesting a rate coefficient of $k_{-3} = 10^{14.34} \exp(1790/RT)$ over the temperature range of 200 to 500°K based on a least-squares treatment of their data. Two more studies have since been published: Hippler and Troe,[25] using a flash-photolytic technique, studied the recombination in the presence of He, Ne, Ar, $N_2$, $CO_2$, $CF_4$, $C_2F_6$, $SiF_4$, and $SF_6$ at 300°K; and Widman and DeGraff,[26] using a similar technique, covered the temperature range of approximately 200 to 373°K for M = He, Ne, and Ar, and 293 to 373°K for M = $N_2$, $SF_6$, $CF_4$, and $CO_2$. The only datum of Hippler and Troe directly comparable with the results of Clyne and Stedman is their Ar measurement, which gives a value of $k_{-3}^{Ar}$ that is 1.7 times that of the previous workers. Widman and DeGraff obtained an expression $k_{-3}^{Ar} = 10^{14.4} \exp(1800/RT)$, giving values slightly larger (by 10–30%) than those of Clyne and Stedman.

**Table 2.**    $2Cl + M \xrightarrow{k_{-3}} Cl_2 + M$

| M | $\log k$ (300°K), $cm^6/mol^2$-sec | Reference |
|---|---|---|
| $Cl_2$ | $15.2 \pm 0.3$ (502°K)[a] | 19 |
| $Cl_2$ | 14.8 (313°K) | 27 |
| $Cl_2$ | 14.5 | 20 |
| $Cl_2$ | 16.13 (313°K) | 21 |
| He | 15.18 (313°K | 21 |
| $Cl_2$ | 16.31 (293°K) | 23 |
| Ar | 15.63 (293°K) | 23 |
| Ar | 15.6 | 24 |
| $Cl_2$ | 16.3 | 24 |
| He | 15.15 | 25 |
| Ne | 15.23 | 25 |
| Ar | 15.88 | 25 |
| $N_2$ | 16.18 | 25 |
| $CO_2$ | 16.75 | 25 |
| $CF_4$ | 16.74 | 25 |
| $C_2F_6$ | 16.32 | 25 |
| $SiF_4$ | 16.76 | 25 |
| $SF_6$ | 16.32 | 25 |
| He | 15.36 | 26 |
| Ne | 15.58 | 26 |
| Ar | 15.73 | 26 |
| $N_2$ | 15.91 | 26 |
| $CO_2$ | 16.28 | 26 |
| $CF_4$ | 16.09 | 26 |
| $SF_6$ | 16.08 | 26 |

[a] As evaluated by Lloyd.[13]

The high- and low-temperature expressions are not consistent with one another, which suggests considerable curvature in the Arrhenius plot. A three-parameter expression that fits both the high- and low-temperature regimes well within experimental uncertainty is $k_{-3}^{Ar} = 10^{16.785} \, T^{-0.88} \exp(1440/RT)$. Combined with an expression for the equilibrium constant computed from JANAF data, $K_{3,-3} = 6.10 \exp(-57,246/RT)$, this gives a dissociation-rate coefficient of $k_{-3}^{Ar} = 10^{17.57} \, T^{-0.88} \exp(-55,810/RT)$.

Lloyd has also tabulated results for the relative efficiencies of $Cl_2$ and Ar as chaperones in dissociation/recombination. There is considerable scatter in the various determinations with $k_3^{Cl_2}/k_3^{Ar}$ varying between 3.6

and 5.8; a ratio of $5 \pm 1$ seems the best choice at present. Jacobs et al.[28] found their shock-tube data best fitted by assuming that $k_3^{Cl} = 10 k_3^{Ar}$, and although there is no a priori justification, we assume the same relative efficiency throughout the temperature range.

Relative efficiencies of other species have been recently measured by Hippler and Troe and by Widman and DeGraff, but not for the chaperones of greatest interest in the laser system, namely HCl, $H_2$, and H. If it is assumed that the primary recombination mechanism for Cl atoms is, as is believed to be the case for I atoms, the radical-molecule mechanism

$$Cl + M \rightarrow ClM^*$$

$$ClM^* + Cl \rightarrow Cl_2 + M$$

then some very qualitative statements can be made regarding the relative efficiencies of H and HCl as chaperones. For example, H atoms would be expected to be somewhat more efficient than Ar because the strong H–Cl bond would favor formation of $ClM^*$, where $M = H$. HCl might be expected to be even less efficient than Ar because of the existence of a potential barrier to formation of the Cl–H–Cl configuration in the potential energy surface.[29] Therefore, we tentatively assume H atoms to be three times as efficient as Ar, and HCl, 0.5 times as efficient. $H_2$ has been reported to be 25 to 60% more efficient than Ar in the case of I recombination[30]; we assume in the case of Cl atom recombination that $k_{H_2}/k_{Ar} = 1.5$ throughout the temperature range. Again, this assumption has no basis in direct evidence and awaits some experimental confirmation.

The results of Hippler and Troe and of Widman and DeGraff on the efficiencies of other chaperones are in considerable disagreement for those species that both groups studied. As there is no consistent trend in the differences between the two groups (e.g., for $M = He$, the Widman and DeGraff rate coefficient is 1.65 times larger than that of Hippler and Troe; whereas for $M = N_2$ it is 0.54 times as large), and inasmuch as there are no obvious experimental oversights in either report, it is difficult to select values of various $k$'s that are reliable to better than a factor of 2. Therefore, it does not seem appropriate at this time to recommend a set of values. The results of the two groups are shown in Table 2.

**F₂.** In recent years, several shock-tube determinations have been made of the rate of $F_2$ dissociation[31–37]

$$F_2 + M \rightleftarrows F + F + M \tag{4}$$

Although the numerical values of the rate coefficient have generally been

**Table 3.** Fluorine Dissociation Rate Coefficient $F_2 + M \xrightarrow{k_4} 2F + M$

$k_4 = A \exp(-E/RT)$

| M | Temperature, °K | log A | E, cal/mol | % $F_2$ | Reference |
|---|---|---|---|---|---|
| Ar | 1330–1580 | 12.85 | 29,800 | 5, 10, 20 | 31[a] |
| Ar | 900–1900 | 10.24 | 14,000 | 3, 6, 15 | 32[b] |
| Ar | 1300–1700 | 12.49 | 27,300 | 5 | 33[c] |
| Ne | 1650–2700 | 12.18 | 23,900 | 0.5, 1 | 34 |
| Ne | 1400–2000 | 13.30 | 35,000 | 0.5 | 35 |
| Ar | 1200–1500 | 13.57 | 31,700 | 10 | 36[d] |
| Ar | 1400–2600 | 13.55 | 34,700 | 10, 20 | 37 |
| $F_2$ | 1400–2600 | 13.99 | 34,800 | 10, 20 | 37 |
|  |  | 12.66 | 28,500 | (Review of data) | 13 |

[a] The tabulated results are for their six runs with 5% $F_2$. In a series of nine runs in 10% $F_2$ in Ar mixtures, Johnson and Britton obtained log $A = 11.54$, $E = 20,680$ cal/mol; in a series of five runs in 20% $F_2$ in Ar, they obtained log $A = 9.61$, $E = 11,025$ cal/mol. They preferred their 5% results as being more reliable.
[b] The results are based primarily on the 12 6% runs. The 3% runs yielded a considerably smaller value for $E$. The scatter in the 15% runs was too large to permit a reliable assignment of $E$.
[c] Results based on data of Refs. 33 and 31. In mixtures of 5% $F_2$, 20% Kr, 75% Ar, Seery and Britton obtained log $A = 13.15$, $E = 21,100$; in 10:20:70 mixtures, they obtained log $A = 11.57$, $E = 19,600$ cal/mol.
[d] Just and Rimpel also expressed their results in the form $k_4 = AT^n \exp(-D/RT)$, obtaining log $A = 11.54$, $n = -1.9$, $D = 36,700$ cal/mol.

in agreement, within the experimental uncertainties of the various determinations, the temperature dependences of the rate coefficient have been noticeably inconsistent. The various results are summarized in Table 3. The extent to which the $F_2$ bond energy exceeds the average experimental activation energy is disconcerting. This discrepancy becomes important when one tries to calculate the rate coefficient for $F + F$ recombination from $K_{4,-4}$ and $k_4$ and extrapolate the results to low temperatures. The problem is compounded by the uncertainty in the bond dissociation energy of fluorine.

Although a critical review of the bond strength of $F_2$ is outside the scope of this paper, it is evident that the possible uncertainty in such a fundamental datum leads to ambiguities in several thermochemical and kinetic values important to this study. In 1968 the National Bureau of Standards committee recommended 37.76 kcal/mol for $D^\circ_{298}(F_2)$; since 1965, the JANAF committee has recommended $37.72 \pm 0.8$ kcal/mol. (The dissociation energy at $0°K$, $D^\circ_0$, in either case is 1.0 kcal/mol smaller.) We use 38 kcal/mol; in view of the uncertainties, a third significant figure seems uncalled for.

With a value of 38 kcal/mol for $D^\circ_{298}(F_2)$, all of the shock-tube values for the activation energy of $F_2$ dissociation in an Ar bath lead to a strongly negative temperature dependence for F-atom recombination, which results in unreasonably large values near 300°K. Therefore, for calculations, we prefer the theoretical value calculated by Shui, Appleton, and Keck,[38] whose results agree numerically with the shock-tube data of Johnson and Britton[31] but lead to an activation energy of 35.1 kcal/mol and a pre-exponential factor of $10^{13.7}$. Until direct measurements near 300°K are available, we use $k_4^{Ar} = 10^{13.7} \exp(-35,100/RT)$. This implies $k_{-4}^{Ar}$ at 300°K is $10^{14.0}$. A preliminary report by Valence et al.[39] suggests $\log k_{-4}^{Ar} \leq 15.5$ at 300°K, which is consistent with the recommended value.

Recent work of Breshears and Bird[37] indicates that $k_4^{F_2}/k_4^{Ar}$ is approximately 2.7 in the temperature range of 1400 to 2600°K. Data on the rate of fluorine dissociation with $M = H$, $H_2$, or F are not available. By analogy with the findings in cases of other homonuclear diatomics, such as $Cl_2$, $Br_2$, $H_2$, $O_2$, and $N_2$, it is expected that F atoms will exhibit a large efficiency relative to Ar; we assume $k_4^F = 10k_4^{Ar}$. H is expected to show enhanced efficiency relative to Ar for reasons given above in the discussion of $Cl_2$ dissociation/recombination, and we assume $k_4^H = 3k_4^{Ar}$.

**$Br_2$ and $I_2$.** The dissociation/recombination of bromine and iodine received great attention in the days before the discovery and development of chemical lasers, and the literature on both of these topics is voluminous. Cohen et al.[40] have tabulated all the bromine-recombination data through 1971, which span the temperature range of 300 to 3000°K, with $M = Ar$, Br, $Br_2$, HBr, and $H_2$. Theoretical work is reviewed by Clarke and Burns.[41] Although iodine recombination received extensive attention in the early days of flash photolysis, interest in the problem has not abated. References to earlier work can be found in several recently published studies.[42-44]

## 1.3. Hydrogen-Halide Dissociation/Recombination

**HCl Dissociation/Recombination.** Five shock-tube studies of the dissociation of HCl have been published[45-49]; all report activation energies

**Table 4.** $HCl + Ar \underset{k_{-5}}{\overset{k_5}{\rightleftharpoons}} H + Cl + Ar$

| Temperature, °K | % HCl | $k_5(T) = AT^n \exp(-E/RT)$ | | | $k_5(4000)^a$ | $k_5(300)^b$ | $k_{-5}(300)^c$ | Reference |
| | | log $A$ | $n$ | $E$, cal/mol | | | | |
| --- | --- | --- | --- | --- | --- | --- | --- | --- |
| 3300–5400 | 1, 2 | 11.28 | 0.5 | 69,700 | $1.89 \times 10^9$ | $5.5 \times 10^{-39}$ | $2.34 \times 10^{36}$ | 45 |
| 2800–4600 | 1, 2 | 12.82 | 0 | 70,000 | $1.02 \times 10^9$ | $6.8 \times 10^{-39}$ | $2.9 \times 10^{36}$ | 46 |
| | | 21.83 | −2 | 102,200 | $1.1 \times 10^9$ | $2.8 \times 10^{-58}$ | $1.2 \times 10^{17}$ | 46 |
| 2900–4000 | 1, 2 | 13.62 | 0 | 81,000 | $1.57 \times 10^9$ | $4.0 \times 10^{-46}$ | $1.7 \times 10^{29}$ | 47 |
| 3500–7000 | 5, 10, 20 | 13.68 | 0 | 82,700 | $1.44 \times 10^9$ | $2.6 \times 10^{-47}$ | $1.1 \times 10^{28}$ | 49 |

[a] Calculated from analytic expression for $k_5$.
[b] Extrapolated from analytic expression for $k_5$.
[c] $k_{-5} = \dfrac{k_5}{K_{5,-5}}$, calculated.

20 to 30 kcal/mol smaller than the H–Cl bond dissociation energy. The four measurements of $k_5^{Ar}$ for the reaction

$$HCl + M \rightleftharpoons H + Cl + M \qquad (5)$$

are displayed in Table 4. Throughout the temperature range of the experiments, the disagreement among the four studies[45-47,49] with M = Ar is less than a factor of 2. Because of the low activation energies obtained, all four results, when extrapolated to room temperature and converted through the use of the equilibrium constant $K_{5,-5}$ to recombination rate coefficients, yield numbers that are too large by many orders of magnitude. Jacobs et al.,[46] in their study, reported $k_5$ in two forms, one using the best fit of 70 kcal/mol for the apparent activation energy, the other using the bond dissociation energy for the activation energy and a $T^{-2}$ pre-exponential temperature dependence. The latter form yields a room-temperature value of $k_{-5}$ of $1.2 \times 10^{17}$ cm$^6$/mol$^2$-sec, which is probably too large by a factor of 30 or so. In a follow-up study, Giedt and Jacobs[48] examined the dissociation rate in 100% HCl at temperature down to 1600°K. These data were well fitted by the form of $k_5$ with the 70 kcal/mol activation energy, but not by the other expression. In order to give values more nearly correct at temperatures near 300°K, we suggest using the second expression of Jacobs et al., but with the exponential energy increased from 102.2 to 104 kcal/mol. It might be noted that two recent results published since the analysis of Jacobs et al. prompt a re-examination of their conclusions. The first is that the current best values for the two exchange reaction-rate coefficients, H + Cl$_2$ and Cl + H$_2$, are nearly a factor of 2 larger than the values that Jacobs et al. used in their analysis (see the following discussion). The second is that Breshears and Bird,[49] in their studies of the decomposition of pure HCl, were able to measure $k_5^{HCl}$ and found it to be significantly larger than $k_5^{Ar}$.

(Giedt and Jacobs,[48] using 100% HCl, could detect no increase in the efficiency of HCl.) The effect of faster exchange rates would be to raise the value Jacobs et al. calculated for $k_5^{HCl}$ by about 10%; the effect of enhanced $k_5^{HCl}$ would lower it by about the same extent. Consequently, the effects tend to cancel, and we see no reason to alter their results over the temperature range of their shock-tube experiments. This leaves the small discrepancy at high temperatures between Jacobs et al.[46] and the other three sets of workers unresolved, but as that regime is outside the range of interest of all laser applications, we do not explore the matter further here.

Of the five sets of experiments on HCl dissociation, three[45–47] were confined to mixtures of 2% HCl or less in Ar; consequently, the relative efficiencies of HCl itself or H and Cl atoms could not be determined. Breshears and Bird[10] worked with mixtures of up to 100% HCl and did obtain a value for the HCl efficiency. Since their measurements were confined to the initial dissociation rate, they could draw no conclusions regarding atom efficiencies. They found HCl to be approximately three to eight times more efficient than Ar, with the difference in efficiencies increasing with temperature. However, the 100% HCl studies were confined to the temperature regime of 3000 to 4000°K, so there is considerable uncertainty in the relative efficiencies in the neighborhood of 300°K. Clearly, the Breshears and Bird value for $k_5^{HCl}$ cannot be extrapolated to low temperatures. Therefore, for simplicity, we sacrifice the detailed information on the high-temperature relative-rate coefficient and assume that $k_5^{HCl} = 5k_5^{Ar}$ throughout the temperature range. This leads to values for $k_5^{HCl}$ that are within 50% of those obtained by Breshears and Bird, but differ somewhat more from the conclusions of Giedt and Jacobs.

There are very few data on the relative efficiencies of atomic species in case of heteronuclear recombination. The radical-molecule model mentioned earlier would lead one to expect greater efficiencies by H and Cl atoms in facilitating H–Cl recombination. $Cl_2$ might also be more efficient than Ar because of the metastable intermediate $Cl_3$ formed. In the absence of experimental guidelines, we assume $k_5^H = k_5^{Cl} = 5k_5^{Ar}$ and $k_5^{Cl_2} = 3k_5^{Ar}$. Other species are assumed to have the same efficiency as Ar.

**HF and DF Dissociation.** The rate of HF dissociation

$$HF + M \rightarrow H + F + M \tag{6}$$

was first measured by Jacobs et al.[50] in the shock tube at temperatures of 3800 to 5300°K. They reported a rate coefficient for Ar as chaperone in two forms: $k_6^{Ar} = 10^{19.05} T^{-1} \exp(-D_0/RT)$ or $k_6^{Ar} = 10^{22.71} T^{-2} \exp(-D_0/RT)$ where $D_0$, the dissociation energy at 0°K from

$v = 0$, is currently taken as 135.1 kcal/mol. (At the time of their study, a slightly smaller value of 134.1 kcal/mol was accepted.) Subsequently, Blauer[51] performed similar shock-tube experiments over a temperature range of 3700 to 6100°K and obtained a rate coefficient smaller by about a factor of 2, with a pre-exponential factor of $10^{18.67} T^{-1}$. The two studies have been reviewed briefly by Armstrong and Holmes[52] and in detail by Brown,[53] who concluded there was no basis for preferring one result over the other. In this paper, we arbitrarily recommend the value reported by Jacobs et al. with the $T^{-1}$ temperature dependence in the pre-exponential; the experiment with $T^{-2}$ temperature dependence extrapolates to unreasonably large values at room temperature.

The dissociation energy of HF is subject to some uncertainty. Here, we take $D_0(\text{HF})$ for the ground state molecule to be 135.1 kcal/mol, calculated from heats of formation of $-65.1$, 18.4, and 51.6 kcal/mol for HF, F, and H, respectively. Thus

$$k_6^{\text{Ar}} = 1.1 \times 10^{19} \, T^{-1} \exp\left(\frac{135{,}100}{RT}\right)$$

There are no data available, either experimental or theoretical, on chaperones other than Ar. Following our reasoning in the case of HCl, we assume that $k_6^{\text{HF}} = k_6^{\text{H}} = k_6^{\text{F}} = 5k_6^{\text{Ar}}$ and that all other species have the same efficiency as Ar.

The dissociation of DF has not been studied. It seems reasonable to assume the same rate as for HF, except for a slightly increased activation energy corresponding to the increased bond dissociation energy of DF relative to HF, that is, $k_7^{\text{Ar}} = 10^{19.05} \, T^{-1} \exp\left(-D_0/RT\right)$, where Reaction (7) is

$$\text{DF} + \text{M} \rightarrow \text{D} + \text{F} + \text{M} \tag{7}$$

and $D_0$ is taken to be 136.7 kcal/mol. Following our arguments in the HF case, we assume that $k_7^{\text{DF}} = k_7^{\text{D}} = k_7^{\text{F}} = 5k_7^{\text{Ar}}$ and that other species have the same efficiency as Ar.

**HBr and HI Dissociation.** Although these reactions have been studied for half a century, and constituted some of the early classical-kinetic studies of Bodenstein and his co-workers, it was only a few years ago that the existence of a reaction path involving a termolecular process,

$$2\text{X} + \text{H}_2 \rightarrow 2\text{HX}$$

where X is either Br or I, was demonstrated.[54,55] In the case of iodine, this path dominated at low temperatures in the thermal decomposition of HI. In the case of Br, the termolecular path is important only in photodecomposition. Presumably, under appropriate conditions, the dissociation of the halide HX proceeds through the reverse of the foregoing reaction. Consideration of the relative energetics of the F and Cl systems indicates that, although the analogous process may be occurring, it can never compete measurably with the usual chain reaction. At high temperatures, of course, the dissociation of HBr or HI proceeds by the simple one-step process as in the HF and HCl cases. Armstrong and Holmes[52] have reviewed briefly the literature to 1968; since then, three shock-tube studies on HBr dissociation have appeared,[40,56,57] as well as a major paper on room-temperature $H_2$–$Br_2$ photolysis.[55]

## 2. PUMPING REACTIONS

Although the recombination reactions discussed in Section I may be producing HX molecules in upper vibrational states, the reactions that are depended upon for the production of vibrationally excited HX are the atom transfer, or metathesis, reactions of the form

$$X + H_2 \rightarrow HX + H$$

and

$$H + X_2 \rightarrow HX + X$$

where X is any of the halogen atoms and H can be replaced by D. The maximum vibrational level of HX obtainable in one of the metathesis reactions depends on the exothermicity of the reaction, the internal energy of the reactants, and the relative translational energy of the reactants. For all practical purposes, especially if reactants are in equilibrium conditions, it can be assumed that

$$E_{v_{max}} \leq -\Delta H + E_{act}$$

where $\Delta H$ is the enthalpy of reaction, $E_{act}$ is the activation energy, and $E_{v_{max}}$ is the energy above the vibrational ground state of the maximum level attainable. These quantities are tabulated in Table 5, using the vibrational energies of Table 6.

The HF system is unique in that both atom transfer reactions are exothermic, giving rise to lasers that operate only on the "cold" (i.e., $F + H_2$) reaction, only on the "hot" reaction ($H + F_2$), or on the chain reaction (both). For all the other halogens, the analogous cold reaction is endothermic and can form hydrogen halide in the ground state only. This statement presumes that the reactant species are in thermal equilibrium; vibrationally excited $H_2$ reacting with Cl, Br, or I is sufficiently energetic to produce vibrationally excited HX, and one day a way may be found to

**Table 5.** Heats of Reaction, Activation Energies, and Maximum Vibrational Levels Energetically Possible for Reactions of Halogens with Hydrogen (Deuterium)[a]

| X, H(D) | X + H$_2$(D$_2$) | | | H(D) + X$_2$ | | |
|---------|-------------------|-------------------|------------|----------------|-------------------|------------|
|         | $-\Delta H$,[a] cal/mol | $E_{act}$, cal/mol | $v_{max}$ | $-\Delta H$,[a] cal/mol | $E_{act}$, cal/mol | $v_{max}$ |
| F, H    | 31,700 | 1,600 | 3 | 97,900 | 2,400 | 11 |
| F, D    | 30,600 | 1,200 | 4 | 99,300 | ~2,400 | 15 |
| Cl, H   | -1,097 | 5,260 | 0 | 45,240 | 2,520 | 6 |
| Cl, D   | -1,770 | 5,480[b] | 0 | 46,390 | ~2,500 | ~9 |
| Br, H   | -16,650 | 18,400 | 0 | 41,460 | 1,820 | 6 |
| I, H    | -32,860 | 34,100[c] | 0 | 35,190 | 0[b] | 6 |

[a] For all species in lowest vibrational and rotational states. Activation energy data from text, except as noted.
[b] Ref. 59.
[c] Ref. 58.

take advantage of this in designing a better laser system. Nevertheless, the reverse reaction between HX($v$) and H must be considered in a laser system because it can represent an important loss mechanism for the excited levels. The rate coefficient for HX(0) + H can be calculated from the equilibrium constant and the experimentally determined rate for the forward reaction (i.e., X + H$_2$). In the event that neither measurements nor theoretical values are available for HX($v$ > 0) + H, it would seem reasonable to assume that the pre-exponential factors for HX($v$) + H are the same as for HX(0) + H. The activation energies may be as small as zero and will probably be no larger than the activation energy for the

**Table 6.** Vibrational Energy Levels

| Vibrational level $v$ | $E_v - E_0$, cal/mol | | | | | | | | |
|---------|------|------|------|------|------|------|------|------|------|
|         | HF | DF | HCl | DCl | HBr | DBr | HI | H$_2$ | D$_2$ |
| 1  | 11,327 | 8,310 | 8,250 | 5,980 | 7,310 | 5,260 | 6,370 | 11,897 | 8,563 |
| 2  | 22,161 | 16,360 | 16.210 | 11,800 | 14,370 | 10,390 | 12,520 | 23,121 | 16,800 |
| 3  | 32,517 | 24,150 | 23,870 | 17,480 | 21,170 | 15,390 | 18,440 | 33,687 | 24,748 |
| 4  | 42,406 | 31,690 | 31,240 | 23,000 | 27,710 | 20,260 | 24,120 | 43,607 | 32,449 |
| 5  | 51,839 | 38,980 | 38,330 | 28,370 | 33,990 | 25,000 | 29,560 | 52,900 | 39,960 |
| 6  | 60,824 | 46,030 | 45,140 | 33,590 | 40,010 | 29,600 | 34,760 | 61,570 | 47,320 |
| 7  | 69,369 | 52,850 | 51,670 | 38,660 | 45,770 | 34,080 | 39,710 | 69,640 | 54,620 |
| 8  | 77,477 | 59,420 | 57,930 | 43,580 | 51,280 | 38,430 | 44,420 | 77,120 | 61,920 |
| 9  | 85,149 | 65,770 | 63,920 | 48,350 | 56,520 | 42,640 | 48,880 | 84,020 | 69,300 |
| 10 | 92,878 |       | 69,640 | 52,970 | 51,510 | 46,720 | 53,080 | 90,360 | 76,840 |

reaction between $HX(0) + H$. In this discussion we are fortunate in that recent theoretical work provides guidelines to the rate coefficients, but in general there is no experimental corroboration.

The key factor in the use of metathesis reactions as the basis of a chemical laser is that such reactions, if exothermic, deposit a significant fraction of the liberated energy in the vibrational mode of the newly formed chemical bond. Furthermore, and more importantly, the relative rates of formation of various vibrational levels of the product molecule often increase with $v$, so that population inversions are obtained.

The general problem of the distribution of the energy of reaction among the vibrational levels of product molecules for reactions of the type $A + BC \rightarrow AB + C$ has been the subject of an extensive series of studies by Polanyi and Wong.[60] Their conclusions relevant to this work can be summarized as follows: For an exothermic reaction, translational energy in the reactants is more effective than vibrational energy in promoting reaction. Furthermore, the excess energy of reaction is divided between translational energy of products and vibrational energy of the newly formed bond, with an average of 60%, and rarely less than 40% or more than 80%, going into the new bond.

A very crude approximation to the relative pumping rates into individual vibrational levels can be obtained by assuming a triangular distribution function that peaks at that vibrational level whose energy corresponds to two-thirds of the exothermicity. The highest level populated is the highest level energetically permitted; the lowest level populated is the one whose energy corresponds to one-third of the exothermicity.[61]

## 2.1. Pumping Reactions in the $H_2$–$Cl_2$ System

$Cl + H_2 \xrightarrow{8} HCl + H$.   The early studies of this reaction[62–64] were reviewed by Fettis and Knox,[59] who showed that the three sets of results fell on "an exceptionally good Arrhenius plot" yielding a rate coefficient of $k_8 = 10^{13.92} \exp(-5480/RT)$. The experimental data spanned the temperature range of 273 to 1071°K. Additional evidence was cited to support the argument that the activation energy must be no lower than the 5480 cal/mol obtained. The results of three more recent studies[65–67] differ slightly from the Fettis and Knox derived value.

Westenberg and de Haas,[65] using ESR to measure Cl atom concentrations, studied the reaction in a flow system over the temperature range of 261 to 456°K in both forward and reverse directions; the rate coefficients obtained were, respectively, $k_8 = 10^{13.08} \exp(-4300/RT)$ and $k_{-8} = 10^{13.36} \exp(-3500/RT)$. The ratio of $k_8/k_{-8}$ was shown to be a factor of 2 to 3 smaller than the true equilibrium constant, a phenomenon that the authors attributed to perturbation by fast reaction of the distribution of reactants over internal energy states. The value of $k_8$ obtained by these workers is approximately 40% larger than the Fettis and Knox expressions at the low end of their temperature range and a factor of about 2 smaller at the high end.

Clyne and Stedman[66] measured the rate of the back reaction, the removal of H atoms by HCl, in a discharge flow system between 195 and 373°K and found the rate coefficient in their temperature range to be well fitted by the expression $k_{-8} = 3.5 \times 10^{11} \, T^{1/2} \exp(-2900/RT)$. When combined with the equilibrium constant calculated from JANAF data, $K_{8,-8} = k_8/k_{-8} = 5.87 \, T^{-0.163} \exp(-1200/RT)$, this yields a forward-rate coefficient of $k_8 = 2.05 \times 10^{12} \, T^{0.337} \exp(-4100/RT)$. The value they obtained for the $k_{-8}$ is about 1.6 times that of Westenberg and de Haas; their $k_8$ disagrees with the Fettis and Knox expression for $k_8$ at low temperatures by a factor of more than 4.

Benson et al.[67] studied the reaction in the forward direction at 478–610°K using ICl as a thermal source of Cl atoms, thus permitting suppression of any possible Cl chain reactions by eliminating the presence of $Cl_2$ and by using $I_2$ molecules as radical scavengers. Their results indicate a $k_8$ about 30% smaller than the Fettis and Knox expression in the temperature range of their measurements. Benson and co-workers reanalyzed the earlier data of Ashmore and Chanmugam and of Steiner and Rideal using modern thermochemical values where appropriate, and found the revised values to agree quite closely with their own determination.

If we accept the Benson et al. result of $k_8 = 4.8 \times 10^{13} \exp(-5260/RT)$ as the best determination for the forward rate, it is apparent that the rate calculated from the measurements on the backward rate by Westenberg and de Haas and by Clyne and Stedman are increasingly too large at low temperatures. Several theoretical discussions[68–70] have been published, but none appears to have reconciled the discrepancy. Recently, Galante and Gislason[71] have re-examined the problem and concluded that both flow-tube measurements of the backward reaction were complicated by the occurrence of fast H + Cl recombination on the walls; since H atoms were measured in both cases, this would give a faster apparent reaction rate. If the wall recombination is very fast compared to the reaction being measured, namely that between H + HCl, then the apparent rate coefficient for the latter process would be twice its true value. This, Galante and Gislason felt, could account reasonably well for the experimental discrepancies. This seems to make sense, especially when one considers the temperature dependence of the discrepancy between the forward and backward measurements. If wall recombination is producing an effect, it should be most noticeable at low temperatures, when the rate of H+HCl is slower and the wall recombination is probably faster. At high temperatures, the opposite is true of both reactions, so the discrepancy should be less, as indeed it is. Thus, at this time, it seems best to assume that the normal relationship exists between $k_8$ and $k_{-8}$ and to use the value derived by Benson et al.[67] for the former.

**H+Cl₂.** Until recently, there were no direct measurements of the rate coefficient of this reaction. All evaluations were based on studies of the ratio of rate coefficients for the pair of competitive H-atom reactions

$$H+HCl \rightarrow H_2+Cl \qquad (-8)$$

and

$$H+Cl_2 \rightarrow HCl+Cl \qquad (9)$$

This ratio had been determined experimentally by Klein and Wolfsberg[72] and was then calculated theoretically by Wilkins.[73] The former workers found that $k_{-8}/k_9$ was given by $0.143 \exp(-1540/RT)$ in the temperature range of 273 to 335°K. Wilkin's calculations were in substantial agreement with this result, differing only in that a smaller temperature dependence (slightly under 1 kcal/mol) was obtained. A second competitive study of the same reaction system by Davidow et al.[74] produced results for $k_{-8}/k_9$ that were slightly smaller by about 10 to 20% than those of Klein and Wolfsberg in their temperature range, but the measurement at 196°K suggests a sharp increase in the difference between the two activation energies, $E_{-8}$ and $E_9$. However, the system used by Davidow et al. was rather complex, and, until more direct confirmation for the low-temperature behavior is obtained, we prefer to accept the results of Klein and Wolfsberg as valid even when extrapolated outside the temperature range of their measurements.

The first direct measurement of $k_9$ was reported in 1969 by Albright et al.[75] The work was performed in a flow system, with H atoms formed in a high-frequency discharge through $H_2$, all reactants and products being monitored by a mass spectrometer. Special care was taken to avoid wall reactions. These workers reported $k_9$ to be $3.7 \times 10^{14} \exp(-1800/RT)$ in the temperature range of 298 to 565°K, suggesting a steric factor of nearly unity. In order to compare this result with the measurements of Klein and Wolfsberg, it is necessary to assume some value for $k_8$. For this purpose, we use the value of $k_8 = 4.8 \times 10^{13} \exp(-5260/RT)$ obtained by Benson et al.[67] and the equilibrium constant $K_{8,-8} = 5.87 \, T^{-0.163} \exp(-1200/RT)$ calculated from JANAF data. These values, combined with the ratio for $k_{-8}/k_9$ obtained by Klein and Wolfsberg, give $k_9 = 5.7 \times 10^{13} \, T^{0.163} \exp(-2520/RT)$. This yields values considerably smaller than those of Albright et al. (by a factor of almost 9 at 300°K and almost 5 at 500°K).

Two other experimental values can be compared with the foregoing results. Stedman et al.,[76] using a coupled discharge-flow system and a time-of-flight mass spectrometer, measured the rates of both H and D atoms with $Cl_2$ at 300°K, and for the former found a rate coefficient of $2.1 \times 10^{13}$ cm³/mol-sec, which agrees with the Albright value. Stedman et al. obtained a value for the $D + Cl_2$ rate coefficient almost three times smaller than their $H + Cl_2$ value.

Jacobs et al.[28] studied the effect of the reverse reaction on the dissociation of HCl in the temperature range of 3500–5200°K by adding large quantities of $Cl_2$ to an HCl–Ar mix in the shock tube. They found they could match the experimental results with computer profiles by assuming $k_9 = 10^{13.85} \, T^{0.163} \exp(-2480/RT)$. This agrees fairly well with the extrapolated rate coefficient obtained by Klein and Wolfsberg, using the value obtained by Benson et al. for $k_8$. The results of Albright et al. extrapolated to this high-temperature regime yield a value for $k$ about four times larger than the shock-tube values. However, it is probably unfair to judge the merits of the low-temperature results on the basis of the agreement or lack of agreement with shock-tube results.

Thus it appears that there are discrepancies here. The possible implications are: (1) The results of Albright and Stedman are too large by a considerable margin; (2) the ratios of $k_{-8}/k_9$ obtained by Klein and Wolfsberg, by Davidow et al., and by Wilkins are all too small by a corresponding factor; (3) the value of $k_{-8}$ used to derive $k_9$ from the relative-rate ratio is too small, If, instead of the results of Benson et al. for $k_8$, we had used the direct measurements of $k_{-8}$ by Westenberg and de Haas,[65] the discrepancy between the two sets of results would be decreased to within a factor of 2 to 2.5. However, we have already indicated reasons for regarding the Westenberg and de Haas measurements as too large by a factor of about 2.

We, therefore, at this time favor the results obtained from the Klein and Wolfsberg experiments coupled with the expression of Benson et al., but we recognize that this reaction merits further examination.

*Production of Vibrationally Excited HCl.* The vibrational distribution of product HCl in the $H + Cl_2$ reaction has been studied extensively by Polanyi and co-workers over a period of nearly a decade.[77-82] The basic technique remains the same: H atoms and $Cl_2$ molecules are mixed in a flow tube and the relative intensities of the various HCl emission lines are measured. These line intensities are then related to populations by means of Einstein coefficients. Two techniques can be applied. In the "arrested-relaxation" method, emission measurements are made before the vibrationally excited species have a chance to relax. In the "measured-relaxation" technique, the populations are extrapolated to zero-flow distance to correct for population changes caused by collisional (and to a lesser extent, radiative) processes. This is not a trivial problem and considerable effort has been expended by Polanyi and his colleagues to minimize the effects of very rapid vibrational energy transfer processes. Five separate sets of distribution numbers (i.e., relative rates into specific vibrational levels) have been published. Because these experiments give only relative distributions and no value for the total rate coefficient, and because the ground vibrational level is not visible in emission, there is no comparable result for the formation of HCl(0) in the reaction. Absorption measurements,[82] however, indicate that very little if any HCl(0) is produced directly, so it is assumed with some confidence that only HCl($v$) where $v = 1$ through 6 can be formed. HCl(6) is the highest possible level obtainable in the direct reaction. Any higher vibrational levels observed must be due to $V - V$ energy-transfer reaction.

The various sets of relative distribution numbers are shown in Table 7. In each case they have been normalized so that the rate coefficients sum to unity. Presumably, the latest set of values of Polanyi obtained by the arrested-relaxation method, is the most reliable, and we prefer those results here. Renormalized to sum to unity, these relative distribution numbers are $f_1:f_2:f_3:f_4:f_5:f_6 = 0.12:0.42:0.39:0.04:0.02:0.002$ ($f_v$ is the fraction of product molecule formed in the $v$th vibrational level). These numbers should, strictly speaking, be recomputed using the Einstein coefficients of Herbelin and Emanuel.[84] However, the present experimental uncertainties do not warrant the correction.

**Table 7.** Relative Pumping Rates for $H + Cl_2 \xrightarrow{k_9(v)} HCl(v) + Cl$; $k_9(v) = f(v) \cdot k_9(\text{Total})$

| | | | $f(v)$ | | | | |
|---|---|---|---|---|---|---|---|
| $v = 0$ | 1 | 2 | 3 | 4 | 5 | 6 | Reference |
| 0.76 | 0.11 | 0.076 | 0.038 | 0.011 | 0.0015 | 0.00015 | 79 |
| — | 0.13 | 0.48 | 0.36 | <0.02 | — | — | 80[a] |
| — | 0.13 | 0.26 | 0.52 | 0.10 | 0.01 | 0.001 | 81[a] |
| — | 0.14 | 0.28 | 0.47 | 0.09 | 0.014 | 0.0014 | 82 |
| | 0.12 | 0.42 | 0.39 | 0.04 | 0.02 | 0.002 | 83 |

[a] Method of measured relaxation.

## 2.2.  Pumping Reactions in the $H_2(D_2)$–$F_2$ Systems

**$F + H_2(D_2)$.**  The early literature offers very few experimental results on reactions of F atoms. Most of the early entries for Reaction 10 in Table 8

$$F + H_2 \rightarrow HF + H \tag{10}$$

and

$$F + D_2 \rightarrow DF + D \tag{11}$$

are either estimates or theoretical calculations. An examination of the table reveals that the results cluster about two values for the activation energy: 2.5 and 1.6 kcal/mol. The larger value gives a room-temperature rate coefficient of about $10^{12}$ cm$^3$/mol-sec; the smaller gives a value about 10 times larger. The best study leading to the low activation energy is that of Homann et al.,[96] who produced F atoms in a flow system by the reaction $N + NF_2 \rightarrow N_2 + 2F$ and monitored the products mass spectrometrically. The most careful determination giving a high activation energy is that of Foon and Reid,[102] who studied the competition of several hydrocarbons and $H_2$ for F atoms by collecting products and unspent reagents and analyzing them by gas chromatography. It is difficult to choose between the two experiments. The latter experiment suffers from the possible complication of chain reaction caused by the presence of both F and $F_2$. Also, since no absolute rates are measured, an assumption about the activation energy for F-atom reaction with the standard-reference hydrocarbon $C_2H_6$ is required. The former experiment is subject to interference from reactions of unreacted N or $NF_2$ and uncertainties in the reliability of the ClNO titration for measuring F-atom concentrations.[108] Most laser calculations that have been made in the HF laser system to date have used the smaller activation energy, and agreement with experimental data has been reasonable, but the effect of increasing the activation energy $E_{10}$ is complex: Lowering the value of $k_{10}$ by about an order of magnitude would decrease the concentration of H atoms proportionately, making $[H] = [F]$. This would make F atoms relatively more, and H atoms less, important as deactivators. Nevertheless, this important rate coefficient must be regarded at present as subject to an uncertainty of an order of magnitude, and some better evaluation, preferably by a technique that is clearly free of chain-reaction complications, is called for.

Three experimental studies designed to determine the relative rates of F atoms with $H_2$ and $D_2$, that is, $k_{10}/k_{11}$ have been published.[97,106,109] The three agree at or near room temperature in finding $k_{10}/k_{11} = 1.8$ to 2.0, although theoretical studies[100,110,111] find values of 1.65 and 1.59.

**Table 8.** Rate Coefficient for $F + H_2 \xrightarrow{k_{10}} HF + H$

| Temperature, °K | $k_4 = A \exp(-E/RT)$ A, cm³/mol-sec | E, kcal/mol | log $k_4$, 300°K | Technique | Reference |
|---|---|---|---|---|---|
| — | $3 \times 10^{13}$ | 8.0 | >7.55 | Estimate | 85 |
| — | $9.3 \times 10^{13}$ | 1.71 | 12.72 | Relative to F+CH₄ | 86[a] |
| — | $2.5 \times 10^{10}\,T$ | 8.0 | 6.9 | Estimate | 87 |
| — | $5 \times 10^{12}$ | 5.7 | 8.55 | Estimate | 88 |
| 3800–5300 | $6.8 \times 10^{13}\,T^{-0.3}$ | 3.09 | 10.84 | Shock-tube, computer fit | 50 |
| 298–2500 | $7.8 \times 10^{11}\,T^{0.69}$ | 2.5 | 11.78 | BEBO theory | 89, 90 |
| 395–435 | $1 \times 10^{12}$ | ~4.0 | ~9.1 | Thermal reaction of H₂, F₂, and O₂ | 91[b] |
| — | $7.8 \times 10^{11}\,T^{0.7}$ | 3.0 | 11.43 | Data evaluation | 92 |
| 298–2500 | $1 \times 10^{12}\,T^{0.67}$ | 2.6 | 11.76 | BEBO theory | 93 |
| 293 | — | — | 13.3 | Flow and mass spectroscopy | 94, 95, 75[b] |
| 300–400 | $1.6 \times 10^{14}$ | 1.6 | 13.0 | Flow and titration | 96 |
| 173–293 | $8.3 \times 10^{13}$ ($1.6 \times 10^{14}$) | 2.29 (2.97) | 12.19 (12.0) | Relative to F+D₂, F+CH₄ (using results of Ref. 102) | 97[c] |
| 250–500 | $1.3 \times 10^{14}$ | (1.6) | 12.96 | Absolute rate theory | 98, 99[d] |
| 300–1000 | $1.3 \times 10^{14}$ | 2.34 | 12.4 | Classical trajectory calculation | 100 |
| 300 | — | — | 12.6 | H₂, H, F₂ fast flow and EPR | 101 |
| 253–348 | $2.9 \times 10^{14}$ | 2.47 | 12.63 | Relative to F plus various RH | 102[e] |
| 300 | — | — | 13.58 | Flash photolysis and chemical laser | 105 |
| ~300 | — | — | 12.73 (12.56) | Relative to F+CH₄, F+C₂F₂ (using results of Ref. 102) | 106[f] |
| ~298 | — | — | 13.0 | | 107 |

[a] These results were quoted by Fettis and Knox from a personal communication from Mercer and Pritchard. The F+CH₄ rate was assumed. Later, Fettis and Knox[59] revised the value of the A factor to log A = 14.07.

[b] In their experiments, Levy and Copeland obtained an approximate value for the ratio of rate coefficients for F+H₂ and F+O₂+M. Assuming the rate of the latter reaction to be the same as that of H+O₂+M, they obtained a value for $k_{10}$ at 405°K. The value of the A factor was assumed to be $10^{12}$, thus enabling them to calculate an activation energy of 7 kcal/mol. However, the value of the rate coefficient for the H+O₂+M reaction that they assumed is too small by a factor of approximately 50. Therefore, their assumed preexponential factor for $k_{10}$ of $10^{12}$ would imply an activation energy of approximately 4 kcal/mol.

[c] The results of Ref. 97 have been recomputed, ignoring the data at 77°K, which introduce curvature in the Arrhenius plot. The value in parentheses is obtained by using 1850 cal/mol for the activation energy of F+CH₄ from Ref. 102, rather than 1210 cal/mol of Refs. 59 and 86.

[d] The experimental value for E is used in estimating the overlap integral needed in making the calculations. A slightly different temperature dependence was obtained when calculations as high as 6000°K were included in the data fit. Reference 99 also gives the results of classical trajectory calculations.

[e] The A factor of Ref. 102 has been increased by a factor of 6 to be consistent with Ref. 59, since Foon and Reid took $A_{F+C_2H_6}$ to be $1 \times 10^{13}$ rather than $6 \times 10^{13}$. However, it has been recently suggested[103,104] that the latter value is too large by a factor of 2 to 3. This would mean that the A factors of Refs. 86, 97, and 102 should all be lowered by this factor.

[f] Williams and Rowland determined atom-transfer rates from H₂ and CH₄ relative to C₂H₂. We use their ratio of H₂:CH₄ transfer combined with the previously reported F+CH₄ rate of Ref. 59. (The value in parentheses uses the F+CH₄ rate of Ref. 102.)

However, the temperature dependence of the ratio is subject to disagreement: We compute, from the data of Kapralova et al.,[97] $E_{10}-E_{11}=$ 146 cal/mol between 293 and 173°K, and almost no temperature dependence between 173 and 77°K; while Persky[109] found $E_{10}-E_{11}=$ $-370$ cal/mol between 163 and 417°K. As the technique of Kapralova et al. was rather indirect, we prefer the more recent results of Persky, who found $k_{10}/k_{11}=1.04\exp(370/RT)$. Combining this with the value of Homann et al. for $k_{10}$ of $10^{14.2}\exp(-1600/RT)$ gives $k_{11}=$ $10^{14.2}\exp(-1970/RT)$. This compares fairly well with two theoretical calculations: Wilkins[112] obtained approximately $10^{13.5}\exp(-1200/RT)$ for $k_{11}$, and Jaffe and Anderson[100] calculated $10^{13.8}\exp(-2160/RT)$.

Three laboratories have studied the product distribution of Reaction (10). Pimentel's group[113–115] has examined the reaction by measuring gains of appropriate spectral lines in a chemical laser. Polanyi and co-workers[116–118] and Jonathan and co-workers[119,120] have used a discharge flow system and made measurements in a nonlasing medium. The various results are displayed in Table 9. It can be seen that the three groups have converged on the value of $k_{10}(v=1)/k_{10}(v=2)=0.3$. The two groups that have measured $k_{10}(v=3)$ and $k_{10}(v=2)$ are now in substantial agreement there as well, with the probable value for $k_{10}(v=3)/k_{10}(v=2)$ of 0.48 to 0.5. There has been one recent report on the experimentally measured temperature dependence of the relative pumping rates. Coombe and

**Table 9.** Relative Pumping Rates for $F+H_2 \xrightarrow{k_{10}(v)} HF(v)+H$

| | $k_{10}(v)$ at 300°K | | | | |
|---|---|---|---|---|---|
| $v=0$ | $v=1$ | $v=2$ | $v=3$ | Technique | Reference |
| — | 1.5 | 1 | — | Laser experiment | 113 |
| — | 0.18 | 1 | $<1.33$ | Laser experiment ($T=539$°K) | 114 |
| — | ~0.3 | 1 | — | Laser experiment | 115 |
| — | $\lesssim 0.29$ | 1 | $\gtrsim 0.47$ | Discharge flow | 116 |
| — | 0.31 | 1 | 0.48 | Discharge flow | 117 |
| — | 0.31 | 1 | 0.47 | Discharge flow | 118 |
| — | 0.29 | 1 | 0.76 | Discharge flow | 119 |
| — | 0.30 | 1 | 0.5 | Discharge flow | 120 |
| ~ | 0.15 | 1 | 0.49 | Semiempirical calculation/ Monte Carlo | 99 |
| 0 | 0.53 | 1 | 0.18 | Semiclassical calculation/ Monte Carlo | 100 |
| 0 | 1.28 | 1 | 0.0023 | Variational theory/ Monte Carlo | 121 |

Pimentel,[122] using laser techniques, found $k_{10}(v=2)/k_{10}(v=1) = 2.14 \exp(254/RT)$ and $k_{10}(v=3)/k_{10}(v=2) = 0.39 \exp(117/RT)$.

Because HF concentrations are determined by emission measurements, the formation rate of $HF(v=0)$ has not been ascertained in any of the experiments. In this connection, recent Monte Carlo computer calculations by Wilkins[99] are of interest. His results suggested that $k_{10}(v=0) = 0$ and that the activation energies for $v = 1$, 2, and 3 are, respectively, 1500, 1230, and 1250 cal/mol. This implies a difference in activation energies between $k_{10}(2)$ and $k_{10}(1)$ of 270 cal/mol, in excellent agreement with experiment.[122] As the differences in activation energies are rather slight, we prefer at this time to assume, for computational convenience, the same temperature dependence for all of the $k_{10}$'s, so that each $k_{10}(v)$ can be expressed as $f(v)k_{10\,\text{total}}$, where $f$ is temperature independent. We accept Wilkins' result that negligible pumping in the zeroth vibrational level takes place.

Very recently, Coombe and Pimentel[123] have reported the effect of $H_2$ rotational energy on the vibrational energy distribution in Reaction (10). They found a greater tendency to inversion, especially on the $2 \to 1$ transition when the $H_2$ was para-enriched (85% para).

Anderson[124] has published in brief the results of a classical-trajectory study of the reaction between H and $HF(v)$ for $v = 0$ through 5, showing the threshold translational energy required for reaction. Wilkins,[125] with a Monte Carlo computer technique, has calculated the rates for reactions of the form $H + HF(v) \to F + H_2(v')$, for $v = 3$, 4, 5, and 6 and $v' = 0$, 1, and 2. His results suggest that all of these reactions proceed with activation energies of approximately 500 to 600 cal/mol and that at high temperatures total probability for reaction of a given $v$ into all possible $v'$ is about 0.1 to 0.2; the rate coefficients are tabulated in Table 10. His results can be simplified and summarized approximately (within 20–30%) by the following expressions for various $k_{-10}(v, v')$:

$$k_{-10}(6, v') = f_6(v') \times 1.9 \times 10^{13} \exp\left(\frac{-580}{RT}\right)$$

$$f_6(0) = 0.22, \quad f_6(1) = 0.22, \quad f_6(2) = 0.56$$

$$k_{-10}(5, v') = f_5(v') \times 1.1 \times 10^{13} \exp\left(\frac{-510}{RT}\right)$$

$$f_5(0) = 0.36, \quad f_5(1) = 0.64, \quad f_5(2) = 0$$

$$k_{-10}(4, v') = f_4(v') \times 7.4 \times 10^{12} \exp\left(\frac{-460}{RT}\right)$$

$$f_4(0) = 0.5, \quad f_4(1) = 0.5, \quad f_4(2) = 0$$

$$k_{-10}(3, v') = f_3(v') \times 1.62 \times 10^{13} \, T^{-0.01} \exp\left(\frac{-835}{RT}\right)$$

**Table 10.** Rate Coefficients for H+

$$HF(v) \xrightarrow{k_{-10}(v,v')} H_2(v') + F^a$$

| | | | $k = AT^n e^{-E/RT}$ | |
|---|---|---|---|---|
| $v$ | $v'$ | $\log A$ | $n$ | $E$, cal/mol |
| 6 | 0 | 12.66 | −0.03 | 570 |
| | 1 | 12.41 | 0.085 | 540 |
| | 2 | 13.02 | 0.01 | 565 |
| 5 | 0 | 12.54 | 0.015 | 510 |
| | 1 | 12.92 | −0.005 | 575 |
| 4 | 0 | 12.82 | −0.44 | 570 |
| | 1 | 12.10 | 0.13 | 380 |
| 3 | 0 | 13.21 | −0.01 | 835 |

$^a$ From Wilkins.[125]

and

$$f_3(0) = 1$$

Several determinations of the vibrational energy distribution of Reaction (11) have been published and are listed in Table 11. The molecular-beam experimental measurements of Schafer et al.[126] are for back scattering only and are inconsistent with other data. The three remaining measurements[114,117,118] are in approximate agreement, particularly for $k_{11}(2)$ and $k_{11}(3)$. Kerber et al.[3] have compared the pumping distributions for Reactions (10) and (11) graphically. The new data of Polanyi and Woodall[118] (published since the review of Ref. 3) are consistent with the trends of earlier work, and we recommend these results here. These data should be corrected for the revision in Einstein coefficients discussed by Herbelin and Emanuel,[84] but the numerical change is very slight (within experimental uncertainty), and the results tabulated in Table 11 are uncorrected.

**Table 11.** Relative Pumping Rates for $F + D_2 \xrightarrow{k_{11}(v)} DF(v) + D$, $k_{11}(v) = g_{11}(v) k_{11}$ (Total)

| $g_{11}(0)$ | $g_{11}(1)$ | $g_{11}(2)$ | $g_{11}(3)$ | $g_{11}(4)$ | Reference | Technique |
|---|---|---|---|---|---|---|
| — | — | >0.22 | >0.36 | 0.42 | 114 | Equal-gain chemical laser |
| 0.009 | 0.014 | 0.037 | 0.21 | <0.73 | 126 | Crossed molecular beam |
| — | — | 0.29 | 0.41 | 0.3 | 117 | Discharge flow |
| — | 0.11 | 0.25 | 0.38 | 0.27 | 118 | Discharge flow |

**H(D)+F₂.** In comparison with the previous reaction, experimental work on the reactions

$$H + F_2 \rightarrow HF + F \tag{12}$$

and

$$D + F_2 \rightarrow DF + F \tag{13}$$

has been meager. The data for $k_{12}$, both experimental and theoretical, are tabulated in Table 12. There has been only one experimental determination of $k_{12}$ over a temperature range sufficient to permit calculation of Arrhenius rate parameters. Albright et al.[75,94] obtained a value of $k_{12} = 1.2 \times 10^{14} \exp(-2400/RT)$ in a discharge flow system at temperatures of 294 to 565°K, using mass spectrometry to monitor species. However, as was noted earlier, their determination of the rate coefficient for $H + Cl_2$, obtained at the same time, is larger than some other evaluations by an order of magnitude at 300°K, so their value of $k_{12}$ is somewhat subject to "guilt by association." The measurement of Rabideau et al.[101] was made in an $H_2$–$F_2$ system in which both Reactions (10) and (12) were occurring. Observed F- and H-atom concentrations were matched by computer calculations; the best fit was obtained with their resultant values of $k_{10}$ and $k_{12}$, which possibly are subject to greater uncertainty than their quoted 25% ($k_{10}$) and 8% ($k_{12}$). Thus another determination of $k_{12}$ is needed, but in its absence we used the value of Albright et al. Wilkins[129] obtained a value of $k_{12}$ of $10^{15.86} \, T^{0.614} \exp(-2840/RT)$, or $\sim 10^{14.04} \exp(-2450/RT)$ between 200 and 500°K, by three-dimensional classical Monte Carlo trajectory calculations.

No measurements for the analogous reaction (13) with deuterium atoms have been reported. Wilkins[129] reported a value of $k_{13} = 10^{15.46} \, T^{-0.55} \exp(-2840/RT)$ between 200 and 1000°K or approximately

**Table 12.** Rate Coefficients for $H + F_2 \xrightarrow{k_{12}} HF + F$

| $T$ | $k_{12} = A \exp(-E/RT)$ $A$, cm³/mol-sec | $E$, kcal/mol | $k$, 300°K | Technique | Reference |
|---|---|---|---|---|---|
| — | $>3 \times 10^{13}$ | 4.2 | $>2.6 \times 10^{10}$ | Estimate | 85 |
| — | $5.3 \times 10^{12} \, T^{1/2}$ | 4.0 | $1.1 \times 10^{12}$ | Estimate | 87 |
| 298–2500 | $2.1 \times 10^{12} \, T^{0.67}$ | 1.5 | $7.7 \times 10^{12}$ | BEBO theory | 127 |
| 298 | | $\sim 1.5^a$ | $1.8 \times 10^{12}$ (288°K) | Photochem reactor | 128 |
| 294–565 | $1.2 \times 10^{14}$ | 2.4 | $2.14 \times 10^{12}$ | Flow/mass spectroscopy | 75, 94 |
| 250–500 | $9.4 \times 10^{13}$ | 2.4 | $1.66 \times 10^{12}$ | Absolute rate theory | 99 |
| 300 | | | $2.5 \times 10^{12}$ | Fast flow/EPR | 101 |

$^a$ The value of $k_{12}$ was determined from an experimentally measured ratio of rates of reaction for $H + F_2$ to $H + O_2 + N_2$. Assuming a value of $9.2 \times 10^{15}$ for the latter rate coefficient, Levy and Copeland calculated the value of $k_{12}$ given in the table. Then, assuming a pre-exponential value of log $A = 9.1$, they calculated the activation energy shown in the table.

$10^{13.83}$ exp $(-2460/RT)$ between 200 and 500°K. A comparison of other pairs of reactions of H and D atoms with the same reactant leads one to expect the activation energy for $k_{13}$ to be a few hundred calories smaller than that for $k_{12}$. However, Wilkins' calculated $H + F_2$ rate in the same paper is in excellent agreement with the experimental results of Albright et al.,[75,94] and, since the latter are the best data available for Reaction (10), there is no good reason for not accepting the calculated $D + F_2$ rate.

The energy distribution of Reaction (12) has been studied in two laboratories. Jonathan et al.[130–132] have reported successively improved measurements of the pumping distribution in a discharge-flow system. Polanyi and Sloan,[133] used a somewhat different discharge-flow technique, have also published distribution numbers. Both studies report measurements of emission from $v = 1$ through 9; Jonathan's latest work includes data on the relative rate of formation of the $v = 0$ level monitored by absorption techniques. Jonathan used the measured-relaxation method, while Polanyi and Sloan used the arrested-relaxation technique. In principle, the latter should be the more direct technique, but in practice either procedure requires great care. In both laboratories the emission intensities were converted to population densities by means of Einstein coefficients calculated by the method of Cashion.[134] Herbelin and Emanuel[84] have recalculated the Einstein coefficients and have adjusted the data of Jonathan et al., and of Polanyi and Sloan, to produce revised sets of rate coefficients. The original and the corrected results are shown in Table 13. The $g(v)$ shown are the relative $k$'s normalized to sum to unity. Thus $k_{12}(v) = g(v)k_{12}$(Total). In the original work the $k$'s were normalized by arbitrarily setting the largest value, $k_{12}(v_{max}) = 1$.

Wilkins' classical Monte Carlo trajectory calculations[129] of the distribution of vibrationally excited HF-product molecules in Reaction (12) indicate no molecules are formed in the levels with $v \le 2$ or $\ge 7$. The $g(v)$

**Table 13.** Relative Pumping Rates for $H + F_2 \xrightarrow{k_{12}(v)} HF(v) + F$ $k_{12}(v) = g_{12}(v)k_{12}$ (Total)

| $g_{12}(0)$ | $g_{12}(1)$ | $g_{12}(2)$ | $g_{12}(3)$ | $g_{12}(4)$ | $g_{12}(5)$ | $g_{12}(6)$ | $g_{12}(7)$ | $g_{12}(8)$ | $g_{12}(9)$ | $g_{12}(10)$ | Reference |
|---|---|---|---|---|---|---|---|---|---|---|---|
| — | 0.06 | 0.08 | 0.14 | 0.18 | 0.28 | 0.26 | — | — | — | — | 130 |
| — | 0.05 | 0.09 | 0.14 | 0.17 | 0.24 | 0.23 | <0.07 | — | — | — | 131 |
| 0.012 | 0.026 | 0.032 | 0.038 | 0.13 | 0.26 | 0.29 | 0.13 | 0.059 | 0.012 | 0.012 | 132 |
| 0.009 | 0.021 | 0.029 | 0.030 | 0.134 | 0.30 | 0.40 | 0.046 | 0.02 | 0.004 | 0.004 | 132[a] |
| 0.03 | 0.034 | 0.037 | 0.070 | 0.10 | 0.22 | 0.28 | 0.113 | 0.073 | 0.045 | — | 133 |
| 0.025 | 0.03 | 0.038 | 0.062 | 0.11 | 0.28 | 0.36 | 0.044 | 0.028 | 0.017 | — | 133[a] |

[a] Corrected with Einstein coefficients of Herbelin and Emanuel.[84]

obtained (at 300°K) from his calculations are $g(3)$ through $g(6) =$ 0.10:0.23:0.43:0.25. This suggests that the low vibrational levels observed in the experiments are formed by V–V processes and from the cold reaction (10); the very high vibrational levels are populated by V–V processes, which are expected to be very rapid for large $v$. If it is, therefore, assumed that reaction takes place into levels $v = 4$ through 8 only, and Polanyi and Sloan's experimental results (computed with the Einstein coefficients of Herbelin and Emanuel) are renormalized so that $\sum_{v=3}^{6} g_{12}(v) = 1$, then we obtain $g_{12}(3):g_{12}(4):g_{12}(5):g_{12}(6) =$ 0.07:0.14:0.35:0.44. We expect that these numbers may be near the values that are eventually agreed upon, and we recommend them for computational purposes.

No experimental data on the pumping distribution of Reaction (13) are available. Kerber et al.[3] arrived at a set of distribution numbers by comparison with the analogous reaction (12). This is a reasonable approach in the absence of experimental data: Using such a comparison to the $H + F_2$ pumping distribution measured by Polyanyi and Sloan and recomputed with the Einstein coefficients of Herbelin and Emanuel, we obtain values for $g_{13}(5)$ through $g_{13}(9)$ of 0.10:0.19:0.29:0.38:0.04. Wilkins' theoretical-trajectory calculations indicate that only $v = 5$ through 9 in the product DF molecules are formed. His temperature-dependent rate coefficients, evaluated over the temperature range of 200 to 1000°K, are:

$$k_{13}(5) = 10^{14.41}\, T^{-0.43} \exp\left(\frac{-2830}{RT}\right)$$

$$k_{13}(6) = 10^{14.1}\, T^{-0.37} \exp\left(\frac{-2770}{RT}\right)$$

$$k_{13}(7) = 10^{14.97}\, T^{-0.54} \exp\left(\frac{-3160}{RT}\right)$$

$$k_{13}(8) = 10^{15.40}\, T^{-0.74} \exp\left(\frac{-2750}{RT}\right)$$

and

$$k_{13}(9) = 10^{16.98}\, T^{-1.32} \exp\left(\frac{-4570}{RT}\right)$$

These results give room-temperature distribution numbers of approximately $g_{13}(5)$ through $g_{13}(9) = 0.2:0.15:0.23:0.39:0.02$, which are in reasonable agreement with the numbers derived from the comparison with the $H + F_2$ distributions.

## 2.3.  Pumping Reactions in the $H_2(D_2)/Br_2$ and $H_2(D_2)/I_2$ Systems

The overall rates of reaction in the $H_2/Br_2$ system:

$$H + Br_2 \rightarrow HBr + Br \tag{14}$$

and

$$Br + H_2 \rightarrow HBr + H \tag{15}$$

have been reviewed in detail by Cohen, Giedt, and Jacobs.[40] Fettis and Knox[59] have reviewed the kinetics of Br atoms with both $H_2$ and $D_2$. The kinetics of the $H_2/I_2$ system, long thought to be dominated by a bimolecular four-center reaction, were re-examined by Sullivan[58] in recent years, and the rate coefficient for the iodine-atom reaction

$$I + H_2 \rightarrow HI + H \tag{16}$$

was obtained. In an extended series of papers over the past decade, Polanyi and his co-workers have examined the vibrational-energy distributions of $H + Br_2$[78,81,82,134] and $H + HBr$.[78,135] Cadman and Polanyi[136] studied the reactions of $H + HI$ and $D + HI$, both of which form electronically excited iodine atoms. The possibility of vibrationally excited $H_2(D_2)$ being formed as well could not be excluded.


## 3.  ENERGY-TRANSFER REACTIONS

### 3.1.  Vibrational–Translational (Rotational) Energy Transfer

Vibrational–translational (rotational) energy-transfer processes drive the vibrational populations toward equilibrium with the translational degrees of freedom with a resulting loss of potential power in a laser system. Parameter variations in theoretical modeling studies[1,137-139] show the complex nature of the interaction between the physical- and chemical-rate processes in chemical lasers. However, laser performance basically depends on the relative rates of production of excited states and their removal by V–R,T or V–V processes.

The relaxation processes for the hydrogen halides have posed a challenge to both experimentalists and theoreticians. The hydrogen halides provide a sensitive test of any theory of vibrational energy transfer because of their large rotational velocities, their strong intermolecular attractive forces, and the large effect of isotopic substitution. Theoretical predictions based on the SSH V–T theory,[140] Moore's V–R theory,[141] or Millikan and White's semiempirical correlation[142] do not reproduce the experimental data for the hydrogen halides. A V–R model developed by

Shin[143,144] predicts results in reasonable agreement with the hydrogen halide self-relaxation data. Classical-trajectory calculations, discussed in Wilkins' chapter of this book, have been performed in only a few cases. Because of the difficulties of the theoretical calculations, only experimental data exist for most of the relaxation processes. For some of the processes, there are neither experimental data nor theoretical calculations.

The V–R,T deactivation of the hydrogen halides can be described by

$$HX(v)+M \xrightleftharpoons{k_{v,v-\Delta v}} HX(v-\Delta v)+M+\Delta E(v, \Delta v) \qquad (17)$$

where $k_{v,v-\Delta v}$ is the rate coefficient of the reaction, $v$ represents the initial vibrational level of HX, and $\Delta v$ is the number of quanta transferred to R-T energy. Relaxation-rate coefficients have been measured for the self-deactivation of the hydrogen halides and for their deactivation by other hydrogen halides, other diatomic molecules, and some atoms and polyatomic molecules. Rates for their deactivation by atomic radicals have been measured in a few cases, but the uncertainties of these determinations are large. Most of the experimental measurements were made for $v = \Delta v = 1$; only a few studies have reported rates for the higher vibrational levels.

The complexity of Shin's theory and the specific nature of the classical trajectory calculations preclude any general conclusions on the $v$ and $\Delta v$ dependence of the reaction rates. Since the deactivations of the upper vibrational levels affect the power of a laser, kinetic models of the laser medium must include these upper levels. The harmonic oscillator (HO) model[145] predicts the rate coefficients for the higher vibrational levels to be given by $k_{v,v-1} = vk_{1,0}$ for $\Delta v = 1$ and predicts the rates of multiquantum transitions, $\Delta v > 1$, to be negligible compared to single quantum transitions. The hydrogen halides are, however, far from ideal harmonic oscillators, and the HO model cannot be expected to give a precise description of the relaxation processes. The extent to which anharmonicity changes the $v$ dependence of the HO-predicted rates depends on the collision model assumed. It is not clear whether the $v$ dependence of $k$ increases or decreases because of anharmonicity. The use of the Landau–Teller approximation for the calculation of the relaxation rates in the HO model results in a negligible probability for multiquantum exchanges. This approximation is based on a consideration only of the strong repulsive intermolecular forces and may break down completely where the collision partners exert strong attractive forces on each other. Collisions of the hydrogen halides with themselves, H atoms, and halogen atoms probably do involve strong attractive forces. Trajectory calculations have been performed for a few deactivation processes, and these results are used as a guide for estimating the higher-level rate coefficients of those few processes. In this review, most extrapolations of experimental data for the deactivation rates to higher vibrational levels will be based on the HO model for lack of any other information. This assumption leads to very large uncertainties in the rates for the upper levels. For example, $v^3$ dependence, as has been suggested for HF–HF V–V transfer, would differ from a $v$ dependence by a factor of nine for the $v = 3$ level. The $v$ dependence as well as the multiquantum transitions remain the largest uncertainties in kinetic models of chemical lasers, and will certainly need revision when critical experiments can be devised to measure them.

Most experimental investigations result in the measurement of a relaxation time $\tau$ at a given pressure $p$ and temperature $T$. To convert this relaxation time to a rate coefficient $k$ (cm³/mol-sec), one must make certain assumptions about the nature of the relaxation process. According to the HO model,

$$p\tau = \frac{RT}{k_{10} - k_{01}} = \frac{RT}{k_{10}[1 - \exp(-\Delta E/RT)]} \tag{18}$$

where $k_{10}$ is the rate coefficient for the deactivation of the $v = 1$ level to $v = 0$; $k_{01}$ is the rate coefficient for the backward reaction; $\Delta E$ is the exothermicity of the reaction; and $R$ is the universal gas constant.

In the following paragraphs, we will describe the results of some of the hydrogen–halide deactivation studies. Since HF(DF) and HCl lasers have been studied and modeled most extensively, there is a large body of experimental data available for these molecules, which we have collected in Tables 14 through 19. We have fitted an analytical expression to the results of each experimental study where the deactivation reactions were studied over a range of temperatures. These expressions fit the data with an accuracy of 10 to 15%.

**Self-Relaxation of the Hydrogen Halides (M = HX).** Self-relaxation of the hydrogen halides has been studied at the higher temperatures (above 800°K) in shock tubes and at lower temperatures by laser-induced fluorescence studies. The laser-induced fluorescence technique has been used mostly at room temperature, although some studies in heated or cooled cells or behind reflected shock waves have yielded data over a wide range of temperatures. The relaxation of HF has been studied more extensively than any of the other hydrogen halides, with reported rates between 295 and 5000°K.

Shock-tube studies of Solomon et al.,[146] Bott and Cohen,[147] Just and Rimpel,[148] and Vasil'ev et al.[149] gave values of $p\tau$ for HF–HF relaxation at 1400°K between 0.195 and 0.245 $\mu$sec-atm. At 4000°K, Solomon et al.

**Table 14.** Experimental Rates for V–R,T Relaxation of HF($v = 1$) by Various Collision Partners HF($v = 1$) + M $\xrightleftharpoons{k,(P_T)^{-1}}$ HF($v = 0$) + M

| M | Temperature, °K | $P_{T_{HF-M}}$, $\mu$sec-atm | k, cm³/mol-sec | Reference |
|---|---|---|---|---|
| HF | 350 | 0.014 | $2 \times 10^{12}$ | 150 |
| | 1400–4100 | $6.3 \times 10^{-4} \exp(64/T^{1/3})$ | $6.6 \times 10^2 \, T^{2.85}$ | 146 |
| | 1350–4000 | $1.0 \times 10^{-2} \exp(34.4/T^{1/3})$ | $1.1 \times 10^5 \, T^{2.12}$ | 147 |
| | 350 | 0.025 | $1.1 \times 10^{12}$ | 151 |
| | 294 | 0.015 | $1.6 \times 10^{12}$ | 152 |
| | 295–1000 | $11 \exp(-41/T^{1/3})$ | $(3 \times 10^{14} \, T^{-1} + 1.7 \times 10^5 \, T^2)$ | 153 |
| | 290 | 0.025 | $9.5 \times 10^{11}$ | 154, 224 |
| | 350–750 | $1.9 \exp(-29/T^{1/3})$ | $(2.4 \times 10^{14} \, T^{-1} + 7.1 \times 10^8 \, T)$ | 155 |
| | 295–1000 | $8.0 \exp(-38.9/T^{1/3})$ | $(3 \times 10^{14} \, T^{-1} + 3.5 \times 10^5 \, T^2)$ | 156, 157 |
| | 297–670 | $5.7 \exp(-39/T^{1/3})$ | $(4.2 \times 10^{14} \, T^{-1} + 3.5 \times 10^5 \, T^2)$ | 158 |
| | 1500–5000 | $5.7 \times 10^{-3} \exp(42.0/T^{1/3})$ | $2.8 \times 10^4 \, T^{2.3}$ | 149 |
| | 600–2400 | $(5.5 \times 10^6 \, T^{-2} + 4.4 \times 10^{-3} \, T)^{-1}$ | $(4.4 \times 10^{14} \, T^{-1} + 4 \times 10^5 \, T^2)$ | 159 |
| | 1400–2600 | $0.028 \exp(21.4/T^{1/3})$ | $1.6 \times 10^6 \, T^{1.78}$ | 148 |
| Ar | 1350–4000 | $1.6 \times 10^{-3} \exp(112/T^{1/3})$ | $8.5 \times 10^{-4} \, T^{4.01}$ | 147 |
| | 1500–5000 | $8.5 \times 10^{-3} \exp(89.47/T^{1/3})$ | $0.08 \, T^{3.43}$ | 149 |
| | 800–2400 | $[90 \exp(-85/T^{1/3}) + 10^{-2}]^{-1}$ | $1.7 \, T^{3.05}$ | 159 |
| | 350 | $>4$ | $<8 \times 10^9$ | 150 |
| | 295 | $>22$ | $<1.1 \times 10^9$ | 156 |
| He | 1350–4000 | $1.5 \times 10^{-4} \exp(133/T^{1/3})$ | $5.5 \times 10^{-5} \, T^{4.46}$ | 147 |
| | 295 | $>22$ | $<1.1 \times 10^9$ | 156 |
| F | 1400–4100 | $3.5 \times 10^{-5} \exp(64/T^{1/3})$ | $1.2 \times 10^4 \, T^{2.85}$ | 146 |
| | 1900–3300 | $2.5 \times 10^{-6} \, T$ | $3.2 \times 10^{13}$ | 194 |
| | 2545 | 0.026 | $8 \times 10^{12}$ | 196 |
| | 1500–2400 | $7.9 \times 10^{-4} \exp(46/T^{1/3})$ | $6.8 \times 10^4 \, T^{2.40}$ | 159 |
| $F_2$ | 350 | $>1$ | $<2.5 \times 10^{10}$ | 150 |
| | 350 | $>13$ | $<2 \times 10^9$ | 155 |
| $H_2$ | 295–610 | $5.1 \times 10^3 \, T^{-1.28}$ | $1.6 \times 10^4 \, T^{2.28}$ | 206 |
| | 1400–4000 | $2.5 \times 10^{-2} \exp(34.4/T^{1/3})$ | $4.4 \times 10^4 \, T^{2.12}$ | 209 |
| H | 295 | $\geq 0.04$ | $\leq 6.3 \times 10^{11}$ | 203 |

measured somewhat faster rates and Just and Rimpel measured somewhat slower rates than those found by Bott and Cohen and by Vasil'ev et al.

The laser-induced fluorescence technique has been used to measure HF relaxation times between 295 and 2400°K.[150–159] We have fitted the results of these various determinations with functions describing the temperature dependences and list them in Table 14. The measured relaxation times are in agreement within a factor of 1.6 at any temperature between 295 and 1000°K, with the exception of the early measurement of Airey and Fried[150] at 350°K, which is somewhat faster. The high-temperature data combined with those at low temperature show that the HF–HF V–R,T relaxation times must have a maximum near 1200°K.[153–160] This maximum was confirmed by Blair and co-workers[159] in an experimental study covering the range between 600 and 2400°K. The phenomenon of a relaxation-time maximum has been observed or

**Table 15.** Experimental Rates of V–R,T Relaxation of DF($v=1$) by Various Collision Partners DF($v=1$)+ M $\xrightarrow{k_{1(p\tau)^{-1}}}$ DF($v=0$)+M

| M | Temperature, °K | $P\tau_{DF-M}$, μsec-atm | $k$, cm³/mol-sec | Reference |
|---|---|---|---|---|
| DF | 295–5000 | $[1.32\times10^6 T^{-2}+4.7\times10^2 \exp(-58.8/T^{1/3})]^{-1}$ | $1.2\times10^{14} T^{-1}+1.4\times10^2 T^{2.96}$ | 160, 163, 164 |
| | 1500–4000 | $2.5\times10^{-3}\exp(56.0/T^{1/3})$ | $6.8\times10^2 T^{2.76}$ | 162 |
| | 1500–5000 | $2.2\times10^{-3}\exp(57.25/T^{1/3})$ | $5\times10^2 T^{2.80}$ | 149 |
| | 350 | 0.066 | $4.4\times10^{11}$ | 151 |
| | 300–940 | $(1.2\times10^{11} T^{-4}+2.5)^{-1}$ | $(9.5\times10^{18} T^{-3}+2\times10^8 T)$ | 156, 157 |
| | 295–670 | $(1.35\times10^{11} T^{-4}+4.0)^{-1}$ | $(1.1\times10^{19} T^{-3}+3.2\times10^8 T)$ | 158 |
| HF | 294–580 | $4.4\times10^{-7} T^2$ | $1.8\times10^{14} T^{-1}$ | 156, 157 |
| | 295–750 | $[2.3\times10^6 T^{-2}+3.7\times10^2 \exp(-54.8/T^{1/3})]^{-1}$ | $(1.9\times10^{14} T^{-1}+1.35\times10^2 T^3)$ | 160 |
| | 300, 350 | 0.029, 0.040 | $8.5\times10^{11}, 7.2\times10^{11}$ | 181 |
| | 300 | 0.019 | $1.3\times10^{12}$ | 183 |
| Ar | 1600–5000 | $7\times10^{-4}\exp(128.6/T^{1/3})$ | $1.1\times10^{-5} T^{4.53}$ | 163 |
| | 1500–5000 | $0.014\exp(78.04/T^{1/3})$ | $0.43 T^{3.27}$ | 149 |
| He | 295 | >43 | $<5.6\times10^8$ | 156 |
| | 295 | >43 | $<5.6\times10^8$ | 156 |
| F | 1900–3000 | $5.5\times10^{-6} T$ | $1.5\times10^{13}$ | 163 |
| | 2000–3000 | $6.3\times10^{-4}\exp(36/T^{1/3})$ | $7.6\times10^5 T^{2.23}$ | 195 |
| $H_2$ | 295 | 0.3 | $8.1\times10^{10}$ | 156 |
| $D_2$ | 295–4000 | $[52\exp(-35/T^{1/3})+1.6\times10^4 T^{-2}]^{-1}$ | $9\times10^{11} T^{-1}+2.7\times10^4 T^{2.23}$ | 163, 206 |
| | 295–430 | $1.3\times10^6 T^{-2}$ | $55 T^3$ | 206 |

**Table 16.** Experimental Rates of V–R,T Relaxation of HCl ($v = 1$) in the Presence of Various Molecules $HCl(v = 1) + M \overset{k_v(P_T)^{-1}}{\rightleftharpoons} HCl(v = 0) + M$

| M | Temperature, °K | $P_{T_{HCl-M}}$, $\mu$sec-atm | $k$, cm³/mol-sec | Reference |
|---|---|---|---|---|
| HCl | 1000–2100 | $1.3 \times 10^{-3} \exp(64/T^{1/3})$ | $74.2\ T^{3.05}$ | 169 |
|  | 1100–2100 | $4.9 \times 10^{-4} \exp(79/T^{1/3})$ | $2.0\ T^{3.5}$ | 173 |
|  | 295 | 1.5 | $1.6 \times 10^{10}$ | 174 |
|  | 144–584 | $(10^{9.27}\ T^{-4} + 10^{-2.97}\ T)^{-1}$ | $1.55 \times 10^{17}\ T^{-3} + 8.7 \times 10^4\ T^2$ | 175 |
| Ar | 1100–2100 | $0.022 \exp(79/T^{1/3})$ | $0.046\ T^{3.5}$ | 173 |
|  | 1300–2000 | $3.3 \times 10^{-4} \exp(133.9/T^{1/3})$ | $4.4 \times 10^{-7}\ T^5$ | 184 |
|  | 296–700 | $[5.6 \times 10^2 \exp(-110/T^{1/3}) + 4 \times 10^{-5}]^{-1}$ | $7.9 \times 10^5 + 3.2 \times 10^{-8}\ T^{5.5}$ | 186 |
| He | 1100–1600 | $5.4 \times 10^{-4} \exp(93.6/T^{1/3})$ | $0.013\ T^4$ | 184 |
|  | 296 | 658 | $3.7 \times 10^7$ | 176 |
|  | 296–700 | $[8.5 \times 10^3 \exp(-110/T^{1/3}) + 6 \times 10^{-4}]^{-1}$ | $1.2 \times 10^7 + 4.8 \times 10^{-7}\ T^{5.5}$ | 186 |
| Cl | R.T. | $3.8 \times 10^{-3}$ | $6.5 \times 10^{12}$ | 188 |
|  | R.T. | $4.1 \times 10^{-2}$ | $5.9 \times 10^{11}$ | 189 |
| Cl₂ | R.T. | 7.3 | $3.3 \times 10^9$ | 188 |
|  | R.T. | 0.5 | $4.8 \times 10^{10}$ | 189 |
|  | 295 | 0.59 | $4.1 \times 10^{10}$ | 190 |
| H₂ | 295 | 7.7 | $3.2 \times 10^9$ | 174 |
| N₂[a] | 295 | 1.51 | $1.6 \times 10^{10}$ | 226 |

[a] $N_2$ relaxes $HCl(v = 1)$ by both V–V and V–R,T processes.

predicted for other diatomic molecules[161] but at much lower temperatures.

Similar studies have resulted in relaxation times for DF between 295 and 5000°K. Three shock-tube studies[149,162,163] reported DF relaxation times between 1400 and 4000°K that are within 20% agreement. The shock-tube data of Bott and Cohen[163] at 800°K overlapped and agreed with their measurements[164] at 800°K obtained with the laser-induced fluorescence technique; these measurements are in turn in substantial agreement with those of Hinchen[157] between 295 and 800°K. Lucht and Cool[158] measured rates in this temperature regime that are faster by a factor of 1.2 to 1.5. Between 295 and 1400°K, the DF relaxation times are approximately 2.5 times longer than those of HF. The data of Vasilév et al.[149] and Bott and Cohen[163] show the relaxation times of HF and DF to be about the same at 4000°K.

Shin has developed a model[143,144] for vibrational relaxation based on a rotation-averaged oscillator and a rigid rotor; the two hydrogen atoms interact strongly in close-in collisions to transfer vibrational and rotational energies. His predictions[143,165] for HCl, HBr, and HI relaxation at 700°K and above are in substantial agreement with experimental data. Applying the same theory to HF–HF and DF–DF V–R energy transfer, Shin[144,166] has found it necessary to include strong dipole–dipole attraction to reproduce the experimental data between 300 and 4000°K.

**Table 17.** Vibrational Relaxation of the Hydrogen Halides at Room Temperature

| Molecule 1 | Molecule 2 | $k_{21}$,[a] $(\mu\text{sec-Torr})^{-1}$ | $P_{21}$[c] | $k_{vv}+k_{12}$,[d] $(\mu\text{sec-Torr})^{-1}$ | $P_{vv}+P_{12}$[c] | Reference |
|---|---|---|---|---|---|---|
| HF | HF | $(5.6\pm0.5)\times10^{-2}$ | $11\times10^{-3}$ | | | 153[e] |
| HF | DF | $(3.5\pm0.5)\times10^{-2}$ | $7.0\times10^{-3}$ | $(7.7\pm0.4)\times10^{-2}$ | $1.5\times10^{-2}$ | 180[e] |
| HF | HCl | $(1.5\pm0.2)\times10^{-2}$ | $2.5\times10^{-3}$ | $(1.7\pm0.1)\times10^{-2}$ | $2.8\times10^{-3}$ | 180[f] |
| HF | HBr | $(0.9\pm0.1)\times10^{-2}$ | $1.6\times10^{-3}$ | $(7.5\pm1.0)\times10^{-3}$ | $1.3\times10^{-3}$ | 180[f] |
| DF | DF | $(2.1\pm0.2)\times10^{-2}$ | $4.2\times10^{-3}$ | | | 164[e] |
| DF | HCl | $(1.6\pm0.3)\times10^{-2\,b}$ | $2.7\times10^{-3}$ | $(4.0\pm0.4)\times10^{-1}$ | $6.8\times10^{-2}$ | 182 |
| DF | HBr | $(1.0\pm0.2)\times10^{-2}$ | $1.8\times10^{-3}$ | $(7.1\pm0.7)\times10^{-2}$ | $1.26\times10^{-4}$ | 182 |
| DF | DBr | $(2.3\pm0.7)\times10^{-2}$ | $4.1\times10^{-3}$ | $(5.4\pm1.0)\times10^{-3}$ | $9.5\times10^{-4}$ | 182 |
| HCl | HCl | $(8.3\pm0.8)\times10^{-4}$ | $1.27\times10^{-4}$ | | | 174, 175 |
| HCl | DCl | $(5.75\pm0.6)\times10^{-4}$ | $8.9\times10^{-5}$ | $(3.25\pm0.2)\times10^{-3}$ | $5.0\times10^{-4}$ | 174, 176 |
| HCl | HBr | $(1.31\pm0.14)\times10^{-3}$ | $2.3\times10^{-4}$ | $(3.4\pm0.3)\times10^{-2}$ | $6.0\times10^{-3}$ | 177, 225 |
| DCl | DCl | $(2.5^{+0.3}_{-0.9})\times10^{-4}$ | $3.9\times10^{-5}$ | | | 174, 176 |
| HBr | HBr | $(5.7\pm0.5)\times10^{-4}$ | $1.23\times10^{-4}$ | | | 177 |
| HBr | DBr | $(3.5\pm0.6)\times10^{-4}$ | $7.5\times10^{-5}$ | $(1.96\pm0.2)\times10^{-3}$ | $4.22\times10^{-4}$ | 178 |
| DBr | DBr | $(1.7\pm0.3)\times10^{-4}$ | $3.66\times10^{-5}$ | | | 178 |
| HI | HI | $(3.75\pm0.5)\times10^{-4}$ | $7.0\times10^{-5}$ | | | 179 |

[a] $k_{21}$ is the deactivation rate of Molecule 2 by Molecule 1; $k_{12}$ is the deactivation rate of Molecule 1 by Molecule 2.

[b] $k_{21}+0.88k_{12}=(1.6\pm0.3)\times10^{-2}$ $(\mu\text{sec-Torr})^{-1}$.

[c] $P_{21}=k_{21}/k_{\text{gas kin.}}$, $P_{12}=k_{12}/k_{\text{gas kin.}}$. At $T=295°\text{K}$, $k_{\text{gas kin.}}=2.57\times d^2/\bar{m}^{1/2}$ $(\mu\text{sec-Torr})^{-1}$ with $d=(d_{\text{HX}}+d_{\text{HY}})/2$ in Å, and $\bar{m}$ the reduced collision mass in a.m.u. Collision diameters of 2.5, 3.3, 3.5, and 4.1 Å have been used for HF(DF), HCl(DCl), HBr(DBr), and HI, respectively.

[d] $k_{vv}$ = rate of V–V energy transfer from Molecule 1 to Molecule 2 where Molecule 1 changes from $v=1$ to $v=0$ and Molecule 2 changes from $v=0$ to $v=1$. Value listed for DF–HCl is $k_{vv}$ and not $k_{vv}+k_{12}$.

[e] See Tables 14 and 15 for values from other investigations.

[f] Somewhat faster rates were reported in Ref. 181.

**Table 18.** Experimental Rates for V–V Energy Transfer from HF($v = 1$) to Other Collision Partners $AB$, $\text{HF}(v = 1) + AB(v = 0) \xrightleftharpoons{k_i(P_T)^{-1}} \text{HF}(v = 0) + AB(v = 1, 0)^b$

| $AB$ | Temperature, °K | $P_{T_{\text{HF–AB}}}$, $\mu$sec-atm | $k$, cm³/mol-sec | Reference |
|------|-----------------|------------------------------------|------------------|-----------|
| DF | 295–1050 | $(1.3 \times 10^9\ T^{-3} + 7.6)^{-1}$ | $1.1 \times 10^{17}\ T^{-2} + 6.3 \times 10^8\ T$ | 180 |
| | 300, 350 | $9.8 \times 10^{-3},\ 1.33 \times 10^{-2}$ | $2.5 \times 10^{12},\ 2.1 \times 10^{12}$ | 181, 202 |
| | 295 | 0.02 | $1.2 \times 10^{12}$ | 156 |
| $H_2{}^a$ | 294 | 0.021 | $1.1 \times 10^{12}$ | 152 |
| | 295–1000 | $9.1 \times 10^{-5}\ T$ | $8.3 \times 10^{11}$ | 180 |
| | 295 | 0.021 | $1.1 \times 10^{12}$ | 156 |
| $N_2$ | 294 | 10.5 | $2.3 \times 10^9$ | 152 |
| | 295 | $6.6^{+6.6}_{-2.2}$ | $4 \times 10^9$ | 155 |
| | 1500–3000 | $10^7\ T^{-2}$ | $8.7\ T^3$ | 207 |
| | 295–1600 | $(30\ T^{-1} + 1.4 \times 10^{-7}\ T^2)^{-1}$ | $2.4 \times 10^9 + 11\ T^3$ | 180 |
| $O_2$ | 295 | 3.76 | $6.5 \times 10^9$ | 229 |
| | 295–800 | $(6.9\ T^{-1} + 1.2 \times 10^{-7}\ T^2)^{-1}$ | $5.6 \times 10^8 + 9.8\ T^3$ | 180 |

[a] Values calculated for $H_2(v = 1) + HF(v = 0) \rightleftarrows H_2(v = 0) + HF(v = 1)$.
[b] The measurements contain V–R,T contributions as well as V–V.

Trajectory calculations by Berend and Thommarson for HF–HF and DF–DF deactivation gave good agreement with the experimental data when a sufficient number of $J$ levels were considered.[167] Their initial calculations[168] were based on fewer $J$ levels than were necessary to describe adequately the relaxation at low temperatures. They observed that below 1000°K the deactivation resulted from the collision of HF($v = 1$) with HF($v = 0$) and the formation of internally excited (HF)$_2$ dimers. The excited dimer lives long enough to allow a redistribution of energy so that neither of the HF molecules is vibrationally excited when the dimer dissociates.

**Table 19.** Experimental Rates for V–V Energy Transfer from DF($v = 1$) to Other Chaperones, $\text{DF}(v = 1) + AB(v = 0) \xrightleftharpoons{k_i(P_T)^{-1}} \text{DF}(v = 0) + AB(v = 1)$

| $AB$ | Temperature, °K | $P_{T_{\text{DF–AB}}}$, $\mu$sec-atm | $k$, cm³/mol-sec | Reference |
|------|-----------------|------------------------------------|------------------|-----------|
| $D_2$ | 295 | $0.043^a$ | $5.6 \times 10^{11}$ | 156 |
| | 295–750 | $1.5 \times 10^{-4}\ T$ | $5.4 \times 10^{11}$ | 160, 182 |
| $N_2$ | 295 | 0.66 | $3.6 \times 10^{10}$ | 156 |
| | 295–1100 | $(5 \times 10^4\ T^{-2} + 4.0 \times 10^{-4}\ T)^{-1}$ | $4.1 \times 10^{12}\ T^{-1} + 3.2 \times 10^4\ T^2$ | 182 |
| $O_2$ | 295–890 | $(17\ T^{-1} + 8.5 \times 10^{-11}\ T^3)^{-1}$ | $1.3 \times 10^9 + 6.9 \times 10^{-3}\ T^4$ | 182 |

[a] Values calculated for exothermic direction, that is, $D_2(1) + DF(0) \rightarrow D_2(0) + DF(1)$.

**Table 20.** Recommended Rate Coefficients for $H_2$–$F_2$ System

| Reaction number | Reaction | Rate coefficients ($cm^3$, mole, sec, cal units) | Notes |
|---|---|---|---|
| 1 | $F_2 + M_1 \rightleftarrows 2F + M_1$ | $k_1 = 5 \times 10^{13} A_M \exp(-35,100/RT)$ | $A_M = 1$ for all species except F, $F_2$, H; $A_{F_2} = 2.4$, $A_F = 10$, $A_H = 3$<br>$M_2 = $ all species except H, $H_2$ |
| 2a<br>2b<br>2c | $H_2(0) + M_2 \rightleftarrows 2H + M_2$<br>$H_2(0) + H_2 \rightleftarrows 2H + H_2$<br>$H_2(0) + H \rightleftarrows 2H + H$ | $k_{-2a} = 6.2 \times 10^{17} T^{-0.95}$<br>$k_{-2b} = 9.4 \times 10^{16} T^{-0.61}$<br>$k_{-2c} = 20 k_{-2a}$ | |
| 3 | $HF(v) + M_3 \rightleftarrows H + F + M_3$ | $k_3(v) = \dfrac{1.2}{(n+1)} \times 10^{19} T^{-1} A_M \exp \times [(-135,100 + E_v - E_0)/RT]$ | $A_M = 1$ for species except H,F, HF; $A_H = A_F = A_{HF} = 5$; $v = 0, \ldots, n$ |
| 4 | $F + H_2(0) \rightleftarrows HF(v) + H$ | $k_4 = g(v) \times 1.6 \times 10^{14} \exp(-1,600/RT)$ | $v = 1, 2, 3;\ g(1) = 0.17,\ g(2) = 0.55$<br>$g(3) = 0.28$ |
| 5 | $H + F_2 \rightleftarrows HF(v) + F$ | $k_5 = g(v) \times 1.2 \times 10^{14} \exp(-2,400/RT)$ | $g(v) = 0,\ v = 0, 1, 2;\ g(3) = 0.08$,<br>$g(4) = 0.13,\ g(5) = 0.35,\ g(6) = 0.44$ |
| 6<br>7<br>8 | $HF(v) + HF \rightleftarrows HF(v-1) + HF$<br>$HF(v) + H_2 \rightleftarrows HF(v-1) + H_2$<br>$HF(v-1) + H_2(1) \rightleftarrows HF(v) + H_2(v'-1)$ | $k_{6v} = v(3 \times 10^{14} T^{-1} + 3.5 \times 10^4 T^{2.26})$<br>$k_7 = 1.6 \times 10^4 T^{2.28} v$<br>$k_8 = 8.3 \times 10^{11} g(v)$ | $v = 1 \ldots$<br>$v = 1 \ldots$<br>$g(v) = 1, 3.2, 10, 23, 46, 90$ for $v = 1$ to 6, respectively |
| 9<br>10<br>11 | $HF(v) + M_9 \rightleftarrows HF(v-1) + M_9$<br>$HF(v) + N_2(0) \rightleftarrows HF(v-1) + N_2(1)$<br>$HF(v) + H \rightleftarrows HF(v') + H$ | $k_9 = 7.7 \times 10^{-7} A_M T^5 v$<br>$k_{10} = v(11 T^3 + 2.4 \times 10^9)$<br>$k_{11} = A(v)\ 1.8 \times 10^{13} \exp(-700/RT)$ | $A_{Ar} = A_{F_2} = 1;\ A_{He} = 2$<br>$v = 1 \ldots$<br>$0 \leq v' < v,\ v = 1 \ldots A(1) = 0.1$, $A(v > 1) = 1$ |
| 12<br>13 | $HF(v) + F \rightleftarrows HF(v-1) + F$<br>$HF(v) + HF(v') \rightleftarrows HF(v+1)$<br>$\quad + HF(v'-1)$ | $k_{12} = 1.6 \times 10^{13} v \exp(-2700/RT)$<br>$k_{13} = 1.6 \times 10^{15} v(v'+1) T^{-1}$<br>$k_{13} = 8 \times 10^{16} T^{-1}$ | $v = 1 \ldots$<br>$v = 1, \ldots, 6;\ v' \geq v + 2;\ v'(v+1) \leq 40$<br>$v'(v+1) > 40$ |
| 14a<br>14b<br>15 | $H_2(v) + M \rightleftarrows H_2(v-1) + M$<br>$H_2(v) + H \rightleftarrows H_2(v-1) + H$<br>$HF(v) + DF(v') \rightleftarrows HF(v-1)$<br>$\quad + DF(v'+1)$ | $k_{14a} = 2.5 \times 10^{-4} T^{4.3} A_M$<br>$k_{14b} = 2 \times 10^{13} \exp(-2720/RT)$<br>$k_{15} = (1.1 \times 10^{17} T^{-2} + 6.3 \times 10^8 T) v$ | $A_{He} = 2,\ A_{Ar} = A_{F_2} = 1$<br>$v = 1 \ldots$<br>$v = 1, \ldots, v' = 0, \ldots, 6$ |
| 16a<br>16b<br>16c | $HF(4) + H \rightleftarrows H_2(v) + F$<br>$HF(5) + H \rightleftarrows H_2(v) + F$<br>$HF(6) + H \rightleftarrows H_2(v) + F$ | $k_{16a} = g(v)\ 7.4 \times 10^{12} \exp(-460/RT)$<br>$k_{16b} = g(v)\ 1.1 \times 10^{13} \exp(-510/RT)$<br>$k_{16c} = g(v)\ 1.9 \times 10^{13} \exp(-580/RT)$ | $g(0) = g(1) = 0.5$<br>$g(0) = 0.36,\ g(1) = 0.64$<br>$g(0) = 0.22,\ g(1) = 0.22$, $g(2) = 0.56$ |

Both HF and DF self-relaxation rates have been described by two-term expressions[160] that fit the experimental data over the temperature range of 295 to 4000°K. At temperatures below 1000°K the low-temperature term dominates over the high-temperature term and $p\tau$ is proportional to $T^2$. The probability of HF self-deactivation is $1 \times 10^{-2}$ at 300°K and extrapolates to $2 \times 10^{-2}$ at 200°K and $6 \times 10^{-2}$ at 100°K. We recommend rate coefficients for HF and DF in Tables 20 and 21 that are two-term expressions. The high-temperature term has been chosen for computational ease to be of the form of $T^n$ instead of the usual Landau–Teller $\exp(B/T)^{1/3}$ dependence. We have set $k_{v,v-1} = vk_{1,0}$ for lack of better information. Allowing multiquantum transitions and assuming $k_{v,v'} = k_{1,0}$ for $v \geq v' \geq 0$ would result in almost the same deactivation rate for a given $v$ level, but the total vibrational energy of the system would be removed at a faster rate. It may yet be found that multiquantum transitions do occur, but at present there is no experimental evidence one way or the other.

Breshears and Bird[169–171] and Kiefer et al.[172] have used a densitometric technique to make shock-tube measurements of the self-relaxation of HCl, HBr, and HI at temperatures from 700 to 2100°K. Bowman and Seery[173] measured the relaxation times of HCl between 1100 and 2100°K by monitoring the rise time of the infrared emission behind shock waves. They obtained results in substantial agreement with those of Breshears and Bird.[169] Other investigators have used the laser-induced fluorescence technique to measure room-temperature relaxation rates for HCl,[174,175] DCl,[176] HBr,[175,177] DBr,[178] and HI.[179] Zittel and Moore[175] extended the measurements for HCl and HBr to temperatures between 144 and 584°K. Relaxation times for HBr obtained with a vacuum uv flash-photolysis technique[161] were shorter than those of Zittel and Moore[175] and Chen[177] by a factor of 2.8. Otherwise, the measurements seem consistent with each other and reproducible. The relaxation times of HCl and HBr are maximum at ~400°K compared with the 1200°K maximum for HF and DF self-relaxation. The higher temperature turn-around for HF(DF) results in its relaxation rate being faster than those of HCl and HBr at 300°K by about a factor of 100. The self-relaxation times of DCl and DBr are approximately three times longer than those of HCl and HBr, respectively, at room temperature. Similarly, the DF–DF relaxation times are ~2.5 times longer than those of HF. A two term expression describing HCl self-relaxation rates between 144 and 1000°K is listed in Table 22.

**Relaxation of the Hydrogen Halides by Other Hydrogen Halides.** Table 17 lists for easy comparison the room-temperature relaxation rates for

**Table 21.** Recommended Rate Coefficients for $D_2$–$F_2$ System

| Reaction number | Reaction | Rate coefficient (cm³, mole, sec, cal units) | Notes |
|---|---|---|---|
| 1 | $F_2 + M \rightleftarrows 2F + M$ | $k_1 = 5 \times 10^{13}\, A_M \exp(-35{,}100/RT)$ | $A_M = 1$ for all species except F, $F_2$, H; $A_{F_2} = 2.4$ $A_F = 10$, $A_H = 3$ |
| 2a | $D_2 + M_2 \rightleftarrows 2D + M_2$ | $k_{-2a} = 10^{18}\, T^{-1}$ | $M_2 =$ all species except D, $D_2$ |
| 2b | $D_2 + D_2 \rightleftarrows 2D + D_2$ | $k_{-2b} = 10^{17}\, T^{-0.67}$ | |
| 2c | $D_2 + D \rightleftarrows 2D + D$ | $k_{-2c} = 3 \times 10^{17}\, T^{-0.5}$ | |
| 3 | $DF(v) + M \rightleftarrows D + F + M$ | $k_{3,v} = \dfrac{1.2}{n+1} \times 10^{19}\, T^{-1} A_M \exp[(-136{,}700 + E_v - E_0)/RT]$ | $A_M = 1$ for all species except DF, D, F; $A_D = A_F = A_{DF} = 5$ |
| 4a | $F + D_2(0) \rightleftarrows DF(v) + D$ | $k_4 = g(v) \times 1.6 \times 10^{14} \exp(-1970/RT)$ | $v = 1, 2, 3, 4;\ g(1) = 0.11,\ g(2) = 0.25,\ g(3) = 0.38,\ g(4) = 0.27;$ |
| 4b | $F + D_2(v') \rightleftarrows DF(v) + D$ | $k_{-4} = 10^{13} \exp(-500/RT)$ | $v' \le v - 4$ |
| 5 | $D + F_2 \rightleftarrows DF(v) + F$ | $k_5(v) = g(v) \times 6.8 \times 10^{13} \exp(-2460/RT)$ | $g(v) = 0$ for $v = 1, 2, 3, 4;\ g(5) = 0.07,\ g(6) = 0.15,\ g(7) = 0.25,\ g(8) = 0.40,\ g(9) = 0.12$ (see text for more detailed expressions) |
| 6 | $DF(v) + DF \rightleftarrows DF(v-1) + DF$ | $k_6(v) = v(1.2 \times 10^{14}\, T^{-1} + 1.4 \times 10^2\, T^{2.96})$ | $v = 1 \ldots$ |
| 7 | $DF(v) + D_2 \rightleftarrows DF(v-1) + D_2$ | $k_7 = 55\, vT^3$ | $v = 1 \ldots$ |
| 8 | $DF(v-1) + D_2(v') \rightleftarrows DF(v) + D_2(v'-1)$ | $k_8 = 5.4 \times 10^{11}\, g(v)$ | $g(v) = 1,\ 3.2,\ 10,\ 22,\ 46,\ 90,$ for $v = 1$ to 6, respectively |
| 9 | $DF(v) + M_9 \rightleftarrows DF(v-1) + M_9$ | $k_9 = 7.8 \times 10^{-7}\, vT^5\, A_M$ | $A_{Ar} = A_{F_2} = 1;\ A_{He} = 2$ |
| 10 | $DF(v) + N_2(0) \rightleftarrows DF(v-1) + N_2(1)$ | $k_{10} = v(4.1 \times 10^{12}\, T^{-1} + 3.2 \times 10^4\, T^2)$ | $v = 1 \ldots$ |
| 11 | $DF(v) + D \rightleftarrows DF(v') + D$ | $k_{11} = g(v) \times 2 \times 10^{15}\, T^{-0.8} \exp(-2200/RT)$ | $v' < v,\ v = 1, \ldots,\ g(1) = 0.1,\ g(v > 1) = 1.0$ |
| 12 | $DF(v) + F \rightleftarrows DF(v-1) + F$ | $k_{12} = 1.6 \times 10^{13}\, v \exp(-3380/RT)$ | $v = 1 \ldots$ |
| 13 | $DF(v) + DF(v') \rightleftarrows DF(v+1) + DF(v'-1)$ | $k_{13} = 3.3 \times 10^{15}\, v'(v+1)T^{-1}$ $k_{13} = 8 \times 10^{16}\, T^{-1}$ | $v = 1, \ldots, 9;\ 1 \le v' \le 3$ $v = 1, \ldots, 9;\ v \ge v' \ge 3$ |
| 14 | $D_2(v) + M \rightleftarrows D_2(v-1) + M$ | $k_{14} = 9 \times 10^{14}\, T^{4.3}\, A_M$ | $A_{D_2} = 1;\ A_F = A_{F_2} = A_{Ar} = 0.2$ |
| 15 | $DF(v) + HF \rightleftarrows DF(v-1) + HF$ | $k_{15} = v(1.9 \times 10^{14}\, T^{-1} + 1.35 \times 10^2\, T^3)$ | $v = 1 \ldots$ |

**Table 22.** Recommended Rate Coefficients for $H_2$–$Cl_2$ System

| Reaction number | Reaction | Rate coefficients $(cm^3,$ mole, sec, cal units) | Notes |
|---|---|---|---|
| 1 | $Cl_2 + M_1 \rightleftarrows 2\,Cl + M_1$ | $k_{-1} = 6.1 \times 10^{16}\,A_M\,T^{-0.88}\,\exp(1440/RT)$ | $A_M = 1$, for all M except $Cl_2$, Cl, H, HCl, $H_2$; $A_{Cl_2} = 5$, $A_{Cl} = 10$, $A_H = 3$, $A_{HCl} = 0.5$, $A_{H_2} = 1.5$ |
| 2a | $H_2(0) + M_2 \rightleftarrows 2\,H + M_2$ | $k_{-2a} = 6.2 \times 10^{17}\,T^{-0.95}$ | All species except H, $H_2$ |
| 2b | $H_2(0) + H_2 \rightleftarrows 2\,H + H_2$ | $k_{-2b} = 9.4 \times 10^{16}\,T^{-0.61}$ | |
| 2c | $H_2(0) + H \rightleftarrows 2\,H + H$ | $k_{-2c} = 20\,k_{-2a}$ | |
| 3 | $HCl(0) + M_3 \rightleftarrows H + Cl + M_3$ | $k_3 = 6.8 \times 10^{21}\,T^{-2}\,\exp(-104{,}000/RT)$ | $A_{Cl_2} = 3$, $A_H = A_{Cl} = A_{HCl} = 5$, $A_M = 1$ for all other M. |
| 4 | $Cl + H_2 \rightleftarrows HCl(v) + H$ | $k_4(v) = 4.8 \times 10^{13}\,\exp[(-5260 - E_v + E_0)/RT]$ | See Table 6 for values of $E_v - E_0$ |
| 5 | $H + Cl_2 \rightleftarrows HCl(v) + Cl$ | $k_5 = 5.7 \times 10^{13}\,g(v)\,T^{0.163}\,\exp(-2520/RT)$ | $g(v) = 0.12, 0.42, 0.39, 0.04, 0.02, 0.002$ for $v = 1$ to $6$, respectively |
| 6 | $HCl(v) + HCl \rightleftarrows HCl(v-1) + HCl$ | $k_6 = v(1.6 \times 10^{17}\,T^{-\frac{3}{2}} + 8.7 \times 10^4\,T^2)$ | $v = 1, \ldots, 6$ |
| 7 | $HCl(v) + H_2 \rightleftarrows HCl(v-1) + H_2$ | $k_7 = 9.6 \times 10^3\,v\,T^{2.23}$ | $v = 1, \ldots, 6$ |
| 8 | $HCl(v) + Cl_2 \rightleftarrows HCl(v-1) + Cl_2$ | $k_8 = 5 \times 10^4\,v\,T^2$ | $v = 1, \ldots, 6$ |
| 9 | $HCl(v) + H \rightleftarrows HCl(v') + H$ | $k_9 = 5 \times 10^{13}\,\exp(-650/RT)$ | $v = 1, \ldots, 6,\ v' < v$ |
| 10 | $HCl(v) + Cl \rightleftarrows HCl(v') + Cl$ | $k_{10} = 2 \times 10^{13}\,\exp(-700/RT)$ | $v = 1, \ldots, 6,\ v' < v$ |
| 11 | $HCl(v) + M_{11} \rightleftarrows HCl(v-1) + M_{11}$ | $k_{11} = 7.8 \times 10^{-7}\,v\,T^5\,A_M$ | $A_{Ar} = 1$, $A_{He} = 15$ |
| 12 | $HCl(v) + HCl(v') \rightleftarrows HCl(v+1) + HCl(v'-1)$ | $k_{12} = 4 \times 10^{14}\,v'(v+1)\,T^{-1}$ | $v = 1, \ldots;\ 1 \leq v' \leq v$ |

the hydrogen halides in the presence of other hydrogen halides as well as their self-deactivation rates. These rates are all for the first vibrational level. There is either a single measurement or two measurements in close agreement except for some deactivation rates involving HF or DF. The three measurements of DF–HF relaxation agree within a factor of 1.3 while factors of 1.3 and 1.8 separate the two values for the HCl–HF and HBr–HF relaxation rates, respectively.

The probabilities of deactivation of the hydrogen halides by HF and DF range between $1.6 \times 10^{-3}$ and $11 \times 10^{-3}$ per collision if HF self-deactivation is included, while the deactivations by the hydrogen halides other than HF(DF) have much lower probabilities of $3.66 \times 10^{-5}$ to $2.3 \times 10^{-4}$. The probability of HBr($v = 1$) deactivation by HF is greater than that of HF($v = 1$) by HBr, since Table 17 shows that $P_{\mathrm{HBr-HF}} = 1.6 \times 10^{-3} > (P_{vv} + P_{\mathrm{HF-HBr}}) = 1.3 \times 10^{-3}$, where $P_{vv}$ is the probability of V–V transfer from HF($v = 1$) to HBr in a single collision. Similarly, $P_{\mathrm{DBr-DF}} > (P_{vv} + P_{\mathrm{DF-DBr}})$ and $P_{\mathrm{HCl-HF}} \approx (P_{vv} + P_{\mathrm{HF-HCl}})$ (See Ref. 180). These data show that $P_{\mathrm{HF-HF}} > P_{\mathrm{HX-HF}} > P_{\mathrm{HF-HX}}$ for X = Cl or Br, and that $P_{\mathrm{DF-DF}} > P_{\mathrm{DBr-DF}} > P_{\mathrm{DF-DBr}}$. It appears that if $P_{\mathrm{HX-HX}} > P_{\mathrm{HY-HY}}$ for a pair of hydrogen halides, then $P_{\mathrm{HY-HX}} > P_{\mathrm{HX-HY}}$. This empirical rule permits one to estimate the values of unknown rate coefficients.

Chen and Moore proposed three V–R deactivation models[174] for the hydrogen halides and concluded that a considerable fraction, perhaps almost all, of the vibrational energy of the initially excited molecule is converted into rotational energy of the same molecule. Their model I (Fig. 8 of Ref. 174) predicts the deactivation rates of a vibrationally excited molecule HX($v = 1$) by HY and DY to be the same. This prediction agrees with the data for HCl($v = 1$) and HBr($v = 1$) deactivation by HF(DF) if we assume on the basis of the preceding empirical rule that $P_{\mathrm{HCl-DF}} > P_{\mathrm{DF-HCl}}$ so that $k_{\mathrm{HCl-DF}} \approx (1.6 \pm 0.3) \times 10^{-2} \, (\mu\text{sec-Torr})^{-1}$. Their Model I predictions do not agree quite so well with the results for HF–DF, HCl–DCl, or HBr–DBr relaxations, where in each case the experimental data show $P_{\mathrm{DX-HX}} > P_{\mathrm{DX-DX}}$. Chen and Moore[174] suggested that a small contribution from Models II or III could account for this. One might conclude that Model I dominates the other two models more completely when the collision partners contain different halogens than when they both contain the same halogen, but there is no good theoretical basis for this.

Relaxation times of DF($v = 1$) in the presence of HF have been measured[157,160,181] at temperatures between 295 and 750°K. The data fall between the relaxation times of HF–HF and DF–DF and can be described with the same $T^2$ temperature dependence. Blauer and co-workers[162] observed fast V–V coupling in mixtures of HF and DF at 1600 to 3000°K but could not extract a numerical value for the rate coefficient for this process. Relaxation rates for $HCl(v = 1) - HF$,[180,181] $HCl(v = 1) - DF$,[182] and $HBr(v = 1) - HF$[181] at temperatures higher than 300°K have inverse-temperature dependences as do the self-deactivation rates of HF and DF. No temperature-dependence data are available for hydrogen-halide relaxation by hydrogen halides except for HF(DF) and self-relaxation of HBr and HCl. Experimental results of Airey and Smith[183] indicate that HF deactivates $DF(v = 2)$ approximately 1.30 times as fast as it deactivates $DF(v = 1)$. We suggest scaling the rates as $k_{v,v-1} = vk_{1,0}$ until more extensive data are available.

**V–R,T Deactivation of the Hydrogen Halides by Atoms and Other Molecules.** Seery[184] measured the relaxation of $HCl(v = 1)$ by rare gases in high-temperature shock-tube experiments and determined the relaxation-rate dependence on the mass of the collision partners Kr, Ar, He, and Ne. He found He to be the most efficient and Kr the least. In a previous study, Bowman and Seery[173] had measured somewhat faster relaxation times for HCl–Ar than did Seery.[184] The rare gases are very inefficient deactivators of the hydrogen halides at room temperature, and only a few measurements have been reported. The relaxation times of HCl and DCl by He were found[176] to be $660 \pm 100$ and $610 \pm 60$ $\mu$sec-atm, respectively. A measurement[185] of HBr relaxation by He gave 120 $\mu$sec-atm. Relaxation-time measurements[186] between 295 and 700°K for HCl and DCl in the presence of $^4$He and Ar are approximately linear on a Landau–Teller plot of $\ln P\tau$ versus $T^{-1/3}$. Extrapolations of these HCl–Ar data agree with the mean of the two shock-tube measurements for HCl–Ar relaxation. Although He deactivates HCl and DCl 15 times faster than does Ar, the relaxation times have the same temperature dependence between 295 and 2000°K.

Two laboratories have reported relaxation times for HF–Ar relaxation[147,149] and DF–Ar relaxation[149,163] between 1500 and 5000°K. The HF–Ar data are in substantial agreement with each other and also with the mean of the HCl–Ar data, which is something of a surprise. The DF–Ar relaxation times of Bott and Cohen[163] and Vasil'ev et al.[149] were in agreement at 4600°K but differed by a factor of 4.3 at 1500°K. It was reported by Steele and Moore[186] that Ar deactivated HCl and DCl at very similar rates. Interestingly enough, the geometric mean of the two sets of DF–Ar relaxation times of Bott and Cohen[163] and Vasil'ev et al.[149] were whereas, the HCl data[184] show He to be 15 times more efficient than Ar, Bott and Cohen[147] found He to be only a factor of 2 more efficient than Ar for deactivating HF. Blair et al.[159] measured somewhat faster HF–Ar relaxation times between 800 and 2400°K than the overlapping and extrapolated data of Bott and Cohen[147] and Vasil'ev et al.[149] Their data also show some curvature on the Landau–Teller plot, while the HCl–Ar data do not. The curvature could be real or due to the effects of impurities, such as $H_2O$ or residual HF, which become more important at low temperatures.

The SSH V–T calculations do not adequately predict the data for the rare-gas deactivations of the hydrogen halides.[163,176] Trajectory calculations[187] for the relaxation of HF by Ar agree within a factor of 3 with the shock-tube data but do not have as steep a temperature dependence. Berend and Thommarson[187] concluded that a reduction of the range parameter $\alpha_{Ar-H}$ by 20 to 30% would lead to much better agreement with the shock-tube data. Such a variation is well within the range of uncertainty of $\alpha$. Variation of 20 to 30% in the range parameter does not substantially improve the SSH predictions. Steele and Moore[186] made approximate calculations for the relaxation of HCl(DCl) by Ar and He, and concluded from the qualitative fit to their data that both rotational and translational degrees of freedom play important roles.

We have used in Tables 20, 21, and 22 the same rates for deactivation of HF, DF, and HCl by Ar. This seems appropriate considering the shock-tube data and lack of HF(DF) data at low temperatures. In most systems these deactivations should not play a large role anyway.

Halogen and hydrogen atoms are much more efficient deactivators of the hydrogen halides than are the rare gases. However, the measurement of these rates requires the production and measurement of atomic species, with attendant complications. Three teams of workers[188-190] have reported room-temperature rates for deactivation of HCl by $Cl_2$; two of these reported rates for HCl–Cl deactivation. These data are listed in Table 16. Craig and Moore[188] reported a value for $k_{HCl-Cl}$ that is faster by a factor of 10 than that of Ridley and Smith[189] and a value for $k_{HCl-Cl_2}$ slower by a factor of 10. The large spread in experimental values makes a choice of rate coefficients difficult. On the basis of preliminary experimental results in our laboratory, we have chosen rate coefficients for HCl($v = 1$) deactivation by $Cl_2$ that agree more closely with the measurements of Craig and Moore and that are proportional to $T^2$. Ridley and Smith interpreted their HCl–Cl results in terms of an exchange reaction, where the incident Cl atom abstracts the H atom from the excited HCl and forms a ground-state HCl. If this is true, the differences in rate coefficients for the several vibrational levels of HCl can be ascribed to differences in activation energy of the abstraction process. Thompson[191] with classical trajectories calculated the overall rate of energy transfer for this process, and found that, indeed, it could explain the upper vibrational-level data of Ridley and Smith. Thommarson and Berend[192] performed quasiclassical-trajectory calculations exploring the effect of different barrier heights. Their calculated vibrational-excitation rate coefficients increased with decreasing barrier height, but the overall values were lower than either of the two experimental values. A third trajectory calculation by Smith and Wood[193] results in deactivation rates for $HCl(v = 1, 2; J = 3)$ by Cl that are slower by a factor of 3 than those of Ridley and Smith. Their calculated rates had activation energies of 1.8 and 1.0 kcal/mol for the first and second levels, respectively.

It is difficult to choose between the divergent values of Ridley and Smith on the one hand and Craig and Moore on the other in selecting a reasonable rate coefficient for HCl vibrational relaxation by Cl atoms. Experimental uncertainties are echoed in theoretical uncertainties. It is not known with certainty whether there is an attractive well or a potential barrier in the Cl–HCl potential energy surface. A barrier would suggest deactivation rates that increase with vibrational energy of the reactant HCl. On the other hand, a well would imply rates that are possibly independent of $v$, or perhaps even decreasing slightly with $v$. Ridley and Smith found deactivation rates that increased with $v$, which is consistent with the barrier picture. Craig and Moore found a value for $k_{1,0}$ that is an order of magnitude faster than that of Ridley and Smith and is more indicative of a well than of a barrier. If there is a well in the surface, then $k_{2,1}$ and $k_{3,2}$, had Craig and Moore measured them, would probably not differ very much from the measurements of Ridley and Smith. In other words, the upper vibrational rates of Ridley and Smith are not necessarily to be questioned because of the discrepancy in the $k_{1,0}$ values. In Table 22 we have used the value of $k_{1,0}$ derived by Craig and Moore and set the higher $k$'s to the same value, recognizing the considerable uncertainty that now exists in all the values.

Bott and Cohen[194] and Blauer et al.[162] have measured the deactivation of $HF(v = 1)$ by F atoms produced by dissociating $F_2$ in one case[162] and $SF_6$ in the other case[194] at temperatures between 1900 and 3500°K in shock tubes. They reported rate coefficients of $5 \times 10^{13}$ cm$^3$/mol-sec[162] and $3 \times 10^{13}$ cm$^3$/mol-sec[194] at 2500°K. Similar studies[163,195] were reported for DF–F relaxation rates. Blauer and Solomon[196] also measured by a different technique the rate for $HF(v = 1)$ and $HF(v = 2)$ relaxation by F atoms at 2545°K. They monitored the formation and decay of $HF(v = 1)$ and $HF(v = 2)$ in the reaction of F with HCl behind an incident shock wave. Their estimate of $8 \times 10^{12}$ cm$^3$/mol-sec for the deactivation of $HF(v = 1)$ is slower by a factor of 6 than both their previously measured value[162] and their estimate for $HF(v = 2)$ deactivation by F. This slow-rate coefficient agrees with the value obtained by Blair et al.[159] with a laser-induced fluorescence behind a reflected shock wave in mixtures of

HF and $F_2$. Of the four measurements, the faster two were obtained for conditions for which the underpopulated first vibrational level was relaxing toward an equilibrium concentration while the slower rates were measured for an overpopulated HF($v = 1$) concentration relaxing toward a lower equilibrium concentration. Thompson[197] performed trajectory calculations and obtained rates that extrapolate to $8 \times 10^{12}$ cm³/mol-sec at 2500°K in agreement with the lower of the two sets of experimental numbers; however, the laser-induced fluorescence data between 1500 and 2500°K have a much steeper temperature dependence. Wilkins[198] calculated a lower value of $2 \times 10^{12}$ at 2500°K with a different potential-energy surface. Shin[199] assumed an inverse-power potential and calculated the transition probability for the de-excitation of HF($v = 1$) by F. His calculations extrapolate to $9 \times 10^{11}$ cm³/mol-sec at 2500°K; however, his rate coefficient increases below 800°K to about $2 \times 10^{13}$ cm³/mol-sec at 300°K. Neither Wilkins nor Thompson calculated such an increase. For laser-modeling purposes, we have chosen $K_{1,0} = 10^{13.2} \exp(-2700/RT)$ for deactivation of HF by F atoms, which has the temperature dependence obtained by Wilkins between 300 and 1000°K but has a value of $9 \times 10^{12}$ cm³/mol-sec at 2500°K, a compromise between theoretical and experimental results. The theoretical rates[197,198] for DF($v = 1$) deactivation by F are comparable to the HF($v = 1$)–F rates at 2500°K and are a factor of 2 slower at 300°K. The experimental data, however, show the DF($v = 1$)–F deactivation rate to be slower than that of HF($v = 1$)–F by a factor of 2 at 2500°K. We suggest in Table 21 for DF($v = 1$) that $k_{1,0} = 10^{13.2} \exp(-3380/RT)$ cm³/mol-sec.

Wilkins' theoretical calculations[198] show that multiquantum transitions contribute significantly to the deactivation of the vibrational levels higher than 2. Both Wilkins and Thompson found that the rate coefficients for single-quantum deactivation HF($v$) to be between $v$ and $v^2$ times as fast as the deactivation of HF($v = 1$). We have listed rates in Tables 20 and 21 that are proportional to $v$. In most cases HF deactivation by F atoms will have a minimal effect on laser performance compared to other mechanisms. The rate coefficient for HF($v = 1$) deactivation has a steep temperature dependence and decreases to $\sim 10^{11.2}$ cm³/mol-sec at room temperature. Both HF and H atoms are usually more important deactivators. If F atoms are found to contribute significantly to the deactivation in specific lasers, it may be desirable to include reactions involving multiquantum transitions.

In contrast to HCl–$Cl_2$ and HBr–$Br_2$ relaxation, $F_2$ does not deactivate HF($v = 1$) efficiently. Fried et al.[155] measured the deactivation rate of HF($v = 1$) by $F_2$ to be less than $2 \times 10^9$ cm³/mol-sec at 350°K. Donovan et al.[200] found that both $Br_2$ and Br deactivate HBr($v = 1$) equally fast with a deactivation probability of about 0.01, or about 27 times faster than HBr itself. Therefore, the slow $HF(v = 1)-F_2$ rate is somewhat unexpected. For Tables 20 and 21, we have assumed $F_2$ to have the same efficiency as Ar.

The deactivation of HF(DF) by H atoms has been studied by classical-trajectory calculations.[201,202] The deactivation can occur by (1) the loss of vibrational energy to translation or rotation (nonreactive), (2) an exchange reaction in which the incident H atom abstracts the F atom from the excited HF leaving the vibrational energy as kinetic energy in the liberated H atom (reaction), and (3) for levels of $v \geq 3$, the incident H atom can react with $HF(v)$ forming $H_2$ and F. Wilkins[125] calculated the deactivation rates for the third case, $H + HF(v) \rightarrow H_2(v') + F$, $v \geq 3$, with the same LEPS surface he used for the pumping reaction, $F + H_2$. The deactivation results have been presented in Section 2.2 on pumping reactions in the $H_2(D_2)-F_2$ systems. Using this same surface, Wilkins calculated for both reactive and nonreactive collisions the rates for $H + HF(v) \rightarrow H + HF(v')$. Rates obtained with the surface for the pumping reactions were in good agreement with experimental data, although this is no guarantee that the surface correctly describes the part of the collision important for $H + HF(v) = H + HF(v')$. Wilkins' results[202] show that in contrast to the HO model the probabilities of multiquantum transitions from a given vibrational level are within a factor of 3 of those for single-quantum transitions. When the reactive and nonreactive processes are added together for $H + HF(v) \rightarrow H + HF(v')$, the resulting rate coefficients can be described within a factor of 2 by $k = 10^{13.25 \pm 0.2} \exp(-700 \pm 20/RT)$. Kwok and Wilkins[203] have interpreted flow-tube data at 300°K assuming single-quantum transitions. They estimated values of $(0.7 \pm 0.4) \times 10^{12}$, $9 \times 10^{12}$, and $1.4 \times 10^{13}$ cm$^3$/mol-sec for the deactivation of the $v = 1$, 2, and 3 levels, respectively. These compare with Wilkins' calculated values of $2.5 \times 10^{12}$, $1.0 \times 10^{13}$, and $1.7 \times 10^{13}$ for the same levels. We have listed in Table 20 deactivation rates for $HF(v \geq 2)$ by H that are close to Wilkins' results.[202] A slower rate coefficient for $HF(v = 1)-H$ deactivation which is more in agreement with the experimental datum has been suggested. Wilkins[204] has made similar calculations for the deactivation of $DF(v)$ by D. In the absence of experimental data, we have used Wilkins' calculated rates for $v \geq 2$ and have reduced his calculated rate for $DF(v = 1)-D$ deactivation by the same proportion as we reduced his $HF(v = 1)-H$ rate to agree with the experimental value. The deactivation of $HF(v)$ by H plays a large role in degrading the performance of certain HF lasers if these rates are accurate for the upper levels. There is a definite need for further experimental study of these processes to confirm the fast deactivation rates of $HF(v)$ and $DF(v)$ by H and D.

Smith and Wood[193] have calculated rates for H-atom deactivation of $HCl(v = 1$ and 2) in a classical-trajectory study. However, the calculations were performed for a single $J = 3$ rotational level and can be taken only as a guide. Although their value of $8 \times 10^{10}$ cm$^3$/mol-sec at room temperature is approximately five times as fast as HCl self-deactivation, it is much slower than the value of $(3.9 \pm 1.3) \times 10^{12}$ cm$^3$/mol-sec measured by Arnoldi and Wolfrum[205] in a laser-induced fluorescence experiment. Such a fast rate coefficient should not be very temperature dependent, and we have listed $k_{v,v-1} = 1.4v \times 10^{14} \exp(-1000/RT)$ in Table 22. Additional experimental or theoretical studies will be required to establish the precise $v$ and $T$ dependence.

The V–R,T relaxation times of $DF(v = 1)$ in the presence of $H_2$ have been measured by Bott[206] between 295 and 750°K in a combined shock-tube laser-induced fluorescence experiment. The results agree with extrapolated results of a shock-tube study[163] in the range of 900 to 3000°K. Bott also reported[206] rates for HF–$H_2$ and DF–$D_2$ from 295 to 550 K; however, the precision of the data is much lower because of the V–V coupling of the molecules. The room temperature V–R,T data for the other hydrogen halides in the presence of $H_2(D_2)$ have been listed in Ref. 206. Although not all of the possible combinations have been studied, it appears that the deactivation rate of the hydrogen halides by $H_2(D_2)$ is roughly independent of the identity of the halogen atom. Although hydrogen deactivates the hydrogen-halide molecules much more rapidly than it deactivates itself, the isotopic substitution of D for H atoms appears to have a similar effect in hydrogen-halide deactivation by hydrogen as in hydrogen self-deactivation. For lack of any better information, the temperature dependence of the relaxation rates of $DF(v = 1)$ by $H_2$ can be used as a guide for the deactivation rates of other hydrogen halides by $H_2(D_2)$.

Breshears and Bird[171] have measured the deactivation rates of HI and DI by $N_2$ at temperatures in the range of 1000 to 2700°K and 1200 to 2000°K, respectively, and obtained relaxation times of 1 to 10 $\mu$sec-atm. They concluded that HI(DI) deactivated $N_2$ too slowly to be measurable. Sentman and Solomon,[207] on the other hand, concluded that deactivation of $N_2$ by HF is fast while deactivation of HF by $N_2$ is negligible at temperatures from 1500 to 3000°K. The rates for $HF(v = 1)$ relaxation in the presence of $N_2$ have been measured at room temperature[152,156] and between 295 and 1500°K[180] with the laser-induced fluorescence technique. These rates contain both V–V and V–R,T contributions but the V–V contribution is expected to dominate. The V $\rightarrow$ V energy exchange requires a smaller amount of energy to be converted to rotational (translational) degrees of freedom and probably is much faster than the V–R,T deactivation of $HF(v=1)$ by $N_2$. Chen[208] observed double exponential decays in the relaxation of $HBr(v=1)$–$N_2$ mixtures and estimated the V–R,T relaxation rate of $HBr(v=1)$ by $N_2$ to be less than 3% of its V–V relaxation rate.

## 3.2. V–V Energy Transfer

Vibrational–vibrational energy-transfer data for the hydrogen halides in the temperature range of interest consist almost entirely of experimental results. However, there are a few theoretical calculations that should be mentioned. Sentman[210] has calculated classical energy-transfer rates at high temperatures for

$$HX(v=1)+AB(v=0) \rightleftarrows HX(v=0)+AB(v=1) \qquad (19)$$

where X represents Cl or F and AB is $N_2$, $H_2$, $D_2$, or several of the hydrogen halides. Dillon and Stephenson[211] calculated rate coefficients for HF, DF, and HCl V–V energy transfer to $CO_2$ using a model that does not rely on the Born approximation and that incorporates curved classical trajectories. They obtained excellent agreement with HCl–$CO_2$ experimental data[212] between 298 and 510°K. Sharma et al.[213] have used the Sharma–Brau theory to make calculations for V–V exchanges of $HCl(v=2)$ with $HCl(v=0)$ and $HBr(v=2)$ with $HBr(v=0)$. On the other hand, experimental data have been obtained for a large number of room-temperature exchanges involving the first vibrational level and a smaller number of exchanges involving the higher vibrational levels. Moore[214] has described the experimental techniques for measuring V–V energy-transfer rates, and has listed data for several V–V exchanges. In many of the experiments, the small contribution of V–R,T deactivation cannot be separated from the deactivation by V–V transfer, so that the sum of the two rates has been obtained. The $v$ dependence of the V–V rates in most cases has to be estimated for lack of experimental information. We have assumed $k_{v,v-1} = vk_{1,0}$ for this dependence based on the HO model in the absence of other data, although we expect this will prove to be a poor rule in the case of hydrogen halides when experimental results become available.

**$HX(v)+HX(0) \rightleftarrows HX(v-1)+HX(1)$.** The hydrogen-halide molecules formed in a pumping reaction have an initial vibrational distribution that is rearranged by rapid V–V energy exchanges. These exchanges are fast because only small amounts of excess energy must be converted to R–T energy. For HF/DF lasers, the redistribution of energy in the vibrational levels of the same molecule has little effect[139] on total power but affects the spectral output. In HCl lasers[1] the total power can be affected by V–V transfers. The HO model predicts that only single quantum, $\Delta v = 1$, transitions are important, and the exchange can be described by

$$HX(v) + HX(v') \underset{k_{v,v-1}^{v',v'+1}}{\overset{k_{v,v-1}^{v',v'+1}}{\rightleftharpoons}} HX(v-1) + HX(v'+1) \qquad (20)$$

with rate coefficients related by ·

$$k_{v,v-1}^{v',v'+1} = v(v'+1)k_{1,0}^{0,1} \qquad (21)$$

The validity of this model is subject to the same uncertainties previously discussed for V–T reactions. As the hydrogen halides are very anharmonic, the actual relationship among the various $k$'s may be considerably different, and, in fact, $k$ may decrease rather than increase with $v$. However, the relationship is obeyed reasonably well for V–V transfer between NO and $N_2$,[215] which are more nearly harmonic.

Several groups have reported experimental results for Reaction 20 with $v' = 0$, $v = 2$, and $HX = HF$,[153,154,183] $DF$,[164] $HCl$,[216–219] and $HBr$,[220,221] at room temperature. There is good agreement among overlapping studies for each molecule. HCl and HBr have similar V–V rates, just as they have similar V–R,T rates, and their V–V rates are slower than those for HF(DF) by a factor of 5 to 10. Studies of the rate coefficients for reactions involving HCl and HBr[222] and DF[223] show inverse temperature dependences characteristic of attractive collisions. In all three cases, the probabilities for the exothermic transfer vary as $T^{-3/2}$, the same temperature dependence as that measured for HF(DF) self-relaxation in this temperature range.[160]

Osgood et al.[224] obtained measurements for the V–V exchange of HF($v$) with HF(0) for $2 \le v \le 4$ in experiments in which one to three vibrational levels were simultaneously pumped in fixed ratios by the absorption of laser radiation. The exchange rates were deduced from the rise times of the fluorescence from the vibrational level above the highest level being pumped. They obtained rate coefficients for the exothermic exchanges

$$HF(1) + HF(v') \underset{k_{1,0}^{v',v'+1}}{\overset{k_{1,0}^{v',v'+1}}{\rightleftharpoons}} HF(0) + HF(v'+1) \qquad (22)$$

in ratios of

$$k_{1,0}^{3,4} : k_{1,0}^{2,3} : k_{1,0}^{1,2} : : 8.8 : 3.7 : 1$$

which differ considerably from the HO model predictions of $2 : 1.5 : 1$. In an earlier experiment, Airey and Smith[183] measured these rates and the next one in the sequence in a steady-state fluorescence-quenching experiment and obtained rates in the ratio of $44 : 13.5 : 2.4 : 1$. The fastest two rates both exceed the gas-kinetic collision rate. In both sets of experiments, the dependence on $v$ is much stronger than the HO prediction. Neither study yields a direct measurement of the rate coefficients but depends either on computer modeling or properly adjusted initial vibrational-level concentrations. Possible multiquantum transitions might effect these two experiments in different ways, and, therefore, present another difficulty for the data reduction. The deduced V–V rate coefficients are also sensitive to the choice of HF–HF V–T relaxation rates for upper vibrational levels. Osgood et al. assume the HO type of relationship, that is, $k_{v,v-1} = vk_{1,0}$; a stronger $v$ dependence, say $k_{v,v-1} = v^2k_{1,0}$, will weaken the $v$ dependence of the V–V rate. If the published analyses are accepted, a $v^3$ dependence would be inferred for scaling the V–V rates with $v$. However, when these fast rates for upper levels are used in modeling HF lasers, they lead to greater emission from high levels ($v > 4$) than is observed and less emission from low levels. This argues against the validity of such rapid V–V rates for $k_{1,0}^{3,4}$ and $k_{1,0}^{4,5}$, which exceed gas kinetic, not to mention the unlikelihood of a $v^3$ extrapolation to larger $v$ levels being correct. The rates cannot continue to increase with $v$, and may, in fact, reach a maximum and then decrease. On the other hand, a more complicated set of reactions, including multiquantum transitions, may be required to explain both laser-emission and laser-induced fluorescence experiments in a consistent manner. Thus, at present, there are insufficient data to resolve the difficulties of scaling these V–V and V–R,T processes to the higher levels. We have set $k_{v,v-1}^{v',v+1} = v(v'+1)k_{1,0}^{0,1}$ in Tables 20 and 21. However, the numbers should be used with caution.

Gorshkov et al.[216] estimated HCl V→V rates from gain measurements in an HCl laser. They measured $k_{1,0}^{1,2} = (1.5 - 3) \times 10^{12}$ cm$^3$/mol-sec which is consistent with later measurements,[217-219] $k_{1,0}^{2,3} = (0.64 - 1.28) \times 10^{12}$, and $k_{2,1}^{2,3} = (4.5 - 9) \times 10^{12}$ cm$^3$/mol-sec.

Ridley and Smith[225] examined the vibrational relaxation of HCl produced by the reaction of Cl atoms, formed in a discharge through $Cl_2$, with HI. They measured the quenching of HCl emission in the presence of added HCl (in the ground vibrational state) and thus obtained quenching rates for both HCl(2) and HCl(3). Since the rates, thus obtained, were over an order of magnitude faster than the published rate of HCl(1) V–T relaxation by HCl, they attributed the observed quenching to V–V processes:

$$HCl(2) + HCl(0) \rightarrow HCl(1) + HCl(1)$$
$$HCl(3) + HCl(0) \rightarrow HCl(2) + HCl(1)$$

obtaining the rate coefficients $8.4 \times 10^{11}$ and $1.1 \times 10^{12}$ cm$^3$/mol-sec, respectively, for the reactions. If the contributions of the V–T processes to the relaxation are indeed negligible, as they assume, and, furthermore, only the one single-quantum transfer process is occurring, then the exothermic rate coefficients are $k_{1;0}^{1;2} = 1.4 \times 10^{12}$ and $k_{1;0}^{2;3} = 2.9 \times 10^{12}$. The first of these values is about a factor of 2 slower than the measurements reported previously,[102–105] all of which are in reasonably good agreement with each other. Therefore, it is difficult to assess the accuracy of the value for $k_{1;0}^{2;3}$. It should also be noted that Ridley and Smith deduced $k_{1;0}^{2;3}$ to be about twice as fast as $k_{1;0}^{1;2}$; whereas, Gorshkov et al.[216] found it to be less than half as fast. In Table 22 we use the HO rule for evaluating $k$'s for larger $v$'s.

**The V–V Exchange Between Two Hydrogen Halides.**   Table 17 shows the results of a number of room-temperature measurements of V–V transfer rates from one hydrogen halide to another. These exchanges involve the de-excitation of the first vibrational level of one and the excitation of the first level of the second molecule. The maximum spread in values obtained by different investigators is the factor of 2 in the case of HF–DF. Data for the other exchanges are in closer agreement. In general, the probability of energy exchange for the exothermic process

$$HX(v = 1) + HY(v = 0) \rightleftarrows HX(v = 0) + HY(v = 1) - \Delta E \qquad (23)$$

decreases as the energy defect $\Delta E$ [which must be converted to rotational (translational) energy] increases. Published plots of probability versus $\Delta E$ show this dependency for initially excited HF[180] and HCl.[226] The temperature dependences of the rate coefficients for exothermic transfers from HF and DF have been measured[180–182] between 295 and 750°K. The rate coefficients ($cm^3$/mol-sec) for the exothermic transfer have negative temperature dependences of about $T^{-1}$ near 300°K. They become nearly temperature independent near 700°K, above which the rates appear to increase. Again, the $v$ dependence can only be estimated, probably lying between the HO prediction of $k(v)$ proportional to $v$ and the stronger $v$ dependence previously described for HF–HF results of $k(v)$ proportional to $v^3$. We suggest using the HO predicted $v$ dependence.

**The V–V Exchange Between Hydrogen Halides and Homonuclear Diatomic Molecules.**   Chen[208] and Bott and Cohen[180] have plotted the probabilities for exothermic energy exchange in collisions of hydrogen halides with homonuclear diatomics at room temperature. They are 10 to 100 times less probable than the energy exchange between two hydrogen halides having a comparable energy defect. The temperature dependences have been measured in only a few cases for transfers involving HF,[180,207] DF,[182] and HI and DI.[171] Bott and Cohen[180] and Bott[182] show that between 300 and 800°K the rate coefficients are independent of temperature for nearly resonant transfers and are proportional to $T$ for the transfers having larger energy discrepancies. For example, for V–V transfer from $H_2(1)$ to $HF(0)$, they found $k = 8.3 \times 10^{11}$ cm$^3$/mol-sec, and for $D_2$–DF, they reported[160,182] $k = 5 \times 10^{11}$, both values being temperature independent. The data[171] for $N_2$ exchange with HI and DI show a similar temperature behavior between 1000 and 2800°K.

Kwok[227] has studied the deactivation of $HF(v)$ by $H_2$ in a discharge-flow system at 300°K and obtained the decay rates for $v = 1$ to 5. If only single-quantum transitions are assumed and V–T processes can be neglected, then exothermic-rate coefficients can be calculated for the reactions

$$HF(v-1) + H_2(1) \xrightleftharpoons{k_{1,0}^{v-1,v}} H_2(0) + HF(v) \tag{24}$$

and his data indicate $k_{1,0}^{1,2}/k_{1,0}^{0,1} = 3.2$, $k_{1,0}^{2,3}/k_{1,0}^{0,1} = 10$, $k_{1,0}^{3,4}/k_{1,0}^{0,1} = 22$, and $k_{1,0}^{4,5}/k_{1,0}^{0,1} = 46$. The analysis becomes more complicated when multiquantum transitions are allowed and more equations have to be included. Nevertheless, the data indicate that the $v$ dependence of the rate coefficients $k_{1,0}^{v-1,v}$ is close to or slightly larger than $v^2$. Similar results of $k_{1,0}^{1,2}/k_{1,0}^{0,1} = 4$ were obtained by Anlauf et al.[228] We have tentatively left out the possibility of multiquantum transitions and scaled the rates according to Kwok's results. In the tables we have assumed that the exchange rates are independent of the vibrational level of $H_2$. The same scaling is assumed in the DF–$D_2$ case.

**The V–V Transfer from the Hydrogen Halides to Heteronuclear Diatomics and Polyatomics.** For some laser systems this category of hydrogen-halide deactivations can dominate other processes, since it includes V–V transfer to molecules supplying either H(D) or the halide for the pumping reaction, molecules produced in a precombustor, and impurities such as $H_2O$. On the other hand, this exchange reaction has been used to advantage in the case of HF, DF, and HCl transfer to $CO_2$ to produce lasing on excited $CO_2$. (See the chapter by Cool.)

Room-temperature data have been reported for V–V transfer to CO and NO from HF(DF),[180,182,220] HCl(DCl),[176,226] HBr(DBr),[176,208] and DI.[176] These transfer rates are much faster than the transfer rates to the homonuclear diatomics such as $N_2$, although the energy discrepancies are similar. Bott[182] reported weak temperature dependences for the rate coefficients of HF(DF) transfer to CO and NO.

The very rapid deactivation of HF(DF) by $H_2O(D_2O)$ has been measured at room temperature[152] and at high temperatures.[230] At room temperature, $H_2O$ relaxes HCl with a probability[174] of 0.1 and HBr with a probability[185] of only $4.5 \times 10^{-4}$. Deactivation studies[228,231] of HF by parafinic hydrocarbons, which are of interest as H-atom sources, indicate the transfer rate increases with the size of the $C_nH_{2n+2}$ molecules. The deactivation rate of $HF(v = 1)$ by $CH_4$ is comparable to the HF self-relaxation rate, and, therefore, such molecules can contribute significantly to the deactivation of $HF(v = 1)$. Anlauf et al.[228] found HF(2) to be deactivated about four times faster by these hydrocarbons than HF(1). Rates have also been measured[226,232] for HCl, DCl transfer to $CH_4$ and $CD_4$.

The rates for V–V exchange of $HF(v = 1)$ with a wide range of fluorine containing molecules have a wide range of values, from $1.1 \times 10^5$ (sec-Torr)$^{-1}$ for $ClF_3$[231] down to $90 \pm 75$ (sec-Torr)$^{-1}$ for $SF_6$.[155] Molecules such as $CBrF_3$, $CF_4$, and $SiF_4$ deactivate $HF(v = 1)$ inefficiently with rates of $(4 \pm 1) \times 10^2$ (sec-Torr)$^{-1}$.

## NOTE ADDED IN PROOF

In the time that has elapsed since the completion of the manuscript version of this chapter, considerable work has been carried out in the three principal chemical laser systems discussed: $H_2$–$F_2$, $D_2$–$F_2$, and $H_2$–$Cl_2$. Some of the uncertainties discussed in this review have been resolved; we consider it, therefore, appropriate to draw attention to these reports. In all cases but one the new work is in the area of energy transfer processes; the exception is a report on the three-body recombination of F atoms. Ganguli and Kaufman[233] measured the room temperature process in the presence of Ar as chaperone and obtained an approximate value of $\log k_{-4}^{Ar} \cong 13.5$, which is smaller than the value recommended in Tables 20 and 21 by about a factor of 3.

Arnoldi and Wolfrum[234] measured the vibrational relaxation of HCl in the presence of H atoms at 300°K and obtained a rate coefficient of $4 \times 10^{12}$, which is about a factor of 4 slower than the value given in Table 22. The important uncertainty in the deactivation rate of HCl(1) by Cl atoms seems to have been resolved by the recent remeasurements in Moore's laboratory; a result of $5.1 \times 10^{12}$ cm/mole-sec was obtained,[235] substantiating the faster of the two rates previously discussed.

Heidner and Bott[236] studied the deactivation of HF(1) by H atoms and reported a rate coefficient of $1.4 \times 10^{11}$ cm/mole-sec, somewhat smaller than the initial trajectory calculations had suggested; this can be interpreted as indicating a larger barrier in the potential energy surface than the 1.5 kcal/mol that Wilkins had used in his calculations.

Considerable attention has focused on the deactivation of the upper vibrational levels of HF. Cohen and Bott[237] reported arguments leading to the conclusion that the 2–1 V–T deactivation rate is six to seven times larger than the 1–0 rate, which is considerably larger than the harmonic oscillator assumption would imply. A revised conclusion about the $v$-dependence of the V–T rates will affect the deduced V–V rates as well, because most experimental data reflect situations in which the total disappearance of a given HF($v$) level is monitored, with both V–V and V–T processes contributing. Experimental data for $v = 2$, 3, 4, 5 have been reported and discussed by Kwok and Wilkins.[238] One interpretation of the data presented to date is that the V–T rates scale up sharply with $v$ (i.e., approximately as $v^{2.5}$) at least for the first few levels and that the V–V rates are fairly independent of $v$.

Finally, the HF–HF V–T deactivation (of the $v = 1$ level) has been studied down to 200°K by Lucht and Cool.[239] Below about 700°K, the relaxation rate coefficient varies as $T^{-1.35}$.

Some of the material in this review has been updated and expanded and published in separate technical reports.[240,241]

## REFERENCES

1. N. Cohen, T. A. Jacobs, G. Emanuel, and R. L. Wilkins, Int. J. Chem. Kinet. **1,** 551 (1969); **2,** 339 (1970).

2. N. Cohen, "A Review of Rate Coefficients for Reactions in the $H_2$–$F_2$ Laser System," Report No. TR-0172(2779)-2 (The Aerospace Corp., El Segundo, California, 1971); "A Review of Rate Coefficients for Reactions in the $H_2$–$F_2$ Laser System," Report No. TR-0073(3430)-9 (The Aerospace Corp., El Segundo, California, 1972).

3. R. L. Kerber, N. Cohen, and G. Emanuel, IEEE J. Quant. Electron. **QE-9,** 94 (1973).

4. D. L. Baulch, D. D. Drysdale, D. G. Horne, and A. C. Lloyd, *Evaluated Kinetic Data for High Temperature Reactions, Vol. 1* (Butterworths, London, 1972), p. 261.

5. T. A. Jacobs, R. R. Giedt, and N. Cohen, J. Chem. Phys. **47,** 54 (1967).

6. D. O. Ham, D. W. Trainor, and F. Kaufman, J. Chem. Phys. **53,** 4395 (1970).

7. D. W. Trainor, D. O. Ham, and F. Kaufman, J. Chem. Phys. **58,** 4599 (1973).

8. J. E. Bennett and D. R. Blackmore, Proc. Roy. Soc. London Ser. A **305,** 553 (1968).

9. T. A. Jacobs, R. R. Giedt, and N. Cohen, J. Chem. Phys. **48,** 947 (1968).

10. J. P. Rink, J. Chem. Phys. **36,** 1398 (1962).

11. E. A. Sutton, J. Chem. Phys. **36,** 2923 (1962).

12. I. Amdur, J. Chem. Soc. **57,** 856 (1935).

13. A. C. Lloyd, Int. J. Chem. Kinet. **3,** 39 (1971).

14. H. Hiraoka and R. Hardwick, J. Chem. Phys. **36,** 1715 (1962).

15. R. W. Diesen and W. J. Felmlee, J. Chem. Phys. **39,** 2115 (1963).

16. T. A. Jacobs and R. R. Giedt, J. Chem. Phys. **39,** 749 (1963).

17. M. Van Thiel, D. J. Seery, and D. Britton, J. Phys. Chem. **69,** 834 (1965).

18. R. A. Carabetta and H. B. Palmer, J. Chem. Phys. **46,** 1333 (1967); **47,** 2202 (1967).

19. G. Chiltz, R. Eckling, P. Goldfinger, G. Huybrechts, G. Martens, and G. Simoens, Bull Soc. Chim. Belg. **71,** 747 (1962).

20. J. W. Linnett and M. H. Booth, Nature **199,** 1181 (1963).

21. L. W. Bader and E. A. Ogryzlo, Nature **201,** 491 (1964).

22. E. Hutton, Nature **203,** 835 (1964).

23. E. Hutton and M. Wright, Trans. Faraday Soc. **61,** 78 (1965).

24. M. A. A. Clyne and D. H. Stedman, Trans. Faraday Soc. **64,** 2698 (1968).

25. H. Hippler and J. Troe, Chem. Phys. Lett. **19,** 607 (1973).

26. R. P. Widman and B. A. DeGraff, J. Phys. Chem. **77,** 1325 (1973).

27. P. B. Ayscough, A. S. Cocker, F. S. Dainton, and S. Hirst, Trans. Faraday Soc. **58,** 295 (1962).

28. T. A. Jacobs, R. R. Giedt, and N. Cohen, J. Chem. Phys. **49,** 1271 (1968).

29. F. S. Klein and A. Persky, J. Chem. Phys. **59,** 2775 (1973).

30. R. Engleman, Jr. and N. R. Davidson, J. Amer. Chem. Soc. **82,** 4770 (1960).

31. C. D. Johnson and D. Britton, J. Phys. Chem. **68,** 3032 (1964).

32. R. L. Oglukian, "Determination of the Dissociation Rate of Fluorine," AFRPL-TR-65-152 (Air Force Rocket Propulsion Laboratory, 1965).

33. D. J. Seery and D. Britton, J. Phys. Chem. **70,** 4074 (1966).

34. R. W. Diesen, J. Chem. Phys. **44,** 3662 (1966).

35. R. W. Diesen, J. Phys. Chem. **72,** 108 (1968).

36. Th. Just and G. Rimpel, DLR FB 70-02, Deutsche Luft- und Raumfahrt, Porz-Wahn, Germany (1970).

37. W. D. Breshears and P. F. Bird, J. Chem. Phys. **58,** 1576 (1973).

38. V. H. Shui, J. P. Appleton, and J. C. Keck, *Fourteenth Symposium (International) Combustion* (Pittsburgh Combustion Institute, Pittsburgh, 1971), p. 21.

39. W. Valance, B. Birang, and D. I. MacLean, "Measurement of Fluorine Atom Concentrations and Recombination Rates by ESR Spectroscopy," FRK-116 (Boston College, Boston, Massachusetts 1971).

40. N. Cohen, R. R. Giedt, and T. A. Jacobs, Int. J. Chem. Kinet. **5,** 425 (1973).

41. A. G. Clarke and G. Burns, J. Chem. Phys. **56,** 4636 (1972).

42. W. H. Wong and G. Burns, J. Chem. Phys. **58,** 4459 (1973).

43. H. Hippler, K. Luther, and J. Troe, Chem. Phys. Lett. **16,** 174 (1972).

44. H. M. Maier and F. W. Lampe, J. Chem. Phys. **77,** 430 (1973).

45. E. S. Fishburne, J. Chem. Phys. **45,** 4053 (1966).

46. T. A. Jacobs, N. Cohen, and R. R. Giedt, J. Chem. Phys. **46,** 1958 (1967).

47. D. J. Seery and C. T. Bowman, J. Chem. Phys. **48,** 4314 (1968).

48. R. R. Giedt and T. A. Jacobs, J. Chem. Phys. **55,** 4144 (1971).

49. W. D. Breshears and P. F. Bird, J. Chem. Phys. **56,** 5347 (1972).

50. T. A. Jacobs, R. R. Giedt, and N. Cohen, J. Chem. Phys. **43,** 3688 (1965).

51. J. A. Blauer, J. Phys. Chem. **72,** 79 (1968).

52. D. A. Armstrong and J. L. Holmes, Comp. Chem. Kinet. **4,** 143 (1972).

53. R. L. Brown, " A Brief Review of Experimental Determinations of the Rate of Thermal Decomposition of HF in Shock Tubes," NBS Report 10-635 (Washington, D.C.).

54. J. H. Sullivan, J. Chem. Phys. **46,** 73 (1967).

55. J. H. Sullivan, J. Chem. Phys. **49,** 1155 (1968).

56. R. R. Giedt, N. Cohen, and T. A. Jacobs, J. Chem. Phys. **50,** 5374 (1969).

57. K. Westberg and E. F. Greene, J. Chem. Phys. **56,** 2713 (1972).

58. J. H. Sullivan, J. Chem. Phys. **30,** 1292 (1959); **30,** 1577 (1959); **36,** 1925 (1962).

59. G. C. Fettis and J. H. Knox, *Progress in Reaction Kinetics, Vol. 2,* G. Porter, Ed. (Pergamon, London, 1964).

60. J. C. Polanyi and W. H. Wong, J. Chem. Phys. **51,** 1439, 1451 (1969), and the references therein.

61. J. M. Herbelin, private communication. For another approach to this problem, see the chapter in this book by Ben-Shaul and Hofacker.

62. H. Steiner and E. K. Rideal, Proc. Roy. Soc. London Ser. A **173,** 503 (1939).

63. W. H. Rodebush and W. C. Klingelhoefer, J. Amer. Chem. Soc. **55,** 130 (1933).

64. P. G. Ashmore and J. Chanmugam, Trans. Faraday Soc. **49,** 254 (1953).

65. A. A. Westenberg and N. de Haas, J. Chem. Phys. **48,** 4405 (1968).

66. M. A. A. Clyne and D. H. Stedman, Trans. Faraday Soc. **62,** 2164 (1966).

67. S. W. Benson, F. R. Cruickshank, and R. Shaw, Int. J. Chem. Kinet. **1,** 29 (1969).

68. J. H. Sullivan, J. Chem. Phys. **51,** 2288 (1969).

69. N. S. Snider, J. Chem. Phys. **53,** 4116 (1970).

70. A. A. Westenberg, J. Chem. Phys. **53,** 4117 (1970).

71. J. J. Galante and E. A. Gislason, Chem. Phys. Lett. **18,** 231 (1973).

72. F. S. Klein and M. Wolfsberg, J. Chem. Phys. **34,** 1494 (1961).

73. R. L. Wilkins, J. Chem. Phys. **42,** 806 (1965).

74. R. S. Davidow, R. A. Lee, and D. A. Armstrong, J. Chem. Phys. **45,** 3364 (1966).

75. *a* R. G. Albright, A. F. Dodonov, G. K. Lavrovskaya, I. I. Morosov, and V. L. Tal'rose, J. Chem. Phys. **50,** 3632 (1969).

   *b* Also in V. L. Tal'rose, G. K. Lavrovskaya, A. F. Dodonov, and I. I. Morosov, *Recent Developments in Mass Spectroscopy, Proceedings of the International Conference in Mass Spectroscopy,* Kyoto, K. Ogata and T. Hayakawa, eds. (University Park Press, Baltimore, 1969), p. 1022.

76. D. H. Stedman, D. Steffenson, and H. Niki, Chem. Phys. Lett. **7,** 173 (1970).

77. J. K. Cashion and J. C. Polanyi, Proc. Roy. Soc. London Ser. A **258,** 529 (1960).

78. P. E. Charters and J. C. Polanyi, Disc. Faraday Soc. **33,** 107 (1962).

79. J. R. Airey, R. R. Getty, J. C. Polanyi, and D. R. Snelling, J. Chem. Phys. **41,** 3255 (1964).

80. K. G. Anlauf, D. H. Maylotte, P. D. Pacey, and J. C. Polanyi, Phys. Lett. **24A,** 208 (1967).

81. K. G. Anlauf, P. J. Kuntz, D. H. Maylotte, P. D. Pacey, and J. C. Polanyi, Disc. Faraday Soc. **44,** 183 (1967).

82. P. D. Pacey and J. C. Polanyi, Appl. Opt. **10,** 1725 (1971).

83. K. G. Anlauf, D. S. Horne, R. G. Macdonald, J. C. Polanyi, and K. B. Woodall, J. Chem. Phys. **57,** 1561 (1972).

84. J. M. Herbelin and G. Emanuel, J. Chem. Phys. **60,** 689 (1974).

85. S. S. Penner, J. Franklin Inst. **249,** 441 (1950), as quoted in G. Bahn, Pyrodynamics **4,** 371 (1966).

86. G. C. Fettis, J. H. Knox, and A. F. Trotman-Dickenson, J. Chem. Soc. 1064 (1960).

87. R. Dunlap and R. Hermsen, *Nonequilibrium Flow in Nozzles,* UTC 2032-FR (United Aircraft Corp., Hartford, Connecticut 1963).

88. K. A. Wilde, AIAA J. **2,** 374 (1964).

89. S. W. Mayer, J. B. Szabo, and L. Schieler, "Computed High-Temperature Rate Constants for H-Atom Transfer Involving the Light Elements," Report No. TDR-496(9210-02)-4 (The Aerospace Corp., El Segundo, California, 1965).

90. S. W. Mayer, L. Schieler, and H. S. Johnston, J. Chem. Phys. **45,** 385 (1966).

91. J. B. Levy and B. K. W. Copeland, J. Phys. Chem. **69,** 408 (1965).

92. R. Tunder, S. W. Mayer, E. Cook, and L. Schieler, "Compilation of Reaction Rate Data for Non-equilibrium Performance and Reentry Calculation Programs," Report No. TR-1001(9210-02)-1 (The Aerospace Corp., El Segundo, California, 1967).

93. S. W. Mayer and L. Schieler, J. Phys. Chem. **72,** 236 (1968).

94. A. F. Dodonov, E. B. Gordon, G. K. Lavrovskaya, I. I. Morosov, A. N. Ponomarev, and V. L. Tal'rose, "Mass-Spectrometric and Laser Determinations of Rate Constants for Elementary Reactions Characteristic of Hydrogen-Halogen Laser," reported at International Symposium on Chemical Lasers, Moscow, September 1969. As quoted by J. R. Airey, Int. J. Chem. Kinet. **2,** 65 (1970); also quoted in Ref. 75.

95. A. F. Dodonov, G. K. Lavrovskaya, I. I. Morosov, and V. L. Tal'roze, Dokl. Akad. Nauk SSSR **198,** 622 (1971) (p. 440 in English translation).

96. K. H. Homann, W. C. Solomon, J. Warnatz, H. Gg. Wagner, and C. Zetzsch, Ber. Bunsenges. phys. Chem. **74,** 585 (1970).

97 G. A. Kapralova, A. L. Margolin, and A. M. Chaikin, Kinet. Katal. **11,** 810 (1970) (p. 669 in English edition).

98. R. L. Wilkins, "Absolute Rates of the Reactions $H_2 + F$ and $F_2 + H$," Report No. TR-0059(6753-20)-1 (The Aerospace Corp., El Segundo, California, 1971).

99. R. L. Wilkins, J. Chem. Phys. **57,** 912 (1972).

100. R. L. Jaffe and J. B. Anderson, J. Chem. Phys. **54,** 2224 (1971); **56,** 682 (1972).

101. S. W. Rabideau, H. G. Hecht, and W. B. Lewis, J. Magn. Resonance **6,** 384 (1972).

102. R. Foon and G. P. Reid, Trans. Faraday Soc. **67,** 3513 (1971).

103. D. G. Truhlar, J. Chem. Phys. **56,** 3189 (1972).

104. J. T. Muckerman and M. D. Newton, J. Chem. Phys. **56,** 3191 (1972).

105. K. L. Kompa and J. Wanner, Chem. Phys. Lett. **12,** 560 (1972).

106. R. L. Williams and F. S. Rowland, J. Phys. Chem. **77,** 301 (1973).

107. J. F. Hon, A. Axworthy, and G. Schneider, "Advanced Fuels for Chemical Lasers," Report R-9297 (Rocketdyne Division, Rockwell International, Canoga Park, California, 1973).

108. M. A. A. Clyne, D. J. McKenney, and R. F. Walker, Can. J. Chem. **51,** 3596 (1973).

109. A. Persky, J. Chem. Phys. **59,** 3612 (1973).

110. J. T. Muckerman, J. Chem. Phys. **54,** 1155 (1971).

111. J. T. Muckerman, J. Chem. Phys. **56,** 2997 (1972).

112. R. L. Wilkins, Mol. Phys. **28,** 21 (1974).

113. K. L. Kompa and G. C. Pimentel, Ber. Bunsenges. phys. Chem. **72,** 1067 (1968).

114. J. H. Parker and G. C. Pimentel, J. Chem. Phys. **51,** 91 (1969).

115. O. D. Krogh and G. C. Pimentel, J. Chem. Phys. **56,** 969 (1972).

116. J. C. Polanyi and D. C. Tardy, J. Chem. Phys. **51,** 5717 (1969).

117. K. G. Anlauf, P. E. Charters, D. S. Horne, R. G. MacDonald, D. H. Maylotte, J. C. Polanyi, W. J. Skrlac, D. C. Tardy, and K. B. Woodall, J. Chem. Phys. **53,** 4091 (1970).

118. J. C. Polanyi and K. B. Woodall, J. Chem. Phys. **57,** 1574 (1972).

119. N. Jonathan, C. M. Melliar-Smith, and D. H. Slater, Mol. Phys. **20,** 93 (1971).

120. N. Jonathan, C. M. Melliar-Smith, S. Okuda, D. H. Slater, and D. Timlin, Mol. Phys. **22,** 561 (1971).

121. R. L. Jaffe, J. M. Henry, and J. B. Anderson, J. Chem. Phys. **59,** 1128 (1973).

122. R. D. Coombe and G. C. Pimentel, J. Chem. Phys. **59,** 251 (1973).

123. R. D. Coombe and G. C. Pimentel, J. Chem. Phys. **59,** 1535 (1973).

124. J. B. Anderson, J. Chem. Phys. **52,** 3849 (1970).

125. R. L. Wilkins, J. Chem. Phys. **58,** 3038 (1973).

126. T. P. Schafer, P. E. Siska, J. M. Parson, F. P. Tully, Y. C. Wong, and Y. T. Lee, J. Chem. Phys. **53,** 3385 (1970).

127. S. W. Mayer, L. Schieler, and H. S. Johnston, *Eleventh Symposium* (*International*) *on Combustion* (Pittsburgh Combustion Institute, Pittsburgh, 1967), p. 837.

128. J. F. Levy and B. W. K. Copeland, J. Phys. Chem. **72,** 3168 (1968).

129. R. L. Wilkins, J. Chem. Phys. **58,** 2326 (1973).

130. N. Jonathan, C. M. Melliar-Smith, and D. H. Slater, J. Chem. Phys. **53,** 4396 (1970).

131. N. Jonathan, C. M. Melliar-Smith, D. Timlin, and D. H. Slater, Appl. Opt. **10,** 1821 (1971).

132. N. Jonathan, S. Okuda, and D. Timlin, Mol. Phys. **24,** 1143 (1972).

133. J. C. Polanyi and J. J. Sloan, J. Chem. Phys. **57,** 4988 (1972).

134. K. Cashion, J. Mol. Spectrosc. **10,** 182 (1963).

135. J. R. Airey, P. D. Pacey, and J. C. Polanyi, *Eleventh Symposium* (*International*) *on Combustion* (Pittsburgh Combustion Institute, Pittsburgh, 1967), p. 85.

136. P. Cadman and J. C. Polanyi, J. Phys. Chem. **72,** 3715 (1968).

137. J. R. Airey, J. Chem. Phys. **52,** 156 (1970).

138. R. L. Kerber, G. Emanuel, and J. S. Whittier, Appl. Opt. **11,** 1112 (1972).

139. G. Emanuel and J. S. Whittier, Appl. Opt. **11,** 2047 (1972).

140. R. N. Schwartz, A. I. Slawsky, and K. F. Herzfeld, J. Chem. Phys. **20,** 1591 (1952).

141. C. B. Moore, J. Chem. Phys. **43,** 2979 (1965).

142. R. C. Millikan and D. R. White, J. Chem. Phys. **39,** 3209 (1963).

143. H. K. Shin, J. Phys. Chem. **75,** 1079 (1971).

144. H. K. Shin, J. Chem. Phys. **59,** 897 (1973).

145. K. F. Herzfeld and T. A. Litovitz *Absorption and Dispersion of Ultrasonic Waves* (Academic, New York, 1959), p. 86.

146. W. C. Solomon, J. A. Blauer, F. C. Jaye, and J. G. Hnat, Int. J. Chem. Kinet. **3,** 215 (1971).

147. J. F. Bott and N. Cohen, J. Chem. Phys. **55,** 3698 (1971).

148. T. Just and G. Rimpel, "Studies of HF Relaxation and on the Reaction of $H_2$ with $F_2$ Behind Shocks," presented at 3rd Conference on Chemical and Molecular Lasers, St. Louis, Missouri, May 1972; see also Fig. 2 of Ref. 162.

149. G. K. Vasil'ev, E. F. Makarov, V. G. Papin, V. L. Tal'roze, Zh. Eksp. Teor. Fiz. **64,** 2046 (1973). (Russian).

150. J. R. Airey and S. S. Fried, Chem. Phys. Lett. **8,** 23 (1971).

151. R. R. Stephens and T. A. Cool, J. Chem. Phys. **56,** 5863 (1972).

152. J. K. Hancock and W. H. Green, J. Chem. Phys. **57,** 4515 (1972).

153. J. F. Bott, J. Chem. Phys. **57,** 96 (1972).

154. R. M. Osgood, Jr., A. Javan, and P. B. Sackett, Appl. Phys. Lett. **20,** 269 (1972).

155. S. S. Fried, J. Wilson, and R. L. Taylor, IEEE J. Quan. Electron. **QE-9,** 59 (1973).

156. J. J. Hinchen, J. Chem. Phys. **59,** 233 (1973).

157. J. J. Hinchen, J. Chem. Phys. **59,** 2224 (1973).

158. R. A. Lucht and T. A. Cool, J. Chem. Phys. **60,** 1026 (1974).

159. L. S. Blair, W. D. Breshears, and G. L. Schott, J. Chem. Phys. **59,** 1582 (1973).

160. J. F. Bott and N. Cohen, *Recent Developments in Shock Tube Research, Proceedings of the Ninth International Shock Tube Symposium, Stanford University, July 16–19, 1973,* D. Bershader and W. Griffith, Eds. (Stanford Univ. Press, Stanford, California, 1973), p. 259.

161. R. J. Donovan, D. Husain, and C. D. Stevenson, Trans. Faraday Soc. **66,** 2148 (1970).

162. J. A. Blauer, W. C. Solomon, and T. W. Owens, Int. J. Chem. Kinet. **4,** 293 (1972).

163. J. F. Bott and N. Cohen, J. Chem. Phys. **58,** 934 (1973).

164. J. F. Bott and N. Cohen, J. Chem. Phys. **59,** 447 (1973).

165. H. K. Shin, Chem. Phys. Lett. **6,** 494 (1970).

166. H. K. Shin, Chem. Phys. Lett. **10,** 81 (1971).

167. G. C. Berend and R. L. Thommarson, private communication.

168. G. C. Berend and R. L. Thommarson, J. Chem. Phys. **58,** 3203 (1973).

169. W. D. Breshears and P. F. Bird, J. Chem. Phys. **50,** 333 (1969).

170. W. D. Breshears and P. F. Bird, J. Chem. Phys. **52,** 999 (1970).

171. W. D. Breshears and P. F. Bird, J. Chem. Phys. **54,** 2698 (1971).

172. J. H. Kiefer, W. D. Breshears, and P. F. Bird, J. Chem. Phys. **50,** 3641 (1969).

173. C. T. Bowman and D. J. Seery, J. Chem. Phys. **50,** 1904 (1969).

174. H.-L. Chen and C. B. Moore, J. Chem. Phys. **54,** 4072 (1971).

175. P. F. Zittel and C. B. Moore, J. Chem. Phys. **59,** 6636 (1973).

176. P. F. Zittel and C. B. Moore, J. Chem. Phys. **58,** 2922 (1973).

177. H.-L. Chen, J. Chem. Phys. **55,** 5551 (1971).

178. Y.-D. Chen and H.-L. Chen, J. Chem. Phys. **56,** 3315 (1972).

179. H.-L. Chen, J. C. Stephenson, and C. B. Moore, Chem. Phys. Lett. **2,** 593 (1968).

180. J. F. Bott and N. Cohen, J. Chem. Phys. **58,** 4539 (1973).

181. J. L. Ahl and T. A. Cool, J. Chem. Phys. **58,** 5540 (1973).

182. J. F. Bott, J. Chem. Phys. **60,** 427 (1974).

183. J. R. Airey and I. W. M. Smith, J. Chem. Phys. **57,** 1669 (1972).

184. D. J. Seery, J. Chem. Phys. **58,** 1796 (1973).

185. B. M. Hopkins and H.-L. Chen, J. Chem. Phys. **59,** 1495 (1973).

186. R. V. Steele, Jr. and C. B. Moore, J. Chem. Phys. **60,** 2794 (1974).

187. G. C. Berend and R. L. Thommarson, J. Chem. Phys. **58,** 3454 (1973).

188. N. C. Craig and C. B. Moore, J. Phys. Chem. **75,** 1622 (1971).

189. B. A. Ridley and I. W. M. Smith, Chem. Phys. Lett. **9,** 457 (1971).

190. R. L. Johnson, M. J. Perona, and D. W. Setser, J. Chem. Phys. **52,** 6372 (1970). See their Ref. 22.

191. D. L. Thompson, J. Chem. Phys. **56,** 3570 (1972).

192. R. L. Thommarson and G. C. Berend, Int. J. Chem. Kinet. **5,** 629 (1973).

193. I. W. M. Smith and P. M. Wood, Mol. Phys. **25,** 441 (1973).

194. J. F. Bott and N. Cohen, J. Chem. Phys. **55,** 5124 (1971).

195. J. A. Blauer and W. C. Solomon, *14th Symposium (International) on Combustion* (State College, Pennsylvania, 1972), p. 189.

196. J. A. Blauer and W. C. Solomon, Int. J. Chem. Kinet. **5,** 553 (1973).

197. D. L. Thompson, J. Chem. Phys. **57,** 4164 (1972).

198. R. L. Wilkins, J. Chem. Phys. **59,** 698 (1973).

199. H. K. Shin, Chem. Phys. Lett. **14,** 64 (1972).

200. R. J. Donovan, D. Husain, and C. D. Stevenson, Nature **227,** 602 (1970).

201. D. L. Thompson, J. Chem. Phys. **57,** 4170 (1972).

202. R. L. Wilkins, J. Chem. Phys. **58,** 3038 (1973).

203. M. A. Kwok and R. L. Wilkins, J. Chem. Phys. **60,** 2189 (1974).

204. R. L. Wilkins, Mol. Phys. **29,** 555 (1975).

205. D. Arnoldi and J. Wolfrum, Chem. Phys. Lett. **24,** 234 (1974).

206. J. F. Bott, J. Chem. Phys. **61,** 2530 (1974).

207. L. H. Sentman and W. C. Solomon, J. Chem. Phys. **59,** 89 (1973).

208. H.-L. Chen, J. Chem. Phys. **55,** 5557 (1971).

209. J. F. Bott and N. Cohen, "Vibrational Relaxation of HF in the Presence of $H_2$," Report No. TR-0073(3451)-1 (The Aerospace Corp., El Segundo, California, 1972).

210. L. H. Sentman, Chem. Phys. Lett. **18,** 493 (1973).

211. T. A. Dillon and J. C. Stephenson, J. Chem. Phys. **58,** 2056 (1973).

212. J. C. Stephenson, J. Finzi, and C. B. Moore, J. Chem. Phys. **56,** 5214 (1972).

213. R. D. Sharma, H.-L. Chen, and A. Szöke, J. Chem. Phys. **58,** 3519 (1973).

214. C. B. Moore, Adv. Chem. Phys. **23,** 41 (1973).

215. A. B. Callear and I. W. M. Smith, Trans. Faraday Soc. **59,** 1735 (1963).

216. V. I. Gorshkov, V. V. Gromer, V. I. Igoshin, E. L. Keshelev, E. P. Markin, and A. N. Oraevsky, Appl. Opt. **10,** 1781 (1971).

217.  M. M. Hopkins and H. I. Chen, Chem. Phys. Lett. **17,** 500 (1972).

218.  I. Burak, Y. Noter, A. M. Ronn, and A. Szöke, Chem. Phys. Lett. **17,** 345 (1972).

219.  S. R. Leone and C. B. Moore, Chem. Phys. Lett. **19,** 340 (1973).

220.  I. Burak, Y. Noter, A. M. Ronn, and A. Szöke, Chem. Phys. Lett. **16,** 306 (1972).

221.  B. M. Hopkins and H.-L. Chen, Chem. Phys. Lett. **17,** 500 (1972).

222.  Y. Noter, I. Burak, and A. Szöke, J. Chem. Phys. **59,** 970 (1973).

223.  J. F. Bott, Chem. Phys. Lett. **23,** 335 (1973).

224.  R. M. Osgood, Jr., A. Javan, and P. B. Sackett, J. Chem. Phys. **60,** 1464 (1973).

225.  B. A. Ridley and I. W. M. Smith, J. Chem. Soc. Faraday Trans. **68,** 123 (1972).

226.  H.-L. Chen and C. B. Moore, J. Chem. Phys. **54,** 4080 (1971).

227.  M. A. Kwok, presented at 3rd Conference on Chemical Lasers, St. Louis, Missouri, May 1972.

228.  P. G. Anlauf, P. H. Dawson, and J. A. Herman, J. Chem. Phys. **58,** 5354 (1973).

229.  W. H. Green and J. K. Hancock, IEEE J. Quan. Electron. **QE-9,** 50 (1973).

230.  J. A. Blauer, W. C. Solomon, L. H. Sentman, and T. W. Owens, J. Chem. Phys. **57,** 3277 (1972).

231.  J. K. Hancock and W. H. Green, J. Chem. Phys. **59,** 6350 (1973).

232.  P. F. Zittel and C. B. Moore, J. Chem. Phys. **58,** 2004 (1974).

233.  P. S. Ganguli and M. Kaufman, *Chem. Phys. Lett.* **25,** 221 (1974).

234.  D. Arnoldi and J. Wolfrum, *Chem. Phys. Lett.* **24,** 234 (1974).

235.  R. G. MacDonald, C. B. Moore, I. W. M. Smith and F. J. Wodarczyk, *J. Chem. Phys.* **62,** 2934 (1975).

236.  R. F. Heidner, III, and J. F. Bott, *J. Chem. Phys.* **63,** 1810 (1975).

237.  N. Cohen and J. F. Bott, *Appl. Optics* **15,** 28 (1976).

238.  M. A. Kwok and R. L. Wilkins, *J. Chem. Phys.* **63,** 2453 (1975).

239.  R. A. Lucht and T. A. Cool, *J. Chem. Phys.* **60,** 1026 (1974).

240.  N. Cohen and J. F. Bott, "A Review of Rate Coefficients in the $H_2$–$Cl_2$ Chemical Laser System," *Aerospace* TR-0075(5530)-7 (1975).

241.  N. Cohen and J. F. Bott, "A Review of Rate Coefficients in the $H_2$–$F_2$ Chemical Laser System," *Aerospace* TR-0076(6603)-2 (1976).

CHAPTER **3**

# OPTICAL ASPECTS OF CHEMICAL LASERS

## R. A. CHODZKO

The Aerospace Corporation

Los Angeles, California

## A. N. CHESTER

Hughes Research Laboratories

Malibu, California

Very few optics and mode-control problems are unique to chemical lasers. However, chemical lasers possess many characteristics that present genuine challenges to optical technology. Among these are

1. High optical gain, leading to superradiance and other parasitic effects

2. Complicated, nonequilibrium excitation and de-excitation kinetic processes.

3. Population cascade and competition effects, with simultaneous laser operation on many separate but interconnected transitions.

4. Operation on both partial and total population inversions.

5. Spurious absorption, caused by spent laser gases remaining within the optical cavity.

6. Gain media with large characteristic Fresnel numbers.

7. Spatial and temporal variations in both gain and refractive index.

8. High average-beam-power densities, at wavelengths where many high-power optical components are not yet available.

These effects tend to call for a highly sophisticated approach to mode control and optics design whenever a chemical laser is required to exhibit well-controlled output and good beam quality. However, because a comprehensive systematic treatment of this complex problem does not exist at this time, our discussion of the optical aspects of chemical lasers will focus on particular effects that can be isolated for theoretical or experimental study. Moreover, since some of the important effects can be more easily studied on other types of lasers, we will refer to experiments performed with gas dynamical or electrical lasers as well as chemical lasers.

We begin by discussing one of the overriding concerns of chemical-laser technology today: obtaining good beam quality. In Section 1, the physical principles underlying optical-resonator design are explained as they relate to good output beam quality, and the principal types of optical resonators that are used with chemical lasers are described.

Our best present means for achieving good beam quality at large Fresnel numbers, the unstable resonator, requires more extensive treatment. Section 2 summarizes our theoretical understanding of unstable resonators, both empty and containing an active gain medium. Experimental measurements of chemical-laser beam quality are described in Section 3, including measurement techniques, the performance of both unstable resonators and oscillator–amplifier optical configurations, and the use of grating resonators for simultaneous wavelength and mode control.

# 1. PHYSICAL PRINCIPLES OF OPTICAL RESONATORS

## 1.1. Key Optical Considerations in Chemical Lasers

The optical resonator of a laser oscillator has two important functions:

1. To provide optical feedback to maintain laser oscillation; this temporal coherence gives the output beam its high monochromaticity.

2. To control the optical phase across the output beam front; this spatial coherence leads to low beam divergence, or high focused intensities.

Maximum temporal coherence is called for in applications such as wideband optical communications, interferometry, spectroscopy, metrology, and Doppler radar. Maximum spatial coherence (good beam quality) is needed for long-distance propagation, or when focusing a beam to maximum intensity, as in welding, surgery, or materials processing. Related aspects of laser output control include wavelength selection in multiline lasers and the general area of modulation and pulse waveform control.

Since spatial coherence, or good beam quality, has been the overriding concern in chemical-laser optical-resonator work, most of this chapter is devoted to exploring how narrow-beam divergence is obtained from high-power chemical lasers. The same basic principles apply to all high-energy laser devices, but we will give special emphasis to results obtained on chemical lasers when those are available. We begin by discussing the types of optical resonators and how they determine the spatial coherence of the laser radiation.

## 1.2. Stable Resonators and Beam Quality

In a high-gain laser, the spontaneous emission can be amplified to high intensity with a single pass through the gain medium, so that a superradiant source of stimulated laser emission can be designed without feedback mirrors. However, superradiant output generally has poor phase coherence, yielding impaired beam quality. Thus, a superradiant optical design is usually limited to lasers with pulse lengths not much longer than

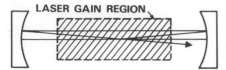

**Fig. 1.** Ray paths in a stable laser resonator.

the transient time of light through the medium, so that insufficient time is available to build up a resonator mode.

Most lasers utilize feedback rather than a superradiant design to improve the spatial and temporal coherence of the output. A laser oscillator, with mirrors providing optical feedback, may be used either alone or in combination with one or more stages of laser amplification. In the latter case, the term "MOPA" is frequently used to describe the combination of a laser master oscillator with a laser power amplifier. However, in both the oscillator and MOPA geometries, it is the oscillator itself that primarily determines the output beam quality.

Several types of optical resonators have traditionally been used around a laser medium to provide optical feedback (Fig. 1).[1] Most laser resonators are "open," that is, mirrors are provided only on two opposing sides of the laser medium. The earliest open resonators, using flat, parallel mirrors, quickly gave way to stable resonators. As indicated in Fig. 1, in a stable resonator the curvatures of the mirrors (usually both concave) are chosen so that an optical ray that wanders away from the optical axis of the resonator is returned to it by the focusing action of the mirrors. This greatly reduces the diffraction losses of plane-parallel resonators because of lateral leakage of radiation.

In recent years optical waveguides have been used to form enclosed or partially enclosed resonators[2-4] for some small-volume lasers with output power of a few watts or less.[5] However, high-power and high-energy lasers involve large-volume gain media, which at this time do not appear to be compatible with the waveguiding technique when good beam quality is desired.

The more widely used open resonators can be plotted on the resonator stability diagram[6,7] of Fig. 2, according to how paraxial rays propagate within the resonator. The g parameters describing resonator stability are defined by

$$g_i = 1 - \frac{L}{R_i} \tag{1}$$

where $L$ is the mirror separation and $R_i$ is the radius of curvature of mirror number $i$ (positive for concave mirrors). For mirror curvatures lying in the shaded region, the resonator is stable, in the sense that

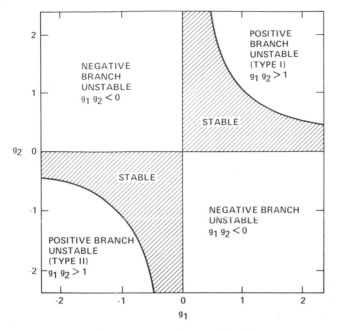

**Fig. 2.** Resonator stability diagram.

paraxial rays are confined near the optical axis. The plane-parallel resonator appears at the point ($g_1 = 1$, $g_2 = 1$), and the various unstable regions will be discussed later.

The spatial coherence controlled by the resonator determines the output beam quality. Beam quality may be measured by the divergence of the beam after propagating through a long distance (Fig. 3a). The far-field pattern of the laser is that brightness distribution $B(\theta_x, \theta_y)$ (measured in $W/sr$) to which the beam converges as it propagates. This propagation is assumed to occur in a vacuum or some other homogeneous isotropic medium.

Beam quality can also be measured by the intensity pattern that is produced in the focal plane of a lens through which the beam passes (Fig. 3b). Both of these definitions of laser beam quality are essentially equivalent, since the brightness pattern $B(\theta_x, \theta_y)$ in the far field is related to the intensity $I(x, y)$ measured in the focal plane of the lens through the relation

$$B(\theta_x, \theta_y) = f^2 I(f\theta_x, f\theta_y) \tag{2}$$

for a position in the focal plane $x = f\theta_x$, $y = f\theta_y$. In fact, even when the

(a) FAR-FIELD BRIGHTNESS

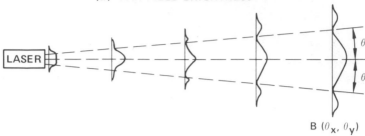

(b) FOCAL SPOT INTENSITY

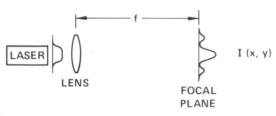

**Fig. 3.** Two equivalent measures of spatial coherence or beam quality. The intensity distribution of output radiation is plotted at various distances from the laser source.

application of interest calls for maximum far-field brightness, focal-spot intensity is commonly measured to determine beam quality.

The beam divergence in the far field, or equivalently, the maximum focal spot intensity, is determined by the intensity and optical phase of the beam as shown schematically in Fig. 4. As we know, a beam of diameter $D$ having uniform intensity and phase in the near field will diffract to produce the well-known Airy pattern[8] in the far field. On axis, the Huygens wavelets from different parts of the beam front all add up in phase to produce an intensity maximum. Away from the axis, differences in optical-path length cause the wavelets to arrive out of phase with one another, and their mutual interference leads to a smaller total intensity. The angular diameter of this localized far-field pattern is $2.44\lambda/D$, measured at the first null in intensity.

Now consider what happens when the near-field intensity is still uniform, but the phase varies irregularly across the beam front, as in Fig. 4b. There is no position in the far field at which the Huygens wavelets all arrive in phase. Since there will be both positive and negative interference between wavelets arriving from different parts of the beam front, the intensities tend to add incoherently. This produces a reduced far-field intensity, one which is more independent of position with respect to the optical axis. Thus, in this case, we obtain a fairly large beam divergence leading to a broad far-field pattern, which may have some irregularities but which has no strong narrow central peak.

A disturbance in the near-field intensity has much less impact upon beam quality. As indicated in Fig. 4c, a beam with uniform phase but irregular intensity yields a far-field pattern that still has a large amount of energy concentrated in a narrow central peak. The effect of the nonuniform intensity is to increase the amount of energy present in the normally weak side lobes surrounding the central intensity maximum.

It is true that, in extreme cases, the near-field intensity can be so nonuniform that large amounts of energy may be delivered to the off-axis side lobes, yielding poor overall divergence. However, in practice, laser-resonator calculations usually show that disturbances in phase are the

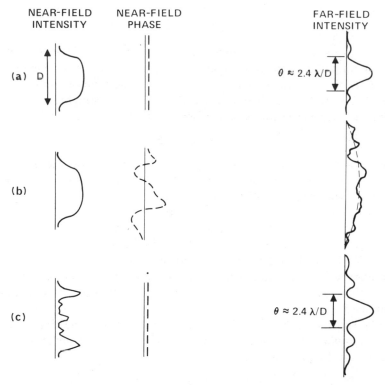

**Fig. 4.** Effect of intensity and phase distribution on the far-field intensity pattern of a laser beam.

crucial determinant of output beam quality, rather than intensity nonuniformities.

To determine how the resonator affects output beam quality, it is necessary to analyze the optical field distribution, both in intensity and phase. An initial amplitude and phase distribution for the laser radiation $\psi(r)$ is assumed to be incident upon one of the resonator mirrors. This field distribution is then propagated for one complete round trip through the resonator, arriving back at its initial position. This propagation, which we represent by the operator $\hat{H}$, is computed using the Fresnel–Kirchhoff integral[9] or some other solution of Maxwell's equations.

The optical fields can be analyzed in terms of the steady-state field distributions that can be sustained within an empty resonator. These can be found through the requirement that the field distribution obtained after one round trip through the resonator be identical to the initial assumed field distribution, apart from an overall multiplicative complex constant. A field distribution that obeys this condition, written schematically as

$$\hat{H}\psi(r) = \gamma\psi(r) \tag{3}$$

is a normal mode of the resonator, in the sense that once it is excited it will maintain the same functional form for all time. The eigenvalue $\gamma$ represents the loss in total energy and the optical-phase shift that this mode encounters on each round trip through the resonator.

More explicitly, the complex number $\gamma$ may be written in polar form as

$$\gamma = C \exp(i\phi) \tag{4}$$

where $\mathcal{R} = C^2$ is the round-trip power reflectivity of the resonator, $\delta = 1 - C^2$ is the output coupling fraction or fractional loss per round trip, and $\phi$ is the excess optical-phase shift per round trip through the resonator, compared with free-space propagation through a distance equal to twice the mirror separation. The phase $\phi$ determines the exact oscillation frequency or frequencies within the gain linewidth of the laser. Since we are dealing with empty resonators at the moment, there will always be some energy loss per round trip; in a laser, the gain of the laser medium offsets this loss so that a mode of constant energy content can be maintained.

To be specific, if the Fresnel–Kirchhoff integral or some similar propagation formula is used, the resonator modes are found through the solution of an integral equation of the form

$$\gamma\psi(r) = \int dS \, K(r, r') \, \psi(r') \tag{5}$$

where the integration takes place over one mirror of the resonator.

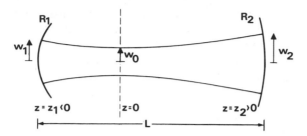

**Fig. 5.** Transverse-mode parameters in a stable resonator.

The eigenfunctions $\psi(r)$ of the integral equation (5) represent the various transverse modes of the optical resonator. Since these modes are mutually orthogonal, they maintain themselves independently of one another once they are initially excited. The optical field distribution within the resonator can be expressed as a suitable sum of these transverse mode functions. If more than one axial or longitudinal mode is present, distinguished by different oscillating frequencies, then each axial-mode frequency may contain a different mixture of the transverse-mode distribution functions.

As an example, let us consider the transverse modes of a stable resonator containing radiation of wavelength $\lambda$, as originally obtained by approximate solution of Eq. (5) by Boyd and Gordon.[10] In the solution it is found that there is one location along the optical axis, as shown by the dotted line in Fig. 5, where all of the transverse modes have uniform phase across the beam front. If we measure the $z$-coordinate from this position and the $x$- and $y$-coordinates at right angles to the optical axis, the electric-field distribution of transverse-mode number $nm$ is given by the expression

$$E_{nm}(x, y, z) = (2^{m+n-1} m! \, n! \, \pi)^{-1/2} w^{-1} H_n\left(\frac{x\sqrt{2}}{w}\right)$$

$$\times H_m\left(\frac{y\sqrt{2}}{w}\right) \exp\left[\frac{-(x^2+y^2)}{w^2}\right]$$

$$\times \exp\left[ikz + i(n+m+1)\tan^{-1}\left(\frac{z}{z_0}\right) - ik\left(\frac{x^2+y^2}{2R}\right)\right] \quad (6)$$

where the confocal parameter $z_0$ (sometimes denoted $b$) is defined by

$$z_0 = \frac{\pi w_0^2}{\lambda} \quad (7)$$

and

$$k = \frac{2\pi}{\lambda} \quad (8)$$

The basic transverse-coordinate dependence consists of Hermite polynomials of $x$ and of $y$, multiplied by the Gaussian functions $\exp(-x^2/w^2)$ and $\exp(-y^2/w^2)$. In addition, there is a complex-phase term involving the coordinate $z$ and the wave-front radius of curvature $R$, which varies in a simple way with $z$:

$$R = \frac{(z^2 + z_0^2)}{z} \tag{9}$$

The characteristic beam radius $w$ obeys

$$w = w_0 \left[ 1 + \left( \frac{z}{z_0} \right)^2 \right]^{1/2} \tag{10}$$

having its minimum value $w_0$ at the beam waist, $z = 0$.

The wavefront curvatures match the mirror curvature at the mirror positions $z = z_1$ and $z = z_2$:

$$-R_1 = \frac{(z_1^2 + z_0^2)}{z_1} \tag{11}$$

and

$$R_2 = \frac{(z_2^2 + z_0^2)}{z_2} \tag{12}$$

[The negative sign in Eq. (11) results from our convention of positive radii of curvature for concave mirrors.] Eliminating $z_0$ from Eqs. (11) and (12) and noting that

$$-z_1 + z_2 = L$$

the mirror separation, we find that

$$\frac{-z_1}{L} = \frac{L - R_1}{2L - R_1 - R_2} = \frac{g_2(1 - g_1)}{g_1 + g_2 - 2g_1 g_2} \tag{13}$$

and

$$\frac{z_2}{L} = \frac{L - R_1}{2L - R_1 - R_2} = \frac{g_1(1 - g_2)}{g_1 + g_2 - 2g_1 g_2} \tag{14}$$

$z_0$, and hence $w_0$, may then be found from Eq. (11) or (12).

At $z = 0$, the transverse-mode functions are real, and the most general transverse mode $nm$ may be written

$$E_{nm} = U_n(x) U_m(y) \tag{15}$$

The $U$'s are simple Hermite–Gaussian functions and have the functional form shown in Fig. 6. These functions are in fact identical with the

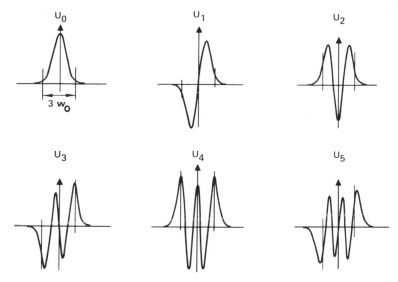

**Fig. 6.** Lowest-order Hermite–Gaussian functions $U_n(x)$.

wavefunctions of the simple harmonic oscillator in quantum mechanics.[11,12]

Figure 6 illustrates how transverse-mode selection and consequently good beam quality can come about. The vertical lines are drawn on each of the $U_n$ graphs at the position $x = \pm 1.5 w_0$. Therefore, if the laser gain medium were contained in a tube or cylinder of diameter $3w_0$, it is apparent from Fig. 6 that the transverse mode $U_0$ would suffer the least aperturing loss at the edges of the gain medium, since very little of its intensity extends beyond the vertical lines. Thus, if the limiting aperture of the resonator at the position of the beam waist has diameter $3w_0$, the mode $U_0$ will have less loss than the higher order transverse modes. This lowest-loss or fundamental mode will grow the fastest from spontaneous emission noise, and may sufficiently saturate the gain medium that no higher-order mode will have sufficient gain to overcome the aperturing losses. Note that the higher-order modes all have one or more changes in phase because their amplitudes change sign; the higher-order modes, therefore, will have a wider beam divergence than the fundamental mode, according to the previous discussion of Fig. 4. Thus this aperturing, which leads to single-transverse mode operation, also yields good beam quality.

Note in Fig. 6 that if we increase the diameter of the gain medium, perhaps to obtain more output power, without increasing the fundamental mode diameter, then many of the different lower-order transverse modes

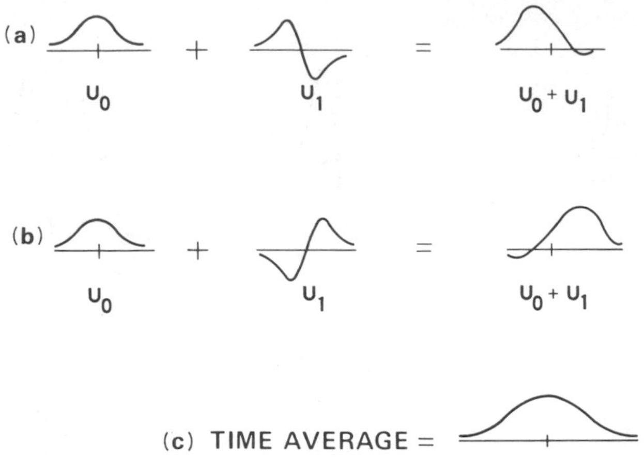

Fig. 7.  The blurring of a time-averaged far-field pattern caused by multiple transverse modes.

will have almost the same aperturing losses. This can permit several modes to oscillate simultaneously, as we shall see shortly.

Before we examine multimode operation, let us point out how beam quality is adversely affected by multiple modes. Suppose that only the $U_0$ and $U_1$ modes were present in the laser output, as suggested in Fig. 7. The far-field intensity patterns of these modes, which have the same functional form as the near-field patterns shown in Fig. 6, may at one time add up as shown in Fig. 7a to produce a skewed off-center intensity distribution. However, in general, these different transverse modes will oscillate with slightly different optical frequencies.[10] Thus at some later time, $U_0$ will have returned to its original optical phase; whereas, $U_1$ will be out of phase compared with its earlier configuration, as indicated in Fig. 7b. Therefore, at this later time, the total laser output will be skewed to the opposite side of the optical axis in the far field. Since the transverse modes go in and out of phase with respect to each other with a characteristic beat frequency that is typically tens of megahertz, the time-averaged far-field pattern will be broader than the instantaneous far-field pattern. As a result, the average far-field pattern is broader than that which would be observed if only the $U_0$ mode were present. In general, the larger the number of transverse modes that are present, the broader the beam divergence in the far field will be. Since the outer radius of the $Q$th mode is approximately $Q^{1/2} w_0$, a useful rule of thumb is that the presence of $Q$ different transverse-mode orders in both $x$- and $y$-directions produces a far-field pattern with approximately $Q$ times the total area, so that the on-axis intensity in the far field is about $Q$ times lower than if the beam were diffraction-limited.

Now we must discuss how the presence of the gain medium can lead to multiple transverse-mode operation. If the gain is exactly uniform throughout the laser medium, the fundamental mode $U_0$ always has the least aperturing loss and, therefore, will tend to dominate. However, the presence of a saturable gain medium can lead to simultaneous oscillation on a number of transverse modes, as shown schematically in Fig. 8. For example, a laser might have the small signal gain profile indicated in Fig. 8a, with a constant value $G_0$ (gain per unit distance) throughout most of the laser medium, decreasing to zero near the boundaries. Suppose that the fundamental 00 mode has an aperturing loss $G_t$, so that an average gain greater than $G_t$ is sufficient to cause that mode to oscillate. If $G_0 > G_t$, the 00 mode will turn on, contributing the Gaussian field distribution indicated in Fig. 8b. The fundamental mode increases in

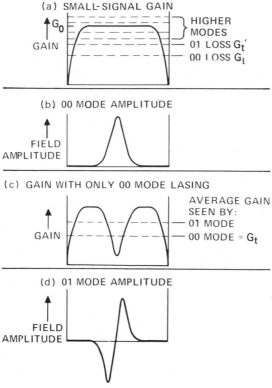

**Fig. 8.** The onset of multimode operation caused by high saturable gain.

intensity on every pass through the medium until the gain for that mode is decreased to the value $G_t$, to satisfy the gain-equals-loss steady-state lasing condition. The saturation of the gain medium depends upon the local-field intensity of the mode; once the fundamental mode reaches its equilibrium intensity, the remaining saturated gain of the medium will have a profile similar to that in Fig. 8c. The dotted line at $G_t$ in Fig. 8c suggests that the indicated saturated-gain profile, averaged with a weight function equal to the field intensity of the fundamental mode, yields an average saturated-gain value equal to the loss $G_t$.

Let us now note that the next higher-order mode, the 01 mode, has a different spatial distribution of intensity, as shown in Fig. 8d. Since the 01 mode has intensity concentrated farther from the center-of-the-gain medium, it samples more of the unsaturated higher-gain medium closer to the gain boundaries. Thus as indicated in Fig. 8c, the average gain seen by the 01 mode under saturated conditions, in general, will exceed the average gain $G_t$ of the 00 mode. If this average gain of the 01 mode happens to exceed the 01 mode loss $G'_t$, the 01 mode will also begin oscillating, even in the presence of the 00 mode. Thus we can see that if the boundaries of the gain medium are far away so that there is little difference in aperturing losses for the various transverse modes, or if the gain is much higher than the threshold value of the fundamental mode, it is possible for several different transverse modes to oscillate simultaneously. As previously seen, this leads to increased beam divergence or poorer beam quality.

Let us consider how this multimode degradation in beam quality constrains laser-operating parameters with a specific numerical example. Assume a stable optical resonator with rectangular geometry and give the laser medium a small spatially uniform homogeneously saturable gain $G_0$. Assume that

$$G_0 > G_t \tag{16}$$

where, as before, $G_t$ is the loss of the fundamental 00 mode, so that the gain is sufficient to make the fundamental mode oscillate. Then it can be shown that with the 00 mode lasing, the next higher Hermite–Gaussian mode, the 01 mode, has an average gain

$$G = \tfrac{1}{2}(G_0 + G_t) \tag{17}$$

Thus the 01 mode also lases simultaneously with the 00 mode provided that

$$\tfrac{1}{2}(G_0 + G_t) > G'_t \tag{18}$$

where $G'_t$ is the loss of the 01 mode. Thus, to obtain single-mode

(a) LARGE-RADII MIRRORS

(b) CONCAVE-CONVEX RESONATORS

**Fig. 9.** Methods for increasing mode volume in stable resonators. (*a*) Increase in mirror radii of curvature, and (*b*) use of concave–convex stable-mirror configurations.

operation on only the 00 mode, the small-signal gain of the medium is constrained to lie in a certain range given by the equation

$$1 < \frac{G_0}{G_t} \leq \frac{2G_t'}{G_t} - 1 \tag{19}$$

Normally, we would like $G_0/G_t \gg 1$ so that the laser operates well above threshold and power extraction is efficient. However, single-mode operation then requires that $G_t' \gg G_t$, which, as we saw in connection with Fig. 6, can only be achieved if the laser medium is not much larger than the fundamental mode size, so that the fundamental mode has much lower aperturing losses than the high-order modes. This limits the amount of active medium volume that can be used if good beam quality is to be obtained.

Unfortunately, high-output energy from a laser requires a large active-gain medium. Therefore, let us look at the techniques that are available for increasing the size of the fundamental resonator mode until it occupies a large-volume laser medium, thereby assuring efficient single transverse-mode operation and good beam quality.

In a conventional stable resonator the fundamental mode diameter can be increased through the use of larger-radii mirrors (Fig. 9a). However, as the mode volume increases, the requirements on mirror figure, gain medium uniformity, and mirror alignment increase very rapidly. A useful descriptive parameter is the Fresnel number $N_F$, defined as

$$N_F = \frac{a^2}{\lambda L} \tag{20}$$

where $2a$ is the gain medium or resonator-mirror diameter, $\lambda$ is the optical wavelength, and $L$ is the mirror separation. In terms of this parameter, it is found that the mode volume of the fundamental mode of the stable resonator can be increased as a practical matter only up to Fresnel numbers of ~10 because of the limitations just mentioned.

There is a variation of the stable resonator, which can be used at Fresnel numbers as high as 100, that is indicated schematically in Fig. 9b. In this case, a combination of a concave and a convex mirror is chosen that yields a resonator still lying within the stable region of Fig. 2.[13,14] With proper choice of mirror curvatures, the mode maintained between these mirrors can be identical with the mode that would be obtained in a limited portion near one end of a much longer stable resonator whose effective Fresnel number is only ~10. The limitations of this technique for increasing fundamental mode volume are the stringent requirements on mirror figure (and probably also on medium uniformity) required as the Fresnel number is increased.[15]

Thus our efforts to increase the mode volume of stable resonators lead to a limitation on the optical Fresnel number that can be used. How does this translate to a limitation on active-medium volume and, as a consequence, on laser-output energy? This is shown in Table 1, which lists the mirror separation required to achieve various Fresnel numbers with an active laser-mode volume of various sizes. For example, in Table 1, we see that for a laser with 0.001 liters of active volume, Fresnel numbers between 1 and 100 can readily be achieved with convenient cavity lengths. A reasonable cavity length requires a higher and higher Fresnel number as the laser volume increases, until with a 1000-liter volume, even a Fresnel number as high as 100 results in an unwieldy cavity length

**Table 1.** Laser-Cavity Length Requirements for Various Optical Fresnel Numbers ($\lambda = 3$ $\mu$m)

| Resonator mode volume, liters | Resonator Fresnel number $a^2/\lambda L$, $L$ in cm | | |
|---|---|---|---|
| | 1 | 10 | 100 |
| 0.001 | 10.3 | 3.3 | 1.0 |
| 0.010 | 32.6 | 10.3 | 3.3 |
| 0.100 | 103 | 32.6 | 10.3 |
| 1 | 326 | 103 | 32.6 |
| 10 | 1030 | 326 | 103 |
| 100 | 3257 | 1030 | 326 |
| 1000 | 10301 | 3257 | 1036 |

of $\sim 10$ m. The examples given refer to the specific wavelength $\lambda = 3$ $\mu$m, but similar considerations apply at other wavelengths.

## 1.3. Unstable Resonators

In our discussion of the stable resonator, both in its conventional config-uration and in the large-mode concave–convex version, it was necessary to limit our considerations to Fresnel numbers less than about 100. However, it has been found that through the operation of resonators in the unstable region of Fig. 2, limitations on Fresnel number for single transverse-mode operation can be overcome for most practical purposes. An accurate evaluation of the performance of various resonators requires detailed calculations using electromagnetic wave-propagation theory, as reported in the work of many authors and discussed in Section 2. However, the following heuristic arguments give some insight into why certain resonators yield uniphase laser operation and others do not.

When the Fresnel number of a laser resonator is only about unity, the beam expands significantly on each round-trip propagation through the resonator, so that energy travels laterally across the beam front. This means that diffraction in effect couples all parts of the gain medium together. This is suggested in Fig. 10*a* by a shaded region of high diffractive cross-coupling, essentially filling the dotted boundaries of the laser-gain medium. Since the existing radiation in the resonator locally stimulates the emission of the laser radiation, this cross-coupling between all parts of the gain medium means that all of the stimulating inversion centers can be phase-locked together and will be phase-coherent with respect to one another. Thus the output beam can be uniphase, or at least phase-coherent, across the beam front.

Let us now consider what happens when the gain medium is expanded laterally within the same stable resonator, so that the Fresnel number is now much larger than unity (Fig. 10*b*). In a ray optics approximation, there are now other closed-beam paths not lying strictly along the optical axis, which can exist within the laser medium and be supported by the resonator mirrors. These other paths would correspond to higher-order transverse modes in a more exact treatment including diffraction.

These various transverse modes extract energy from different parts of the laser medium and have little spatial overlap. Since the Fresnel number is large, ray optics is a good approximation, and we can neglect diffractive coupling between these modes. Thus, these different transverse modes oscillate independently of one another and lead to an output-beam front that is not phase coherent.

In contrast to this picture of a stable resonator operating at a large

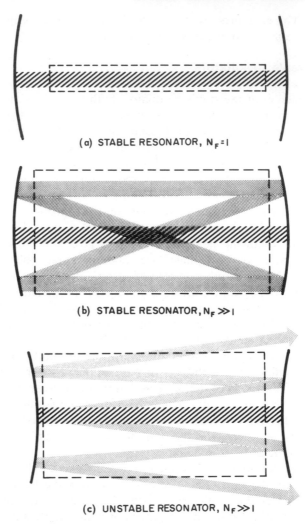

(a) STABLE RESONATOR, $N_F = 1$

(b) STABLE RESONATOR, $N_F \gg 1$

(c) UNSTABLE RESONATOR, $N_F \gg 1$

**Fig. 10.** Regions of strong cross-coupling in various resonator geometries, in various ranges of Fresnel number.

Fresnel number, consider what happens when we use convex mirrors to form an unstable resonator as in Fig. 10c. Now there is still a central region near the optical axis, of such a radius that its Fresnel number is near unity, within which all the laser radiation is closely coupled together by diffraction. The laser emission produced by this central region is, therefore, phase coherent. However, because the convex mirrors magnify the beam continually as it propagates through the optical cavity, this central uniphase mode spreads out to fill the entire laser medium, eventually overflowing around one mirror to form the laser output as in the figure. Thus the unstable resonator can be thought of as consisting of a low Fresnel-number oscillator region, surrounded by a multipass amplifier. Since phase control of the output beam occurs within the central portion of the resonator, a high-quality output beam can be maintained. However, since the principal amplification of the output occurs in the outer regions and these are not limited in usable Fresnel number, good phase control of the beam can be obtained even with a large Fresnel-number medium. The unstable laser resonator was first studied extensively by Siegman;[16] it is very widely used today in high-energy lasers, as discussed in Section 2.

Some typical types of unstable resonators are shown in Fig. 11. The symmetric unstable resonator (Fig. 11a), with laser output overflowing around both end mirrors, is rarely used in practice because multiple output beams usually are not desired. The asymmetric, unstable resonators (Fig. 11b) are useful in particular situations, for example, when a diffraction grating is used in place of one cavity mirror to perform wavelength selection. The most commonly used unstable resonator is the confocal geometry (Fig. 11c), which in the geometric optics approximation yields a collimated output beam.

The output beams from these types of unstable resonators have the form of an annulus, with a zero-intensity central region produced by the part of the beam that is blocked by the output mirror. Such an output beam might be optimum for use with a Cassegrain telescope or focusing optics, but with a conventional telescope designed to accommodate an unobscured beam this annular intensity distribution departs from the ideal uniform intensity, which yields minimum far-field beamwidth. The hole rapidly fills in as the beam propagates, but the annular beam still retains more power in its side lobes than does a uniform circular beam. The far-field pattern produced by a uniform annular beam is well known[17] and depends critically on the relative size of the hole compared with the total beam diameter. However, in high-power lasers, where the obscuration commonly amounts to less than 50% of the unobscured beam area, it is frequently found that the annular beam departs so little in beam quality from the uniform beam that other factors, such as medium uniformity, are more important than uniform intensity when optimization of laser performance is attempted. For those cases in which the annulus is so thin that beam quality is impaired, as with unstable resonators with an output coupling of 30% or less, a telescope incorporating conical lenses or mirrors can be used to produce a modified output beam that does not have a hole.[18] Alternatively, a modified unstable resonator may be

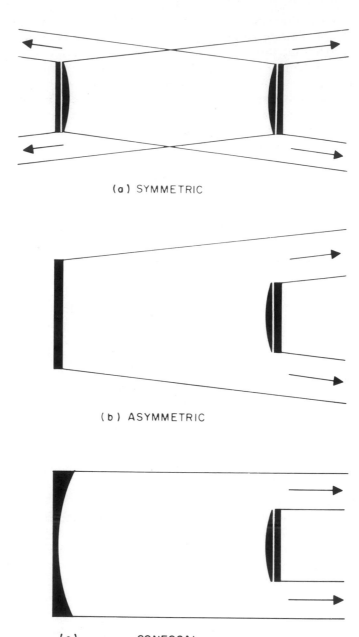

(a) SYMMETRIC

(b) ASYMMETRIC

(c)     CONFOCAL
(COLLIMATED OUTPUT BEAM)

**Fig. 11.** Types of unstable laser resonators.

utilized that produces a more uniform output beam, such as the continuously coupled diffraction-grating resonator reported by Chodzko et al.[19]

The wave-front patterns for the modes in unstable resonators can be derived very simply in the geometric optics approximation and are shown in Fig. 12, which is adapted from Ref. 16. In the type I positive-branch unstable resonator (Fig. 12a), the beam magnifies on each round trip through the resonator; this is the usual type of unstable resonator used. However, there are two additional types, shown in Figs. 12b and c. In these resonators, the beam magnifies on each round trip through the resonator, but it is inverted either once (in the negative-branch resonator) or twice (in the type II positive-branch resonator). Thus these types of resonators have a focal point inside the optical cavity at which the laser radiation is concentrated. The various types of unstable resonators can be distinguished by their locations in the resonator stability diagram of Fig. 2.

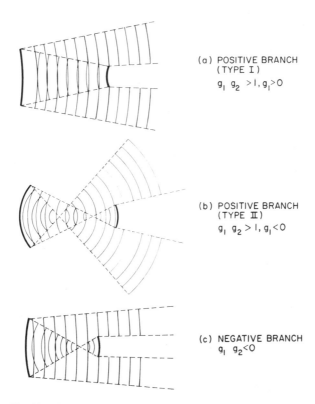

(a) POSITIVE BRANCH
(TYPE I)
$g_1 g_2 > 1, g_1 > 0$

(b) POSITIVE BRANCH
(TYPE II)
$g_1 g_2 > 1, g_1 < 0$

(c) NEGATIVE BRANCH
$g_1 g_2 < 0$

**Fig. 12.** Geometric wave-front patterns in unstable resonators.

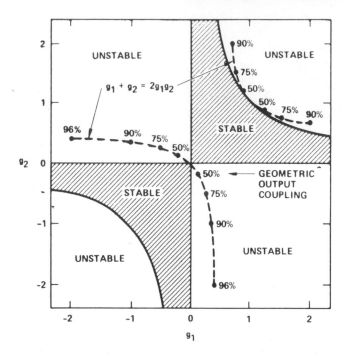

**Fig. 13.** Confocal unstable resonators and their output couplings plotted on the resonator-stability diagram.

The important special case of the confocal unstable resonator satisfies the condition[20]

$$g_1 + g_2 = 2g_1g_2 \qquad (21)$$

These resonators, and their fractional output coupling per round trip, are plotted on the stability diagram in Fig. 13. Note that, as soon as a resonator departs even slightly from the stable region, the output coupling is extremely high; thus unstable resonators tend to require high laser gain per pass in order to oscillate.

We have explained heuristically, in connection with Fig. 10, why unstable resonators can be expected to provide good spatial coherence or good beam quality even at large Fresnel numbers. However, no simple closed expression, such as Eq. (6), is presently known that describes the mode properties of unstable optical resonators. In addition, since unstable resonators are used with high-gain laser configurations in which the small-signal gain per pass may be 100% or more, the presence of the gain medium is not always a minor perturbation on the performance of an unstable resonator. In contrast, since stable resonators are typically used with gains per pass of 20% or less, the transverse-mode patterns of stable resonators are not greatly distorted by the presence of the gain medium.

In Section 2 we will first discuss the ideal performance of unstable resonators in the absence of a gain medium. Then, the effect of cavity perturbations, such as gain, mirror tilt, and refractive-index disturbances, will be noted. It will be found that unstable-resonator beam quality and output power are both typically sensitive to all of these effects. We will then survey the experimental measurements of beam quality using unstable laser resonators, including the effects of wavelength selection in chemical-laser experiments. Comparative beam quality measurements using master oscillator power amplifier (MOPA) optical configurations will also be discussed.

## 2. THEORETICAL RESULTS

In 2.1 through 2.4, the current understanding of the empty-cavity modes of an unstable resonator will be discussed in more detail. These results apply to either pulsed or cw chemical lasers. The effect of the active medium of a cw chemical laser on the empty-cavity modes will be discussed in Section 2.5.

### 2.1. Geometric Solution for an Unstable Resonator

The gross features of unstable-cavity performance can be understood from the simple geometric model discussed first by Siegman,[16] which also predicts the wave-front patterns previously shown in Fig. 12. A schematic diagram of the resonator geometry is shown in Fig. 14. It is hypothesized that the unstable cavity will support a geometric mode consisting of a spherical wave of uniform amplitude. The requirement that the wave front be self-reproducing leads to the simultaneous solution of two

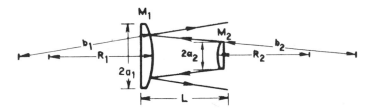

**Fig. 14.** Unstable-resonator geometry with single-ended output.

paraxial imaging equations:

$$\frac{1}{b_1} = \frac{1}{b_2+L} - \frac{2}{R_1} \tag{22}$$

and

$$\frac{1}{b_2} = \frac{1}{b_1+L} - \frac{2}{R_2} \tag{23}$$

where $R_1/2$ and $R_2/2$ are the focal lengths associated with mirrors $M_1$ and $M_2$, respectively, and $L$ is the mirror separation. The virtual centers of the unstable-resonator modes in Fig. 12 are at distances $b_1$ and $b_2$ from the mirrors. Thus, if an arbitrary wave front (e.g., a plane wave) is launched within the equivalent sequence of lenses, $b_1$ and $b_2$ are the steady-state wave-front curvatures at $M_1$ and $M_2$ obtained after a large number of passes. The values of $b_1$ and $b_2$ determined from Eqs. (22) and (23) are given by

$$b_1 = \frac{L[1/g_1 - 1 + (1 - 1/g_1 g_2)^{1/2}]}{(2 - 1/g_1 - 1/g_2)} \tag{24}$$

and

$$b_2 = \frac{L[1/g_2 - 1 + (1 - 1/g_1 g_2)^{1/2}]}{(2 - 1/g_1 - 1/g_2)} \tag{25}$$

where

$$g_1 = 1 - \frac{L}{R_1}, \qquad g_2 = 1 - \frac{L}{R_2}$$

The round-trip transverse geometric magnification is given by

$$M = \left(1 + \frac{L}{b_1}\right)\left(1 + \frac{L}{b_2}\right) \tag{26}$$

assuming $b_1$ and $b_2$ are positive. The output-coupling fraction is

$$\delta_s = 1 - \frac{1}{M^2} \tag{27}$$

for spherical mirrors, and

$$\delta_c = 1 - \frac{1}{M} \tag{28}$$

for cylindrical mirrors. Thus a small-diameter $[\sim(\lambda L)^{1/2}]$ geometric mode will form along the axis and expand radially because of the transverse magnification. The expansion will continue for the single-ended output cavity in Fig. 14 until the radiation reaches the boundary of the smallest mirror $M_2$ of diameter $2a_2$. The output beam of diameter $D_F = M(2a_2)$ is then obtained in the last two passes. The total number of passes is given

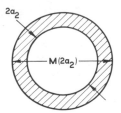

(a) $M_2$ WITH CIRCULAR BOUNDARY, CIRCULAR ANNULUS

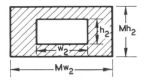

(b) $M_2$ WITH RECTANGULAR BOUNDARY, RECTANGULAR ANNULUS

**Fig. 15.** Geometric solution for near-field intensity distribution of unstable cavity-output beam. (a) $M_2$ with circular boundary and (b) $M_2$ with rectangular boundary.

by

$$n_p = \frac{\ln (D_F/D_i)}{\ln (M)} \tag{29}$$

within the geometric approximation where $D_F$ is the final beam diameter and $D_i$ is the initial beam diameter $(\lambda L)^{1/2}$. [This assumes $D_F = D_i(M)^{n_p}$.]

Thus for a spherical cavity mirror, the geometry of the output beam is determined by the shape and size of the small mirror $M_2$. If $M_2$ has a circular boundary centered on the optical axis, a circular annular beam is produced. If the $M_2$ boundary is rectangular, the rectangular beam shown in Fig. 15 is produced. The rectangular-mode geometry can sometimes be more efficient at energy extraction than the circular, if it provides a better match to the shape of the active laser medium (a supersonic diffusion-type cw chemical laser[21] generally has a rectangular active medium, for instance).

As an example, consider the confocal unstable cavity[20,22] determined by the condition $R_1 + R_2 = 2L$ or $g_1 + g_2 = 2g_1g_2$. If $g_1g_2 > 1(R_1 > 0$ and $R_2 < 0$, for instance), we have a positive branch confocal resonator as seen in the stability diagram of Fig. 13. This implies

$$b_1 = \infty \tag{30}$$

and

$$b_2 = \frac{-R_2}{2} \tag{31}$$

The virtual center position is then positive, which means it lies outside the resonator. If $g_1g_2<0$ ($R_1>0$, $R_2>0$), we have a negative-branch confocal unstable resonator. The virtual center is then negative, which implies that it is inside the resonator. Since the cavity radiation is focused at this location, negative-branch unstable resonators are frequently avoided because of the high-energy densities produced. The magnification for the positive-branch confocal unstable cavity is

$$M = 1 - \frac{2L}{R_2} \tag{32}$$

Another example is an unstable resonator consisting of a flat and a convex mirror of radius $R$. This is applicable to the single-line unstable cavity.[19] The virtual centers are given by

$$b_1 = L\left(1 - \frac{R}{L}\right)^{1/2} \tag{33}$$

and

$$b_2 = L\left[\left(1 - \frac{R}{L}\right)^{1/2} - 1\right] \tag{34}$$

and the magnification

$$M = \frac{(1 - R/L)^{1/2} + 1}{(1 - R/L)^{1/2} - 1} \tag{35}$$

The geometric analysis of the empty-cavity modes of an unstable resonator is a useful theory that gives an approximate value to the magnification $M$ and an accurate determination of the radii of curvature of the emerging wave fronts $b_1$ and $b_2$. The foregoing geometric analysis can clearly be extended to include more optical elements within the unstable resonator, an unstable resonator with an internal lens, for instance, but the formulas for the virtual centers become more cumbersome.

## 2.2.  Diffraction Analysis—Uniform Reflectivity

The empty-cavity modes of an unstable resonator within the scalar wave approximation can be determined from Kirchhoff's formulation of Huygen's principle. Thus within the Fresnel diffraction[23] approximation and under the assumption of rectangular symmetry, the integral equations are given by

$$\gamma_x^{(2)} U_x^{(2)}(x_2) = \left(\frac{i}{\lambda L}\right)^{1/2} \int_{-a_1}^{a_1} U^{(1)}(x_1')$$

$$\times \exp\left[-\frac{ik}{2L}(g_1 x_1'^2 + g_2 x_2^2 - 2x_1' x_2)\right] dx_1' \qquad (36)$$

$$\gamma_x^{(1)} U_x^{(1)}(x_1) = \left(\frac{i}{\lambda L}\right)^{1/2} \int_{-a_2}^{a_2} U^{(2)}(x_2')$$

$$\times \exp\left[-\frac{ik}{2L}(g_1 x_1^2 + g_2 x_2'^2 - 2x_1 x_2')\right] dx_2' \qquad (37)$$

where $U_x^{(1)}(x_1)$ is the field amplitude on mirror (1) at coordinate $x_1$, $\gamma_x^{(1)}$ is the eigenvalue of the integral equation, which determines the diffraction losses and the frequencies of the eigenmodes, $g_i = 1 - (L/R_i)$ is defined as before for the specific cavity geometry, and $2a_1$ is the width of mirror (1) in the $x_1$ dimension (see Fig. 14). Identical equations are to be solved in the $y_1$ transverse coordinate orthogonal to $x_1$, and the field amplitudes of the modes of the resonator are the products of the $x$- and $y$-field distributions so obtained. The total eigenvalues of the resonator modes are given by

$$(\gamma_x^{(1)} \gamma_x^{(2)})_m (\gamma_y^{(1)} \gamma_y^{(2)})_n = \gamma_m \gamma_n \qquad (38)$$

where $m$ and $n$ are integers defining the transverse-mode order of the resonator; $m = n = 0$ corresponds to the lowest-loss or fundamental mode. The fractional diffraction loss per round trip within the resonator is given by

$$\delta_{mn} = 1 - |\gamma_m \gamma_n|^2 \qquad (39)$$

and the laser-oscillation frequency for the $mn$ mode is

$$\nu_{qmn} = \nu_0\left[q + 1 + \frac{1}{\pi} \arg(\gamma_m \gamma_n)\right] \qquad (40)$$

where $q$ is the longitudinal-mode order and $\nu_0 = c/2L$ is the fundamental-standing-wave beat frequency.

Equations (36) and (37) can be combined into a single integral equation in nondimensional form given by

$$\gamma_m U^{(2)}(X_2) = i(NN_0)^{1/2} \int_{-1}^{1} \exp\{i\pi N_0(g_1 X_1'^2 | g_2\alpha^2 X_2^2 \quad 2\alpha X_1' X_2)\} dX_1'$$

$$\times \int_{-1}^{1} \exp\{i\pi N_0(g_1\alpha^2 X_1'^2 + g_2 X_2'^2$$

$$- 2\alpha X_1' X_2')\} U_x^{(2)}(X_2') dX_2' \qquad (41)$$

where $X_2 = x_2/a_2$, $N = a_2^2/\lambda L$ is the Fresnel number computed from the small-mirror diameter, $N_0 = a_1^2/\lambda L$ is the Fresnel number based on the large mirror, and $\alpha = a_2/a_1$ is the mirror-diameter ratio.

For an unstable resonator that is edge coupled,[19] there is a small mirror (2) and a larger mirror (1). If the aperture of mirror (1) is large compared with mirror (2), which is often the case, the approximation $a_1 \to \infty$ can be made, and Eqs. (36) and (37) can be combined into a single integral equation given by

$$\gamma_x^{(1)} \gamma_x^{(2)} U^{(2)}(x_2) = \int_{-a_2}^{a_2} K(x_2, x_2') U^{(1)}(x_2') dx_1' \qquad (42)$$

where

$$K(x_2, x_2') = \left(\frac{i}{2\lambda L g_1}\right)^{1/2} \exp -\left\{\frac{i\pi}{2\lambda L g_1}\left[(2g_1 g_2 - 1)(x_2^2 + x_2'^2) - 2x_2 x_2'\right]\right\} \qquad (43)$$

Equation (42) is of the same form[24,25] as the integral equations (36) and (37) for a single pass through a symmetric unstable resonator with mirror parameter $g = 2g_1 g_2 - 1$. Thus a solution for a symmetric-strip unstable resonator gives the solution for the asymmetric-strip unstable resonator. This is an important equivalence relation that reduces the number of numerical calculations required.

The symmetric unstable resonator has $R_1 = R_2 = R$, $a_1 = a_2 = a$ and

$$g = \frac{(M+1)^2 + (M-1)^2}{(M+1)^2 - (M-1)^2} \qquad (44)$$

A new parameter defined as the equivalent Fresnel number is

$$N_{eq} = \frac{a^2}{2\lambda L}\left(M - \frac{1}{M}\right) = N(g^2 - 1)^{1/2} \qquad (45)$$

where $N = a^2/\lambda L$ is the Fresnel number based on either mirror diameter. The single-pass integral equation for the symmetric unstable resonator in nondimensional form is

$$\gamma u(X) = (iN)^{1/2} \int_{-1}^{1} \exp\{i\pi N[g(X^2 + X'^2) - 2XX']\} u(X') dX' \qquad (46)$$

where $X = x/a$. If the transformation

$$u(X) = v(X) \exp(-i\pi N_{eq} X^2) \qquad (47)$$

is applied to Eq. (46), one obtains

$$\gamma_n v_n(X) = (iN)^{1/2} \int_{-1}^{1} \exp\left[i\pi NM\left(X' - \frac{X}{M}\right)^2\right] v_n(X') dX' \qquad (48)$$

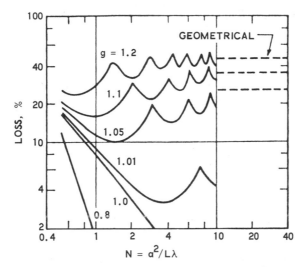

**Fig. 16.** Comparison of loss per reflection versus Fresnel number $N$ for the symmetric-strip case as obtained from the exact numerical calculations (solid lines) and the geometric analysis (dashed lines).[26]

which is the general integral equation to be solved for the unstable-resonator empty-cavity eigenmodes.

It should be pointed out again that Eqs. (36), (37), and (48) correspond to a two-dimensional solution or the case of rectangular mirrors. For an empty-cavity solution, the $x$–$z$- and $y$–$z$-directions correspond to two separable two-dimensional problems, each with its own characteristic Fresnel number. When the symmetry is spoiled by either the gain medium or refractive-index disturbances, the $x$–$z$- and $y$–$z$-solutions are no longer independent.

Equation (48) was first studied extensively by Siegman and Arathoon[26]; the solution for the diffraction losses is shown in Fig. 16. Thus, there is an oscillatory variation in the diffraction loss per transit as a function of Fresnel number $N$. The diffraction loss per pass approaches the geometric value, and the amplitudes of the oscillations are reduced in the large Fresnel-number limit. The same results, plotted in Fig. 17 as a function of the equivalent Fresnel number, show that the relative maxima occur at integral values of $N_{eq}$. It will be shown[27] in 2.3 that $N_{eq}$ plays a role in determining diffraction effects caused by the mirror-edge boundary. Figure 18 shows the radius of curvature of the emerging wave front calculated numerically from Eq. (48), compared with the geometric model of Siegman for the lowest-order mode. Thus the geometric model quite accurately predicts the phase distribution of the lowest-order unstable-resonator mode.

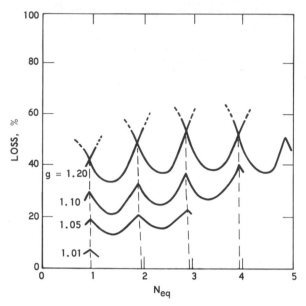

**Fig. 17.** A replot of Fig. 16 in terms of the equivalent Fresnel number $N_{eq}$. The dashed curves correspond to higher-order modes.[26]

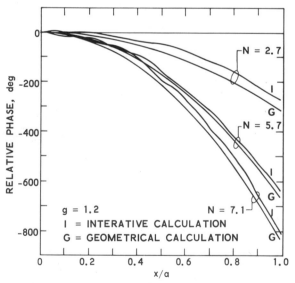

**Fig. 18.** Phase variation relative to the mirror surface of the lowest-order mode versus radius[26] as obtained from the geometric analysis $G$ and the exact iterative computer solution $I$ for three typical values of Fresnel number $N$.

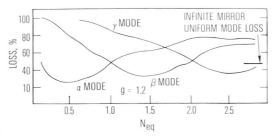

**Fig. 19.** Loss per reflection for even symmetric $\alpha$, $\beta$, $\gamma$ modes as a function of equivalent Fresnel number for $g = 1.2$.[28]

The oscillatory variation in the loss per transit of the unstable resonator shown in Fig 16 suggests a complicated transverse-mode structure. Normally, transverse modes are defined and ordered according to their loss per transit with $1 - |\gamma_0|^2 < 1 - |\gamma_1|^2 < 1 - |\gamma_2|^2$, and so on. Sanderson and Streifer[28] found that this ordering process is ambiguous for the modes in an unstable resonator as opposed to previous results[10,23,29] obtained for stable and Fabry-Perot cavities. Figure 19 shows the diffraction loss per pass calculated from eq. (48) (rectangular-strip case) versus $N_{eq}$. The $\alpha$, $\beta$, $\gamma$, correspond to transverse modes that have increasing value of $N_{eq}$ at the first loss minimum. This notation for labeling the modes ($\alpha$, $\beta$, $\gamma$, . . .) follows the continuous variation of their complex eigenvalues rather than arranging them in order of their relative loss at any particular value of $N_{eq}$; there is a unique correspondence between these two methods of labeling that extends through mode crossings. After the first loss minimum, there is a complicated oscillatory loss variation with $N_{eq}$ resulting in mode crossing points for integral values of $N_{eq}$. One might expect that this implies multitransverse-mode operation for the unstable cavity and, hence, poor beam quality. At half-integral equivalent Fresnel numbers, on the other hand, there is no mode crossing and, as a result, perhaps improved mode selection. Figure 20 shows how the various modes are tracked through the crossing points by tracing their complex eigenvalues. The real versus imaginary parts of $\gamma_\alpha$, $\gamma_\beta$, and so on are plotted as a function of Fresnel number $N$, and the curves are continuous functions of $N$. Siegman and Miller[24] obtained the solution to the unstable-resonator integral equation for the case of circular-mirror boundaries. They found results similar to the rectangular case, a periodic variation in loss with $N_{eq}$, and mode crossing occurring at integral values of $N_{eq}$.

The foregoing diffraction calculations were obtained from numerical computer solutions to the appropriate integral equations. The various numerical techniques applied have limitations in the maximum value of $N$

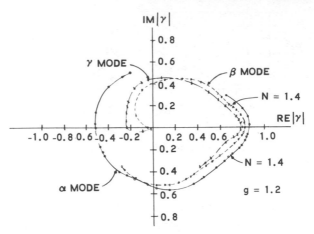

**Fig. 20.** Complex eigenvalue loci for the even symmetric $\alpha$, $\beta$, $\gamma$ modes with $g = 1.2$ and $0.16 \leq N \leq 5.1$. Arrows indicate the direction of increasing $N$.[28]

for which solutions can be obtained. Recently, Horwitz[25] has obtained solutions to Eq. (48) by an asymptotic expansion technique valid in the limit of large values of $N_{eq}$. These results should approach the geometric-optic solution in the limit $N_{eq} \to \infty$. Horwitz found that only the lowest-loss symmetric mode approaches the geometric-optics limit and the higher-order modes do not. Figure 21 shows the diffraction loss for higher-order modes in a region of large values of $N_{eq}$, from numerical calculations of Ref. 25 for a $M = 1.9$ confocal cavity with rectangular mirrors. Note the "cusping" or mode-separation[25,28] phenomena where a single symmetric mode oscillates about the geometric limit for sufficiently large $N_{eq}$. Such a mode separation was not observed by Siegman and Miller for the case of circular mirrors. Thus, the diffraction calculations

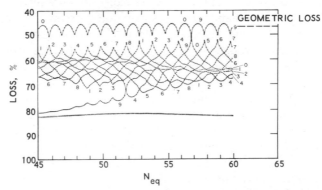

**Fig. 21.** Loss per reflection for the even symmetric modes as a function of equivalent Fresnel number for an $M = 1.9$ confocal unstable cavity with rectangular mirrors.[25]

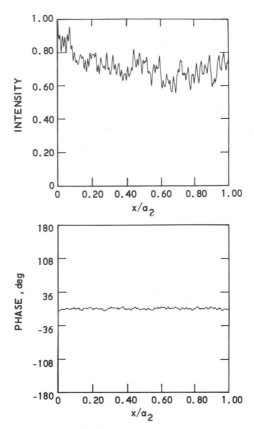

**Fig. 22.** Intensity and phase distribution on the small mirror of the fundamental mode corresponding to Fig. 21 with $N_{eq} = 49.4$.[25]

indicate a difference in the mode selectivity between square aperture mirrors and circular aperture mirrors. This could be a result of the circular nature of the Fresnel zones and be related to the edge effects to be discussed. The amplitude and phase distributions associated with Fig. 21 ($m = 0$ mode) at $N_{eq} = 49.4$ are shown in Fig. 22, indicating a nonuniform amplitude distribution, but a uniform phase distribution.

The mode-crossing behavior described previously indicates that single transverse-mode operation may not occur easily in an unstable resonator. The question of interest, however, is not whether some particular transverse mode dominates, but whether there is a unique amplitude and uniform phase distribution that reproduces itself after many passes through the resonator. The unstable-resonator problem was addressed in this way by Rensch and Chester.[30] The dominant mode for a confocal

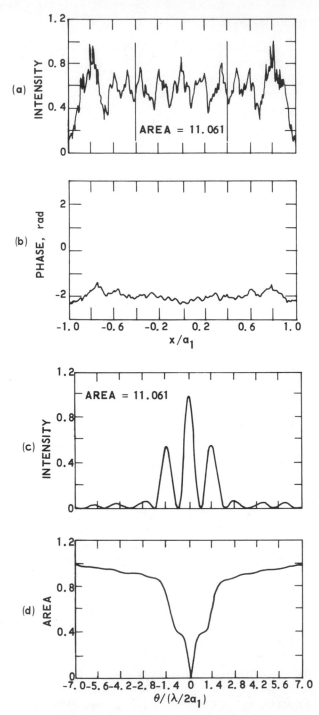

**Fig. 23.** Beam-quality calculations for the empty-cavity fundamental mode of a positive-branch confocal unstable cavity with $M = 2.5$ and tube Fresnel number $N_0 = M^2 N = 60.$[30]

unstable resonator (rectangular mirrors) was determined by the Fox and Li[23] iteration technique. Single transverse-mode ouput was verified by a lack of mode beating. When mode beating occurs, a fixed steady-state intensity and phase profile is not obtained after a large number of iterations. Figure 23 is a plot for a typical calculation of Ref. 30 showing the near-field intensity and phase distribution and the far-field intensity and integrated intensity distribution for $N_0 = 60$, $M = 2.5$. Thus although there is nonuniform amplitude distribution, the phase distribution is sufficiently uniform to produce a far-field diffraction pattern with a well-defined side-lobe structure; a large fraction of the total power is within the main lobe ($\sim 40\%$).

Various numerical techniques have been used to solve Eq. (41) or (48) for the empty-cavity unstable-resonator eigenmodes. The problem is that many solutions of interest correspond to large Fresnel numbers ($N$ or $N_0$). For large $N$, the integrands in Eq. (41) or (48) have high-frequency oscillations that lead to inaccuracies in the numerical integration for a finite number of sample points. Sanderson and Streifer[31] have made a comparison of the various numerical methods. The integral equations can be solved iteratively, as first done by Fox and Li,[23] where an assumed amplitude and phase distribution is launched within the equivalent lens sequence and the process of mode build-up is calculated numerically. The iterative method was used by Siegman and Arrathoon[26] and Rensch and Chester[30] to calculate the unstable-cavity modes.

Equations (41) and (48) are homogeneous Fredholm integral equations of the second kind and constitute an eigenvalue problem. This problem can be cast in matrix form for a given numerical-integration scheme, and the (complex) eigenvalues can be determined by a similarity transformation. In the matrix method, the complex eigenfunctions for the fundamental and many higher-order modes are computed simultaneously, but the process of mode build-up is not determined. This technique was developed by Sanderson and Streifer.[28] The maximum Fresnel number that can be handled by the matrix technique is determined by the maximum size of the matrix that can be inverted on available computers for a reasonable computing time. The size of the matrix required for a given accuracy depends on the numerical integration scheme, that is, trapezoidal rule, Simpson's rule or Gaussian quadrature. Sanderson and Streifer[31] found that Gaussian quadrature was the optimal numerical integration scheme. The standard Gaussian quadrature technique has unequal sampling-point spacings, which are determined by the roots of Legendre polynomials. Unstable-resonator modes for Fresnel numbers ($N_0$) up to about 10 can be conveniently calculated[32] with a CDC 6600 computer.

Bullock et al.[32] have recently extended the Gauss–Legendre matrix-diagonalization technique to larger Fresnel numbers. The integrand in Eq. (41) or (48) can be separated into two factors. One represents the Green's function of the source. In the case of an unstable resonator, the more oscillatory contribution to the integrand comes from the Green's function factor rather than the source distribution factor. A numerical integration scheme, which removes the structure of the Green's function term from the sampling requirement and allows for the evaluation of the mode structure of resonators at large Fresnel number, has been derived by Bullock et al. A brief description of the scheme, which is termed the modified Gauss–Legendre (MGL) method, is given here.

Consider integrals of the form

$$\mathcal{I} = \int_{-1}^{1} \exp\left[i(\beta_2 x^2 + \beta_1 x)\right] f(x)\, dx \tag{49}$$

The numerical-integration formula for Eq. (49) may be represented by

$$\mathcal{I} = \sum_{k=1}^{n} w_k f(\xi_k^{(n)}) \tag{50}$$

where the integration weights are given by

$$w_k = \frac{1}{n P_{n-1}(\xi_k^{(n)}) P_n'(\xi_k^{(n)})} \sum_{m=0}^{n-1} (2m+1) P_m(\xi_k^{(n)}) L_m \tag{51}$$

and where

$$L_m = \int_{-1}^{1} \exp\left[i(\beta_2 x^2 + \beta_1 x)\right] P_m(x)\, dx \tag{52}$$

Here $\beta_1$ and $\beta_2$ are real numbers that relate to the Fresnel-number and mirror-curvature parameters of a cylindrical-strip resonator. The $P_m(x)$ are the Legendre polynomials, and the $\xi_k^{(n)}$ are the $n$ roots of $P_n(x)$. Calculation of the integration weights with this method then requires the sampling points $\xi_k^{(n)}$ to be sufficiently dense to resolve only the field structure $f(x)$. In practical applications, the evaluation of the $L_m$ can be difficult for large $m$. Through the use of the recursion relations for the Legendre polynomials, a recursion relation for the $L_m$ may be derived:

$$\frac{n}{2n-1} L_n = (2n-3)\left[\frac{1}{(2n-1)(2n-5)} + \frac{i}{2\beta_2}\right] L_{n-2}$$

$$-\frac{\beta_1}{2\beta_2}\left[L_{n-1} - L_{n-3}\right] + \frac{n-3}{2n-5} L_{n-4} \tag{53}$$

for $n \geq 3$ and $\beta_2 \neq 0$. For the recursion process to be initiated, the values of $L_0$, $L_1$, and $L_2$ are required. By direct calculation, we find

$$L_0 = \left(\frac{\pi}{2\beta_2}\right)^{1/2} \exp\left[\frac{i\beta_1^2}{4\beta_2}\right]\left\{E\left[\left(\frac{2\beta_2}{\pi}\right)^{1/2}\left(1+\frac{\beta_1}{2\beta_2}\right)\right]\right.$$

$$\left. + E\left[\left(\frac{2\beta_2}{\pi}\right)^{1/2}\left(1-\frac{\beta_1}{2\beta_2}\right)\right]\right\}$$

$$L_1 = \frac{1}{2i\beta_2}\{\exp\left[i(\beta_2+\beta_1)\right] - \exp\left[i(\beta_2-\beta_1)\right] - i\beta_1 L_0\}$$

$$L_2 = \frac{3}{4i\beta_2}\{\exp\left[i(\beta_2+\beta_1)\right] + \exp\left[i(\beta_2-\beta_1)\right] - L_0\} - \frac{3\beta_1}{4\beta_2}L_1 - \frac{1}{2}L_0$$

(54)

where the function $E(x)$ is the complex Fresnel integral

$$E(x) = \int_0^x \exp\left(i\frac{\pi}{2}t^2\right)dt \tag{55}$$

Figure 24 shows the results of Bullock et al., where 64 sampling points were used to approximate the Green's function in the MGL numerical

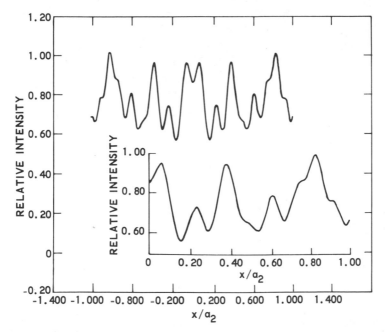

**Fig. 24.** Intensity distribution on the small mirror of a positive-branch confocal unstable resonator at a Fresnel number $N = 10$ and $M = 2.0$. The insert shows the calculation of Horwitz.[32]

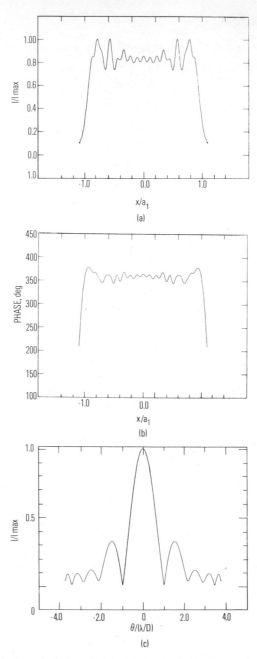

**Fig. 25.** Beam-quality calculations for the empty-cavity fundamental mode of a positive-branch confocal unstable cavity with $M = 3.16$ and $N_0 = 60$.[32] (a) Near-field intensity distribution (on scraper). (b) near-field phase distribution (on scraper), and (c) far-field intensity distribution.

scheme. The intensity distribution across the small mirror is plotted for a confocal unstable resonator with a Fresnel number $N$ of 10 (tube Fresnel number of 40) and compared with the results of Horwitz.[25] The comparison for the previously discussed Fresnel number is excellent. At higher values of $N$, however, the comparison was poorer because 64 sample points were not sufficient to approximate the high-frequency fluctuations in the Green's function. These high-frequency fluctuations were not deemed important since, as will be discussed, they are caused by edge effects that will not exist in a practical active resonator. Figure 25 shows the near-field amplitude and phase distribution (on the output-coupling mirror) and the far-field intensity distribution for a confocal unstable resonator calculated by Bullock et al. The calculation again assumes 64 sampling points and can be compared approximately with the results of Rensch and Chester in Fig. 23. Although the results are similar, the calculations of Rensch and Chester show more fine structure than those of Bullock et al., which possibly is due to the limitations of the MGL technique.

## 2.3.  Diffraction Analysis—Nonuniform Reflectivity

The complex mode-crossing behavior of the unstable resonator was first explained by Anan'ev and Sherstobitov.[27] The effect is shown to be a result of diffraction of the geometric eigenmode at the finite edge of the mirror boundaries. Figure 26 shows a symmetric unstable resonator with mirror radii of curvature $R$ and a geometric wave front of radius

$$b = \frac{LM}{M-1} \tag{56}$$

at the mirror surface. The radiation striking the mirror boundary is reflected through an angle $2\theta_c$ as it leaves the resonator. A certain

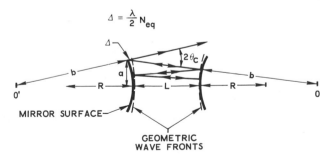

**Fig. 26.**  Symmetric unstable resonator showing edge-effect parameters.

fraction of this radiation can be transmitted back into the resonator by diffraction along the rays denoted by double arrows. This edge wave can influence the mode of the unstable cavity. A diffraction zone $h_d$ is defined near the mirror boundary by

$$\frac{\lambda}{h_d} = 2\theta_c \qquad (57)$$

Zones of larger width than $h_d$ cannot be diffracted back into the resonator. The angle of reflection $\theta_c$ is related to the cavity geometry by

$$\theta_c = \frac{\Delta}{a} \qquad (58)$$

where $\Delta$ is the deviation of the geometric wave front from the mirror edge and $a$ is the mirror radius. But $\Delta$ is related to the equivalent Fresnel number by

$$\Delta = \frac{\lambda}{2} N_{eq} \qquad (59)$$

since $N_{eq}$ is the number of half-period zones on the mirror surface as viewed from the virtual center 0. Hence

$$h_d = \frac{a}{N_{eq}} \qquad (60)$$

For a symmetric unstable cavity

$$N_{eq} = \frac{N}{2}\left(M - \frac{1}{M}\right) \qquad (61)$$

and for a confocal unstable cavity

$$N_{eq} = \frac{N}{2}(M - 1) \qquad (62)$$

For a typical confocal unstable cavity with $a_2 = 2.5$ cm, $M = 1.5$, $\lambda = 4 \times 10^{-4}$ cm, and $L = 1$ m, one gets $h_d = 0.026a = 0.65$ mm. If the reflectivity of the mirror is attenuated over this narrow region, the edge-wave amplitude will be reduced and the output beam should closely approach the geometric solution, a spherical wave of uniform amplitude.

One method of reducing the mirror-edge effects is to taper the reflectivity distribution. For the case of a Gaussian reflectivity distribution and rectangular mirrors, Anan'ev and Sherstobitov[27] obtained an exact solution to the unstable-cavity diffraction integral equation. The integral equation to be solved for a symmetric resonator is

$$\gamma_m U_m(x) = \left(\frac{ik}{2\pi L}\right)^{1/2} \int_{-\infty}^{\infty} \exp\left\{-\frac{2x'^2}{a^2} + ik\left[\frac{(x-x')^2}{2L} - \frac{(x+x')^2}{2R}\right]\right\} U_m(x') \, dx'$$

(63)

where we assume a reflectivity distribution $\rho(x) = \exp(-2x^2/a^2)$ and where $a$ is the mirror scale. The solution is given by

$$U_m(x) = \exp\left[(1 - i\pi N'_{eq})\left(\frac{x}{a}\right)^2\right] H_m\left[2(i\pi N'_{eq})^{1/2} \frac{x}{a}\right]$$

(64)

where

$$N'_{eq} = \frac{N'}{2}\left(M' - \frac{1}{M'}\right)$$

$$M' = 1 + \frac{L}{R'} + \left[\frac{2L}{R'} + \left(\frac{L}{R'}\right)^2\right]^{1/2}$$

and

$$\frac{ik}{2R'} = \frac{ik}{2R} + \frac{1}{a^2}$$

with eigenvalues

$$\gamma_m = \left(\frac{1}{M'}\right)^{m+1/2}$$

(65)

The resonators of interest for high-power applications typically have large values of the Fresnel number $N$ and the equivalent Fresnel number $N_{eq}$. For these conditions, the complex equivalent Fresnel number $N_{eq}$ is given approximately by

$$N'_{eq} \cong N_{eq} - \frac{(i/\pi)(M^2 + 1)}{M^2 - 1}$$

(66)

and the eigenfunctions are given by

$$U_m(x) = \exp\left[-\left(\frac{x}{a}\right)^2\left(i\pi N_{eq} + \frac{2}{M^2 - 1}\right)\right]$$

$$\times H_m\left\{(2\pi N_{eq})^{1/2}\left[1 - \frac{\epsilon}{2} + i\left(1 + \frac{\epsilon}{2}\right)\right]\frac{x}{a}\right\}$$

(67)

where

$$\epsilon = \frac{M^2 + 1}{\pi N_{eq}(M^2 - 1)}, \qquad (\epsilon < 1)$$

The round-trip power losses are

$$\delta_m = 1 - |\gamma_m|^2 = 1 - M^{-(2m+1)}\left(1 + \frac{1}{\pi^2 N_{eq}^2}\right)^{-m+1/2} \tag{68}$$

and the eigenfrequencies are

$$\nu_m = \frac{1}{2\pi}\frac{c}{L}\left(\pi q + \frac{m+1/2}{\pi N_{eq}}\right) \tag{69}$$

The solution is a Hermite–Gaussian polynominal, as in the stable-cavity solution. However, the argument of the polynomial is complex because of the losses introduced by the variable reflectivity mirrors. Analytic solutions of this type were obtained previously by Bergstein[33] and Streifer[34] for uniform-reflectivity unstable-cavity mirrors with rectangular geometry, but the importance of the edge effects was not pointed out. Zucker[35] obtained analogous Laguerre–Gaussian polynominal solutions for the case of circular mirrors, and their application and physical interpretation in terms of unstable-resonator modes was pointed out by Chester.[36,37]

The parameters of Anan'ev's solution are the magnification $M$, the Equivalent Fresnel number $N_{eq}$, and the mirror scale $a$. The spot size is given by $a(M^2 - 1)^{1/2}$. Figure 27 shows the intensity distribution of the fundamental and higher-order transverse modes given by Eq. (67). Figure

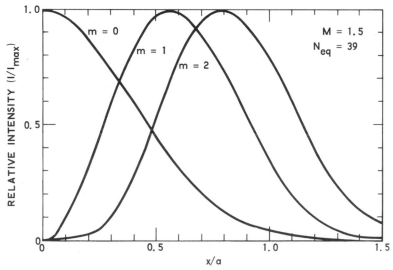

**Fig. 27.** Intensity distribution from the analytic solution for the transverse modes of a symmetric unstable cavity with Gaussian mirrors.[27]

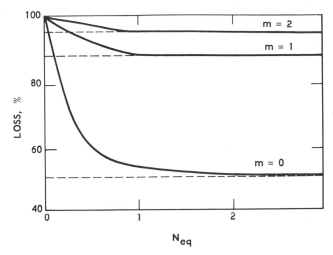

**Fig. 28.** Loss per reflection versus $N_{eq}$ from analytic solution for the transverse modes of a symmetric unstable cavity with Gaussian mirrors.[27]

28 shows the losses (cylindrical mirrors) versus $N_{eq}$ for the various transverse modes. Thus the intensity distribution is now uniform, showing none of the rippled structure in Figs. 22 and 23. The higher-order modes show increasing off-axis intensity and, hence, will have larger losses than the fundamental $m = 0$ mode for the Gaussian reflectivity mirrors. Notice that neither the mode-crossing behavior nor the oscillatory variation in loss with $N_{eq}$ are observed. There is a high degree of mode discrimination, with mode losses asymptotically approaching $1 - (1/M^{2m+1})$ for large $N_{eq}$. The $\exp[i\pi N_{eq}(x/a)^2]$ phase factor in Eq. (67) is just the expected phase distribution across the mirror surface for the geometric wave front shown in Fig. 26. Equation (67) shows that the curvature of the emerging wave front corresponds to the geometric solution for the $m = 0$ and $m = 1$ modes but not for the higher-order modes. The above mode-loss discrimination gives good assurance that the $m = 0$ fundamental Gaussian mode will dominate. The eigenfrequencies for the various transverse modes are seen from Eq. (69) to depend on $N_{eq}$. Figure 29 shows that the frequency separation in megahertz between adjacent transverse modes is small for large $N_{eq}$. For the previous confocal-cavity example this separation is 0.4 MHz, which is small compared with a typical Doppler width ($\sim 200$ MHz). Thus the above mode-loss discrimination is required to ensure single frequency operation for an unstable resonator with perfectly smoothed edges. This close frequency spacing of the transverse modes of an unstable resonator has also been observed by Sanderson and Streifer[28] for uniform reflectivity mirrors.

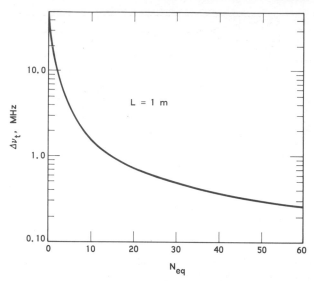

**Fig. 29.** Frequency spacing between transverse modes versus $N_{eq}$ from analytic solution for a symmetric unstable cavity with Gaussian mirrors.[27]

Sherstobitov and Vinokurov[38] made numerical calculations on the effect of tapering the reflectivity in the edge region $h_d$ (defined in Eq. (57)), where diffraction dominates the unstable-resonator modes. Thus Eq. (63) was solved for a linearly tapered reflectivity profile where $\rho(x) =$ constant for $x \leq a - h_d$ and $\rho(x)$ decreases linearly to zero for $a - h_d \leq x \leq a$ and $h_d = a/2N_{eq}$. Figure 30 shows the diffraction loss per

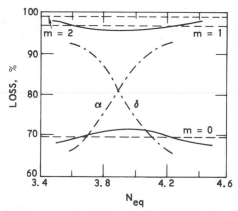

**Fig. 30.** Loss per reflection versus $N_{eq}$ from analytic solution with Gaussian mirror (dashed curve), numerical solution with a linearly tapered reflectivity distribution near mirror edge (solid curve), and numerical solution with perfectly sharp mirror edges (dash dotted curve). Symmetric unstable resonator with $M = 3.3$, even symmetric modes.[38]

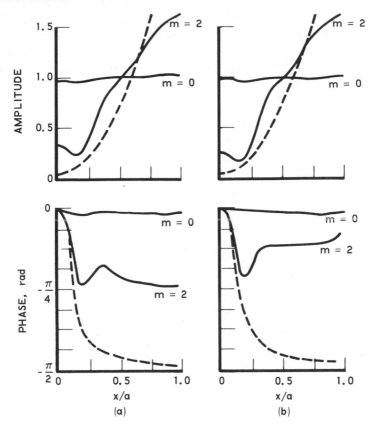

**Fig. 31.** Amplitude and phase distribution for a symmetric unstable resonator with a linearly tapered reflectivity distribution near the mirror edge. The solid curve is the numerical solution ($M = 3.3$), and the dashed curve is the analytic solution for Gaussian mirrors.[38] (a) $N_{eq} = 3.89$, and (b) $N_{eq} = 4.04$.

pass of the $\alpha$ and $\delta$ even symmetric modes in the vicinity of a mode-crossing point at $N_{eq} = 4.0$ and $M = 3.3$, from Ref. 28. Also, the losses for the linearly tapered reflectivity profile as well as the $1 - (1/M^{2m+1})$ losses for Gaussian reflectivity mirrors are plotted (dotted lines). Thus tapering in the edge wave region $h_d$ lifts the unstable resonator mode degeneracy as expected. Figure 31 shows the corresponding calculated amplitude and phase distribution for the linearly tapered profile for $N_{eq} = 4.04$. The $m = 0$ mode has the uniform amplitude and phase distribution, corresponding to the geometric solution, and the $m = 2$ mode approaches the solution of Eq. (67). In conclusion, an unstable resonator with a large number of Fresnel zones will produce the geometric mode if the reflected wave in the outer Fresnel-zone region is attenuated. This reflected wave from the outer two Fresnel zones can be attenuated in several ways, such as tapering the reflectivity, slightly misaligning the resonator so that the optical axis is shifted by one Fresnel zone so that phases of the edge waves cancel, or fabricating the mirror boundary so that it has a roughness scale of two Fresnel zones.

Further studies of the effect of tapering the amplitude and phase of the reflected-edge wave on the transverse modes of an unstable resonator have been conducted by McAllister et al.[39] They point out that the theory of the edge wave is derived from standard scalar-diffraction theory in Born and Wolf.[8] This formulation is obtained by expressing the field amplitude resulting from an illuminated aperture as the sum of a geometric optics term plus a diffraction term, the latter of which is expressed as a line integral around the aperture boundary.

Thus Kirchhoff's approximation to Huygens' principle gives

$$U(\bar{P}) = \frac{1}{4\pi} \iint_A U_0 \left[ \frac{1}{r} e^{ikr} \frac{\partial}{\partial n} \left( \frac{1}{s} e^{iks} \right) - \frac{1}{s} e^{iks} \frac{\partial}{\partial n} \left( \frac{1}{r} e^{ikr} \right) \right] dA' \tag{70}$$

where $A$ is an aperture illuminated by a spherical wave from $\bar{P}_0$, $r$ is the distance from $\bar{P}_0$ to a point on $A$, $\bar{P}$ is the field point, $s$ is the distance from $\bar{P}$ to $A$, and $\partial/\partial n$ is the derivative normal to the surface. Then it can be shown that

$$U(\bar{P}) = U_G(\bar{P}) + U_D(\bar{P}) \tag{71}$$

where

$$U_G(\bar{P}) = \frac{1}{R} e^{ikR} \tag{72}$$

which corresponds to a geometric wave, and

$$U_D(\bar{P}) = -\frac{1}{2\pi R} e^{ikR} \oint_\Gamma U_0 \frac{\exp\left[ i\pi N_{eq}(\xi) \right] dl}{\xi} \tag{73}$$

which corresponds to a diffraction wave formed at the mirror boundary; $R = \bar{P}_0\bar{P}$, $\Gamma$ is the aperture (mirror) boundary, and $\xi$ is the distance of the cavity axis from mirror boundary $\Gamma$. Equation (73) applies to the case of an aligned symmetric unstable resonator where

$$N_{eq}(\xi) = \frac{\xi^2}{2\lambda L} \left( M - \frac{1}{M} \right) \tag{74}$$

To account for an amplitude or phase variation of the reflected wave with $\xi$, Eq. (73) can be written assuming circular mirrors of radius $\xi$ as

$$U_D(\bar{P}) = -\frac{1}{R} e^{ikR} \int_0^\xi \frac{dU}{d\xi'} \exp\left[ i2\pi N_{eq} \left( \frac{\xi'}{a} \right)^2 \right] d\xi' \tag{75}$$

Note that $N_{eq}$ occurs as a phase factor in Eq. (75) and the edge-wave amplitude is predicted to have a periodic dependence. Equation (75) can be used to estimate the effect of various reflectivity profiles on the edge-wave amplitude and to optimize the taper so that a geometric mode is produced by the resonator. For the case of a parabolic reflectivity profile, one assumes

$$U(\zeta) = U_0 \text{ for } 0 \leq \zeta \leq \zeta_0 \tag{76}$$

$$U(\zeta) = \frac{U_0(1-\zeta^2)}{(1-\zeta_0^2)} \text{ for } \zeta_0 < \zeta \leq 1 \tag{77}$$

where $\zeta = \xi/a$ and $\zeta_0$ denotes the position where taper begins. After substituting Eq. (77) into Eq. (75), one gets

$$U_D(\bar{P}) = \frac{2U_0}{(1-\zeta_0^2)R} \int_{\zeta_0}^{1} \exp(i2\pi N_{eq}\zeta^2)\, \zeta \, d\zeta \tag{78}$$

$$= \frac{U_0}{R} \exp[ikR + i\pi N_{eq}(1+\zeta_0^2)] \frac{\sin \bar{X}}{\bar{X}} \tag{79}$$

where $\bar{X} = \pi N_{eq}(1-\zeta_0^2)$. The minima occur for

$$N_{eq}(1-\zeta_0^2) = m, \qquad m = 1, 2, \ldots \tag{80}$$

and the secondary maxima asymptotically approach

$$N_{eq}(1-\zeta_0^2) \cong (m+\tfrac{1}{2}), \qquad m = 1, 2, \ldots \tag{81}$$

For the previous confocal unstable cavity example, $N_{eq} = 39$, and the first maxima occur at $\zeta_0 = [1-(1.43/N_{eq})]^{1/2} = 0.985$ with the edge-wave intensity reduced to 5% of its maximum value. Note that the minima imply $U_D = 0$ or that there is no edge wave, that is, the geometric solution applies exactly. From Eq. (78), radial zones can be defined over which the phase term under the integral changes by $\pi$, or $\bar{X}$ in Eq. (79) changes by $\pi/2$. Thus, the minima at $\bar{X} = m\pi$ correspond to two zones or equivalently to a change in $N_{eq}$ of unity. McAllister et al.[39] also solved numerically for the transverse modes of a symmetric unstable cavity with circular mirrors and a parabolic reflectivity taper using the Prony technique developed by Siegman and Miller.[24] Figure 32 shows the calculated loss per pass versus $\zeta_0$, the normalized radius where taper begins, for the case of $M = 2$ and $N_{eq} = 3.5$. The value of the phase term in Eq. (78), corresponding to $\zeta_0$, is also shown. The results in Fig. 32 are consistent with Eq. (79) in that the geometric loss of $\approx 75\%$ output coupling is predicted for an even number of radial zones [$N_{eq}(1-\zeta_0^2) = m$, $m = 1, 2, \ldots$]. The secondary maxima approach the geometric loss for increasing taper, and any taper is better than a uniform reflectivity mirror ($\zeta_0 = 1$).

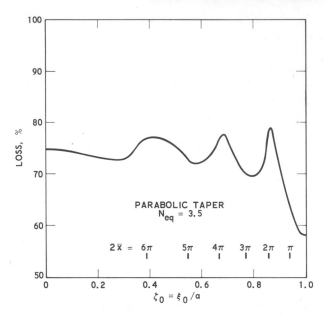

**Fig. 32.** Loss per reflection versus normalized radius $\zeta_0$, where taper begins for a symmetric unstable resonator with parabolic amplitude reflectivity tapered mirrors. The numerical calculations for the fundamental mode were made by the Prony technique for an $M = 2$, $N_{eq} = 3.5$ cavity. The loss is also plotted as a function of the parameter $\bar{X} = \pi N_{eq}(1 - \zeta_0^2)$ for comparison with the analytic solution for the diffracted edge wave.[39]

The formulation of Eq. (75) was also applied by McAllister et al.[39] to study analytically the effect of a tapered phase distribution on the edge-wave amplitude. This is equivalent to beveling the mirror or changing the mirror curvature over at least two radial equivalent Fresnel zones. It was found that tapering the phase has an effect similar to tapering the reflectivity in that the unstable cavity losses approach the geometric value. The solution again depends on $\zeta_0$, the radial position where taper begins, in a periodic manner as the radial zones alternately add or cancel one another. The solution shows that in order to obtain zero diffracted field the mirror edges must be ground off rather than built up.

An application of Eq. (73) in Ref. 39 was to determine the effect of mirror aperture shaping on the amplitude of the diffracted edge wave. An optimum mirror boundary path $\Gamma$ was found that produces a zero-amplitude edge wave. If the mirror shape is given by

$$dl = \left(\frac{\xi}{a}\right)^2 d\xi \qquad (82)$$

then Eq. (73) reduces to

$$U_D = U_0 \frac{e^{ikR}}{R} \int_1^{\zeta_M} \exp\left(i2\pi N_{eq}\zeta^2\right) \zeta \, d\zeta \tag{83}$$

where $\zeta_M$ denotes the maximum radius on the mirror boundary. In the same manner as in Eq. (79), $\zeta_M$ is selected by the condition

$$N_{eq}(\zeta_M^2 - 1) = m, \quad m = 1, 2, \ldots \tag{84}$$

to make $U_D = 0$. From Eq. (82), it can be shown that the cylindrical coordinate $(\zeta, \theta)$ representation of the optimum boundary shape is given by the transcendental equation

$$\cos\left[2\theta - (\zeta^4 - 1)^{1/2}\right] = \frac{1}{\zeta^2} \tag{85}$$

The optimum boundary is shown in Fig. 33, with the solid curve corresponding to the solution to Eq. (85) for $0 \le \theta \le \pi/4$. There are many possible solutions, provided the segments cover an even number of equivalent Fresnel radial zones so that cancellation of the diffracted edge

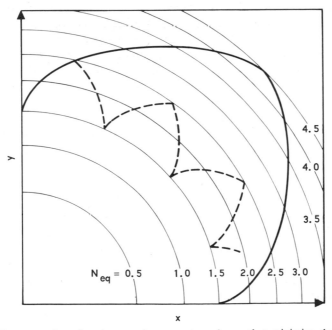

**Fig. 33.** Two examples of optimum mirror aperture shapes that minimize the diffracted edge wave in a symmetric unstable resonator. Solution is obtained from an analytic solution for the edge-wave amplitude.[39]

wave can be achieved. One may assume that slightly roughening the mirror boundary over two zones would achieve similar results.

From the preceding discussion, it is seen that the amplitude of the edge wave for an unstable resonator can be reduced in a variety of ways. Fresnel-zone averaging can be achieved by the use of rectangular instead of circular mirrors, by cavity-mirror tilt that translates the cavity axis one or more zones, and by variations in the saturable gain across the mode volume. Except for the case of an aligned resonator with ideal sharp-edged circular mirrors and a uniform-gain distribution, there appears to be no advantage in designing an unstable cavity for half integral values of $N_{eq}$.[40,41]

## 2.4. Effects of Mirror Tilt on the Modes of an Unstable Resonator

In practice, laser-resonator mirrors are always subject to environmentally induced vibration and misalignment, and consequently unstable-resonator beam quality can be highly sensitive in these effects. However, there are other reasons to study mirror-tilt effects. A mirror tilt, introducing as it does a linear optical phase shift across the beam front within the resonator, is the simplest example of a phase disturbance within the resonator. Other sources of phase disturbance, such as mirror astigmatism, mirror distortion induced by laser-beam power loading, laser-medium density variations caused by shock waves, and anomalous dispersion, all can occur in high-power laser devices. Useful insights can be gained into these complex effects through the study of the sensitivity of resonator performance to such simple effects as mirror tilt.

The dramatic effects of mirror tilt in unstable resonators were analyzed by Sanderson and Streifer[42] and Krupke and Sooy[20] using geometric optics, and were strikingly verified by the latter in their classic paper[20] on unstable-resonator performance. The central result of these analyses may be explained by reference to Fig. 34, in which both mirrors of an unstable resonator have been tilted through an angle $\beta$ with respect to the original optical axis.

The effect of this mirror tilt is to rotate the optical axis of the resonator (defined as the line that intersects both mirrors at normal incidence). However, the angle of rotation $\phi$ of the optical axis depends on the magnification $M$, and one finds by simple geometry that

$$\phi = 2\beta \frac{M+1}{M-1} \tag{86}$$

It should be noted that several different conventions are in use for

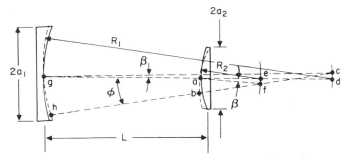

**Fig. 34.** Unstable resonator, showing the effect of mirror-tilt angle $\beta$ on the optical-axis tilt angle $\phi$ within the geometric-optics approximation.[20]

describing both the mirror-tilt parameter and the change in optical axis, depending on the relative simultaneous tilts assigned to each of the two resonator mirrors, as summarized in Table 2.

The geometric analysis presumes that the laser-output beam is parallel to this new optical axis, so that $\phi$ is also the angle through which the output beam is steered. This turns out to be true for small mirror-tilt angles, as verified by both experimental measurements and by the more accurate diffraction calculations described later. Thus mirror misalignment is expected to produce beam steering, and mirror vibration leads to scanning of the output beam and a consequent increase in time-averaged beam divergence. The constraints on mirror alignment stability that result are simply summarized in Ref. 37. Equation (86) predicts that very large beam-steering results from small mirror-tilt angles when the magnification $M$ approaches unity. This strong alignment sensitivity has been observed experimentally, and is a principal reason why the unstable

**Table 2.** Comparison Between Various Conventions Describing Mirror-Tilt Effects

| References | Tilt angle of output mirror | Tilt angle of back mirror | Resulting change in direction of optical axis |
|---|---|---|---|
| This article, 30 | $\beta$ | $\beta$ | $\dfrac{2\beta\,(M+1)}{(M-1)}$ |
| 20 | — | $\theta$ | $\dfrac{2\theta\,M}{(M-1)}$ |
| 36, 37 | — | $\chi$ | $\dfrac{2\chi\,M}{(M-1)}$ |
| 42 | $\eta/a$ | $\eta/a$ | 0 |

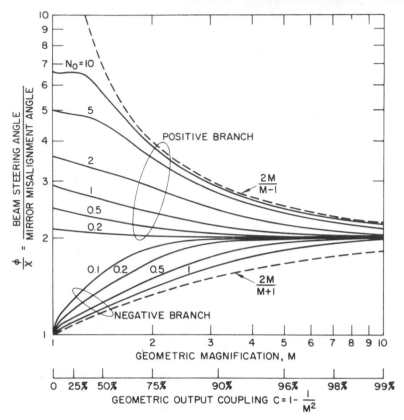

**Fig. 35.** Magnification of mirror-misalignment effects by an unstable resonator.[37] The ratio of far-field beam-steering angle $\phi$ to mirror-misalignment angle $\chi$ is plotted as a function of the geometric magnification $M$ of the resonator for various values of outer Fresnel number $N_0$. All values calculated by perturbation expansion for the lowest-loss mode in a Gaussian-mirror confocal unstable resonator by sampling the field distribution circulating within the resonator. The dotted lines show the results of the Krupke–Sooy[20] geometric analysis of the optical axis of a tilted resonator.[27]

resonators usually used have $M \gtrsim 1.5$. This relatively high magnification, of course, means that the output coupling is high, and thus, a laser-medium geometry yielding relatively high total gain must be employed.

Since the beam-steering effect described in Eq. (86) and Table 2 is obtained by a geometric-optics analysis, it would be expected to be accurate only at large Fresnel numbers. The exactly soluble circularly symmetric case using Gaussian mirrors was solved by Chester,[36,37] who used perturbation theory to analyze the effect of tilting the back mirror through angle $\chi$. The resulting beam steering $\phi$, plotted in Fig. 35 as the

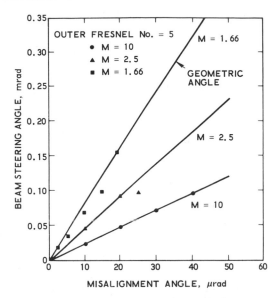

**Fig. 36.** Beam-steering angle $\phi$ versus convex-mirror misalignment angle $\beta$ for various values of the confocal unstable-cavity magnification $M$. The solid lines correspond to the geometric theory, and the data points correspond to numerical calculations from diffraction theory.[30] Outer Fresnel number $N_0 = 5$.

ratio $\phi/\chi$, shows that in a Gaussian resonator with $M > 1.5$, the geometric approximation is valid whenever the outer Fresnel number exceeds

$$N_0 = \frac{a_1^2}{\lambda L} = M^2 N = 1 \tag{87}$$

for negative-branch confocal resonators or when $N_0$ exceeds 10 for positive-branch confocal resonators.

Such an exact solution is not available for sharp-edged mirrors at finite Fresnel numbers. However, diffraction calculations by Rensch and Chester[30] on rectangular-mirror confocal resonators, summarized in Figs. 36 and 37 show that for $N_0 \geq 5$ the geometric-optics approximation gives a fairly good estimate of beam-steering effects. However, an examination of the actual mode shapes, calculated, including diffraction effects, reveals other effects in addition to beam steering. Figures 38 and 39 show the near-field intensity and phase distributions for a particular unstable-resonator case as the tilt angle $\beta$ is varied. In addition to beam steering produced by the linear phase shift resulting from the mirror tilt, it is found that as the tilt angle increases there is a drop in output power and a distortion of the fundamental mode amplitude and phase, which causes increased beam divergence in the far field. The phase irregularities

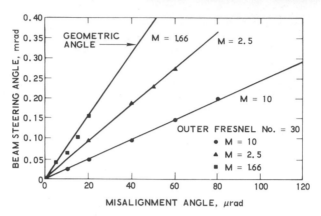

**Fig. 37.**  See description in Fig. 36, $N_0 = 30$.

beginning to appear in Fig. 39 herald the onset of multimode operation as the tilt increases still further.

These calculations show that, although the laser output initially consists of a single transverse mode, the power and quality of that mode progressively suffer as the mirror-tilt angle increases, until finally it breaks into several simultaneously oscillating modes exhibiting relatively poor overall beam divergence.

The degree of mirror misalignment that eventually leads to serious degradation in the output beam as a result of multimode operation can be estimated with empirical rules developed by Rensch.[30] Basically, the important requirement is to keep the optical axis $\overline{bh}$ in Fig. 34 from walking off the small mirror. It was found[30] that one can obtain single transverse-mode operation even in the presence of small-signal round-trip gains as high as 21 dB ($2G_0L = 5$) if $\overline{ab}/a_2 \leq 0.5$. In terms of the magnification $M$ and mirror spacing $L$ of a confocal positive-branch unstable resonator, this means that single-mode operation is ensured provided that

$$\frac{\overline{ab}}{a_2} = \frac{4(M+1)L\beta}{(M-1)^2 \, a_2} \leq \frac{1}{2} \tag{88}$$

This is apparently one of the few instances in which a crucial property of unstable resonators depends upon a parameter other than the magnification or the Fresnel number. However, as partially pointed out in Ref. 30, in most applications beam steering and loss of far-field intensity rather than multimode operation will probably be the limiting factors that determine acceptable mirror vibration or misalignment angles. As shown by the experimental measurements in Fig. 14 of Ref. 20 and by the

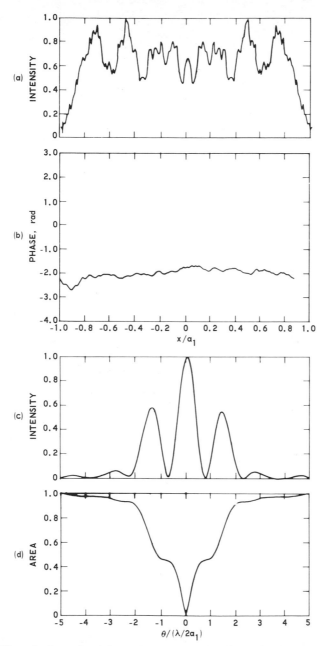

**Fig. 38.** Effect of mirror-tilt angle on the beam quality of a confocal unstable cavity with sharp-edged mirrors. Numerical calculations assume an outer Fresnel number $N_0 = 30$, $M = 2.5$ and zero convex mirror-tilt angle $\beta$, (a) near-field intensity distribution on scraper mirror, (b) near-field phase distribution, (c) far-field intensity distribution, and (d) far-field power distribution.

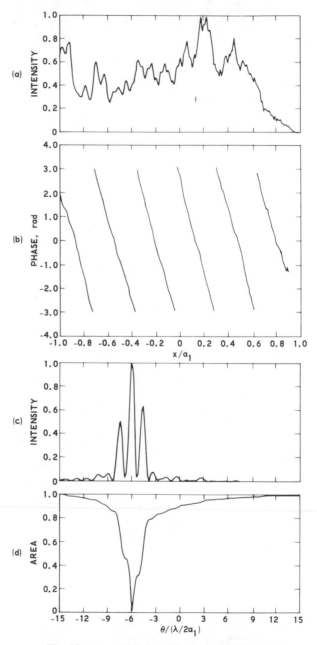

**Fig. 39.** See description in Fig. 38, $\beta = 60\ \mu$rad.

theoretical results in Fig. 16 of Ref. 30, the reduction in output power alone can limit mirror-misalignment tolerances to smaller values, except in some special cases (for example, $N_0 = 30$, $M = 1.66$ in Ref. 14) where multimode operation occurs before the output power drops.

## 2.5. Laser-Medium Effects and More-Complex Phase-Disturbance Effects

The laser medium introduces additional complex phase and amplitude disturbances into the unstable-resonator output beam besides the mirror-tilt effects previously discussed. The active medium can have a gain distribution and a refractive-index distribution that will change the modes from their empty-cavity values. Once a saturable-gain medium of finite length is inserted in the cavity, the calculation of the modes becomes a nonlinear problem that is mathematically intractable.[43] The approach used originally by Statz and Tang[44] and subsequently by Fox and Li[43] was to assume that the saturable gain is concentrated in a thin sheet near one of the cavity mirrors. This allows one to decouple the linear-wave propagation between the mirrors from the nonlinear propagation through the gain medium. For an unstable resonator, the gain-sheet approximation should be valid provided that the beam does not geometrically expand appreciably throughout the active medium (sufficiently small $M$). Rensch and Chester[30] have done a study of the limitations of the gain-sheet approximation by comparing mode calculations that utilize a single saturable gain-sheet model with those that utilize a distributed multiple gain-sheet ($\sim 20$ sheets) model. From numerical calculations, they found that the single-sheet approximation broke down when $M \geq 3$ for a gain length equal to the cavity length. Rensch and Chester[30] have made calculations of confocal unstable-resonator modes that assume a uniform saturable-gain medium and compared them with the empty-cavity modes for small-signal gains up to ten times threshold. They found, as one might expect, that the effect of saturable gain was to smooth the intensity distribution of the fundamental mode so that the high-intensity regions were decreased and the low-intensity regions were increased. The phase distribution was not appreciably affected and remained uniphase.

The only chemical laser for which an active unstable-resonator theoretical model has been developed thus far is the supersonic diffusion cw HF/DF device. As a result of the dominant processes, that is, supersonic diffusion mixing, reaction, and deactivation kinetics of a multilevel system and multispectral lasing, the HF and DF gain medium cannot be adequately described in terms of a fixed small-signal gain or gain distribution, or a fixed saturation parameter or parameter distribution. Instead, a highly interactive model must be constructed in which mixing, kinetics, and stimulated emission must be tied in with resonator characterization in order to be able to determine the degree of transverse-mode control and of power-extraction efficiency to be attained with a given resonator configuration. Two approaches to this problem are currently being pursued: (1) Mirels[45] has developed a geometric model for unstable-cavity modes coupled with a simplified mixing-kinetics model, and (2) Bullock et al.[32] have developed a physical-optics model for the unstable-cavity modes coupled with a detailed chemical-kinetics model and a simplified turbulent diffusion-mixing length model. Both approaches apply the gain-sheet approximation, and neither considers medium refractive index effects on the cavity modes.

The work of Mirels has a simplified single-line kinetics model, illustrated in Fig. 40, which characterizes the cw HF/DF active medium by a single chemical-pumping rate $k_F$, a single collisional-deactivation rate $k_{cd}$, and an arbitrary stimulated emission rate given by $BI$ where $B$ is the Einstein $B$ coefficient for the transition and $I$ is the local laser intensity. The mixing model assumes a flame-sheet approximation where the flame-sheet position is given by $\bar{y}_f(\bar{x}) = C_m \bar{x}^\nu$, where $\nu = 1$ for turbulent mixing and $\nu = 1/2$ for laminar mixing. Thus, the reactants are assumed to be premixed but nonreacting until they reach the flame sheet defined by $\bar{y}_f = \bar{y}_f(\bar{x})$ for $\bar{x} < \bar{x}_D$ and $\bar{y}_f = \bar{w}$ for $\bar{x} > \bar{x}_D$. Here, $\bar{w}$ is the nozzle-exit half width, and $\bar{x}_D$ is the diffusion distance and denotes the streamwise station where the flame sheet reaches the center line of the helium–fluorine flow.

The interaction of the unstable resonator and a cw diffusion chemical laser is analysed by Mirels for the aforementioned active medium model with a geometric-optics approximation that is valid for large Fresnel numbers. Both cylindrical $(j = 1)$ and spherical $(j = 2)$ optics are considered. The gain sheet is assumed to be adjacent to one of the resonator mirrors. Laser performance is found to depend on the parameters, $M$, $K_1 = k_F/k_{cd}$, $\rho_D = \bar{x}_D \bar{u}/k_{cd}$, $\rho_c = \bar{x}_c \bar{u}/k_{cd}$, $\rho_e = \bar{x}_e \bar{u}/k_{cd}$, $\mathscr{A}$, and $\gamma = \bar{w}\sigma [F]_0 \eta_{sc}$, where $M$ is the cavity magnification, $\bar{x}_D = \bar{w}/c_m$, $\bar{u}$ is the mean flow velocity, $\bar{x}_c$ is the centerline of cavity axis with respect to nozzle exit, $\bar{x}_e$ is the streamwise scale of unstable-cavity geometric mode, $\mathscr{A}$ is the mirror absorptivity, $\sigma$ is the cross-section for stimulated emission, $[F]_0$ is the initial F-atom concentration, and $\eta_{sc}$ is the number of nozzle semichannels. Analytic solutions were obtained for the limiting cases $(M-1) \ll 1$ and $M \gg 1$. In the former case, the laser medium almost completely saturated and the resonator efficiency is high. In the latter case, the radiation incident on the gain region originates near the optical axis and is uniform.

For the case $(M-1) = \epsilon \ll 1$, $K_1 \to \infty$, and turbulent mixing, Fig. 41

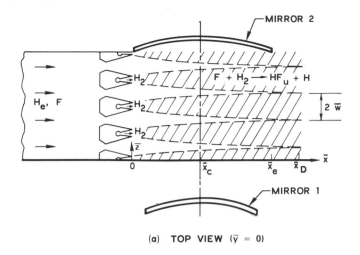

**(a)  TOP VIEW ($\bar{y} = 0$)**

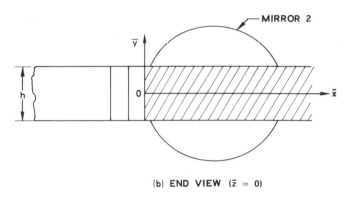

**(b)  END VIEW ($\bar{z} = 0$)**

**Fig. 40.** Schematic diagram of a diffusion-type cw HF chemical laser with an unstable resonator.[45] Model calculates the near-field intensity distribution and power-on gain distribution in the plane of symmetry ($\bar{y} = 0$) within the geometric-optics approximation.

shows the variation of resonator efficiency $\eta_r$ with optical-cavity location $\rho_c$. It should be noted that this solution will not be valid for $\epsilon \to 0$ since diffraction effects will dominate. The present results neglect terms of order $\epsilon^2$ compared with unity. It is hoped, therefore, that there will be a range of values of $\epsilon$, where this geometric model applies (e.g., $\epsilon \approx 0.3$). The resonator efficiency is defined by $\eta_r = P_0/P_S$, where $P_0$ is the net resonator output power and $P_S$ is the putput power that would be obtained if the gain medium were completely saturated. Note that $\eta_r \rightarrow 1$ as $M \rightarrow 1$ and $\eta_r$ decreases linearly with $\epsilon$. In the limit $\epsilon \ll 1$, the gain region occupies $0 \leq \rho \leq 1$, and the length of the resonator, to obtain maximum output, is $\rho_e = 1$. Figure 41 shows that the optimum resonator efficiency is obtained as $\rho_c \rightarrow 0$. Note from Fig. 40 that $\rho_c \neq 0.5$ corresponds to an unsymmetrical resonator; $\rho_c = 0$ corresponds to the position of peak power-on gain. Hence, these results are consistent with the general observation, in Ref. 45, that the output power is a maximum when $\rho_c$ is located at the position of maximum power-on gain (all other parameters being constant). The fact that higher efficiency $\eta_r$ is obtained for $j = 1$ than for $j = 2$ is reasonable from physical considerations.

It may be concluded that the preceding simplified geometric model[45] is useful in that it identified the important physical parameters affecting the power-extraction efficiency and gain saturation. It does not, however, give information on the phase distribution of the output or beam quality. It could be extended to examine the effect of medium refractive index on beam quality from a geometric-optics point of view. The relative validity of the geometric-optics approach versus the physical optics one should also be examined theoretically and, possibly, experimentally.

Bullock et al.[21,32] have developed a numerical computer code, known as BLAZER[21] that models a cw HF/DF diffusion-limited unstable resonator. BLAZER uses the reaction–deactivation kinetics treatment of RESALE 1[74] that pertains to cold-reaction devices in HF/DF chemical lasers. The principal phenomena modeled in BLAZER that differ from RESALE are extraction of optical power from the gain medium by the unstable optical resonator via stimulated emission and a two-stream mixing corresponding to turbulent diffusion. BLAZER contains numerous compromises with reality that are necessary for an efficient and economical numerical analysis of chemical-laser performance. The two-stream mixing-model used in BLAZER is indicated in Fig. 42. The fuel and oxidant mixed volume grows linearly in the direction normal to the flow until both the fuel ($H_2/D_2$) and oxidizer (F, He) streams are fully mixed at the mixing length $\bar{x}_D$. A two-dimensional slit-nozzle injector is modeled so that the mixed volume grows only along the direction parallel to the resonator-optic axis. The temperature within the mixed zone is considered to vary along the flow axis only, that is, optic-axis temperature variations in the mixed volume are not treated. Both the fuel and oxidant streams are treated as though they have a common initial temperature, defined as the initiation temperature, which is an important parameter in the model. This temperature was treated as a parameter to be determined by independent methods, that is, the chemiluminescence temperature-measurement technique.

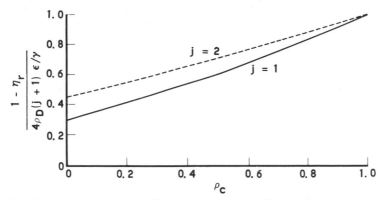

**Fig. 41.** Variation of resonator-efficiency parameters with normalized location of the optical axis $\rho_c$ from the simplified geometric model in the limit $M - 1 = \epsilon \ll 1$.[45] The resonator efficiency $\eta_r = P_0/P_s$ is a measure of the degree of saturation of the chemical-laser active medium, where $P_0$ is the output power and $P_s$ is the maximum power in the saturated limit.

A three-element resonator was modeled in BLAZER (Fig. 43) and consisted of a back mirror, a front mirror, and a scraper (output-coupling) mirror. Both stable and unstable configurations involving spherically curved mirrors were included. In the stable case, the scraper was eliminated and a two-element resonator resulted.

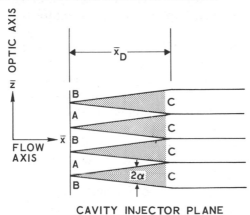

**A**   FLUORINE STREAM CONTAINING $[F]°$, $[DF]°$, $[He]°$
**B**   HYDROGEN STREAM CONTAINING $[H_2]°$
**C**   MIXED VOLUME CONTAINING $[F]$, $[H_2]$, $[HF]$, $[DF]°$, $[He]°$
$\bar{x}_D$   MIXING LENGTH

**Fig. 42.** Mixing model used in the BLAZER computer code. The model[32] assumes a linear growth with $\bar{x}$ or turbulent mixing and a two-dimensional slit nozzle.

The single gain sheet approximation was applied with the sheet placed adjacent to the back mirror. The one-way gain through the infinitesimally thin sheet was assumed to be equivalent to the gain per pass through the resonator. The round-trip propagation through the resonator was accomplished using the Kirchhoff–Huygens integral scalar-wave approximation to propagation in a cylindrical-strip resonator. The cylindrical-strip calculation was performed in both the streamwise and cross-flow directions to obtain the spherical-mirror solution, assuming rectangular mirrors. The modified Gauss–Legendre numerical-integration scheme described previously was used, with the number of sampling points set equal to 64.

The solution to the active cw HF/DF resonator model was found by the process of iteration. The kinetic-rate equations and the wave-propagation equations were solved simultaneously to find a self-consistent set of functions for the number-density distribution $N_{v,j}(x)$ and the electric-field distribution $E_{v,j}(x)$. The BLAZER code solves alternately by first assuming a set of initial fields in order to solve the rate equations; then, in turn, it generates a new set of fields from the original fields by propagating the initial fields round trip through the resonator and through the gain distribution determined by the solution to the rate equations. The process of iteration continues until self-consistent fields, vibrational–rotational level concentrations, temperatures, gains, and so on, are found. Convergence is defined to occur when the total integrated optical power in each band changes by less than some specified percent for two successive iterations.

The output data of the BLAZER code includes the gain, number density, and field distributions for each HF/DF vibrational–rotational transition. The near-field intensity and phase distribution per transition on the scraper mirror are also calculated as well as the single and multiline far-field intensity distributions of the annular output beam. Good comparison of the BLAZER model calculations with experimental near-field data has been obtained as discussed in the next section on experimental results. The code can be used to predict the total output power and, hence, the degree of gain saturation for a specific nozzle–injector–resonator design, and to predict the far-field beam quality. Single transverse-mode operation is assumed if an unchanging near-field intensity and phase distribution is obtained after a sufficient number of iterations. Multitransverse-mode operation is assumed if a steady-state solution is not obtained, giving evidence for mode beating. Bullock et al. have applied the BLAZER code to a variety of cw HF/DF nozzle designs and unstable-resonator designs. The BLAZER code predicts multitransverse-mode output for unstable-cavity magnifications below a certain value for a given size device, that is, a specific gain length. As the

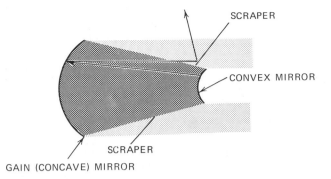

**Fig. 43.** Three-element resonator model used in the BLAZER computer code.[32] The model typically assumes a confocal unstable-resonator geometry.

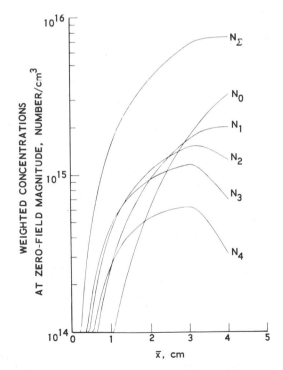

**Fig. 44.** Calculated integrated-number densities of excited species versus distance downstream of nozzle-exit plane $\bar{x}$ for a cw DF laser from the BLAZER computer code.[21] The number-density distributions for various vibrational DF transitions are shown for a typical slit nozzle (CL-II).

nozzle gain increases, the minimum magnification for single transverse-mode output is predicted to increase. This mode-beating phenomenon predicted by the computer code is believed to be associated with the asymmetric-gain distribution typically observed in cw chemical lasers, but no experimental data are available yet to confirm the predictions.

Typical predictions obtained with the BLAZER code are shown in Figs. 44 and 45. Figure 44 shows the calculated number-density distributions for various DF vibrational transitions. Thus the BLAZER code predicts significant variations in the magnitude and position of the number densities with vibrational quantum number $v$ for a given transition. Figure 45 shows the corresponding multiline near-field intensity and single-line phase distribution at the position of the scraper mirror, predicted for an edge-coupled confocal unstable cavity with a geometric magnification $M = 1.75$. The BLAZER code predicts a uniform phase output beam (nearly a plane wave) and, consequently, good beam quality for a multiline supersonic diffusion cw HF/DF unstable resonator. The effect of the active medium is an asymmetry in the near-field intensity distribution in the streamwise dimension, caused by the nonuniform gain profiles.

## 3.  LASER-OPTICS EXPERIMENTS

### 3.1.  Introduction

The term "laser optics" in this section refers to the effect of the optical-cavity design on the output radiation of a laser. The optical-resonator design largely determines the transverse-mode selection or phase coherence of the output beam, the gain saturation of the active medium, the selection of lines, and the frequency (longitudinal modes) of the output. The term "beam quality" refers to transverse-mode selection or the phase coherence of the output beam, with good beam quality referring to a beam with minimum divergence. A given device will have diffraction-limited beam quality if the amplitude and phase distribution are both sufficiently uniform that the beam divergence is as small as permitted by diffraction. The present work describes a series of laser-optics experiments conducted primarily with a diffusion-limited cw HF/DF chemical laser.

As indicated in the previous section on theory, the near-field intensity distribution of an active unstable resonator depends strongly on the nature of the gain region; whereas, the phase front tends to remain uniform for negligible refractive-index variations. The gain distribution for the various vibrational–rotational transitions of a cw diffusion chemical laser depends on the detailed kinetic-rate processes and the nozzle design. Thus, a chemical-laser fundamental transverse mode for an unstable cavity can only be defined by the active medium being modeled in detail. The geometric radius of curvature of this phase front for an unstable resonator is given by Siegman's geometric model as discussed in Section 2.

The selection of a resonator design depends on the available small-signal gain and the transverse scale of the active medium. For a cw diffusion chemical laser, the transverse-mode scale is determined by the nozzle height and the streamwise dimension of the gain zone. Table 3 shows the positive small-signal gain in the streamwise dimension for typical cw chemical lasers. The table is based on experimental data[46,47] and theoretical estimates. The associated active medium Fresnel numbers based on these distances are also listed, taking the gain length as given in the table as the lateral cavity dimension and assuming a 1-m mirror separation. Typical peak small-signal gain values for the various devices are also listed in Table 3. Thus the Fresnel number associated with the dc discharge subsonic diffusion cw HF/DF laser is on the order of unity, implying that a stable cavity could be adequate to provide single transverse-mode output. The subsonic diffusion cw DF–$CO_2$ transfer chemical laser,[48] on the other hand, with its 10-cm gain zone, has a medium Fresnel number of 250 and requires an unstable resonator for

**Table 3.** Medium Fresnel Numbers for Typical cw Diffusion Chemical Lasers

| Type of device | Maximum small-signal gain, %/cm | Extent of positive gain along flow, cm | Medium Fresnel number | Resonator |
|---|---|---|---|---|
| Subsonic diffusion cw HF/DF laser (Ref. 47) | DF—1 HF—3.0 | 0.3 | DF—1.0 HF—1.33 | Stable |
| Supersonic diffusion cw HF/DF laser | | | | |
| (a) Aerospace standard slit nozzle (Ref. 46) | DF—3.5 HF—10.0 | 5.0 | DF—150 HF—200 | Unstable |
| (b) UARL matrix hole nozzle (Ref. 49) | DF—7.0 HF—20.0 | 1.0 | DF—6.2 HF—8.3 | Unstable |
| Supersonic diffusion DF–$CO_2$ transfer chemical laser (Ref. 48) | $CO_2$—3.0 | 10.0 | 250 | Unstable |

single transverse-mode output. The supersonic diffusion cw HF/DF laser has a positive-gain zone that can vary from 1 cm for the United Aircraft Research Laboratory (UARL) matrix-hole nozzle design[49] to 5 cm for the Aerospace standard-slit nozzle.[50] The associated Fresnel numbers vary between 10 and 200, which suggests the use of an unstable cavity.

In addition to the medium Fresnel number, the available small-signal gain of the active medium must be considered when choosing the cavity parameters. Table 3 shows that the small-signal gain of various chemical lasers varies from about 1 to 20%/cm. Note that the small-signal gain in DF is about one third the gain in HF, primarily because of the difference in Einstein coefficients.[51] In the geometric optics limit, the threshold gain of the resonator[52] is given by

$$G_t = \frac{1}{2l} \ln \left( \frac{M^2}{\mathcal{R}_1 \mathcal{R}_2} \right) \qquad (89)$$

where $\mathcal{R}_1$ and $\mathcal{R}_2$ are mirror reflectivities, $M$ is the cavity magnification ($M = 1.0$ for a stable cavity), and $l$ is the length of the active medium parallel to the optical axis. Thus, for a specific $l$, there is a cavity magnification $M$ that yields a desirable threshold gain $G_t$. It is desirable for the medium to be well saturated, with $G_t$ a small fraction of the unsaturated gain $G_0$, since the relative efficiency of laser-power extraction is approximately proportional to $1 - G_t/G_0$. However, a very low value of $G_t$ requires values of $M$ close to 1, which, as previously discussed, leads to a resonator with high sensitivity to misalignment and other internal disturbances. Thus the optimum value of $G_t$, and hence $M$, can only be determined as a best compromise between these opposing factors.

The experimental results discussed in this section were obtained by several authors. In the early multiline supersonic diffusion cw HF/DF laser experiments, Spencer et al.[53] made closed-cavity power measurements. A confocal positive-branch unstable resonator was applied to the same device by Mansell and Love,[54] who measured near-field output power and obtained Plexiglas near-field burn patterns. Near diffraction-limited beam quality was first measured on a multiline confocal unstable cavity on the same type of device by Wisner et al.[49] Further extensive beam-quality measurements on a multiline cw HF/DF confocal unstable cavity were made by Chodzko[55] and Hook.[56] The first demonstration of diffraction-limited beam quality from a cw chemical laser was reported by Chodzko et al.,[19] who used a single-line nonconfocal unstable cavity on a supersonic diffusion cw HF device. More recently, Hinchen[47] has demonstrated single-transverse mode, single-frequency operation with a stable cavity on a small subsonic diffusion cw HF/DF laser.

## 3.2. Supersonic Diffusion cw HF/DF Multiline Laser

**Stable Closed-Cavity Power.** As indicated in Table 3, the medium Fresnel number for a typical supersonic diffusion cw HF/DF laser (e.g., the Aerospace standard-slit nozzle[50]) is about 200, and one expects multitransverse-mode operation for a stable resonator. Nevertheless, there is a very useful application for a stable resonator, namely the closed-cavity power-measurement technique developed by Spencer et al.[53]

The technique, illustrated in Fig. 46, consists of a highly stable resonator to minimize diffraction losses from the cavity. The optical axis of the resonator is located a distance $\bar{x}_c$ downstream of the nozzle exit plane. The mode volume for this cavity, by symmetry, extends a distance $2\bar{x}_c$ downstream. Through the use of a flat and a concave mirror, this mode volume can be conveniently varied by tilting the concave mirror. The maximum mode volume is determined by the mirror dimensions and the nozzle-exit plane overlap as indicated in Fig. 46. If one makes the cavity mirrors highly reflecting, the radiation flux inside the cavity will build up to a very high level and saturate the active medium over a distance $2\bar{x}_c$. Application of Eq. (89) for a typical mirror reflectivities ($\mathcal{R}_1 = \mathcal{R}_2 = 0.99$ to 0.98) and $l = 18$ cm[50] gives a threshold gain of 0.05 to 0.10%/cm. Thus

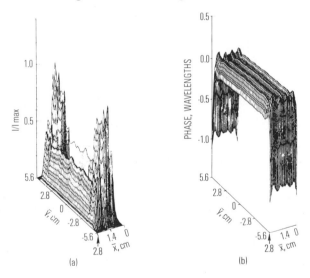

(a)  (b)

**Fig. 45.** The cw DF laser-beam output from an $M = 1.75$ confocal unstable resonator as predicted by the BLAZER computer code.[21] The maximum outer Fresnel number was $N_0 = 450$ and $\bar{x}_c = 1.25$ cm. (a) Multiline near-field intensity distribution on the scraper mirror, and (b) singleline [$P_2(8)$] near-field phase distribution on the scraper mirror.

vibrational–rotational transitions with small-signal gain values of a fraction of a percent per centimeter will be saturated over a distance $2\bar{x}_c$.

If one measures the power absorbed in the finite reflectivity mirrors by standard calorimetric techniques, it can be shown in the limit of high mirror reflectivity, that this is the maximum power available from the active medium. The analysis assumes negligible scattering losses, that is, all of the flux incident on the mirrors is either specularly reflected or absorbed. For the case of homogeneous broadening of a single spectral line, the gain-saturation law[57] is given by

$$G(\nu) = G_0(\nu)\left(1 + \frac{I_\nu}{I_s(\nu)}\right)^{-1} \tag{90}$$

If one follows the simple two-level, premixed kinetic model given by Yariv[58] the parameters are given by

$$G_0(\nu) = \Delta\mathcal{N}_0\sigma(\nu)(\text{cm}^{-1}) \tag{91}$$

$$I_s(\nu) = \frac{h_P\nu}{(\tau_1 + \tau_2)\sigma(\nu)} \, (\text{W cm}^{-2}) \tag{92}$$

where $G_0(\nu)$ is the small-signal gain, $I_s(\nu)$ is the saturation intensity, and $\sigma(\nu)$ is the cross-section for stimulated emission:

$$\sigma(\nu) = \lambda^2 A_{21} \frac{E(\nu)}{8\pi} \, (\text{cm}^2)$$

$\Delta\mathcal{N}_0$ is the small signal inversion:

$$\Delta\mathcal{N}_0 = k_{F_2}\tau_2 - k_{F_1}\tau_1 \, (\text{cm}^{-3})$$

$k_{F_i}$ is the pumping rate ($\text{cm}^{-3} \, \text{sec}^{-1}$) into level $i$ ($i = 1, 2$); $\bar{\tau}_i$ is the characteristic collisional deactivation time (sec) level $i$ ($i = 1, 2$); $A_{21}$ is the Einstein coefficient for spontaneous emission ($\text{sec}^{-1}$); $E(\nu)$ is the normalized line-shape factor (sec); $\lambda = c/\nu$ is the wavelength (cm); and $h_P = $ Planck's constant. Thus $I_s(\nu)$ defines the photon flux that reduces the inversion by a factor of 2. In this derivation, it is assumed that spontaneous emission can be neglected compared with collisional deactivation, as is the case for most molecular lasers.

Following the development of Rigrod,[57] the flux absorbed in both mirrors $I_{\mathcal{A}}$ is given by

$$\frac{I_{\mathcal{A}}}{I_s} = \mathcal{A}\left(G_0 l + \frac{\ln Z}{\ln(1-Z)}\right)$$

where

$$Z = (1 - \mathcal{A})\left(1 - \frac{\mathcal{T}}{1 - \mathcal{A}}\right)^{1/2} \tag{93}$$

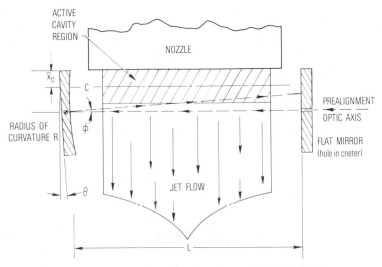

**Fig. 46.** Schematic of closed-cavity power-measurement technique.[53] The method measures the saturated power (power absorbed in mirrors) from the active medium a distance $2\bar{x}_c$ downstream of the nozzle-exit plane as shown in the shaded region. The variation in the optical-axis location $\bar{x}_c$ is obtained by the tilting of the concave mirror of the stable cavity.

$\mathscr{A}$ is the absorptivity of both mirrors, $\mathscr{T}$ is the transmissivity of one mirror, and $l$ is the length of the active medium. A plot of the absorbed flux versus mirror absorptivity for two values of transmissivity and different values of $G_0 l$ is shown in Fig. 47. The conclusion is that, for finite $\mathscr{T}$ the absorbed flux approaches zero as $\mathscr{A}$ approaches zero; however, for zero $\mathscr{T}$ as $\mathscr{A}$ approaches zero, the absorbed flux approaches a finite upper limit given by $I_{\mathscr{A}_{max}} = G_0 l I_s$ or a maximum power of approximately $P_{\mathscr{A}_{max}} = G_0 I_s V = P_{cc}$, where $V$ is the volume of the active medium. In the limit of small absorptivity, Eq. (93) reduces to

$$\frac{I_{\mathscr{A}}}{I_s} = G_0 l \left[ 1 - \frac{\mathscr{A}}{G_0 l} \right] \quad \text{for} \quad \frac{\mathscr{A}}{G_0 l} \ll 1$$

It is seen from Fig. 47 that one must have very high reflectivity mirrors to extract a large fraction of the closed-cavity power from regions of low small-signal gain. For a high-$Q$ stable resonator, the internal-cavity flux reaches very high values, and, according to Eq. (90), this implies an energy per unit volume produced by the active region of $GI = G_0 I_s$ for $I \gg I_s$. Thus $P_{cc} = G_0 I_s V$ is an upper limit to the power from the active medium. In terms of the kinetic rates, this is given by

$$P_{cc} = \frac{(k_{F_2}\tau_2 - k_{F_1}\tau_1)}{(\tau_1 + \tau_2)} h_P \nu V$$

This reduces to $P_{cc} = k_{F_2} h_P \nu V$ for the case of a "good" laser, where $\tau_1 \ll \tau_2$. Thus within the assumptions of the foregoing simplified model, the closed-cavity power gives all of the power pumped into the upper level, which is the maximum available power.

The effect of finite mixing rates in a cw chemical laser clearly complicates the problem, since the spatial distribution of gain and absorption will be different for each spectral line. It is believed that, in the limit of perfect mirror reflectivity, the closed-cavity power is an upper bound in power for current practical cw HF/DF lasers, but this has not yet been proven mathematically. Experimentally, no output power has been measured from either stable or unstable resonators that is greater than the closed-cavity power. However, it should be noted that the closed-cavity measurement technique is subject to possible errors if incoherent radiation or hot-gas circulation carries significant amounts of heat to the mirrors in addition to the laser radiation absorbed. In addition, the high reflectivity of the cavity may permit the participation of parasitic modes that use stray reflections from walls or other surfaces to extract laser energy from regions outside of the nominal cavity-mode volume. Thus, application of the closed-cavity technique for power measurements requires some care in practice.

As a rough interpretation of the closed-cavity power versus $\bar{x}_c$ variation, one may assume that the resonator is extracting power from the small-signal gain averaged over all lines a distance $2\bar{x}_c$ downstream of the nozzle-exit plane. This average-gain profile will have a rapidly changing high-gain region, possibly a slowly changing low-gain region, and a region of absorption. If the closed-cavity mode volume is extended a distance $2\bar{x}_c$ over the positive average-gain region, the maximum available power will be extracted from the active region. If $2\bar{x}_c$ is extended still further into the absorbing region, the multimode radiation will couple into these regions, and, consequently, the closed-cavity power will be reduced. Figure 48 shows a typical closed-cavity power versus $\bar{x}_c$ curve measured by Spencer et al.[50] with the Aerospace standard slit nozzle and either $N_2$ or He as diluents. Some information on the average multiline gain profiles can be extracted from Fig. 48. Thus for He diluent, the high-gain region extends a distance $2\bar{x}_c$ of ~3.7 cm downstream of the nozzle exit, and there is a low-gain region extending from 3.7 to 5.6 cm before one gets into the absorbing regions. For $N_2$, on the other hand, the low-gain region is much reduced, and the high-gain region extends the same distance (3.7 cm) downstream. Notice that, since the closed-cavity technique completely saturates the high-gain region, no information on the position or magnitude of the peak gain can be obtained.

With proper precautions, the closed-cavity power-measurement tech-

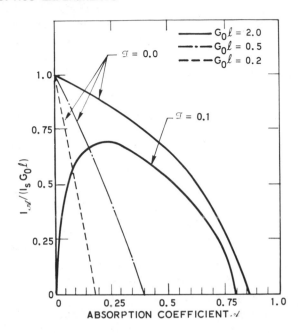

**Fig. 47.** Flux absorbed in both cavity mirrors versus absorption coefficient $\mathcal{A}$ for various values of small-signal gain $G_0 l$ and mirror transmissivity $\mathcal{T}$. Maximum absorbed flux occurs for $\mathcal{T} = \mathcal{A} = 0$, which is approximated in closed-cavity method.

nique appears to be a useful method to determine the upper bound on the output power for a given cw chemical laser. These data are scalable, that is, one can measure the closed-cavity power for a small nozzle (with a sufficiently low $G_t$) and extrapolate the power for a larger nozzle by scaling the area (height times nozzle width). The data can be used to check experimentally how well one is saturating the medium with an output-coupled resonator such as an unstable cavity.

### Edge-Coupled Unstable-Cavity Experiments

*Output Power.* It is of interest to determine whether a multiline unstable cavity can efficiently saturate the cw HF/DF active medium. This is a complicated question because the peak small-signal gain varies over a wide range (from 2 to 10%/cm for HF for the Aerospace standard slit nozzle)[46] for the various lines and the streamwise-gain distribution for each line is different because of cascading and $J$ shifting (see chapter by Emmanuel). Theoretical models, such as the BLAZER code developed by Bullock et al.,[32] can predict unstable-cavity gain saturation but rely on the accuracy of the mixing and kinetics model in correctly predicting the number-density distribution of excited species. It is important, therefore, to check experimentally the validity of the theories and measure independently the degree of saturation.

Mansell et al.[54] and Wisner et al.[49] made some early measurements of unstable-resonator power-extraction efficiency. Wisner et al. compared the output power from the confocal unstable resonator with the multimode stable-resonator output for nearly optimum coupling ($\cong 30\%$). They found that more than half the multimode power could be extracted by the unstable cavity under conditions of good beam quality. This result is inconclusive, however, since the unstable cavity mode geometry was not optimized with respect to the gain profile and the stable-cavity output-coupled power is less than the more fully saturated closed-cavity power. More recently, Hook[56] has reported about 75% of the closed-cavity power extracted from a cw HF confocal unstable cavity.

To determine the optimum unstable-cavity mode geometry on a cw chemical laser, a new apparatus was developed by Chodzko et al.[59] The apparatus, sketched in Fig. 49, consists of a calorimeter for measurement of the output power of an edge-coupled cw unstable resonator as a

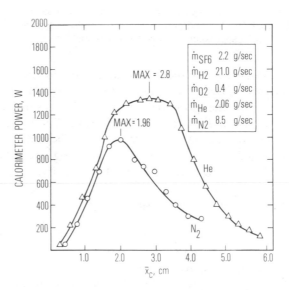

**Fig. 48.**  Closed-cavity power versus $\bar{x}_c$ as obtained from arc-driven supersonic diffusion cw HF laser with a 36-slit standard nozzle.[50] Measurements were made with both He and $N_2$ diluents and yield information on the extent of the positive-gain zone as well as the maximum power available.

function of mode geometry. The calorimeter consists of an absorbing copper-plate assembly, which replaces the output-coupling mirror of the unstable resonator. A rectangular hole in the plate of width $w$ and height $h$ is nominally centered with and orthogonal to the cavity axis. The radiation intercepted by the water-cooled plate provides the unstable-cavity output power while containing the radiation within the cavity. The absorbing plate is constructed from four separate (labeled $T$, $B$, $U$, and $D$) movable rectangular leaves. When the location of these leaves is varied, the height and width of the unstable-cavity mode can be changed independently for a fixed-cavity magnification. Furthermore, through the measurement of power in each leaf, the distribution of power from the active medium upstream and downstream of the optical axes can be inferred. Such a device, termed a "variable-aperture calorimeter-absorbing scraper" (VACAS), has been applied to a cw HF/DF confocal unstable resonator.[59] The output power versus mode geometry and $\bar{x}_c$ was measured for different nozzle designs, resonator magnifications, and cavity pressures in both HF and DF.[60]

The values of $\bar{x}_c$ given in the following discussion refer to the prealigned distance of the cavity-optical axis from the nozzle-exit plane, which is also the distance of the center line of the rectangular hole in the VACAS from the exit plane. In the course of the VACAS experiments,[60] it was found that this initial prealignment did not yield the maximum output power. A possible interpretation is that the maximum output power was obtained by the translation of the optical axis of the unstable cavity to the region of maximum gain. This translation was accomplished by tilting the cavity mirrors in the streamwise plane and optimizing the output power. Thus the data in Fig. 50 correspond to a fixed value of $\bar{x}_c$, which has been translated slightly upstream of the prealigned value of 1.25 cm (to ∼1.0 cm) for optimum power. The optimum was obtained with the VACAS leaves initially set for a large value of $w$ and $h$, that is, $w \simeq \bar{x}_c/M$ and $h = H_N/M$, where $H_N$ is the nozzle height and $\bar{x}_c$ is the prealigned value. The optimization procedure required for maximum power may imply that an asymmetric mode of the unstable cavity is most efficient, probably because of the asymmetric gain distribution, which is peaked near the nozzle exit plane for this fast-mixing nozzle. A theoretical discussion of asymmetric modes is given in Ref. 45, and the results depend on the degree of saturation of the active medium. Further experimental evidence for asymmetric unstable-resonator modes is given in Ref. 60 from the near-field intensity distribution of the output beam of a cw DF confocal unstable cavity.

Figure 50 shows some typical VACAS data[60] obtained with DF through the use of an Aerospace $3.8 \times 23$-cm axisymmetric nozzle array,

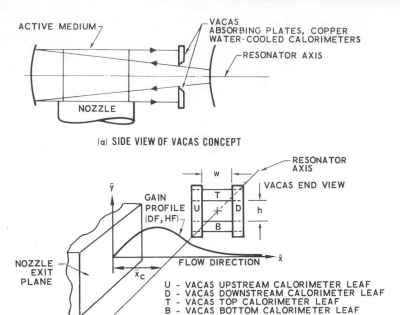

(a) SIDE VIEW OF VACAS CONCEPT

(b) ISOMETRIC VIEW

U - VACAS UPSTREAM CALORIMETER LEAF
D - VACAS DOWNSTREAM CALORIMETER LEAF
T - VACAS TOP CALORIMETER LEAF
B - VACAS BOTTOM CALORIMETER LEAF

**Fig. 49.** The VACAS technique for measuring the output power from an edge-coupled unstable resonator.[59] The internal-cavity calorimeter consists of a flat absorbing plate with a variable rectangular-hole geometry of width $w$ and height $h$ such that the output power in a rectangular mode of width $Mw$ and height $Mh$ is obtained. The output-power distribution on the upstream ($U$), downstream ($D$), top ($T$), and bottom ($B$) calorimeter leaves is also measured.

and $M = 1.5$ confocal unstable resonator, a cavity pressure of 8 Torr, and a value of $\bar{x}_c = 1.25$ cm. The output power is plotted as a function of both beam height ($Mh$) or beam width ($Mw$). Figure 50 shows the two characteristic dimensions given by the nozzle-exit plane and the nozzle height. Thus the maximum power for a fixed $\bar{x}_c$ and beam height is obtained by the opening of the beam width up to the nozzle-exit plane limit. At this maximum beam width, the maximum power is obtained at a beam height nearly equal to the nozzle height. Thus, at $\bar{x}_c \cong 1.25$ cm, the nozzle flow has not expanded significantly past the nozzle height. Notice that the power distribution in the VACAS leaves at the optimum power point gives 34% of the power on the upstream leaf and 21% of the power on the downstream leaf, which indicates that the Aerospace axisymmetric nozzle[60] array is a fast-mixing device ($\bar{x}_c \cong 1.25$ cm). The VACAS measurements were also made with the same $M = 1.5$ cavity and HF lasing at $\bar{x}_c = 2.5$ cm. An optimum power of 11.0 kW was obtained at 2.5-Torr cavity pressure with the same nozzle. This is to be compared with an optimum closed-cavity power of 12.0 kW giving a 91% power-extraction efficiency for the multiline confocal unstable cavity. The maximum closed-cavity power obtained with DF was ~10 kW for the flow conditions corresponding to Fig. 50, indicating that the $M = 1.5$ unstable resonator did not saturate the DF medium because of its reduced small-signal gain compared with HF.

In conclusion, efficient saturation of the active medium has been demonstrated for a multiline cw HF confocal unstable cavity with $l \cong$ 20 cm gain length. A new experimental technique (VACAS) for determining the optimum-mode geometry for an edge-coupled unstable resonator was developed and applied to a supersonic diffusion cw HF/DF laser. For the case of an HF active medium and a specific nozzle (an Aerospace axisymmetric nozzle)[60] over 90% of the closed-cavity power was measured with a confocal unstable cavity. The optimum-mode geometry for a fixed $\bar{x}_c$ appears to correspond to a beam width of $2\bar{x}_c$ for the unstable cavity. Saturation in DF is, as expected, more difficult to achieve than in HF because of the lower small-signal gain of DF. Comparison of available theories with the above conclusions should be possible but has not yet been carried out.

*Beam Quality.* The first demonstration of near diffraction-limited beam quality with a multiline confocal unstable cavity, and a cw HF/DF chemical-laser medium was reported by Wisner et al.[49] of the United Aircraft Research Laboratories. Attempts made earlier at other laboratories, such as The Aerospace Corporation and TRW Systems, Inc., were unsuccessful because of such problems as thermal blooming caused by absorption of the HF laser beam by atmospheric water vapor and thermal distortion of the cavity mirrors. In Ref. 49, thermal distortion of the cavity mirrors at power levels of up to 2 kW was avoided by the use of multilayered dielectric-coated mirrors of high reflectivity. The cavity mirrors were dielectric coated for maximum reflectivity ($\geq$99.8%) at 2.8 $\mu$m with a silicon substrate.

Figure 51 is a schematic of the beam-quality measurement apparatus used by Wisner et al.[49,61] Beam-quality measurements had to be made rapidly since the combustion-driven supersonic diffusion cw HF/DF laser had a blow-down vacuum system with a 15 to 20 sec run time. The beam quality was measured in terms of the fraction of total power transmitted through an aperture at the focal plane. The reflected component from a NaCl wedge was imaged with a 1-m focal-length mirror onto an integrated irradiance analyzer at the focal plane. This device, developed by Wey and Wisner,[61] consists of a rotating disc with a series of graduated

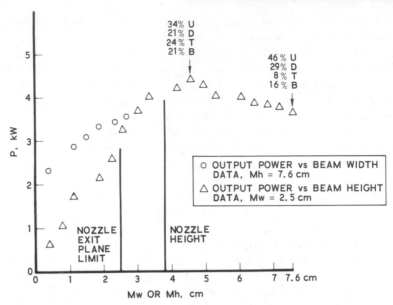

**Fig. 50.** The VACAS output power versus beam width ($M$w) or beam height ($M$h) for a cw DF chemical laser.[61] The data corresponds to a $3.8 \times 23$ cm axisymmetric nozzle array, with an $M = 1.5$ confocal unstable cavity located at a nominal $\bar{x}_c = 1.25$ cm, and flow conditions corresponding to $\sim 10$ kW of closed-cavity power. The VACAS leaves were opened from a narrow slit up to the limit determined by the position of the nozzle-exit plane for a fixed beam height of 7.6 cm. The power distribution on the VACAS leaves at the optimum output-power geometry is also shown.

apertures at constant radius and a fast response power meter that measures the power transmitted through a given aperture. The resulting oscilloscope display is a plot of normalized power subtended by the apertures. A large aperture (diameter $\cong 30\ F\lambda/D$ where $F$ is the focal length and $D$ is the aperture diameter) is used as the 100% reference. Typical rotational frequencies for the disc were 20 rps, giving a complete far-field "power-in-the-bucket" experimental curve in 50 msec.

The cw HF laser used by Wisner et al.[49] in their beam-quality experiments had a $1.25 \times 20$-cm matrix-hole nozzle design. It was observed that the maximum unstable-cavity output power was obtained with a mode diameter of $\sim 1$ cm and a cavity magnification of 2 ($G_t \cong 3.5\%$/cm). This implies a peak small-signal gain of $\sim 20\%$/cm and a fast mixing rate.

Figures 52 and 53 show some typical beam quality data obtained at two different confocal unstable cavity magnification values ($M = 2, 3$) with the integrated irradiance analyzer by Wisner et al.[49] The theoretical curves correspond to a uniformly illuminated annulus of constant phase, where the values of $M$ are assumed to be calculated from the cavity geometry within the geometric approximation. Unfortunately, no corresponding

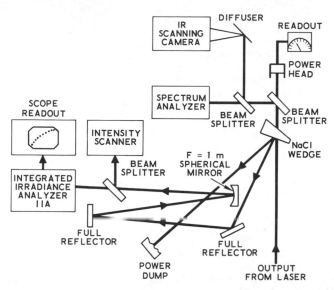

**Fig. 51.** Schematic of beam-quality measurement apparatus.[49] An integrated irradiance analyzer IIA was used to rapidly measure the far-field power transmitted through various apertures at the focal plane.

near-field data were given in their report[49] for comparison with the assumed near field. Thus at 1160-W output power (Fig. 52), 20% of the power was measured in the diffraction-limited main lobe, which is 40% of the theoretical limit. At 300-W output power (Fig. 53), 50% of the power was measured in the main lobe, which is ~80% of the theoretical limit. The cause of the apparent variation of beam quality with power has not been established and could be a result of thermal distortion of the cavity mirrors. Further data showing the output-power stability in time over the beam-quality scanning time (~50 msec) would be desirable.

Chodzko[55] has made beam quality measurements on a multiline cw HF confocal unstable cavity. The arc-driven cw HF laser is identical with that described in Ref. 50, that is, the Aerospace standard slit nozzle. Figure 54 is a schematic diagram of the apparatus. The multiline output beam from the $M = 1.414$ confocal unstable resonator (2.2-cm beam diameter) transmitted through a $CaF_2$ wedge beam splitter into a power meter. The first surface reflection off the wedge goes to a 1.5-m focal-length mirror. A rotating mirror sweeps the focal spot, whose main lobe has a diameter of ~0.5 mm, past a 0.05-mm diameter scanning aperture. A $CaF_2$ lens in conjunction with a beam splitter simultaneously images the scanning aperture onto the entrance slit of a monochromator and an InSb detector. The output signals displayed on a dual-trace oscilloscope simultaneously give the far-field intensity distribution of the multiline sum and any

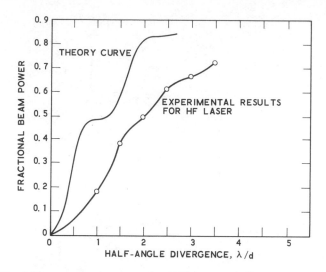

**Fig. 52.** Far-field power fraction versus beam-divergence half angle as measured with IIA on a cw HF chemical laser.[49] The solid curve corresponds to a uniformly illuminated annulus of constant phase and the data points correspond to an $M = 2.0$ confocal unstable cavity, an output power of $P_0 = 1160$ W, and the matrix-hole nozzle design.

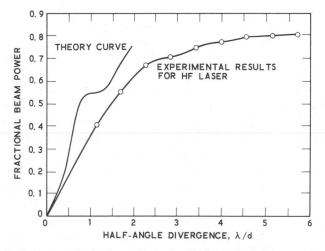

**Fig. 53.** Far-field power fraction versus beam-divergence half angle as measured with the IIA on a cw HF chemical laser.[49] The solid curve corresponds to a uniformly illuminated annulus of constant phase and the data points correspond to an $M = 3.0$ confocal unstable cavity, an output power of $P_0 = 300$ W, and the matrix-hole nozzle design.

single-line component beam. The scanner, therefore, gives the spectrally resolved stream-wise far-field intensity distribution. The apparatus in Fig. 54 also can measure the fraction of power transmitted through an aperture at the focal plane with the rotating mirror in a fixed orientation, and a power meter placed behind the aperture. The multiline near-field intensity distribution is obtained with a ZnS thermal image screen 30 cm from the laser exit plane. The entire apparatus is placed in a dry box whose relative humidity is <1% to eliminate thermal blooming effects. At power levels ≤100 W, it was found that there was no variation in beam quality with power, while above 100 W the beam quality degraded with power because of thermal distortion of the water-cooled, BeCu cavity mirrors.

Figure 55 shows the experimental data obtained by Chodzko[55] at 100-W output power with the apparatus of Fig. 54. The multiline near-field intensity distribution is asymmetric in the streamwise direction, because of the nonuniform-gain profile and possibly because of upstream saturation causing a depletion in gain on the downstream side. A complex azimuthal mode of the unstable resonator is suggested by the irregular shape of the pattern. The time-averaged power-in-the-bucket curve shows that ~25% of the power lies within the diffraction-limited main lobe, compared with a theoretical value of 40% for a uniformly illumi

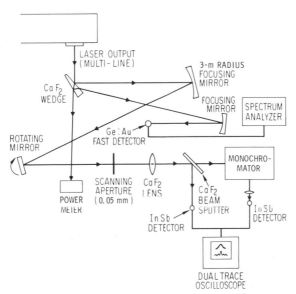

**Fig. 54.** Schematic of beam-quality measurement apparatus.[55] The spectrally resolved far-field intensity distribution of a multiline output beam is made by simultaneously imaging the far-field spot on a monochromator slit and directly on an InSb detector.

nated annulus of constant phase ($M = 1.414$). The spectrally resolved far-field intensity distribution showing the $P_2(4)$ beam and the simultaneous multiline beam is also shown. There is a well-defined side-lobe structure, and the main lobe is diverging at the diffraction-limited angle. The relative height of the main-lobe peak compared with the side lobes is less (by about a factor of 3), however, than that for a uniformly illuminated annulus of constant phase. This is qualitatively consistent with the power-in-the-bucket data, which indicates that about one-half the theoretical power is inside the main lobe. It is seen that the main lobe of the $P_2(4)$ single-line component beam is coincident with the main lobe of the multiline beam. This implies that the wave fronts of the various single-line component beams are essentially parallel. The relative distribution of power in the side lobes of the different single-line component beams are seen to be similar but not identical. This means that the near-field intensity distribution of each transition is different due to the different gain distributions for each transition. These results clearly depend on the details of the mixing process.

As a final consideration in this section, a comparison between an available theory (BLAZER code by Bullock et al.) and experimental data will be made. Hook[56] of TRW Systems, Inc., has applied a multiline confocal unstable resonator ($M = 2$, rectangular hole in the coupling mirror) to a combustion-driven supersonic diffusion cw DF laser. Figure 56 shows a Plexiglas near-field burn pattern obtained 9 m from the laser-exit plane. A theoretical prediction of the multiline DF near-field intensity distribution is plotted for comparison. There is clearly good qualitative agreement between experiment and theory. The side view of the burn pattern (Fig. 56$a$), with the $\bar{y}$-axis measured at right angles to both the gas flow and the optical axis, shows a double-peak structure determined by diffraction effects. Since the small-signal gain distribution is relatively uniform across the nozzle, the solution is similar to an empty cavity mode (see Fig. 25, for instance). The side view of the burn pattern (Fig. 56$b$), with the $\bar{x}$-axis measured parallel to the gas flow and perpendicular to the optical axis, shows a peak on the upstream side caused by the nonuniform gain distribution, and diffraction effects are still observable in the secondary maxima.

To summarize these types of measurements, approximately twice diffraction-limited performance has been obtained from a multiline supersonic diffusion cw HF/DF laser with confocal unstable cavities. Between 50 and 80% of the theoretical power in the diffraction-limited main lobe has been measured for various nozzle designs. In addition to measuring the fraction of power transmitted through an aperture at the focal plane, beam-quality data have been obtained in terms of the spectrally resolved

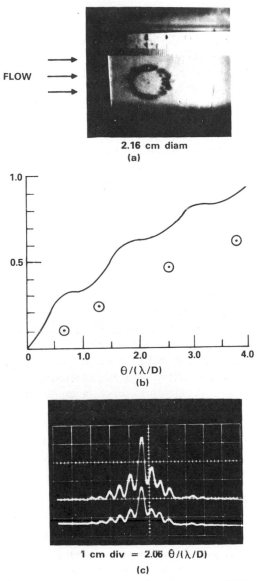

**FLOW**

**2.16 cm diam**

**(a)**

θ/(λ/D)

**(b)**

**1 cm div = 2.06 θ/(λ/D)**

**(c)**

**Fig. 55.** Beam-quality measurements for a multiline cw HF chemical laser.[55] The confocal unstable resonator had a magnification $M = 1.414$, $N_0 = 43$, the optical axis was located at $\bar{x}_c = 1.25$ cm, a standard 36-slit nozzle was used, and the output power $P_0 = 100$ W. (a) Near-field intensity distribution observed on ZnS thermal image screen (1 m from source), (b) far-field power fraction versus beam-divergence half angle, and (c) spectrally resolved far-field intensity distribution: upper trace—$P_2(4)$ component beam and lower trace— multiline sum beam ($\sim$10 lines).

far-field intensity distribution. The data showed a well-defined side-lobe structure with the main-lobe diverging at the diffraction-limited angle. The wave fronts of the single-line component beams that constituted the multiline sum were parallel. The intensity distribution of the side lobes was different for each V–R transition, probably because of the different gain distribution for each line. An experimental near-field intensity distribution[56] has been compared with the active resonator theory of Bullock et al.,[21,32] and good qualitative agreement has been obtained.

### 3.3. Supersonic Diffusion cw HF Selected-Line Lasers

**Single-Line Unstable Cavity.**    Demonstration of diffraction-limited beam quality on a cw chemical laser was first reported by Chodzko et al.[19] with a single-line nonconfocal unstable resonator. Single-line operation is of interest, for example, because of atmospheric propagation considerations, since most HF wavelengths are strongly absorbed by water vapor in the atmosphere. Some $P_2(J)$ HF lines, however, are not absorbed sufficiently to produce thermal-blooming effects over laboratory distances. In particular, the $P_2(8)$ HF line at 2.911 $\mu$m has a calculated 15% transmission through a vertical atmosphere at midsummer latitude.[62] In addition to single-line output, the transverse-mode selection and arbitrary-mode volume generally characteristic of unstable resonators are also desirable. Beam quality measurements using a single-line cw HF unstable cavity should be easier to interpret than multiline data, since there will be a single streamwise gain profile and, hence, a single near-field pattern that can be related to the far-field intensity distribution.

Figure 57 is a schematic diagram of the single-line optical cavity designs used by Chodzko et al.[19] in their cw HF laser-beam quality measurements. In both configurations of Fig. 57, the cavity consists of a convex mirror and a plane-diffraction grating.

It should be noted here that a plane-diffraction grating is equivalent to a flat mirror provided that the aberrations introduced by the grating in the output beam are not too large. For a stable cavity incorporating a diffraction grating in the Littrow configuration, the beam waist occurs at the grating location, and, hence, a plane wave is incident on the grating, which implies no aberrations in the first-order reflected beam. In a nonconfocal unstable cavity incorporating a flat and convex mirror, a diverging spherical wave is incident on the grating, and aberrations are introduced. This may be seen from the grating equation [Eq. (96)], which implies that the first-order reflected beam does not reflect like a flat mirror for off-axis rays. It can be shown that the aberration or deviation from an ideal spherical wave front $\Delta\Phi$ in the first-order reflected beam

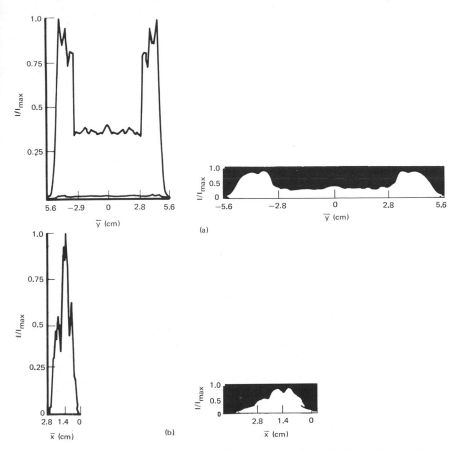

**Fig. 56.** Comparison of theoretical BLAZER computer-code predictions with experimental near-field intensity distribution for a multiline cw DF chemical laser.[21] The near-field intensity distribution was obtained from a Plexiglas burn pattern 9 m from the output coupling mirror that had a $5 \times 1.25$-cm rectangular hole. The edge-coupled confocal unstable cavity had a magnification $M = 2$, a maximum outer Fresnel number of $N_0 = 450$, and $\bar{x}_c = 1.25$ cm. The effect of diffraction because of the 9-m propagation path was taken into account in the numerical calculations.

depends on the diffraction-grating groove spacing, the wavelength $\lambda$, the cavity length $L$, the magnification $M$, and the beam diameter $D_B$. For example, a 600-line/mm grating with $\lambda = 2.8$ $\mu$m, $L = 1$ m, $M = 1.5$, and $D_B = 5$ cm produces an aberration of $\Delta\Phi \cong \lambda/10$. This astigmatic aberration is reduced to $\Delta\Phi \cong \lambda/20$ for a 300-line/mm grating.

Since, the diffraction grating is equivalent to a flat mirror, the resonator of Fig. 57 is simply a nonconfocal unstable cavity. The cavity in Fig. 57a

**Table 4.**   Single-Line Non-confocal Unstable-Cavity Geometries

| Cavity | Coupling | Cavity length, cm | Grating, line/mm | Convex mirror radius, m | Coupling mirror, cm dia. hole | Magnification |
|---|---|---|---|---|---|---|
| Fig. 57$a$ Edge | | 92 | 600 | −13.7 | 0.95 | 1.67 |
| Fig. 57$b$ Continuous | | 92 | 600 | −22.7 | none | 1.49 |

is termed "edge-coupled" since it extracts power in an annular beam with a 45° output coupling mirror that has a hole drilled parallel to the optical axis, as was previously used by Krupke and Sooy.[20] Since the diffraction grating will not have 100% efficiency, not all of the radiation is reflected into the first order along the optical axis. A fraction of the power will be lost as a zero-order beam that is absorbed in a heat sink. The direction of the output beam is independent of the line selected by the grating in this configuration. In contrast, the unstable cavity in Fig. 57$b$ is termed "continuously coupled" and extracts power out through the zeroth order. The resonator in this case produces a relatively uniform intensity pattern in the near field; the size of the output beam is determined primarily by the limiting aperture of the cavity (usually the ruled area of the grating) and the extent of the active region transverse to the optical axis. Table 4 gives the unstable-cavity mirror geometries used in these experiments.

Table 5 shows the single-line power levels and the corresponding optimized gas-flow conditions of the arc-driven cw HF laser[50] and the continuously coupled cavity (Fig. 56$b$) used by Chodzko et al.[19] About 10% of the closed-cavity power was extracted on a single strong line.

**Table 5.**   Operating Conditions of cw HF Laser for Single-Line Unstable-Cavity Experiments

| Cavity | Wavelength, $\mu$m | Power, W | Cavity pressure, Torr | Plenum pressure, psi | $\dot{m}_{SF_6}$, g/sec | $\dot{m}_{H_2}$, g/sec | $\dot{m}_{He}$, g/sec | $\dot{m}_{O_2}$, g/sec |
|---|---|---|---|---|---|---|---|---|
| Closed cavity[a] | Multiline | 900 | 5 | 16 | 2.0 | 0.9 | 2.6 | 0.8 |
| Single line (Fig. 57$b$) | 2.727, $P_2(3)$ | 60 | 5 | 16 | 2.0 | 0.9 | 2.6 | 0.8 |
| | 2.760, $P_2(4)$ | 100 | 5 | 16 | 2.0 | 0.9 | 2.6 | 0.8 |
| | 2.795, $P_2(5)$ | 100 | 5 | 16 | 2.0 | 0.9 | 2.6 | 0.8 |
| | 2.832, $P_2(6)$ | 60 | 5 | 16 | 2.0 | 0.9 | 2.6 | 0.8 |
| | 2.871, $P_2(7)$ | 40 | 5 | 16 | 2.8 | 0.5 | 1.6 | 1.1 |
| | 2.911, $P_2(8)$ | 10 | 5 | 16 | 2.1 | 0.25 | 1.0 | 0.9 |

[a] Closed cavity with 5-cm internal aperture.
Flow rates for closed-power level are given for comparison with rates that optimize the power (unapertured zeroth-order beam) on a single line.

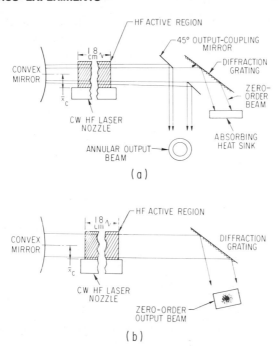

**Fig. 57.** Schematic diagram of single-line nonconfocal unstable-cavity arrangements.[19] (a) Edge-coupled cavity (annular output beam) and (b) continuously coupled cavity (uniform output beam).

Note that the optimum-flow conditions for one line are not the same as another, since higher $J$ transitions can be emphasized with the increase of the rotational temperature. The small fraction of closed-cavity power measured in the output beam is partly due to lack of gain saturation. The threshold gain computed from Eq. 89 is 3.3%/cm when a 70% grating efficiency is assumed, and, consequently, high-gain saturation was not achieved for the individual transitions.

Figure 58 shows the beam-quality data obtained with the edge-coupled single-line unstable cavity (Fig. 57a) at a 20-W output power on the $P_2(5)$ transition. The near-field intensity distribution (obtained from a ZnS thermal-image screen) is similar to the multiline near-field pattern in Fig. 55. There is again an asymmetry in the streamwise direction, which is probably caused by the asymmetric gain profile, and irregularities, which suggest the presence of higher-order azimuthal modes of the unstable cavity. The measured magnification of ~1.5 is less than the geometrical value of 1.67, possibly because of the diffraction effects discussed by Siegman.[63] The theoretical comparison for the beam quality is based on a uniformly illuminated annulus of constant phase with the measured magnification ($M = 1.5$). The power transmitted through an aperture at the focal plane shows ~50% of the theoretical power in the main lobe. This far-field degradation is due at least in part to the nonuniform near-field intensity pattern. The streamwise scan of the far-field intensity distribution shows well-defined side lobes and the main lobe diverging at the diffraction-limited angle, based on the measured beam diameter.

As an independent measure of beam quality, a spectrum analysis of the intensity fluctuations of the output was made. The apparatus (see Fig. 54) could measure the frequency spectrum from 0 to 15 MHz. It was found that when the cavity mirrors were tuned, the spectral width around zero-center frequency could be reduced to 1 kHz, which corresponds to the expected fluctuations in the arc current. For this narrow spectral width, the near-field pattern appeared as in Fig. 58. High-frequency fluctuations up to 15 MHz were observed when the cavity mirrors were detuned from the annular near-field pattern in Fig. 58. Recent theoretical work of Bullock[21] predicts ~10 MHz spacing between the fundamental and the second higher-order transverse mode for an active cw HF unstable resonator. It is concluded that the near-field pattern in Fig. 58 corresponds to the fundamental transverse mode of the active edge-coupled single-line unstable cavity.

The annular beam produced by the edge-coupled configuration (Fig. 57a) has the disadvantage of having a large fraction of the power within the side lobes, as indicated in Fig. 58. This could be improved by the increasing of the magnification $M$ of the unstable resonator if one has sufficiently high small-signal gain. In many cases, however, the gain is limited, and it is desirable to have a large fraction of power in the main lobe independent of $M$. The continuously coupled single-line unstable cavity (Fig. 57b) accomplishes this by producing a more nearly homogeneous near-field beam pattern.

Figure 59 shows the beam-quality data obtained with the continuously coupled cavity (Fig. 57b) at an output power of 10 W on the $P_2(8)$ HF line. The near-field intensity distribution of the zeroth-order beam is seen to be more uniform and diffuse than that of the edge-coupled beam (Fig. 58), with the overall size determined by the boundaries of the active-gain region. The limiting aperture of the cavity was 3.8 cm from the nozzle exit plane to the grating boundary. Even though there is radiation spreading outward because of magnification, the high-intensity regions are confined to the high-gain regions. The irregular shape of the near field is presumably related to a nonuniform gain distribution of the specific nozzle used. In order to compare the beam-quality data with a known theoretical solution, the beam was passed through an external circular of 1.25-cm

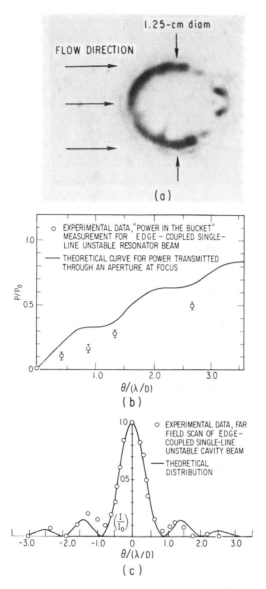

**Fig. 58.** Beam-quality measurements for a single-line edge-coupled cw HF unstable resonator.[19] The cavity had a magnification $M = 1.67$, grating efficiency $\cong 70\%$, $N_0 \cong 15$, the optical axis was located at $\bar{x}_c = 2.5$ cm, a standard 36-slit nozzle[50] was used, the output power $P_0 \cong 20$ W, and a $P_2(5)$ transition was selected. A spectrum analysis indicated single transverse-mode operation. (*a*) Near-field intensity distribution observed on ZnS thermal image screen (1 m from source), (*b*) far-field power fraction versus beam-divergence half angle, and (*c*) far-field intensity versus beam-divergence half angle from streamwise scan obtained at focus.

**181**

diameter, reducing the power by a factor of 2. When the beam was focused, the power transmitted through another aperture was seen to follow quite closely the theoretical curve for a uniformly illuminated circular disc of constant phase. In Fig. 59 essentially all of the theoretical power (84%) is within the main lobe. The far-field intensity distribution at focus also agrees with the theory, with the main-lobe beam divergence close to the diffraction limit. Note that the relative height of the side lobe as compared with the main lobe is also approximately correct. A two-dimensional scan was obtained 165 m from the source (which was a 0.8-cm diameter parallel beam) and, hence, well into the far field. The horizontal- and vertical-beam widths at half intensity were about 0.39 and 0.42 mrad, respectively, to be compared with a theoretical value of 0.36 mrad.

A spectral analysis of the intensity fluctuations of the output beam was made as described previously. A narrow spectral distribution a few kilohertz wide about zero-center frequency was observed, with no fluctuations up to 15 MHz. Tilting the convex mirror in the streamwise direction did not affect the spectral distribution. This is expected for the continuously coupled cavity, since there is no coupling mirror with a circular hole to define a symmetry axis for the transverse modes. It is suggested that the optical axis can be translated anywhere within the gain region (above threshold) and still produce a phase-coherent output beam without mode beats.

It is concluded that the supersonic diffusion cw HF single-line unstable-resonator experiments by Chodzko et al.[19] demonstrate diffraction-limited performance for a specific device and flow conditions. Since the length of the active region was 18 cm, the cavity pressure was 5 Torr, and the gas density in the cavity was $\sim 10^{-6}$ cm$^{-3}$, the calculated deviation of the phase front per pass through the active medium is no more than $\sim \lambda/50$. Depending on the assumed number of passes through the active medium for the unstable cavity, it appears that the density variations in the present scale device are not sufficient to degrade the beam quality. A mechanism for enhancing the refractive index variations, such as anomalous dispersion, could modify this conclusion, but the excellent comparison between experimental data and theory obtained with the continuously coupled single-line unstable cavity is not consistent with this. It should be noted, however, that the maximum Fresnel number for which coherent, diffraction-limited performance has been shown (Fig. 59) is $\sim 15$ for the apertured (1.25-cm diameter) zeroth-order beam.

The edge-coupled single-line unstable cavity data (Fig. 58) indicate that, even with single transverse-mode output, there can be nonuniformities in the near-field intensity distribution, presumably because of the nonuniform gain profile. In further experiments a two-dimensional near-field intensity distribution can be Fourier-transformed to provide a theoretical comparison with the data assuming a uniform phase front. The near field can be obtained quantitatively by observing it on a diffuse screen with an infrared camera as done before by Freiberg et al.[64] Active resonator computer models such as those developed by Bullock et al.[32] can also be applied and compared with experiment.

**Selected-Multiline Experiments.** The cw HF single-line unstable cavity experiments demonstrated a maximum of ~10% of the closed-cavity power in a single transverse mode. This power extraction efficiency can be improved by increasing the small-signal gain or by decreasing the cavity magnification $M$ so that better gain saturation is achieved. Theoretical calculations indicate (see chapter by Emmanuel), however, that multiwavelength cascading between the HF/DF vibrational levels is also required for efficient power extraction. To obtain multiline operation to enhance output power, but to exclude lines with poor atmospheric transmission, special types of multiline resonators have been used. Chodzko[65] has considered a selected-multiline unstable resonator. Angelbeck et al.[66] have considered a selected-multiline stable oscillator in conjunction with parallel staged, phase-controlled amplifiers to produce high-power coherent output.

References 65 and 66 use a diffraction grating for multiline selection as will be discussed. An intracavity prism may also be used for line selection, but it is difficult to obtain sufficient dispersion to ensure selected-line output for a high-gain cw HF/DF device with a prism. An intracavity absorption cell may also be used to eliminate undesirable lines. Water-vapor absorption cells, such as those developed by Bhaumik,[67] or simply an open-air cavity, may be used to weaken the HF ($v = 1$ to $0$) lines, for instance. It should be noted, however, that for high-power oscillator applications, the diffraction grating is scalable since it can be cooled, while the prism probably is not because of beam-quality degradation by localized self-heating effects. The intracavity absorption cell may be scalable if one uses a flowing gas to eliminate thermal blooming,[68] but no such work has been reported in connection with cw chemical lasers.

The selected-multiline unstable resonators discussed by Chodzko[65] are illustrated in Fig. 60, and the principle of operation is shown in Fig. 61. The angles of incidence $\bar{i}$ and reflection $\bar{r}$ for a reflection-type diffraction grating are related by[69]

$$\sin \bar{i} + \sin \bar{r} = \frac{\bar{n}\lambda}{d} \qquad (94)$$

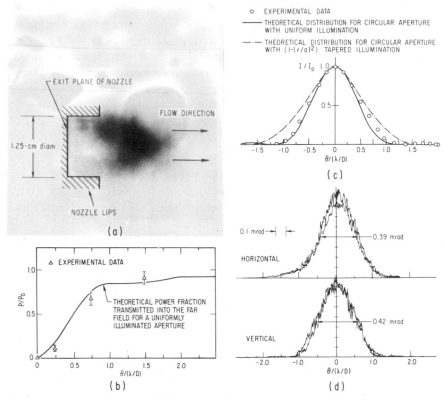

**Fig. 59.** Beam-quality measurements for a single-line continuously coupled cw HF unstable resonator.[19] The cavity had a magnification $M = 1.5$, grating efficiency $\cong 70\%$, $N_0 \cong 15$, the optical axis was located at $\bar{x}_c = 2.5$ cm, a standard slit nozzle was used, the output power $P_0 \cong 10$ W, and a $P_2(8)$ transition was selected. A spectral analysis indicated single transverse-mode operation. (a) Near-field intensity distribution observed on ZnS thermal image screen (1 m from source), (b) far-field power fraction versus beam-divergence half angle, (c) far-field intensity versus beam-divergence half angle (streamwise scan obtained at focus), and (d) far-field intensity versus beam-divergence half angle (obtained 165 m from a 0.8-cm diameter parallel-beam source).

where $\bar{n}$ is the order of the diffraction pattern, $\lambda$ is the wavelength, and $d$ is the groove spacing. For single-line operation, the grating is oriented at the Littrow angle $\bar{i}_L$ so that the first-order radiation is reflected back on itself ($\bar{i} = \bar{r}$), which gives

$$\sin \bar{i}_L = \frac{\lambda}{2d} \qquad (95)$$

The groove spacing $d$ of the diffraction grating is limited by the condition $0.5\lambda \le d \le 1.5\lambda$ in order to ensure that only the zeroth- and first-order reflections exist.[70]

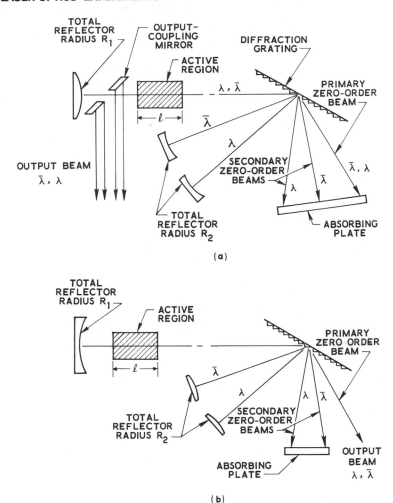

**Fig. 60.** Off-Littrow configuration applied to a confocal selected-multiline unstable resonator.[67] For an off-Littrow cavity, secondary zeroth-order beams exist at each selected wavelength in addition to a multiwavelength primary zeroth-order beam. For a confocal cavity (parallel output beam) $R_1 + R_2 = 2L$ where $L$ is the total distance between mirrors 1 and 2. (*a*) Edge-coupled cavity (annular output beam) and (*b*) continuously coupled cavity (uniform output beam).

If the incident radiation $I_{\bar{i}}$ is incident at an angle $\bar{i}$ different from $\bar{i}_L$ (off-Littrow configuration), then a first-order beam $I_{\bar{r}_1}$ will be reflected at an angle

$$\sin \bar{r}_1 = \frac{\lambda}{d} - \sin \bar{i} \qquad (96)$$

and a zeroth-order beam $I_{\bar{r}_0}$ will be reflected at an angle

$$\bar{r}_0 = -\bar{i} \tag{97}$$

If a secondary-feedback mirror is applied to $I_{\bar{r}_1}$, then a fraction will be reflected in the first order along the direction of the incident beam $\bar{i}$, a fraction will be reflected as a primary zeroth-order beam in the direction $\bar{r}_0 = -\bar{i}$, and a fraction will be reflected as a secondary zeroth-order beam in a direction $\bar{r}_0 = -\bar{r}_1$. Note that the secondary feedback mirror implies two reflections off the grating per round-trip pass. Thus, the off-Littrow configuration, applied to radiation of wavelength $\lambda$ and $\bar{\lambda}$ as shown in Fig. 60, provides two three-element resonators. These three-element cavities can be independently length tuned so that the $\lambda$ and $\bar{\lambda}$ transitions are both oscillating on line center.

Figures 60$a$ and $b$ show the off-Littrow configuration applied to an edge-coupled and a continuously coupled unstable cavity, respectively. For the edge-coupled configuration (Fig. 60$a$), the output could be annular and contains both $\lambda$ and $\bar{\lambda}$ radiation. The primary and secondary zeroth-order beams represent wasted power. The direction of the output beam is independent of the lines selected. For the continuously coupled case (Fig. 60$b$), the output is the primary zeroth-order beam, which contains $\lambda$ and $\bar{\lambda}$ radiation. The secondary zeroth-order beams of wavelengths $\lambda$ and $\bar{\lambda}$, respectively, represent wasted power for most applications. The direction of the output beam for a fixed angle of incidence $\bar{i}$ is independent of the lines selected.

For an ideal diffraction grating in the Littrow configuration, when no scattering or reflection losses are assumed, the round-trip reflectivity and transmissivity are $\eta$ and $1 - \eta$, respectively, where $\eta$ is the diffraction efficiency of the grating. In the off-Littrow configuration, on the other hand, the round-trip reflectivity is $\eta^2$, since there are two reflections per round trip (this assumes that the efficiency is independent of $\bar{i}$ over the wavelength range of interest). Table 6 is a summary of the round-trip diffraction grating reflectivities, transmissivities, and absorptivities for the off-Littrow resonators shown in Fig. 60 when an ideal grating is assumed.

**Table 6.** Comparison of Effective Reflectivities, Transmissivities, and Absorptivities of Continuously Coupled and Edge-Coupled Configurations Shown in Fig. 60

| Configuration | $\mathscr{R}$ | $\mathscr{T}$ | $\mathscr{A}$ |
|---|---|---|---|
| Edge coupled | $\eta^2$ | 0 | $1 - \eta^2$ |
| Continuously coupled | $\eta^2$ | $1 - \eta$ | $\eta(1 - \eta)$ |

Since the off-Littrow configuration forms a three-element cavity, there is more flexibility in cavity design than in the single-line Littrow configuration, which is necessarily a nonconfocal unstable cavity. As indicated in Fig. 60, both the edge-coupled and the continuously-coupled cavities can be confocal unstable (positive or negative branch), provided they satisfy the condition $R_1 + R_2 = 2L$, where $R_1$, $R_2$ are the mirror radii and $L$ is the total cavity length. In addition to producing a parallel output beam, the confocal cavity has more easily fabricated, shorter-radii mirrors for the same magnification than the nonconfocal Littrow configuration. It should be noted that it is possible to operate one wavelength in the Littrow configuration and one wavelength in the off-Littrow configuration simultaneously, as discussed by Osgood et al.[71] This configuration is of little use for an unstable resonator, however, since, in general, the magnification depends on $L$, which will be different for the Littrow resonator than the off-Littrow one. Hence, in general, the virtual centers of the unstable resonator or the radii of curvature of the output wave fronts will be different for the two wavelengths; since the two components cannot be simultaneously collimated, the effective beam quality will be poor.

Table 6 shows that, for the edge-coupled case (Fig. 60a), $\eta = 1$ gives the maximum output power. For the continuously coupled cavity (Fig. 60b), there is an optimum $\eta$ that depends on the small-signal gain $G_0l$ and the magnification $M$. The optimum can be calculated within the geometric-optics limit from the theory of Mirels and Batdorf,[52] which extends Rigrod's[57] theory to an unstable resonator. This theory relates the unstable-cavity center-line intensity to the gain and losses, assuming either homogeneous or inhomogeneous broadening and, hence determines the degree of gain saturation. The foregoing theory in conjunction with Table 6 allows calculation of the primary and secondary zeroth-order beam center-line output intensities as functions of $\eta$. These results are plotted in Fig. 62 for the case of homogeneous broadening, an $M = 1.5$ confocal unstable cavity, and two values of small-signal gain. Thus, the optimum grating efficiency for maximum output intensity $I_{out}$ decreases for increasing gain, approaching low efficiencies for sufficiently high gain. The secondary zeroth-order beam output intensity $I_{lost}$, on the other hand, is reduced for low-grating efficiencies in comparison with the primary zeroth-order beam. Also, there is an $\eta$ so that two independent laser lines with different high gain could achieve nearly optimum output and small secondary zeroth-order beam losses. Thus the continuously coupled off-Littrow configuration may be useful for a high-gain device. Similar conclusions can be drawn for an inhomogeneously broadened line.

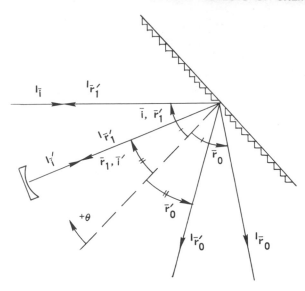

**Fig. 61.** Angles and intensities associated with a specific wavelength in the off-Littrow configuration.[67]

Preliminary experiments were conducted by Chodzko[65] on a supersonic diffusion cw HF laser with an unstable resonator in the off-Littrow configuration. Selected-multiline output was successfully shown for both edge-coupled and continuously coupled cavities. Both the primary and secondary zeroth-order beams were observed. The cavity consisted of two flat secondary-feedback mirrors, a 51-m radius convex mirror, and a 2.5-m cavity length, giving a magnification of $M = 1.55$. The diffraction grating was a water-cooled 600-line/mm interferometrically ruled master with a 2.8 mirror blaze angle. The measured efficiency at 2.795 $\mu$m ($E$ vector perpendicular to the grooves) was $\eta = 0.8$, 10% in the zeroth-order and 10% lost because of scattering and absorption. The cavity was set for simultaneous oscillation on the $P_2(5)$ and $P_1(6)$ HF transitions. From the continuously coupled cavity, 25, 3, and 33 W were measured with the $P_2(5)$ feedback mirror alone, the $P_1(6)$ mirror alone, and both the $P_1(6)$ and $P_2(5)$ feedback mirrors, respectively. The flow conditions corresponded to 1 kW of closed-cavity power as in Table 5. The observed-power enhancement with both lines oscillating indicates some cascading effect between the lines. From the measured grating losses, the power loss because of secondary zeroth-order beams was 26 W, and the power lost because of grating scattering and absorption was 60 W. The spectral content of the two-line output beam was identified with a

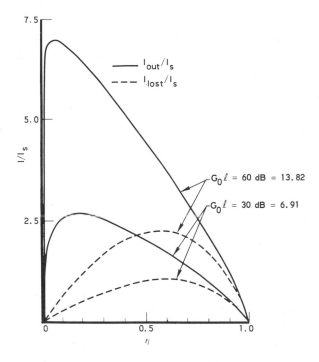

**Fig. 62.** Primary zeroth-order beam-output intensity ($I_{out}$) and secondary zeroth-order beam-loss intensity ($I_{lost}$) versus diffraction-grating efficiency for two values of small-signal gain.[67] Calculation applies to a continuously coupled cavity (Fig. 60b), is a generalization of Rigrod theory[57] applied to unstable cavities within the geometric-optics approximation, and yields the centerline intensity. A homogeneously broadened line and an $M = 1.5$ confocal unstable cavity are assumed.

monochromator as $P_2(5)$ and $P_1(6)$ at 2.795 and 2.707 $\mu$m, respectively. The near-field and far-field intensity distribution was similar to Fig. 59, indicating diffraction-limited output. The low-output power in comparison with the closed-cavity power was due in part to a lack of gain saturation, since the cavity threshold gain was nearly 4%/cm.

Figure 63 is a schematic of a parallel-staged MOPA chemical-laser configuration currently being investigated by Angelbeck et al.[66] of United Aircraft Research Laboratories. The objective is to build a phased array of cw HF/DF parallel-staged amplifiers. Parallel staging could be useful for high-power scaling since the undesirable superradiance effects that may occur in an equivalent long-gain length resonator may be partially suppressed. Furthermore, with a selected-line master oscillator, phase control can be accomplished (in principle) so that the output beams from the amplifiers will be in phase, thereby enhancing the on-axis brightness in comparison with independent beams. The master oscillator indicated in Fig. 63 consists of a three selected-line stable resonator (three vibrational levels in cascade), where a diffraction grating in the off-Littrow configuration described previously is used for line selection. An electronic feedback system, in conjunction with piezoelectric elements, drives cavity mirrors so that line-center operation is maintained on each line.

The three-line output beam from the master oscillator is split into two components (one for each amplifier) with a beam splitter. The first split beam propagates directly through the first amplifier, while the second split beam is further decomposed into the three spectral lines with a dispersive element (a diffraction grating). The individually dispersed beams are then recombined with a second diffraction grating after reflection by three flat total reflectors with piezoelectric drivers. The optical path for each wavelength of the second split beam, therefore, can be varied with respect to the first split master-oscillator beam. The recombined beam is then propagated through the second amplifier, and the two amplified beams provide the combined output.

Phase control is provided by means of a phase measurement system that determines the relative phase of the two amplified beams at each wavelength at a single point on the wave front. A relative phase difference generates an error signal and an electronic feedback loop gives a signal to the piezoelectric drivers for phase correction. To measure the relative phase, a small sample of each of the output beams is analyzed by means of an optical heterodyne technique. Thus a local oscillator signal is generated by splitting a small fraction of the beam from amplifier number two and upshifting the frequency by 40 MHz with an acousto-optic modulator (germanium substrate with a $LiNbO_3$ transducer). This local oscillator beam is then expanded and combined by means of a beam splitter with small samples of the two output beams. These two combined beams are then dispersed with a diffraction grating, and two phase-sensitive detectors for each wavelength measure the beat-frequency signals, which are related to the phase difference between the two output beams at that wavelength. The phase-sensitive detectors consist of fast InAs detectors combined with a large bandwidth ($\sim100$ MHz) electronic phase-measuring circuit. The actual frequency response of the electronic phase correction system is limited by the piezoelectric mirror mounts to $\sim100$ Hz.

It should be noted also that the two output beams must be accurately parallel for the system to perform as a phased array. The wave fronts must be parallel to some small fraction of the diffraction-limited beam-divergence angle, which depends on the beam diameter. For example, this angle is $\sim120$ $\mu$rad for $\lambda = 2.9$ $\mu$m and $D = 2.5$ cm. An autoalignment system, not indicated in Fig. 63, maintains parallel output to within 10 $\mu$rad, with a 100-Hz bandwidth. The system consists of a He–Ne laser autocollimator that follows an optical path nearly identical to the infrared beam. The He–Ne beam is divided into two paths with a visible beam splitter at the same position as the first infrared beam splitter near the exit plane of the master oscillator. The He–Ne beam is processed through the optical system up to the exit plane of the chemical laser amplifiers, where a reference flat reflects both beams back to the autocollimator. The autocollimator produces error signals that drive servo-controlled pointing mirrors until the He–Ne beams, and, hence, the two amplified cw HF/DF output beams are parallel.

Preliminary experiments have been performed by Angelbeck et al.[66] with the MOPA configuration described previously. Preliminary to the chemical-laser experiments, an argon-ion laser was applied to the electronic-feedback phase-measurement system and the improvement in relative phase stability was observed. The argon-ion laser single-line (0.5147 $\mu$m) beam was split into two optical paths, and a 3.5$\lambda$ peak-to-peak periodic disturbance was introduced into one optical path. At 80 Hz, a factor of 14 decrease in the relative path length was obtained with a closed feedback loop in comparison with open loop operation. At 20 Hz, an absolute-phase stability of $\lambda/8$ was observed, which corresponds to about $\lambda/30$ at 3 $\mu$m. Preliminary MOPA experiments were made with an arc-driven cw HF master oscillator and a combustion-driven cw HF amplifier. The nozzle in the amplifier and the oscillator was of the matrix-hole design discussed previously.[49] The amplifier was operated at flow conditions corresponding to ~800 W of closed-cavity power. The master oscillator beam made a triple pass through the 20-cm active region as indicated in Fig. 64. Table 7 shows the amplification obtained on various HF spectral lines with a 3-mm diameter Gaussian input beam. Thus a maximum of ~50 W has been extracted from the amplifier on a single line, and average gains of the order of 10%/cm were measured. Evidence of saturation effects are seen for the $P_1(4)$ transition, as the gain decreases with increasing intensity.

Further experiments will be conducted to determine if the power-extraction efficiency, presently <10% of the closed-cavity power, can be increased for some type of MOPA configuration. The problem is clearly complicated by the fact that the amplifier small-signal gain profiles for the various lines have different regions of positive gain and absorption. The oscillator-beam diameter and position relative to the nozzle-exit plane may be different to optimize the power extraction for each line. In addition, cascading effects between vibrational levels will also affect the gain profiles and need further investigation.

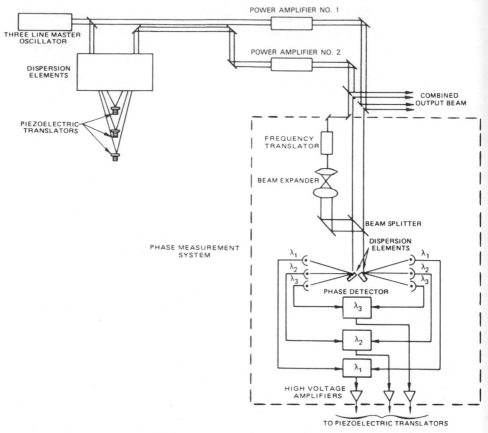

**Fig. 63.** Schematic of a parallel-staged MOPA cw chemical-laser concept.[68] The output from a three-selected line oscillator (using the off-Littrow configuration described previously) is amplified by two independent power amplifiers. The relative phase of the two amplified beams is measured by an optical-heterodyne technique and phase corrections are made such that the two output beams are in phase to yield increased far-field intensity.

In summary, several techniques are presently available for multiline selection in a chemical laser. A diffraction grating, a prism, or an internal-cavity absorption cell may be used. The diffraction grating looks attractive for high-power oscillator applications since it can be water-cooled. One multiline-selection technique using a grating is the off-Littrow configuration. The grating oriented at an angle different from the Littrow angle, a primary mirror, and a secondary feedback form a three-element cavity for each selected wavelength. The off-Littrow configuration is not as efficient as the Littrow configuration because there are two reflections off the grating per round-trip pass, which wastes laser

**Table 7.** MOPA Results[a]

| Transition | $\lambda$, $\mu$m | $P_0$, W | $P$, W | $P/P_0$ | $\langle G \rangle$, cm$^{-1}$ |
|---|---|---|---|---|---|
| $P_2(3)$ | 2.727 | 1.12 | 35 | 31.3 | 0.057 |
| $P_2(4)$ | 2.760 | 0.70 | 44 | 62.9 | 0.069 |
| $P_2(5)$ | 2.795 | 0.88 | 51 | 58.0 | 0.068 |
| $P_2(6)$ | 2.832 | 1.16 | 47 | 40.5 | 0.062 |
| $P_1(3)$ | 2.608 | 0.80 | 53 | 66.3 | 0.070 |
| $P_1(4)$ | 2.639 | 0.13 | 43 | 331 | 0.097 |
|  |  | 0.72 | 52 | 72.2 | 0.071 |
| $P_1(5)$ | 2.673 | 0.08 | 41 | 513 | 0.104 |

[a] Amplifier conditions: Total combustor gas flow—0.384 (mol/sec); $F_2 : D_2$ mole ratio $= 2.5$; combustor He mole fraction $= 80\%$; $H_2 : F_2$ mole ratio $= 5.0$. Optical configuration: 60-cm triple-pass delta; 3.0-mm beam diameter ($P_0$ is the initial power of oscillator).

energy in secondary zeroth-order beams at each selected wavelength in addition to the primary zeroth-order beam at all wavelengths. High-power coherent output from a supersonic diffusion HF/DF laser can be obtained, in principle, using the off-Littrow cavity as a stable oscillator in an oscillator–amplifier configuration or as an unstable oscillator. The off-Littrow selected-multiline unstable cavity can be output-coupled in two ways: by edge coupling or by continuous coupling. It was shown theoretically that, in the limit of high gain, both methods of coupling can efficiently extract radiation from the active medium. For the case of edge coupling, a grating of high efficiency is required ($\eta \to 1$), while for continuous coupling a low-efficiency grating with small scattering and absorption losses is sufficient. Preliminary experiments on a cw HF laser were performed that proved the feasibility of the selected-multiline off-Littrow unstable cavity. For the case of the cw HF/DF oscillator–amplifier configuration, a new technique for phase control has been developed that permits phased arrays of amplifiers for increased far-field brightness. It remains to be shown, however, that either a selected-multiline oscillator–amplifier or a selected-multiline unstable resonator can extract a large fraction of the available cw HF/DF multiline closed-cavity power.

## 3.4. Subsonic Diffusion cw HF/DF Lasers

In addition to the high-power chemical lasers described previously, there is a need for low-power sources for a variety of applications, such as small-signal gain measurements on larger cw and pulsed HF/DF devices, atmospheric-absorption measurements, aerosol-scattering measurements, chemical kinetics, and molecular-spectroscopy studies. Several investigators[47,72,73] have developed small cw HF/DF lasers. Cool et al.[72] developed a longitudinal-flow laser, with $H_2$ injected along the flow direction and F atoms provided by an rf discharge. Glaze and Linford[73] then developed a higher-gain transverse-flow device excited by an rf discharge. Hinchen[47] has recently developed a transverse flow laser where F atoms are provided by a dc discharge. The device developed by Hinchen has stable output power and is simple to construct; detailed laser-optics measurements have been made. The techniques for mode and frequency control in small lasers are well established, and we will summarize results obtained with Hinchen's laser to illustrate this general class of device.

In the Hinchen laser a mixture of $SF_6$ and He was fed into a dc discharge operated at 12 kV and 100 mA, where F atoms were formed. The F atoms flow into a narrow channel 3-mm high, and $H_2/D_2$ was injected transverse to the flow in the optical-cavity region so that HF*/DF* was formed in a narrow region of ~3-mm width along the optical axis. The narrow gain region was consistent with the typical subsonic $10^4$ cm/sec flow velocity maintained by a small 15 liter/sec vacuum pump. Further details in this device will be found in the chapter by Gross and Spencer.

The short 10-cm gain length permitted a short cavity length, so that a single longitudinal mode lay under the Doppler-broadened line. A diffraction grating (600 lines/mm for HF and 300 lines/mm for DF) was used to select a single line in a stable cavity configuration. Power was coupled out of a 2-m radius partially transmitting mirror. The cavity-mirror spacing was established by Invar rods, and the mirror was mounted on a piezoelectric crystal so that the longitudinal-mode frequency could be scanned across the individual molecular line. The small-signal gain of this device has been measured with a second cw HF probe laser of identical design. The peak-line center gain on a $P_1(4)$ HF transition was found to be about 3%/cm with a 4-mm wide positive-gain region. As indicated in Table 1, a medium Fresnel number near unity implies that most of the power could be extracted in a single transverse mode with a stable cavity. The output-power level developed on the gain of the various transitions and varied between 0.1 and 1 W on a single HF line. The beam profile of the $P_1(4)$ HF beam was measured 70 cm from the exit plane by sweeping the beam past a 0.1-mm diameter detector. The data shown in Fig. 65 have a Gaussian shape and 1.2-mm beam width (FWHM), which is near the calculated value for the $TEM_{00}$ mode.

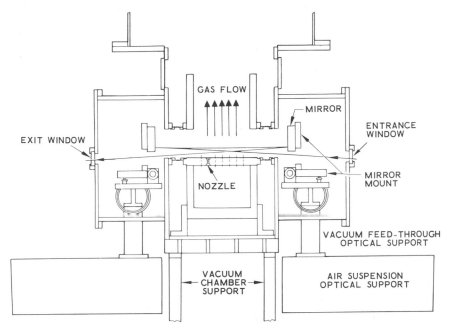

**Fig. 64.** Schematic of triple pass, single-line MOPA experimental configuration as applied to a cw HF laser.[68] The oscillator consisted of an arc-driven supersonic diffusion cw HF device with a stable resonator yielding a Gaussian input beam. The amplifier was a combustion-driven device utilizing the matrix-hole nozzle design. Care had to be taken that the temperature in the active region of the oscillator matched that of the amplifier so that there was positive gain for the various HF input transitions.

A cavity length of 28.6 cm was selected, which gives a longitudinal-mode spacing of $c/2L = 525$ MHz, significantly greater than the Doppler-line width of 380 MHz calculated from the measured-gas temperature of 475°K. Figure 66 shows the $P_1(4)$ HF power output as the piezoelectric drive changed the cavity length through successive longitudinal modes; the vanishing power for certain cavity lengths confirms the single longitudinal-mode operation.

A spectral analysis of the intensity fluctuations up to 700 MHz was also carried out. As indicated in Fig. 67, a beat frequency at 60 MHz was observed. This beat frequency was eliminated with a 3-mm intracavity aperture, and no other fluctuations up to 700 MHz were observed. Thus it is concluded that the laser cavity is oscillating on a single-transverse and single-longitudinal mode. The amplitude stability of the output power of the HF laser was also measured, and a short-term stability of ±1.5% was measured at center frequency. Thus the measurements show that Hinchen[47] has developed a very simple, small, single frequency, stable cw

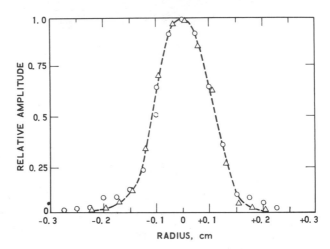

○  EXPERIMENTAL, NO INTERNAL
   APERTURE

— — —  GAUSSIAN BEAM THEORY

△  EXPERIMENTAL, WITH
   INTERNAL APERTURE

**Fig. 65.** Near-field intensity versus radius for small, single-line, single-frequency cw HF/DF laser.[47] For this dc discharge subsonic flow device, the output beam closely matches a $TEM_{00}$ Gaussian with a stable cavity.

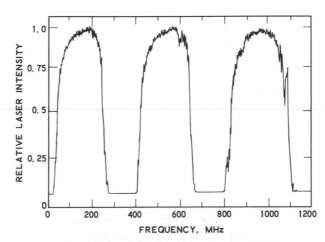

**Fig. 66.** Output power versus longitudinal-mode frequency for a small, single-line, single-frequency cw HF/DF laser.[47] Frequency variation was obtained by varying the length of the stable cavity with a piezoelectric drive on the mirror mount.

**196**

HF/DF probe laser, which will have a variety of applications as a laboratory diagnostic tool.

## 4. SUMMARY AND CONCLUSIONS

This chapter has dealt in some detail with a variety of theoretical and experimental aspects of laser-mode control. Here we intend to place this material in perspective and, where possible, to draw more general conclusions based on our present state of knowledge.

We began by discussing the physical principles underlying laser-mode control. Laser-beam divergence and laser focal-spot size, both equivalent descriptions of the laser-beam quality, can be made close to the ideal case permitted by the laws of diffraction if the optical phase can be maintained constant across the laser-beam front. The optical resonator, or cavity, performs this phase control through the use of diffracted radiation to phase-couple all of the individual radiating molecules.

At low Fresnel numbers, diffraction can distribute laser radiation throughout the laser medium. Under these circumstances, a stable resonator, which uses concave mirrors to help confine off-axis cavity radia-

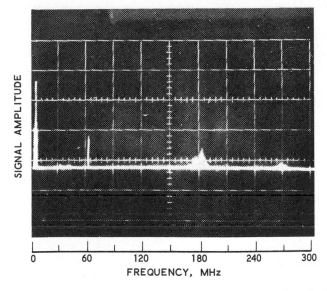

**Fig. 67.** Spectral analysis of intensity fluctuations for a small, single-line, single-frequency cw HF/DF laser.[47] This data corresponds to the circle data points in Fig. 65 where there is no internal-cavity aperture. Through the application of an internal-aperture mode, beats could be eliminated up to 300 MHz, indicating single transverse-mode operation.

tion, produces a diffraction-limited, phase-coherent output beam. The radiation within the cavity corresponds to the lowest-loss solution, or mode, of those electromagnetic field patterns that are allowed by the geometry of the mirrors used.

At the larger Fresnel numbers encountered with high-power lasers, and especially when the optical gain of the laser is high, diffraction alone is not generally strong enough to phase-lock all parts of the laser medium together. In this case, an unstable resonator, with convex mirrors, can be used. Instead of confining the radiation near the optical axis, the unstable resonator purposely magnifies it on every pass through the medium, until the radiation originally emitted near the center of the resonator fills the entire laser medium. In this way, the geometry of the cavity is used to phase-couple all parts of the laser medium together, and a phase-coherent output beam can once again be obtained.

A geometric optics analysis in Section 2.1 allowed us to predict the main features of unstable resonator performance: wave-front patterns, output-coupling fraction, beam diameter, and degree of beam collimation. However, since geometric optics neglects diffraction, it cannot predict under which conditions a phase-coherent diffraction-limited output beam will be obtained. In Section 2.2 we examined the Fresnel–Kirchhoff integral equations describing radiation within the unstable resonator, from which are obtained the eigenmodes (possible radiation patterns) and eigenvalues (output coupling and oscillating frequency) of the radiation field.

The various modes were seen to have losses that vary periodically as a function of a newly defined quantity, the equivalent Fresnel number $N_{eq}$. Near integral values of $N_{eq}$, there are generally two modes with equally low losses. To avoid the loss of phase coherence generally obtained when several different transverse modes simultaneously oscillate, it is deemed preferable to operate instead at half-integral values of $N_{eq}$, where there is a single mode of lowest loss that can hopefully dominate the output. The various numerical methods available for calculating the eigenmodes are exemplified by the modified Gauss–Legendre integration method discussed in the text.

As discussed in Section 2.3, the complex behavior of the unstable-resonator modes as a function of cavity parameters turns out to be due to the edge wave, radiation-diffracted back into the cavity from the edges of the output-coupling mirror. Anan'ev's exact solution for a cavity whose output mirror has a Gaussian reflectivity profile not only illustrates what happens when the edge wave is suppressed by smoothing the edge, but also provides explicit formulas for the exact coupling losses and oscillating frequencies for this various transverse modes of a particular unstable-resonator geometry. Further analyses of the edge wave from other types of output mirror, with changes in reflectivity, curvature, or boundary shape near the edge of the mirror, show that almost any departure from a sharp-edged circular output mirror can be adequate, particularly at large Fresnel numbers, to assure single-mode operation at almost any value of $N_{eq}$. In particular, square or rectangular resonator mirrors suffer much less from potential multimode operation caused by the edge wave, because their edges cross many Fresnel zones of the cavity mirror, and edge contributions from different parts of the mirror boundary, therefore, tend to cancel one another within the resonator.

Not only these intentional mirror adjustments affect the performance of the unstable resonator. Section 2.4 described how small amounts of mirror misalignment in an unstable resonator cavity can lead to significant beam steering in the output. Thus mirror vibration can cause a blurring of the far-field or focal-spot pattern. Exact solutions were given for the unstable resonator, with Gaussian reflectivity and typical numerical calculations presented for conventional unstable resonators. An empirical rule was given for estimating the maximum amount of mirror misalignment permitted if the resonator is to operate single mode, with relatively good output-beam quality.

The complications introduced by the laser-gain medium itself are more troublesome and were discussed in Section 2.5. Results were given for two simplified models of the cw chemical-laser operation: Mirels' flame-sheet mixing model, coupled with a geometric-optics cavity model, and Bullock et al.'s turbulent diffusion mixing model, coupled with multilevel chemical kinetics and optical-diffraction effects. In both cases, the intensity distribution of radiation within the resonator was shown to be strongly dependent on the details of the laser-medium gain distribution, and no longer determined solely by the resonator itself. The more detailed model by Bullock et al. yields gain profiles and near-field intensity patterns in reasonable agreement with experiment, and predicts some gain and cavity conditions under which multiple transverse-mode operation might be encountered.

Using this theoretical material as a basis for understanding, we then examined the experimental-mode control work thus far performed on chemical-laser devices. We began in Section 3.2 by discussing the cw HF and DF supersonic diffusion lasers, operated without any attempt to control the output wavelengths. The closed-cavity technique of measuring maximum-laser output power was described, in which all of the laser power is absorbed in the cavity mirrors and measured calorimetrically. The VACAS device, which permits the mode volume of an unstable resonator to be matched to that of the laser medium, yields an optimized output power of 11.0 kW compared with 12.0 kW for the closed-cavity case, indicating that the optimized unstable cavity was capable of extracting over 90% of the available laser power.

The far-field intensity pattern obtained from such a chemical laser without wavelength selection generally has a diffraction-limited central lobe containing between 50 and 80% of the theoretical power and several side lobes. The beam divergence may be characterized as generally about twice the diffraction limit, in terms of the far-field angle, including a given fraction of the beam power as in Figs. 52, 53, and 55 of Section 3.

Wavelength selection was added to the basic unstable-resonator cavity by the replacement of one of the cavity mirrors with a diffraction grating, as described in Section 3.3. By using the grating in the continuously coupled resonator geometry, Chodzko et al. demonstrated essentially diffraction-limited single-line output from the cw HF laser, with output approximately equal to 10% of the closed-cavity power. The resonator was changed to permit simultaneous oscillation ón several selected spectral lines, but because of cavity losses improved total output power was not obtained. An alternate approach to efficient power extraction with wavelength control is the parallel-staged oscillator–amplifier configuration studied by Angelbeck et al. However, at present, the multiline power extracted using this technique is still less even than the 10% achieved with Chodzko's resonator.

We may summarize these experimental results as follows. It appears that the unstable resonator is generally quite effective in extracting energy from the HF/DF laser medium. However, the output beams from present devices, probably because of the nonuniformity and saturation characteristics of the gain medium, tend to have a beam divergence about twice that of an ideal diffraction-limited beam. If wavelength control is introduced into the resonator, beam qualities seem to improve considerably, but only at the sacrifice of some 90% of the laser-output power. Multiple-wavelength optical configurations have been developed to increase the wavelength-controlled output available, but so far these have not achieved higher extraction efficiencies. Finally, Section 3.4 described mode-control considerations in low-power, low-Fresnel number chemical lasers.

In general, we may summarize our state of knowledge as follows. Most work to date on optical control of chemical lasers has involved the use of unstable resonators to obtain good beam quality from high-power devices. Unstable resonators themselves are well understood, but a flowing, mixing chemical-laser medium adds complications and distortions to the laser-radiation field, which at present are beyond our capability to model in detail. Nevertheless, significant advances have been made in approximate computer models of cw chemical-laser devices. Far too little experimental work has yet been reported, and at this time we do not know how to obtain simultaneously full output power, diffraction-limited beam quality, and wavelength selection on any high-power chemical laser.

A useful way to bridge the gap between theory and experiment, in order to learn how to improve present devices, might be to use measured values of small-signal gain to check kinetics-mixing codes (as in Fig. 31), and measured values of saturated gain during lasing as input to cavity-mode calculations to check numerical-resonator models. Up to this time, the lack of this sort of integrated experimental-theoretical effort in the chemical-laser field has been mainly responsible for our limited progress in achieving full optical control of these devices. However, considerable work is still going on, and it can be expected that new results will be forthcoming even before these words reach print.

# REFERENCES

1. For a general discussion of laser resonators, see, for example: (a) H. Kogelnik, "Modes in Optical Resonators," in *Lasers, Vol. 1*, A. K. Levine, Ed. (Dekker, New York, 1966), pp. 295 347; (b) A. Maitland and M. II. Dunn, *Laser Physics* (North-Holland, Amsterdam, 1969), pp. 93–152.

2. E. A. J. Marcatili and R. A. Schmeltzer, Bell Syst. Tech. J. **43**, 1783 (1964).

3. H. Steffen and F. K. Kneubuhl, Phys. Lett. **A27,** 612 (1968); P. Schwaller, H. Steffen, J. F. Moser, and F. K. Kneubuhl, Appl. Opt. **6,** 827 (1967).

4. P. W. Smith, Appl. Phys. Lett. **19,** 132 (1971).

5. R. L. Abrams and W. B. Bridges, IEEE J. Quant. Electron. **QE-9,** 940 (1973).

6. G. D. Boyd and H. Kogelnik, Bell Syst. Tech. J. **41,** 1347 (1962).

7. A. G. Fox and T. Li, Proc. IEEE **51,** 80 (1963).

8. See, for example, M. Born and E. Wolf, *Principles of Optics*, third ed. (Pergamon, New York, 1965), p. 396.

9. M. Born and E. Wolf, *Principles of Optics*, third ed. (Pergamon, New York, 1965), p. 380.

10. G. D. Boyd and J. P. Gordon, Bell Syst. Tech. J. **40,** 489 (1961).

11. E. Merzbacher, *Quantum Mechanics* (Wiley, New York, 1961), p. 57.

12. L. Pauling and E. B. Wilson, Jr., *Introduction to Quantum Mechanics* (McGraw-Hill, New York, 1935), p. 75.

13. K. H. Wrolstad, P. V. Avizonis, and D. A. Holmes, J. Phys. **E-4,** 143 (1971).

14. R. B. Chesler and D. Maydan, J. Appl. Phys. **43,** 2254 (1972).

15. The concave–convex mirror configuration was used at large Fresnel numbers in a ring-resonator geometry by G. M. Janney, IEEE International Electron Devices Meetings, Washington, D.C., October 1970 and October 1971.

16. A. E. Siegman, Proc. IEEE **53,** 277 (1965).

17. M. Born and E. Wolf, *Principles of Optics*, third ed. (Pergamon, New York, 1965), pp. 416–417.

18. W. N. Peters and A. M. Ledger, Appl. Opt. **9,** 1435 (1970).

19. R. A. Chodzko, H. Mirels, F. S. Roehrs, and R. J. Pederson, IEEE J. Quant. Electron. **QE-9,** 523 (1973). Diffraction-grating coupled unstable resonators, both ring and linear, were first considered in detail by G. M. Janney in "Advanced Mode Control and High Power Optics Technology Program, Interim Technical Report, Contract F29601-71-C0101 (Hughes Research Laboratories, Malibu, California, 12 Nov. 1971), pp. 106–110.

20. W. F. Krupke and W. R. Sooy, IEEE J. Quant. Electron. **QE-5,** 575 (1969).

21. D. Bullock, TRW Systems Group, private communication.

22. Yu. A. Anan'ev, G. N. Venokurov, L. V. Koval'chuck, N. A. Sventsitskaya, and V. E. Sherstobitov, Sov. Phys. JETP **31,** 420 (1970).

23. A. G. Fox and T. Li, Bell Syst. Tech. J. **40,** 453 (1961).

24. A. E. Siegman and H. Y. Miller, Appl. Opt. **9,** 2729 (1970).

25. P. Horwitz, J. Opt. Soc. Amer. **63,** 1528 (1973).

26. A. E. Siegman and R. W. Arrathoon, IEEE J. Quantum Electron. **QE-3,** 156 (1967).

27. Yu. A. Anan'ev and V. E. Sherstobitov, Sov. J. Quantum Electron. **QE-3,** 156 (1967).

28. R. L. Sanderson and W. Streifer, Appl. Opt. **8,** 2129 (1969).

29. H. Kogelnik and T. Li, Proc. IEEE **54,** No. 10, 1312 (1966).

30. D. B. Rensch and A. N. Chester, Appl. Opt. **12,** 997 (1973).

31. R. L. Sanderson and W. Streifer, Appl. Opt. **8,** 131 (1969).

32. D. L. Bullock, R. J. Wagner, and R. S. Lipkis, "New Unstable Resonator Program," Report 00173-2-006447, TRW Systems Group (Los Angeles, California, 30 Nov. 1973).

33. L. Bergstein, Appl. Opt. **7,** 495 (1968).

34. W. Streifer, IEEE J. Quant. Electron. **QE-4,** 229 (1968).

35. H. Zucker, Bell Syst. Tech. J. **49,** 2349 (1970).

36. A. N. Chester, Appl. Opt. **11,** 2584 (1972).

37. A. N. Chester, IEEE J. Quant. Electron. **QE-9,** 209 (1973).

38. V. E. Sherstobitov and G. N. Vinokurov, Sov. J. Quant. Electron. **2,** 224 (1972).

39. G. L. McAllister, W. H. Steirer, and W. B. Lacina, IEEE J. Quant. Electron. **QE-10,** 346 (1974).

40. Yu. A. Anan'ev, Sov. J. Quant. Electron. **1,** 565 (1972).

41. A. E. Siegman, Appl. Opt. **13,** 353 (1974).

42. R. L. Sanderson and W. Streifer, Appl. Opt. **8,** 2241 (1962).

43. A. G. Fox and T. Li, IEEE J. Quant. Electron. **QE-2,** 774 (1966).

44. H. Statz and C. L. Tang, J. Appl. Phys. **36,** 1816 (1965).

45. H. Mirels, AIAA Journal, **13,** 785 (1975).

46. R. A. Chodzko, D. J. Spencer, and H. Mirels, IEEE J. Quant. Electron. **QE-9,** 550 (1973).

47. J. J. Hinchen, J. Appl. Phys. **45,** 1818 (1974).

48. T. A. Cool, IEEE J. Quant. Electron. **QE-1,** 72 (1973).

49. G. R. Wisner, G. Palma, and M. Foster, "Laser Beam Quality Study Report," Report

UARL M911239-23 (United Aircraft Research Laboratories, East Hartford, Connecticut, Sept. 1973).

50.  D. J. Spencer, H. Mirels, and D. A. Durran, J. Appl. Phys. **43**, 1151 (1972).

51.  J. M. Herbelin and G. Emanuel, J. Chem. Phys. **60**, 689 (1974).

52.  H. Mirels and S. B. Batdorf, Appl. Opt. **11**, 2384 (1972).

53.  D. J. Spencer, D. A. Durran, and H. A. Bixler, Appl. Phys. Lett. **20**, 164 (1972).

54.  D. N. Mansell, J. A. Love, and W. L. Snell, IEEE J. Quant. Electron. **QE-7**, 177 (1971).

55.  R. A. Chodzko, The Aerospace Corp., unpublished work.

56.  D. Hook, TRW Systems Group, private communication.

57.  W. W. Rigrod, J. Appl. Phys. **36**, 2487 (1965).

58.  A. Yariv, *Quantum Electronics* (Wiley, New York, 1968), pp. 263–267.

59.  R. A. Chodzko, S. Mason, R. R. Giedt, and D. A. Durran, "Variable Aperture Calorimeter for an Unstable Resonator," to be published in Appl. Opt.

60.  R. A. Chodzko, E. J. Cross, R. R. Giedt, M. A. Kwok, and D. H. Ross, "Experimental Studies of a cw HF/DF Confocal Unstable Resonator," Report TR-0076(6605)-1 (The Aerospace Corp., Los Angeles, California, May 1976).

61.  R. Wey and G. R. Wisner, "An Integrated Irradiance Analyzer for Real Time Laser Beam Diagnostics," Report UARL148 (United Aircraft Research Laboratories, East Hartford, Connecticut, Oct. 1972).

62.  D. J. Spencer, The Aerospace Corp., private communication.

63.  A. E. Siegman, "Stabilizing Output with Unstable Resonators," *Laser Focus* (May 1971), pp. 27–42.

64.  R. J. Freiberg, P. P. Chenausky, and C. J. Buczek, IEEE J. Quant. Electron. **QE-8**, 882 (1972).

65.  R. A. Chodzko, Appl. Opt. **13**, 2321 (1974).

66.  A. W. Angelbeck, L. R. Boedeker, R. O. Decker, M. C. Foster, E. Hasselmark, N. L. Krascella, L. Lynds, S. N. Mapes, D. G. McMahon, R. A. Meizner, G. E. Palma, T. J. Sadowski, E. W. Vinje, G. R. Wisner, and T. B. Milan, "Investigations of Master Oscillator Power Amplifier Systems for Chemical Lasers," TR-73-156 (United Aircraft Research Laboratories, East Hartford, Connecticut, April 1974).

67.  M. L. Bhaumik, Appl. Phys. Lett. **20**, 342 (1972).

68.  F. G. Gebhardt and D. C. Smith, Appl. Phys. Lett. **14**, 52 (1969).

69.  R. S. Longhurst, *Geometrical and Physical Optics* (Wiley, New York, 1967), pp. 242–243.

70.  T. M. Hard, Appl. Opt. **9**, 1825 (1970).

71.  R. M. Osgood, P. B. Sackett, and A. Javan, J. Chem. Phys. **60**, 1464 (1974).

72.  T. A. Cool, R. R. Stephens, and J. A. Shirley, J. Appl. Phys. **41**, 4038 (1970).

73.  J. A. Glaze and G. J. Linford, Rev. Sci. Instrum. **44**, 600 (1973).

74.  W. D. Adams, E. B. Turner, J. F. Holt, D. G. Sutton, and H. Mirels, "The Resale Chemical Laser Computer Program," Report TR-0075 (5530)-5 (The Aerospace Corp., Los Angeles, California, February 1975).

CHAPTER **4**

# CONTINUOUS-WAVE HYDROGEN–HALIDE LASERS

## R. W. F. GROSS

## D. J. SPENCER

**The Aerospace Corporation**
**El Segundo, California**

The goal of all chemical lasers is the efficient conversion of chemical energy into coherent radiation of high power, if possible with a negligible addition of nonchemical energy. High power calls for reactions with large energy release, and good conversion efficiencies require that a large fraction of this energy be channeled through optically excited states of the reaction product. The reaction of fluorine with hydrogen has such features: the heat of reaction is large, 14.3 kJ/g of fluorine, and ~60% of this exothermicity appears initially in vibrational excitation of the HF molecules formed, the remainder being lost in rotation and translation. It is, therefore, quite easy to obtain vibrational inversions of HF by this reaction, as the early discovery of the HF laser shows.

Unfortunately, however, the favorable initial energy distribution in HF is rapidly destroyed by unusually fast collisional-deactivation processes, which result in short lifetimes of the excited molecules and in poor efficiencies. Rapid deactivation of the higher-vibrational states of HF is a particular problem, since two thirds of the exothermicity of the $H_2$–$F_2$ chain reaction is deposited by the "hot" branch into these upper excited states. Deactivation in the form of V–R, T collisions is also responsible for the rapid heating of the laser medium. Since cw HF lasers operate in general on partial V–R inversions, heating of the laser medium results in a reduction of the gain and the spatial extent of the medium.

Efficient conversion of the reaction exothermicity into coherent radiation requires the judicious control of all contributing factors. The production rates should be fast compared to deactivation. The temperature of the active medium has to be kept low, and the photon flux in the cavity high and spatially uniform. These factors are especially important in the design of efficient high-power lasers.

Simultaneous control of all parameters is difficult to achieve for pulse lasers but relatively easy for continuous lasers. In the flowing medium of a continuous laser, gas-dynamical techniques can be used to optimize the production of excited states, to cool the gas, and finally to remove the expended molecules and the waste heat from the cavity. This is why efficient extraction of high powers from chemical reactions was first obtained with continuous, gas-dynamically controlled HF lasers. Gas-dynamical techniques also make it possible to initiate the HF reaction by thermal dissociation of fluorine, a process that is the key to purely chemical lasers of very high power.

As early as 1967, a number of people began to consider gas-dynamical control of chemical lasers.[1–4] It was realized that, if the reaction of hydrogen with chlorine[4] or fluorine[2,3] could be made to take place at constant pressure and in a supersonic flow, one should be able to control its explosive nature, to produce a suitable laser medium, and to convert a sizable fraction of the exothermicity of the reaction into coherent radiation. It was also realized that it would be important to carefully define the initial concentration of F atoms. A supersonic free jet appeared to offer the desired conditions, since it combined a fast flow with the possibility of expansion necessary for constant pressure operation. The high-pressure high-temperature stagnation conditions that would produce this supersonic jet could also be used to produce the necessary F atoms by thermal dissociation.

The most important problem was to mix the reactants fast enough and without prereaction. Previous supersonic-combustion work had two solutions to offer[4]: (1) a free-jet "flame" induced by a standing shock and (2) a supersonic-diffusion "flame." Conceptually, the first technique seemed to have a number of advantages: The gases could be premixed, the start of the reaction would be spatially well defined by the shock, and shock-tube experiments had successfully demonstrated lasing behind shock waves.[2] The second technique required a very fast mixing process right in the active medium, which, it was feared, would create severe optical disturbances. Shock-induced concepts were, therefore, tried first, but for various reasons were never successful. In 1969 two independent groups[3,5] almost simultaneously observed lasing from a supersonic-diffusion flame, and this concept eventually led to the high-power high-efficiency HF/DF lasers of today.

Simultaneously, a different approach was taken by Cool et al.,[6] who used the fast-energy transfer between DF and $CO_2$ to extract the energy from chemically produced DF. Since the deactivation of $CO_2$ is slow by comparison with DF, and the $CO_2$ transition does not terminate on the ground-state, gas-dynamical, and kinetic requirements for this laser are less stringent. Following the first observation of continuous lasing in HF, several other successful hydrogen–halide cw lasers were discovered. The reactions producing the excited hydrogen–halide molecules in all these lasers are similar; in all cases except one, halogen atoms react with hydrogen molecules. The process by which the halogen atoms are produced differs and is the main distinction between the various lasers. The halogen-dissociation technique to a large degree also defines the required flow system. We shall, therefore, arrange this review according to the different techniques of halogen-atom generation.

Thermal dissociation of various fluorine compounds in combination with a fast adiabatic expansion to supersonic speeds has proved the most successful approach. The reason is that thermal dissociation under high-temperature equilibrium conditions permits the production of large concentrations of F atoms in the most efficient way. A number of different thermal drivers have been used, arc heaters, shock tubes, regenerative and resistance heaters, combustors, and finally chemical reactions. The parameters of these lasers have been collected in Table 1.

## Table 1

| Name, Year, Reference | Reaction, Main Flow, Device | Maximum Flow (g/sec) | Maximum Power (W) | Mirror Size, Cavity Type, Output Mode | Nozzle Dimensions (cm) | Chemical Efficiency (%) | Specific Power (W/g/sec) | Power Input (kW) | Pressure Cavity (Torr) | Pressure Plenum (atm) | Plenum Temp. (K) |
|---|---|---|---|---|---|---|---|---|---|---|---|
| Gross et al., 1969, (2) | $F_2 + H_2$, $F_2O + H_2 + Ar$, Shock tube detonation wave laser | (Shock speed 1.8-2.0 mm/μsec) | 10 (peak power) | 2.5 cm ∅, Stable, Single hole transverse | 2.5 × 17.6 (cavity) | | | None | 18 ($p_2$) | | 2000 ($T_2$) |
| Airey and McKay, 1969, (3) | $F + HCl$, $F_2 + Ar$, Supersonic shock tube driven | | $8 \times 10^{-4}$ | 1.0 cm ∅, Stable, Single hole transverse | 1.0 × 15 (1 slit) | | | None | 8-19 | 0.7-2.3 | 1050 |
| Spencer et al., 1969, (5) | $F + H_2$, $SF_6 + N_2$, Supersonic arc heater | $N_2 = 15$, $H_2 = 2$, $SF_6 = 0.54$ | 1 | 4.5 cm ∅, Stable, Single hole transverse | 0.95 × 17.6 (1 slit) | | 0.057 | 20 | 10 | 3 | 2000 |
| 1970, (15) | $F + H_2$, $D_2$, $SF_6 + N_2$, Supersonic arc heater | $N_2 = 7.4$, $H_2 = 1.0$, $D_2 = 1.6$, $SF_2 = 1.7$ | HF: 600 DF: 450 | 4.5 cm ∅, Stable, Multi-hole transverse | 0.95 × 17.6 (36 slits) | 12 | 60 | 30 | 4.7 | 1.7 | 2300 |
| 1973, (23) | $F + H_2$, $D_2$, $SF_6 + He + O_2$, Supersonic arc heater | $He = 6.9$, $H_2 = 1.0$, $O_2 = 0.9$, $SF_6 = 3.3$ | 2 000 | 9.6 × 9.6 cm, Stable, Closed cavity transverse | 1.27 × 17.6 (36 slits) | 9.8 | 243 | 35-50 | 4.5 | 1.3 | 2000 |

| Reference | System/Method | Flow rates | | Cavity | Nozzle | | | | | | Power |
|---|---|---|---|---|---|---|---|---|---|---|---|
| Giedt 1973, (31) | F + H₂, D₂, F₂ + He, Supersonic arc heater | He = 20, H₂ = 10, F₂ = 15 | 15 500 | 9.6 × 18.3 cm, Stable, Closed cavity transverse | 1.9 × 22.9 (50 slits) | 14.7 | 344 | 312 | 2.7 | 6.5 | 1340 |
| Morsell 1971, (32) | F + H₂, F₂ + He, Supersonic resistive heater | He = 2.0, H₂ = 0.33, F₂ = 1.3 | 148 | 5.0 cm Ø, Stable, Germanium flat transverse | 1 × 15 (1 slit) | 2.5 | 40 | 2.4 | 4.1 | 0.30 | 2020 |
| Meinzer et al., 1970, (38) | F + D₂, H₂, F₂ + He, Supersonic combustion heater | F₂ = 0.38, D₂ = 0.04, H₂ = 0.02 (combustor) | 10 | Stable, Single hole transverse | 1 × 10 (1 slit) | | | None | 3.8 | 5 | 2200 |
| Shirley et al., 1971, (36) | F + H₂, D₂, F₂ + He + NO, Subsonic driven by chemical reaction | He = 0.2-0.6, H₂ = 0.01, F₂ = 0.76, NO = 0.03 | 7.2 | Stable, Germanium flat transverse | 1 × 15 (1 slit) | 0.5 | | None | 8.0 | | 400-700 |
| Blaszuk et al., 1973, (39) | F + D₂, H₂, F₂ + He, Supersonic combustion heater | 37 (total) | HF: 3 000 | Stable, Silicon flat, transverse | 1.05 × 45.5 (matrix of 3600 holes) | | 162 | None | 7.5 | 1.1 | 1300 |
| | | | DF: 1 740 | Unstable, transverse | 1.05 × 45.5 (matrix of 3600 holes) | | 52 | None | 8.2 | 1.5 | 1300 |

**Table 2**

| Name, Year, Reference | Reaction, Device, Main Flow | Discharge Type | Maximum Power (mW) | Laser Transitions | Coupling Mode | Pump Capacity (ℓ/sec) | Cavity Pressure (Torr) | Flow Rates (mmol/sec) | Channel Cross Section (cm) | Remarks |
|---|---|---|---|---|---|---|---|---|---|---|
| Naegeli and Ultee, 1970, (43) | Cl + HI → HCl, Subsonic mixing laser, $Cl_2$ + He | High voltage + microwave 60 Hz + 2450 MHz | 0.47 | 2-1 | Single hole | 235 | 2 | Not measured | 0.5 × 15 | |
| Cool et al., 1970, (41) | F + $H_2$ → HF<br>H + $F_2$ → HF<br>F + HI → HF<br>F + $D_2$ → DF<br>D + $F_2$ → DF<br>F + DI → DF<br>H + $Cl_2$ → HCl<br>Cl + HI → HCl<br>H + HBr → HBr<br>Br + HI → HBr,<br>Near sonic longitudinal mixing laser,<br>$F_2$, $Cl_2$, $Br_2$<br>$H_2$, $D_2$, He | 27 MHz rf discharge | 53<br>50<br>4<br>83<br>80<br>2<br>1<br>12<br>None<br>None | 2-1<br>2-1<br>3-2, 2-1<br>3-2, 2-1<br>3-2<br>3-2<br>1-0<br>2-1, 1-0<br>---<br>--- | 5%<br>5%<br>0.5%<br>2%<br>2%<br>0.5%<br>0.5%<br>--<br>--<br>couplers | 200 | 24<br>27<br>17<br>21<br>32<br>19<br>45<br>21<br>--<br>-- | He   $H_2$, $D_2$   $F_2$, $Cl_2$<br>18   0.18   0.10<br>21   0.28   0.22<br>11   0.06   0.11<br>16   0.18   0.17<br>25   0.37   0.45<br>14   0.12   0.13<br>37   2.00   0.28<br>18   0.05   0.04<br>----   ----<br>----   ---- | | Narrow bore tube, cavity axis along flow axis, very detailed study |
| Hinchen and Banas, 1970, (51) | F + $H_2$ → HF, F + $D_2$ → DF, Subsonic mixing laser, $SF_6$ + $N_2$, He | High voltage dc discharge, 1500-6000 W | HF: 5 500<br>DF: 3 000 | 2-1, 1-0<br>3-2 to 1-0 | Single hole | 235 | 5.5 | He = 10<br>$N_2$ = 20<br>$H_2$ = 3<br>$SF_6$ = 3-20 | 1.25 × 30 | |
| Buczek et al., 1970, (47) | F + $H_2$ → HF, F + $D_2$ → DF, Subsonic premixed, $SF_6$ + $H_2$, $D_2$ + He | High voltage dc discharge, transverse to flow, magnetically stabilized | HF: 800<br>DF: 30 | 2-1, 1-0<br>3-2, 2-1 | 20% coupler<br>2.5% coupler | >300 | 8.5 | He = 13.6<br>$H_2$ = 2.6<br>$D_2$ = 2.6<br>$SF_6$ = 8.5 | 1.6 × 3.0 | |
| Glaze et al., 1971, (44) | Cl + HI → HCl, Near sonic mixing laser, $F_2$ + He | 21 MHz rf discharge, 50-100 W | 70 | 2-1, 1-0 | 4.8% $MgF_2$ flat coupler | 41 | 3.0 | He = 3.0<br>$Cl_2$ = 0.2<br>HI = 0.25 | 0.5 × 3.0 | Multi-slit nozzle, gain measurements |

| Reference | Reaction system | Excitation | Output | Transitions | Coupler | | | Gas composition | Dimensions | Comments |
|---|---|---|---|---|---|---|---|---|---|---|
| Glaze, 1971, (50) | F + H₂ → HF, F + D₂ → DF, Near sonic mixing laser, F₂ + He | 21 MHz rf discharge, 50–100 W | HF: 300  DF: 120 | 2-1, 1-0  3-2 | 7% MgF₂ flat coupler  5% MgF₂ flat coupler | 41 | 1.0 | He = 2.0, H₂ = 0.1, D₂ = 0.1, F₂ = 0.3 | 0.5 × 3.0 | Multi-slit nozzle, detailed gain measurements |
| Stephen and Cool, 1971, (49) | F + H₂ → HF, F + D₂ → DF, Near sonic mixing laser, SF₆ + He | 27 MHz rf discharge, 2500 W | HF: 1 500  DF: 500 | 2-1, 1-0  3-2 to 1-0 | 5% coupler  2% coupler | 180  180 | 9.8  4.7 | He = 11, H₂ = 0.78, SF₆ = 2.1  He = 11, D₂ = 1.4, SF₆ = 0.26 | 3.0 × 4.5 | Single slit, sonic wall injection |
| Rosen et al., 1973, (42) | F + H₂ → HF, F + HF → HF, F + CH₄ → HF, F + HBr → HF, F + HCl → HF, Cl + HI → HCl, Subsonic mixing laser, F₂, Cl₂ + He | 2450 MHz microwave discharge, 1000 W | 7 500, 2 300, 1 500, 2 000, 1 700, 1 650, 100 | 3-2 to 1-0, 5-4 to 1-0, 3-2 to 1-0, 4-3 to 1-0, 3-2 to 1-0, 4-3 to 1-0, 2-1 to 1-0 | 10%, 10%, 5%, 5%, 2%, 5%, 0.5% couplers | 840 | 6.0 | He  H₂,HI  F₂,Cl₂ / 64 2.9 0.60 / 67 0.39 0.59 / 67 0.89 0.49 / 70 0.34 0.49 / 70 0.53 0.86 / 56 0.73 2.3 / 70 1.5 2.1 | 0.7 × 10.0 | Multi-tube injector, detailed spectroscopy of all reactions, chemiluminescence studied |
| Glaze and Lindford, 1973, (45) | F + H₂ → HF, Subsonic mixing laser, F₂ + He | 8 MHz rf discharge, 1300 W | 10 000 | 3-2 to 1-0 | 25% MgF₂ flat coupler | 94.6 | 3.0 | He = 20, H₂ = 2, F₂ = 1 | 1.0 × 14.0 | Multi-slit nozzle |
| Hinchen, 1974, (46) | F + H₂ → HF, F + D₂ → DF, Subsonic mixing laser, SF₆ + He | High voltage dc discharge, 1300 W | HF: 1 300  DF: | 2-1, 1-0  3-2 to 1-0 | 25% coupler  Single hole | 14  14 | 10–15  10–15 | He = 5.8, H₂ = 0.8, O₂ = 0.5, SF₆ = 2.4  He = 2.5, D₂ = 1.0, SF₆ = 3.0 | 0.3 × 10.0 | Very stable single mode output, gain line measurements |
| Proch et al., 1974, (52) | F + H₂ → HF, Subsonic mixing laser, SF₆ + Ar | High voltage dc discharge, 2 × 3.6 kW | 40 000 | 2-1, 1-0 | 10% Dielectric coupler | 1250 | 2.7 | Ar = 89.0, SF₆ = 9.0, H₂ = 31.0 | 1.5 × 40 | Tunable by varying temperature of medium |

A second class of cw lasers uses electrical discharges to produce the halogen atoms. This can usually be done only at low gas pressures and is a relatively inefficient process. Electrically driven hydrogen–halide lasers are, therefore, inherently low-power, low-efficiency lasers. Their virtue, however, lies in their small size, low cost, high-spectral line output, and high output stability. Table 2 chronologically lists all such cw lasers published until early 1974.

We shall first discuss the thermally driven lasers. Their prototype, the arc-driven supersonic-diffusion laser, will be described in some detail, including its technology.

## 1. THERMALLY DRIVEN LASERS

In Table 1 we have collected relevant information on various thermally driven hydrogen–halide lasers. They are all HF/DF lasers, and all operate on the reaction of F atoms with $H_2$, the so-called "cold" reaction. As an important predecessor, we also included the shock-tube laser of Gross et al.[2] The first column lists the references, followed in the second column by the type of reaction, the type of gas dynamics, and the heating method. The next two columns list the maximum reported mass flows, the composition of the laser-gas input, and the maximum power obtained with this gas flow. In some experiments, power was measured intracavity; this closed-cavity technique is discussed in detail in the chapter by Chester and Chodzko. Another column gives information about the cavity, and the following one about the size of the nozzle and, therefore, about the size of the active-laser medium. The chemical efficiency will be defined presently. Specific power, the power produced by a total gas flow of 1 g/sec, plays a similar role as the energy produced per liter in pulse lasers. The listing is completed by the power input into the heater, the pressure in the heater and the cavity, and the temperature in the heater.

We shall first review the historically important early attempts to thermally produce lasing in HF with shock tubes and subsonic-diffusion flames. This will be followed by a detailed technical description of the arc-heated supersonic HF laser and its performance. The rather scanty experimental information on HF/DF combustion lasers and other thermal-dissociation lasers will be reviewed thereafter. We shall also discuss a number of unpublished devices and experiments that we consider important.

## 1.1 Shock-Tube and HF Flame Lasers

The very first attempt to obtain continuous lasing from the reaction of $F_2$ with $H_2$ was made by Spinnler and Kittle[1] at the Rohm and Haas Co.

Laboratories in Huntsville, Alabama, during the years 1967 to 1969. The only report of their preliminary results was given at the second Conference on Chemical and Molecular Lasers in St. Louis in 1969.[1] Based on their experience with hydrogen–halide rocket-propulsion systems, Spinnler and Kittle built a simple tubular reactor, 3.5 cm diameter and 50-cm long with $H_2$ and $F_2$ injector holes facing each other along its length. The tube was evacuated, and the reactants were admitted simultaneously by quick-opening valves. Brewster windows at the tube's ends and a hemispherical cavity completed the laser. Spinnler and Kittle observed pulses of 10-$\mu$sec duration from the $P(8)$ and $P(5)$ transitions of the 2–1 band of HF. The cavity pressure had reached a few Torr at the time of the laser emission.

From such experiments, it appeared that a simple subsonic-diffusion flame did not provide sufficient control of the short inversion zones because of the very rapid, exothermic reaction of $F_2$ and $H_2$, and a search for more sophisticated gas-dynamical techniques began at three laboratories: AVCO Everett Research Laboratory, The Aerospace Corporation, and Cornell University. Workers at Aerospace and AVCO decided to use gas-dynamical techniques based on their experience with $CO_2$ gas-dynamical lasers (GDL).[7] Cool, at Cornell, was the first to recognize that the DF–$CO_2$ transfer reaction[6,8] can be used to make a subsonic cw chemically pumped $CO_2$ laser possible. The work on transfer lasers will be discussed in the chapter by Cool.

To explore the feasibility of a gas-dynamically controlled and thermally initiated chemical laser, both Gross et al.[2] and Airey and McKay[3] used shock tubes. In the first investigation a shock wave was generated in mixtures of $F_2O$ and $H_2$, highly diluted with argon. The shock strength had been adjusted so that the shock speed exceeded the critical Chapman–Jouguet detonation-wave speed. Such a shock wave initiates a stable deflagration wave, and the reaction speed is controlled by the pressure and temperature behind the shock. Using a transverse cavity, the authors observed laser pulses of $\sim$10-$\mu$sec duration from vibrationally excited HF. The inversion was found to be restricted to a narrow nonequilibrium reaction zone following the shock wave. These experiments showed that laser action can be obtained from the reaction of $H_2$ with F atoms, even at temperatures in excess of 2000°K, but also that the control of the reaction zone was exceedingly critical. Similar results were obtained in unpublished work by the same workers from mixtures of $SF_6$ and $H_2$. A detailed analysis of the chemical kinetics of these experiments failed at that time because of the many unknown rate processes.

Airey and McKay[3] at AVCO came one step closer to a chemical cw laser and in fact demonstrated cw laser action for the first time. They used a shock tube to heat a mixture of argon and fluorine in the reflected shock region to temperatures sufficiently high to dissociate a sizable fraction of the $F_2$. The gas mixture was then expanded in a short supersonic nozzle. The expansion cooled the gas so drastically that the F atoms had no time to recombine. The result was a low-pressure low-temperature gas jet with a nonequilibrium concentration of F atoms frozen at the stagnation temperature. The second reactant, required for the chemical laser, was introduced into the jet by a mixing device added to the nozzle. In Airey's experiments this mixing injector (Fig. 1) was an ingenious device some-

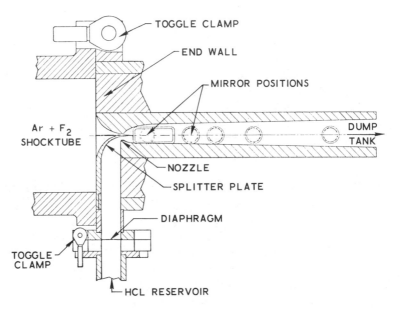

**Fig. 1.**   End-wall and nozzle of shock-tube laser of Airey and McKay.[3]

what reminiscent of constructions by M. C. Escher.[9] HCl was injected from a second tube through a corrugated splitter plate and rapidly mixed with the expanded F–Ar mixture. The reaction of F atoms with HCl takes place as soon as the gases mix; it produces excited HF, which is rapidly carried through the transverse-cavity region by the supersonic-jet flow. Two internally mounted mirrors formed the cavity across the 15.3 cm wide nozzle. Laser power was coupled from the cavity by a hole of 0.1-mm diameter in one of the mirrors. The authors observed laser pulses of 800 $\mu$W that lasted for the full 1.8-msec testing time of the shock tube. The operating conditions are listed in Table 1.

## 1.2   Arc-Heated Supersonic-Diffusion Lasers

The group at The Aerospace Corporation was the first to build and operate a cw HF laser that could be run for arbitrary lengths of time.[5] A high-enthalpy, arc-heated supersonic wind tunnel was modified for these experiments. This facility had already been used as a conventional arc-driven $CO_2$–GDL producing powers of 300 W.[10] The first successful HF laser experiments used the nozzle of the $CO_2$ laser, a simple single-slit two-dimensional rapid-expansion nozzle 17.6-cm wide and 0.95-cm high at its exit. Two hydrogen injectors, 0.6-cm diameter copper tubes with a row of small holes, mounted one above and one below the nozzle exit (Fig. 2), were added to the nozzle.

$SF_6$ was chosen as the fluorine carrier, since it is a nontoxic, noncorrosive gas that could be handled safely in large quantities. The high temperature, which can be obtained with the arc heater, made it possible to dissociate $SF_6$ in the plenum chamber. A gas-mixing plenum was added between the arc chamber and the nozzle (Fig. 2) for the addition of the $SF_6$.

To protect the arc's cathode, nitrogen was blown through the arc heater. More nitrogen added with the $SF_6$ into the mixing plenum served as a convenient diluent that allowed control of the plenum temperature and, most importantly, the gas dynamics of the nozzle and the kinetics in the supersonic jet. Without diluent the nozzle would have had to be

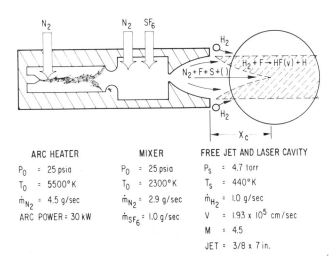

**Fig. 2.**   Schematic of arc heater and nozzle of first supersonic diffusion laser of Spencer et al.[5]

**Fig. 3.** Visible emission (photographic negative) from the reaction zone of first supersonic diffusion laser. The nozzle exit is on the left, and hydrogen is injected from above and below. The reaction zones are clearly visible.

operated at unreasonably low pressure, and the heat released from the reaction would have been many times the total thermal enthalpy of the flow. As we shall see later, such a large relative heat release will destroy the supersonic flow in the nozzle and the jet. By varying the amount of diluent added to the mixing plenum, the temperature in the plenum could be adjusted to produce the desired degree of dissociation of $SF_6$ and mass flow of F atoms.

The equilibrium gas mixture of $N_2$, $SF_6$, F, and other dissociation products of $SF_6$ are expanded through the nozzle sufficiently rapidly that the gas has no time to equilibrate its chemical composition to the low temperatures reached in the adiabatic expansion. The chemical composition of the gas is frozen near its plenum value.

The hydrogen injected from the two tubes form a shear layer enveloping the free jet issuing from the nozzle. Hydrogen diffuses very rapidly into the jet's interior, where it reacts with the free F atoms to produce the desired vibrationally excited HF. Figure 3 shows a side-view photograph of the visible vibrational overtone emission from HF formed in the reaction zone. There are two regions of excited HF both above and below the jet axis. They correspond to the areas of hydrogen diffusion; the reaction under the given conditions is rate limited by diffusion or mixing.

The cavity was arranged transverse to the flow and consisted of two 4.5-cm diameter mirrors, one with a hole for coupling power out. A laser power of a few watts was observed in the earliest experiments.[5]

**Laser Technology.**  During these early experiments, it soon became apparent that the chemical reaction producing the excited HF was limited by diffusion and that fast mixing was the key to higher powers and better efficiencies. Despite its early shortcomings, the approach showed promise. In an extensive research and developmental effort, the laser was studied and improved technically to its present performance. Since this work has become the basis for all other supersonic cw HF lasers, we shall describe it here in more detail.

*Power Supply, Arc Heater, Vacuum, and Cooling Systems.* For laboratory experiments, the use of $SF_6$ avoids the dangers connected with the handling of molecular fluorine. The high dissociation temperature of $SF_6$ requires an arc heater. The arc heater has, however, the advantage that it allows one to vary the plenum temperature nearly independent of the plenum pressure and composition. This makes an arc a very attractive tool for parametric studies of the cold reaction. Obviously, $F_2$ instead of $SF_6$ has to be used for studies of the full $F_2$–$H_2$ chain reaction. Such experiments were later undertaken by The Aerospace Corporation in cooperation with the Air Force Rocket Propulsion Laboratory.

The Aerospace Corporation arc heater was powered by a rectifier, which can produce 160 kW at 400 V and 400 A. The arc heater (Fig. 4) consisted of a thoriated tungsten cathode and an annular-ring anode made from copper. About 70% of the input power was delivered to the gas; the remainder was lost to the cooling water. The diluent gas was injected into the arc chamber from behind the cathode and left at high speed through an anode hole. This prevents back diffusion of F atoms into the arc chamber, which would otherwise severely damage the cathode.

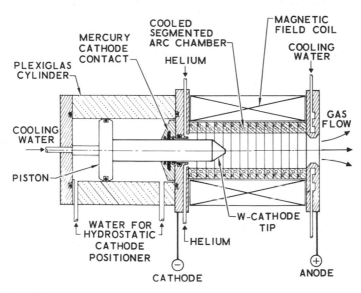

**Fig. 4.** Technical details of the arc heater of the Aerospace supersonic diffusion lasers.

Immediately downstream of the anode, more diluent gas was injected through a number of small peripheral holes. This reduced the temperature of the hot gas issuing from the anode hole from 5500°K to the desired plenum value. Injection at this point also helped to uniformly mix the hot gas throughout the plenum. In this way it was possible to control the temperature in the plenum and, therefore, the degree of dissociation of $SF_6$, whereas the arc was operated at a mass flow that was near its optimum efficiency. $SF_6$ was injected into the plenum further downstream through another series of small holes in the wall. A constant-area transition-section changed the circular cross-section of the plenum first to a square, and then to a elongated rectangle, which ends in the nozzle. The transit time of the gas through the plenum is typically a few milliseconds, sufficient for both good mixing and good thermal and chemical equilibration.

Arc efficiency, plenum pressure, plenum temperature, and, hence, the degree of dissociation and the F atom concentration were all inferred from a measurement of the hot and cold plenum pressure under constant mass flow. Since the gas leaves the plenum through choked sonic orifices in the nozzle, the ratio of the pressures measured in the diluent flow before and after the arc is struck, defines the energy addition to the gas and, therefore, its temperature: $T_{hot}/300 = (P_{hot}/P_{300})^2$. It is assumed that the energy put into the gas by the heater remains the same when $SF_6$ is added later. The equilibrium temperature and the degree of dissociation are calculated from the measurement with an equilibrium computer program (NEST). This method is convenient, but necessarily approximate, as it neglects changes of the boundary layers in the throat region and thus assumes that the nozzle-throat area remains constant. It also neglects variations in the heat transfer from the gas to the plenum walls and the nozzle and changes in the chemical composition in the nozzle-entrance flow.

The arc-plenum chambers, as well as the nozzle, are water cooled, and high-power laser runs of 2 hr have been made with this machine without difficulties.

The vacuum pump consists of a Roots–Connersville blower, model 20x60 RGS-HV, backed by nine Stokes Model 412H mechanical pumps. The combined capacity of this rather large vacuum system is 16,000 cfm at pressures up to 15 Torr. The vacuum pump's exhaust is passed through a scrubber to reduce atmospheric pollution, but the gases entering the pump are not processed, and most of the HF produced by the reaction is absorbed by the oil of the mechanical pumps. This pump system has ingested, during the past 5.5 yr about 50 lb HF/wk with no mechanical failure or corrosion problems in the pumps. The pump oil is changed every 75 hr of operation, only twice as frequently as under normal noncorrosive service.


*Nozzles.* The purpose of the cw chemical-laser nozzle is threefold: (1) to expand and accelerate the gas flow to supersonic velocities, (2) to cool the gas adiabatically to low temperatures and freeze the F-atom concentration at or near its value in the plenum chamber, and (3) to provide supersonic mixing of the oxidizer-diluent mixture with the fuel ($H_2$) in the jet. To accomplish the last two requirements, the nozzle should be small and short, and a minimum length or Busemann nozzle[11] is used for convenience.

These nozzles are decidedly different from wind-tunnel or rocket-propulsion nozzles, in which the main object is to produce a gas flow that is in equilibrium and in which all energy introduced by the heater is recovered as kinetic energy and not frozen in the internal energy modes of the molecules. Rapid expansion or minimum length nozzles are also employed in GDL.[7] In contrast to GDL nozzles, which actually produce the nonequilibrium vibrational states, the nozzle of a chemical laser generates only a nonequilibrium chemical composition. Because of the long atom recombination times, a nonequilibrium chemical composition can be achieved much more easily and efficiently than a vibrational nonequilibrium. In addition, the high velocity of the supersonic gas jet is an important part of the chemical laser. The very rapid deactivation of the excited HF, and, therefore, the very short lifetimes of the lasing species, results in short inversion zones. Unless the reacting molecules are transported through the cavity rapidly, diffraction and deactivation losses will dominate the active medium. Most importantly, however, the nozzle in a chemical laser has to mix the fuel, $H_2$ or $D_2$, rapidly and thoroughly into the supersonic flow. These considerations and the experience with the early nozzle led to a second chemical-laser nozzle,[12] an array of 36 slit-nozzles placed vertically and side-by-side. Figure 5 shows a schematic drawing and Fig. 6a a photograph of this 36-slit nozzle bank. The 36 contoured nozzles were integrally machined from a single copper block. At the trailing edges of each of the interspaced vanes, one sees the $H_2$ injectors, 1.5-mm diameter stainless-steel capillaries with four 0.343-mm holes. Each injector tube fits into an individual O-ring sealed hole in the copper frame above and below the nozzle bank. Through these holes, hydrogen is supplied from both ends to the capillary tubes. Each of the vanes is water cooled by an integral bore near its leading edge; a channel in the top frame supplies and another in the bottom frame drains the cooling water. The majority of experimental results were obtained with

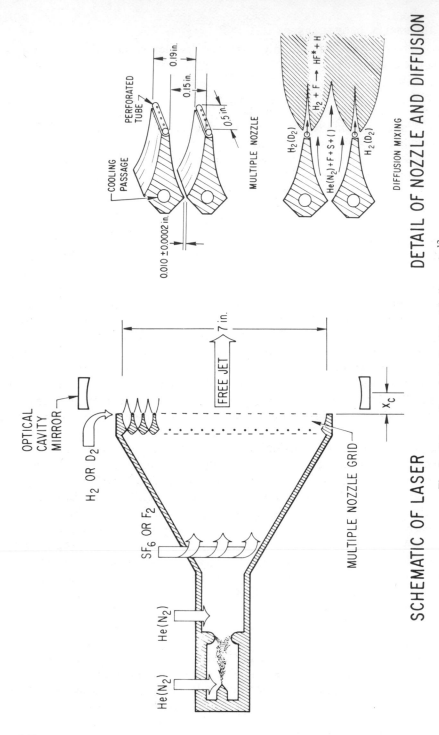

**Fig. 5.** Schematic and details of the 36-slit nozzle.[12]

DETAIL OF NOZZLE AND DIFFUSION

PERFORATED TUBE

0.19 in.

0.15 in.

0.5 in.

MULTIPLE NOZZLE

COOLING PASSAGE

0.010 ±0.0002 in.

$H_2(D_2)$

$He(N_2) + F + S + ( )$

$H_2(D_2)$

$H_2 + F \rightarrow HF^* + H$

DIFFUSION MIXING

OPTICAL CAVITY MIRROR

$H_2$ OR $D_2$

FREE JET

7 in.

$x_c$

$SF_6$ OR $F_2$

MULTIPLE NOZZLE GRID

$He(N_2)$

$He(N_2)$

SCHEMATIC OF LASER

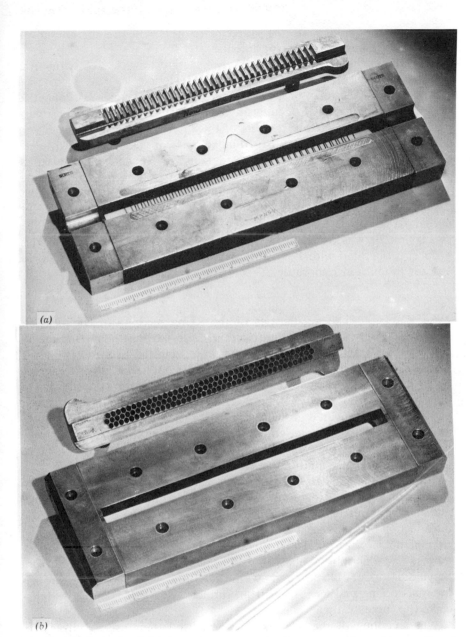

**Fig. 6.** Nozzle banks of the Aerospace supersonic diffusion laser. The nozzles are viewed from their downstream sides and fit into the support block from behind: (*a*) Standard 36-slit nozzle bank, with the injectors visible in the support block (note the individually contoured nozzles); (*b*) Axisymmetric matrix nozzle. Hydrogen is ejected through the star-like hole pattern between the nozzle exits.

this nozzle; it has become the standard-reference nozzle for all subsequent supersonic HF laser experiments.

To further increase the mixing of the reactant flows, an axisymmetric matrix-nozzle, shown in Fig. 6b, was developed at The Aerospace Corporation in the form of a closely packed array of small axisymmetric supersonic nozzles. Hydrogen is injected through the triangular interspaces created by the close packing of the circular nozzles. In this design the main jets are almost fully surrounded by hydrogen, and the distance the hydrogen has to diffuse is considerably shortened compared to a slit-nozzle array of equal area-ratio. Two such nozzles were built: one of $1.25 \times 17.8$-cm area with a total of 73 individual nozzles of 0.423-cm exit diameter and sonic orifices of 0.013-cm diameter, and a second one of considerably larger size with 655 nozzles in a $3.8 \times 22.8$-cm array. The smaller nozzle was tested in the Aerospace arc laser,[13] and both nozzles were also operated with $F_2$ in the MESA laser facility at the Air Force Rocket Propulsion Laboratory.[14]

*Laser Cavities.* The pecularities of the lasing medium in chemical cw lasers necessitate special attention to the cavity arrangement. Because of the very fast reactions generating and destroying the excited HF, lasing starts very close to the nozzle-exit plane and then extends downstream for up to 15 cm. There is a severe variation of small-signal gain and power density along the flow direction. Downstream of the lasing regime, vibrationally deactivated ground-state HF actually absorbs the laser radiation. In HF, a host of rotational lines from three vibrational levels shows gain, but not at the same time nor at the same point in the flow; lasing is, therefore, inherently multifrequency.

Originally a stable cavity with 4.5-cm diameter mirrors was used for all experiments with nitrogen as a diluent.[12,15] Later, when the active medium was stretched beyond the 4.5-cm length of these mirrors by the introduction of helium as a diluent, 9.6-cm square mirrors were used. The circular mirrors were held by a simple friction support, and the square mirrors had elaborate gimbals (Fig. 7). The mirrors were machined from beryllium–copper blocks and were carefully water cooled through internal-cooling passages. Their front surfaces were gold- or silver–MgF$_2$-coated to increase their reflectivity.

Water cooling of the mirrors also permitted their use as intracavity calorimeters, a great convenience in the diagonsis of the laser medium, since the technical problems of coupling large powers out of the cavity could be separated from the investigation of the laser medium itself. Thermocouples measured the temperature rise of the cooling water while

**Fig. 7.** The 9.6-cm square mirror with its gimbal. The water channels in the mirror are clearly visible, and so are the water-supply tubes on top of the support structure. The micrometer adjustment is remotely controlled by an electric motor.

the water mass–flow was determined by a flow rotameter. This technique is discussed in the chapter by Chester and Chodzko.

When necessary, power could be coupled out of the stable cavity[16] with a multiple-hole coupler, a flat metal mirror with an array of small coupling holes drilled through its face. The coupler was, of course, also water cooled; Fig. 8 shows such an output coupler.[17] The size of the holes and their density were arranged so as to produce diffraction-controlled coupling over the whole mirror area.

In a medium that changes from very high gain to absorption in a few centimeters, spatial scanning of the power distribution is very important, if only to enable one to locate the cavity axis at the maximum power position. This was accomplished with a spatial scanning cavity.[18] A large, square, flat output coupler, or mirror in the closed-cavity experiments, faces a spherical mirror. The flat mirror is fixed, and the spherical mirror can be rotated around an axis that is parallel to the upstream edge of the

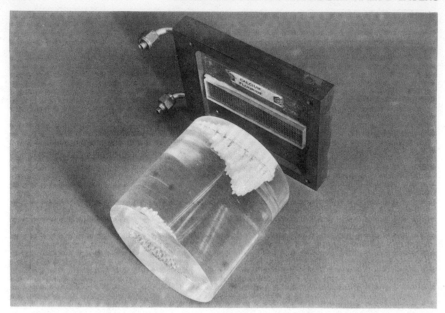

**Fig. 8.** Multihole output coupler seen from its back. Also shown is a lucite block with near-field burn pattern.

mirror and perpendicular to both the flow and the cavity axes. Rotation of the spherical mirror will move the effective cavity axis along the flow direction; only a certain symmetrical strip of the mirrors on both sides of the cavity axis will return the photons onto themselves, and no masking of the unused mirror areas is required. In this fashion, the power distribution along the flow direction can most easily be investigated.

It was found important to prevent expended HF from recirculating into the cavity region, the mirror extensions, and the free jet. Ground-state HF not only absorbs power and reduces the gain, but it is also an efficient collisional deactivator of excited HF. The mirror extensions, which were used to recess the sensitive optical surfaces away from the corrosive, high-temperature jet, were flushed with nitrogen. Failure to flush decreased the laser power. To prevent reaspiration of HF into the jet, a jet-catcher—a simple rectangular water-cooled duct of slightly larger cross-sectional area than the jet—was installed 15-cm downstream of the nozzle exit. At the same time, the jet-catcher acted as a supersonic diffusor, providing gas recompression by a factor of about 2.

Figure 9 shows a photograph of the complete laser machine. On the left, one can see the cylindrical arc housing. The big rectangular box is the cavity chamber containing the nozzle, jet, and jet-catcher. One cavity

**Fig. 9.** General view of the Aerospace supersonic HF diffusion laser. The laser is shown with an unstable cavity attached to it; the 45° edge coupler is housed in the odd-shaped extension in the right foreground. The arc is an earlier version with a manually operated cathode.

extension and a mirror gimbal is visible in the middle foreground. The large vertical tube is the vacuum duct leading to the pumps; it also houses a heat exchanger.

**Laser-Performance Studies.** During the past years, a large body of experimental data has been collected with this machine, and the laser performance has been improved from an initial multimode low-power output of a few watts to a diffraction-limited output of several kilowatts. This improvement was obtained by judicious optimization of all parameters without enlarging the basic dimensions of the nozzle, pump capacity, or arc heater. Table 1 lists the accomplishments of this work until 1974.

The early studies of the laser in which nitrogen was used as diluent and either HF or DF as the active medium were reported in Refs. 6, 12, 15, and 19. In 1970 Spencer et al.[15] reviewed this work in a paper first presented at the International Quantum Electronics Conference, Kyoto (1970). At that time the device produced a multimode power of 600 W in

HF and 450 W in DF.[15] The introduction of $O_2$, which improved the dissociation of $SF_6$ and raised the power to 950 W, was reported by Mirels and Spencer in 1971.[20] During that same year, helium was introduced as a diluent, and the performance went up to 1700 W; this work was reported by Spencer et al.[17] The same paper also discusses the improved DF-laser performance. Spectroscopic studies of the laser output were reported by Kwok et al.[19] and of the medium, by Kwok and Wilkens.[21] The scanning cavity was discussed by Spencer et al.[18] Detailed gain measurements were obtained by Chodzko et al.[22] A number of papers are devoted to the investigation of the flow field in the jet and the nozzle.[23–26] The most comprehensive discussion is the one by Spencer and Varwig.[23] The medium was studied through its infrared emission by Kwok and Cross[27] and in the visible by Varwig.[24]

It is obvious that we shall have to restrict ourselves to the review of the important findings of this voluminous work. In doing this, the reader has to be reminded that we are concerned with engineering studies that are trying to optimize a large number of interconnected parameters with the goal of improving the power output and the efficiency of the laser as a thermodynamic machine. In this single-minded quest, the scientific reasons for an improvement are not always understood, and the explanations are often hypothetical. Only extensive computer modeling, guided by the experimental findings, can eventually shed light on the influence of the various parameters. The intuition of a good research engineer, therefore, is the most powerful guide.

Before we continue, it is necessary to define laser power and efficiency. In most of the experiments, power was measured in the closed cavity. The question of how this power can be extracted in a well-defined laser beam is discussed in the chapter by Chester and Chodzko. In a number of experiments, power was coupled out of the cavity by multihole mirrors to study power distributions in the near field. This was done fully disregarding far-field beam quality, which is poor for this mode of coupling. Power measurements in the near field were made either with a calorimeter or the lucite block technique, if the spatial distribution was studied.

Polished blocks of lucite or Plexiglas were found to be excellent integrating high-power sensors.[15] Lucite has a high absorption coefficient at wavelengths beyond 2 $\mu$m, a low sublimation temperature of 260°C, and a low heat conductivity of $10^{-4}$ cal/cm-sec-°C. Collimated radiation falling on a lucite surface will, therefore, ablate the material, which leaves a three-dimensional image of the two-dimensional power distribution in the beam. This beautifully simple technique can even be made quantitative. Figure 8 shows a typical example: the power distribution in the near-field beam from a multihole-output coupler.

There are various efficiencies one can define; the one used in this work is the chemical efficiency $\eta$, which relates the radiative power $P$ produced by the laser to the power theoretically available from the reaction of $H_2$ with all F atoms produced by the heater. Expressed in terms of the molar flow rate of F atoms through the nozzle $\dot{n}_F$(mol/sec) and $P$(kW),

$$\eta = \frac{P}{133\,\dot{n}_F}$$

The rationale for this choice of an efficiency, which disregards the large power put into the heater, is, of course, the argument that the arc heater is merely a convenience for these experiments and that we can replace it with a combustion device that will not require external power.

*Experiments with $N_2$ Diluent.* A comprehensive review by Spencer et al.[15] of the experimental work prior to 1970 collects all information from the early experiments using nitrogen as diluent and the basic 36-slit-nozzle. The variation of total laser power as a function of $SF_6$ and $H_2$ mass flows was measured for the first time with the closed-cavity calorimeter mirror system. Figure 10 shows the power variation with the

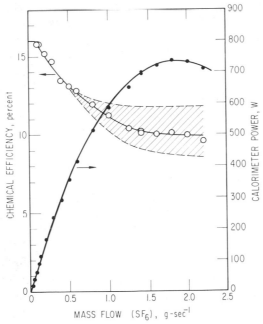

**Fig. 10.** Chemical efficiency and laser power-output as a function of $SF_6$ flow rate in $N_2$ diluent.[20]

cavity positioned at the maximum power location downstream from the nozzle, and a constant power input of 30 kW into the arc. At low $SF_6$ flow rates, the $SF_6$ is fully dissociated, and the laser-output power is proportional to the $SF_6$ mass flow. At higher $SF_6$ flow rates into the plenum, laser-output power decreases. This is largely due to a drop in plenum temperature. As more and more gas is added into the plenum, the temperature decreases, the dissociation of $SF_6$ decreases, and, therefore, so does the concentration of F atoms. A similar dependence of output power on the injected $H_2$ flow was found. The loss of power for high hydrogen flow rates appears to be due to the increased deactivation of HF by $H_2$ and to possible gas-dynamical disturbances by the greatly increased $H_2$ flow. At peak power, the mass flow of $SF_6$ was 1.8 g/sec, which corresponds to a fluorine molar flow of 0.056 mol/sec, calculated for the plenum conditions of the experiment. The hydrogen molar flow rate at optimum was 0.5 mol/sec, which is approximately nine times the stoichiometric value needed for complete reaction of all F atoms. This large, excess hydrogen flow is required in order to produce sufficiently fast mixing of the hydrogen and fluorine in the jet, which shows once again that the reaction of F atoms with $H_2$ is diffusion controlled under the given conditions.

The variation of chemical efficiency as a function of the $SF_6$ mass-flow rate is shown in Fig. 10. Power in this series of experiments was measured in a 4.5-cm diameter closed cavity with mirror absorptivities of 2%. A peak power of 738 W was found at the optimum flow rates.

It is seen that the efficiency decreases with increasing $SF_6$ flow rate and increasing laser power from 16% to $\approx$10%. Since the chemical efficiency is calculated from the flow rate of F atoms, not $SF_6$, this decrease in efficiency can be explained by a combination of increased losses of excited HF molecules because of collisional deactivation and fluid-dynamical effects.

During the early experiments, 4.5-cm cavity mirrors were used. Power was taken out of the cavity with multiple hole couplers. Through the use of couplers with different ratios of open to reflecting areas, the optimum coupling fraction was determined.[15] Figure 11 shows the variation of output power as a function of the coupling fraction. The results are compared with a calculation based on Rigrod's theory.[16] The curve shows a broad maximum between 20 and 50% output coupling. When the mirror absorption losses of 2% per surface are taken into account, an effective integrated zero-power gain coefficient $g_0$ of almost 0.08 cm$^{-1}$ can be calculated from the theory. This large gain is an indication of the high concentration of excited HF molecules generated in the device.

The extent of the lasing medium along the jet was first investigated[15]

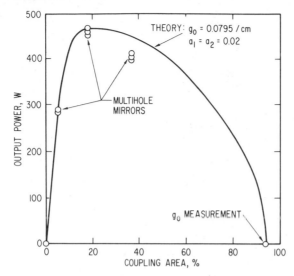

**Fig. 11.** Output power as a function of the open coupling area of a multihole output coupler. Mirror absorption $a_1 = a_2 = 0.02$. Theoretical curve is based on Rigrod.[26] Nitrogen as diluent.[5]

with a Fabry–Perot cavity, an output coupler of 18.75% transmission, and the lucite-block technique; Fig. 12 shows a typical example. One can see that the peak of the power distribution is very close to the nozzle exit and that, thereafter, power falls to zero in ≤10-cm. This shows the very high degree of nonuniformity of the medium along the flow direction that is typical of this type of laser.

Equilibrium calculations showed that, at optimum-power conditions, $SF_6$ is only 75% dissociated in the plenum.[15] This means that a large amount of fluorine remains tied up in $SF_6$, $SF_4$, and lower dissociation products, and is lost for the formation of active HF. Moreover, large quantities of sulfur produced by the arc heater tended to coat all cooled surfaces. Therefore, oxygen was added to the plenum in the expectation that it would react with the sulfur and lower sulfur fluorides, increasing the concentration of F atoms and reducing the sulfur-deposition problem. The addition of $O_2$ resulted in a 20 to 30% improvement in power[20] as shown in Fig. 13.

*Experiments with He Diluent.* These early experiments led to the use of helium as a diluent instead of nitrogen. Helium offers two advantages:[17] a better arc-heater efficiency and larger pressure ratios in the same nozzle. The improved arc-heater performance is probably due to a combination

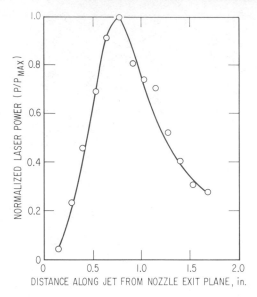

**Fig. 12.** Normalized power output as a function of distance from the nozzle in nitrogen diluent.[20]

of factors: production of helium metastable atoms, absence of useless energy stored in the vibrational states of $N_2$, and better matching of arc and power supply may all contribute. The increased adiabatic expansion ratios are, of course, due to the higher specific-heat ratio and the lower molecular mass of helium as compared to nitrogen and are independent of the mode of heat addition to the plenum. Figure 14 compares a nitrogen and helium flow through the same supersonic nozzle.[17] Jet-exit temperatures and pressures are both lower, and the jet velocity is higher in helium. This results in a faster transport of the gases through the cavity, an increased inversion length, and decreased collisional deactivation.

Figure 15 compares the power and efficiency of the same laser configuration operated once with He and then with $N_2$, both times with an optimum addition of $O_2$.[17] Peak-power output in helium as a function of the $SF_6$ flow rate is found to be almost twice that for nitrogen, at very nearly the same efficiency. The peak-power output of the laser reached 1750 W.

The higher flow speeds in the jet also extend the power distribution along the jet axis by almost a factor of 2. This extension required the use of larger mirrors, but reduced mirror loading considerably. Figure 16 shows[17] the axial power variation in $N_2$ and He, determined with the

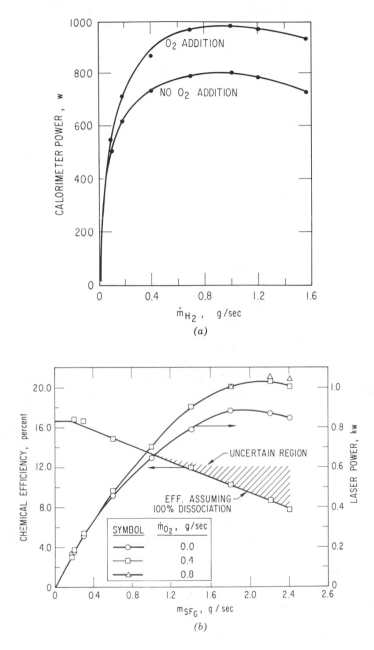

**Fig. 13.** Improvement of output power and efficiency due to the addition of $O_2$ to the mixing plenum (*a*) as a function of the hydrogen mass-flow rate and (*b*) as a function of the $SF_6$ mass-flow rate.[20]

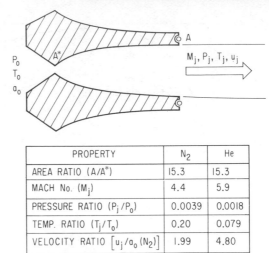

| PROPERTY | $N_2$ | He |
|---|---|---|
| AREA RATIO $(A/A^*)$ | 15.3 | 15.3 |
| MACH No. $(M_j)$ | 4.4 | 5.9 |
| PRESSURE RATIO $(P_j/P_0)$ | 0.0039 | 0.0018 |
| TEMP. RATIO $(T_j/T_0)$ | 0.20 | 0.079 |
| VELOCITY RATIO $[u_j/a_0 (N_2)]$ | 1.99 | 4.80 |

**Fig. 14.**  Comparison of nitrogen and helium diluent in the same supersonic nozzle.[17]

large, tiltable rectangular-mirror system. The point of maximum power has moved from 1.95 cm in nitrogen to 2.75 cm in helium.

*DF-Laser Experiments.*   Atmospheric transmission of an HF-laser beam is poor[28] because of the profuse water-absorption bands in the wavelength region between 2.5 and 3.0 $\mu$m (Fig. 17). DF-laser emission in the 3.6 to 4.0 $\mu$m regime is, by contrast, absorbed only very little; for this reason, a strong interest in the DF laser existed from the very beginning. Early experiments[15] in which $D_2$ was injected into the jet in place of $H_2$ showed that in nitrogen a DF laser produced only 75% of the power of the HF counterpart. Later experiments[17] with helium as diluent showed, however, that the same power could be extracted from either medium. The explanation is most likely that DF loses vibrational energy in V–V exchange collision with $N_2$, especially the higher, 3–2 laser transitions which are closest to the first vibrational level of $N_2$ at 4.3 $\mu$m. Because of the larger number of V–R transitions that lase in DF-lasers, power is distributed over a larger number of states than in HF. In addition, the stimulated emission cross-section of DF is smaller than that of HF, therefore, DF requires high power densities in the cavity for efficient power extraction. If the DF laser is operated under unsaturated-power conditions, it will produce only $\approx 0.1$ of the power of the corresponding HF laser, even in a helium carrier.[17]

*$H_2$–$D_2$ Injection into the Plenum.*   In a combustion driven HF or DF laser the heating of the plenum gases to high temperatures is most

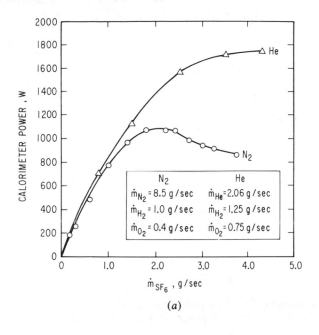

(a)

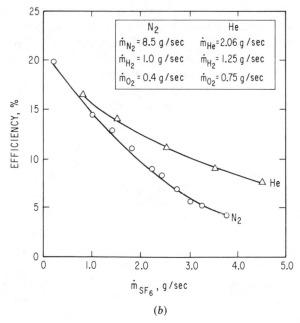

(b)

**Fig. 15.** Comparison of nitrogen- and helium-diluent operation of the same laser config-uration: (a) closed-cavity power and (b) efficiency as functions of the SF$_6$ flow rate.[17]

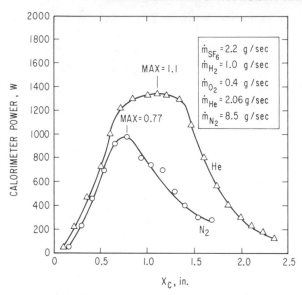

**Fig. 16.** Comparison of nitrogen- and helium-diluent operation of the same laser configuration, variation of closed-cavity power along the flow direction $x_c$.[17]

conveniently done by burning deuterium or hydrogen in an excess of fluorine in the plenum. Such a combustion laser has been described by Meinzer[29] and will be reviewed later. The combustion will produce DF or HF in the plenum, and the question arises how severely these gases will interfere with the excited HF or DF in the jet. To investigate this problem $D_2$ and $H_2$ were added to the plenum gases.[20] Figure 18 shows the results of the experiments for all four permutations. The straight lines of the

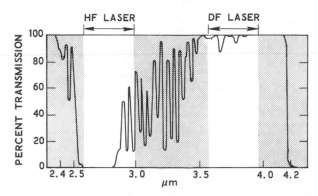

**Fig. 17.** Absorption of the atmosphere in the wavelength region of the HF and the DF lasers.

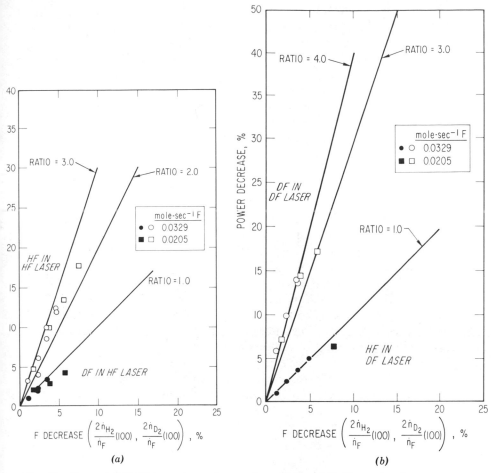

**Fig. 18.** Loss of HF/DF laser power due to the addition of $H_2/D_2$ to the plenum as a function of the loss of F atoms due to the combustion of $F_2$ with $H_2/D_2$ in the plenum. The straight lines have slopes of 1, 2, 3, and 4: (*a*) for an HF laser and (*b*) for a DF laser.[20]

figures are lines of slope 1, 2, 3, and 4; they indicate that a partial molar concentration of 0.1 HF in a DF laser and DF in an HF laser did not decrease the output power significantly, while the addition of HF to an HF laser decreased the power by a factor of 2.5, and the addition of DF to a DF laser reduced the DF laser-power by a factor of 4. These experiments showed that low concentrations of DF will not interfere with HF lasing.

*Spectral Investigations.* The spectrum of both the HF and DF laser output were measured and reported in several papers,[15,17,19,21] and are shown in Fig. 19. As one would have expected, the number of lasing lines is larger in helium than in nitrogen for both lasers. The appearance of two laser lines from the $v = 3 \rightarrow v = 2$ band of HF in helium is interesting. This is due to the lower rotational temperature in helium, where the jet temperature was 400°K compared to 600°K in nitrogen. The lower temperature results in a higher partial inversion of the medium. In contrast to HF pulse lasers, cw HF lasers generally operate on partial V–R inversions; only very close to the exit of a supersonic mixing-nozzle have total inversions been observed[21] in cw-laser media. This is also the reason for the absence of $R$-branch transitions in the output spectrum of cw HF lasers, in full agreement with the results of numerical-model calculations.[30]

The question of rotational equilibrium in the active medium of continuous HF lasers is still unresolved. Observations of the chemiluminescent emission from the medium under nonlasing conditions[21] show that the rotational distributions are apparently in equilibrium at the translational temperature. Such measurements have, however, to be regarded with caution. Because of the low light intensities of the chemiluminescence, the measurements require extensive time and spatial averaging, which could easily mask local short-time deviations from equilibrium. No meaningful measurements of the rotational distributions in the active medium in a cavity under lasing conditions have yet been reported.

*Small-Signal Gain Measurements.* The small-signal gain of the laser medium was measured by Chodzko et al.,[22] who used a high repetition

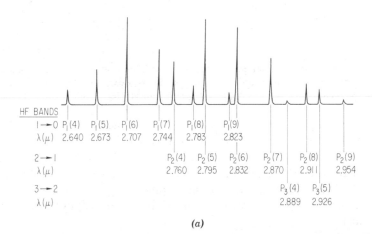

*(a)*

**Fig. 19.** Spectral distribution of the laser output ($a$) of an HF laser and ($b$) of a DF laser. Both lasers were operated with helium diluent.[21]

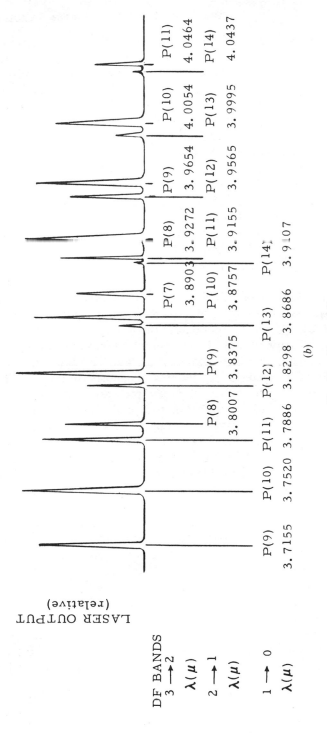

**Fig. 19.** (Continued)

LASER OUTPUT
(relative)

DF BANDS

3 → 2
λ(μ)

2 → 1
λ(μ)

1 → 0
λ(μ)

|       |        | P(9)   | P(8)   | P(7)   | P(8)   | P(9)   | P(10)  | P(11)  |
|       |        | 3.8007 | 3.8375 | 3.8903 | 3.9272 | 3.9654 | 4.0054 | 4.0464 |
|       | P(10)  | P(11)  | P(9)   | P(10)  | P(11)  | P(12)  | P(13)  | P(14)  |
|       | 3.7520 | 3.7886 | 3.8375 | 3.8757 | 3.9155 | 3.9565 | 3.9995 | 4.0437 |
| P(9)  | P(10)  | P(11)  | P(12)  | P(13)  | P(14)  |        |        |        |
| 3.7155| 3.7520 | 3.7886 | 3.8298 | 3.8686 | 3.9107 |        |        |        |

237

rate, pulsed probe laser to determine $g_0$ for a number of rotational lines at various positions along the jet axis. Chodzko found a gain coefficient $g_0$ of $0.1\,cm^{-1}$ on the $P_2(3)$ transition and $0.07\,cm^{-1}$ on the $P_1(6)$ transition of HF. The variation of small-signal gain with the rotational quantum number $J$ indicated a rotational temperature of 350°K and a vibrational temperature of 28,000°K at a distance of 0.4 cm from the nozzle exit. Further downstream the rotational temperature increased as both gain and vibrational temperature decreased, a result in agreement with the chemiluminescent measurements by Kwok and Wilkens.[21] Small-signal gain measurements in the multiline HF laser with a single-line probe laser should be applied with caution. Cascading from higher vibrational levels to lower ones and stimulated emission pumping of lower laser transitions under saturated conditions will alter the inversion densities of the levels markedly in a cavity. Small-signal gains, therefore, may have only limited use in predicting the output from a cavity.

**The Flow Field in the Laser Cavity.**   In their chapter on the gas dynamics of chemical lasers, Grohs and Emanuel have discussed various theoretical attempts to describe and to understand the gas dynamics of the flow field in the laser cavity. Scant experimental information on the flow phenomena in the jet has been obtained to data. This is in part due to the difficulties in using standard gas dynamic diagnostic techniques. The low density in the jet, because of the low pressure and high stagnation temperature, has made interferometric and Schlieren photography difficult except for nonreacting, low stagnation temperature flows.[31] The high stagnation temperatures and the chemical reactivity of the gases also make the use of standard pitot pressure and temperature probes exceedingly difficult. All of these problems are compounded by the spatial and temporal resolution required by the small dimensions of the individual nozzles and the fast mixing processes.

Figure 20 schematically illustrates the various gas-dynamical phenomena occurring at a nozzle exit.[23] The rather thick boundary layers on the nozzle wall have a tendency to separate, giving rise to two shock waves, which intersect in the flow outside of the nozzle. Underneath the separated flow, a recirculation region develops that permits hydrogen to diffuse upstream. The upstream diffusion of hydrogen will lead to pre-reaction in the nozzle, generating not only very undesirable ground-state HF but also sizable amounts of heat. The heat released in the boundary layer will further aggravate the separation problem. Most of the reaction heat is released, of course, in the reaction zone downstream, leading to a lateral expansion of the individual jets and a crowding of the parallel jet flows. As the concentrations of the reactants are increased, these

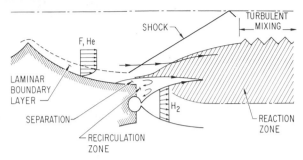

**Fig. 20.** Flow phenomena in the nozzle and an individual jet of a supersonic diffusion laser. Only the lower half of one two-dimensional slit is shown.[23]

phenomena will become more severe, the final result being a choking of the nozzle flow, an effect that is also summarily referred to as thermal blockage.[23]

*Flow Visualization Techniques.* Fortunately, the medium provides us with a rather unique clue to its condition; it is highly chemiluminescent, both in the infrared as well as in the visible. The first chemiluminescent investigations of the reaction regime were made with an infrared vidicon camera in the wavelength regime between 2 and 3 $\mu$m.[27] This camera permitted the observation of the spatial distribution of spontaneous radiation from the excited HF V–R levels. The television-type pictures showed for the first time the extent and the structure of the individual reaction zones. The jet does, however, also radiate in the visible region of the spectrum. Plate 1 shows the flow behind a 17-slit nozzle photographed from above.[24] The nozzle contour has been schematically drawn in on the lower margin; the flow direction is upward with the main flow issuing from the open nozzle exits. The core regions of the four center jets is illuminated by scattered light from the arc. The bright regions between the jets, glowing green-yellow, are due to HF overtone emission, that is, transitions to the ground state from the third, fourth, and fifth vibrational levels of the excited HF. The yellow-green regions indicate, therefore, the reaction zones. Unreacted hydrogen does not radiate; it is injected in the dark spaces between the jets. Clearly, the laser medium is highly nonuniform not only in its chemical composition, but also gas-dynamically. The waviness visible in the reaction-zone boundaries is due to periodic shock waves, which are the typical diamond patterns of supersonic free jets.

*Pressure and Temperature Measurements.* Pitot pressure and total-temperature traverses of the nozzle exit flow (Fig. 21) confirm this

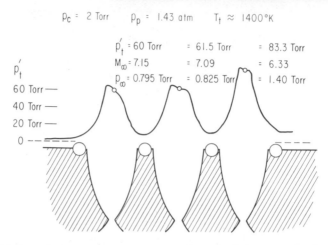

**Fig. 21.** Horizontal pitot-pressure scan across the exit of three nozzle slits. $M_\infty$ is the freestream Mach number, $p_t$ is the pitot pressure, $p_c$ is the cavity ambient pressure, $p_p$ is the plenum stagnation pressure, $p_\infty$ is the free-stream static pressure, and $T_t$ is the total plenum temperature.[23]

picture.[23,24,26,30] They showed, moreover, that the hot helium flow in the nozzles was almost completely viscous, which means that the boundary layers occupied almost the full nozzle area. Theoretical calculations of the nozzle flow, taking viscous effects into account, are only in partial agreement with the experimental observations, especially at low pressures. At the time of writing these problems were still under active investigation; the reader is referred to the original literature and the chapter by Grohs and Emanuel for a further discussion.

The dependence of the laser-output power on the cavity pressure $p_c$ and the plenum pressure $p_p$ was investigated by Spencer and Varwig.[23] Figure 22 shows the results of their experiments. Power increases as the cavity pressure is reduced at constant plenum pressure. This effect is due to improved mixing at lower pressures. Spencer and Varwig in their original paper surmised that there was an optimum cavity pressure and that it was approximately given by the ratio $p_c/p_p$ equal to its matched value calculated on the geometrical area ratio of the nozzle. This interpretation was largely based on the observed maximum of the experimental results at $p_p = 38$ psia. Current thinking is that there is no optimum pressure ratio $p_c/p_p$; the power simply increases with decreasing pressure, and nozzle mismatch has less influence on power than improved mixing. This observation has a strong influence on attempts to recompress the cavity gases to atmospheric pressures by the use of a downstream

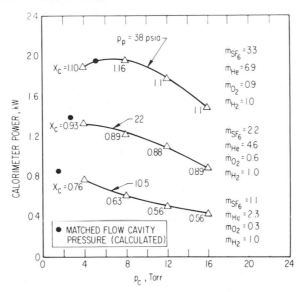

**Fig. 22.** Closed-cavity power as a function of cavity ambient pressure $p_c$ with $p_p$, the plenum pressure as the parameter and helium as the diluent.

diffusor, a subject discussed in detail in the chapter by Grohs and Emanuel.

*Thermal Blockage.* One of the most dramatic gas-dynamical effects is thermal blockage.[23,30] As both the $SF_6$ mass flow into the plenum and the power to the arc were increased, laser power dropped sharply above a certain $SF_6$ flow rate. To investigate this effect, 24 slits of a standard 36-slit nozzle were plugged leaving only the 12 central slits open. Closed-cavity power produced by this nozzle was then measured for increasing $SF_6$ flow rates, carefully keeping other flow parameters constant. Figure 23 shows the results of this experiment;[23] up to a value of 1.5 g/sec of $SF_6$ flow, total power increased roughly proportional to the $SF_6$ flow rate; at 3.4 g/sec, however, the maximum power dropped by almost a factor of 2. The nozzle was then replugged such that only every third slit was open. A large space separated the individual jets, allowing for sideways expansion (base relief) of the jets. All hydrogen injectors were kept in operation to prevent dead-gas regions from forming behind the plugged nozzles. Again closed-cavity power was measured, and the results are also shown in Fig. 23. There was little change in power for low $SF_6$ flow rates. At the highest $SF_6$ flow, the effect is, however, most dramatic; instead of 530 W measured with the unrelieved nozzle, 1080 W were now obtained. The simplest explanation of this effect is that it is due to thermal blockage. When

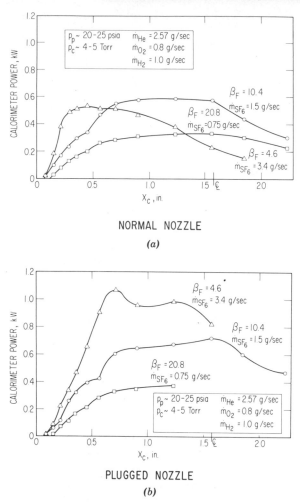

**Fig. 23.** Variation of power for various ratios $\beta_F$ of F atom mass flow to diluent mass flow obtained by varying the $SF_6$ injected into the plenum. (a) Normal spacing of individual nozzles: 12 central nozzles open, 12 nozzles on either side plugged, and (b) Plugged nozzle: every third nozzle open, all others plugged; $\beta_F = \dot{n}_{He}/\dot{n}_F$.[23]

large quantities of fluorine are burned in the jet, large amounts of heat are released. The resulting temperature rise will produce a large expansion of the gases in the jet, especially at low diluent concentrations. If the crowding of the individual jets is so large that they cannot expand sideways, the pressure in the jet will rise. In the extreme case this process can lead to the formation of shock waves and thermal choking. But even a small rise of the pressure $p_c$ in the jet will reduce mixing and power. "Base relief," that is, spacing the nozzles apart, will help reduce the problem, as will "shrouds" above and below the jet, which have the additional advantage of aiding pressure recovery. This subject is discussed in detail in the chapter by Grohs and Emanuel.

**Project MESA Experiments.**   A series of optimization experiments were also undertaken by The Aerospace Corporation with molecular fluorine instead of sulfur hexafluoride, at the remote desert site of the Air Force Rocket Propulsion Laboratory.[14,31] A large steam-ejector system allowed mass flows of >50 g/sec at pressures of ≈2 Torr in the cavity to be pumped through the laser device. This permitted two larger nozzle arrays to be tested: A slit nozzle with 50 slits for fluorine and 51 secondary slits for hydrogen of overall size of $1.9 \times 22.9$ cm; and an axisymmetric nozzle with 655 individual jets of overall size of $3.8 \times 22.8$ cm. Plenum pressures were increased to 10 atm. A typical high-power point taken from MESA Test 601[31] is listed in Table 1.

## 1.3.  Resistance-Heater Driven Lasers

$SF_6$ has an average bond strength of 60 kcal/mol; the dissociation of this molecule requires temperatures in excess of 2000°K at 1 atm, which in a continuous device can only be achieved by an electric arc. $F_2$, on the other hand, has a bond strength of 35 kcal/mol and is nearly completely dissociated at 1400°K and 1 atm. This makes it possible to use a low-temperature heater, for example, a resistively heated pebble-bed heater or a combustor, to generate the F atoms needed for the laser. Such a resistive heating scheme was used by Morsell[32] in unpublished work at the Boeing Aerospace Center in Kent, Washington, during 1970 and 1971. The following review of this work is based on the unpublished notes, which Dr. Morsell kindly supplied to the authors.

The flow facility used by Morsell was a blow-down tunnel capable of a total gas flow of about 3 g/sec for a few seconds. The short test times make elaborate water cooling of nozzles and tunnel walls unnecessary and permit the use of a regenerative pebble-bed heater. In contrast to shock-tube experiments, this technique results in enough time for gas mixing and achievement of steady-state conditions in the laser chamber.[33]

Morsell used two different regenerative heaters: The first consisted of a 120 cm long by 1.9-cm diameter nickel tube containing four 6-mm diameter nickel tubes; a second more successful device was a 1.9-cm diameter, 35-cm long graphite tube filled with 3-mm diameter graphite

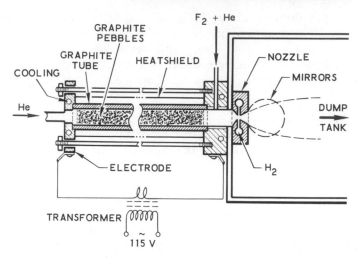

**Fig. 24.** Resistance-heater driven supersonic diffusion laser of Morsell.[32] The laser was operated as a blow-down tunnel.

pellets. Both heaters could be heated by passing current from a low voltage 3 kW, 60-Hz transformer directly through the tubes. Figure 24 shows a diagram of the apparatus. Following the plenum chamber was a simple supersonic nozzle producing a Mach 3 flow. $H_2$ or $D_2$ was injected through a series of 78 holes of 0.02-cm diameter at the nozzle throat. A transverse stable cavity with either hole coupling or a germanium flat coupler of 48% transmission was used to extract power.

With this device 23.6 W of multimode, multiline power was obtained at a flow of 0.58 mol/sec of He, 0.03 mol/sec of $F_2$ through the nickel-tube heater, and 0.166 mol/sec $H_2$ injected at the nozzle throat. The gas temperature at the heater exit was found to be 1130°K; however, losses to the cold-nozzle walls resulted in a considerably lower stagnation temperature. Morsell then changed to the graphite heater, which permitted helium to be heated to 2000°K; of course, in absence of $F_2$, which was added together with more He downstream of the heater. Working at about the same molar flow rates as before, however, at a stagnation temperature of 1930°K, a power of 150 W was extracted through the Ge flat. In a series of three experiments the laser was also operated with $F_2:D_2:He:CO_2$ mixtures of $0.03:0.044:0.468:0.336$ mol/sec and gave a power output of 500 W at the 10.6 $\mu$m $CO_2$ laser transition.

Another unpublished resistive heater designed for continuous operation was built by Gross at The Aerospace Corporation. It was based on a design developed at NASA.[34,35] Current was passed through a set of 20

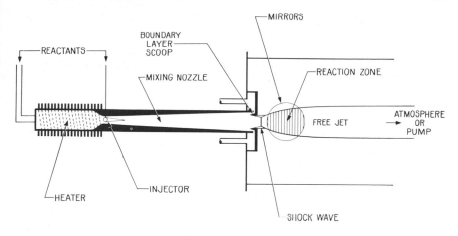

**Fig. 25.** Resistance-heater driven shock-initiated HF laser of Gross (see text).

heavy tungsten wire-mesh elements placed in series, one behind the other in a carefully water-cooled duct of $2.5 \times 10$ cm cross-section. With a power input of 94 kW into the heater, he was able to heat continuously 76 g/sec of argon at 2 atm to 2000°K. The theory and scaling laws of this type of heater have been discussed extensively by Siegel.[35]

The heater was used in an attempt to obtain lasing in HF behind a standing shock wave. The experiments were unsuccessful and never published. Since the idea was an outgrowth of shock-tube experiments at The Aerospace Corporation[2] and has also been suggested by others,[4] we believe a discussion nevertheless justified. Figure 25 shows a schematic of the device. The heater was followed by a relatively long supersonic mixing nozzle with an exit area of $5 \times 10$ cm. A short distance downstream of the throat, a bank of supersonic injector nozzles was installed. The exit area of the mixing nozzle was surrounded by a boundary-layer scoop, which was also intended to serve as a shock holder. Premixed hydrogen and helium were passed through the heater at a pressure of 2 atm. The mixture reached a stagnation temperature of 1200°K. In passing through the first part of the mixing nozzle, the hydrogen–helium mixture was supposed to be adiabatically cooled to near 200°K. At this point molecular fluorine was injected. The three gases then proceeded to mix in the remaining part of the nozzle and were finally ejected in a free jet through a standing-shock system. Just as in the shock tube, the standing normal Mach shock recompressed the gas to a pressure slightly higher than the ambient pressure and a temperature close to stagnation temperature. This should have lead to a rapid dissociation of $F_2$ and a free-burning subsonic flame behind the shock. We had expected to find an inversion zone following the shock in full analogy to the shock-tube experiments.

No lasing was observed in the transverse cavity. Using the IR vidicon camera,[27] we did however, find a shock-induced HF flame of the expected shape. Failure to lase was probably due to a number of factors; the most important being extensive prereaction in the long mixing nozzle. The nozzle had been designed on the basis of inviscid calculations, and the boundary layers turned out to be much thicker than expected. Pressure traverses along the nozzle axis detected a strong shock system produced by the injector, which lead to boundary-layer separation. The combination of shock and boundary layers probably resulted in early ignition of the reaction inside the nozzle. The shock-tube experiments had shown that at the high temperatures behind the shock waves the inversion zone was only a few millimeters thick. Lasing at these temperatures is, in fact, possible only on total vibrational inversions. Schlieren photographs of the standing shock system showed that the Mach shock was small and slightly curved. If there was an inversion zone as thin as the one observed in the shock tube, the shock curvature would have prevented a straight line of sight through this zone. Possibly a shorter nozzle with thinner boundary layers and a more effective boundary-layer control system could have produced a larger, planar shock wave, less prereaction, and a better chance for lasing. The success of the diffusion laser, however, made such further work superfluous.

## 1.4.  Combustion and Chemically Driven Lasers

Electrical heating of the fluorine-bearing gas introduces inefficiencies into the overall laser system, and large power supplies are required for high-power cw devices. In addition, there is always the engineer's dream of a "true" chemical laser that needs "only gas bottles for operation" and no "alien" power. Of course, these gas bottles are only another, if convenient, energy storage, and it should not be forgotten that in an honest closed-cycle energy balance, this energy has been put into the gases by the manufacturer. Such considerations do, in fact, become important if one considers the use of chemical lasers, for example, in fusion applications. Nevertheless, "true" chemical lasers seem to have a large fascination. Obviously, the HF laser can be operated entirely independent of an electrical heater by replacing the heater with a combustor. The only requirement is that the combustion products do not interfere with the lasing gases by deactivation or direct reaction.

The first laser of this class was reported by Cool[6] who used the reaction of NO with $F_2$ to generate F atoms for the purely chemical operation of a

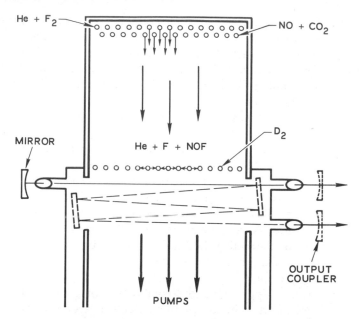

**Fig. 26.** HF/DF laser driven by the chemical reaction of NO with $F_2$ of Shirley et al.[36]

DF–$CO_2$ transfer laser; in a later paper with his co-workers, Shirley et al.[36] described the application of the same technique to an HF–DF laser. The mechanisms of the NO+$F_2$ reaction have been discussed in the chapter by Cool and will not be repeated here. Earlier attempts to obtain lasing in HF or DF with this reaction in a longitudinal-cavity laser[6] had failed, and a fast subsonic flow system with a transverse cavity was built to repeat the experiments. Figure 26 shows the apparatus.

He and $F_2$ were injected through a bank of tubes with minute holes near the rear of the device; NO entered through a second row of tubes a few centimeters downstream. The gases mixed and reacted while traveling 59 cm down the channel at which point $H_2$ or $D_2$ was injected through another row of 63 tubes of 0.125 cm diameter each with three holes of 0.016 cm diameter. In this bank of tubes, the injected holes faced each other, and the injection was transverse to the flow. The transverse dimension of the channel was 15 cm.

A folded five-path transverse optical cavity permitted the extraction of laser power from a 6-cm long region starting as close as 0.5 cm to the $H_2/D_2$ injector bank. Since in the HF/DF experiments the inversion was found to extend only $\approx 1.5$ cm downstream of the injector, the multiple path cavity was not used. A maximum laser output power of 7.2 W for HF and 7.0 W for DF was measured. Table 1 lists the experimental conditions. Detailed spectroscopic measurements were also reported both in multiline and in single-line operation. Only $P$-branch transitions could be observed with an intracavity diffraction grating, an indication that the medium was partially inverted. Lasing was strongest on lines with rotational quantum number $J = 7$, 8, 9, 10 in agreement with gas temperatures of 400 to 700°K measured with a thermocouple. A maximum-gain coefficient of 0.02 cm$^{-1}$ was determined for the strongest line with the aid of a small probe-laser.

In the light of more recent investigations[37] of the kinetics of the NO reactions with $F_2$ and HF, it appears that deactivation of HF by NO and NOF is slow and that this does not explain the lower laser efficiencies. A more likely reason is the low initial concentration of F atoms generated by the NO technique. This leads to slow chemical pump-reactions and low HF-production rates. As a result of this and the small dimensions of the laser, saturation of the active medium was probably not reached in the cavity. A more detailed discussion of this kind of laser operation is given in the chapter by Cool.

A more powerful approach is the combustion of deuterium or hydrogen with excess fluorine in the plenum. Such a laser was first demonstrated by Meinzer[29] at the United Aircraft Laboratories in Hartford, Connecticut. The advantage of this combustion method is that, because of the very high exothermicity of the reaction, combustor temperatures of 4000°K can be achieved, and the combustion can be run very lean, in a large excess of fluorine. Equilibrium calculations[38] showed that for a molar ratio of $F_2$ to $D_2$ of 2.5 at 1 atm and a combustion temperature of 2500°K, more than 50% of the fluorine was dissociated. Figure 27 shows schematically Meinzer's laser.[38] The plenum mixture is further diluted with helium and then expanded through a fast supersonic nozzle, which

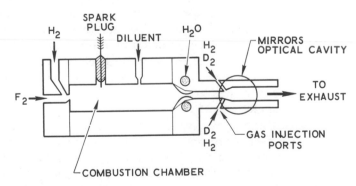

**Fig. 27.** HF supersonic diffusion laser driven by the combustion of $D_2$ with $F_2$ in the plenum of Meinzer et al.[38]

freezes the F-atom concentration. Hydrogen is injected into the flow downstream, and a transverse cavity extracts power from the laser medium. The pertinent operational parameters of the experiment are listed in Table 1. A power of 10 W was obtained with a single-hole coupler.

An improved combustion-driven laser has recently been reported by Blaszuk et al.[39] The combustor has 18 injectors spaced 2.5 cm apart along the center line of the back plate of a $45.5 \times 10.2 \times 2.5$-cm box. Each injector consisted of 26 closely packed tubes of 0.16-cm diameter. Gaseous fluorine is supplied through one half of these tubes, whereas the other half carries a mixture of helium and hydrogen or deuterium. The side of the box opposite to the combustion injectors consists of a $45.5 \times 10.5$-cm matrix of 1800 axisymmetric supersonic nozzles alternated with 1800 hydrogen-injector nozzles. The individual nozzles had a diameter of 0.57 cm. The supersonic jet is housed in a duct of $45.5 \times 3.8$ cm, which, after a length of 6 cm, opens up into a diffusor of 20 cm length and $2°$ half angle. Limited testing of this device produced powers of up to 3 kW in HF and 1850 W in DF. The maximum operational conditions of these experiments are listed in Table 1. Some of these experiments also employed an unstable cavity with a 50% effective output coupling.

Work on another combustor driven HF/DF laser is in progress at TRW Systems, Inc. This work is discussed in the chapter by Grohs and Emanuel. Figure 28 shows schematically the TRW combustor and laser.[40] The oxidizer ($F_2$) and the fuel ($D_2$) are injected from a manifold of small

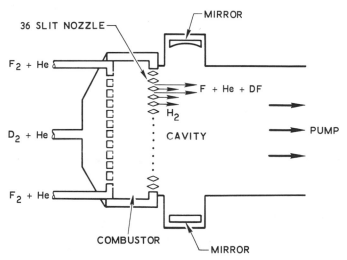

**Fig. 28.** DF-combustor driven HF laser developed at TRW Systems, Inc.[40]

holes in the rear wall of the plenum chamber. The gas temperature in the combustor could be varied between 1200 and 2200°K by changing the ratios of oxidizer, fuel, and helium diluent. The nozzle used in the machine was a standard Aerospace 36-slit nozzle. Closed cavity powers of 1 to 4 kW have been obtained in this laser at pressures, flow rates, and efficiencies comparable to the arc-heated laser.

## 2.  ELECTRICAL-DISCHARGE DRIVEN LASERS

In this section we shall review all cw hydrogen–halide lasers in which the halogen atoms are produced by a direct electrical discharge rather than by thermal equilibrium dissociation. These lasers have an inherently low power output, and they are relatively inefficient. Moreover, they do not appear to be scalable to high powers. They are, however, very useful laboratory devices. Being small they do not require large pumps or power supplies and will operate at subsonic flow speeds. Most importantly, however, they are eminently suitable as probe lasers for gain measurements and as laser light sources for laser-induced chemistry experiments. They have proved invaluable tools for the investigation of the basic chemical kinetics of chemical lasers.

It seems, therefore, that we should review this work from the aspect of someone looking for a laser light source that produces a large number of lines in the 2 to 4-$\mu$m regime at powers of a few watts, with only moderate beam requirements but with high temporal stability. Presumably this potential user is not too concerned about efficiency or high specific power, but would like to know how much he has to invest into pumps, power supplies, and sophistication of nozzle designs. This point of view is reflected in Table 2, which lists all such lasers reported until 1974 in chronological order.

### 2.1.  Laser-Medium Studies

Most of the publications are largely device oriented; the papers by Cool et al.[41] and Rosen et al.,[42] however, require a more detailed review since they contain two of the most comprehensive studies of the medium of a chemical laser.

Figure 29 shows a schematic drawing of Cool's longitudinal discharge laser.[6,41] An electrodeless discharge in one arm produces the atoms that are mixed with the other reactant in the laser tube, and the reaction takes place in a narrow bore inside a Teflon block. The axis of the cavity

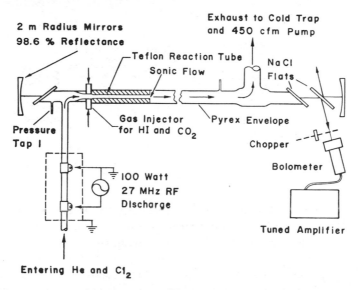

**Fig. 29.** First discharge driven chemical laser of Cool et al.[41] with longitudinal cavity axis.

coincides with the axis of the flow tube. Table 2 summarizes the experiments and results obtained with this first discharge-driven chemical cw laser. Comparison with the other discharge lasers shows that a longitudinal laser is a very inefficient device even at fairly high pump speeds and carefully optimized cavity coupling. The reason is, of course, the presence of ground-state hydrogen–halide molecules in the optical path, which are produced by the very fast deactivation processes in the cavity. Nevertheless, this early paper[41] reports the only research in which initiation with both halogen as well as with hydrogen atoms was studied; also, lasing was investigated for a large number of hydrogen–halide reactions. This very detailed work is still the best introduction to many aspects and problems of hydrogen–halide cw lasers.

Cool and his students[42] in 1973 repeated the systematic investigation of a large number of reactions with improved discharge equipment and an improved laser (Fig. 30). $F_2$ or $Cl_2$ in a helium carrier was dissociated by a high-power microwave discharge. A Litton model L-5001 magnetron was attached to a tunable C-band waveguide crossing the discharge tube under a $10°$ angle. This interesting arrangement allowed easy matching of the microwave cavity to the 2.5-cm diameter quartz discharge tube. The magnetron developed 1 kW at 2450 MHz. A transition section expanded the flow into the $0.7 \times 10$-cm laser section. Secondary reactants were injected through a row of 40 stainless-steel tubes of 0.8-mm diameter

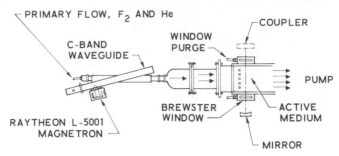

**Fig. 30.** Microwave-discharge driven HF/DF laser of Rosen et al.[42]

with seven small holes each. The flow velocity in the cavity was subsonic at $\approx 350$ m/sec, and the static pressure was 6 Torr. The cavity was arranged transverse to the flow, and calcium-fluoride Brewster-angle windows were used. Germanium flats with dielectric coatings of different reflectivity served as output couplers. The active medium could be scanned with the cavity over 17 cm downstream from the place of injection.

The investigated reactions and their maximum power output are listed in Table 2. The authors presented a very complete study of the laser-power distribution along the flow axis at optimum-output coupling, the spectral output of the laser, and the distribution of the chemiluminescent radiation from the various gas combinations. The main emphasis of the work is on the investigation of the chemical kinetics of the active medium by spectroscopic means.

The relative populations of the vibrational levels of HF along the flow axis were obtained from detailed spectra of the chemiluminescent radiation from the gas in absence of the cavity. Plotting the reduced intensities of the radiation from the various rotational levels, the authors showed that there was rotational equilibrium in the flow at a temperature of $370°K$ in good agreement with the translational gas temperature of $375°K$ measured with a thermocouple. The number of molecules in a vibrational level was obtained by extrapolation of the rotational distribution to $J = 0$. Figure 31 shows the results for the $F + H_2$ reaction.[42] The important conclusion from the energy distribution of the vibrational populations is that there is no total inversion anywhere in the investigated flow regime; the deactivation processes fully dominate the relaxation of the vibrational populations produced by the pumping reactions. This is in contrast to pulsed HF lasers where total inversions are observed in the initial stages of the reaction. Similar recent investigations by Kwok and Wilkens[21] in fast-mixing supersonic cw HF lasers have shown that total inversions can

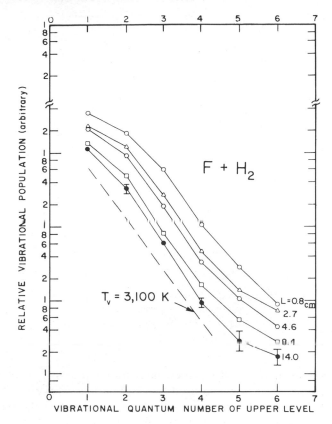

**Fig. 31.** Variation of population densities with vibrational quantum number in a subsonic HF laser. Chemiluminescence measurements in absence of a cavity by Rosen et al.[42]

exist also in these devices near the nozzle exit, if the flow speed is high, the mixing is fast, and the T–R temperatures are low. Figure 32 shows such a case taken from Kwok and Wilkens' work.[21] The reason for the distributions observed by Rosen et al.,[42] therefore, has to be the slow mixing and possible prereaction upstream of the injectors; the pumping reactions are rate limited by the mixing process. This observation once more stresses the need for the detailed analysis of the reactive mixing processes in these lasers.

## 2.2. Probe Lasers and Laser Light-Sources

The majority of the remaining papers is concerned with a variety of HF/DF lasers and, in two cases, with the HCl laser. Naegeli and Ultee[43]

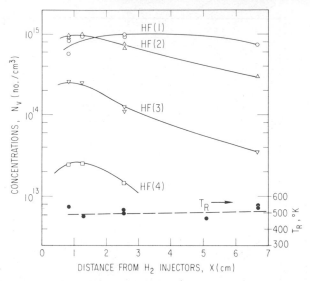

**Fig. 32.** Variation of population densities along the flow direction $x_c$ in a supersonic HF laser. Chemiluminescence measurements of Kwok and Wilkens.[21] Laser operation conditions: $SF_6 = 4.3$ g/sec, $H_2 = 1.0$ g/sec, $O_2 = 0.8$ g/sec, $He = 2.0$ g/sec; $p_p = 3.5$ atm, $p_c = 3$ Torr, $T_p = 1800°K$.

first observed lasing from HCl in a cw laser pumped by the reaction of Cl with HI. This reaction was also used by Cool and co-workers[41,42] and by Glaze et al.[44] to obtain lasing from HCl. The reason for the use of HI instead of $H_2$ is, of course, that the reaction of $H_2$ with Cl atoms is endothermic by 1.1 kcal/mol and is, therefore, unable to populate even the lowest vibrational level of HCl. The reaction of HI with Cl has an exothermicity of 32 kcal/mol and should produce HCl in up to the third vibrational level. This has been observed by Rosen et al.,[42] but the other investigations show lasing only from the stronger transitions from the second vibrational level, which is undoubtedly due to slow flow, slow mixing, and poor saturation of the medium.

In the investigations of HF lasers, both $F_2$ and $SF_6$ have been used as a fluorine source in the discharge. For the direct dissociation of $F_2$, electrodeless discharges have to be used; $SF_6$, on the other hand, can be dissociated by a simple high-voltage dc discharge. There seems to be little difference in laser performance between $SF_6$ and $F_2$ operation. The differences in power output and electrical efficiency observed by Glaze and Lindford[45] and Hinchen[46] are due to the larger size of the active medium, the larger flow rates, and possibly the better mixing in Glaze's laser. The ease of handling $SF_6$ and the much simpler dc discharge

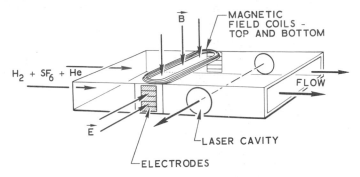

**Fig. 33.** Magnetically stabilized dc discharge technique of Buczek et al.[47]

techniques make $SF_6$ the best fluorine source from a practical point of view. Helium was used as diluent in most studies, because it resulted in a better discharge than nitrogen or argon, and because its higher sound speed allowed larger velocities in the laser cavity.

Many discharge techniques have been tried. The most unusual one is the unique technique of Buczek et al.[47] shown in Fig. 33. A 5-kV dc discharge is held in place transverse to the flow by a weak magnetic field perpendicular to both flow and current. The flow velocity was 70 m/sec at a pressure of 8.5 Torr in a 30-cm wide flow channel. The transverse cavity is located a few centimeters downstream of the discharge. This arrangement permitted premixing of all reactants, which eliminates problems of injection and mixing encountered in the other devices. This is the only work investigating transverse discharges in chemical cw lasers. The relatively successful results of this study should encourage further experimentation with such devices and also with discharges of the multiple-pin variety, which have proved so useful in pulse-laser applications.

Again, the laser performances obtained with the various discharge techniques are similar, and the choice between them can be decided by the equipment and the experience available to the researcher. Microwave supplies appear somewhat easier to match to the impedance of the discharge than do rf transmitters, especially if one uses the $C$-band waveguide described by Rosen et al.[42] The simplest method, however, is a straightforward high-voltage dc discharge. Hinchen[46] developed a most effective electrode arrangement for this purpose that gives exceptionally stable laser performance not available with the other methods. Figure 34 shows a laser built by Spencer,[48] based on Hinchen's design; the cathode consists of a number of pins extending $\approx 10$ cm into the gas. Each pin is separately ballasted with a resistor to decouple the various arcs. Helium and $SF_6$ is injected from behind the pin cathode through a loose packing

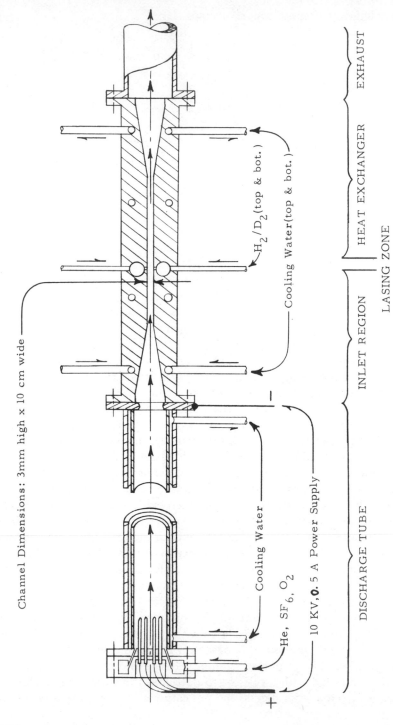

Channel Dimensions: 3mm high x 10 cm wide

$H_2/D_2$(top & bot.)

Cooling Water(top & bot.)

Cooling Water

He, $SF_6$, $O_2$

10 KV, 0.5 A Power Supply

EXHAUST

HEAT EXCHANGER

LASING ZONE

INLET REGION

DISCHARGE TUBE

**Fig. 34.** Multiple-pin dc discharge laser of Spencer[48] and Hinchen.[46]

of glass wool. The glass wool was found to be important in diffusing the flow uniformly over the entire tube's cross-section. With this electrode design, the discharge tube could be filled nearly uniformly with a large number of discharge filaments.

Since most of the power input to the discharge is dissipated as heat, water cooling of the discharge tube is necessary. This should also help to reduce wall recombination of fluorine atoms. Wall coatings have been variously used to suppress recombination; they seem to be helpful if applied to metal parts in contact with the gas. Teflon coatings[45] or halocarbon waxes seem to be most practical. Unfortunately, there are no conclusive investigations of wall recombination of F atoms, and such work would be very helpful.

The design of the mixing nozzle has the most crucial influence on the laser performance. It is probably not fortuitous that the laser of Glaze and Lindford,[45] which has the most sophisticated nozzle, also produces the highest power and seems to be most efficient. Their nozzle, shown in Fig. 35, was similar to the supersonic mixing nozzles. The nozzle covers an area of $1 \times 14$ cm and consists of a bank of 47 individual ducts of

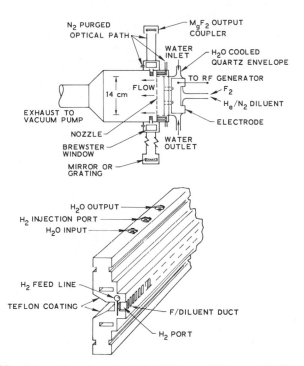

**Fig. 35.** Schematic and nozzle of the HF laser of Glaze and Lindford.[45]

$0.15 \times 1$-cm cross-section interspersed with 46 injector ducts of $0.1 \times 1$-cm cross-section. The wall thickness between the ducts is 0.05 cm. The nozzle is water cooled, and all upstream surfaces are Teflon coated. The high gas speed in the nozzle also prevents upstream diffusion of hydrogen, which could be a problem in subsonic lasers.

Another less elaborate nozzle design, which gave a good performance, was first described by Stephens and Cool[49] and later modified by Hinchen.[46] Cool's original nozzle was a straight rectangular duct. In Hinchen's version (Fig. 34) the duct opened up after a throat section. In both versions hydrogen was injected at high pressure through a row of small holes of 0.033-mm diameter in the top and bottom walls of the nozzle. The row of minute supersonic injection jets served the dual purpose of enhancing rapid mixing and of acting as an effective gas-dynamical displacement throat. This type of injection will, however, permit upstream diffusion of hydrogen unless the main flow speed is sufficiently high. It is important, therefore, that the cross-sectional area of the duct or throat be kept as small as upstream pressures will permit at the chosen mass flow rate.

Since all these lasers operate on partial inversions, the translational gas temperature should be kept as low as possible. No investigations of efficient cooling of the gas flow either by a heat exchanger upstream of the nozzle or by adiabatic cooling through supersonic expansion of the main flow have been reported. Either method should result in an improvement of laser performance; a supersonic expansion should, moreover, also give higher flow speeds in the cavity.

As in the large supersonic cw lasers, care should be exercised to prevent the recirculation of ground-state HF. Dead-gas regions, such as the extensions for Brewster windows, should be flushed with copious amounts of diluent, and the exit from the cavity section should continue as directly as possible into the pump line.

Gain measurements have been reported by a number of authors.[44,46,49–51] Corresponding to the lower concentrations of F atoms and, therefore, excited HF molecules, small-signal gain coefficients are lower than in the supersonic lasers; measured values of $g_0$ are of the order of 0.01 to $0.04 \, \text{cm}^{-1}$. Hinchen[46] also measured the gain line width in HF to be 300 MHz at 10 to 15 Torr pressure.

The spectral output characteristics of the various lasers have been measured in great detail and are amply documented in Table 2. Careful studies of the power output as a function of coupling fraction have, however, not been reported. Usually an optimum output coupling has been determined; the most detailed study is that of Rosen et al.[42] In accordance with the low gains and pumping powers, the optimum output coupling is of the order of 5 to 20%. It has to be assumed that none of the reported lasers operates under full saturation of the pumping chemistry. Efficient power extraction from the medium can, therefore, not be expected in these low-power lasers, unless the active medium is greatly lengthened. Under these circumstances, unstable cavities will not be of any advantage over stable cavities. Satisfactory output-beam quality can, however, be obtained by severely limiting the mode volume by an aperture in the cavity.[46]

Recently, Proch et al.[52] published the technical details of an HF laser modeled on the Hinchen and Banas[51] laser. With a multimode maximum output power of 40 W, this laser is the most powerful electrical discharge laser reported to date. $SF_6$ was dissociated in two parallel, 1-m long, dc discharge tubes with an electrical power input of 3.6 kW each. The nozzle was a simple rectangular channel 40-cm wide and 1.5-cm high. Hydrogen was injected through 16 small holes in a $2.5 \times 6$-mm tube of elliptical cross-section placed at the center of the channel just upstream of the cavity section. The system was pumped by two roots pumps in series with a total capacity of 1250 liter/sec. It is this large pump capacity that is largely responsible for the power produced by the laser. Because of the central injection, two distinct inversion regimes were observed: one above and one below the centerline of the channel. Despite the fact that the stable cavity is not excited in the $TEM_{00}$ mode longtime stablility was better than 5%. Proch et al. also showed that they could "tune" the laser emission by varying the temperature of the active medium. In this way, lasing was observed on all rotational levels between $J = 4$ and $J = 10$ on both the $P_{1-0}$ and $P_{2-1}$ bands.

## ACKNOWLEDGMENTS

The authors have received help and encouragement from a number of their colleagues. Foremost is R. Oglukian of the Air Force Weapons Laboratory in Albuquerque, New Mexico, who undertook the thankless task of reviewing the chapter and who, from his long and intimate knowledge of the subject, suggested an uncounted number of improvements and corrections. The chapter was also read by G. Emanuel, L. Wilson, and R. Giedt. J. Hinchen, D. Proch, T. A. Jacobs, and R. Giedt made preprints of their work available to the authors, and L. Morsell and D. McClure gave permission to use unpublished work performed at the Boeing Aerospace Center, Kent, Washington. We express our thanks to all these contributors.

## REFERENCES

1. J. F. Spinnler and P. A. Kittle, "Hydrogen Fluoride Chemical Laser—A Demonstration of Pure Chemical Pumping," presented at Second Conference on Chemical and Molecular Lasers, St Louis, May 22–24, 1969.

2. R. W. F. Gross, R. R. Giedt, and T. A. Jacobs, J. Chem. Phys. **51,** 1250 (1969).

3. J. R. Airey and S. F. McKay, App. Phys. Lett. **15,** 401 (1969); J. R. Airey, S. S. Fried, and S. F. McKay, Final Report Contr. DAH01-68-2144 (AVCO Everett Research Laboratory, Everett, Masachusetts, Jan. 1970).

4. J. R. Bowen and K. A. Overholtser, Astronautica Acta **14,** 475 (1969).

5. D. J. Spencer, T. A. Jacobs, H. Mirels, and R. W. F. Gross, Int. J. Chem. Kinet. **1,** 493 (1969); also App. Phys. Lett. **16,** 235 (1970).

6. T. A. Cool, R. R. Stephens, and T. J. Falk, Int. J. Chem. Kinet. **1,** 295 (1970).

7. E. T. Gerry, IEEE Spectrum **7,** 51 (1970); N. G. Basov, V. G. Mikhaylov, A. N. Oraevskii, and V. A. Sheglov, Zh. Tekh. Fiz. **38,** 2031 (1968).

8. R. W. F. Gross, J. Chem. Phys. **50,** 1889 (1969).

9. M. C. Escher, "The Graphic Work of M. C. Escher" (Meredith, New York, 1967).

10. D. J. Spencer, unpublished results, see also Ref. 48.

11. L. Prandtl and A. Buseman, "Stodola Festschrift" (Zürich, 1929) for design purposes; K. Foelsch, J. Aero. Sci. **16,** 160 (1949); H. Shames and F. L. Seashore, NACA Report: RM E8J12 (Dec. 1948).

12. D. J. Spencer, H. Mirels, and T. A. Jacobs, App. Phys. Lett. **16,** 384 (1970).

13. D. A. Durran and D. J. Spencer, "Axisymmetrical Mixing Nozzle for Supersonic Diffusion Laser," Report TR-0059 (6756-02)-1, (The Aerospace Corp., Los Angeles, California, July 15, 1970).

14. R. R. Giedt, J. F. Bott, G. Emanuel, M. A. Kwok, H. Mirels, D. J. Spencer, and W. S. King, "Project Mesa Data Summary 1970," Report TR-0172(2779)-3 (The Aerospace Corp., Los Angeles, California, Nov. 5, 1971).

15. D. J. Spencer, H. Mirels, and T. A. Jacobs, Opto-Electronics **2,** 155 (1970).

16. W. W. Rigrod, J. App. Phys. **36,** 2487 (1965).

17. D. J. Spencer, D. A. Durran, and H. Mirels, Appl. Phys. Lett. **43,** 1151 (1972).

18. D. J. Spencer, D. A. Durran, and H. A. Bixler, Appl. Phys. Lett. **20,** 164 (1972).

19. M. A. Kwok, R. R. Giedt, and R. W. F. Gross, Appl. Phys. Lett. **16,** 386 (1970).

20. H. Mirels and D. J. Spencer, IEEE J. Quant. Electron. **QE-7,** 501 (1971).

21. M. A. Kwok and R. L. Wilkins, J. Chem. Phys. **60,** 2189 (1974); also M. A. Kwok, D. J. Spencer, and R. W. F. Gross, J. Appl. Phys. **45,** 3500 (1974).

22. R. A. Chodzko, D. J. Spencer, and H. Mirels, IEEE J. Quant. Electron. **QE-9,** 550 (1973).

23. D. J. Spencer and R. L. Varwig, AIAA J. **11,** 1000 (1973).

24. R. L. Varwig, "Photographic Observations of a cw HF Chemical Laser Reacting Flow Field," Report TR-0074(4240-10)-6 (The Aerospace Corp., Los Angeles, California, Aug. 15, 1973).

25. R. L. Varwig and M. A. Kwok, AIAA J. **12,** 208 (1974).

26. D. A. Durran and R. L. Varwig, "Performance of Triple Slit Nozzle for Chemical Laser Applications," Report TR-0073(4240-10)-2, (The Aerospace Corp., Los Angeles, California, Sept. 28, 1973).

27. M. A. Kwok and E. F. Cross, Opt. Eng. **11,** 131 (1972).

28. R. A. McClatchey and J. E. A. Selby, "Atmospheric Attenuation of HF and DF Laser Radiation," Report AFCRL-72-0312 (Air Force Cambridge Research Laboratories, L. G. Hanscom Field, Bedford, Massachusetts, May 23, 1972).

29. R. A. Meinzer, Int. J. Chem. Kinet. **2,** 335 (1970).

30. R. L. Varwig, "Chemical Laser Nozzle Flow Diagnostics," Report TR-172(2779)-1 (The Aerospace Corp., Los Angeles, California, Feb. 15, 1972).

31. R. R. Giedt, "MESA II 1971 Data Review," Report TR-0073(3435)-1 (The Aerospace Corp., Los Angeles, California, March 30, 1973); also MESA, Test 601 Field Data Report (The Aerospace Corp., Los Angeles, California, Sept. 7, 1973).

32. L. Morsell, unpublished results. The authors are also indebted to D. McClure of the Boeing Aerospace Center, Kent, Washington, for permission to use L. Morsell's data.

33. A blow down tunnel had also been used for experiments with a $CO_2$ transfer laser by A. N. Oraevskii at the Lebedev Institute, Moscow, USSR.

34. B. L. Siegel, NASA Technical Memorandum TM X-1466 (NASA, Nov. 1967).

35. B. L. Siegel, NASA Technical Note TN D-4514 (NASA, April 1968).

36. J. A. Shirley, R. N. Sileo, R. R. Stephens, and T. A. Cool, "Purely Chemical Laser Operation in the HF, DF, HF-$CO_2$, and DF-$CO_2$ Systems," presented at the AIAA 9th Aerospace Science Meeting, New York (Jan. 1971).

37. See the chapter by T. A. Cool.

38. R. A. Meinzer, R. H. Hall, J. Van Bogaerde, and B. R. Bronfin, "CW Combustion-Mixing Chemical Laser," paper presented at the 6th International Quantum Electrodynamics Conference, Kyoto, Sept. 1970.

39. P. Blaszuk, W. Burwell, J. Davis, M. Foster, L. Lynds, R. Meinzer, G. Palma, E. Shulman, D. Verbridge, and G. Wisner, "Laser Source Development," Final Report, UARL M 911239-22-3, Volume 3, Test Results (United Aircraft Research Laboratories, East Hartford, Connecticut, Sept. 1973).

40. F. N. Mastrup, J. E. Broadwell, J. Miller, and T. A. Jacobs, private communication of unpublished results, the writers are indebted to T. A. Jacobs of TRW Systems, Inc., Redondo Beach, California, for permission to use the figures and quote the data.

41. T. A. Cool, R. R. Stephens, and J. A. Shirley, J. Appl. Phys. **41,** 4038 (1970).

42. D. I. Rosen, R. N. Sileo, and T. A. Cool, IEEE J. Quant. Electron. **QE-9,** 163 (1973).

43. D. W. Naegeli and C. J. Ultee, Chem. Phys. Lett. **6,** 121 (1970).

44. J. A. Glaze, J. Finzi, and W. F. Krupke, Appl. Phys. Lett. **18,** 173 (1971).

45. J. A. Glaze and G. J. Lindford, Rev. Sci. Instrum. **44,** 600 (1973).

46. J. J. Hinchen, J. App. Phys. **45,** 1818 (1974).

47. C. J. Buczek, R. J. Freiberg, J. J. Hinchen, P. P. Chenausky, and R. J. Wayne, Appl. Phys. Lett. **17,** 514 (1970).

48. D. J. Spencer, J. A. Beggs, and H. Mirels, "Small Scale cw HF-DF Chemical Laser,"

Report TR-0076(6940)-6 (The Aerospace Corp., Los Angeles, California, April 1976).

49. R. R. Stephens and T. A. Cool, Rev. Sci. Instrum. **42,** 1489 (1971).
50. J. A. Glaze, Appl. Phys. Lett. **19,** 135 (1971).
51. J. J. Hinchen and C. M. Banas, Appl. Phys. Lett. **17,** 386 (1970).
52. D. Proch, J. Wanner, and H. Pummer, Report IPP IV/73 (Institut für Plasmaphysik, Garching, Germany, 1974).

CHAPTER **5**

# GAS DYNAMICS OF SUPERSONIC MIXING LASERS

## G. GROHS

## G. EMANUEL

**TRW Systems Group, Redondo Beach, California**

**263**

Continuous-wave operation of a chemical laser was first demonstrated[1-3] in 1969. This type of laser typically produces inversion by supersonically mixing initially separate oxidizer and fuel streams inside the optical cavity. Since the efficiency depends strongly on the rate of mixing, a large number of small nozzles (half of which contain the oxidizer, the other half the fuel) are generally used for rapid mixing. Once the reactants are mixed, a fast highly exothermic pumping reaction generates vibrationally excited product species. Collisional deactivation by V–T and V–V reactions competes with stimulated emission. Consequently, the optical, kinetic, and gas-dynamical processes are coupled. With this in mind, a review and extension are presented of the gas-dynamical literature applicable to such lasers.

Discussion is limited to a mixing laser in which inversion is provided by the "cold" $F + H_2 \rightarrow HF(v) + H$ or $F + D_2 \rightarrow DF(v) + D$ reactions. This laser has achieved good chemical efficiency[4] (~15%) and has been theoretically studied.[5,6] In laboratory experiments an electric arc, requiring substantial external power, is used as a thermal source for the dissociation of $F_2$ or $SF_6$, which provides the atomic fluorine. The thermal source, however, can also be provided by the combustion[7] of $D_2 + F_2$ for an HF laser, or of $H_2 + F_2$ for a DF laser. In this case no external power supply is required; the lasers operates completely on chemical energy.

A combustor-driven supersonic HF mixing laser requires the solution of a number of gas-dynamical problems. These include the determination of appropriate thermodynamic conditions for the dissociation of fluorine in the oxidizer plenum. The effects of boundary-layer growth on the highly viscous expansion through the oxidizer and fuel nozzles have to be taken into account. In the cavity, we must evaluate the rate of mixing as influenced by the nozzle boundary layers; by the presence of shock waves, vorticity, and turbulence; and by heat addition from the exothermic reactions. The rate of mixing must then be related to the lasing process. In addition, it is essential to establish a high velocity to remove the spent, absorbing gas from the cavity, and to maintain a cavity temperature and pressure suitable for efficient lasing. We also consider to what extent the supersonic flow, downstream of the cavity, can recover pressure in a diffuser.

As discussed in Ref. 8, cw lasing of HF has also been demonstrated in a subsonic mixing laser. We have not considered this configuration because of its low efficiency and lack of pressure-recovery potential.

Since the pulsed HF laser utilizes premixed reactants, one may wonder why a cw HF premixed laser[9] has received little attention. In this regard, we briefly mention a number of aspects, peculiar to pulsed operation, that cannot readily be achieved in a cw configuration. The premixing itself is done with great care, frequently at a reduced temperature,[10] with the reactants introduced alternately[11] into a mixing vessel. Thermodynamic conditions in the vessel, the feed lines from the vessel, and the laser tube itself must be maintained in the stable regime for the $H_2$–$F_2$ mixture.[11] This calls for a rather slow flow between vessel and tube, with no large changes in pressure. In the tube, the pumping reactions are triggered by the partial dissociation of $F_2$ with either a flashlamp or a discharge.[9–11] By comparison, the mixing process itself is the "trigger" for the cw laser.

For the reader's convenience, the text is subdivided into three major sections that can be read independently. Section 1 discusses the combustor-driven operation with emphasis on engineering estimates of flow conditions, Section 2 reviews the physics of mixing, and Section 3 presents scaling and performance estimates for the HF cold-reaction laser.

In Section 1, an overview is presented of the combustor-driven cold-reaction HF (or DF) laser. The discussion follows the flow starting with the combustor plenum, through the nozzles and the cavity, and terminating with the diffuser. Mixing and heat addition in the cavity are treated in the simplest possible manner with a one-dimensional description.[12,13] Estimates are obtained for the temperature, pressure, velocity, and Mach number in the cavity, and for the effect of mixing and reaction on these parameters. This section thus provides background material for the discussion in Sections 2 and 3 where the mixing and lasing processes are considered in greater detail. Another objective is to discuss pressure recovery and the choice of oxidizer diluent, since both are related to the laser power and efficiency.

Section 2 discusses the mixing process in the cavity, primarily from the point of view of a single mixing layer. The experimental and theoretical literature is reviewed for applicable results on shear layers and wakes with laminar, transitional, and turbulent flow. The flow field at the cavity inlet and its effect on mixing and reaction are described. Because of its importance for supersonic reactive mixing, we also discuss current theoretical studies of turbulent mixing.

In the last section a theoretical model for the power and efficiency is given that combines the analyses of Mirels et al.[5] and Broadwell.[6] Lasing with finite-rate chemistry is considered in the laminar or turbulent mixing layer between parallel uniform oxidizer and fuel jets of finite width. The physical significance of the dimensionless scaling parameters is discussed. Scaling laws predicted by this model are then used to correlate experimental laser-performance data obtained with both arc and combustor plenums.

## 1.  COMBUSTOR-DRIVEN HF cw LASER

### 1.1.  Introductory Remarks

We limit the presentation to the HF cold-reaction laser, where the $F + H_2$ reaction provides excited HF. All results, however, carry over directly to the DF cold-reaction laser. The thermal energy needed to dissociate molecular fluorine is provided by combustion of $F_2 + D_2$ in a plenum or combustor. (A DF laser utilizes $F_2 + H_2$ combustion. The reason for the isotope switch is discussed in Section 1.3.) As shown in Section 3, the combustor-produced DF (or HF for a DF laser) is an important deactivator in the laser cavity. The use of an arc as the thermal-energy source, therefore, results in a more efficient laser. For commercial applications, however, such as welding, cutting, and laser-induced chemistry, a combustor offers the practical advantage of operating from bottled gases.

Section 1.2 provides an overview of a combustor-driven laser. Subsequent subsections then discuss the combustor chamber, nozzle bank, cavity, and diffuser in the same sequence as encountered by the gas.

### 1.2.  Overview

An exploded-view photograph of a combustor-driven laser is shown in Fig. 1. At the upstream end, we have a rectangular combustor constructed of nickel or copper for compatibility with fluorine. Diluent mixed with $D_2$ enters the combustor through a set of small holes (0.074-cm diam.) situated on a raised post, as shown in Fig. 2. Additional diluent plus $F_2$ enter through the remaining holes (0.061-cm diam.), which are inclined so that the $F_2$ and $D_2$ jets impinge on each other. The $D_2 + F_2$ combustion is hypergolic, and equilibrium is rapidly achieved. As a consequence, the combustor is only 5-cm long. Because of this design, the combustor performance[14] is insensitive to the fraction of diluent injected with the $D_2$ or $F_2$. The injector plate, where the reactants enter, is not water cooled; the side surfaces of the combustor are water cooled.

The hot gas from the combustor, containing the dissociated fluorine,

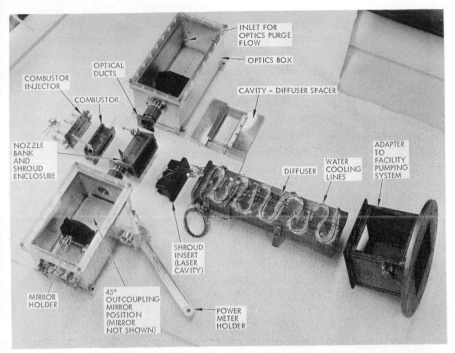

**Fig. 1.** Exploded view of combustion-driven laser.

flows through an array of small nozzles. These are either of two-dimensional or axisymmetric design, and do not differ from those used for arc-driven operation.

In most laboratory experiments, the gas from the nozzle bank enters the cavity as free jets. Since recirculating gas is detrimental to laser performance because of deactivation and absorption by ground-state HF, a provision for purging (generally with $N_2$) the stagnant gas surrounding the free jet is important. This purge flow can be entirely eliminated if the jet is enclosed by metal shrouds, as shown in Fig. 1. As discussed in Section 1.6, shrouds also contribute to improved pressure recovery.

Openings for the laser beam must, of course, be provided on both sides of the jet. To minimize the presence of absorbing gas in the beam path, purge tubes (often called "optical ears") are attached to the side walls. Purge gas, at a small fraction of the flow rate required for purging a free-jet cavity, enters the tubes near the mirrors and flows toward the jet, where it is entrained. This purge flow also protects the mirrors from the corrosive gases in the cavity. Figure 3 shows the shroud surfaces viewed through an optical ear. Also visible is the exit of a two-dimensional nozzle bank.

**Fig. 2.** Combustor injector.

Downstream of the shrouds, and joining them smoothly, is a diffuser of rectangular design. Its function is to convert as much as possible of the stagnation pressure of the inlet supersonic flow into static pressure at the exit, where the flow is subsonic. The diffuser-exit pressure is referred to as the recovered pressure.

Like the shrouds, a diffuser is not an essential feature of a cw chemical laser. It has the same function as the "jet catcher" discussed elsewhere, that is, to increase the pump-inlet pressure. Although diffusers are a standard component of supersonic wind tunnels, their inlet conditions when used with a chemical-laser cavity are sufficiently different that further discussion is warranted.

**Fig. 3.** Shrouds and nozzle bank viewed through an optical duct.

## 1.3. Combustor

**General Equations for Combustor Operation.** Three gases are injected into the combustor: $D_2$, $F_2$, and diluent (He or $N_2$). The function of the combustor is to produce heat to dissociate fluorine. Hence, excess fluorine is used, and the $D_2$ fuel is completely consumed by reaction with $F_2$:

$$D_2 + F_2 \rightarrow 2\,DF, \qquad \Delta H = -130 \text{ kcal/mol } F_2$$

After combustion, the hot-gas mixture consists of F, $F_2$, DF, and added diluent. The thermal energy generated by this combustion is $130 \times \dot{N}_{D_2}$ kcal/sec, where $\dot{N}_i$ is the molar flow rate of species $i$. Part of this energy is lost by heat transfer $\dot{Q}_c$ to the combustor walls, part is used to heat the gas mixture to its final stagnation temperature $T_0$, and the remainder dissociates some or all of the excess fluorine. With $\dot{m}_i$ denoting the mass flow rate of species $i$, we characterize the gas after combustion by

$$\alpha = \frac{\dot{m}_F}{\dot{m}_F + \dot{m}_{F_2}}, \qquad \psi = \frac{\dot{N}_{DF} + \dot{N}_{dil}}{0.5\dot{N}_F + \dot{N}_{F_2}} \tag{1}$$

where $\dot{m}_i = W_i \dot{N}_i$, and $W_i$ is the molecular weight of species $i$. The parameter $\alpha$ is the degree of $F_2$ dissociation, or the F-atom mass fraction, while $\psi$ is the molar ratio of diluent, including combustor-produced DF, to the unburned excess fluorine considered as $F_2$.

Since deuterium is expensive, one would prefer to use $H_2 + F_2$ combustion as the thermal-energy source. Experiments,[4] however, have shown HF-laser efficiency to be very poor in this case. There are two reasons for this. First, the HF(0) concentration, from the combustor, exceeds by more than an order of magnitude the HF(1) concentration produced in the cavity. The $1 \rightarrow 0$ (i.e., $v = 1 \rightarrow v = 0$) band is thus absorbing, and lasing cannot occur in this band. Second, HF(0) is a much faster deactivator of HF($v$), by V–V reactions, than is DF(0). The particular reaction that dominates is HF(2) + HF(0) $\rightarrow$ 2 HF(1), which adversely affects lasing on the $2 \rightarrow 1$ band. For the same reasons, poor efficiency is observed[4] when $D_2 + F_2$ combustion is utilized for a DF laser. Consequently, an HF laser is driven by a DF combustor and vice versa.

The concentration of F atoms in the combustor, in equilibrium with $F_2$,

$$F_2 \rightleftharpoons 2 F, \qquad \Delta H = 37.8 \text{ kcal/mol } F_2$$

is determined by the temperature-dependent equilibrium constant $K_p$. The partial pressures are related to $K_p$ by

$$K_p = \frac{p_F^2}{p_{F_2}} \qquad (2)$$

where $p_i$ is the partial pressure (atm) of species $i$. The equilibrium constant for fluorine dissociation is given by

$$K_p(T_0) = 6.32 \times 10^3 \, T_0^{0.8} \exp\left(\frac{-3.76 \times 10^4}{RT_0}\right) \qquad (3)$$

where $R(= 1.987 \text{ cal/mol-°K})$ is the universal gas constant, and $T_0$ is the combustor gas temperature.

We determine the degree of dissociation $\alpha$ in terms of $\psi$ and the combustor pressure $p_0$. Since $\dot{N}_i$ is proportional to $p_i$ and $\dot{m}_i = W_i \dot{N}_i$, Eqs. (1) can be written in terms of partial pressures as

$$\alpha = \frac{p_F}{p_F + 2p_{F_2}}, \qquad \psi = \frac{p_{DF} + p_{dil}}{0.5 p_F + p_{F_2}} \qquad (4)$$

The combustor pressure is the sum of the partial pressures

$$p_0 = p_F + p_{F_2} + p_{DF} + p_{dil} \qquad (5)$$

By combining Eqs. (2), (4), and (5), we obtain the required relation

$$\alpha = \frac{1}{2a} \{[\psi^2 + 4a(1 + \psi)]^{1/2} - \psi\} \qquad (6)$$

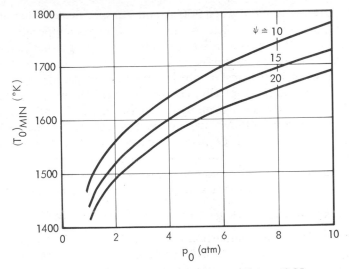

**Fig. 4.** Minimum combustor temperature for $\alpha \geq 0.95$.

where $a$ is

$$a = 1 + \frac{4p_0}{K_p(T_0)} \tag{7}$$

and $K_p$ is given by Eq. (3).

Most of the unburned fluorine leaving the combustor should be atomic for cold-reaction operation, and $\alpha$ should be close to unity. Figure 4 shows the minimum combustor temperature $(T_0)_{min}$ needed for $\alpha \geq 0.95$ as a function of $p_0$ for $\psi = 10$, 15, and 20. Observe that as $\psi$ decreases or $p_0$ increases, $(T_0)_{min}$ increases.

Our primary interest is in cold-reaction operation; we, thus, assume $T_0 \geq (T_0)_{min}$. For this case, we determine injected and combustor-exit flow rates in terms of $T_0$, $p_0$, $\psi$, the combustor heat-transfer rate $\dot{Q}_c$ (kcal/sec) and the nozzle bank throat area $A^*$ (cm²). A prime superscript is used to denote the flow rates injected into the combustor. With $\dot{N}_{F_2}$ equal to zero, combustor-exit flow rates are given by

$$\dot{N}_{dil} = \dot{N}'_{dil}, \qquad \dot{N}_{DF} = 2\,\dot{N}'_{D_2}, \qquad \dot{N}_F = 2(\dot{N}'_{F_2} - \dot{N}'_{D_2}) \tag{8}$$

and $\psi$ now is

$$\psi = \frac{2(\dot{N}_{DF} + \dot{N}_{dil})}{\dot{N}_F} \tag{9}$$

The energy equation has the form

$$65\dot{N}_{DF} = \dot{Q}_c + 18.9\dot{N}_F + 10^{-3}(T_0 - 300)(\bar{C}_{p\,dil}\dot{N}_{dil} + \bar{C}_{pDF}\dot{N}_{DF} + \bar{C}_{pF}\dot{N}_F) \tag{10}$$

where $\bar{C}_{pi}$ are average (between $300°K$ and $T_0$) specific heats (cal/mol-°K) of species $i$, and where we have assumed a $300°K$ gas temperature before injection. The left-hand term is the combustion energy (kcal/sec), while the $18.9\dot{N}_F$ term is the energy needed for $F_2$ dissociation.

The combustor heat-transfer rate is a function of $p_0$, $T_0$, gas composition, and surface area. Typically, combustors operate with $\alpha \approx 0.95$ and $\psi \gtrsim 10$. Consequently, $T_0$ is given by the curves in Fig. 4, and temperature and composition are no longer independent variables. For the combustor described in Section 1.2, $\dot{Q}_c$ can be expressed as a fraction of the combustion energy, where this fraction depends on $p_0$ and the water-cooled surface area $A_c$ of the combustor. Experimental heat-transfer data for a $5 \times 5 \times 20$-cm$^3$ combustor with $D_2$-$F_2$-He reactants yield[14]

$$\dot{Q}_c = 6.52 \times 10^{-4} A_c \, p_0^{-0.2} \, (65\dot{N}_{DF}) = 0.163 p_0^{-0.2} \, (65\dot{N}_{DF}) \qquad (11)$$

and about 16% of the combustion energy is lost by heat transfer at low pressure ($\sim 1$ atm), while 10% is lost at high pressure ($\sim 10$ atm).

The combustor mass-flow rate (g/sec) is given by the choked-nozzle flow equation[15]

$$\dot{m} = 111 \left( \frac{2}{\gamma+1} \right)^{(\gamma+1)/2(\gamma-1)} \left( \frac{\gamma W}{T_0} \right)^{1/2} p_0 A^* \qquad (12)$$

The mass-flow rate, molecular weight $W$, and ratio of specific heats $\gamma$ of the combustion products are given by

$$\dot{m} = W_{dil}\dot{N}'_{dil} + 4\dot{N}'_{D_2} + 38\dot{N}'_{F_2} \qquad (13)$$

$$W = \dot{m}(\dot{N}_{dil} + \dot{N}_{DF} + \dot{N}_F)^{-1} \qquad (14)$$

and

$$\gamma = \frac{C_{pdil}\dot{N}_{dil} + C_{pDF}\dot{N}_{DF} + C_{pF}\dot{N}_F}{(C_{pdil} - R)\dot{N}_{dil} + (C_{pDF} - R)\dot{N}_{DF} + (C_{pF} - R)\dot{N}_F} \qquad (15)$$

Table 1. Specific Heat Values for Combustor Gases ($T_0 \approx 1700°K$)

| Species | $\bar{C}_{pi}$, cal/mol-°K | $C_{pi}$, cal/mol-°K |
|---------|---------------------------|----------------------|
| F       | 5.10                      | 5.01                 |
| DF      | 7.60                      | 8.30                 |
| He      | 4.97                      | 4.97                 |
| $N_2$   | 7.80                      | 8.46                 |

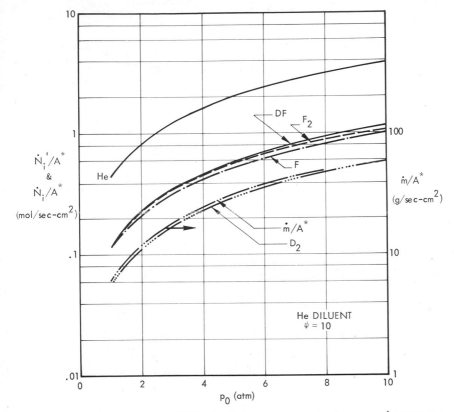

**Fig. 5.** Injected molar flow rates $N'(F_2–D_2–He)$, combustor molar flow rates $\dot{N}(F–DF–He)$, and total mass flow rate $\dot{m}$. Combustor temperature is given by Fig. 4, and flow rates are per unit throat area $A^*$, $\psi = 10$.

The foregoing molar specific heats are evaluated at the combustor temperature $T_0$. Since the specific heats vary slowly with temperature, we evaluate them at 1700°K as shown in Table 1.

**He Versus $N_2$ Diluent.** It is a simple matter to solve Eqs. (9) through (15) on a computer, given $p_0$, $T_0$, $\psi$, and the type of diluent, He or $N_2$. Figures 5 through 8 show flow rates per unit nozzle-throat area for the $\psi = 10$ and $\psi = 20$ curves of Fig. 4, where $\alpha = 0.95$. Note that $\dot{N}_{He} = \dot{N}'_{He}$ and that $\dot{N}_{N_2} = \dot{N}'_{N_2}$.

Comparing $N_2$ versus He diluent operation at the same $\psi$, $T_0$, and $p_0$ shows that $\dot{m}/A^*$ is ~1.6 times larger with $N_2$, as would be expected from its higher molecular weight. Despite this, the F-atom molar flow rate $\dot{N}_F/A^*$ is ~1.8 times larger with He diluent. As shown in Section 3, laser

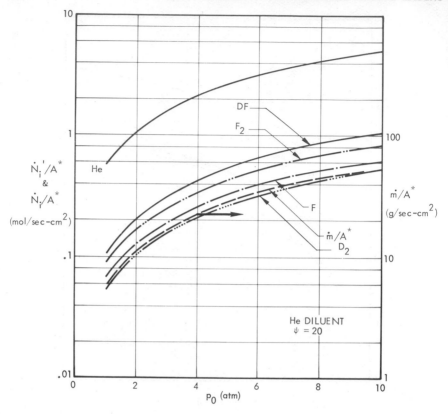

**Fig. 6.** Injected molar flow rates $\dot{N}'$ ($F_2$–$D_2$–He), combustor molar flow rates $\dot{N}$ (F–DF–He), and total mass flow rate $\dot{m}$. Combustor temperature is given by Fig. 4, and flow rates are per unit throat area $A^*$, $\psi = 20$.

power increases with the F-atom flow rate. For a given nozzle bank, we, therefore, expect more power with He diluent.

Operation with $N_2$ differs from that with He in the specific heat ratio $\gamma$ and the mixture molecular weight. For all considered $p_0$ and $\psi$ values, $\gamma$ for He diluent operation ranges from 1.55 to 1.57, whereas for $N_2$ diluent it ranges from 1.32 to 1.34. Consequently, for the same $p_0$, $T_0$, and effective nozzle-area ratio, the nozzle-exit static temperature and pressure are substantially higher with $N_2$ diluent. Similarly, the molecular weight for He diluent operation ranges from 7.8 to 9.8; whereas, for $N_2$, it ranges from 24.8 to 25.9. The higher molecular weight reduces both the nozzle-exit velocity and the rate of laminar diffusion in the cavity.

Within the context of the mixing model presented in Section 3, laser

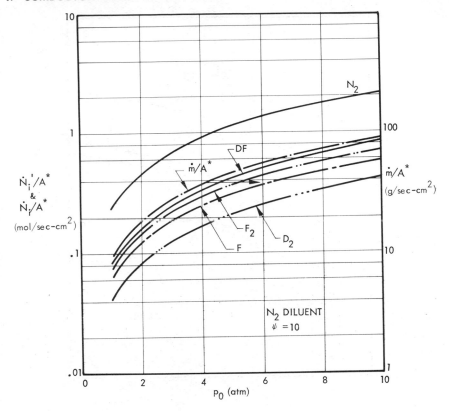

**Fig. 7.** Injected molar flow rates $\dot{N}'$ ($F_2$–$D_2$–$N_2$), combustor molar flow rates $\dot{N}$ (F–DF–$N_2$), and total mass flow rate $\dot{m}$. Combustor temperature is given by Fig. 4, and flow rates are per unit throat area $A^*$, $\psi = 10$.

power is proportional to $\dot{N}_F/\dot{N}_{DF}$.* This molar ratio decreases slightly with increasing $p_0$, while it decreases more substantially with increasing $\psi$. The ratio is also lower with $N_2$ diluent than with He diluent. For example, at $p_0 = 10$ atm and $\psi = 20$, ($\dot{N}_F/\dot{N}_{DF}$) equals 0.435 for $N_2$, but equals 0.575 for He. At $p_0 = 1$ atm and $\psi = 10$, both He and $N_2$ give a $\dot{N}_F/\dot{N}_{DF}$ ratio near unity. Since $N_2$ has a larger heat capacity than He, more combustion is needed with $N_2$ diluent to dissociate the same amount of fluorine. Consequently, a large fraction of the diluent is DF. Thus, at $p_0 = 1$ atm and $\psi = 10$, we have $\dot{N}_{N_2}/\dot{N}_{DF} = 2.85$ and $\dot{N}_{He}/\dot{N}_{DF} = 3.74$.

* As shown in Section 3, this is only one of the factors that determines power. Our discussion here assumes laser performance is in the high-pressure regime, and that combustor produced DF is the principal collisional deactivator.

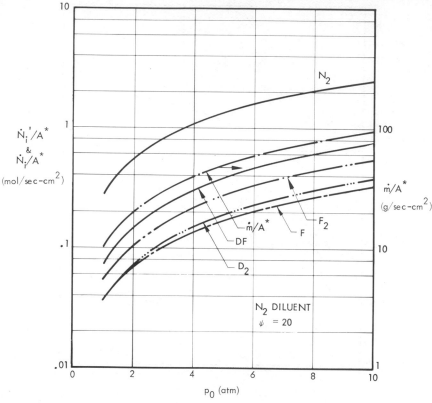

**Fig. 8.** Injected molar flow rates $\dot{N}'$ ($F_2$–$D_2$–$N_2$), combustor molar flow rates $\dot{N}$ (F–DF–$N_2$), and total mass-flow rate $\dot{m}$. Combustor temperature is given by Fig. 4, and flow rates are per unit throat area $A^*$, $\psi = 20$.

A variety of diluent dependent factors have been examined. In terms of laser power, all of them favor He-diluent operation. For a given nozzle bank, the peak experimental[16] power with He diluent is about a factor of 2 larger than with $N_2$ diluent.

## 1.4. Nozzle Bank

**General Remarks.** The nozzle bank performs a number of critical functions required for efficient lasing. First and foremost, it provides the fluorine oxidizer and hydrogen fuel jets, which, upon mixing in the cavity, react to generate vibrationally excited HF. Rapid mixing, important for good efficiency, is assisted by using a large number of small nozzles. These nozzles are designed to produce supersonic flow at a temperature and pressure substantially lower than that in the combustor. As discussed in Section 3, a temperature in the cavity of about 400°K and a pressure of a few Torr are required for efficient lasing. Another function, for cold reaction operation, is to "freeze" chemically the F-atom mole fraction at or near its combustor value. The combustor nozzles thus establish the appropriate pressure, temperature, and composition for the all-important pumping reactions in the laser cavity. Finally, the large cavity-inlet velocity, provided by the supersonic nozzles, stretches the lasing-zone length and provides the potential for pressure recovery.

Figure 9 shows two different nozzle arrays, both of which are two-dimensional. The one[17] shown in Fig. 9a has a contoured fluorine-oxidizer nozzle and slit injection for the hydrogen-fuel gas. This nozzle has been used extensively for both arc[16] and combustor-driven[14] experiments. Shown in Fig. 9b is a similar nozzle array, except that the fuel enters through supersonic nozzles, and both the oxidizer and fuel nozzles are wedge shaped. Straight cones also are often used for axisymmetric nozzles.[17]

Because the boundary-layer thickness at the nozzle exit is a substantial fraction of the exit dimension, contoured nozzles are designed by first

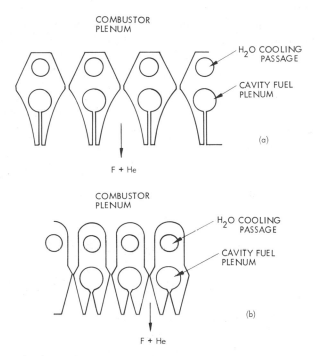

**Fig. 9.** Schematic of two-dimensional supersonic nozzle arrays. (a) Slit fuel injection and (b) supersonic fuel injection.

calculating an inviscid core flow by the method of characteristics. Subsequently, a boundary-layer calculation is performed, and the displacement thickness is added to the core flow to obtain the final nozzle contour. The pressure distribution used for the boundary-layer calculation, of course, must match the distribution at the outer edge of the core flow. A parallel shock-free nozzle-exit flow can be provided by contoured nozzles, but for only one combustor pressure and temperature. The reason is that the Reynolds number changes when plenum conditions change, thereby altering the boundary-layer growth rate from that of the design calculation.

The heat-transfer rate from the combustor gases to the subsonic and transonic parts of the nozzles is substantial. This is a consequence of the sizable stagnation region per nozzle blade, the amount of wetted surface area, and the high mass flux in the transonic region. Typically, the energy lost to the nozzle blades exceeds that lost to the combustor walls by a factor of 2. Because of the high heat-transfer rates, the convergent part of the nozzle blades is water cooled, as shown in Fig. 9. The fuel flow also provides some blade cooling downstream of the throats. However, the combustor-gas heat-transfer rate there is sharply reduced compared to its value upstream of the throat.

There is the possibility that some F atoms recombine in the gas phase, because the static temperature inside the nozzles decreases rapidly. This recombination is a three-body process and, thus, very pressure dependent. Whether it occurs depends on the combustor pressure and on the residence time of the flow inside the convergent part of the nozzles. Because of the small distance (~0.3 cm) from nozzle inlet to throat, this residence time is generally very short, and three-body recombination is slow even at pressures of several atmospheres. Downstream of the throat, residence times are even shorter and pressures lower. Calculations assuming a fully catalytic surface[18] indicate, however, that a significant amount (as much as 20%) of fluorine may recombine at the wall, principally upstream of the throat. This affects the F and $F_2$ concentration profile across the nozzle-exit plane, especially in the boundary layer.

**Nozzle-Flow Analysis.**  Because of their small size (see Table 2), the nozzles are characterized by an unusually small Reynolds number, which results in a rapid boundary-layer growth rate. Consequently, the laminar boundary layers are quite thick at the exit of both fuel and oxidizer nozzles. A convenient Reynolds number for viscous nozzle flow can be defined by

$$\mathrm{Re} = \frac{\rho_0(2H_0)^{1/2}h^*}{2\mu_0} \tag{16}$$

**Table 2.** Dimensions for a Two-Dimensional Nozzle Bank of the Type Shown in Fig. 9b

|  | Oxidizer | Fuel |
|---|---|---|
| Throat width, cm | 0.0102 | 0.00762 |
| Throat-to-exit distance, cm | 0.335 | 0.175 |
| Area ratio | 18.75 | 13.33 |

where $h^*$ is the throat half width, or the throat radius for an axisymmetric nozzle, $H$ is the enthalpy, $\rho$ is the density, $\mu$ is the viscosity, and the subscript 0 refers to stagnation conditions. With $\mu = cT^\omega$ and the equation of state, we obtain in cgs units (except for $p_0$, which is in atmospheres)

$$\text{Re} = 78.5 \left( \frac{\gamma W}{\gamma - 1} \right)^{1/2} \frac{p_0 h^*}{c T_0^{\omega + 0.5}} \tag{17}$$

One method for determining the viscous flow inside a given nozzle is to solve the Navier–Stokes equations, simplified by neglecting the transverse pressure gradient, for the complete nozzle-flow field.[19] Another approach[20] is to estimate the displacement thickness distribution, calculate the pressure gradient using a one-dimensional approximation for the core flow, solve the boundary-layer equations for a new displacement thickness, and iterate until the pressure distribution along the nozzle has converged. Both approaches yield results in good agreement with low Reynolds-number experimental data.[20,21]

Standard measurements during laser operation include injected-combustor flow rates, combustor pressure, and heat-transfer losses. These measurements allow combustor plenum conditions to be determined, as outlined in Section 1.3. A knowledge of the nozzle geometry then is sufficient for computing, by one of the previously mentioned methods, the nozzle-flow field. Average values for quantities such as static temperature, stagnation pressure, velocity, and so on, are obtained at the nozzle-exit plane by conserving fluxes of mass, momentum, and energy. These integrated fluxes relate directly to the usual boundary-layer thickness parameters, such as displacement thickness $\delta^*$ and momentum thickness $\theta$.

Laser-test data[14] for an $H_2$–$F_2$–He combustor using the two-dimensional nozzle bank shown in Fig. 9b were processed by the foregoing procedure, with an inviscid core-boundary layer calculation for the nozzle. In these tests, the combustor pressure varied from 1 to 7.25 atm. Table 2 provides the pertinent nozzle-geometry parameters, and Fig. 10 shows the results of the computations. In Fig. 10a, $h_o$ is the exit half

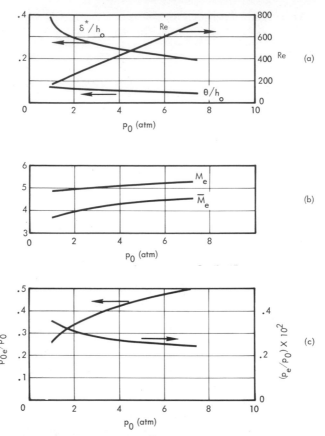

**Fig. 10.** Nozzle-exit conditions versus combustor pressure for an $H_2$–$F_2$–He combustor. Nozzle geometry is given in Table 2. (a) Re, $\delta^*/h_o$, and $\theta/h_o$ versus combustor pressure, (b) $M_e$ and $\overline{M}_e$ versus combustor pressure, and (c) static and stagnation-pressure ratios versus combustor pressure.

width of the oxidizer nozzle, which from Table 2 is 0.0956 cm. Both $\delta^*$ and $\theta$ are proportional to $\mathrm{Re}^{-1/2}$, as expected from laminar boundary-layer theory,[22] with $\theta$ approximately one fifth of $\delta^*$. The large value of $\delta^*/h_o$, between 0.2 and 0.4, shows that the boundary layer fills much of the nozzle-exit plane. The effective nozzle-area ratio is, thus, substantially reduced from its geometric value. For the cases under consideration, the velocity boundary-layer thickness is about three times larger than $\delta^*$. Since $(\delta^*/h_o) = 0.33$ at about 1.25 atm combustor pressure, the boundary layers from the two side walls are merged at the nozzle exit when $p_0 \leq 1.25$ atm.

In these experiments the Reynolds number, Eq. (16), ranged from 160 at 1 atm pressure to 700 at 7.25 atm (Fig. 10a). At the lower Reynolds numbers (~200), the boundary-layer displacement thickness at the nozzle's throat is no longer negligible.[19] As the Reynolds number decreases, the increasing displacement thickness at the throat can cause the effective nozzle-area ratio to increase. Since this effect does not appear in Fig. 10, viscous effects are underestimated at the lowest Reynolds numbers.

A nozzle-wall boundary layer separates if the ambient pressure sufficiently exceeds the nozzle-exit static pressure. For a large nozzle, where the boundary layer is turbulent, separation begins when the ambient pressure exceeds the exit pressure by a factor ranging from 2.5 to 5.[23] For a single, low Reynolds-number nozzle, however, separation can occur at a pressure ratio as low as 1.3.[23] The extent of separation in the oxidizer nozzles, particularly when the fuel is slit injected, and its effect on the mixing process are not known at this time (see Section 2).

Figure 10b shows the centerline, or core, exit Mach number $M_e$ and the average exit Mach number $\bar{M}_e$. The latter is based on the flux averaging method previously discussed and is given by

$$\bar{M}_e = \left(1 - \frac{\delta^* + \theta}{h_o}\right)^{1/2} M_e \tag{18}$$

The appreciable difference between the two Mach numbers is a direct result of the large value of $(\delta^* + \theta)/h_o$. At high Reynolds numbers, of course, the Mach numbers become independent of $p_0$, and this trend is evident in the figure. At low values of $p_0$, however, $\bar{M}_e$ shows considerable variation.

Figure 10c shows the ratio of the static exit pressure $p_e$ (uniform across the flow) to $p_0$ and the ratio of the average exit-stagnation pressure $\overline{p_{0e}}$ to $p_0$. These ratios are determined by the isentropic formulas

$$\frac{p_e}{p_0} = \left(1 + \frac{\gamma - 1}{2} M_e^2\right)^{-\gamma/(\gamma-1)} \tag{19}$$

and

$$\frac{\overline{p_{0e}}}{p_0} = \frac{p_e}{p_0} \frac{\overline{p_{0e}}}{p_e} = \left(\frac{p_e}{p_0}\right)\left(1 + \frac{\gamma - 1}{2} \bar{M}_e^2\right)^{\gamma/(\gamma-1)} \tag{20}$$

The nozzle-exit static pressure ranges from 2.66 to 13.8 Torr (at the higher combustor pressure). Of importance is the large loss of stagnation pressure in the nozzle because of viscous dissipation. At the nozzle exit, the average stagnation pressure ranges from $(1/4)p_0$ when $p_0 = 1$ atm to $(1/2)p_0$ when $p_0 = 7.25$ atm. This large loss in stagnation pressure adversely affects pressure recovery. From the point of view of pressure recovery, best results are obtained at the highest combustor pressure.

## 1.5. Cavity

We determine in the simplest manner possible the effect of the mixing and heat-addition processes on the temperature, pressure, and Mach number (also see Section 2). Evaluation of laser power is covered in Section 3. The topic of base relief is first discussed to help clarify the assumptions and logic of the subsequent analysis.

**Base Relief.** For simplicity, the discussion is limited to two-dimensional nozzles that produce alternate supersonic oxidizer and fuel jets. We first consider the case where the jets have matched static pressures at the nozzle-exit plane and the nozzles have sharp trailing edges (no base relief), as shown in Fig. 11a. The presence of shrouds or the ambient cavity pressure first affects the flow along the Mach waves, Fig. 11b, that originate at the top and bottom edges of the nozzles. Since the oxidizer and fuel jets may have different Mach numbers and ratios of specific heats, the Mach angles can vary from jet to jet. In any event, there is a triangular region downstream of the nozzles where the flow is unaffected by edge conditions. In this region, the flow is approximately parallel, or a constant-area flow.

As shown in Fig. 11a, mixing starts at the trailing edges of the nozzles. Thus heat addition and lasing first occur in the triangular region. The heat generated in this region causes the pressure and temperature to increase and the supersonic Mach number to decrease.[15] Choking will occur if the Mach number in the region decreases to unity. In this situation, a strong shock system forces the upstream flow to readjust.[15]

Shock waves may also be present in the triangular region because of trailing-edge disturbance or unequal nozzle-exit pressures. The heat-addition process interacts with these disturbances to strengthen them (Section 2). In any case, the increase in pressure because of heat addition tends to reduce laser efficiency (Section 3). When choking occurs, efficiency drops sharply,[24] a phenomenon sometimes referred to as "thermal blockage."

If choking occurs in the triangular region, then reducing the ambient pressure or increasing the shroud angle does not appreciably alter the shock system. One method[24] for reducing the pressure rise and preventing choking is to introduce a base region between hydrogen nozzles, or slits, as shown schematically in Fig. 12 and discussed further in Section 2. Immediately downstream of the base, we have a subsonic wake region at or below the ambient pressure. Expansion fans from the wake region interact with the reacting flow in the mixing region to reduce the pressure

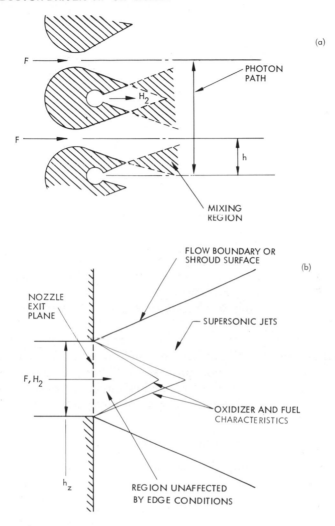

**Fig. 11.** Nozzle schematic without base relief showing mixing regions and regions unaffected by edge conditions. (*a*) Top view of nozzle and (*b*) side view of nozzle.

increase. Base relief has proved to be an effective method for avoiding choking at low values of the diluent ratio $\psi$.[24]

An alternate approach is to increase $\psi$, that is, the amount of diluent in the flow. The additional diluent absorbs part of the added heat, resulting in a smaller pressure rise. Both techniques are utilized, the choice being an arbitrary one. On the one hand, base relief increases the size of the nozzle bank and decreases the potential for pressure recovery (Section

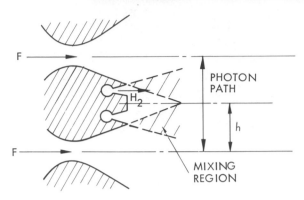

**Fig. 12.**   Nozzle schematic with base relief.

1.6). On the other hand, increasing $\psi$ increases the flow rate, assuming $\dot{N}_F$ is kept fixed to maintain laser power. With regard to laser efficiency, experimental testing has not yet shown one approach to be more advantageous than the other. For instance, at low $\psi$, base relief improves efficiency, but increasing $\psi$ without base relief similarly improves efficiency.

**Analytical Approach.**   We present a simple one-dimensional model for the cavity mixing and reaction processes. Within the context of such a model, we have the choice of assuming either a constant-area or a constant-pressure flow. The latter assumption requires that the following conditions be met:

1.   The flow is a free jet in the cavity.

2.   The nozzle-exit pressures for the fluorine and hydrogen jets are approximately matched.

3.   The correct amount of base relief is utilized.

Free-jet experiments using nozzles without base relief are often performed with an adjustable ambient pressure. (This pressure is controlled either by increasing the cavity purge flow or by changing the opening of a downstream exhaust valve.) Although generally referred to as a constant-pressure jet, in fact the pressure is constant only on the jet's surface. It is indeed possible to increase the ambient pressure to the extent that the jet has approximately a constant cross-sectional area. In this case, the variable pressure inside the jet differs appreciably from ambient.

The constant-area flow assumption does not require matched nozzle-exit pressures for the oxidizer and fuel jets. Furthermore, the base-relief configuration, Fig. 12, can also be treated within the constant-area approximation, provided a pressure can be assigned to the base region between the fuel nozzles. The constant-area assumption is valid when the shrouds plus the boundary-layer displacement thicknesses produce a parallel flow, or when a sufficiently high ambient pressure is utilized. The shroud-angle condition is often approximately achieved in practice. Shroud half angles as small as 5° have been used without an adverse effect on laser efficiency.[14]

Space limitations do not permit discussion of both the constant-area and constant-pressure assumptions. We treat the constant-area case without base relief because it is less restrictive and results in a higher recovered pressure.[12]

We also assume that all specific-heat ratios are temperature independent, and that parallel fuel and oxidizer flows enter the cavity at sonic or supersonic speeds. As before, all unreacted fluorine is assumed to be dissociated. Viscous and heat-transfer losses to the shrouds are not included. These losses generally are small.

In a constant-area flow (or one that is constant-pressure), the equations of motion that conserve mass, momentum, and energy, and the equation of state are all algebraic. An exact solution is thus readily obtained that connects different uniform states. In the subsequent analysis, we are concerned with the following states: the oxidizer and fuel jets at the nozzle-exit plane, the mixed but unreacted flow, and, finally, the mixed and reacted flow. An important advantage of this approach is that the solution is independent of the details of the intervening processes.[25] For example, the mixed state is the same for laminar and turbulent mixing and is independent of the mixing-layer growth rate. Similarly, the state after heat addition is independent of the rate of reaction or the specific steps involved in the kinetics. Furthermore, the mixing and heat-addition processes can be analyzed separately, although in reality they occur simultaneously. When the two solutions are connected, they represent an exact solution to the governing equations irrespective of the detailed simultaneous processes. The final result will be shown to depend on only three dimensionless parameters: a ratio of specific heats, a mixing parameter, and a heat addition parameter.

**Mixed Solution.** We now present the constant-area solution for the nonreacting mixing of two parallel streams, denoted by single and double primes, respectively. A subscript 0 denotes stagnation conditions, a subscript 1 denotes the fully mixed flow, and cgs units are used throughout, exept for pressure, which is in atmospheric units.

Some readily determined values for the mixed flow (area $A$, mass $\dot{m}$, molar flow rates $\dot{N}$, molecular weight $W$, ratio of specific heats $\gamma$, and stagnation temperature $T_0$) are given by:

$$A_1 = A' + A'' \tag{21}$$

$$\dot{m}_1 = (\rho u A)_1 = (W\dot{N})' + (W\dot{N})'' \tag{22}$$

$$\dot{N}_1 = \dot{N}' + \dot{N}'' \tag{23}$$

$$W_1 = \left(\frac{\dot{m}}{\dot{N}}\right)_1 \tag{24}$$

$$\gamma_1 = \frac{\{[\gamma/(\gamma-1)]\dot{N}\}' + \{[\gamma/(\gamma-1)]\dot{N}\}''}{\{[1/(\gamma-1)]\dot{N}\}' + \{[1/(\gamma-1)]\dot{N}\}''} \tag{25}$$

$$\left(\frac{\gamma}{\gamma-1}\dot{N}T_0\right)_1 = \left(\frac{\gamma}{\gamma-1}\dot{N}T_0\right)' + \left(\frac{\gamma}{\gamma-1}\dot{N}T_0\right)'' \tag{26}$$

where $u$ is the velocity. The equations of energy and impulse (or momentum) are

$$T_{01} = T_1 + \frac{[(\gamma-1)Wu^2/\gamma]_1}{2\times 4.184\times 10^7\times R} \tag{27}$$

$$J_1 = J' + J'' \tag{28}$$

where the impulse function $J$ is

$$J = 1.013\times 10^6 pA + \dot{m}u$$
$$= 1.013\times 10^6 pA(1+\gamma M^2) \tag{29}$$

The solution to the foregoing equations can be expressed in terms of $\gamma_1$ and a dimensionless mixing parameter $Z_m$:

$$Z_m = 1 - \left\{2\times 4.184\times 10^7 R\left[\left(\frac{\gamma+1}{\gamma}\right)\frac{T_0}{W}\left(\frac{\dot{m}}{J}\right)^2\right]_1\right\} \tag{30}$$

Actually, a variety of constant area mixing solutions is possible. They are categorized as follows:[26]

1.  When $Z_m < 0$, no solution exists, since the flow chokes.[25]

2.  When $0 \le Z_m \le \gamma_1^{-2}$, two solutions occur, one subsonic and the other supersonic. The supersonic solution may not satisfy the second law of thermodynamics, in which case only the subsonic solution is valid. A check on the change in entropy should be performed, particularly when $M_1$ is large compared to unity, which is when the second law is likely to be violated. The subsonic solution always satisfies the second law and can be obtained from the supersonic solution by the normal shock relations.

3. When $Z_m > \gamma_1^{-2}$, only the subsonic solution occurs. This solution satisfies the second law.

The solution that experimentally occurs in a laser cavity is supersonic, and we, hereafter, assume this to be the case, that is, $M_1 \geq 1$.

The physical meaning of $Z_m$ becomes clear when $M_1$ is large compared to unity. We then have from Eqs. (29) that $(J/\dot{m})_1 \cong u_1$. The maximum possible velocity of the mixed flow is

$$u_{max} = \left[2 \times 4.184 \times 10^7 R\left(\frac{\gamma}{\gamma-1}\frac{T_0}{W}\right)_1\right]^{1/2} \tag{31}$$

Thus $Z_m$ has the form

$$Z_m \cong 1 - \left(\frac{\gamma^2-1}{\gamma^2}\right)_1 \left(\frac{u_{max}}{u_1}\right)^2 \tag{32}$$

and is a measure of the energy ratio $(u_1/u_{max})^2$.

After some algebra, the governing equations provide the solution:

$$M_1^2 = \frac{1+Z_m^{1/2}}{1-\gamma_1 Z_m^{1/2}} \tag{33}$$

$$\frac{u_1}{u_r} = \left[\frac{(\gamma+1)M^2}{1+\gamma M^2}\right]_1 \tag{34}$$

$$\frac{T_1}{T_{01}} = \left(1+\frac{\gamma-1}{2}M^2\right)_1^{-1} \tag{35}$$

$$\frac{p_1}{p_r} = \left[\frac{1+\gamma M^2}{(\gamma+1)M^2\{1+[(\gamma-1)/2]M^2\}}\right]_1 \tag{36}$$

$$\frac{p_{01}}{p_{0r}} = \left[\frac{\{1+[(\gamma-1)/2]M^2\}^{\gamma/(\gamma-1)}}{1+\gamma M^2}\right]_1 \tag{37}$$

where the reference quantities are

$$u_r = \left(\frac{\gamma}{\gamma+1}\frac{J}{\dot{m}}\right)_1, \quad p_r = \left(\mathcal{R}\frac{\gamma+1}{\gamma}\frac{T_0}{W}\frac{\dot{m}^2}{JA}\right)_1, \quad p_{0r} = \left(\frac{J}{1.013 \times 10^6 A}\right)_1, \tag{38}$$

and $\mathcal{R}$ (= 82.06 atm-cm$^3$/mol-°K) is the universal gas constant.

Figure 13 presents $M_1$, $u_1/u_r$, $p_1/p_r$, $T_1/T_{01}$, and $p_{01}/p_{0r}$ versus $Z_m$ for $\gamma_1 = 1.4$. (The actual value for $\gamma_1$ is close to 1.4 because of the large $H_2$-molar flow rate.) We observe that the stagnation-pressure ratio and $M_1$ increase rapidly when $Z_m$ exceeds 0.3. In this region, $(\gamma M^2)_1$ considerably exceeds unity and Eqs. (29) indicate that $J_1 \cong (\dot{m}u)_1$. Combining this with Eq. (38) shows that $p_{0r}$ varies linearly with $(\dot{m}/A)_1$, since $u_1$ has

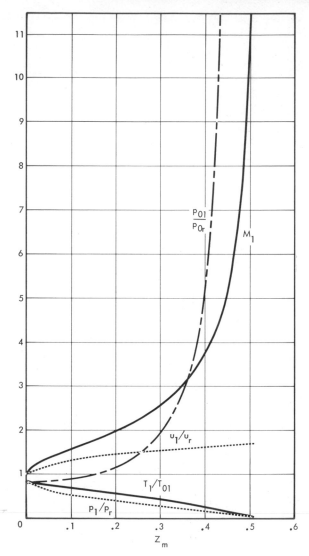

**Fig. 13.** $M_1$, $p_1/p_r$, $p_{01}/p_{0r}$, $T_1/T_{01}$, and $u_1/u_r$ after mixing versus $Z_m$ for $\gamma = 1.4$.

little variation when $Z_m \geq 0.3$ (Fig. 13). The significance of this result will become apparent in Section 1.6. Figure 13 also shows that a low mixed static-temperature and pressure result when $Z_m \geq 0.3$.

**Heat-Addition Solution.** The effect of heat addition on the constant-area flow by the pumping reaction of $F + H_2$ and the removal of laser power is now examined. We assume that the fast-pumping reaction goes to completion, the V–V and V–T reactions go to equilibrium, and the ratio of specific heats remains constant at its initial value $\gamma_1$. This last assumption is justified when changes in composition and temperature are not extreme, as is the case when large amounts of diluent, including excess fuel, are present. It is not necessary to assume that the molecular weight and total molar flow rate are unchanged, since these quantities are constant anyway during binary reactions, such as the V–V, V–T, and pumping reactions.

We previously defined in Eq. (9) the molar ratio $\psi$ of diluent to fluorine flow. In a similar manner, we define for the cavity (hydrogen) fuel flow

$$R_L = \frac{\dot{N}_{H_2}}{0.5\dot{N}_F + \dot{N}_{F_2}} \cong \frac{2\,\dot{N}_{H_2}}{\dot{N}_F} \tag{39}$$

Optimal experimental values for $R_L$ are in the range of 8 to 20, which means that the fuel molar-flow rate is four to ten times larger than the F-atom molar flow rate. Most of the fuel, therefore, does not react and thus constitutes additional diluent. A large molar fraction ($\sim 50\%$) of the injected fuel can be replaced by He without any appreciable change in laser power.[27] This indicates that a large fuel flow is required for gas-dynamic rather than kinetic reasons, such as increasing the $F + H_2$ pumping rate. The gas-dynamical phenomenon involved centers on enhanced mixing, and possibly cooling effects caused by added $H_2$ (or He) acting as diluent. Note, however, that an increasing fuel flow tends to compress the oxidizer jet and results in a higher oxidizer static temperature and pressure.

Examination of Eqs. (9) and (39) shows that the initial value of $\dot{N}/\dot{N}_F$, where $\dot{N}$ is the (constant) total molar flow rate, can be written as

$$\frac{\dot{N}}{\dot{N}_F} = \frac{\dot{N}_{DF} + \dot{N}_{dil} + \dot{N}_{H_2} + \dot{N}_F}{\dot{N}_F} = \frac{\psi + R_L + 2}{2} \tag{40}$$

Thus the energy $Q_{\Delta H}$ added by the pumping reaction (kcal/g) is given by

$$Q_{\Delta H} = \frac{|\Delta H|\,\dot{N}_F}{\dot{m}} = \frac{\dot{N}_F}{\dot{N}}\frac{\dot{N}}{\dot{m}}|\Delta H| = \frac{2}{W_1}\frac{|\Delta H|}{\psi + R_L + 2} \tag{41}$$

We define the chemical efficiency of the laser $\eta$ in terms of the exothermicity $|\Delta H|$ of the $F + H_2$ pumping reaction:

$$\eta = \frac{P}{4.184 \times 10^3 \dot{N}_F |\Delta H|} \tag{42}$$

where $P$ is laser power in watts. By combining Eqs. (40) and (42), the power can be written in terms of efficiency as

$$P = 4.184 \times 10^3 \frac{2}{W_1} \frac{|\Delta H|}{\psi + R_L + 2} \dot{m}\eta \tag{43}$$

If we denote flow conditions after the heat addition and lasing processes are complete by a subscript 2, the energy equation has the form

$$R\left(\frac{\gamma}{\gamma-1}\right)_1 \frac{T_{02}}{W_1} = R\left(\frac{\gamma}{\gamma-1} \frac{T_0}{W}\right)_1 - \frac{P}{4.184\dot{m}} + 10^3 Q_{\Delta H} \tag{44}$$

By combining Eqs. (41), (43), and (44), we obtain for the stagnation-temperature ratio

$$Z_{\Delta H} = \frac{T_{02}}{T_{01}} = 1 + 2 \times 10^3 \left(\frac{\gamma-1}{\gamma}\right)_1 \frac{|\Delta H|}{\psi + R_L + 2} \frac{1-\eta}{RT_{01}} \tag{45}$$

which characterizes the heat-addition and lasing processes.

We now refer to Ref. 15 for the rest of the solution. The Mach number $M_2$ is first determined by solving

$$g(M_2) = \frac{g(M_1)}{Z_{\Delta H}} = \frac{g_1}{Z_{\Delta H}} \tag{46}$$

where $g$ is defined by

$$g(M) = \frac{(1+\gamma_1 M^2)^2}{M^2\{1 + [(\gamma_1-1)/2]M^2\}} \tag{47}$$

The supersonic solution of Eq. (46) is readily shown to be

$$M_2^2 = \frac{g_1 - 2\gamma_1 Z_{\Delta H} + [g_1^2 - 2(\gamma_1+1)g_1 Z_{\Delta H}]^{1/2}}{2\gamma_1^2 Z_{\Delta H} - (\gamma_1-1)g_1} \tag{48}$$

and other quantities of interest are given by

$$\frac{T_2}{T_1} = \left(\frac{M_2}{M_1}\right)^2 \left(\frac{1+\gamma_1 M_1^2}{1+\gamma_1 M_2^2}\right)^2 \tag{49}$$

$$\frac{p_2}{p_1} = \frac{1+\gamma_1 M_1^2}{1+\gamma_1 M_2^2} \tag{50}$$

$$\frac{p_{02}}{p_2} = \left(1 + \frac{\gamma_1-1}{2} M_2^2\right)^{\gamma_1/(\gamma_1-1)} \tag{51}$$

**Combined Solution.** The mixing and heat-addition solutions are easily combined* into results that depend only on the dimensionless parameters $\gamma_1$, $Z_m$, and $Z_{\Delta H}$. Figures 14 through 17 show $M_2$, $T_2/T_{01}$, $p_2/p_r$, and

---

* For example, $(T_2/T_{01}) = (T_2/T_1)(T_1/T_{01})$.

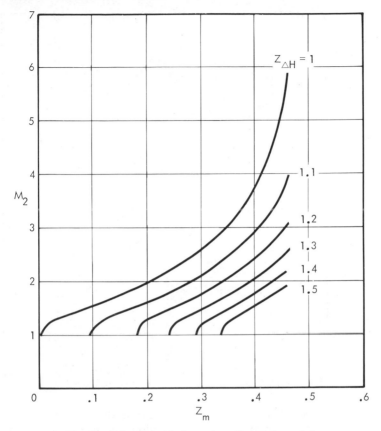

**Fig. 14.** $M_2$ versus $Z_m$ for various $Z_{\Delta H}$ and $\gamma_1 = 1.4$.

$p_{02}/p_{0r}$ versus $Z_m$ for various values of the stagnation temperature ratio $Z_{\Delta H}$, assuming $\gamma_1 = 1.4$.

The foregoing mixed solution is represented by the $Z_{\Delta H} = 1$ curves. Thus, the effect of heat addition on the Mach number is given by the difference $[(M_2)_{Z_{\Delta H}=1}] - (M_2)_{Z_{\Delta H}}$ at a fixed value of $Z_m$. Choking occurs when $M_2 = 1$ or, from Eq. (48), when $Z_{\Delta H} = (\frac{1}{2})g_1/(\gamma_1 + 1)$. This result, when combined with Eq. (45), yields the minimum value for $\psi + R_L$ needed to avoid choking.

Figure 14 shows the well-known result that a supersonic Mach number decreases with heat addition, with choking occurring abruptly, since the curves are vertical at $M_2 = 1$. Figure 15 shows the expected increase in temperature with heat addition. The ratio $T_2/T_{01}$ can exceed unity, since the normalization uses the stagnation temperature before heat addition. Figures 16 and 17 show the static and stagnation pressures, respectively,

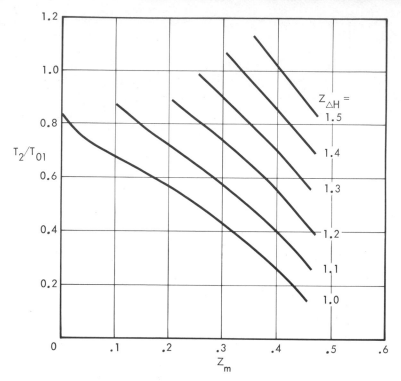

**Fig. 15.** $T_2/T_{01}$ versus $Z_m$ for various $Z_{\Delta H}$ and $\gamma_1 = 1.4$.

which vary in opposite directions with heat addition. The decrease in stagnation pressure is particularly severe when the mixing parameter $Z_m$ exceeds 0.35.

The principal methods for adjusting $Z_m$ and $Z_{\Delta H}$ are through the oxidizer-nozzle area ratio and $\psi + R_L$, respectively. The area ratio affects $Z_m$ through $(J/\dot{m})$, Eq. (30). For example, consider the use of a large-area ratio nozzle, which would result in a large value for $Z_m$. Such a nozzle, however, has either a very small throat and thus a thick boundary layer at the exit, or a large exit dimension and thus fewer mixing regions. Consequently, a large-area ratio nozzle results in either a large stagnation-pressure loss in the nozzle, or poor laser efficiency stemming from the slower mixing. To a limited extent, other parameters in $Z_m$ can be varied. Thus, the combustor temperature can be increased above the value needed for complete fluorine dissociation, thereby increasing $T_{01}$ in $Z_m$. This implies a larger DF/F ratio, however, and increased deactivation in the cavity. From Eq. (45), we see that increasing $\psi + R_L$ decreases $Z_{\Delta H}$. The reduced reactive heating makes possible the use of low-angle shrouds, which, in turn, improves pressure recovery, and is one of the reasons for operating with large amounts of diluent.

We illustrate the use of Figs. 14 through 17 by considering typical values for a cold reaction HF laser with He as diluent. These values are $\gamma_1 = 1.4$, $W_1 = 6$ g/mol, $|\Delta H| = 31.6$ kcal/mol, $T_{01} = 10^{3\circ}$K, $(J/\dot{m})_1 = 2.7 \times 10^5$ cm/sec, $(J/A)_1 = 1.9 \times 10^5$ g/cm-sec$^2$, $\psi = 17$, $R_L = 8$, and $\eta = 0.1$, and result in $Z_m = 0.35$ [Eq. (30)], $Z_{\Delta H} = 1.3$ [Eq. (45)], and reference pressures of $p_r = 6.11 \times 10^{-2}$ atm and $p_{0r} = 0.187$ atm. Table 3 summarizes conditions after mixing and after reaction as obtained from the figures. The heat addition drives the Mach number down to 1.75, which is rather close to choking, and increases the static temperature by 450°K. Lasing occurs during this temperature increase, but may terminate before all the fluorine is utilized (see Section 3) and, thus, before the 800°K temperature is achieved. A constant-area cavity experiences an adverse pressure gradient, as exemplified by the static pressures in the table. The increase in static pressure contributes significantly toward the recovered pressure. As shown in the table, this is accompanied by a sharply decreased stagnation pressure.

It is pertinent at this time to consider the effect of chain-reaction operation where $\alpha \ll 1$, and both the $F + H_2$ and $H + F_2$ reactions produce excited HF. The heat release is 64.8 kcal/mol of HF, compared to

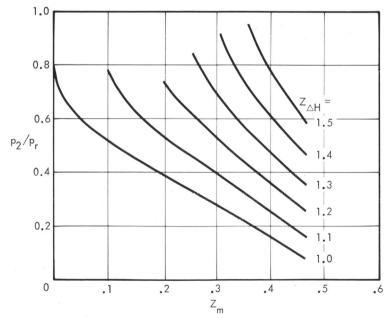

**Fig. 16.** $p_2/p_r$ versus $Z_m$ for various $Z_{\Delta H}$ and $\gamma_1 = 1.4$.

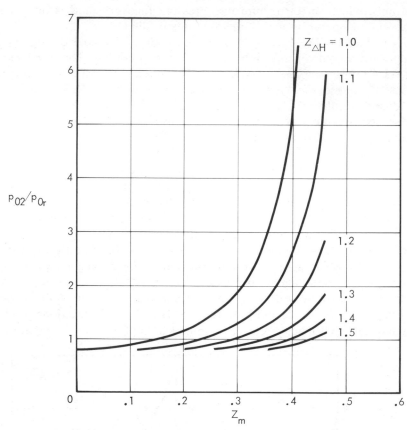

**Fig. 17.** $p_{02}/p_{0r}$ versus $Z_m$ for various $Z_{\Delta H}$ and $\gamma_1 = 1.4$.

**Table 3.** Flow Conditions after Mixing, State 1, and after Heat Addition, State 2

|  | State 1 | State 2 |
| --- | --- | --- |
| M | 3.0 | 1.75 |
| $T$, °K | 350 | 800 |
| $T_0$, °K | 1000 | 1300 |
| $p$, atm | $1.35 \times 10^{-2}$ | $3.50 \times 10^{-2}$ |
| $p_0$, atm | 0.533 | 0.187 |

31.6 kcal/mol for cold-reaction operation. The higher exothermicity produces higher cavity temperatures and possibly choking. In fact, the foregoing example with 64.8 kcal/mol for $|\Delta H|$ results in a choked flow. Furthermore, laser efficiency for chain operation has proved to be less than that obtainable from cold-reaction operation.[27] It is for these reasons that only cold-reaction operation has been considered in this article.

## 1.6. Diffuser

One important factor that sets a chemical-laser diffuser apart from standard diffusers is the low entrance Reynolds number, which results in thick boundary layers. The low Reynolds number, Eq. (17), is a consequence of four contributing factors:

1.  Low entrance static pressure ($\sim 3.5 \times 10^{-2}$ atm),

2.  Low molecular-weight gas ($\sim 6$ g/mol),

3.  High entrance static temperature ($\sim 800°$K), and

4.  Small entrance height ($\sim 5$ cm).

Items 1 through 3 result in a low density. Item 4 stems from the elongated rectangular cross-section of the cavity.

The simplest diffuser used to date has been a constant-area duct of approximately 10:1 length to height ratio, followed by a 6° half-angle subsonic diffuser.[14] The diffuser inlet mates directly and smoothly to the ends of the cavity shrouds. The shock waves inside the diffuser interact with the wall boundary layers causing them to separate, which leads to additional upstream disturbances. It is important, therefore, to isolate the laser-cavity flow from these disturbances. This is accomplished by extending the shrouds a short distance past the region of lasing and by providing a small positive half angle ($\sim 5°$) for the shrouds. Without these precautions, the shock system developed by the diffuser can extend into the cavity region, causing a severe reduction in laser power.[14]

As a consequence of the isolation requirement, diffuser performance cannot be entirely separated from the shroud configuration, which may, in fact, be considered as part of the diffuser. As discussed shortly, optimum recovered pressure occurs when the shrouds have a minimum positive angle, that is, the diffuser inlet area is a minimum. The minimum angle consistent with good laser efficiency depends on the height of the nozzle bank, on the boundary-layer growth rate, and on the magnitude of the adverse-pressure gradient caused by the heat addition. Thus low $\psi$ operation, or highly exothermic chain-reaction operation, requires a large shroud angle or base relief to offset the severe adverse-pressure gradient.

For a low Reynolds-number straight-duct diffuser, the common normal-shock diffuser model[28] is particularly appropriate. The stagnation pressure downstream of a normal shock, designated by $p_{03}$, is, therefore, a good first-order upper estimate of the static pressure at the exit of the subsonic diffuser. In our formulation the recovered pressure is given by

$$\frac{p_{03}}{p_{0r}} = \frac{\gamma_1 + 1}{2} \frac{M_2^2}{1 + \gamma_1 M_2^2} \left[ \frac{(\gamma_1 + 1)^2 M_2^2}{4\gamma_1 M_2^2 - 2(\gamma_1 - 1)} \right]^{1/(\gamma_1 - 1)} \tag{52}$$

and is shown in Fig. 18. The most important conclusion from the figure is the insensitivity of $p_{03}/p_{0r}$, which varies from 0.8 to 0.9, to either $Z_m$ or $Z_{\Delta H}$, provided the flow does not choke. The large loss in stagnation pressure caused by heat addition, when $Z_m \geq 0.35$, is compensated by the static pressure increase $p_2/p_1$, and by the weak normal shock (since $M_2$ approaches unity) with its small stagnation-pressure loss. Because of this

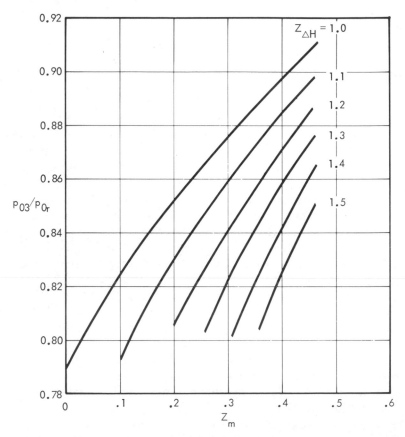

**Fig. 18.**  $p_{03}/p_{0r}$ versus $Z_m$ for various $Z_{\Delta H}$ and $\gamma_1 = 1.4$.

insensitivity, the most effective method for increasing the recovered pressure is to increase $p_{0r}$, which is given by Eq. (38).

The recovered pressure, thus, largely depends on $p_{0r}$. As shown in Section 1.5, $p_{0r}$ is proportional to $(\dot{m}/A)_1$ when $Z_m \gtrsim 0.3$. [Here, $(\dot{m}/A)_1$ refers to the flow rate at the diffuser entrance.] Consequently, the recovered pressure scales linearly with $(\dot{m}/A)_1$, a relation confirmed by diffuser testing[14] with cavity shrouds.

The foregoing implies that base relief, which would increase $A_1$, has an adverse effect on pressure recovery. To understand this conclusion, recall that a nozzle with base relief allows the flow to expand to a lower static pressure and a higher Mach number. With a higher entrance Mach number, the normal-shock stagnation-pressure loss in the diffuser increases, and the recovered pressure is reduced.

The constant-pressure solution, without base relief, results in a jet whose cross-sectional area increases in the cavity because of the heat addition. The situation is analogous to base relief, since the inlet area of a diffuser must exceed the jet's initial cross-section. The percent increase in area depends, of course, on factors such as $\psi + R_L$. Again, the Mach number of the jet at the inlet of the diffuser is higher than for a constant-area flow, and the recovered pressure is reduced.

Applying Fig. 18 to the example given in Table 3, we obtain an estimated recovered pressure $p_{03}$ of 0.157 atm. The overall compression ratio measured from the start of the cavity to the diffuser exit is $(p_{03}/p_1) = 11.6$. Of this ratio, 2.59 occurs across the cavity, while 4.48 occurs across the diffuser.

A number of attempts[14] have been made to improve diffuser performance for a chemical laser by using either a second throat with a small contraction ratio or a thin splitter plate in the constant-area duct section. In all cases, the diffuser failed to start, and a shock system established itself inside the cavity. This behavior is probably a result of the low entrance Reynolds number.

Exothermic reactions may also occur inside the diffuser. The most important of these is the highly exothermic (104 kcal/mol $H_2$) recombination of H atoms that are produced in the cavity by the pumping reaction. If this heat addition occurred within a straight-duct diffuser, it would generally choke the flow, a phenomenon easily detected because of the strong shocks that would then occur upstream of the diffuser entrance. This has not been observed in any experiment to date. Gas-phase recombination requires three-body collisions, and its rate is apparently negligible at pressures and velocities typical of the flow in the diffuser. The recombination most likely occurs on the catalytic walls with the exothermicity contributing to the wall heat transfer.

## 2.  THE FLOW FIELD IN A MIXING-LASER CAVITY

Collisional deactivation processes in HF or DF cw lasers are nearly as fast as the pumping reactions that populate excited vibrational levels, so that pumping must be initiated within the optical cavity. Hence, the reactants are kept separate until they reach the cavity inlet, where pumping is initiated by mixing between adjacent supersonic jets of fuel and oxidizer. The distributions of excited species and optical gain are thus intimately coupled to the mixing process, which is inherently viscous and two-dimensional.

In Section 2.1, we describe the gross features of the flow field observed in laser cavities to provide a frame of reference for the subsequent discussion. Section 2.2 combines experimental and theoretical results on shear layers and wakes into a description of the mixing layer. The flow field at the cavity inlet and its effect on mixing and reaction are discussed in Section 2.3. Section 2.4 reviews current theoretical studies of turbulent mixing, in view of their applicability to supersonic reactive mixing in the cavity.

### 2.1.  Flow Configuration in Laser Cavities, Overview

**Boundary and Initial Conditions.**    As discussed in Section 1 and illustrated by Fig. 19, the reactants typically enter the laser cavity from alternate narrow two-dimensional oxidizer and fuel nozzles (e.g., see Fig.

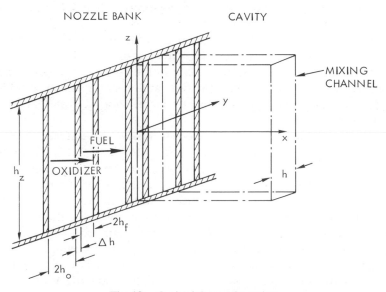

**Fig. 19.**  Cavity-inlet configuration.

9). To a good approximation, changes in the $z$-direction can be neglected compared with those in the $x$–$y$-plane spanned by the mean-flow direction and the optical axis (see Fig. 11$a$). Mach numbers are sufficiently high such that the Mach waves carrying perturbations from the nozzle-end walls or shrouds (Fig. 11$b$) reach the jet axes only far downstream. Because of the large number of nozzles stacked side-by-side in a nozzle bank, the flow pattern resulting from mixing and reaction of the fuel and oxidizer jets is periodic in $y$, with symmetry conditions imposed at the nozzle centerlines. One need, therefore, consider only one channel of width $h = h_o + h_f + \Delta h$, where $h_o$ and $h_f$ are half the widths of an oxidizer and fuel nozzle, respectively, and where $\Delta h$ is the width of the dividing wall or land area between these nozzles at the cavity inlet.

Table 4 gives free-stream conditions as well as nozzle and boundary-layer scale lengths for the exit of the nozzle bank shown in Fig. 9$b$, with dimensions given in Table 2. These conditions are representative of a combustor-driven HF cold-reaction laser[14] and were calculated by the methods of Section 1.4. In this case, mixing occurs between two supersonic jets, one containing atomic fluorine, DF from the combustion reaction, and helium as diluent, and the other supplying pure $H_2$ at a molar flux considerably higher than that of F. Exit pressures in both nozzles are of the order of a few Torr and ideally are matched. The jet

**Table 4.** Typical Free-Stream Conditions and Scale Lengths at Cavity Inlet[14] (nozzle bank of Fig. 9$b$, HF cold-reaction laser, DF combustor, He diluent)

| Parameter | Symbol | Units | Oxidizer jet[a] | Fuel jet[a] |
|---|---|---|---|---|
| Mach no. | $M$ | — | 4.9(0)–5.2(0) | 3.6(0)–4.1(0) |
| Unit Reynolds no. | $Re/l$ | $cm^{-1}$ | 4.7(3)–1.2(4) | 2.8(3)–9.2(3) |
| Pressure | $p$ | Torr | 3.3(0)–9.3(0) | 3.3(0)–9.3(0) |
| Velocity | $u$ | cm/sec | 2.3(5)–2.7(5) | 2.6(5)–2.7(5) |
| Temperature | $T$ | °K | 1.7(2)–1.9(2) | 9.0(1)–8.1(1) |
| Density | $\rho$ | $g/cm^3$ | 3.0(−6)–7.1(−6) | 1.2(−6)–4.0(−6) |
| Mole fractions | $Y_F$ | — | 0.09–0.17 | — |
| | $Y_{DF}$ | — | 0.15–0.18 | — |
| | $Y_{He}$ | — | 0.76–0.65 | — |
| | $Y_{H_2}$ | — | — | 1.0 |
| Displacement thickness | $\delta^*$ | cm | 3.2(−2)–2.2(−2) | 2.3(−2)–1.6(−2) |
| Momentum thickness | $\theta$ | cm | 6.9(−3)–4.8(−3) | 1.1(−3)–7.0(−4) |
| Nozzle half width | $h_o$, $h_f$ | cm | 9.5(−2) | 5.1(−2) |
| Width of land area | $\Delta h$ | cm | 2.5(−2) | |
| Nozzle height | $h_z$ | cm | 2.5(0) | |

[a] 4.7(3) read as $4.7 \times 10^3$.

velocities given in Table 4 do not differ greatly, but both temperature and density are appreciably lower in the fuel than in the oxidizer jet, so that the momentum carried by the $H_2$ jet is smaller.

Because of the low pressures and small-nozzle dimensions, Reynolds numbers are much lower than usually encountered in large-scale supersonic flow facilities, that is, smaller than 3000 when based on the nozzle exit widths $2h_o$ or $2h_f$. As a result, the jets are laminar, and initial turbulence levels are negligible. Boundary layers in the fuel and oxidizer nozzles are generally thick, and at the lower pressures may fill the entire nozzle exit. The effect of the boundary-layer momentum deficits on the development of the mixing layer cannot be neglected. The inner part of the boundary layers is subsonic, permitting upstream propagation of disturbances from the trailing edge of the dividing wall. Gas temperatures in this region should be close to the wall temperature of about 400°K (Section 1.4). Surface recombination in the oxidizer nozzle may deplete F and add $F_2$ near the wall, which would lead to some $H + F_2$ reaction in the early mixing layer.

Fuel and oxidizer jets issuing from actual laser nozzles are rarely uniform and parallel, or matched in pressure. For example, the nozzles shown in Fig. 9b have straight-ramp supersonic sections, so that fuel and oxidizer tend to impinge on each other. If this misalignment persists into the outer supersonic boundary layers, shocks will form downstream of the nozzle exit that turn the two flows parallel. Such recompression shocks also appear after expansion into the void created by a finite land area between the nozzles. Slit or spray-bar injection of hydrogen, see Figs. 9a and 20, result in sonic orifice flow at exit pressures greatly exceeding that in the oxidizer nozzle. The fuel jet then achieves supersonic speeds while pluming out into the lower-pressure oxidizer flow, and pressures are gradually equalized by a complex pattern of expansion and shock waves.

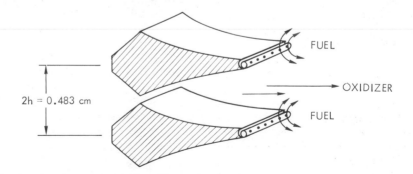

**Fig. 20.**   Spray-bar injector nozzle.[4]

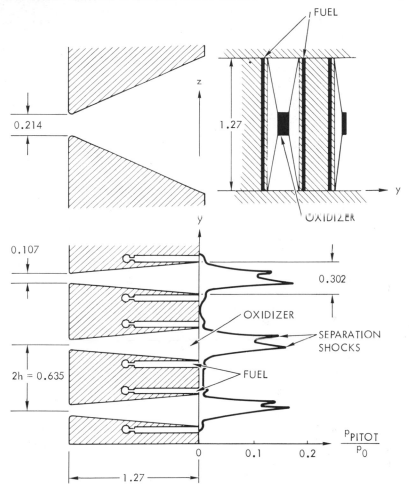

**Fig. 21.** Configuration of base-relief nozzle, pitot-pressure profile at nozzle exit (all dimensions in centimeters and $p_0$ is the oxidizer plenum pressure).[29]

These waves are reflected at the nozzle centerlines and then interact with the mixing layer. Thus both transverse- and axial-pressure gradients occur in the mixing region.

The only published detailed flow-field measurements in a chemically active laser cavity are those of Ref. 29. The nozzle bank investigated, see Fig. 21, is somewhat unusual. The oxidizer nozzle expands in both the $y$- and $z$-directions, and the hydrogen is injected through two sonic slits separated by a large land area (to provide base pressure relief, see Section 1.5. The large spikes of the pitot pressure in the oxidizer-nozzle outlet,

Fig. 21, indicate shock waves that originate further upstream in the nozzle.[29] A plausible explanation is that, because of excess pressure of the hydrogen jet, the thick boundary layer has separated from the wall of the oxidizer nozzle. The constriction of the supersonic nozzle flow by the separated boundary layers then results in separation shocks.

Separated flow regions between the fuel and oxidizer streams can be detrimental to lasing, since $H_2$ and F can react to equilibrium in the low-speed recirculating flow. The vibrationally equilibrated HF absorbs laser radiation, and when it diffuses into the downstream lasing zone, it is an efficient deactivator of excited HF.

**Mixing and Reaction.** When the fuel and oxidizer streams first meet downstream of the nozzle lip, the concentration, velocity, and temperature profiles or their first $y$-derivatives are discontinuous at the interface, as schematically shown in Fig. 22$a$. The species profiles shown indicate the effect of wall recombination of fluorine. Diffusion, viscous dissipation, and thermal conduction smooth out these discontinuities over a mixing zone that grows into the adjacent fuel and oxidizer streams, Fig. 22$b$. The profile shapes and the growth rate depend on the initial profiles as well as on whether the mixing is laminar, transitional, or turbulent, a question that has not yet been resolved for the flow in laser cavities.

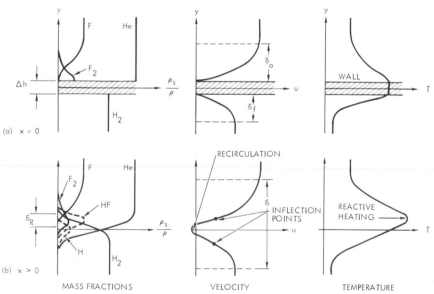

**Fig. 22.** Initial profiles and profiles in the early mixing layer ($\Delta h$ thickness of dividing wall, $\delta_o$, $\delta_f$ boundary-layer thicknesses, $\delta$ thickness of viscous mixing layer, $\delta_R$ of reaction zone, $\rho_s$ species density).

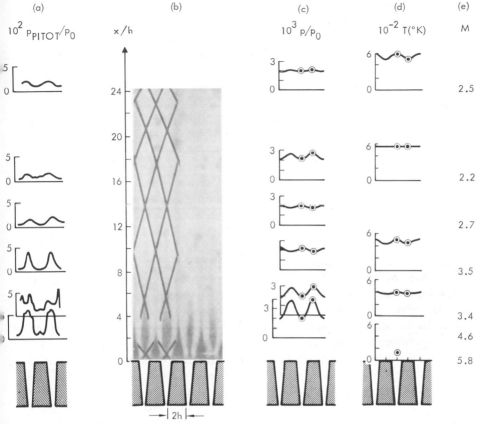

**Fig. 23.** Flowfield in a cavity with base relief.[29] (a) Pitot pressure/plenum pressure; (b) reaction zones, shock-wave pattern; (c) static pressure/plenum pressure; (d) R–T temperature; and (e) average Mach number.

Fuel and oxidizer entrained into the mixing layer diffuse toward each other and are consumed by the pumping reaction that occurs in a narrower reaction zone. There, the excited HF spontaneously emits infrared line radiation, with overtones (e.g., $v = 3 \to 0$) in the near ir that can be photographed.[29,30] The insert in Fig. 23b shows the luminous regions downstream of the nozzle bank of Fig. 21.[29] Because of overpressure of the $H_2$ jets, the reaction zones first appear in the exit of the oxidizer nozzles, where the boundary layers are separated. The separation shocks, tentatively sketched in for the left two-nozzle elements, are rapidly cancelled by the expansion of the flow into the low-pressure nonradiating base region between adjacent fuel-injection slits. The base pressure relief widens the unmixed oxidizer-core flow and deflects the luminous zones to the fuel side where they nearly coalesce at $x/h \simeq 3$. Further downstream, the oxidizer core again contracts and the fuel jet expands ($x/h \simeq 8$). Around $x/h \simeq 10$, there is a second bulge in the oxidizer flow, which ends in a long spike near $x/h = 15$ where the mixing layers appear to merge. The analysis presented in Section 3 predicts that cold-reaction lasing does not continue much beyond this point, since no more fluorine is entrained into the mixing layers after merging.

The apparent deflections of the luminous front and bulges in the nonreacting-core flows can be explained by interactions with shock waves, as indicated in the left half of Fig. 23b. The curvature of the shock results from the independently observed axial decrease in Mach number, and from partial reflection when the shocks interact with the inhomogeneous mixing layers. This interpretation is confirmed by the pitot-pressure traverses (Fig. 23a) and static-pressure measurements (Fig. 23c) shown alongside the chemiluminescence photograph. The pitot-pressure spikes corresponding to the separation shocks in the oxidizer-nozzle exit (Fig. 21) have virtually disappeared at $x/h = 2$, because of the expansion into the base region. At $x/h = 4$, a new system of recompression shocks has developed that aligns the oxidizer and fuel streams. At both stations, a weak peak on the symmetry line of the land area indicates that the viscous shear regions have not completely penetrated the fuel stream. Pitot-pressure profiles at $x/h \geq 8$ do not exhibit large spikes, which implies weakening of the shock system. The mixing layers now appear to gradually grow into the oxidizer stream and to have merged on the fuel side, although on that side they do not radiate because of the excess of deactivating $H_2$. Static pressures at $x/h = 2$ and 4 are considerably lower on the oxidizer than on the fuel side, indicating that base pressure relief has not completely eliminated the original pressure mismatch. At $x/h = 8$, this trend is reversed following the first shock interaction with the mixing layer. A second reversal at $x/h = 16$ indicates that weak wave interactions continue for some distance beyond merging.

This shock system is not solely owing to multiple reflections of separation shocks or of recompression shocks after expansion into the base region. Even in the absence of a land area and of boundary-layer separation, shocks arise from displacement of the supersonic oxidizer or fuel streams by the reactively heated mixing layer.[12] When pumping is initiated by mixing of fuel and oxidizer, a sizable fraction ($\sim$26%) of the heat of reaction goes directly into translational and rotational energy. Since no more than another 20% can be extracted optically, most of the vibrational energy is subsequently also converted to heat by collisional deactivation. This heat addition causes the gas in the mixing layer to expand, and to displace the oxidizer and fuel streams.

In the cavity tests of Fig. 23, reactive heating reinforces the shock/expansion wave system downstream of $x/h = 4$ and explains its persistence to $x/h \geq 16$. Reactive heating causes the gradual rise in temperature by nearly 200°K and the oscillatory decrease in Mach number. The sharp initial drop in Mach number from 5.8 in the exit of the oxidizer nozzle to 3.4 at $x/h = 4$, and the corresponding temperature increase from 149 to 400°K,[29,31] are probably largely owing to the strong separation shocks. (The Mach number and temperature in the nozzle exit are calculated from plenum conditions.)

## 2.2. Structure of the Mixing Layer

Although laminar and turbulent mixing have been studied for decades, only limited information is available on reactive supersonic mixing layers, or supersonic diffusion flames. Experimental techniques for detailed mapping of the mean and fluctuating flow-field properties in such flows, particularly in the low-density high-stagnation temperature corrosive environment of chemical-laser cavities, are still being developed.[32,33] We will discuss some mean-flow measurements, but otherwise rely on non-reactive or incompressible experiments. The analysis of laminar mixing with reactions is relatively straightforward, at least as long as transverse-pressure gradients are small, and comparisons with such calculations will be presented. However, free-mixing layers are inherently unstable and theoretical understanding of transitional and turbulent compressible reactive mixing is far from adequate (Section 2.4).

**Supersonic Diffusion Flames.** Since 1960, a number of experimental and simplified theoretical studies of mixing between coaxial supersonic jets of hydrogen and air have been performed,[34–38] stimulated by the prospect of supersonic combustion ram jets for propulsion in hypersonic flight. References 35, 36, and 38 review this effort. In these experiments pressures and Reynolds numbers have typically been much higher than in a laser cavity, so that upstream boundary layers were less and free-stream turbulence was more important. The near field was found to be dominated by ignition delays in the $H_2 + O_2$ combustion, a phenomenon not observed for the much faster $H_2 + F$ reaction.* The majority of the mean-profile measurements[34,37] apparently have been taken either in the inlet flow or in the turbulent axisymmetric jet after merging of the mixing layer at the axis. Theoretical analyses have emphasized the fully developed turbulent far field. The applicability of these studies to the unmerged laser flow is not immediately obvious.

---

* Finite small-signal gains have been measured immediately downstream of the nozzle exit.[39]

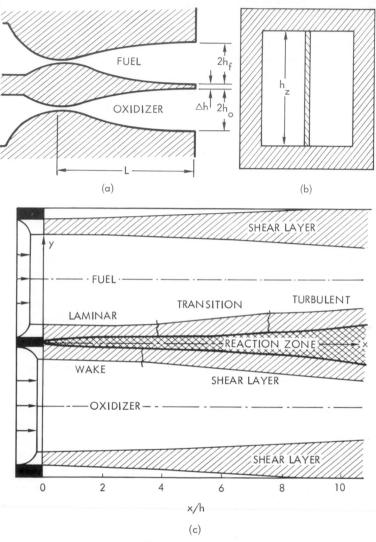

**Fig. 24.** Reactive-flow experiment.[40] (*a*) Nozzle contours, (*b*) nozzle exit, and (*c*) flow field.

However, reasonably detailed mean-flow measurements have been obtained for two-dimensional supersonic mixing and reaction of F and $H_2$.[33,40] Figures 24*a* and *b* show the inlet configuration, and Fig. 24*c* is a schematic of the flow field studied in Ref. 40. Fluorine, diluted with $N_2$, was produced by dissociation of $SF_6$ in an arc plenum and then expanded through a smoothly contoured supersonic nozzle designed for parallel

$M = 4$ exit flow at $p = 10$ Torr. Hydrogen, diluted with $N_2$ and optionally heated by a separate arc, was expanded through an identical nozzle. Mixing and reaction occurred downstream of the lip of a thin dividing wall, in a large cavity whose pressure was carefully matched to the fuel and oxidizer exit pressures. With pressure matching and a high degree of dilution, shocks were weak and were not reflected at the nozzle center-lines in the absence of symmetry constraints. Thus, to first approximation mixing occurred at constant pressure.

Table 5 summarizes nozzle dimensions and operating conditions. Two nozzle designs were used, one for pressures of 1 to 6 Torr (RF–1), and a scaled-down version (RF–2) to extend measurements to the 2 to 13 Torr range. Gas-dynamical conditions, including unit Reynolds numbers, were in the range encountered in laser cavities (Table 4). Compared to laser nozzles, ratios of boundary-layer thicknesses to nozzle-exit widths ranged from a factor of 2 smaller at 13 Torr to slightly larger at 1 Torr, where the oxidizer flow was observed to be fully viscous. In view of the larger nozzles and of adjustable ratios of free-jet densities, velocities, and reactant concentrations, the experiment allowed a more detailed study of the interactions between boundary layers, viscous mixing, and reaction.

Net spontaneous-emission intensity profiles in the 2.4 to 3.2 $\mu$m band

**Table 5.** Operating Conditions and Dimensions for Reactive Flow Nozzles[40]

| | Nozzle[a] | RF-1 Oxidizer[c] | RF-1 Fuel[c] unheated/heated | RF-2 Oxidizer[c] | RF-2 Fuel[c] unheated/heated |
|---|---|---|---|---|---|
| $M$ | —[b] | | 4 | | 4 |
| $Re/l$ | cm$^{-1 b}$ | | 7.1(3) | | 7.1(3) |
| $p$ | Torr | | 1–6 | | 2–13 |
| $u$ | cm/sec | 2.2(5)–1.7(5) | 7.4(4)/1.9(5)–1.8(5)[d] | 2.0(5)–1.8(5) | 7.0(4)/1.7(5)–2.1(5)[d] |
| $T$ | °K | 8.8(2)–5.3(2) | 7.1(1)/6.3(2)–3.3(2)[d] | 9.6(2)–7.5(2) | 7.1(1)/3.9(2)–5.1(2)[d] |
| $F/(N_2 + O_2)$ | — | 0.1–0.2 | | 0.1–0.2 | |
| $H_2/N_2$ | — | | 0.02–0.2 | | 0.03–0.6 |
| $H_2/F$ | — | | 0.59–3.64 | | 0.59–3.64 |
| $\delta^*$ | cm | 6.8(−1)–1.8(−1) | 5.4(−1)–1.8(−1)/ 6.1(−1)–1.5(−1)[d] | 4.2(−1)–1.3(−1) | 3.1(−1)–1.2(−1)/ 3.1(−1)–1.7(−1)[d] |
| $\theta$ | cm | 9.8(−2)–2.5(−2) | 1.4(−2)–5.0(−3)/ 8.6(−2)–2.2(−2)[d] | 6.5(−2)–1.6(−2) | 9.1(−3)–3.0(−3)/ 4.4(−2)–2.4(−2)[d] |
| $h_0, h_f$ | cm | | 1.11 | | 0.80 |
| $\Delta h$ | cm | | 0.15 | | 0.15 |
| $h_z$ | cm | | 6.10 | | 4.32 |
| $L$ | cm | | 7.24 | | 5.11 |

[a] Identical oxidizer and fuel nozzles, see Fig. 24a.
[b] Mach number and unit Reynolds number for design point at $p = 10$ Torr, $T_0 = 2000$°K.
[c] 7.1(3) reads as $7.1 \times 10^3$.
[d] $u$, $T$, $\delta^*$, and $\theta$ ranges given correspond to the pressure range in line three.

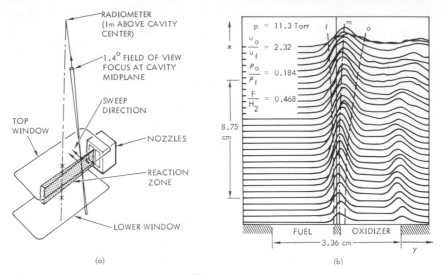

**Fig. 25.** Infrared scanner experiment.[40] (a) Schematic and (b) typical ir scans of reaction zone (maximum intensity at $y_m$, half intensity on fuel side at $y_f$, on oxidizer side at $y_o$).

were obtained with an infrared radiometer that mapped the reaction zone in a rapid sequence of transverse scans (Fig. 25a). The instantaneous field of view was $0.5 \times 0.5$ mm in the $x$–$y$-plane. A typical ir map of the cavity is shown in Fig. 25b. The location $y_m$ of peak intensity and the growth of the reaction zone (half-intensity points $y_f$ and $y_o$) into the fuel and oxidizer streams can be found as functions of distance $x$ from the nozzle exit. The integrated intensity provides a measure of the effectiveness of mixing and pumping in producing excited HF. The outer-shear layer of the oxidizer jet is luminous owing to recirculation of excess $H_2$, so that interference with the mixing layer can be detected. The undisturbed free-mixing layer region extends to $x \simeq 11$ cm at 1 Torr, and to $x \simeq 17$ cm at 13 Torr pressure.

Figure 26 ($p = 1$ Torr) shows the location of the reaction zone, and Fig. 27 ($p = 2.3$ Torr), the integrated intensity as functions of $x$, as observed at various ratios of free-stream reactant concentrations and of free-stream velocities. As the mole ratio $H_2/F$ is increased from 0.6 to 3.7, the reaction zone shifts from the fuel to the oxidizer side of the mixing layer, apparently without significant change in the growth rate (Figs. 26a and b). Yet the net rate of production of excited HF increases (Fig. 27a or b), in agreement with the fact[4] that peak laser power is achieved at $H_2/F$ ratios of O(10). Since both reactants are consumed at the same rate, peak effective pumping rates for an excess of the lighter reactant must be attributable to the mixing process. This phenomenon is presently not understood, but experiments[14] have shown that part of the excess $H_2$ can be replaced by an inert light gas such as helium.

Another surprising result is exhibited by Figs. 26$b$ and $c$ and 27$a$ and $b$. When the fuel jet was heated, the ratio of the free-jet velocities, and indeed also the density ratio, could be reduced to unity. From the literature on compressible nonreactive mixing at high Reynolds numbers, for example, Refs. 41 and 42, we expect that mixing becomes quite slow when the momenta of the two jets are about equal. However, Figs. 26$b$ and $c$ show little effect on the rate of growth of the reaction zone, and comparison of Figs. 27$a$ and $b$ suggests that the integrated intensity is also relatively insensitive to the free-stream momentum ratio. This trend was found to persist to the highest pressures (13 Torr) investigated, where mixing was probably turbulent.

For spectral resolution, point measurements were obtained with an ir spectrometer, with a field of view of $\Delta x = 1$ cm, $\Delta y = 1$ mm, and a position accuracy of $\Delta y = \pm 0.5$ mm. From the measured-line intensities in the $1 \rightarrow 0$ to $4 \rightarrow 3$ vibrational bands, concentrations of the $v = 1$ to 4 levels of HF as well as the R–T temperature have been evaluated. Determination of $v = 0$ populations is more uncertain because of absorption of the $1 \rightarrow 0$ band by recirculating HF.

Additional gas-dynamical data were obtained from pitot-pressure traverses and static-pressure taps. The pitot-pressure profiles confirm the absence of any shock-wave interactions with the mixing layer and allow a direct estimate of its thickness. In the absence of strong static-pressure

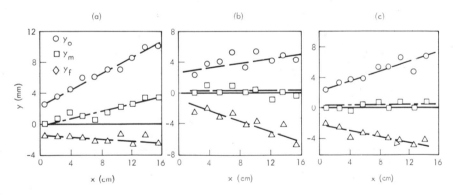

**Fig. 26.** Location of reaction zone, effect of stoichiometry, and velocity ratio (RF-1 nozzle, and $p = 1$ Torr).[40]

| | $u_o/u_f$ | $H_2/F$ | $T_{of}(°K)$ | $T_{oo}(°K)$ |
|---|---|---|---|---|
| (a) | 2.84 | 3.65 | 300 | 2300 |
| (b) | 3.01 | 0.59 | 300 | 2228 |
| (c) | 1.10 | 0.59 | 2140 | 2287 |

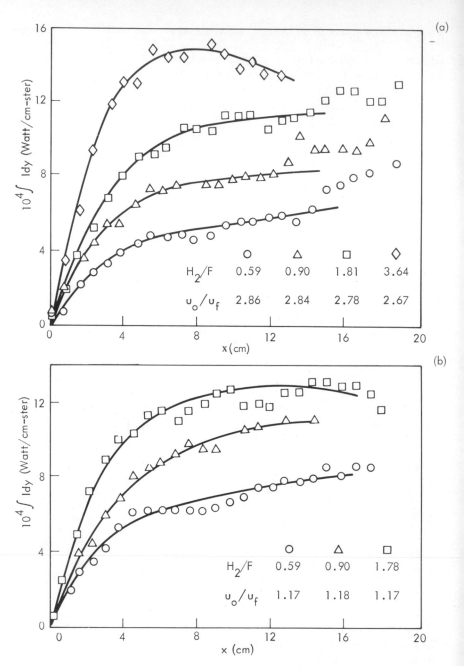

**Fig. 27.** Axial variation of integrated intensity, effect of stoichiometry, and velocity ratio (RF-2 nozzle, and $p = 2.3$ Torr).[40]

| | $T_{0o}(°K)$ | $T_{0f}(°K)$ |
|---|---|---|
| (a) | 2550 | 300 |
| (b) | 2550 | 1600 |

gradients, the ratio of pitot to static pressure is a known function of the ratio of specific heats $\gamma$ and of the Mach number.[29] Thus with an estimate for the transverse variations of $\gamma$ and molecular weight and with the spectroscopically measured temperature distribution, velocity profiles can be calculated for the reaction zone.

Figure 28 provides a synopsis of the pitot-pressure profiles (the width of the viscous region is marked), measured HF(0) to HF(4) concentrations

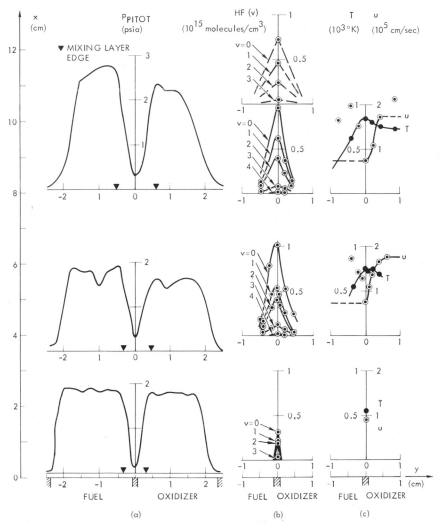

**Fig. 28.** Pitot pressure, excited-HF concentration, temperature, and velocity profiles across mixing layer (RF 1 nozzle, $p \simeq 6$ Torr, $H_2/F = 3.7$, $T_{0f} = 300°K$).[40]

in the reaction zone, and the temperature and estimated velocity distributions for one test condition (RF-1 nozzle, $p \simeq 6$ Torr, $H_2/F = 3.7$, unheated fuel jet). At this pressure, the width of the mixing region is seen to be nearly the same at $x = 3.5$ cm as at the nozzle exit, but has grown noticeably at the $x = 8$-cm station. The initial width agrees well with estimates of the sum $\delta_f + \Delta h + \delta_o$ of the boundary-layer thicknesses in the fuel and oxidizer nozzles plus the thickness of the dividing wall, indicating that the initial development of the mixing region is dominated by the momentum defect in the wake of boundary layers and dividing wall. Imbedded in this viscous region is the chemiluminescent zone where excited HF is produced by the pumping reaction and then diffuses outward. The half-intensity width, $\Delta y_{1/2} = y_o - y_f$, determined from the ir scans, is generally narrower than the viscous region and is in agreement with the species profiles obtained spectroscopically.

The velocity profiles deduced from measured pitot/static pressures and temperatures show a pronounced wake dip. Although the velocity of the cold-fuel jet was only half that of the oxidizer stream, the measurements give about equal velocities on both sides of the wake. The expected[40] shape of the (shear-layer) velocity profiles is indicated in Fig. 28. This discrepancy is attributed in Ref. 40 to transition from laminar to turbulent mixing. Eddies of cold unreacted fuel and oxidizer alternately drift by the spectrometer, which registers only luminescence from the hot flame stretched out between these eddies. Hence measured temperatures can considerably exceed time-averaged values, with which the mean velocities should have been calculated. The profile shapes themselves do not provide a clue as to whether the mixing is laminar, transitional, or turbulent.

An indirect indication of transition is also given by the overall growth rates of the viscous mixing layer and the reaction zone, Fig. 29. In view of the scaling laws predicted by the simplified analysis of Section 3, both width and axial distance have been multiplied by the pressure. Measurements at the lowest pressures (1 to 2 Torr) suggest parabolic growth, those at the highest pressures (12 to 14 Torr) linear growth of the reaction zone. Curve fits give approximately

$$\Delta y_{1/2} = \begin{cases} 4\left(\dfrac{\mu x}{\rho u}\right)^{1/2} \\[2em] 0.08x \end{cases} \quad \text{for} \quad px \begin{cases} < 20 \\[2em] > 100 \end{cases} \quad \text{Torr-cm}$$

with large scatter in the intermediate regime. Infrared scans taken at 13 Torr actually exhibit a fairly well-defined break in the growth rate

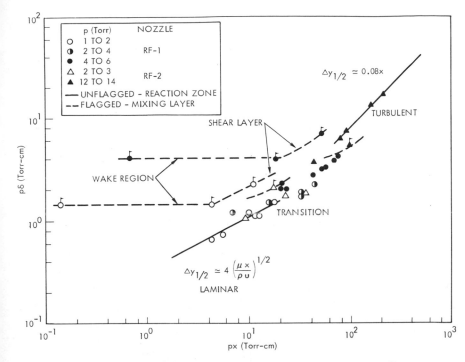

**Fig. 29.** Growth of mixing layer and reaction zone ($T_{0f} = 300°K$).[40]

approximately at $x = 7$ cm, for both heated and unheated hydrogen, and thus is independent of the free-stream velocity ratio.

The few available data for the thickness of the entire viscous region (from pitot-pressure traverses) indicate virtually no growth for the first few nozzle heights downstream of the nozzle exit, and then spreading at a rate comparable to that of the reaction zone. This change in the growth rate is observed at $p = 6$ Torr, as well as at 1 Torr where the reaction zone continues to grow parabolically. Heating of the hydrogen jet has a negligible effect on the width of the viscous region, even at 13 Torr ($px = 100$).

Figure 30 compares the spectroscopically determined $HF(v)$ and temperature profiles with results of a laminar reactive-mixing calculation using the Blottner code.[32] Initial velocity and temperature profiles were obtained from chemically frozen, nozzle boundary-layer calculations (see Section 1.4) for the test conditions of Fig. 28. The width of the dividing wall was neglected, and the pressure was assumed to be constant (6 Torr). The most obvious disagreement is in the $HF(0)$ concentrations, even at

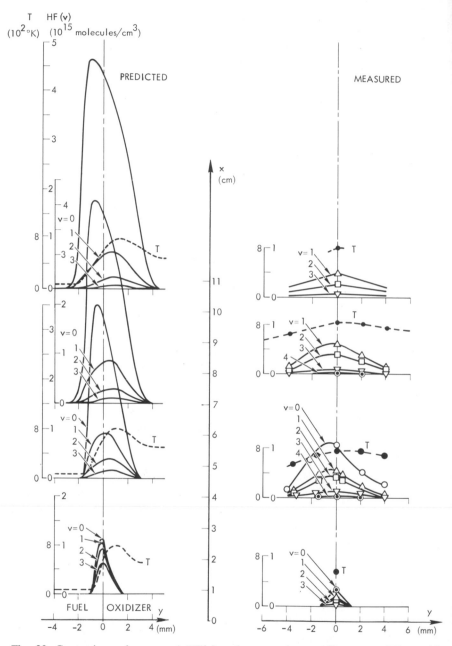

**Fig. 30.** Comparison of measured HF(v) and temperature profiles at $p = 6$ Torr with laminar mixing calculation.[32,40]

314

$x = 4$ cm where the measurements were presumably corrected[40] for absorption by recirculating equilibrated HF. Measured peak concentrations for the higher vibrational levels generally are about a factor of 2 lower than theoretically predicted. Peak temperatures in the reaction zone agree rather well, but the width of both the species and the temperature profiles found from ir scans and spectrometer measurements is about twice that of the computed laminar profiles. Even if the flow was laminar in the 6-Torr experiment, a factor of 2 disagreement between theory and experiment is not unexpected, in view of the uncertainties in rate and transport coefficients. Reference 40 suggests that the measurements are averages over large-scale fluctuations. Although it is not clear whether they were due to an instability in the arc heater, or to transitional fluctuations in the mixing layer itself, the effect would be an apparent widening of the luminous region and reduction in peak radiance.

**Comparison with Nonreactive Mixing in the Laminar and Fully Turbulent Limits.** The growth of the reaction zone at small and large $px$ (or Reynolds numbers) has been compared,[40] respectively, with growth rates of laminar and turbulent nonreactive shear layers at constant pressure.

In the laminar incompressible limit, the first two terms of the Görtler series[43] give a parabolic growth with

$$\Delta y_{1/2} = 2.2 \left( \frac{\mu x}{\rho u} \right)^{1/2} \tag{53}$$

where $U = (u_1 + u_2)/2$, $u_1$ and $u_2$ are the free-stream velocities, and $\Delta y_{1/2} = |y[(u_1 + U)/2] - y[(u_2 + U)/2]|$. Compressible laminar-shear layers are also expected to grow parabolically if free-stream velocities, temperatures, and pressures remain constant.

Fully turbulent shear layers grow linearly in $x$. Experimental data for turbulent shear layers have been reviewed in Ref. 42, which for $M \to 0$ and $\rho_2 = \rho_1$ suggests

$$\Delta y_{1/2} = \frac{x}{\sigma} \left| \frac{u_1 - u_2}{u_1 + u_2} \right|, \qquad \sigma \simeq 12 \tag{54}$$

The density ratio $\rho_2/\rho_1$ appears to have little effect on the growth rate. The Mach-number dependence[42] of $\sigma$ is shown in Fig. 31. Open symbols, which refer to fully developed turbulence at large $\mathrm{Re}(x)$, indicate a significant decrease in the growth rate at high Mach numbers, for example, by a factor of 3 at $M = 4$. But measurements at lower Reynolds numbers (solid symbols), where the turbulent flow field is not yet fully established, show little influence of Mach number.

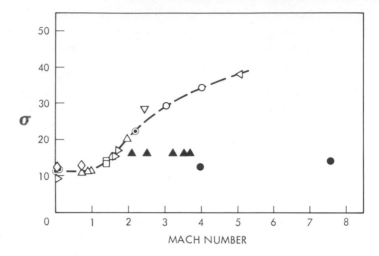

**Fig. 31.** Variation of shear-layer spreading parameter with Mach number.[42,57]

Since the axial growth of the reaction zone is parabolic at low pressures and linear at high pressures, it has been concluded[40] that mixing in the reactive-flow experiments is laminar at $px < 20$ and turbulent (though probably not fully developed) at $px > 100$ Torr-cm, with the data scatter in the intermediate regime attributable to large-scale fluctuations during transition.

The initial width of the viscous region and the central dip in the velocity profiles of Fig. 28, however, suggest that wake rather than shear-layer effects are dominant. Moreover, Eqs. (53) and (54) predict a strong influence of the free-stream velocity difference on shear-layer growth and mixing. Yet the integrated intensity, the widths of both the reaction zone and the viscous region, and the breaks in their growth rates were found in Ref. 40 to be insensitive to free-stream velocity and momentum ratios, even at $px = 100$ Torr-cm. Similar observations have been reported for both supersonic[36] and subsonic[45] $H_2 + O_2$ diffusion flames, as well as some heterogeneous incompressible mixing experiments.[46] This behavior can be understood if the momentum deficit $(\Delta \rho u)_w$ in the wake of upstream boundary layers and base step exceeds the free-stream momentum difference $(\Delta \rho u)_{fs}$, see Fig. 32$a$. Then $(\Delta \rho u)_{fs}$ has little influence on the profiles in the inner wake where the reactants mix or on the maximum shear regions that control transition (discussed later). Ultimately, of course, at a large distance from the nozzle lip, we expect that shear-layer characteristics will dominate if the free-stream momenta are not matched, because the wake momentum deficit is gradually filled out by viscous dissipation, Fig. 32b. But, in the reactive mixing experiments of Ref. 40, this apparently does not happen until $px > 100$ Torr-cm, as shown by the agreement of viscous-layer thicknesses in cold- and hot-fuel tests and by the velocity profiles in Fig. 28c ($px = 48$ Torr-cm).

In view of the possibility that lasing may be controlled by wake rather than by shear-layer mixing, we now examine wake growth rates. A similarity analysis of the laminar incompressible wake of a thin flat plate of wetted length $L$[22] predicts

$$\Delta y_{1/2} = 3.3 \left( \frac{\mu \tilde{x}}{\rho u_\infty} \right)^{1/2} \tag{55a}$$

Here $\tilde{x}$ is measured from the leading edge, $\Delta y_{1/2}$ is the width of the wake region where $(u_\infty - u) > (u_\infty - u_c)/2$, $u_\infty$ is the free-stream velocity, and $u_c$ the centerline velocity. The latter increases as

$$\frac{u_c}{u_\infty} = 1 - 0.37 \left( \frac{\tilde{x}}{L} \right)^{-1/2} \tag{55b}$$

These predictions are valid for $\tilde{x}/L > 3$ or, with $x = \tilde{x} - L$ and the boundary-layer momentum thickness[22] $\theta_L = 0.664(\mu L/\rho u_\infty)^{1/2}$, for $x > 4.5\theta_L$ Re $(\theta_L)$. Closer to the trailing edge, the laminar wake does not grow according to the parabolic similarity law of Eqs. (55), and compressible ($M_\infty = 6$) wakes behind flat plates and slender wedges have shown virtually no growth prior to transition,[47] in agreement with the reactive flow[40] pitot-pressure data.

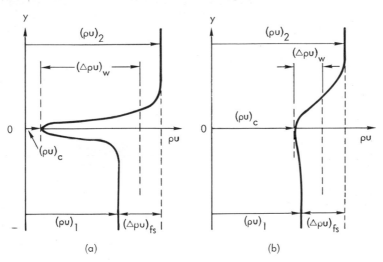

**Fig. 32.** Transition from (a) wake to (b) shear-layer dominated two-stream mixing.

Experiments in turbulent incompressible two-dimensional wakes are reviewed in Ref. 41.* For the small deficit ($u_\infty - u_c \ll u_\infty$) wake at large distances behind a flat plate, a semiempirical analysis[22] predicts self-similar growth according to

$$\Delta y_{1/2} = 0.312(x\theta_L)^{1/2} \tag{56a}$$

$$\frac{u_c}{u_\infty} = 1 - 1.38\left(\frac{x}{\theta_L}\right)^{-1/2} \tag{56b}$$

if the upstream boundary layers are laminar. Note that Eqs. (56) are of the same form as the laminar-wake similarity laws, Eqs. (55). Compressible ($M = 3$) far wake measurements behind a flat plate[48] also exhibit parabolic growth. Indeed, the mean profiles are essentially of the same form as in the incompressible limit if the transverse coordinate is stretched by the transformation $\bar{y} = \int (\rho/\rho_\infty)\, dy$.

Thus, in supersonic wake-like mixing we expect at first no growth, then nonsimilar spreading during transition, and finally an asymptotic approach to parabolic growth in the fully turbulent regime at large $x$. The change in the growth rate of the viscous region observed in the reactive-flow experiments[40] at $p = 1$ and 6 Torr could reflect wake transition, or it could be due to shear-layer effects becoming important in the outer wake. A decision should come from a correlation with wake and shear-layer transition studies, which we review in the next subsection.

**Transition in Nonreactive Shear Layers and Wakes, Implications for Reactive Mixing.** From classical linearized-stability analyses,[49,50] it is known that laminar boundary-layer flow becomes unstable at any Reynolds number as soon as it separates from the wall. This applies to the trailing edge of a splitter plate, as well as to separation induced further upstream by an unfavorable pressure gradient. Amplification of small perturbations is strongest near inflection points of the velocity profile,[51,52] where shear and vorticity reach a maximum. The result is an exponentially growing waviness of the maximum shear line, with preferential amplification at a characteristic frequency proportional to the velocity gradient. These waves steepen and roll up toward the low-speed side into vortices that are convected downstream. Mutual interaction and repeated pairing of the vortices[53] signify the onset of transition and nonlinear amplification, accompanied by transfer of fluctuation energy to harmonics and subharmonics of the natural frequency[51,52] and by a peak in r.m.s. fluctuations and turbulent shear stress $-\rho u' v'$.[52,54] Thereafter, nonlinear

---

* Compressible wake measurements discussed in Ref. 41 are all for axisymmetric configurations.

viscous interactions break up the large vortices into ever smaller turbulent eddies and fill out the turbulent fluctuation spectrum. The turbulent flow field is fully established when the rate of production of large eddies in the mean shear flow is balanced by the rate of viscous dissipation, so that the spectral distribution reaches a steady state.

A typical shear-layer profile, see Fig. 32b, has only one inflection point, and a single row of vortices develops. Figure 33 shows their formation, pairing, and final breakup into random turbulence, as observed[55] in the shear layer between a smoke-seeded subsonic air jet and still air behind a rearward-facing step. From similar experiments, Bradshaw[54] predicts that transition occurs at $x_{tr} \simeq 150\, \theta_L$, and that fully developed self-preserving turbulent flow is established at $x \geq x_{ft} \simeq 1000\, \theta_L$, where $\theta_L$ is the boundary-layer momentum thickness at separation. These points are indicated in Fig. 33. Vortex pairing can be observed near $x_{tr}$, and the large vortices seem to have disappeared at $x_{ft}$. The gradual development of the fluctuation spectrum during incompressible shear-layer transition is shown in Fig. 34. In the instability region of the laminar flow at $x < x_{tr}$, the natural frequency $f_0 = S\,|u_2 - u_1|/\theta_L \simeq 30$ Hz is exponentially amplified. Froude numbers $S$ reported in the literature[50,54] range from 0.009 to 0.014. Around $x_{tr} \simeq 8$ cm, the second and third harmonics are at a peak. But, as the shear layer approaches the fully turbulent state ($x_{ft} \simeq 50$ cm), spectral peaks at $f_0$, $2f_0$, and $3f_0$ are seen to vanish, and the fluctuation spectrum assumes its fully turbulent form. The amplification of subharmonics $f_0/2$ and $3f_0/2$ at $x_{tr} < x < x_{ft}$ appears to be characteristic for shear layers.[51]

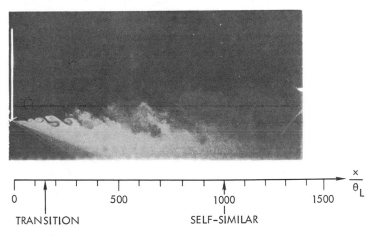

**Fig. 33.** Instability, vortex formation, vortex pairing, and final breakup into turbulence in a smoke-seeded incompressible shear layer ($\theta_L = 5.2 \times 10^{-3}$ cm, $\mathrm{Re}(\theta_L) = 243$).[55]

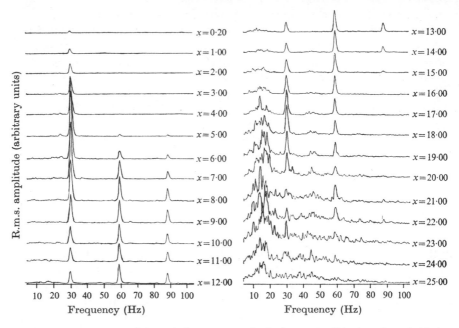

**Fig. 34.** Development of the turbulent spectrum in the incompressible shear layer behind a splitter plate. Natural frequency $f_0 = 30$ Hz artificially excited (axial distances in centimeters, $x_{tr} \simeq 8$ cm, $x_{ft} \simeq 50$ cm). [R. W. Miksad, J. Fluid Mech. **56**, 695 (1972).]

Concerning the structure of the fully turbulent shear flow, there exists a so-far unsettled dispute. References 44, 53, and 56 report the persistence of an ordered pattern of vortices, which by repeated pairing increase their average spacing linearly in $x$ to distances well above the initial natural-instability wavelength, and whose size grows with the shear layer. The spectral peak corresponding to such a vortex pattern would shift closer to zero frequency as the shear layer grows. Since the vortices penetrate the shear layers, this finding is of potential importance for the modeling of entrainment in reactive mixing layers (see Section 2.4). However, no preferred frequencies have been observed in either the incompressible experiments of Ref. 55 or in the supersonic self-preserving shear-layer measurements of Ref. 57.

Figure 35[58] exhibits a dramatic influence of the free-stream density ratio $\rho_2/\rho_1$ on transition in incompressible heterogeneous shear-layer mixing. When $u_2/u_1 < 1$ (zero in the present experiments) and $\rho_2/\rho_1 > 1$ (Fig. 35$a$, He jet into still air), the Froude number $S$ and the transition length $x_{tr}$ are significantly reduced, while for $\rho_2/\rho_1 < 1$ (Fig. 35$b$, air jet into He), the shear layer is stabilized. It is then somewhat surprising that,

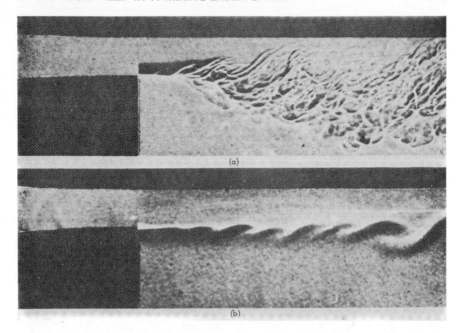

**Fig. 35.** Influence of the density ratio on shear-layer transition behind a rearward facing step. [R. F. Davey and A. Roshko, J. Fluid Mech. *53*, 523 (1972).] (*a*) He jet into air and (*b*) air jet into He.

in the fully turbulent limit [see discussion after Eq. (54)], the growth rate of the mixing layer should be only weakly affected by the density ratio.[44]

Wake-velocity profiles have typically two inflection points, see Fig. 32*a*, and two rows of vortices of opposite sign (Karman vortex street) are shed into the wake. In the most likely configuration,[59] the vortices appear alternately on the two sides of the wake. Figure 36 compares the vortex patterns in transitional wakes (*a*) and shear layers (*b*). For the incompressible wake behind a flat plate, a nonlinear analysis[59] of the amplification of small perturbations at $f_0 = S(u_\infty - u_c)/\theta_L$ predicts transition at $x_{tr} \simeq 110\,\theta_L$,* that is, at about the same distance as in a shear layer.[54] Froude numbers $S$ also fall into the range from 0.009 to 0.014 observed in homogeneous shear layers. It appears that the mechanisms leading to transition at $x_{tr}$ are essentially the same in wakes and shear layers, and that the instability and initial amplification in the high-shear region on one side of the wake is not strongly influenced by that on the other side.

* Note that for typical Re $(\theta_L) = 0(100)$, transition occurs upstream of the range of validity of the similarity solution for laminar wake, Eq. (55), so that the laminar region $x < x_{tr}$ may not be similar.

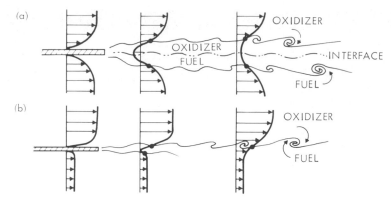

**Fig. 36.**   Initial growth of instabilities and vortex formation. ($a$) Wake and ($b$) shear layer.

Interference between the vortices of the two rows apparently occurs only at $x > x_{tr}$.

Results obtained in the compressible ($M_\infty = 6$) wake of a flat plate[52] of length $L$ are shown in Figs. 37$a$ and $b$. The Froude number of the laminar instability is not affected by the free-stream Mach number. As seen in Fig. 37$a$, fluctuations at $f_0$ are essentially confined to the two off-axis high-shear regions, even after transition at $x_{tr}/L \simeq 4.5$. Fluctuations at the second (Fig. 37$b$) and higher harmonics reach the centerline only after transition. Compressibility stabilizes the wake, and at $M_\infty = 6$ the transition distance referenced to the boundary-layer momentum thickness becomes $x_{tr}/\theta_L \simeq 2000$, in agreement with the nonlinear compressible analysis of Ko.[60] Alber[61] predicts a scaling of $x_{tr}/\theta_L$ with $[1 + (\gamma - 1)M_\infty^2/2]^2$, but Ko's numerical results and the $M_\infty = 6$ measurements suggest a variation with only the first power of the square bracket. In view of the similarities in the pretransition processes of incompressible shear layers and wakes discussed previously, this scaling with $M_\infty$ probably applies to compressible shear layers as well.

These estimates for the transition length are compared in Table 6, with the changes in the growth rates of the viscous region and the reaction zone as observed in Ref. 40 (Fig. 29) at $M_\infty = 4$. An average of the momentum thicknesses of the oxidizer and cold-fuel boundary layers at the nozzle exit has been used as the reference length, and the shear-layer predictions are shown in parenthesis. It appears that the change in the growth rate of the viscous-wake region at $p = 1$ and 6 Torr precedes transition, and is perhaps attributable to the growth of the vortices in the two maximum shear regions. At 13 Torr, the switch from about parabolic to linear growth of the reaction zone occurs somewhat downstream of the predicted $x_{tr}$. At 6 Torr, transition is expected at the last stations where

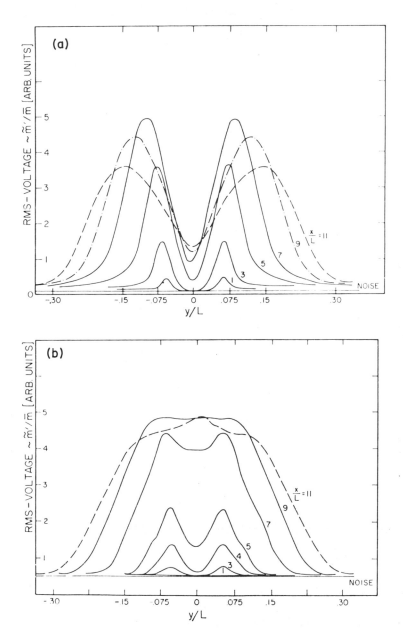

**Fig. 37.** Fluctuation-amplitude distributions across the wake of a flat plate at $M_\infty = 6$ ($L = 1$ cm, $\mathrm{Re}(L) = 9.3 \times 10^4$, $x_{tr} \approx 4.5\,L$). [W. Behrens and D. R. S. Ko, AIAA J. *9*, 851 (1971).] (*a*) $f = f_0$ and (*b*) $f - 2f_0$.

**Table 6.** Observed Break Points in the Growth of Mixing layer and Reaction Zone,[40] and Predictions of Transition in Wakes and Shear Layers

| $p$, torr | $(\theta_o + \theta_f)/2$,[b] cm | Mixing-layer | Reaction zone | Transition[a] | |
| | | $px$, Torr-cm | $px$, Torr-cm | $x$, cm | $px$, Torr-cm |
| --- | --- | --- | --- | --- | --- |
| 1 | 6.1(−2) | ~5 | — | 28(38) | 28(38) |
| 6 | 1.6(−2) | ~20 | — | 7.4(10) | 44(61) |
| 13 | 1.0(−2) | — | 91 | 4.6(6.3) | 60(82) |

[a] Shear-layer estimates in parentheses.
[b] 6.1(−2) reads as $6.1 \times 10^{-2}$.

data were taken, while measurements at 1 Torr terminated well before transition. With $x_{ft} \geq 1000 \, \theta_L$, none of the measurements extended into the fully turbulent regime.

Since boundary layers in laser nozzles are about as thick, relative to nozzle dimensions, as those in the reactive flow experiments, some of the preceeding conclusions apply directly to mixing in laser cavities. In particular, mixing is expected to occur in the unstable laminar or transitional wake of the nozzle boundary layers and the land areas between the fuel and oxidizer nozzles. Reaction occurs in the low-momentum wake core. The width of the imbedded reaction zone is controlled by local concentration gradients, that is, by mass rather than momentum transport, so that it spreads like a shear layer rather than a wake. Turbulent mixing in the reaction zone is triggered by transition in the off-axis high-shear regions of the wake. An increase in the molar ratio of free-stream hydrogen to fluorine to values of $O(10)$ improves the net rate of pumping. One possible explanation is that, as the reaction zone moves into the higher shear region on the oxidizer side, it is rolled up in the vortices and stretched.

**Mixing in an Adverse-Pressure Gradient.** The shear layer and wake studies discussed so far all have been concerned with the special case of constant pressure. As explained in Sections 1.5 and 2.1, the pressure in a laser cavity generally rises in the streamwise direction, because of the displacement effect of the reactively heated mixing layers and the interaction with shock waves.

The effect of an adverse-pressure gradient on nonreactive incompressible two-dimensional turbulent mixing has been investigated by Gartshore[62] for the wake of a thin rectangular-rod transverse to the flow, and by Rebollo[56] for heterogeneous shear-layer mixing of He and $N_2$

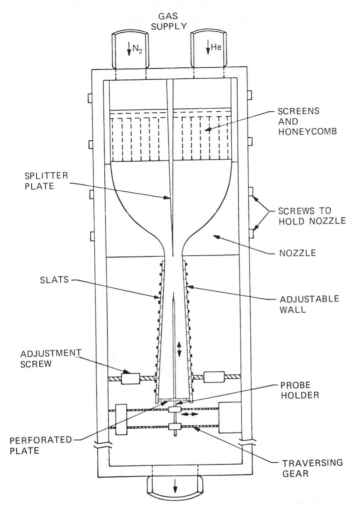

**Fig. 38.** Shear-layer mixing in an adverse pressure gradient, sketch of test facility.[56]

behind a thin splitter plate. In both experiments slats in the walls of the test section (see Fig. 38[56]) were adjusted to produce a power-law pressure rise as required for self-preserving fully turbulent flow.[56,62,63] The velocities then vary as

$$\left.\begin{array}{l} u_1 \sim u_2 \\[1.5em] u_c \sim u_\infty \end{array}\right\} \sim (x - x_0)^\alpha \qquad \begin{array}{l} \text{shear layer} \\[1.5em] \text{wake} \end{array} \qquad (57a)$$

and the viscous region spreads linearly in $x$

$$\Delta y_{1/2} \sim (x - x_0) \tag{57b}*$$

Here $\alpha > 0$ corresponds to an axial decrease of pressure and $\alpha < 0$ to a pressure rise. Since pressures on the two sides of a mixing layer must be matched, an additional constraint is[56]

$$\rho_1 u_1^2 = \rho_2 u_2^2 \tag{57c}$$

The virtual origin $x_0$ for the turbulent region is downstream of the trailing edge $x = 0$ of the body.

Gartshore compares his measurements at $\alpha = -0.316$ and $-0.312$, $(u_\infty - u_c)/u_\infty \approx 0.192$ and $0.239$ to the small deficit $\alpha = 0$ wake data of Ref. 63. In both cases the mean-velocity profiles assume the characteristic Gaussian form

$$\frac{(u_\infty - u)}{(u_\infty - u_c)} = \exp\left[ -\ln 2 \left( \frac{2y}{\Delta y_{1/2}} \right)^2 \right]$$

In the adverse-pressure gradient, the normalized turbulent shear stress $-\overline{\rho u'v'}/\rho(u_\infty - u_c)^2$ is found to be only half that at $\alpha = 0$, and the velocity fluctuations appear to be more nearly isotropic. Prabhu and Narasimha[64] observe an increase in the growth rate and the velocity defect of a fully turbulent wake when passing through a short region with a strong adverse-pressure gradient. As in Bradshaw's experiments,[54] the flow relaxes to a new equilibrium state in 1000 wake-momentum thicknesses, provided that the streamwise velocity gradient remains small compared to $(u_\infty - u_c)/\Delta y_{1/2}$.

For the shear layer, Rebello finds that at $\alpha = -0.18$ the spreading rate is larger by about 60% and the peak nondimensional shear stress $-\overline{\rho u'v'}/\rho_1 u_1^2$ has increased by 70%, compared to $\alpha = 0$ measurements in the same facility.[44,56] Note that the shear stress increases, rather than decreases as in Gartshore's wake. The normalized density fluctuations $(\overline{\rho'^2})^{1/2}/(\rho_2 - \rho_1)$, however, are virtually unaffected, and turbulent mass diffusion $\overline{\rho'v'}/\rho_1 u_1$ increases by only 20%. The pressure gradient seems to affect primarily the velocity field and the momentum transport, rather than the turbulent diffusion of species that control the rate of reaction in a reactive mixing layer. Observed turbulent Schmidt numbers

$$Sc^T = \frac{\overline{\rho u'v'}/(\rho\, \partial u/\partial y)}{\overline{\rho'v'}/(\partial\rho/\partial y)}$$

---

* The parabolically growing small-deficit wake at constant pressure, Eqs. (56), is only approximately self-preserving.

are unexpectedly low, that is, only 0.16 for $\alpha = 0$ and 0.33 for $\alpha = -0.18$ near the dividing streamline. A detailed examination of the density fluctuations reveals excursions of the order of the free-stream density difference, consistent with the persistent, large vortex structure observed in these and the Brown–Roshko experiments.[44] The mixing seems to be more thorough on the light species (He) side of the shear layer.

Shock impingement on laminar boundary layers[65] is known to produce appreciable thickening, separation if the shock is strong, and often transition to turbulent mixing. The reason is that the boundary-layer profile develops an inflection point and becomes more unstable.[22] Since laminar free-shear layer and wake profiles are inherently unstable, here too shock interaction conceivably could trigger transition and enhance mixing. An attempt to determine experimentally the effect on supersonic heterogeneous mixing is described in Ref. 66. A sonic helium jet was injected into a supersonic air stream. A shock holder could be inserted into the test section (see Fig. 39a) to generate an oblique shock impinging on the jet. Principal diagnostic tools included static-pressure taps in the side walls of the test section, a Schlieren system, and photography of chemiluminescence from the reaction of nitric-oxide and atomic-oxygen traces added to the air and helium streams, respectively.

Without the shock holder, the jet was found to grow gradually until intercepted near $x = 7$ cm by the reflection of the weak lip shock ($A - C$ in Fig. 39a). Estimates based on the measured thickness of the turbulent-nozzle boundary layer suggest that the shear-layer transition did not occur until $x > 6$ cm. With the shock holder in place, the flow field changed to that shown in Fig. 39a. Before the oblique shock $D-E$ reaches the jet boundary at $G$, it is intercepted by a new strong recompression shock $B-C$ originating further upstream. The helium jet apparently has become subsonic between $B$ and $G$, allowing upstream propagation of the pressure rise (see Fig. 39b). It seems to have undergone transition near $B$ and spread rapidly. This picture, including the second recompression shock at $H$, is quite similar to that observed during shock-induced separation of an initially laminar boundary layer.[65] Ferri[36] suggests that strong adverse-pressure gradients can result in detached bubbles of recirculating flow imbedded in subsonic or slightly supersonic jets (and indeed also in wake-like mixing layers with strongly exothermic reactions). Although Ref. 66 also refers to the region between $B$ and $H$ as a recirculation bubble, no conclusive evidence of recirculation is presented.

Figure 39b also presents the axial variation of the intensity as obtained from densitometer scans of the chemiluminescence photographs, and transverse-intensity profiles with and without the shock interaction are

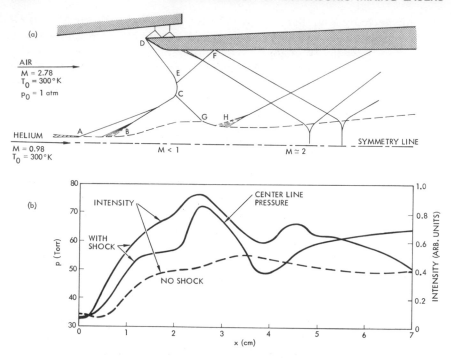

**Fig. 39.** Sonic helium-supersonic air mixing with shock interaction.[66] (a) Flow-field configuration (only upper half is shown) and (b) axial variation of pressure (with shock interaction) and intensity (with and without shock interaction).

compared in Figs. 40a and b. Between points B and G, the intensity has nearly doubled, and so has the width of the luminous jet. This suggests a dramatic enhancement of the mixing rate by the shock interaction.[66] But if the chemical rate rather than mixing is rate limiting, a good portion of the intensity increase could be attributable to the rise in pressure (by a factor of 2) and temperature.

Imbedded recirculation regions will not arise if both streams are sufficiently supersonic and the shock is not too strong, and, in any case, they should be avoided in laser cavities for reasons discussed earlier. The main question is then whether the increase of the mass diffusion rate by shock-induced transition can offset the disadvantages of increased temperature and pressure. Once the flow is fully turbulent, an adverse pressure gradient (whether owing to shock interaction or reactive heating) will not significantly improve reactive mixing, as indicated by Rebollo's experiment.[56]

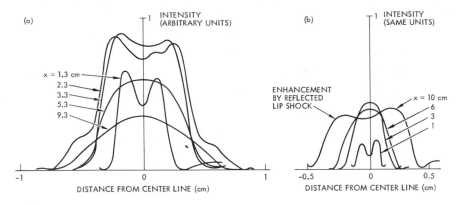

**Fig. 40.** Intensity profiles at various axial stations.[66] (a) With shock interaction and (b) without shock interaction.

## 2.3. Effects of the Cavity-Inlet Configuration

In Section 2.2 we discussed the structure of the mixing layer between parallel supersonic jets of fuel and oxidizer. It should be clear from Section 2.1 that this is an idealization of the actual flow field in mixing-laser cavities. Since a major fraction of the laser power is generated in the first few nozzle scale lengths $h$ from the cavity inlet,[16] we now take a closer look at the complex gas-dynamical interactions in this region, and at their possible effects on reactive mixing and lasing.

**Description of the Inlet Flow Field.**   One way of describing the inlet flow field is to start with the wake formed between two oxidizer streams in the absence of fuel injection, and to consider its perturbation by injection into the base flow region. This approach was suggested by Ferri[36] for the supersonic $H_2 + O_2$ combustion experiments, and by Hayday[67] for cw chemical lasers.

Figure 41 shows the laminar $(Re_\infty(H) = 4.1 \times 10^4)$ near wake of a slender wedge of base height $H$ in a supersonic $(M_\infty = 6)$ air stream, as calculated by Ohrenberger and Baum.[68] Their numerical method couples a finite-difference solution (including transverse-pressure gradients) for the expansion of the viscous boundary layer into the wake, and an integral analysis of the recirculating flow at the base. Uniqueness of the combined solution is determined by smooth passage through saddle-point singularities in the sonic wake neck. The results, which are in good agreement with wind tunnel measurements by Batt,[69] exhibit a corner

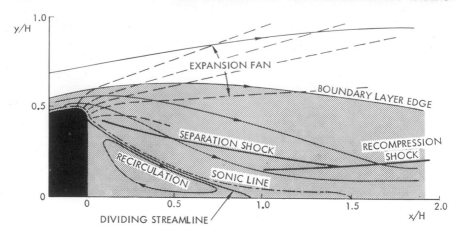

**Fig. 41.** Near wake of a $10°$ half-angle wedge ($M_\infty = 6$, $Re(H) = 4.1 \times 10^4$, $T_w/T_{0_\infty} = 0.19$).[68]

expansion that turns the supersonic outer boundary layer, separation from the base accompanied by a separation (lip) shock, and a stronger recompression (wake) shock that turns the flow parallel to the wake centerline. Entrainment of boundary-layer material into the recirculation region occurs by diffusion across the shear layer that develops along the dividing streamline.

The effects of base injection on the near wake have been studied extensively, if not very systematically, in an attempt to increase the base pressure and reduce the drag of supersonic vehicles. Figure 42 shows some results obtained with injection of $N_2$ through the porous base of a $6°$ half-angle wedge in an $M_\infty = 4$ air stream.[70] In Fig. 42a, contours of constant centerline Mach numbers are plotted as functions of axial distance normalized with the base height $H$, and of the ratio $\lambda$ of the injectant mass flux to the free-stream mass flux intercepted by the wedge. The latter is approximately the mass flux in the boundary layer. As illustrated in Fig. 42b, four regimes can be distinguished. For $0 < \lambda < \lambda_1$, the recirculation region is detached and shrinks with increasing $\lambda$, until at $\lambda = \lambda_1$ it completely disappears. In the regime $\lambda_1 < \lambda < \lambda_2$, the injected gas remains subsonic during the initial expansion and recompression, and only thereafter is it accelerated to supersonic speed by viscous interaction with the free stream. The base pressure is found to rise to a peak as $\lambda \to \lambda_2$. When the mass flux exceeds $\lambda_2$, the injected gas very rapidly overexpands to $M > 1$. Impingement on the high-pressure outer flow causes a recompression shock wave. For $\lambda_2 \leq \lambda \leq \lambda_3$, this shock must be normal at the jet axis, and it isolates the base region so that the base pressure drops drastically. Finally, when a sufficiently high Mach number is reached in the expansion, recompression can be achieved by a weaker oblique shock, and for $\lambda > \lambda_3$ the injectant is supersonic along the entire centerline. Then the base pressure gradually rises again as $\lambda$ is increased. The experiment gives critical values of $\lambda_1 = 0.013$, $\lambda_2 = 0.030$, and $\lambda_3 = 0.053$.

Since the interaction involves both compressibility and viscous or turbulent phenomena (such as exchange of mass, momentum, and thermal energy between injectant and free stream), it cannot be expected to depend on the mass flux ratio $\lambda$ alone. The free-stream Mach and Reynolds numbers are also important. Even without base injection, heat transfer to the base has a singificant influence on one of the saddle-point singularities mentioned previously.[68] In the $\lambda < \lambda_2$ regime, other experiments with base injection have shown that preheating the injectant,[71] or injecting gases with smaller molecular weight and higher heat capacity, such as He[70,72] or $H_2$,[73,74] results in higher base pressures. The recirculation

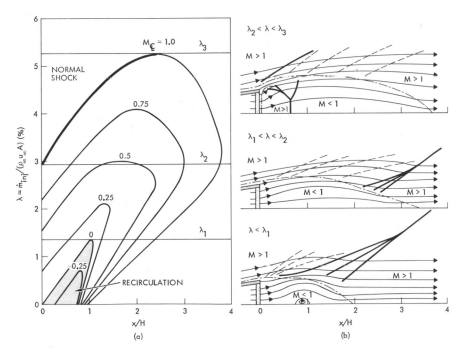

**Fig. 42.** The effect of base injection on the near wake (6° half-angle wedge, adiabatic wall, $M_\infty = 4$, $Re_\infty(H) = 3 \times 10^5$, $N_2$ injection).[70] (a) Contours of constant centerline Mach number as function of normalized axial distance and injected mass flux and (b) schematics of flowfield in three operating regimes ($\rightarrow$ streamline, — shock, -- expansion, -·- sonic line).

region is displaced by the low-density injected gas, the entire base flow is dilated, and both the corner expansion and the recompression shock propagating into the free stream are weakened. Combustion of hydrogen injected near the shoulder was found to have similar effects, and produced base pressures slightly above the free-stream pressure.[74] Simple phenomenological models for these effects have been proposed in Refs. 75 and 76. For $\lambda_2 < \lambda < \lambda_3$, experiments have shown that a wake, though with a widened core and weaker wake shocks, persists at least to $x/H = 8$,[77] and that the base injection delays wake transition.[78] Weiss[79] describes a first-order analytical model, based on experimental growth rates for the wake and the imbedded jet, from which the effect of injection on near wake temperatures can be estimated.

In actual lasers total fuel to oxidizer mass flux ratios are typically around 0.2, for both arc[16] and combustor (Table 4) operation with helium as diluent. Not more than half of the oxidizer mass flux is in the boundary layers. Hence mass-flux ratios $\lambda = \dot{m}_f/\dot{m}_{OB.L.}$ are significantly higher than in most base-injection experiments, and lasers should operate in the $\lambda > \lambda_3$ regime. We expect no detached recirculation bubbles, rapid expansion of the fuel to supersonic speeds, and recompression by oblique shocks. However, slit or supersonic fuel-injection nozzles are usually separated from the oxidizer nozzles by finite land areas, which develop their own recirculation regions and near wakes. Moreover, we recall that the flow is symmetric with respect to the nozzle centerlines, owing to the large number of such nozzle elements combined into a nozzle bank.

Figure 43 shows three likely cavity-inlet flow configurations. In Fig. 43a, both the oxidizer and the fuel are injected through parallel supersonic nozzles (Fig. 9b), their pressures are approximately matched, and the core velocities are of comparable magnitude. Then the oxidizer wake that exists at zero fuel injection is filled out by the fuel jet, and the wakes of the land areas become more important. These wakes look essentially like that of Fig. 41, but are asymmetric because of the differences in density, temperature, and Mach number of the fuel and oxidizer streams. At the low Reynolds numbers, the boundary layers are thicker and the recirculation region is smaller. (The effect of reactive heating is discussed separately.) The expansion and shock waves are reflected at the symmetry lines.

For slit-nozzle injection of fuel into supersonic oxidizer streams, with exit pressures much higher on the fuel than on the oxidizer side (see Fig. 43b), the inlet flow field should resemble a rocket plume in supersonic flight.[80,81] The fuel reaches sonic speed inside the slit nozzle and expands like a supersonic source flow through the corner expansion waves that reflect at the nozzle centerline. When it impinges on the recirculation

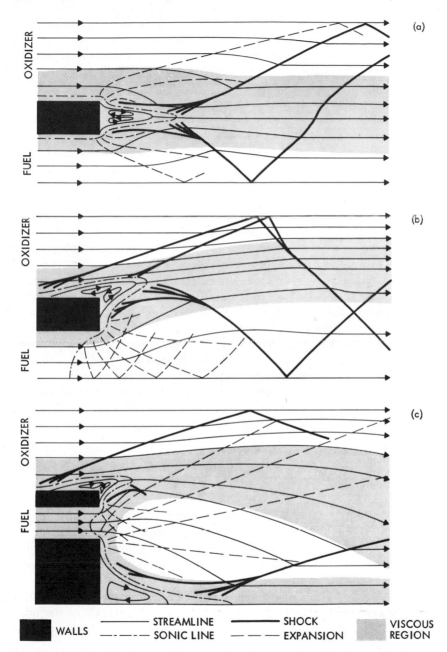

**Fig. 43.** Possible inlet-flow configurations for laser nozzles. (*a*) Supersonic fuel and oxidizer nozzles, matched pressures; (*b*) slit nozzle fuel injection at excess pressure; and (*c*) injector nozzle with base relief.

region of the dividing wall, it is turned by a system of strong oblique lip and wake shocks. Owing to the excess pressure on the fuel side, the recirculation region is pushed into the oxidizer nozzle, whose boundary layer separates from the wall. The oxidizer lip shock becomes a regular separation shock, and is intercepted by a recompression shock originating at the trailing edge of the attached recirculation bubble.

In both of the foregoing cases, fuel and oxidizer enter the recirculation region and react, which is detrimental to lasing (Section 2.1). Reactive heating will expand the recirculating gas and increase the strength of the lip shocks. Similarly, heat addition in the reactive mixing layers results in reaction shocks, which may merge with and strengthen the wake shocks. To counterbalance the increase in temperature and pressure and reduction in Mach number associated with this shock system, base-relief nozzles (Fig. 21) have thin dividing walls between the fuel and oxidizer jets, and introduce a large land area between two fuel-injection slits, as indicated in Fig. 43$c$. This approach, discussed from a different point of view in Section 1.5, is in a sense opposite to that chosen for base-drag reduction, and results in low-base pressures behind the wider land area. Its recirculation region, however, is not accessible to the oxidizer and does not react or absorb radiation. Moreover, as shown by the experiments of Ref. 29, the corner expansion into this base region is apparently sufficient to cancel out the wake shocks between the oxidizer and fuel jets, so that reactive mixing occurs in a favorable pressure gradient until intercepted by the fuel-recompression shock. With this type of nozzle, significant increases of laser power have been obtained at small $x$ and low-diluent concentrations,[82] where reactive heating is most severe.

**Spray-Bar Injection.**   Most of the arc-driven HF and DF laser studies performed at The Aerospace Corporation (e.g., Refs. 4 and 16) used a perforated tube, or spray bar, to inject the fuel streamwise between adjacent oxidizer nozzles (see Fig. 20). To compare the performance of this injector configuration with that of parallel supersonic-jet mixing, it was studied in some detail as part of the reactive flow experiments of Refs. 32 and 40.* For this purpose, both of the RF-2 nozzles (Fig. 24 and Table 5) were run with oxidizer, and a spray bar was mounted at the base of the dividing wall.

Figure 44 shows ir radiometer scans of the cavity, together with pitot-pressure profiles taken 3.8-cm ($x/h = 4.34$) downstream of the nozzle exit. These data were taken at a cavity pressure of 2.3 Torr, with

---

* Similar tests were also conducted at Martin Marietta,[33] but in a short-test duration facility driven with gas from the end wall of a short tube. Mach–Zehnder interferograms were used as primary diagnostics but are not easily interpreted.

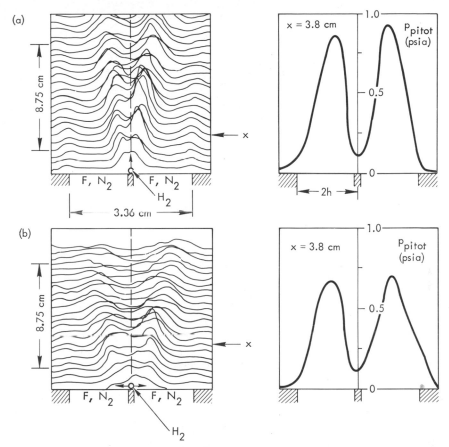

**Fig. 44.** Infrared radiometer scans and pitot-pressure profiles at $x = 3.8$ cm with spray-bar injection of $H_2$ ($H_2/F = 2$, $p = 2.3$ Torr).[32] (*a*) Axial injection and (*b*) transverse injection.

injection at an $H_2/F$ mole ratio of 2. With axial fuel injection, Fig. 44*a*, a single wake forms between the two oxidizer jets, similar to that observed in base-injection experiments.[77] The luminous reaction zones develop in the maximum shear regions on both sides of the wake. Initially, they seem to overlap, but for $x \geq 4$ cm they are clearly separated by a less emitting wake core. As shown in Fig. 45, the core actually becomes absorptive further downstream. One possible explanation is provided by deexcitation of excited HF in collisons with the excess $H_2$ in this region, as indicated by an on-axis peak of HF(0).[32] At 2.3 Torr, the mixing is probably still laminar, particularly if base injection stabilizes the wake.[78] Then the width of the reaction zone is controlled by diffusion. At 13 Torr, large-scale low-frequency gain fluctuations have been observed in the reaction

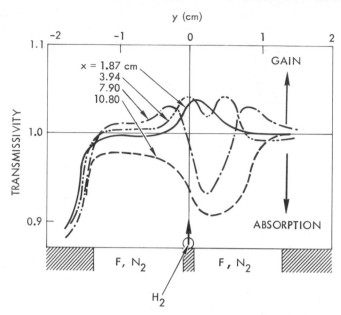

**Fig. 45.** Variation of transmissivity across mixing layer at various axial stations, axial spray-bar injection of $H_2$ ($H_2/F = 2$, $p = 2.3$ Torr).[32]

zones. Whether they were indicative of vortex formation in the unstable high-shear layers or of an arc instability could not be verified.

Partly in the hope that the strong interaction in the inlet flow field would trigger transition in the shear regions and accelerate mixing and reaction, tests were also performed with fuel injection from spray-bar holes at 90° to the oxidizer jets. As seen in Fig. 44b, the wake spreads much faster, and the reaction zones (again in the maximum shear regions) are widened. Figures 46a and b compare the integrated radiated intensity measured for axial and transverse spray-bar injection. Transverse injection is seen to be more effective in producing excited HF, by a factor of 2 at 2.3 Torr, but only by 30% at 13 Torr.

In comparing these results with the untripped parallel-jet mixing experiments discussed in Section 2.2 and included in Fig. 46, it should be remembered that for spray-bar injection both nozzles carried oxidizer, so that the fluorine flux was doubled. On the basis of equal fluorine fluxes, integrated intensities at 2.3 Torr are larger by 50% for parallel, and by a factor of 3 for transverse spray-bar injection. We recall that at this pressure parallel-jet mixing is laminar, so that the impressive factor of 3 increase could be due to transition induced by the strong transverse-jet interaction in the inlet. But at 13 Torr, where the wake between the

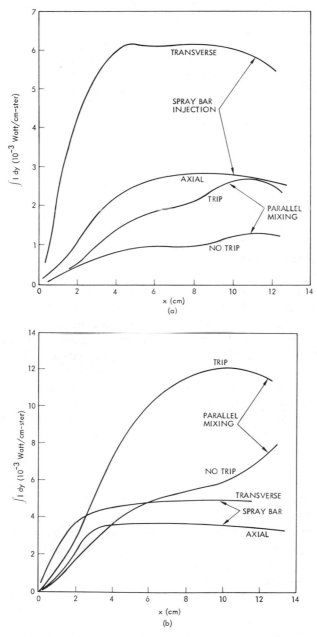

**Fig. 46.** Comparison of axial distributions of integrated spontaneous-emission intensities for axial and transverse spray-bar injection of $H_2$, and for untripped and tripped parallel jet mixing ($H_2/F-2$).[32] (*a*) $p = 2.3$ Torr and (*b*) $p = 13$ Torr.

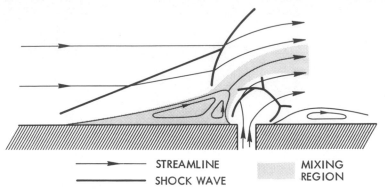

STREAMLINE

SHOCK WAVE

MIXING
REGION

**Fig. 47.**   Flow-field configuration near transverse-injection hole, supersonic free stream.

parallel jets has undergone transition, there is not only no improvement, but actually less output with spray-bar injection, except possibly in the first 2 cm from the cavity inlet.

In the plane of a spray-bar hole, the inlet flow field for parallel injection should look somewhat like that shown in Fig. 43$b$. Essential features of the flow in the neighborhood of a transverse-injection hole are indicated in Fig. 47.[38] Ahead of the hole, the nozzle boundary layer separates from the wall, forming a recirculation bubble with the corresponding oblique separation shock. If the injected jet penetrates the boundary layer, it acts like a blunt body in a supersonic stream, and the recompression shock is like a detached strong bow shock in the nose region. Since it is strongly curved, it introduces vorticity into the separated boundary layer, which could cause transition. Another recirculation region forms downstream of the transverse jet. The separation and recompression shocks on the injectant side merge in a strong normal shock with subsonic flow downstream ("barrel-shock" configuration), similar to that shown in Fig. 42$b$ for the $\lambda_2 < \lambda < \lambda_3$ regime. Substantial reactive heating and absorption will occur in the upstream separation bubble and may affect lasing. Also, it should be noted that only a finite number of holes can be drilled into a spray bar. Thus, the boundary layer is subjected to a periodic three-dimensional perturbation pattern, which sheds a row of longitudinal vortices into the wake. The effect of such vortices on wake transition is not well established. In the last phase of two-dimensional shear-layer transition ($x_{tr} < x < x_{ft}$), longitudinal vortices may play a role.[50,54]

**Shear-Layer Tripping.**   The idea of triggering transition in the wake-shear regions, by tripping both the fuel and oxidizer boundary layers at the trailing edge of the dividing wall, has been examined in Ref. 32. In the present experiments, as in those described in Section 2.2, one of the RF-2 nozzles carried oxidizer, the other fuel. Various devices were tried, the most successful of which was transverse injection of small amounts of diluent ($N_2$) from holes at 90 or 45° angles to the free stream. The effects on the inlet flow field should be much the same as described for transverse-fuel injection, including the generation of three-dimensional vortex patterns. However, the upstream recirculation bubbles are now nonreactive and substantially smaller, and the separation and bow shocks are weaker.

Figure 48 compares ir radiometer scans, and axial distributions of temperature and integrated concentrations in the $v = 0$ to 3 levels of HF,

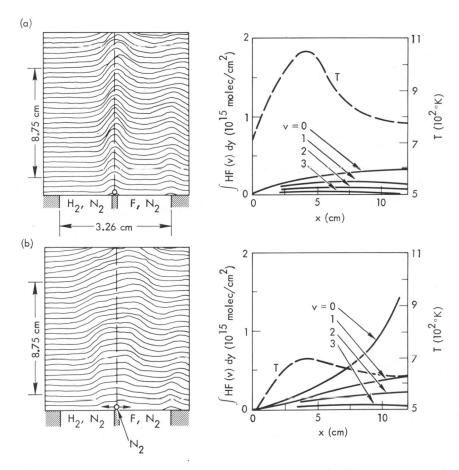

**Fig. 48.** Infrared radiometer scans, and axial distributions of centerline temperature and integrated HF($v$) concentrations for parallel-jet mixing, $p = 2.3$ Torr.[32] ($a$) No trip and ($b$) $N_2$ trip injection ($p_{0\mathrm{uip}} = 100$ psia).

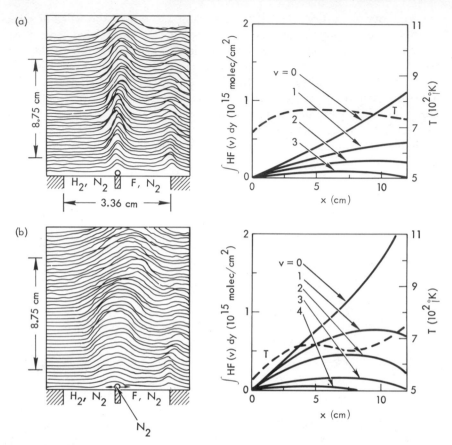

**Fig. 49.** Infrared radiometer scans, and axial distributions of centerline temperature and integrated $HF(v)$ concentrations for parallel-jet mixing, $p = 13$ Torr.[32] (a) No trip and (b) $N_2$ trip injection ($p_{0\text{trip}} = 100$ psia).

for the untripped (a) and tripped (b) mixing layers at 2.3 Torr cavity pressure. Corresponding data at 13 Torr are shown in Fig. 49. Tripping is seen to affect mainly the entrainment of the reactants, and results in faster spreading of the reaction zone with insignificant changes in peak intensity levels. The total $HF(v)$ concentrations integrated across the mixing layer are found to increase by factors of up to 2 or 3 at both low and high pressures. Similar increases are obtained in the spontaneous emission integrated across the mixing layer (see Figs. 46a and b). An additional effect of tripping is cooling of the reaction zone, by nearly 300°K at 2.3 Torr and by about 100°K at 13 Torr, which should be beneficial for lasing.

Reference 32 suggests that tripping induced transition in the mixing layer at 2.3 Torr, somewhere downstream of $x = 3$ cm where the integrated intensity rises above the untripped level (Fig. 46a). However, free-stream Reynolds numbers based on the boundary-layer displacement thickness at the nozzle exit are only of $O(10^3)$ in the reactive flow experiments (Table 5), and of $O(10^2)$ in laser nozzles (Table 4). Experiments by Batt[55,83] at $Re(\delta_L^*) = 1000$ have shown that a wire trip causes large-scale velocity fluctuations in the boundary layer, but leaves the shape of the laminar mean velocity profile unaffected. A wire trip just upstream of a rearward-facing step resulted in faster spreading of the shear layer behind the step. But self-preserving fully turbulent mixing and probably transition occurred at the same $x/\theta_L$ as in the untripped case. Since the boundary layer thickens when tripped, transition requires a longer axial distance than without tripping. For the reactive flow experiments, and even more so for application to the laser cavity, we suspect that the transverse-trip jets may actually only distort the transitional vortex pattern, rather than induce true turbulent mixing.

## 2.4. Analysis of Turbulent, Supersonic, and Reactive Mixing

In the laminar limit numerical solutions of the steady-state Navier–Stokes and species-conservation equations have successfully been obtained for supersonic reactive flows closely related to the cavity flow in mixing lasers. For transitional and turbulent mixing, however, solution of the corresponding time-dependent equations generally exceeds the capacity of even large computers. Modern analyses of turbulent mixing[41,84] are instead based on semiempirical formulations in terms of either time averages (moments) of products of the fluctuating variables or their statistical behavior. Here, we briefly assess more recent attempts to incorporate two effects important in mixing lasers, that is, compressibility and fast exothermic chemical reactions. An equally detailed treatment of transition in mixing layers is presently still out of reach.

**Analysis in Terms of Moments of Turbulent Fluctuations.** For a description of the moment approach to nonreactive turbulent mixing, the reader is referred to review papers by Mellor and Herring[85] and by Bradshaw.[86] Eddy viscosity (first-order closure) models,[87–90] which relate all second- and higher-order moments to local gradients of the mean properties, can describe only the asymptotic state of self-preserving turbulence, $x > x_{ft}$. As seen in Section 2.2, most of the laser power is produced in the regime

$x < x_{ft}$, and thus, second-order closure models are required. In addition to the conservation equations for the mean properties,* these incorporate differential equations for the turbulent kinetic energy $e = \overline{\rho u_i'' u_j''}/2\bar{\rho}$ or the components of the Reynolds stress $\tau_{ij}^T = -\overline{\rho u_i'' u_j''}$, and for a scale length $l$ characterizing the largest turbulent eddies.[92–97]

Models of this type available in 1972 were found[84] to provide no explanation for the decreased shear-layer growth rate at supersonic Mach numbers (Fig. 31). As pointed out by Laufer (p. 687 of Ref. 84), an energy exchange between the mean and turbulent flow field is contained in the pressure work term $\overline{p' \, \partial u_i''/\partial x_j}$ of the mean and turbulent kinetic-energy equations. Oh[98] modeled this term by postulating the formation of shock waves around large eddies that penetrate the high Mach-number region of the mixing layer. The effect of longitudinal pressure gradients has been modeled by Wilcox and Alber.[99] A combination of these two models appears to predict growth rates and profiles in compressible and heterogeneous shear layers.[44,57]

The pumping and deactivation reactions introduce terms of the form $\rho^2 k C_s C_r$ into the species equations and, multiplied by the heat of formation, into the energy equation. Here $k$ is the rate coefficient, and $C_s$ and $C_r$ are the reactant mass fractions. The simplest case, that of isothermal incompressible reactive mixing ($\rho$, $k(T)$ constant), has been discussed by Spalding[100] and Donaldson and Hilst.[101] It was shown that the approximation $\overline{C_s C_r} = \overline{C_s}\,\overline{C_r}$ gives erroneous results for fast reactions. The reason is that the reactants are supplied by molecular diffusion across the boundaries of eddies containing unmixed newly entrained material, and that the rate-limiting step becomes the break-up of the large eddies. Then additional equations for the moments $\overline{C_s' C_r'}, \overline{C_r'^2}$ and $\overline{C_s'^2}$ have to be solved, which in turn introduce third-order moments $\overline{C_s'^2 C_r'}$ and $\overline{C_s' C_r'^2}$ through the reaction term. A tentative closure has been suggested by Hilst.[102]

Laser mixing, however, is compressible, and the reactions are highly exothermic, so that higher-order correlations between $\rho', k'$ and $C_s', C_r'$ must be considered. Furthermore, the rate coefficients depend nonlinearly on the temperature. Moments of this type have not yet been modeled.

A phenomenological approach to the closure problem for fast reactions has been suggested by Marble.[32,103] The Brown–Roshko experiments[44] indicate that material from the two free streams is entrained nearly unmixed in large gulps that penetrate the shear layer. With fast chemical reactions, the coherent interface stretched between these eddies becomes a thin, highly convoluted laminar-diffusion flame. As shown in Ref. 103, stretching of such a flame increases the influx of reactants, but is limited

---

* The notation used here, including mass-centered averages, is explained in Ref. 91.

by the burn-off of narrow protuberances. From these considerations, a differential equation for the mean-flame surface area per unit volume, and hence for the effective rate of reaction, has been derived in Ref. 32.

**Statistical Models for Turbulent Diffusion Flames.**   Existing statistical descriptions of reactive turbulent mixing all consider the case of fast chemical reactions, but neglect the effect of reaction on the turbulence field. This implies that reactive heating is negligible because of either weak exothermicity or a high degree of dilution, or because the flow is incompressible. Toor[104] has shown that under these conditions the statistical behavior of reactant mass fraction and temperature fluctuations can be evaluated in terms of the statistics of an inert species (passive scalar) entrained at one of the mixing-layer edges.

The Brown–Roshko experiments suggest a high degree of unmixedness on the macroscale, so that a passive scalar jump between the values corresponding to the two free streams, but rarely assumes intermediate values. Experiments by Batt[105] for fully developed turbulent shear-layer mixing show a nearly Gaussian passive-scalar probability-density function (PDF) in the core of the layer, indicating thorough mixing. References 106 through 108 suggest a PDF of $\beta$-function form that incorporates both the mixed and unmixed limits. Once the free parameters of the PDF have been evaluated from either experiment or a second-order closure solution for the turbulence field,[105] the mean and r.m.s. concentration profiles for the reactants and reaction products can be determined.

O'Brien[109] discussed the evolution of the PDF from statistically prescribed initial conditions, through a reaction-limited stage followed by a diffusion-limited thin flame regime. He shows that contributions of third-order moments of the PDF are associated with large-amplitude fluctuations across the flame. From these considerations, and a set of inequalities for moments of positive definite scalars, an alternate closure for the correlations $\overline{C_s'^2 C_r'}$ has been derived.[110] Differential equations governing the evolution of the PDF have been formulated[111] for the case of ignition of a premixed turbulent reaction. Although this case is of no interest for chemical lasers, the analysis shows that the appearance of two-point joint probabilities poses at least as severe a closure problem as that in the moment formulation.

In summary, statistical methods currently available cover diffusion-limited reactions with negligible reactive heating, in low Mach number, nearly self-preserving turbulent mixing layers. The advantage of bypassing the closure problem posed by the reaction term is lost when considering a developing turbulent flow, or when reactive heating becomes important at low or moderate degrees of dilution.

## 3. SIMPLE MODELS FOR LASER-PERFORMANCE PREDICTION

### 3.1. Motivation

In this section we derive scaling laws for the optical performance of a cw HF (or DF) laser. Using a simplified description of the mixing-layer dynamics, we deduce the dependence of the axial intensity distribution, power, and efficiency on basic gas-dynamical parameters, such as cavity pressure, temperature, velocity, and reactant and diluent mole fractions.

From Section 1, we recall that the $H_2$ (or $D_2$) fuel is injected through either sonic slits or small supersonic nozzles, generally at exit pressures exceeding that of the oxidizer jet and at flow rates significantly larger than stoichiometric to the fluorine flow. For structural reasons, the dividing wall between the oxidizer and fuel nozzles is truncated, forming a base step of finite height, and the two jets impinge on each other to encourage early mixing. The resulting complex flow field near the nozzle-exit plane, described in Section 2.3, influences the initial development of the mixing and lasing region, and may control transition to turbulent mixing. It is evident from the discussion in Section 2.4 that a consistent detailed analysis of the resulting structure of the mixing zone is presently possible only for laminar mixing, and even in this case requires extensive computational modeling.

Though detailed computational analysis will eventually provide needed insights into the advantages of particular nozzle configurations, it is costly when applied repetitively for scaling-law studies. Instead, an analytical description is outlined in Section 3.2 and evaluated in Section 3.3. The resulting scaling laws for laminar and turbulent mixing are summarized in Section 3.4. Although by necessity they are based on some gross simplifications, comparison with results of more exact analytic and numerical calculations (Section 3.5) and with the measured overall performance of laser cavities (Section 3.6) indicates that essential features of the mixing process have been retained. Theoretically predicted and observed axial-gain distributions are compared in Section 3.7 for the zero-power limit.

### 3.2. Analytic Model for the Mixing Layer

Figure 50 sketches the idealized configuration of the mixing layer adopted in the present model, which follows the analyses of Refs. 5 and 6. Oxidizer and fuel enter the cavity in alternate parallel jets that are uniform in velocity and species distributions, ignoring the nozzle boundary layers. The jet pressures and velocities are presumed matched, so that interaction shocks and shear can be neglected. The dividing wall is thin, and the wake velocity defect, negligible. Within the mixing layer starting at the nozzle lip, the pressure $p$, temperature $T$, and axial velocity $u$ are

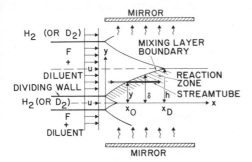

**Fig. 50.** Mixing-layer configuration.

considered constant. Transverse temperature and pressure gradients caused by reactive heating are ignored, since we assume a large amount of diluent.

In actual lasers, the ratio of fuel to oxidizer molar fluxes is large ($R_L \gg 1$), and the fuel jet, narrow. Hence most of the mixing region is dominated by fuel diffusing into the diluent of the adjacent oxidizer jets. The production of excited HF by cold-reaction pumping

$$F + H_2 \xrightarrow{k_p a(v)} HF(v) + H \tag{58}$$

then peaks close to the outer boundary $y_e = \delta(x)$ of the mixing region on the oxidizer side, where $H_2$ is about stoichiometric to F. We assume that pumping and collisional deactivation

$$HF(v) + M_i \xrightarrow{k_{cd,i} a_i(v)} HF(v-1) + M_i \tag{59}$$

are initiated immediately upon entrainment of the fluorine at $y_e = \delta(x)$. The collision partner $M_i$ can be one of the products of the pumping reaction (58), one of the reactants, or the diluent. In a combustor-operated HF laser, the combustion product DF is also an important deactivator. The $k_p$ and $k_{cd,i}$ are the overall forward-rate coefficients (generally large compared to the backward rates); $a(v)$ and $a_i(v)$ are the weights with which the $v$th vibrational level HF($v$) participates in the reactions. For simplicity we neglect V–V exchange between HF molecules. While V–V exchange is important when pumping produces a significant fraction of ground-state molecules (as in the HCl chain laser), it affects the net radiative output of an HF cold reaction laser only when anharmonicity is taken into account.[112]

In the mixing zone, gradients transverse to the flow direction are neglected. The analysis is thus similar to the premixed stream-tube analysis,[112] except that reactions start at the mixing-layer boundary

$y_e = \delta(x)$ rather than at the nozzle exit. Since we are interested in properties integrated across the mixing layer, no attempt is made to resolve species profiles within the layer. The $H_2$ and F (diffusion) fluxes to the flame are equated to the rate of (convective) fluorine entrainment. The only reference to the mixing mechanism is through the rate of growth of $\delta(x)$, for which semiempirical approximations are introduced in Section 3.3.

For simplicity, we adopt a two-level vibrational model for HF, with pumping preferentially into the upper level*:

$$[HF] = [HF_u] + [HF_l], \qquad a(v_u) = 1, \qquad a(v_l) = 0 \tag{60}$$

where $[X]$ is the concentration of species X. Along each stream tube in the mixing layer, the conservation equations for the fluorine, atomic hydrogen, total HF, and the upper vibrational level $HF_u$ become

$$u\frac{d[HF]}{dx} = u\frac{d[H]}{dx} = -u\frac{d[F]}{dx} = k_p[H_2][F] \tag{61}$$

and

$$u\frac{d[HF_u]}{dx} = k_p[H_2][F] - \sum_i k_{cd,i}[M_i][HF_u] - g\frac{I}{\epsilon} \tag{62}$$

where the summation is over all collisional deactivators $M_i$. $I$ is the radiation intensity in the $y$-direction, $\epsilon = h\nu N_A$ is the energy per mole of photons with $N_A$ Avogadro's number, and $g$ is the gain per centimeter given by[112]

$$g = \sigma([HF_u] - e^{-2J\theta}[HF_l])$$
$$\sigma = 2.74 \times 10^{47} \theta_R W_{HF}^{1/2} T^{-3/2} |M|^2 Je^{-J(J-1)\theta} \tag{63}$$

and

$$\theta = \frac{\theta_R}{T}$$

Here $\theta_R = 30.16°K$ is the characteristic rotational temperature of HF, $W_{HF} = 20$ g/mole is its molecular weight, and $|M|^2 = 2.8 \times 10^{-38}$ erg/cm$^3$ is the square of the dipole matrix element. The rotational quantum number $J$ for the lower level of the lasing transition is determined by the transition with the highest chemical efficiency and depends on the rotational temperature $T_R = T$.

No equation has been written down for the $H_2$ concentration, which in the fuel-rich case is controlled by diffusion. Following Refs. 5 and 6, we consider the rates $K_p = k_p[H_2]$ and $K_{cd} = \sum_i k_{cd,i}[M_i]$ as constants. Here, however, $[H_2]$ is not the hydrogen concentration in the undisturbed fuel

---

* Apart from some additional factors involving the rotational quantum number and temperature, the four-level description[6] gives the same results.

jet, but rather in the reaction zone where $[H_2]$ and $[F]$ are approximately stoichiometric. Excess $H_2$ on the fuel side of the mixing layer only pushes the reaction zone toward the oxidizer side. Similarly, the deactivator concentrations $[M_i]$ should be evaluated where deactivation occurs inside the mixing layer.

From Eqs. (61), we obtain for the concentrations

$$[HF] = [H] = [F]_o(1 - e^{-K_p(x-x_0)/u})$$

and

$$[F] = [F]_o e^{-K_p(x-x_0)/u}$$

(64)

where the subscript 0 refers to conditions on a streamline just before it enters the mixing layer and subscripts o and f to the oxidizer and fuel free streams. Note that $x_0$, through $\delta(x_0) = y$, is a function of $y$. With the foregoing results, Eqs. (62) and (63) transform into a relation for the gain

$$\frac{d}{d\zeta}\left(\frac{g}{\sigma[F]_o}\right) + \frac{g}{\sigma[F]_o}\left[1 + (1 + e^{-2J\theta})\frac{\sigma I}{\epsilon K_{cd}}\right] = (K + e^{-2J\theta})e^{-K(\zeta-\zeta_0)} - e^{-2J\theta}$$

(65)

where we have introduced the dimensionless quantities

$$\zeta = \frac{K_{cd}x}{u}, \qquad K = \frac{K_p}{K_{cd}} = \frac{k_p[H_2]}{\sum_i k_{cd,i}[M_i]}$$

(66)

For analytical convenience, we set $e^{-2J\theta} \approx 1$, which is permissible when $T/\theta_R$ is large and lasing occurs at low $J$.

We postulate that the free-mixing layer grows according to the power law

$$\delta = A\zeta^m \quad \text{and hence} \quad \zeta_0 = \left(\frac{y}{A}\right)^{1/m}, \qquad \zeta_D = \left(\frac{h}{A}\right)^{1/m}$$

(67)

The exponent $m$ and coefficient $A$ are determined by the mixing process, and representative choices for these parameters are discussed in Section 3.3. After merging at $\zeta_D = x_D K_{cd}/u$ (Fig. 50), the mixing layer is confined between the centerlines of the fuel and oxidizer jets, $0 \le y \le h$. For the integrated gain

$$G = \begin{cases} \displaystyle\int_0^{\delta(x)} g\,dy \\[1em] \qquad\qquad \text{if } x \lessgtr x_D \\[1em] \displaystyle\int_0^h g\,dy \end{cases}$$

(68)

integration of Eq. (65) across the mixing layer yields the relations

$$\frac{d}{d\zeta}\left(\frac{G}{\sigma[\text{F}]_o}\right) + \frac{G}{\sigma[\text{F}]_o}\left[1 + \frac{2\sigma I}{\epsilon K_{cd}}\right]$$

$$= A\left[mK^{-m}(1+K)e^{-K\zeta}\int_0^{K\zeta} e^z z^{m-1}\, dz - \zeta^m\right] \qquad \text{if } \zeta \le \zeta_D \quad (69a)$$

and

$$\frac{d}{d\zeta}\left(\frac{G}{\sigma[\text{F}]_o}\right) + \frac{G}{\sigma[\text{F}]_o}\left[1 + \frac{2\sigma I}{\epsilon K_{cd}}\right]$$

$$= A\left[mK^{-m}(1+K)e^{-K\zeta}\int_0^{K\zeta_D} e^z z^{m-1}\, dz - \zeta_D^m\right] \qquad \text{if } \zeta > \zeta_D \quad (69b)$$

where the intensity $I$ is a mean value and $g(y_e)$ is assumed zero.

Equations (69) allow performance predictions for both amplifier and oscillator cavities. For an amplifier solution, the reader is referred to Ref. 5. The small-signal gain $G_0(x)$ is obtained in the limit $I = 0$. For an oscillator, the case of primary interest, the reflectivities $r_1$, $r_2$ of the cavity mirrors limit the integrated gain in the lasing zone to a threshold value

$$G_c = -\frac{\ln(r_1 r_2)}{4N} \qquad (70)$$

for which the gain in intensity on one round trip through the cavity balances the mirror losses and the outcoupling. $N$ is the number of oxidizer nozzles in the nozzle bank, each feeding two mixing layers. There is an initial region, $0 \le x \le x_i$, with no lasing, in which the small-signal gain $G_0(x)$ increases until it reaches the threshold. For a saturated cavity $G_c \to 0$ this region is negligible. Throughout the lasing zone, the gain is fixed at $G_c$, and the intensity $I$ is immediately given by Eqs. (69) since the derivative term vanishes. Lasing ends at $x_e$, where $I = 0$. At sufficiently low cavity pressures, merging may occur before lasing terminates, and Eq. (69b) should be used for the region $x_D \le x \le x_e$.

The cavity power distribution

$$P(x) = 2Nh_z \int_0^x GI\, dx \qquad (71)$$

and the chemical efficiency

$$\eta = \frac{P(x_e)}{|\Delta H|\,[\text{F}]_o u 2 N h h_z} = \frac{\displaystyle\int_0^{x_e} GI\, dx}{|\Delta H|\,[\text{F}]_o u h} \qquad (72)$$

follow after integration over $x$. Here $h_z$ is the nozzle height normal to both the flow direction and the optical axis, and $|\Delta H|$ is the heat of reaction (58). The term $[F]_o u 2 N h h_z$ is the total molar flux of atomic fluorine out of the nozzle bank.

## 3.3.  Evaluation for Laminar and Turbulent Mixing

**Mixing-Layer Growth Rates.**   In principle, the coefficient $A$ and exponent $m$ of the power law for the mixing-layer growth, Eq. (67), could be determined by fitting experimental observations, for example, of the growth of the chemiluminescent part of the mixing layer.[30,31] But interpretation of such measurements is difficult because of the complex interaction of boundary layers, wake, and shocks. References 12 and 30 point out the need for corrections for the deflection of the jets by the reaction shocks. Moreover, growth-rate observations alone cannot resolve whether mixing is laminar or turbulent, as will become clear in the subsequent discussion.

Solutions of Eqs. (69) for laminar[5] and turbulent[5,6] mixing are, instead, obtained under the simplifying assumption of self-similar growth prior to merging at $x_D$. This implies the absence of externally imposed scale lengths other than the channel height $2h$. While this assumption is consistent with those made in the beginning of Section 3.2, in actual laser-nozzle configurations, transition and initial perturbations by interaction shocks, nozzle boundary layers, and the base step can result in nonsimilar growth. Then the solutions become valid only asymptotically as these initial perturbations die out downstream of the nozzle lip.

In the absence of shear between the two jets, the growth of a self-similar mixing layer is determined by the balance between molecular diffusion across the layer and entrainment at its boundaries. The corresponding terms in the conservation equations for the entrained species can be estimated by

$$u \frac{\partial [X]}{\partial x} \simeq [X]_e \frac{u_e}{x}, \qquad \frac{\partial}{\partial y}\left(D \frac{\partial [X]}{\partial y}\right) \simeq [X]_e \frac{D_e}{\delta^2} \tag{73}$$

where the subscript e refers to the boundary at which species X is entrained, and where $\delta$ is the layer thickness. In the fuel rich case, and for a highly diluted oxidizer stream, $D$ is the coefficient for diffusion of $H_2$ into the diluent. The balance between the two terms implies the laminar-growth law

$$\delta = A_L \left(\frac{Dx}{u}\right)^{1/2} \quad \text{and hence} \quad m = \tfrac{1}{2}, \quad A = A_L \left(\frac{D}{K_{cd}}\right)^{1/2} \tag{74a}$$

From the corresponding balance between the convection term and the shear-stress of the momentum equation, parabolic growth is also predicted when shear between the jet velocities or the wake of the base step dominate [see Eqs. (53) and (55)]. In the presence of thick laminar boundary layers, with velocity profiles linear near the wall, velocities at the mixing-layer edges at first increase linearly with $\delta$. A slower initial growth rate $\delta \simeq x^{1/3}$ is then expected. But ultimately, wake-like growth, with the sum of the displacement thicknesses of the nozzle boundary layers as scale height, takes over, so that Eq. (74a) is again appropriate.

For fully developed turbulent shear layers, experimental observations suggest a linear growth

$$\delta = A_T x \quad \text{and hence} \quad m = 1, \qquad A = \frac{A_T u}{K_{cd}} \tag{74b}$$

The same behavior is predicted by simple phenomenological models,[22] which replace $D$ by the eddy diffusivity

$$\epsilon = l^2 \left| \frac{\partial u}{\partial y} \right|$$

and postulate that Prandtl's mixing length $l$ is approximately constant across the layer and scales with $\delta$. The presence of shear is implied and indeed provides the mechanism that generates the turbulence in the mixing layer. When shear is due to a velocity difference between the F and $H_2$ jets, we estimate

$$u \frac{\partial [X]}{\partial x} \simeq [X]_e \frac{u_F + u_{H_2}}{2x}, \qquad \frac{\partial}{\partial y} \left( \epsilon \frac{\partial [X]}{\partial y} \right) \simeq [X]_e \left( \frac{l}{\delta} \right)^2 \frac{|u_F - u_{H_2}|}{\delta}$$

With $l/\delta$ constant, we find $\delta \simeq x$. By similar arguments, an initially linear growth is also expected under the influence of the inner profiles of the thick laminar nozzle boundary layers. The growth of a turbulent wake because of a base step, however, is parabolic [see Eq. (56)]. These turbulent growth laws apply only downstream of the initial laminar and transitional regions, when turbulence is fully established. Also, $x$ should be measured from a virtual origin of the turbulent region, located somewhere downstream of the nozzle lip. The presence of shocks or externally imposed disturbances may induce earlier transition and move this origin closer to the nozzle exit (Sections 2.2 and 2.3).

The laser-performance predictions in Sections 3.3 and 3.4 are based on the growth laws given in Eqs. (74). The parameters $A_L$ and $A_T$ are free to be determined experimentally and are expected to depend on the composition of the unmixed reactants and the exothermicity of the reaction, as well as on the two jet velocities. Experiments with nonreacting shear layers, and simple analyses of their internal structure, suggest that $A_L = O(1)$ and $A_T = O(0.1)$.[22]

**Governing Parameters.**    Let us now examine the physical significance of the dimensionless parameters in Eqs. (69). The parameter $K$ is the ratio of the chemical rates for pumping and collisional deactivation. For the most important deactivators, Fig. 51 shows estimates[113] of the ratios of the rate coefficients $k_{cd,i}/k_p$ for HF(2) as functions of temperature. Chemiluminescence measurements[31] in an arc-driven cavity indicate that the excited HF is at a mean translational temperature of 500 to 600°K, and peak temperatures in the pumping zone may be even higher.[40a] In this temperature range, we find that $k_H : k_F : k_{HF} : k_{H_2} : k_p = 1.0 : 0.15 : 0.05 : 0.003 : 1.0$. Now recall that in the pumping zone $[H_2] \simeq [F] \simeq [F]_o/2$, while $[H] \simeq [HF] \simeq [F]_o$. For the fuel rich part of the mixing layer, reasonable estimates are $[H_2] \simeq [H_2]_f/2 \simeq 5[F]_o$, $[F] \ll [F]_o$, and $[H] \simeq [HF] \simeq [F]_o/2$. Thus, with an arc plenum, the dominant deactivator throughout the mixing layer is predicted to be atomic hydrogen. In the

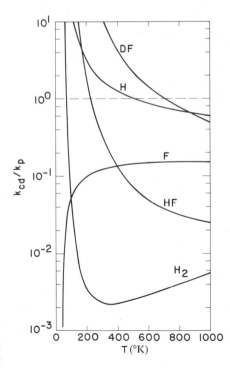

**Fig. 51.**   Rate coefficients as a function of temperature for deactivation of HF(2) normalized by that for pumping HF(2).[113]

pumping zone, where deactivation should peak, we estimate

$$K \simeq \frac{1}{2}\frac{k_{\mathrm{p}}}{k_{\mathrm{H}}} \simeq O(1) \tag{75a}$$

Similar results hold for HF(1), while for HF(3) we expect $K \simeq 0.2$.

In a combustor-driven laser, approximately two DF(0) molecules are produced in the combustor for every fluorine atom (see Section 1.3). In the mixing layer, the DF concentration is, therefore, comparable to $[\mathrm{F}]_{\mathrm{o}}$, and the estimate for $K$ is

$$K \simeq \frac{k_{\mathrm{p}}}{k_{\mathrm{DF}}}\frac{1}{2} = 0.05 \text{ at } 300°\mathrm{K} \text{ to } 1 \text{ at } 1000°\mathrm{K} \tag{75b}$$

Thus DF is the dominant deactivator throughout this temperature range.

The rate coefficients for deactivation by H and DF were not known when Refs. 5 and 6 were written, and both authors placed particular emphasis on the limiting case $K \to \infty$, assuming pumping to be very fast compared to both collisional deactivation and diffusion. If the new rate coefficient estimates hold true,* this limit is of less practical interest.

The product $K\zeta = K_{p}x/u$ can be interpreted as the Damköhler number for pumping, that is, as the ratio of the time needed to mix the reactants freshly entrained at opposite sides of the mixing layer to the time for their consumption by the pumping reaction. This implies that the characteristic time for mixing is comparable to the convection time $\tau_{\mathrm{c}} = x/u$. In the laminar case, mixing is achieved by molecular diffusion, for which Eqs. (73) and (74a) with $A_{\mathrm{L}} = O(1)$ give the time scale $\tau_{\mathrm{D}} \simeq \delta^{2}/D \simeq \tau_{\mathrm{c}}$, and hence $K\zeta \simeq \tau_{\mathrm{D}}/\tau_{\mathrm{p}}$. The turbulent-mixing process is more complicated. The large-scale eddy mixing considered in Section 3.3 redistributes the entrained material in eddies of size $l \simeq \delta$ (see Section 2.4). The pumping reaction, however, can proceed only when the reactants are mixed on the molecular level, and this small-scale mixing is limited by the rate at which the large eddies are broken up. A time constant for the latter process can be estimated[114] as $\tau_{\mathrm{b}} = l/u'$, where $u' \simeq A_{\mathrm{T}}u$ is the r.m.s. of the velocity fluctuations. With the estimate for $\delta$ given in Eq. (74b), we again find that $\tau_{\mathrm{b}} \simeq \tau_{\mathrm{c}}$, and hence $K\zeta = \tau_{\mathrm{b}}/\tau_{\mathrm{p}}$.

When $K\zeta < 1$, the reactants are mixed faster than they are consumed, and excited HF is produced throughout the mixing layer at a rate limited by $K_{\mathrm{p}}$. But when $K\zeta \gg 1$, pumping can proceed only as fast as the reactants are mixed, that is, at an effective averaged rate of the order of $[\mathrm{F}]_{\mathrm{o}}/\tau_{\mathrm{c}}$. In the laminar case, the supply of reactants by diffusion then

---

* Experimental evidence presented in Section 3.5 suggests much slower deactivation by H atoms.

balances consumption by the pumping reaction in a pumping zone of width $\delta_p < \delta$. References 115 and 116 predict $\delta_p/\delta \to (K\zeta)^{-1/3}$. Thus, formation of a thin flame sheet occurs only at very large $K\zeta$. A similar picture may apply instantaneously to turbulent mixing[117] at large $K\zeta$, with the flame stretched along the surfaces of the large eddies. Because of random motion of the flame, the pumping zone appears to be smeared out in the time average, but the averaged rate of pumping is still limited by mixing when $K\zeta \gg 1$.

The dimensionless axial distance $\zeta$ is the Damköhler number for collisional deactivation. When $K\zeta$ is a large and $\zeta$ small, the excited HF molecules diffuse out of the pumping zone before they are deactivated. On the other hand, when both $K\zeta$ and $\zeta$ are large, pumping and deactivation occur only in the pumping zone. For small $K\zeta$, both processes occur throughout most of the mixing layer, and $K\zeta$ and $\zeta$ determine the streamwise extent of the pumping and deactivation regions.

The features just described should be simulated by any credible mixing-layer model. In the derivation of Eqs. (69), fuel-rich conditions and negligible diffusion transverse to the streamlines are assumed. The balance between mixing and the pumping reaction is replaced by one between entrainment and pumping. At small $K\zeta$, this is legitimate, since the growth of the mixing layer is correctly modeled and since, in its interior, pumping is limited by the slower chemical rate rather than by diffusion. The entrainment and mixing rates are proportional to each other, so that the model should also be qualitatively correct at large $K\zeta$, provided that it includes the feature of narrowing of the pumping zone as $K\zeta \to \infty$. In this limit the exponential decay with $K(\zeta - \zeta_0)$ of the fluorine concentration along each streamline, Eq. (64), indicates that something like a flame sheet forms adjacent to the fluorine boundary of the mixing layer.

In addition to the gas-dynamical parameters $K$ and $\zeta$, the following dimensionless optical parameters appear in the solutions:

$$\tilde{G}_c = \frac{G_c}{A\sigma[F]_o}$$

$$\tilde{I} = \frac{I 2\sigma}{\epsilon K_{cd}}$$

$$\tilde{P} = \frac{P}{\epsilon[F]_o u N h_z A} \tag{76}$$

$$\tilde{\eta} = \tilde{P}(\zeta_e) = \frac{\eta 2h |\Delta H|}{\epsilon A}$$

**Table 7.** Oscillator-Performance Predictions

A. Laminar mixing, $m = 1/2$

$$\frac{2K}{K-1} F((\zeta_i)^{1/2}) - \frac{K+1}{K-1} \frac{F((K\zeta_i)^{1/2})}{(K)^{1/2}} - (\zeta_i)^{1/2} = \tilde{G}_c \tag{A1}$$

$$\frac{1+K}{K^{1/2}} F((K\zeta_e)^{1/2}) - (\zeta_e)^{1/2} = \tilde{G}_c \qquad \text{if } \zeta_e \le \zeta_D \tag{A2}$$

$$\frac{1+K}{K^{1/2}} F((K\zeta_D)^{1/2}) e^{-K(\zeta_e - \zeta_D)} - (\zeta_D)^{1/2} = \tilde{G}_c \qquad \text{if } \zeta_e > \zeta_D$$

$$\tilde{I} = \left[ \frac{1+K}{K^{1/2}} F((K\zeta)^{1/2}) - \zeta^{1/2} \right] \frac{1}{\tilde{G}_c} - 1 \qquad \text{if } \zeta \le \zeta_D \tag{A3}$$

$$\tilde{I} = \left[ \frac{1+K}{K^{1/2}} F((K\zeta_D)^{1/2}) e^{-K(\zeta - \zeta_D)} - (\zeta_D)^{1/2} \right] \frac{1}{\tilde{G}_c} - 1 \quad \text{if } \zeta > \zeta_D$$

$$\text{and } \zeta_i \le \zeta \le \zeta_e$$

$$\tilde{P} = \left[ \frac{1+K}{K} \left( \zeta^{1/2} - \frac{F((K\zeta)^{1/2})}{K^{1/2}} \right) - \frac{2}{3} \zeta^{3/2} - \tilde{G}_c \zeta \right]_{\zeta_i}^{\zeta} \qquad \text{if } \zeta \le \zeta_D \tag{A4}$$

$$\tilde{P} = \tilde{P}(\zeta_D) + \frac{1+K}{K} \frac{F((K\zeta_D)^{1/2})}{(K)^{1/2}} (1 - e^{-K(\zeta - \zeta_D)})$$

$$- ((\zeta_D)^{1/2} + \tilde{G}_c)(\zeta - \zeta_D) \qquad \text{if } \zeta > \zeta_D$$

$$\text{and } \zeta_i \le \zeta \le \zeta_e$$

$$\tilde{\eta} = \tilde{P}(\zeta_e) \tag{A5}$$

where $\quad F(x) = e^{-x^2} \displaystyle\int_0^x e^{y^2}\, dy \quad$ (Dawson integral)

$$F(x) \to {}_{1/(2x)}^{x} \quad \text{as} \quad x \to {}_\infty^0$$

$\tilde{G}_c$ is proportional to the ratio of the lasing cavity gain to the peak small-signal gain and must be small compared to unity to achieve saturation. $\tilde{I}$ is the ratio of the collisional to the radiative deactivation time, and $\tilde{I} \gg 1$ implies that most of the population inversion created by pumping is used optically rather than lost by collisional deactivation. In the expression

$$\tilde{P}(\zeta) = \frac{P(x)2\zeta^m}{\epsilon[F]_o u 2N\, \delta h_z}$$

the denominator is the optical power nominally available if for each F atom entrained one photon is emitted, and $P(x)$ is the power actually extracted up to station $x$. $\tilde{P}/(2\zeta^m)$ and $\tilde{\eta}/(2\zeta_e^m)$ are essentially optical efficiencies for a cavity of length $x$ and for the full lasing zone $x_e$, respectively.

**Solutions.** The solutions of Eqs. (69), based on the growth laws (74) for laminar and turbulent mixing layers, are compiled in Tables 7A and B,

**Table 7.**   (Continued)

B. Turbulent mixing, $m = 1$

$$\frac{2K}{K-1} f(\zeta_i) - \frac{K+1}{K-1} \frac{f(K\zeta_i)}{K} - \zeta_i = \tilde{G}_c \tag{B1}$$

$$\frac{1+K}{K} f(K\zeta_e) - \zeta_e = \tilde{G}_c \qquad\qquad \text{if } \zeta_e \le \zeta_D \tag{B2}$$

$$\frac{1+K}{K} f(K\zeta_D) e^{-K(\zeta_e - \zeta_D)} - \zeta_D = \tilde{G}_c \qquad\qquad \text{if } \zeta_e > \zeta_D$$

$$\tilde{I} = \left[ \frac{1+K}{K} f(K\zeta) - \zeta \right] \frac{1}{\tilde{G}_c} - 1 \qquad\qquad \text{if } \zeta \le \zeta_D \tag{B3}$$

$$\tilde{I} = \left[ \frac{1+K}{K} f(K\zeta_D) \, e^{-K(\zeta - \zeta_D)} - \zeta_D \right] \frac{1}{\tilde{G}_c} - 1 \qquad\qquad \begin{array}{l} \text{if } \zeta > \zeta_D \\ \text{and } \zeta_i \le \zeta \le \zeta_e \end{array}$$

$$\tilde{P} = \left[ \frac{1+K}{K} \left( \zeta + \frac{e^{-K\zeta}}{K} \right) - \frac{\zeta^2}{2} - \tilde{G}_c \zeta \right]_{\zeta_i}^{\zeta}$$

$$\tilde{P} = \tilde{P}(\zeta_D) + \frac{1+K}{K} \frac{f(K\zeta_D)}{K} (1 - e^{-K(\zeta - \zeta_D)}) \qquad\qquad \text{if } \zeta \le \zeta_D \tag{B4}$$

$$-(\zeta_D + \tilde{G}_c)(\zeta - \zeta_D) \qquad\qquad \begin{array}{l} \text{if } \zeta > \zeta_D \\ \text{and } \zeta_i \le \zeta \le \zeta_e \end{array}$$

$$\tilde{\eta} = \tilde{P}(\zeta_e) \tag{B5}$$
$$\text{where} \quad f(x) = 1 - e^{-x}$$

respectively. The first two equations determine the start $\zeta_i$ and end $\zeta_e$ of the lasing zone. The third and fourth equations give the normalized intensity and power distributions $\tilde{I}(\zeta)$ and $\tilde{P}(\zeta)$ as functions of the parameters $K$, $\tilde{G}_c$, and $\zeta_D$. The fifth equation determines the efficiency $\tilde{\eta}$. The functions $F(x)$ and $f(x)$ arise from integration of the $\exp(K\zeta_0)$ term in Eq. (65).

For the limiting case of an optically saturated oscillator ($\tilde{G}_c \to 0$), the results are exhibited in Figs. 52$a$ through 54$a$ for the laminar mixing layer, and in Figs. 52$b$ through 54$b$ for turbulent mixing. We note that for $\tilde{G}_c \to 0$ the lasing starts at the nozzle exit ($\zeta_i \to 0$), and radiative deexcitation is fast compared to collisional loss ($\tilde{I} \sim 1/\tilde{G}_c \to \infty$).

Figures 52$a$ and $b$ show the intensity distributions for various $K$, in the form of the normalized local flux $d\tilde{P}/d\zeta = \tilde{I}\tilde{G}_c$, when lasing ends before adjacent layers merge. As pumping becomes faster relative to collisional deactivation, the lasing zone shrinks and the intensity increases, with the peak moving closer to the nozzle exit. As $K \to \infty$, the laminar model

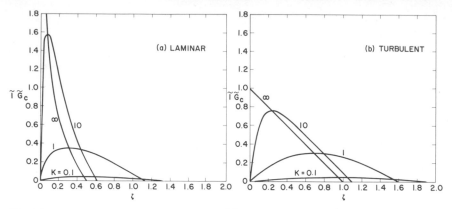

**Fig. 52.** Normalized intensity distributions $\tilde{I}\tilde{G}_c(\zeta)$ at various ratios $K$ of pumping rate to deactivation rate, unmerged mixing layer ($\zeta_D > \zeta_e$). (a) Laminar and (b) turbulent.

predicts $(\tilde{I}\tilde{G}_c)_{max} \to \infty$ and a decay with $(1-2\zeta)/(2\zeta^{1/2})$, while the turbulent prediction is finite and decays as $(1-\zeta)$. In comparing Figs. 52a and b, the reader is reminded that the laminar and turbulent dimensionless variables are normalized differently [see Eqs. (74) and (76)].

Figures 53a and b show the effect on the intensity distribution of the cut-off of the fluorine entrainment when the mixing layers merge at $\zeta_D < \zeta_e$. The distribution at $\zeta \leq \zeta_D$ is unaffected, but beyond $\zeta_D$ it decays exponentially. Consider, for example, that $x_D$ is shifted upstream of $x_e$ by decreasing the nozzle half width $h$, while holding the flow conditions fixed. Then the truncation of the intensity distribution means that less total power can be extracted than when $x_D \geq x_e$. The efficiency can still increase, since the fluorine flux and the available power decrease proportional to $h$.

The chemical efficiency $\eta$, rather than the less significant $\tilde{\eta}$, is plotted as a

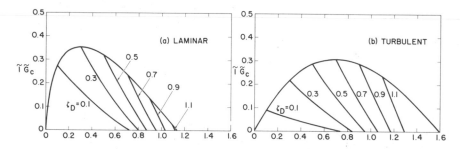

**Fig. 53.** Effect of merging ($\zeta_D < \zeta_e$) on normalized intensity distribution for $K = 1$. (a) Laminar and (b) turbulent.

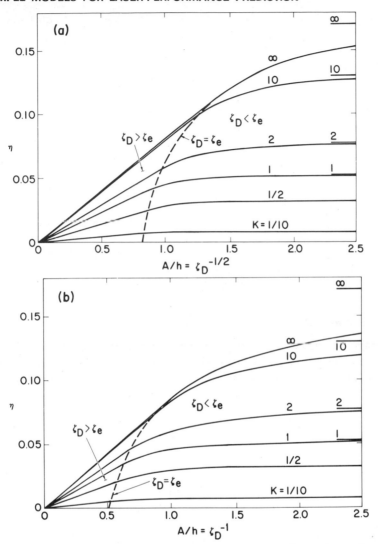

**Fig. 54.** Chemical efficiency $\eta$ as function of scaling parameter $\zeta_D^{-m}$ at various ratios $K$ of pumping rate to deactivation rate where $\zeta_D > \zeta_e$ is the high-pressure regime, and $\zeta_D < \zeta_e$ is the low-pressure regime. (a) Laminar model and (b) turbulent model.

function of $\zeta_D^{-m}$ in Figs. 54a and b. Also shown is the (dashed) curve for $\zeta_D = \zeta_e$. In the unmerged regime, $\zeta_D > \zeta_e$, $\eta$ is seen to be proportional to $\zeta_D^{-m}$, with a slope increasing as $K$ is increased. Here, lasing ends before all fluorine is entrained. Thus, some of the fluorine still mixes and reacts beyond $\zeta_e$, but the pumping rate is insufficient to maintain lasing. When $\zeta_D < \zeta_e$, the efficiency levels off. Highest values of $\eta$ are obtained for merging near the nozzle exit as in a premixed laser. Both the laminar and the turbulent models predict that

$$\lim_{\zeta_D \to 0} \eta = \frac{\epsilon}{2\,|\Delta H|}\left[1 - \frac{\ln(1+K)}{K}\right]$$

Within the two-level approximation, $\epsilon/(2\,|\Delta H|)$, that is, the ratio of the photon energy to the heat of reaction, constitutes an upper bound on the efficiency, reached in the limit $\zeta_D \to 0$ and $K \to \infty$.

The variations of $\eta$ with $\zeta_D^{-1/2}$ for laminar mixing and with $\zeta_D^{-1}$ for turbulent mixing are quite similar. In the next section, we show that in both cases $\zeta_D^{-m} \sim 1/p$, so that Figs. 54 show the inverse pressure dependence of the chemical efficiency. We, therefore, refer to the regions $\zeta_D > \zeta_e$ and $\zeta_D \leq \zeta_e$ as the high- and low-pressure regimes, respectively. Another important observation from Figs. 52 and 53 is that when $K \leq O(1)$, as predicted for the HF laser in Section 3.3, the product $K\zeta_e$ is never much larger than unity. In the lasing zone, pumping is then not truly diffusion limited, and a thin flame sheet does not form.

## 3.4. Scaling Laws

Using the definitions of the dimensionless parameters, Eqs. (66) and (76), and their functional relationships given in Table 7, we can establish scaling laws for the effect of changes in gas-dynamical conditions and nozzle dimensions on the laser performance. Since

$$k_p \simeq f_p(T), \qquad k_{cd} \simeq f_{cd}(T), \qquad [X] \simeq \frac{p}{T}\frac{p_X}{p}, \qquad D \simeq \frac{T^{3/2}}{p}, \qquad \text{and } \sigma \simeq T^{-3/2}$$

$$(77)$$

we find* that the governing parameters scale as

$$\zeta \simeq \frac{xp}{u}\frac{f_{cd}(T)}{T}\frac{p_M}{p}$$

$$\zeta_D^m = \frac{h}{A} \simeq \begin{cases} ph\left\{\left[\dfrac{f_{cd}(T)}{T^{5/2}}\right]\left(\dfrac{p_M}{p}\right)\right\}^{1/2}, & m = \tfrac{1}{2} \quad \text{laminar} \\[2ex] \left(\dfrac{ph}{u}\right)\left[\dfrac{f_{cd}(T)}{T}\right]\left(\dfrac{p_M}{p}\right), & m = 1 \quad \text{turbulent} \end{cases}$$

$$K \simeq \frac{f_p(T)}{f_{cd}(T)}\frac{p_F}{p}\frac{p}{p_M} \qquad (78)$$

* Consistent with earlier approximations, we neglect the exponential dependence on inverse temperature in Eq. (63) for $\sigma$. The empirical coefficients $A_L$ and $A_T$ are considered constant.

$$\tilde{G}_c \approx \frac{1}{N}\frac{p}{p_F}\begin{cases}\left\{[f_{cd}(T)T^{5/2}]\left(\frac{p_M}{p}\right)\right\}^{1/2} & \text{laminar}\\[2ex] [f_{cd}(T)T^{3/2}]\left(\frac{p_M}{p}\right)\left(\frac{1}{u}\right) & \text{turbulent}\end{cases}$$

The relations between the dimensional- and dimensionless-performance parameters can be expressed as

$$I(x) \simeq \tilde{I}(\zeta)p[f_{cd}(T)T^{1/2}]\left(\frac{p_M}{p}\right)$$

$$P(x) \simeq \tilde{P}(\zeta)uNh_z\frac{p_F}{p}\begin{cases}\left\{\left[\frac{f_{cd}(T)}{T^{1/2}}\right]\left(\frac{p_M}{p}\right)\right\}^{-1/2} & \text{laminar}\\[2ex]\left\{f_{cd}(T)\left(\frac{p_M}{p}\right)\left(\frac{1}{u}\right)\right\}^{-1} & \text{turbulent}\end{cases} \tag{79}$$

$$\eta \simeq \frac{\tilde{\eta}}{ph}\begin{cases}\left\{\left[\frac{f_{cd}(T)}{T^{5/2}}\right]\left(\frac{p_M}{p}\right)\right\}^{-1/2} & \text{laminar}\\[2ex]\left\{\left[\frac{f_{cd}(T)}{T}\right]\left(\frac{p_M}{p}\right)\left(\frac{1}{u}\right)\right\}^{-1} & \text{turbulent}\end{cases}$$

In general, this is all we can state on scaling, because the dimensionless quantities $\tilde{I}$, $\tilde{P}$, and $\tilde{\eta}$ are implicit transcendental functions of $\zeta$, $K$, $\tilde{G}_c$, and $\zeta_D$. In particular cases, these functional relations simplify, however, and scaling laws can be given explicitly. In the following discussion we assume saturation and set $\tilde{G}_c = 0$.

### High-Pressure Regime ($\zeta_D > \zeta_e$).

*Scaling with $p$, $u$, $h$, $h_z$.* In the high-pressure regime, $\zeta_D$ does not enter into the dimensionless solutions, and $\zeta_i$, $\zeta_e$, and $\tilde{\eta}$ depend only on $K$ and $\tilde{G}_c$. Since $\tilde{G}_c$ is zero and $K$ is independent of pressure, velocity, or nozzle dimensions, a change from $p$, $u$, $h$, $h_z$ to $p'$, $u'$, $h'$, $h'_z$ (while keeping mole fractions, temperature, and number of nozzles fixed) results in the following scaling:

$$I(x) = I'\left(x\frac{p}{p'}\frac{u'}{u}\right)\frac{p}{p'} \quad \text{laminar and turbulent}$$

$$P(x) = P'\left(x\frac{p}{p'}\frac{u'}{u}\right)\frac{h_z}{h'_z}\begin{cases}\dfrac{u}{u'} & \text{laminar}\\[2ex]\left(\dfrac{u}{u'}\right)^2 & \text{turbulent}\end{cases} \tag{80}$$

$$\eta = \eta'\frac{h'}{h}\frac{p'}{p}\begin{cases}1 & \text{laminar}\\[2ex]\dfrac{u}{u'} & \text{turbulent}\end{cases}$$

Scaling with pressure and nozzle dimensions is the same for laminar and turbulent mixing in this regime. An increase in pressure results in a proportional rise in intensity, compensated by shortening of the lasing zone so that the total power remains constant. Since the net fluorine flux increases with $p$, the efficiency drops as $1/p$. This result is important for chemical-laser operation, because a low-cavity pressure required for good efficiency is detrimental to pressure recovery (Section 1). Because of the two-dimensional cavity configuration, $h_z$ affects power but not efficiency. The power is independent of $h$, but the efficiency varies as $1/h$.

Increasing the velocity stretches the lasing zone without affecting the intensity level. The difference between the laminar and turbulent scaling of $P$ and $\eta$ with $u$ is due to the different mixing mechanisms. The corresponding pressure-scaling laws do not differ, because the laminar-diffusion coefficient $D$ is independent of $u$ but inversely proportional to $p$.

*Scaling with Mole Fractions.*   Scaling with the mole fractions $p_F/p$ and $p_M/p$ differs for arc- and combustor-driven lasers operating in the high-pressure regime. For an arc-driven laser, $p_M = p_H \approx p_F$, so that $K$ is invariant. In a saturated cavity, $\tilde{I}$, $\tilde{P}$, and $\tilde{\eta}$ are constant, and we obtain the relations

$$I(x) = \frac{p_F}{p_F'} I'\left(x\frac{p_F}{p_F'}\right) \qquad \text{laminar and turbulent}$$

$$P(x) = P'\left(x\frac{p_F}{p_F'}\right)\begin{cases} \left(\dfrac{p_F}{p_F'}\right)^{1/2} & \text{laminar} \\ 1 & \text{turbulent} \end{cases}$$

$$\eta = \eta'\begin{cases} \left(\dfrac{p_F}{p_F'}\right)^{-1/2} & \text{laminar} \\ \left(\dfrac{p_F}{p_F'}\right)^{-1} & \text{turbulent} \end{cases}$$

(81a)

With a combustor plenum, $p_M = p_{DF}$ and $p_F$ are essentially independent variables, and thus $K \approx p_F/p_{DF}$ changes. The scaling laws for variations of $p_F/p$ and $p_{DF}/p$ (everything else being held constant) in a combustor-driven saturated cavity becomes

$$I(x) = \frac{p_{DF}}{p_{DF}'}\frac{\tilde{I}(K)}{\tilde{I}(K')} I'\left(x\frac{p_{DF}}{p_{DF}'}\right)\begin{cases} \left(\dfrac{p_{DF}}{p_{DF}'}\right)^{-1/2} & \text{laminar} \\ \end{cases}$$

$$P(x) = \frac{p_F}{p_F'}\frac{\tilde{P}(K)}{\tilde{P}(K')} P'\left(x\frac{p_{DF}}{p_{DF}'}\right)\begin{cases} \\ \left(\dfrac{p_{DF}}{p_{DF}'}\right)^{-1} & \text{turbulent} \end{cases}$$

(81b)

$$\eta = \frac{\bar{\eta}(K)}{\bar{\eta}(K')}\,\eta' \begin{cases} \left(\dfrac{p_{DF}}{p'_{DF}}\right)^{-1/2} & \text{laminar} \\[3ex] \left(\dfrac{p_{DF}}{p'_{DF}}\right)^{-1} & \text{turbulent} \end{cases}$$

The variations of $\tilde{I}$, $\tilde{P}$, and $\bar{\eta}$ with $K$ have to be evaluated from the solutions in Table 7. For $K = O(1)$ we estimate $\tilde{I}\tilde{G}_c \approx K^{0.75}$, $\tilde{P} \approx \tilde{\eta} \approx K^{0.5}$ for laminar mixing, and $\tilde{I}\tilde{G}_c \approx K^{0.6}$, $\tilde{P} \approx \tilde{\eta} \approx K^{0.5}$ for turbulent mixing. As a result, the efficiency also weakly depends on the fluorine-mole fraction and decreases more strongly with $p_{DF}/p$ than when $K \to \infty$.

We have to qualify the mole-fraction scaling laws in two respects. First, unless $p_{H_2}$ is increased together with $p_F$, the mixing layer eventually ceases to be fuel rich. Then the pumping zone moves away from the fluorine stream, and some of the assumptions of the analysis become invalid. Second, as $p_F/p$ is increased toward unity, the diluent mole fraction decreases, and the mixing layer rapidly heats up. As the temperature rises, experiments show that the laser output drops.

*Effect of Temperature and Number of Nozzles.* Temperature scaling is complicated by the variation of both $K$ and $\tilde{G}_c$, and the full functional relations of Table 7 have to be used. In view of the simplifying assumption on the temperature dependence of the gain, the present model is not likely to give dependable predictions for $T$ scaling. Experiments (see Section 3.6) indicate only a weak dependence on the nozzle-exit temperatures of the fuel and oxidizer jets, perhaps because the initial temperature in the mixing layer, in the presence of thick nozzle boundary layers, is largely controlled by the temperature of the nozzle walls. Further downstream, reactive heating, balanced by convective cooling, should dominate temperatures in the mixing layer.

The advantage of many small nozzles is twofold: Eqs. (79) predict that power and efficiency rise with $N \approx 1/h$. In addition $\tilde{G}_c$ decreases with $1/N$, which results in a more highly saturated cavity and a further increase in $P$ and $\eta$.

**Low-Pressure Regime ($\zeta_D \leq \zeta_e$).** When the pressure decreases below a critical value corresponding to $\zeta_D = \zeta_e$, Fig. 54 shows that the efficiency ceases to improve as rapidly as in the high-pressure regime. The critical pressure scales with

$$p_c \approx \begin{cases} \dfrac{1}{h}\left(\dfrac{p_M}{p}\right)^{-1/2} & \text{laminar} \\[3ex] \dfrac{u}{h}\left(\dfrac{p_M}{p}\right)^{-1} & \text{turbulent} \end{cases} \qquad (82)$$

where $p_M = p_H \simeq p_F$ in an arc-driven laser, and $p_M = p_{DF}$ for combustor operation. In the latter case, the contribution of the variation of $\zeta_e^m$ with $K$, and hence with $p_F/p_{DF}$, is found to be insignificant.

In the low-pressure regime $p < p_c$, the transcendental functions cannot be eliminated from the scaling laws. When $\tilde{G}_c = 0$, we obtain from Table 7 the relations

$$
\eta = \frac{\epsilon}{2|\Delta H|}
\begin{cases}
1 - \tfrac{2}{3}\zeta_D - \dfrac{1}{K}\ln\left[(1+K)\dfrac{F((K\zeta_D)^{1/2})}{(K\zeta_D)^{1/2}}\right] & \text{laminar} \\[2ex]
1 - \tfrac{1}{2}\zeta_D - \dfrac{1}{K}\ln\left[(1+K)\dfrac{f(K\zeta_D)}{K\zeta_D}\right] & \text{turbulent}
\end{cases}
\tag{83}
$$

exhibited in Figs. 54. Scaling laws can be read from these figures, if we remember that from Eq. (78)

$$
\zeta_D \simeq
\begin{cases}
(ph)^2\left(\dfrac{p_M}{p}\right) & \text{laminar} \\[2ex]
\dfrac{ph}{u}\left(\dfrac{p_M}{p}\right) & \text{turbulent}
\end{cases}
\tag{84}
$$

and that

$$
P_e \simeq puNhh_z\left(\frac{p_F}{p}\right)\eta
$$

In the limiting case $K \to \infty$ considered in Refs. 5 and 6, the logarithmic terms disappear, and lasing ends at $\zeta_D$ throughout the low-pressure regime. The resulting trend

$$
\eta = \frac{\epsilon}{2|\Delta H|}[1 - \text{const} \cdot (ph)^{1/m}]
$$

is characteristic for the gradual leveling off towards the premixed value of the efficiency as $p \to 0$. This should be compared to the scaling $\eta \simeq (ph)^{-1}$, Eq. (80), found for the high-pressure regime.

## 3.5. Comparison with More Detailed Analyses

To examine the limitations of the foregoing simple model, the predicted scaling laws are compared with results of two more detailed analyses[118,119] of an arc-driven laminar mixing laser. In both analyses, the nozzle boundary layers are ignored, the dividing wall is assumed infinitely thin, and the pressure is assumed uniform in both the transverse and streamwise directions. The four lower HF vibrational levels activated by the cold pumping reaction are considered, with lasing on a single transition between adjacent levels.

Reference 118 provides a closed form solution for the laminar mixing layer in a saturated oscillator cavity, in the limit where pumping and stimulated deexcitation are much faster than diffusion and collisional deactivation. The mixing layer is assumed fuel rich, so that pumping and lasing occur in a thin flame sheet near the mixing-layer boundary that propagates into the fluorine jet. For simplicity, diffusion and collisional deactivation are considered only on the fuel side of the flame sheet, while the fluorine flux to the flame is equated to the rate of convectional entrainment. The error in the fluorine flux is small under fuel rich conditions. But the reaction-product concentrations and the temperature are discontinuous at the flame sheet. Jet temperatures and molecular weights are allowed to differ, the products $\rho^2 D$ and $\rho k_{cd}$ are treated as constant across the layer, and the velocity field is assumed uniform (no shear). Solutions are derived in two steps. First, a similarity solution is obtained neglecting collisional deactivation. Then, a first-order correction for collisional deactivation is calculated by expansion in $\zeta$.

The similarity solution yields a theoretical estimate for the location of the flame sheet and the mixing layer growth as a function of stoichiometry

$$y_f = \delta(x) = A_L \left( \frac{D_f x}{u_f} \right)^{1/2}, \qquad A_L \approx \frac{\rho_f}{\rho_o} \left[ 0.90 + 1.20 \log_{10} \left( \frac{\rho_o [H_2]_f}{\rho_f [F]_o} \right) \right]$$

Here, the subscripts o and f refer to free-stream conditions in the oxidizer and fuel jets, respectively, and $\rho$ is the mixture density. The four-level transitions and the effect of the transverse concentration gradients result in an additional coefficient

$$Z \approx 2.66 \times 10^{-2} \left( \frac{\rho_f}{\rho_o} A_L \right)^{1/2} T_f^{0.2}$$

in the definition of the dimensionless axial distance

$$\zeta = Z \frac{k_{cd,f}[F]_o}{u} x$$

where $k_{cd}$ is the net deactivation rate coefficient for levels $v = 1$ to 3. The number $\phi$ of photons emitted for each HF molecule formed is found to be a weak function of the average temperature

$$\phi = 0.844 \text{ to } 0.783 \text{ for } T = 400 \text{ to } 1000°K$$

Except for the coefficients $Z$ and $2\phi$, the solutions obtained:

$$\left. \begin{array}{l} IG_c = \dfrac{\epsilon[F]_o^2 h k_{cd,f}}{2(\zeta_D)^{1/2}} (2\phi Z) \dfrac{1-2\zeta}{2\zeta^{1/2}} \\[4mm] P(x) = \dfrac{\epsilon[F]_o u N h h_z}{(\zeta_D)^{1/2}} (2\phi) \zeta^{1/2} (1 - \tfrac{2}{3}\zeta) \end{array} \right\} \text{ for } \zeta \leq (\tfrac{1}{2}, \zeta_D)$$

and

$$\eta = \frac{\epsilon}{2\,|\Delta H|}(2\phi)\begin{cases}\frac{1}{3}\left(\frac{2}{\zeta_{D}}\right)^{1/2} & \text{if } \zeta_{D} \begin{array}{c}\geq\\ <\end{array}\frac{1}{2}\\[2mm]1-\frac{2}{3}\zeta_{D}\end{cases}$$

are identical to the predictions of the simplified model, Table 7, in the limit $K \to \infty$, $\tilde{G}_c \to 0$.

In Ref. 119, results of a numerical analysis of a laminar mixing HF amplifier are presented. The problem is simplified by assuming that a constant input intensity $I_0$ is amplified in a single mixing layer, which spreads into the parallel fuel and oxidizer jets. Jet velocities $u_f < u_o$, temperatures $T_f < T_o$, and dimensions $h_f < h_o$ are chosen to approximate those of actual arc-driven lasers. Radiative transitions between the first four vibrational levels are treated as in the code described in Ref. 120. In addition to the cold pumping reaction and V–T transfer, V–V exchange between HF molecules is included. Older estimates for the rate coefficients are used, which indicate that atomic hydrogen is again the dominant deactivator with $K \simeq 1$ to 3. Since $K\zeta$ is not much larger than unity, a flame sheet should not form. This conjecture cannot be checked, however, since no species or gain profiles are given. All calculations are for an unmerged mixing layer.

Solutions are obtained for constant $p$, $(p_F/p)_o$, and four different input intensities, including the small-signal limit $I_0 = 0$. Figure 55 shows the

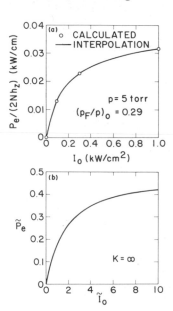

**Fig. 55.** Saturation of a laser amplifier. (a) Total power per mixing layer of unit height as function of input intensity, from numerical HF amplifier calculations;[119] and (b) normalized total power as function of normalized input intensity, from analytical model,[5] $K \to \infty$.

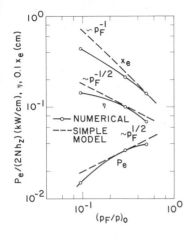

**Fig. 56.** Total power $P_e$, chemical efficiency $\eta$, and lasing length $x_e$ as functions of fluorine mole fraction; comparison of numerical results[119] and simple-model scaling laws.

variation of output power with $I_0$ as calculated, and the prediction of the simple model.[5] Both curves have the same rise toward a maximum power level as the amplifier saturates with $\tilde{I} \simeq I_0/(p_F)_0 \to \infty$.

Scaling with pressure and mole fractions for a saturated amplifier ($\tilde{I} \to \infty$) should be the same as for a saturated oscillator ($\tilde{G}_c \to 0$).[5] Indeed, numerical results for $P(x)$ and $\eta$ obtained at three cavity pressures [holding $\tilde{I}(\gg 1)$, $(p_F/p)_0$, $u$, $h$, and $h_z$ constant] satisfy the scaling laws, Eqs. (80), predicted for the unmerged regime. Also, $G(px)$ is approximately independent of pressure as predicted by the simple model.[5]

The dependence of peak gain, power, efficiency, and lasing length on the fluorine-mole fraction $(p_F/p)_0$—with $\tilde{I}$, $p$ and $[H_2]_f$ held constant—is found to correlate less favorably with the laminar high-pressure scaling laws. Numerical results suggest a rise of the peak small-signal gain $G_{0m}$ with $(p_F/p)_0^{0.6}$ for the $2 \to 1$ transition, and with $(p_F/p)_0^{0.9}$ for the $1 \to 0$ transition, instead of the $(p_F/p)_0^{1/2}$ scaling predicted by Ref. 5. The faster rise of gain for the lower level transition may be attributable to cascading, which is not taken into account by the simple two-level model.

The predicted trends $P_e \simeq (p_F/p)_0^{1/2}$ and $\eta \simeq (p_F/p)_0^{-1/2}$ [see Eq. (81a)] fit the numerical results shown in Fig. 56 only at intermediate fluorine-mole fractions, and the lasing length $x_e$ shrinks less rapidly than with $(p_F/p)_0^{-1}$. Discrepancies at high-fluorine mole fractions are expected, since the temperature in the pumping zone rises as the diluent mole fraction decreases, and since the pumping zone moves away from the fluorine boundary of the mixing layer as the ratio $(p_F)_0/(p_{H_2})_f$ increases. We note that the trends at high $(p_F/p)_0$ indicated by the laminar numerical results for $P_e$ and $\eta$ are similar to the predicted turbulent scaling, which should be kept in mind when interpreting experimental data. At small $(p_F/p)_o$, the sparse numerical data suggest that the efficiency levels off and the power drops nearly linearly with the fluorine mole fraction. This trend is similar to predictions for the low-pressure regime. But the mixing layers did not merge in the present calculations, as confirmed by the $p$ dependence discussed earlier. The $H_2$ flux was held constant and the rate coefficient $k_{H_2}$ used was larger than the estimate shown in Fig. 51. Conceivably, deactivation by $H_2$ in the fuel-rich portion of the mixing layer, and partial absorption of radiation produced in the pumping zone, are becoming increasingly important as $(p_F/p)_o$ decreases.

From the foregoing comparisons, it appears that the simple model correctly predicts the pressure scaling, and that predictions for scaling with partial pressures can be improved by inclusion of all potentially significant deactivators. Incorporation of a multilevel vibrational model should improve absolute predictions. Suggestions for a boundary-layer correction are given in Ref. 31.

## 3.6. Correlation with Observed Laser Performance

Using the scaling laws, we now examine powers and efficiencies observed in cw HF laser experiments with both arc and combustor plenums. Zero-power measurements are discussed separately in Section 3.7.

**Arc-Driven Lasers.** All experimental data to be discussed here were obtained with the standard Aerospace two-dimensional nozzle bank, which uses spray bars at the lips of the oxidizer nozzles (Fig. 20) to inject undiluted $H_2$. Laser power was measured with the closed-cavity configuration.

Figure 57 shows typical variations of laser power with the mass fluxes of $SF_6$ and $H_2$, as observed with $N_2$ as diluent without[4] and with[4,16] $O_2$ addition, as well as for He diluent, which gave the highest power.[16] We will not discuss the improvement of the laser power with addition of $O_2$, or with He instead of $N_2$ as diluent, except to observe that both tend to increase the efficiency of the arc in dissociating $SF_6$, and that the lower molecular weight and higher ratio of specific heats of He lead to a higher velocity and a lower pressure at the exit of the oxidizer nozzle. During each of the three sets of experiments shown in Fig. 57$a$, the plenum pressure and the input power to the arc were approximately held constant. All show at first a nearly linear rise of power with the mass flux of $SF_6$. But as $\dot{m}_{SF_6}$ is increased further, the power reaches a maximum and then begins to fall off. The variation of power with $\dot{m}_{H_2}$, see Fig. 57$b$, is weaker, but also exhibits a maximum, at a mole-flux ratio

$$\frac{\dot{N}_{H_2}}{\dot{N}_F} \geq \frac{W_{SF_6}}{6 W_{H_2}} \frac{\dot{m}_{H_2}}{\dot{m}_{SF_6}} \approx 6.8$$

considerably in excess of stoichiometric.

To uncouple the laser-cavity scaling from variations in the arc performance, the molar flux $\dot{N}_F$ of fluorine out of the plenum is needed, rather than the input of $SF_6$. For the data of curve 1 of Fig. 57$a$, with $N_2$ as diluent and no $O_2$ addition, Ref. 4 gives estimates of $\dot{N}_F$, plenum-mole fraction $(p_F/p)_o$, and temperature $T_o$, and of the oxidizer jet-exit velocity $u_o$ and pressure $p_o$. These estimates are based on mass fluxes and plenum pressures measured before and after the arc is turned on, and on the assumptions that boundary layers in the nozzle throat are negligible, and that heat addition by the arc and losses to plenum and nozzle walls are independent of $\dot{m}_{SF_6}$. The exit conditions $u_o$ and $p_o$ are calculated for an inviscid nozzle flow. Nevertheless, they allow a direct comparison with the constant-pressure scaling predicted by the simple analysis

$$\eta = \eta(\zeta_D^m, K) \qquad \text{(see Fig. 54)}$$
$$P = |\Delta H| \dot{N}_F \eta$$
$$\dot{N}_F \approx \left(\frac{p_F}{p}\right)\left(\frac{u}{T}\right), \qquad K \approx \frac{f_p(T)}{f_{cd}(T)}$$
$$\zeta_D \approx \frac{k_{cd}(T)}{u^{2m-1} T^{4-3m}} \frac{p_F}{p}, \qquad m = \begin{cases} \frac{1}{2} & \text{laminar} \\ 1 & \text{turbulent} \end{cases}$$

where $p_M \approx p_F$ (i.e., collisional deactivation by H, HF, or F) and an optically saturated cavity are assumed. We note that $T$ refers to the temperature in the mixing layer, which is expected to be controlled primarily by reactive heating rather than by the jet exit or plenum temperatures. The velocity $u$ in $\zeta_D$ should be interpreted as that of the fuel jet[118] and is considered constant for the set of experiments. The mole fraction $p_F/p$ is replaced by $(p_F/p)_o$ estimated for the plenum, neglecting any wall recombination of fluorine in the nozzle.

In Figs. 58, the experimental data for curve 1 of Fig. 57$a$ are replotted as a function of the mole fraction $(p_F/p)_o$, ignoring any simultaneous variations in the mixing-layer temperature and velocity. Within experimental scatter, the data appear to be monotonic in $(p_F/p)_o$, which suggests that the roll-off in power in Fig. 57$a$ for $\dot{m}_{SF_6} > 1.8$ g/sec is largely due to incomplete dissociation of $SF_6$ in the arc plenum. Whether this explanation also applies to the stronger roll-off seen in curve 2 of Fig. 57$a$ is not certain. As the ratio of diluent to fluorine decreases, reactive

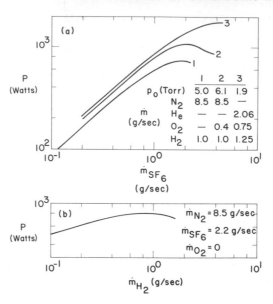

**Fig. 57.** Observed performance of arc-driven HF laser.[4,16] (a) Variation of power with mass flux of $SF_6$ and (b) variation of power with mass flux of $H_2$.

heating increases and should result in smaller population inversions because of higher mixing-layer temperatures, as well as in a rise of pressure within the mixing region (Section 1). In Ref. 82, it is argued that the corresponding increase in the strength of the reaction shocks could induce separation of the nozzle boundary layers, prereaction in the separated regions, and increased absorption of the $1 \rightarrow 0$ band and deactivation by ground-state HF generated there.

The nearly linear rise of power and nearly constant efficiency at small $(p_F/p)_o$, where $SF_6$ is fully dissociated, are reminiscent of the behavior predicted when $\zeta_D$ goes to zero. Surprisingly, efficiencies calculated with the estimated $\bar{N}_F$ are close to those predicted by the simple two-level model for $\zeta_D \rightarrow 0$ and $K \rightarrow \infty$. To check these points, the experimental data are fitted with the predicted variations of $\eta$ with $\zeta_D \approx (p_F/p)_o$ for laminar, Fig. 58a, and for turbulent mixing, Fig. 58b. As suggested by results obtained in four-level calculations[6,118] for the limit $K \rightarrow \infty$, arbitrary scaling factors have been allowed in both $\eta$ and $\zeta_D$ to achieve these fits. It is seen that both the laminar and turbulent scaling laws give excellent approximations to the experimental data. Critical conditions corresponding to $\zeta_D = \zeta_e$ are indicated on the graphs and suggest that most of the data fall into the merged regime. Not enough data points are

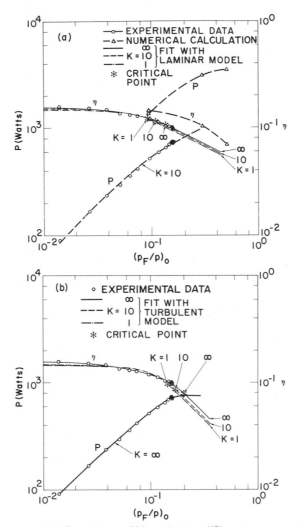

**Fig. 58.** Power and efficiency of arc-driven HF laser[4] as functions of fluorine-mole fraction, comparison of measurement and theory. (*a*) Curve fits of measured data with laminar model at various $K$, comparison with numerical calculation and (*b*) curve fits of measured data with turbulent model at various $K$.

available with $p_F/p > (p_F/p)_{critical}$ to decide whether the experiments follow the laminar $(p_F/p)^{-1/2}$ or turbulent $(p_F/p)^{-1}$ scaling predicted for $\eta$ in the unmerged mixing layer, or the corresponding scaling of power with $(p_F/p)^{1/2}$ or $(p_F/p)^0$.

Curve fits for $K = 1$, 10, and $\infty$ are included in Figs. 58*a* and *b*. Although all approximate the measurements within the experimental data scatter, the overall observed trends appear to be better reproduced when $K$ is assumed to lie between 10 and $\infty$. The measured efficiencies are within 10% of those predicted by the simple two-level analyses for $K = 10$ and $\infty$. The two-level predictions for $K = 1$, however, are lower than experimental efficiencies by a factor of 2.9, which seems more than can be attributed to radiative and collisional cascading. Thus from the curve fits, we conclude that arc-driven lasers with $N_2$ as diluent probably operate at average ratios of pumping to collisional deactivation rates larger than 10, rather than of $O(1)$ as argued on the basis of the rate-coefficient estimates in Fig. 51. We return to this point following the discussion of combustor-driven laser performance.

Also shown in Fig. 58$a$ are the results of the numerical calculations of Ref. 119 for a highly saturated amplifier operating at a similar cavity pressure (5 Torr)—see Section 3.5. Absolute levels of $P$ and $\eta$ should be disregarded, since the degree of saturation of the amplifier cavity is controlled by the somewhat arbitrarily assumed input intensities. But the similarity of the experimental and numerical trends is puzzling, since the calculations suggest that lasing ends well before the laminar mixing layers merge. Indeed, from the numerical results, critical conditions should not occur until $p_F/p$ is of $O(10^{-2})$.

In view of this conflict, it would be helpful to also have measurements of power and efficiency as functions of pressure, at constant mole fractions, since the analytical model and the numerical calculations predict the same pressure scaling in the unmerged regime. Such measurements are reported in Ref. 82 and are shown in Fig. 59. These experiments were performed using He as diluent, and without the addition of $O_2$. The mole fluxes of He, $SF_6$, and $H_2$ were increased proportionally to $p_0$, thereby fixing all mole fractions. Simultaneously, the ambient cavity pressure was regulated by the cavity-purge flow to match the nozzle-exit pressure of the oxidizer jet, and the centerline of the optical cavity was in each case readjusted for maximum power output with the closed cavity configuration.[121] Only the central five of the 36 injector nozzles of the standard

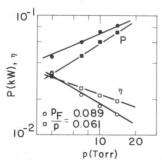

**Fig. 59.** Experimental variation of power and efficiency with pressure at two fluorine mole fractions, arc-driven HF laser.[82]

Aerospace nozzle bank were used in these experiments, to permit operation at higher pressures with the same arc heater.

As a result of the shorter optical path through the active mixing regions, the cavity was probably not saturated, which may explain the lower powers and efficiencies in Fig. 59. If the density of the dominant deactivators (e.g., of H, F, or HF) scales with that of fluorine, the dimensionless integrated gain $\bar{G}_c$ is predicted to vary with $(p_F/p)^{-1/2}$, but to be independent of pressure and hence constant for each of the two sets of experiments shown.

On the average over the pressure range of the experiments, the data suggest that $\eta \simeq p^{-0.4}$, $P \simeq p^{0.6}$ for the mole fraction $p_F/p = 0.061$, and that $\eta \simeq p^{-0.6}$, $P \simeq p^{0.4}$ at $p_F/p = 0.089$. Both the laminar and the turbulent analytical models predict that $\eta \simeq p^{-1}$ and $P \simeq p^0$ throughout the unmerged regime, and that merging results in a gradual approach of the "premixed" limit where $\eta = p^0$ and $P \simeq p^1$. Hence, lasing in an arc-driven cavity is terminated by merging of the mixing layers, and $\zeta_D < \zeta_e$ as suggested by the curve fits in Fig. 58.

The contradiction with the numerical calculations can be resolved, if it is postulated that the dominant deactivation-rate coefficients are smaller by about an order of magnitude than assumed in Ref. 119. This conclusion is also supported by recent H-atom kinetic experiments at The Aerospace Corporation.[113] With this change in $k_{cd}$, merging occurs at a $(p_F/p)_{\text{critical}}$ considerably larger than indicated by the numerical results. Furthermore, the ratio $K \cong k_p/2k_{cd}$ for HF(2) increases to about 10, which makes an average $K > 10$ more believable.

**Combustor-Driven Lasers.** The validity of the scaling laws in Section 3.4 is also demonstrated by the successful correlation of data from a series of 44 tests[14] in a combustor-driven cold-reaction HF laser of the type described in Section 1. In all tests, the combustor reactants were $D_2$ and $F_2$ with He as diluent, and $H_2$ was injected through the small slits of the $2.54 \times 17.8$ cm nozzle bank shown in Fig. 9$a$. In four of these tests, part of the excess $H_2$ was replaced by He. The chemical composition, pressure, temperature, and velocity of the gases entering the cavity were calculated by the method of Section 1.4. Seventy-two two-dimensional mixing regions generated the laser power, which was measured by the closed-cavity method. The primary adjustable parameters for this test series were the combustor plenum pressure and temperature, the parameters $\psi$ and $R_L$ defined in Section 1, and the cavity-shroud angle.

To correlate the data, the following assumptions are made:

1.  The cavity is optically saturated ($\bar{G}_c \to 0$).

2. The dominant collisional-deactivator species is either generated by the pumping reaction, or is a constituent of the oxidizer stream, so that

$$[M] = \frac{\dot{N}_{cd}}{(u_o\, 2Nhh_z)}$$

3. The mixing-layer temperature, and hence the rate coefficients, are assumed constant during the test series.

4. The velocity $u$ in the mixing-layer growth law, and hence in $\zeta_D$, is the velocity $u_f$ of the $H_2$ jet after expansion to the exit pressure of the oxidizer jet.

5. The growth rates $A_L$ or $A_T$ are constant.

6. The mixing layers merge downstream of the lasing zone.

The last assumption will surprise the reader, since we have just argued that arc-driven lasers operate in the merged regime. This working hypothesis, however, is confirmed by the combustor measurements. In any case, merging will show up as a deviation from the high-pressure correlation and should be detectable if $K$ is not too small (see Figs. 54).

Under these assumptions, the analysis predicts

$$\eta = \frac{\epsilon}{2\,|\Delta H|}\, \frac{\tilde{\eta}(K)}{\zeta_D^m}, \qquad m = \begin{cases} \tfrac{1}{2} & \text{laminar} \\ & \text{for the} \qquad\qquad \text{model} \\ 1 & \text{turbulent} \end{cases}$$

Assumptions 2 through 5 result in

$$\zeta_D = x_D\, \frac{k_{cd}}{u_f} \approx \begin{cases} \dfrac{1}{X_{cd}} & \text{laminar} \\[2ex] \dfrac{1}{\xi_{cd}} & \text{turbulent} \end{cases}$$

where

$$X_{cd} = \frac{u_o}{\dot{N}_{cd}p_o}, \qquad \xi_{cd} = \frac{u_o u_f N h_z}{\dot{N}_{cd}}$$

The pressure dependence in the laminar case results from that of the diffusion coefficient. For the ratio $K$ of the pumping to the deactivation rate, we substitute

$$K = \kappa\, \frac{\dot{N}_F}{\dot{N}_{cd}}$$

where the factor $\kappa \simeq k_p/k_{cd}$ is an as yet undetermined adjustable parameter presumed constant for the test series.

In view of the fast H-atom deactivation-rate coefficients predicted by Wilkins,[122] we first try a correlation assuming that the principal deactivator is either H, HF, or F. Then $\dot{N}_{cd}$ is proportional to $\dot{N}_F$, $K$, and hence $\bar{\eta}(K)$ become constant, and

$$\eta \simeq \begin{cases} \chi_F^{1/2} & \text{laminar} \\ \xi_F & \text{turbulent} \end{cases}$$

The result for the turbulent case is shown in Fig. 60, where the data appear to be random.

According to Fig. 51, combustor-produced DF is likely to be a more effective deactivator than is atomic hydrogen. The ratio $\dot{N}_F/\dot{N}_{DF}$ varies with the ratio of $D_2$ and $F_2$ injected into the combustor, and was not held fixed during the test series. Otherwise, the correlation of Fig. 60 should have worked. We must, therefore, allow for variations of $K$ from one test to the next, and unless $K$ is already very large, for variations in $\bar{\eta}(K)$. Under assumptions 1 and 6, Table 7 suggests the following procedure for evaluating $\bar{\eta}$: First, $\zeta_e$ is calculated by inversion of

$$1+K = \begin{cases} \dfrac{(K\zeta_e)^{1/2}}{F((K\zeta_e)^{1/2})} & \text{laminar} \\[2ex] \dfrac{K\zeta_e}{(1-e^{-K\zeta_e})} & \text{turbulent} \end{cases}$$

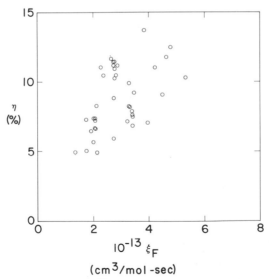

**Fig. 60.** Combustor-driven HF laser,[14] correlation of efficiencies with turbulent model, and F the dominant deactivator.

where $F(x)$ is Dawson's integral.[123] Then $\tilde{\eta}$ is found from

$$\tilde{\eta} = \begin{cases} (\zeta_e)^{1/2}(1 - \tfrac{2}{3}\zeta_e) & \text{laminar} \\[2ex] \zeta_e(1 - \tfrac{1}{2}\zeta_e) & \text{turbulent} \end{cases}$$

Using values for $\kappa$ of 1, 10, and $\infty$, the measured efficiencies are plotted against the laminar scaling parameter $\tilde{\eta}\chi_{DF}^{1/2}$ in Figs. 61a, b, and c, and against $\tilde{\eta}\xi_{DF}$ in Figs. 62a, b, and c.

As for the arc data, correlations with the laminar model are about as good as with the turbulent mixing (or linear mixing-layer growth) assumption, and a decision between the two models cannot be made. The correlations with $\kappa = 1$, Figs. 61a and 62a, still show considerable data

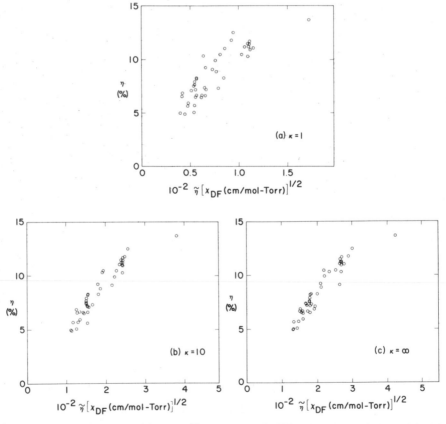

**Fig. 61.** Combustor-driven HF laser,[14] correlation of efficiencies with laminar model, and DF the dominant deactivator.

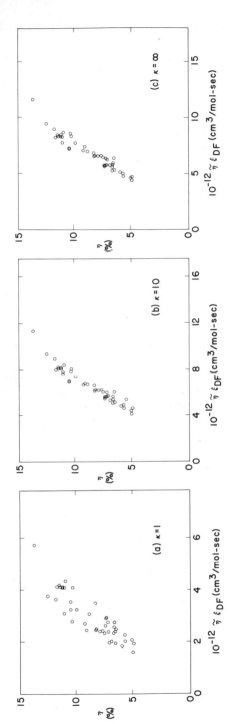

**Fig. 62.** Combustor-driven HF laser,[14] correlation of efficiencies with turbulent model, and DF the dominant deactivator.

scatter, but already exhibit a linear scaling as expected for the high-pressure unmerged regime. The strongest deviation from this trend occurs for the test with the highest efficiency $\eta = 13.7\%$, which may fall into the merged domain. This behavior is more convincingly shown in Figs. 61$b$ and $c$, and 62$b$ and $c$ for $\kappa = 10$ and $\infty$. By comparison with Fig. 60, these correlations are surprisingly successful. It is concluded that for combustor-driven HF lasers, DF produced in the combustor is the dominant deactivator, that averaged ratios of the pumping to the DF deactivation rate are of O(10), and that the majority of the combustor-driven laser data fall into the regime where lasing ends prior to merging of the mixing layers.

The last conclusion is reinforced by Fig. 63, which shows a selection of 13 of the previously mentioned data points, all taken within $\pm 10\%$ of the mole-flux ratios $\dot{N}_F/\dot{N} = 0.12$, $\dot{N}_{DF}/\dot{N}_F = 1.7$, and $\dot{N}_{H_2}/\dot{N}_F = 4.6$, plotted as a function of the inverse of the exit pressure in the oxidizer nozzle. As expected from both the laminar and turbulent models for the unmerged regime, the data suggest a $p_o^{-1}$ dependence of the efficiency. We note that for the present analysis the nozzle flow was corrected for the viscous boundary-layer displacement downstream of the throat, in contrast to Ref. 82 and Fig. 59.

As with the arc-driven laser, the efficiency is apparently insensitive to variations in the jet temperatures and has been found to correlate poorly when corrections for the reaction rates are introduced assuming that the

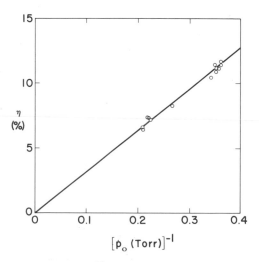

**Fig. 63.** Combustor-driven HF laser,[14] efficiency as function of fluorine-jet pressure, and linear high-pressure scaling.

mixing-layer temperature varies with $T_o$. Some possible explanations have been suggested in Section 3.4.

In the present tests, efficiency is insensitive to $\psi$ and $R_L$ provided they are not too small. When $\psi$ drops below ~12 or $R_L$ is smaller than ~5, the efficiency drops below the data of Figs. 61 and 62. The effect of small $\psi$ is due to a lack of diluent needed to remove the heat added by reaction. Insufficient $H_2$ has a similar effect, since unreacted $H_2$ also works as diluent. As discussed in Section 1, the efficiency is insensitive to the shroud angle, as long as it exceeds about 5°. From subsequent tests with other nozzle banks, it is found that the high-pressure scaling laws also account for the effect of changes in both N and $h_z$.

**Arc Versus Combustor Operation.** The remaining task is to reconcile the findings that both arc- and combustor-driven lasers operate at large $K$, but that in the first case lasing is cut off by merging, while in the second it terminates well before the mixing layers merge. The critical $\zeta_D$ is insensitive to $K$ when $K$ is large. Values of $\zeta_D$ for the arc-driven cavity must, therefore, be about an order of magnitude smaller than for corresponding combustor-driven cavity conditions. Since the nozzle geometries are quite similar,* merging in both cases should occur at about the same axial station $x_D$, and the smaller $\zeta_D$ with the arc can only be explained by a slower deactivation process. The DF mole fractions with the combustor are typically twice the fluorine-mole fraction, and at least comparable to the H and HF mole fractions in the pumping zone. If the H-atom deactivation-rate coefficients are an order of magnitude smaller than predicted[122] while the measured DF deactivation-rate coefficients are accepted, then characteristic deactivation rates with arc operation would indeed be an order of magnitude slower than with combustor-produced DF. Since the pumping rates are the same in both cases, $K_{combustor} \approx K_{arc}/10$. Hence the lasing zone in a combustor-driven cavity should be shorter than with arc operation at the same pressure and mole fractions of F and $H_2$, and power and efficiency should be lower by a factor of 2 or 3. This agrees with observations in the MESA tests.[124]

Figures 64a and b further clarify the essential differences between arc and combustor operation. Here, the arc and combustor driven data are combined on single plots of efficiency versus $\zeta_D$, in the laminar and turbulent interpretation, respectively. Also shown is the ratio of the efficiencies at corresponding operating points of the arc and combustor laser. The poorer performance with a combustor plenum is primarily due to the larger $\zeta_D$, corresponding to unmerged operation and use of only a

---

* Arc and combustor operation with the same nozzle and cavity configuration were compared in the MESA tests.[124]

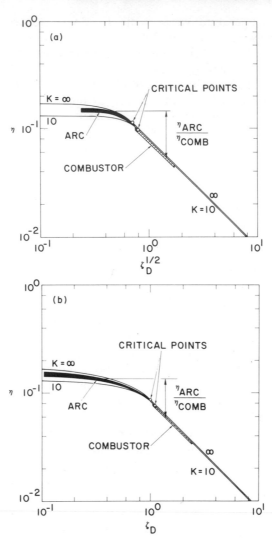

**Fig. 64.** Operating regimes of arc- and combustor-driven HF lasers. (*a*) Laminar model and (*b*) turbulent model.

fraction of the fluorine flux for lasing, or in more physical terms, to the faster collisional deactivation and, hence, a shorter lasing zone.

The last question is why the correlations of Fig. 61 or 62 indicate average values of $K_{combustor} = O(10)$, and hence by implication $K_{arc} = O(10^2)$, which is consistent with the curve fits of Figs. 58. Experimentally determined rate coefficients for pumping and for deactivation by DF give

$K_{combustor} = O(1)$ for the $2 \to 1$ transitions (see Fig. 51). Part of this discrepancy may be attributable to cascading. Typically, about 10% of the net radiative output of a cold-reaction HF laser comes from $3 \to 2$ transitions, 50% from $2 \to 1$ transitions, and 40% from $1 \to 0$ transitions. If we adopt a two-level model, most of the pumping of the $v = 3$ level should be added to that of $v = 2$. The effective rate of deactivation of HF(2) should be diminished by the number of $2 \to 1$ deactivations that lead to subsequent emission in the $1 \to 0$ band. Thus the effective ratio of pumping to deactivation between $v = 2$ and 1 can be appreciably larger than estimated using the rate coefficients for HF(2) pumping and $2 \to 1$ deactivation only.

## 3.7. Predicted and Measured Small-Signal Gain Distributions

We have shown that the simplified mixing-laser analysis is capable of predicting the overall performance of the lasing cavity across a wide range of operating conditions. Determination of axial distributions within the cavity, under given operating conditions, is more demanding on both experiment and theory. The interpretation of measurement of gain, intensity, and power distributions in a lasing cavity is complicated by intercoupling through the cavity optics.[121] This problem is avoided with zero-power measurements taken without mirrors. Such experiments have been performed for the standard Aerospace laser,* and the results have been compared to theoretical calculations.

Reference 31 reports spectrally resolved measurements of the spontaneous ir emission from vibrationally excited HF molecules produced by the cold reaction. The spectrometer slit was arranged such that the measured data are averages over the mixing layers of the central eight nozzles, with a streamwise resolution of about 1.5 mm. From the intensites measured for each of the radiating lines, the concentrations of the vibrationally excited HF molecules ($v = 1$ to 4) and the rotational temperature in the sampled volume have been calculated, and the results for test condition II of Table 8 are shown in Fig. 65. Of particular interest is the small total inversion between the $v = 2$ and 1 populations observed in the first 1.8 cm from the nozzle exit.

Also shown in Fig. 65 are theoretical estimates[31] for the axial distributions of the HF($v$) concentrations and the (average) mixing-layer temperature. The laminar mixing-layer model used for the present estimates is essentially that described in Sections 3.2 and 3.3. However, to better account for the thick-nozzle boundary layers, a parabolic velocity profile

---

* Zero-power measurements in the TRW reactive-flow facility[40a] have been discussed in Section 2. But this facility was never meant to lase.

**Table 8.**   Operating Conditions for Figs. 65 and 67

| Condition | $p_0$, atm | $p_{cav}$, Torr | $\dot{m}_{SF_6}$, g/sec | $\dot{m}_{H_2}$, g/sec | $\dot{m}_{He}$, g/sec | $\dot{m}_{O_2}$, g/sec | $T_0$, °K | Figure | Reference |
|---|---|---|---|---|---|---|---|---|---|
| I | 32 | 3.0 | 4.3 | 1.0 | 2.06 | 0.8 | 1800 | 67a | 31 |
| II | 33 | 3.0 | 4.3 | 1.0 | 2.06 | 0.0 | 1600 | 65, 67b | 31 |
| III | 29 | 3.5 | 4.3 | 1.0 | 2.06 | 0.94 | 2000–2500 | 67a | 128 |

is assumed at the exit of the oxidizer nozzle, as illustrated in Fig. 66. The initial temperature, pressure, and species concentration distributions are assumed uniform, the jets are considered parallel and matched in pressure, and axial pressure gradients are neglected. The parabolic growth of the laminar mixing layers, and the station at which they merge at the centerline of the oxidizer jet, are somewhat arbitrarily matched to the ir emitting zone observed in vidicon scans of the cavity.[125] The simplified chemical and optical model of Sections 3.2 and 3.3 is replaced by a more detailed numerical calculation[126] starting at the mixing-layer boundary.

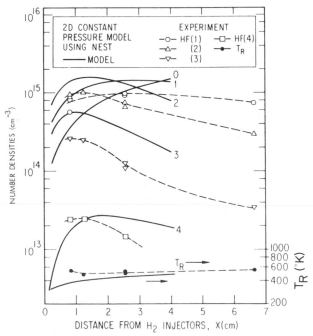

**Fig. 65.**   Arc-driven laser at zero power (condition II), concentrations of vibrational levels of HF as functions of axial distance, chemiluminescence measurements, and predictions by modified mixing-laser model.[31]

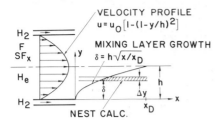

**Fig. 66.** Schematic of flow field for modified mixing-laser model.[31]

Rate coefficients used for the calculations[127] again imply that H is the dominant deactivator, but with a rate coefficient two to three times slower than calculated by Wilkins.[122]

The predictions are seen to exceed the measured concentrations of the lower excited levels, but do reproduce their axial variations. Agreement for HF(4) is poorer. Predictions for the total inversion at small $x$ approximate the measurement considerably better than those of a premixed calculation also presented in Ref. 31. Similar conclusions are reached for a test case run with the $SF_6$ flow reduced to 0.5 g/sec.

Small-signal gain coefficients have been calculated from these excited-state populations and averaged across the mixing layer. Integrated small-signal gains for the $P_2(5)$ and $P_1(6)$ lines have also been measured directly, using a small pulsed HF-laser beam to probe across the mixing layers.[128] Indirect and direct measurements obtained under the similar conditions I and III of Table 8 are compared in Fig. 67a. The agreement of absolute levels is good. Surprisingly, directly measured small-signal gains are found to be finite at the nozzle exit. A possible explanation is that the nozzle boundary layers separate, permitting prereaction in the separation regions.[128]

Figure 67b shows small-signal gains for the $2 \rightarrow 1$ transitions as derived from chemiluminescence measurements[31] for condition II, together with theoretical predictions[128] obtained with the finite difference code of Ref. 119. Calculated-peak gains are in good agreement.* The location of the peak-gain point and the length of the positive-gain region for the $P_2(5)$ line agree only approximately. Since the numerical analysis does not account for the nozzle boundary layers, much less for separation, it cannot predict finite gains at $x = 0$.

The foregoing comparisons suggest that existing theoretical models give good qualitative estimates of the axial distributions, but that the rate coefficients have to be further adjusted and the nozzle boundary layers should be taken into account. The gain, and possibly the total inversion,

---

* The calculation shown in Fig. 67b, which ignored $O_2$ addition to the plenum, is compared in Ref. 128 to the direct measurement for condition III with $O_2$ shown in Fig. 67a.

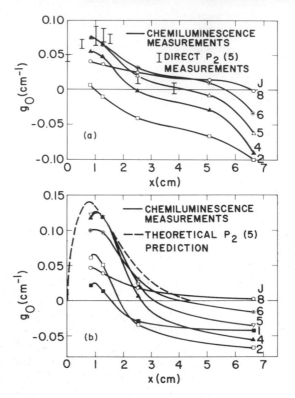

**Fig. 67.** Arc-driven HF laser, small-signal gain distributions for $2 \rightarrow 1$ band. (*a*) With $O_2$ addition, chemiluminescence[31] (condition I) and direct[128] (condition III) measurements and (*b*) without $O_2$ addition (condition II), chemiluminescence measurements[31] and numerical calculations.[128]

at small $x$ may be strongly affected by boundary-layer separation, because of pressure mismatch between the jets and reaction shocks. These effects are not as easily modeled.

## ACKNOWLEDGMENTS

In writing this chapter, the authors have made ample use of the professional experience of their colleagues at both The Aerospace Corporation and the TRW Systems Group. In particular, they thank D. J. Spencer, R. L. Varwig, I. Alber, R. G. Batt, E. Baum, W. Behrens, J. C. Cummings, F. E. Fendell, T. A. Jacobs, and J. T. Ohrenberger for many informative discussions. The authors are grateful to J. Miller and L. A. Hromas of the TRW Systems Group for their encouragement in this effort, and to the editors for their nearly unlimited patience.

## REFERENCES

1. D. J. Spencer, T. A. Jacobs, H. Mirels, and R. W. F. Gross, Int. J. Chem. Kinet, **1,** 493 (1969).
2. T. A. Cool, R. R. Stephens, and T. J. Falk, Int. J. Chem. Kinet. **1,** 495 (1969).
3. J. R. Airey and S. F. McKay, Appl. Phys. Lett. **15,** 401 (1969).
4. H. Mirels and D. J. Spencer, IEEE J. Quant. Electron. **QE-7,** 501 (1971).
5. H. Mirels, R. Hofland, and W. S. King, AIAA J. **11,** 156 (1973).
6. J. E. Broadwell, Appl. Opt. **13,** 962 (1974).
7. R. A. Meinzer, Int. J. Chem. Kinet. **2,** 335 (1970).
8. T. A. Cool, R. R. Stephens, and J. A. Shirley, J. Appl. Phys. **41,** 4038 (1970).
9. C. J. Buczek, R. J. Freiberg, J. J. Hinchen, P. Chenausky, and R. J. Wayne, Appl. Phys. Lett. **17,** 514 (1970).
10. N. R. Greiner, IEEE J. Quant. Electron. **QE-8,** 872 (1972).
11. L. D. Hess, Appl. Phys. Lett. **19,** 1 (1971).
12. W. R. Warren, Jr., "Reacting Flow and Pressure Recovery in HF/DF Chemical Lasers," Technical Report TR-0074 (9240-02)-1 (The Aerospace Corp., Los Angeles, California, 1973).
13. D. A. Russell, "Fluid Mechanics of High Power Grid Nozzle Lasers," AIAA Paper No. 74-223, AIAA 12th Aerospace Science Meeting, Washington, D. C., Jan. 1974.
14. Tests performed at TRW Systems Group under AFWL Contract No. F29601-73-C-0087 (1973).
15. A. H. Shapiro, *Compressible Fluid Flow, Vol. I* (Ronald, New York, 1953).
16. D. J. Spencer, H. Mirels, and D. A. Durran, J. Appl. Phys. **43,** 1151 (1972).
17. D. A. Durran and D. J. Spencer, "Axisymmetric Mixing Nozzle for Supersonic Diffusion Laser," Technical Report TOR-0059 (6756-02)-1 (The Aerospace Corp., Los Angeles, California, 1970).
18. J. E. Ferrell, R. M. Kendall, and H. Tong, "Recombination Effects in Chemical Laser Nozzles," AIAA Paper No. 73-643, 1973.
19. W. J. Rae, AIAA J. **9,** 811 (1971).
20. D. L. Whitfield and C. H. Lewis, J. Spacecraft and Rockets **7,** 462 (1970).
21. D. E. Rothe, AIAA J. **9,** 804 (1971).
22. H. Schlichting, *Boundary Layer Theory*, sixth ed. (McGraw-Hill, New York, 1966).
23. G. A. Hosack, E. C. Curtis, J. C. Hyde, and R. J. Burick, "Aerodynamic Reactive Flow Studies of the HF Laser," Technical Report AFWL-TR-73100 (Rocketdyne Division, Rockwell International, Canoga Park, California, 1973).
24. D. J. Spencer and R. L. Varwig, AIAA J. **11,** 1000 (1973).
25. L. Crocco, "One-Dimensional Treatment of Steady Gas Dynamics," *Fundamentals of Gas Dynamics*, H. W. Emmons, Ed. (Princeton Univ. Press, Princeton, New Jersey, 1958).
26. G. Emanuel, "Feasibility Study of High Energy Ejector Systems," Vol. II, Technical Report 24245-6001-RU-00 (TRW Systems Group, Redondo Beach, California, 1973).
27. "Project MESA Data Summary—1970," Technical Report TR-0172 (2779)-3 (The Aerospace Corp., Los Angeles, California, 1971).

28. J. Lukasiewicz, J. Aero. Sci. **20,** 617 (1953).

29. R. L. Varwig and M. A. Kwok, "Flow Diagnostic Measurements in the Medium of a Continuous Wave HF Chemical Laser," Technical Report TR-0074(4534)-1 (The Aerospace Corp., Los Angeles, California, 1973).

30. R. L. Varwig, "Photographic Observation of CW HF Chemical Laser Reacting Flowfield," "Technical Report TR-0074(4240-10)-6 (The Aerospace Corp., Los Angeles, California, 1973).

31. M. A. Kwok, "Measurement and Analysis of the HF Radiation from a Reacting Supersonic Jet," Technical Report TR-0074(4530)-3 (The Aerospace Corp., Los Angeles, California, 1973).

32. A. B. Witte, J. E. Broadwell, W. L. Shackleford, J. C. Cummings, J. E. Trost, A. S. Whiteman, F. E. Marble, D. R. Crawford, and T. A. Jacobs, "Aerodynamic Reactive Flow Studies of the $H_2/F_2$ Laser—II, "Technical Report AFWL-TR-74-78 (TRW Systems Group, Redondo Beach, California, 1974).

33. S. T. Chapin, "Mixing investigations of $H_2$ and Supersonic $F_2$ Flows," Technical Report MCR-73-37/AFWL-TR-72-242 (Martin Marietta Corporation, Denver, Colorado, 1973.

34. A. Ferri, P. A. Libby, and V. Zakkay, "Theoretical and Experimental Investigation of Supersonic Combustion," Technical Report ARL62-467 (Polytechnic Institute of Brooklyn, Brooklyn, New York, 1962).

35. V. Zakkay and E. Krause, "Mixing Problems with Chemical Reactions," *Supersonic Flow, Chemical Processes and Radiative Transfer*, D. B. Olfe and V. Zakkay, Eds. (Pergamon, Macmillan, New York, 1964), p. 3.

36. A. Ferri, Astronautica Acta **13,** 453 (1968).

37. L. S. Cohen and R. N. Guile, AIAA J. **8,** 1053 (1970).

38. A. Ferri, Ann. Rev. Fluid Mech. **5,** 301 (1973).

39. R. A. Chodzko, D. J. Spencer, and H. Mirels, IEEE J. Quant. Electron. **QE-9,** 550 (1973).

40a. W. L. Shackleford, A. B. Witte, J. E. Broadwell, J. E. Trost, and T. A. Jacobs, "Experimental Studies of Chemically Reactive $(F + H_2)$ Flow in Supersonic Free Jet Mixing Layers," AIAA Paper 73-640, AIAA 6th Fluid and Plasma Dynamics Conference, Palm Springs, California, 1973, AIAA J. **12,** 1009 (1973).

40b. A. B. Witte, J. E. Broadwell, W. L. Shackleford, J. E. Trost, and T. A. Jacobs, "Aerodynamic Reactive Flow Studies of the $H_2/F_2$ Laser," "Technical Report AFWL-TR-72-247 (TRW Systems Group, Redondo Beach, California, 1973).

41. P. T. Harsha, "Free Turbulent Mixing: A Critical Evaluation of Theory and Experiment," Technical Report AEDC-TR-71-36 (Arnold Engineering Development Center, Arnold Air Force Station, Tennessee, 1971).

42. S. F. Birch and J. M. Eggers, "A Critical Review of the Experimental Data for Developed Free Turbulent Shear Layers," *Proceedings of the Conference on Free Turbulent Shear Flows, Vol. I*, Report NASA SP-321 (NASA Langley Research Center, Hampton, Virginia, 1973), p. 11.

43. S. I. Pai, *Viscous Flow Theory, Vol. I, Laminar Flow* (Van Nostrand, Princeton, New Jersey 1956), Section IX, 7.

44a. G. Brown and A. Roshko, "The Effect of Density Differences on the Turbulent Mixing Layer," AGARD Fluid Dynamics Panel Specialist's Meeting on Turbulent Shear Flows, London, England, 1971.

44b. G. L. Brown and A. Roshko, J. Fluid Mech. **64,** 775 (1974).

45. J. H. Kent and R. W. Bilger, *Proceedings of the Fourteenth Symposium (International) on Combustion* (1973), p. 615.

46. L. J. Alpinieri, AIAA J. **2,** 1560 (1964).

47. A. Demetriades, AIAA J. **2,** 245 (1964).

48. A. Demetriades, Phys. Fluids **12,** 24 (1969), **13,** 1672 (1970).

49. H. Sato and K. Kuriki, J. Fluid Mech. **11,** 321 (1961).

50. A. Michalke, "The Instability of Free Shear Layers, a Survey on the State of Art," DLR Mitteilung 70-04 (Deutsche Forschungsund Versuchsanstalt für Luft- und Raumfahrt, Berlin, Germany, 1970).

51. R. W. Miksad, J. Fluid Mech. **56,** 695 (1972).

52. W. Behrens and D. R. S. Ko, AIAA J. **9,** 851 (1971).

53. C. D. Winant and F. K. Browand, J. Fluid Mech. **63,** 237 (1974).

54. P. Bradshaw, J. Fluid Mech. **26,** 225 (1966).

55a. R. Batt, T. Kubota, and J. Laufer, "Experimental Investigation of the Effect of Shear Flow Turbulence on a Chemical Reaction," AIAA Paper 70-721, AIAA Reacting Turbulent Flows Conference, San Diego, California (1970).

55b. R. G. Batt, TRW Systems Group, Redondo Beach, California, private communication.

56. M. R. Rebollo, "Analytical and Experimental Investigation of a Turbulent Mixing Layer of Different Gases in a Pressure Gradient," Ph.D. Thesis, California Institute of Technology, Pasadena, California (1973).

57a. H. Ikawa, "Turbulent Mixing Layer Experiment in Supersonic Flow," Ph.D. Thesis, California Institute of Technology, Pasadena, California (1973).

57b. H. Ikawa and T. Kubota, "An Experimental Investigation of a Two-Dimensional, Self-Similar, Supersonic Turbulent Mixing Layer with Zero Pressure Gradient," AIAA Paper 74-40, AIAA 12th Aerospace Sciences Meeting, Washington, D.C., 1974).

58. R. F. Davey and A. Roshko, J. Fluid Mech. **53,** 523 (1972).

59. D. R. S. Ko, T. Kubota, and L. Lees, J. Fluid Mech. **40,** 315 (1970).

60. D. R. S. Ko, AIAA J. **9,** 1777 (1971).

61. I. E. Alber and L. Lees, AIAA J. **6,** 1343 (1968).

62. I. S. Gartshore, J. Fluid Mech. **30,** 547 (1967).

63. A. A. Townsend, *The Structure of Turbulent Shear Flow*, (Cambridge Univ. Press, Cambridge, England, 1956), Section 5.5.

64. A. Prabhu and R. Narasimha, J. Fluid Mech. **54,** 19 (1972).

65. A. H. Shapiro, *The Dynamics and Thermodynamics of Compressible Fluid Flow*, Vol. II (Ronald, New York, 1954), Section 28.

66. E. Devis, "Shock Wave-Induced Mixing in Parallel Flow," Ph.D. Thesis, Cornell. Univ., Ithaca, New York (1972).

67. A. A. Hayday, "Mixing Phenomena Related to C. W. Chemical Lasers," Technical Report RK-TR-71-16 (Scicom, Inc., Huntsville, Alabama, 1971).

68. J. T. Ohrenberger and E. Baum, AIAA J. **10,** 1165 (1972).

69a. R. G. Batt, "Experimental Investigation of Wakes behind Two-Dimensional Slender Bodies at $M - 6$," Ph.D. Thesis, California Institute of Technology, Pasadena, California (1967).

69b. R. G. Batt and T. Kubota, AIAA J. **6,** 2077 (1968).

70. J. E. Lewis and R. L. Chapkis, "Experimental Investigation of the Effect of Base Injection on the Turbulent Near Wake of a Slender Body at Mach 4.0," Technical Report 06388-6029-R000 (TRW Systems Group, Redondo Beach, California, 1968).

71. J. E. Bowman and W. A. Clayden, AIAA J. **5,** 1924 (1967); **6,** 2029 (1968).

72. A. Watton, "The Fuel-Mixing Mechanism in Diffusion-Type Supersonic Combustion," AIAA Paper 69-32, Seventh Aerospace Sciences Meeting, New York, 1964.

73. H. Fox, V. Zakkay, and R. Sinha, "A Review of Some Problems in Turbulent Mixing," Report NYU-AA-66-63 (New York University, New York, 1966).

74. L. H. Townesend and J. Reid, "Some Effects of Stable Combustion in Wakes Formed in a Supersonic Stream," *Supersonic Flow, Chemical Processes and Radiative Transfer,* D. B. Olfe and V. Zakkay, Eds., (Pergamon, Macmillan, New York, 1968), p. 137.

75. W. L. Chow, AIAA J. **6,** 2422 (1968).

76. R. J. Dixon, R. M. Richardson, and R. H. Page, J. Spacecraft and Rockets **7,** 848 (1970).

77. R. L. Chapkis, J. Fox, L. Hromas, and L. Lees, "An Experimental Investigation of Base Mass Injection on the Laminar Wake behind a 6-Degree Half-Angle Wedge at $M = 4.0$," Technical Report 06388-6009-R000 (TRW Systems Group, Redondo Beach, California, 1967); also AGARD Report CP 19, *Proceedings of the AGARD Conference on Fluid Physics of Hypersonic Wakes* (Ft. Collins, Colorado, 1967).

78. J. E. Lewis and W. Behrens, AIAA J. **7,** 664 (1969).

79. R. F. Weiss, "Near Wake Effects of Base Injection," Research Report 315 (AVCO-Everett Research Labs, Everett, Massachusetts, 1969).

80. R. C. Boger, H. Rosenbaum, and B. L. Reeves, AIAA J. **10,** 300 (1972).

81. L. Walitt, D. C. Wilcox, and C. Y. Liu, "Numerical Study of Plume-Induced Flow Separation," Technical Report ATR-72-33-1 (Applied Theory, Inc., Los Angeles, California, 1973).

82. D. J. Spencer and R. L. Varwig, AIAA J. **11,** 1000 (1973).

83. R. G. Batt, "Some Measurements on the Effect of Tripping the Two-Dimensional Shear Layer," AIAA J. (1975), to be published.

84. M. V. Morkovin (Chairman), "Free Turbulent Shear Flows," Vol. I—Conference Proceedings, Report NASA SP-321 (NASA Langley Research Center, Hampton, Virginia, 1973).

85. G. L. Mellor and H. J. Herring, AIAA J. **11,** 590 (1973).

86. P. Bradshaw, Aeronautical J. **76,** 403 (1972).

87. J. Boussinesq, "Theorie de L'écoulement Tourbillant," Mém. prés. Acad. Sci. **XXIII,** 46 (1877).

88. L. Prandtl, ZAMM **5,** 136 (1925).

89. L. Prandtl, ZAMM **22,** 241 (1942).

90. J. A. Schetz, AIAA J. **6,** 2008 (1968).

91. A. Favre, "Statistical Equations of Turbulent Gases," *Problems of Hydrodynamics and Continuum Mechanics,* I. E. Bloch et al., Eds. (Society of Industrial and Applied Mathematics, Philadelphia, 1969), p. 231.

92. B. J. Daly and F. H. Harlow, Phys. Fluids **13,** 2634 (1970).

93. K. Hanjalic and B. E. Launder, J. Fluid Mech. **52,** 609 (1972).

94. A. K. Varma, R. A. Beddini, R. D. Sullivan, and C. duP. Donaldson, "Application of an Invariant Second-Order Closure Model to Compressible Turbulent Shear Layers," AIAA Paper 74-592, AIAA Seventh Fluid and Plasma Dynamics Conference, Palo Alto, California, 1974.

95. J. C. Rotta, Z. Phys. **129,** 547; **131,** 51 (1951).

96. K. H. Ng and D. B. Spalding, Phys. Fluids **15,** 20 (1972).

97. P. G. Saffman, Proc. Roy. Soc. London Ser. A **317,** 417 (1970).

98. Y. H. Oh, "Analysis of Two-Dimensional Free Turbulent Mixing," AIAA Paper 74-594, AIAA Seventh Fluid and Plasma Dynamics Conference, Palo Alto, California, 1974.

99. D. C. Wilcox and I. E. Alber, *Proceedings of the 1972 Heat Transfer and Fluid Mechanics Institute* (1972), p. 231.

100. D. B. Spalding, *Proceedings of the Thirteenth Symposium (International) on Combustion* (1971), p. 649.

101. C. DuP. Donaldson and G. R. Hilst, *Proceedings of the 1972 Heat Transfer and Fluid Mechanics Institute* (1972), p. 253.

102. G. R. Hilst, "Solutions of the Chemical Kinetic Equations for Initially Inhomogeneous Mixtures," AIAA Paper 73-101, AIAA Eleventh Aerospace Sciences Meeting, Washington, D.C., 1973.

103. G. F. Carrier, F. E. Fendell, and F. E. Marble, "The Effect of Strain on Diffusion Flames," Technical Report TRW-5-PU (TRW Systems Group, Redondo Beach, California, 1973).

104. H. L. Toor, AIChE J. **8,** 70 (1962).

105. I. E. Alber and R. G. Batt, "An Analysis of Diffusion-Limited First and Second Order Chemical Reactions in a Turbulent Shear Layer," AIAA Paper 74-593, AIAA Seventh Fluid and Plasma Dynamics Conference, Palo Alto, California, 1974.

106. W. B. Bush and F. E. Fendell, Acta Astronautica **1,** 645 (1974).

107. W. B. Bush and F. E. Fendell, "On Diffusion Flames in Turbulent Shear Flows— The Two-Step Symmetrical Chain Reaction," Comb. Sci. Tech. (1976), to appear.

108. J. M. Richardson, H. C. Howard, Jr., and R. W. Smith, Jr., *Proceedings of the Fourteenth Symposium (International) on Combustion* (1953), p. 814.

109. E. E. O'Brien, Phys. Fluids **14,** 1326 (1971).

110. C-H. Lin and E. E. O'Brien, Astronautica Acta **17,** 771 (1972).

111. C. Dopazo and E. E. O'Brien, "An Approach to the Autoignition of a Turbulent Flame," presented at the Fourth International Colloqium on Gasdynamics of Explosions and Reactive Systems, La Jolla, California, 1973.

112. G. Emanuel and J. S. Whittier, Appl. Opt. **11,** 2047 (1972).

113. N. Cohen, private communication; see also chapter by N. Cohen and J. Bott.

114. P. M. Chung, AIAA J. **11,** 1040 (1973).

115. A Liñan, "On the Structure of Laminar Diffusion Flames," Aeronautical Engineer Thesis, California Institute of Technology, Pasadena, California (1963).

116. F. E. Fendell, J. Fluid Mech. **21,** 281 (1963).

117a. P. A. Libby, Comb. Sci. Tech. **6,** 23 (1972).

117b. C. H. Gibson and P. A. Libby, Comb. Sci. Tech., **6,** 29 (1972).

118. R. Hofland and H. Mirels, AIAA J. **10,** 420, 1271 (1972).

119. W. S. King and H. Mirels, AIAA J. **10,** 1647 (1972).

120. E. B. Turner, W. D. Adams, and G. Emanuel, J. Comp. Phys. **11,** 15 (1973).

121. D. J. Spencer, D. A. Durran, and H. A. Bixler, Appl. Phys. Lett. **20,** 164 (1972).

122. R. L. Wilkins, J. Chem. Phys. **58,** 2326 (1973).

123. M. Abramowitz and I. A. Stegun, *Handbook of Mathematical Functions* (Dover, New York, 1965).

124. R. R. Giedt, "Experimental Comparison of cw Chemical Laser Gas Heaters and Nozzle Arrays," Fourth Conference on Chemical and Molecular Lasers, St. Louis, Missouri, Oct. 1974.

125. M. A. Kwok and E. F. Cross, Opt. Eng. **11,** 131 (1972).

126. E. B. Turner, G. Emanuel, and R. L. Wilkins, "The NEST Chemistry Computer Program, Vol. I," Technical Report TR-0059 (6240-20)-1 (The Aerospace Corp., Los Angeles, California, 1970).

127. N. Cohen, "A Review of Rate Coefficients for Reactions in the $H_2$–$F_2$ Laser System," Technical Report TR-0172(2779)-2 (The Aerospace Corp., Los Angeles, California, 1971).

128. R. A. Chodzko, D. J. Spencer, and H. Mirels, IEEE J. Quant. Electron. **QE-9,** 550 (1973).

# CHAPTER 6

# PULSED HYDROGEN-HALIDE CHEMICAL LASERS

## STEVEN N. SUCHARD

The Aerospace Corporation,
El Segundo, California

## J. RICHARD AIREY

Naval Research Laboratory,
Washington, D.C.

The discovery in 1965 by Kasper and Pimentel[1] of the first chemical laser stimulated the search for a laser in which the population inversion could be produced by a chain reaction. It was then surmised that such a chemical laser would probably be electrically efficient since only a small amount of energy would be necessary to initiate the chain reaction. In this regard, pulsed HCl chemical lasers have played a very large part in the development of chemical lasers. Not only were they the first chemical laser but they also were the first chemical lasers to operate on a chain reaction.

A short time after reporting the first chemical laser, Kompa and Pimentel[2] reported laser action from the HF molecule initiated by the flash photolysis of $UF_6/H_2$ mixtures, and Deutsch[3] reported laser action initiated by electrical discharge of Freon/$H_2$ mixtures. The first demonstration of laser action from the $H_2/F_2$ chain reaction, however, was not reported until 1969.[4,5] In this case, the chain reaction between the $H_2$ and $F_2$ gases was initiated by both electrical discharges and flash photolysis.

In addition to the HF and HCl molecules and their deuterated analogs, chemically pumped laser action has also been produced from the HBr molecule.[3,6] Because of the lower bond strength of HBr, which reduces the exothermicity of many of its formation reactions below a useful value, it is the least well developed of all the hydrogen halide chemical laser systems.

In addition to extracting energy directly from excited molecules produced in a chemical reaction, Chen, Stephenson, and Moore[7] demonstrated the first chemical transfer laser in 1968. The operation of all chemical-transfer lasers to date is based on the selective production of vibrationally excited species in a chemical reaction followed by a fast transfer of energy to a suitable laser species. By far the most interesting and well-studied transfer system is the $DF/CO_2$ transfer chemical laser in which vibrationally excited $CO_2$ is produced by energy transfer from vibrationally excited DF to ground-state $CO_2$ molecules.

The concern in this chapter is primarily with pulsed hydrogen halide and transfer chemical lasers. In particular, the concern is with initiation techniques, the feasibility of using them as high-power lasers, and an explanation of how they work.

## 1. NON-CHAIN REACTION HF LASERS

Since the first report of the flash-photolysis initiated $UF_6/H_2$ laser by Kompa and Pimentel[2] and the electrical-discharge initiated Freon/$H_2$ laser by Deutsch,[3] there has been a great deal of interest in the pulse-initiated chemical reactions of hydrogen or hydrogen-bearing compounds with fluorine or fluorine-bearing compounds to produce HF lasers. In the case of the non-chain-reaction HF laser, the chemical reaction that produces the laser inversion in the HF molecules is represented by

$$F + H_2 \rightarrow HF(v \leq 3) + H \qquad \Delta H = -31.7 \text{ kcal/mol} \qquad (1)$$

Measurements made of this reaction system have shown that the HF molecules that are formed have an inverted vibrational-level distribution.[8] The relative vibrational level populations have been measured to be $N_1/N_2/N_3 = 0.31/1.00/0.47$. These measurements indicate that a large percentage of the reaction exothermicity goes directly into vibrational excitation of the product molecule, and, consequently, if this energy can be efficiency extracted, the reaction is a good candidate for the high-conversion efficiency needed to convert from chemical to laser energy.

The reactants, method of initiation, spectral content, and time histories for the majority of the reported non-chain-reaction HF lasers, are given in Table 1.

**Table 1.** Pulsed Non-Chain Reaction HF Lasers

| Reaction system[a] | Pressure, Torr | Spectral content | Pulse duration, $\mu$sec | Energy, J | References |
|---|---|---|---|---|---|
| $(CF_4, CBrF_3, CCl_2F_2)$ $+ (H_2, D_2, CH_4, CH_3Cl)/e$ | 0.1–1.2 | $P_1(6)–P_1(15)$ $P_2(2)–P_2(15)$ $P_3(2)–P_3(8)$ | 1 | | 3, 35, 36 |
| $UF_6 + (H_2, D_2, HD)/h\nu$ | 2.25 | $P_2(3)–P_2(8)$ | 10 | $2 \times 10^{-4}$ | 2, 10 |
| $F_2O + H_2/h\nu$ | 100 | | 5 | | 12 |
| $UF_6 + (CH_4, CD_4,$ $C_nH_{2n+2}, CH_3F,$ $CH_2F_2, CHF_3,$ $CH_3Cl, CH_2Cl_2,$ $CHCl_3)/h\nu$ | 2.25 | $P_2(4)–P_2(8)$ | 4–8 | | 37 |
| $CF_3I + CH_3I/h\nu$ | 5–40 | $P_1(3)–P_1(10)$ $P_2(3)–P_2(5)$ | | | 38 |
| $F_2O + H_2/shockwave$ | | | | | 32 |
| $UF_6 + H_2/h\nu$ | 2–5 | | | | 17 |
| $CH_2CHF,$ | | | 3–2 2–1 | | 39 |
| $CH_2CHCl/h\nu$ | | | 1–0 | | |
| $(UF_6, XeF_4, SbF_5, WF_6)$ $+ (H_2, CH_4)/h\nu$ | several | $P_2(2)–P_2(8)$ $P_1(3)–P_1(9)$ | 1 | | 40 |
| $WF_6 + H_2/h\nu$ | several | $P_3(1)–P_3(5)$ $P_2(2)–P_2(8)$ $P_1(3)–P_1(8)$ | 1–2 | | 16 |
| $IF_5 + H_2/h\nu$ | 6.5 | $P_3(4)$ $P_2(3)–P_2(5)$ unidentified lines | 2 | | 41 |

**Table 1.**  (*contd.*)

| Reaction system[a] | Pressure, Torr | Spectral content | Pulse duration, $\mu$sec | Energy, J | References |
|---|---|---|---|---|---|
| $(MoF_6, UF_6)+H_2/h\nu$ | 2–45 | $P_2(4)$–$P_2(7)$ | 7–10 | | 14 |
| $(CHF_2Cl, CHFCl_2,$ $CHF_3, CF_2Cl_2)$ $+H_2/e$ | 2–10 | $P_2(3)$–$P_2(5)$ $P_1(6)$–$P_1(9)$ | | | 42 |
| $SF_6+(H_2, HBr)/e$ | 1–25 | $P_4(4)$ $P_3(6)$–$P_3(9)$ $P_2(3)$–$P_2(9)$ | | | 43 |
| $(CF_3H, CH_2F_2,$ $C_2F_6, SF_6)+H_2/e$ | 1–60 | $P_2(5)$–$P_2(7)$ $P_1(6)$–$P_1(7)$ | 1–40 | | 44 |
| $(BF_3, BCl_3, BBr_e)$ $+H_2O/e$ | | rotational lasing | | | 45 |
| $UF_6+CHCl_3/h\nu$ | | 3–2 2–1 1–0 | | | 18 |
| $N_2F_4+(HCl, CH_4, CH_3F$ $CH_2F_2, CH_3Br, C_2H_6,$ $C_2H_5F, C_2H_5I)/h\nu$ | 1–24 | $P_2(4)$–$P_2(8)$ $P_1(6)$–$P_1(9)$ | 4 | | 46 |
| $N_2F_4+CH_3I/h\nu$ | 12 | $P_2(2)$ $P_1(5)$–$P_1(6)$ | 20 | | 15 |
| $(NF_3, SF_6)+H_2$ $ClN_3/h\nu$ | 12–24 | $P_4(5)$–$P_4(6)$ $P_3(5)$–$P_3(7)$ $P_2(6)$–$P_2(9)$ $P_1(5)$–$P_1(11)$ | | | 33 |
| $N_2F_4+CD_4/h\nu$ | 80 | DF overtone emission | 20 | | 47 |
| $(N_2F_4, NF_3)$ $+(H_2, B_2H_6)/eb$ | 10–250 | $P_3(3)$–$P_3(5)$ $P_2(3)$–$P_2(7)$ $P_1(6)$–$P_1(7)$ | 0.01 | | 9 |
| $O_3+CH_nX_{4-n}/h\nu$ $(X=F, Cl, n=1, 2, 3)$ | 30 | $P_2(4)$–$P_2(6)$ $P_1(6)$–$P_1(10)$ | 2 | | 48 |
| $NF_3+(H_2, C_2H_6)/e$ | several | $P_2(4)$–$P_2(10)$ $P_1(7)$–$P_1(10)$ | | | 49 |
| $(NF_3, N_2F_4)$ $+(H_2, CH_4, C_2H_6, HCl,$ $HBr, natural\ gas)/e$ | 100 | $P_4(3)$–$P_4(4)$ $P_3(3)$–$P_3(7)$ $P_2(3)$–$P_2(9)$ $P_1(6)$–$P_1(9)$ | 0.3 | | 50, 51 |
| $(SF_6, CCl_2F_2, CF_4)$ $+(C_3H_8, C_4H_{10}, C_6H_{14},$ $H_2, CH_4, HI)/e$ | 9–17 | $P_3(4)$–$P_3(7)$ $P_2(3)$–$P_2(8)$ $P_1(4)$–$P_1(8)$ | 2 | | 52 |
| $(SF_6, C_3F_8, C_2F_6, CF_4)$ $+(C_3H_8, CH_4, H_2, C_2H_6,$ $C_4H_{19})/e$ | 30–750 | $P_3(3)$–$P_3(7)$ $P_2(3)$–$P_2(8)$ $P_1(3)$–$P_1(8)$ | 0.1 | | 28 |
| $SF_6+H_2/e$ | 100–250 | vibrational and rotational lasing | 0.3–15 | | 53 |
| $SF_6+H_2/e$ | 30 | $R$-branch lasing | 0.2 | | 29 |
| $SF_6+H_2/e$ | 10–380 | superfluorescence | | | 31 |
| $WF_6+(H_2, D_2, CH_4,$ $C_4H_{10}, HCl)/h\nu$ | | reaction rate measurements | | | 54 |
| $MoF_6+H_2/h\nu$ | | threshold studies | | | 13 |

**Table 1.** (*contd.*)

| Reaction system[a] | Pressure, Torr | Spectral content | Pulse duration, $\mu$sec | Energy, J | References |
|---|---|---|---|---|---|
| $CF_3I+(C_2H_2, C_2H_4,$ $C_2H_6, CH_4)/h\nu$ | | energy partitioning study | | | 55 |
| $CH_3F+C_2H_2O/h\nu$ | | energy partitioning study | | | 56 |
| $CH_2FOCOCH_2F/h\nu$ | | energy partitioning study | | | 57 |
| $CF_3I+HI/h\nu$ | | energy partitioning study | | | 58 |
| $SF_6+H_2(D_2)/e$ | 100–400 | R-branch lasing isotope comparison | 0.2–5 | | 59 |
| $SF_6+(H_2, C_4H_{10})/e$ | 50–400 | $P_3(3)-P_3(6)$ $P_2(3)-P_2(8)$ $P_1(4)-P_1(7)$ | 0.5 | 0.1 | 22 |
| $SF_6+H_2/e$ | 1–60 | $P_3(2)-P_3(9)$ $P_2(3)-P_2(10)$ $P_1(3)-P_1(17)$ | 0.02–3.5 | 1.2 | 23 |
| $SF_6+(H_2, C_2H_6,$ $C_3H_8)/e$ | 50–100 | | 0.03 | 3.5 | 60 |
| $NF_3+(SiH_4, GeH_4,$ $AsH_3, B_2H_6, C_4H_8$ $C_8H_{10}, C_6H_{12}, C_3H_6$ $C_2H_6, CH_4)/e$ | 10–40 | H-donor | 1 | 0.025 | 61 |
| $SF_6+C_3H_8/e$ | 45 | $P_3(3)-P_3(9)$ $P_2(2)-P_2(9)$ $P_1(4)-P_1(9)$ | 0.25 | 7.2W quasi-cw | 27 |
| $SF_6+CH_4/e$ | 10–350 | $P_2(2)-P_2(9)$ $P_1(6)-P_1(8)$ | 0.005 | 0.65 | 24 |
| $SF_6+C_2H_8/eb$ | 140 | | 0.25 | 60 | 34 |
| $SF_6+H_2/e$ | 150 | | 0.20 | 11 | 26 |
| $SF_6+C_3H_8/eb$ | 1000 | | 0.025 | 2500 | 148 |

[a] $h\nu$, flash photolysis; $e$, electrical discharge; $eb$, electron-beam initiation.

## 1.1. Flash-Photolysis Initiation

The production of the fluorine atoms needed to initiate the formation of the HF molecules has been performed by several different techniques. The most important of these techniques is the flash-photolysis and electrical-discharge decomposition of inorganic fluoride molecules such as $UF_6$ and $SF_6$ to produce fluorine atoms. Recently, the use of relativistic electrons from a high-current electron-beam gun has been used for initiating a large volume of reactant gas.[9] The types of hydrogen-bearing molecules used have been even more varied than those of the fluorine-bearing molecules. It appears that almost any molecule that contains a hydrogen atom is a suitable hydrogen source.

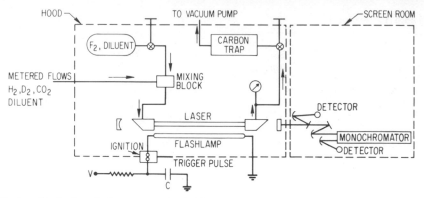

**Fig. 1.** Flash-photolysis apparatus for initiation of both non-chain- and chain-reaction HF lasers. (Adapted from Suchard et al.[11])

A typical experimental apparatus for the flash photolysis-initiated HF laser adopted from Kompa et al.[10] and Suchard et al.[11] is shown schematically in Fig. 1. A low-inductance storage capacitor, charged to 10 to 20 kV, is discharged through a xenon-filled flashlamp that is optically coupled to the laser chamber by an aluminum reflector. The rise time of the light output from the flashlamp is typically 5 to 15 $\mu$sec, depending on the inductance of the flashlamp assembly used. Reactant gases are premixed before their entry into the laser chamber to ensure a spatially uniform production of fluorine atoms for reaction initiation. A typical pulse is shown in Fig. 2. This laser pulse was obtained from a mixture of

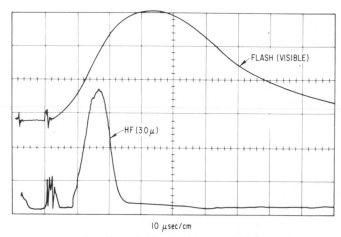

**Fig. 2.** Typical non-chain-reaction HF laser pulse. The upper trace is output of flashlamp (14.7 $\mu$fd at 10 kV); the lower trace is the detector output. Time scale is 10 $\mu$sec/division. Reactant pressures of 40 Torr $H_2$ and 60 Torr $F_2O$. (Adapted from Gross et al.[12])

$H_2/F_2O$ and is adopted from Gross et al.[12] As is evident in this figure, the laser pulse is shorter than the flashlamp duration. The time delay between initiation of the flashlamp and laser threshold is quite variable. This delay feature has been the subject of an experimental, theoretical analysis by Chester and Hess.[13]

There are several disadvantages of using flash photolysis for initiation of a non-chain-reaction HF laser. First, as previously noted, the laser pulse is much shorter than the flashlamp-initiation pulse. Consequently, a large portion of the flashlamp energy is not effective in producing fluorine atoms. Second, and more important, for uniform initiation within the laser chamber, the reactant gases must be optically thin to the photolysis radiation. This implies that only a small percentage of the photolysis radiation can be absorbed by the sample. If the sample pressure is increased so that a larger percentage of the photolysis radiation can be absorbed, the sample becomes optically thick, and the photolysis will not provide uniform reaction initiation. When these two factors are considered, it is easy to see why a flash-initiated non-chain-reaction HF laser has not been electrically efficient.

Conversely, there is a wide variety of fluorine-bearing molecules that, when photolyzed, produce fluorine atoms. These molecules include $UF_6$,[10] $F_2O$,[12] $MoF_6$,[14] and $N_2F_4$.[15] In addition, flash-photolysis initiated lasers offer greater utility because of their ability to be uniformly initiated. Photolysis lasers are amenable to computer modeling, which allows the experimenter the possibility of clearly studying the effects of the various system parameters such as hydrogen source, level of initiation, and effect of diluents on the laser output. Third, actinometric experiments can be performed to measure the photon density deposited into a given laser mixture for a given experimental arrangement.[16]

Numerous experimental measurements have been made on flash-initiated non-chain-reaction HF lasers by Pimentel and co-workers.[17-19] These measurements have been concerned primarily with measurements of the relative rate constants for the production of the HF molecules into various vibrational levels accessible to them.

## 1.2. Electrical-Discharge Initiation

Considerable research has been concerned with the study of the electrical discharge initiation of HF lasers. The basic design of the circuit elements of one of these lasers is shown schematically in Fig. 3, adapted from Kerber et al.[20] Electrical discharge offers a great deal more variety than does flash photolysis in equipment variation and pulse-repetition rates. Many different electrode configurations have been reported for HF lasers,

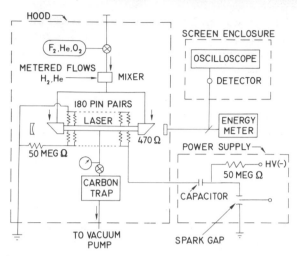

**Fig. 3.** Electrical-discharge apparatus for initiation of both non-chain- and chain-reaction HF lasers. (Adapted from Kerber et al.[20])

for example, helical and double-helical arrays of pins, straight and multiple-pin arrays, and shaped-solid and mesh electrodes.[21-24]

Since the rise time and total duration of an electrical discharge can be made as short as several nanoseconds or as long as milliseconds[25] and the various circuit elements, such as inductance, capacitance, and resistance can be varied over a wide range, experiments can be designed such that there is optimal coupling of the circuit to the reactant medium to ensure maximum energy deposition. By following this procedure of matching the circuit to the reactant gases, Pummer et al,[26] reported laser outputs as high as 11J, with an overall electrical efficiency of 4% when a multiple-pin array was used.

An interesting phenomenon encountered in pin-discharge HF lasers is the double-laser pulse. Figure 4, adopted from Pummer et al.,[26] illustrates the relationship between the laser output pulse and the electrical input. The double-laser pulse phenomenon has been observed over a wide range of pressures and initiation conditions, and, although there is some diversity in the exact time sequencing of the various $v$, $J$ level lasing, there seems to be general agreement that the double pulse is the result of cascading of the HF laser transitions.[25,27,28]

Although electrical efficiencies of 4% have been reported, there are still significant problems that appear to be inherent in electrical discharge-initiated non-chain-reaction HF lasers. The first of these problems is the inability to efficiently couple an electrical discharge to large volumes of reactant gases. As gap separations are increased, arcing, streaks, and

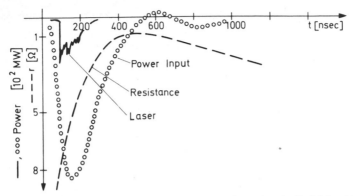

**Fig. 4.** Input/output measurements of non-chain-reaction, electrically initiated HF laser. Reactant pressures of 144 Torr $SF_6$ and 6 Torr $H_2$. The plot shows laser output pulse and electrical input on same power scale. (Adapted from Pummer et al.[25])

other discharge nonuniformities are observed,[21,22] all of which decrease the amount of energy that is coupled into the reactant mixture.

In addition, as has been experienced by almost every experimenter who had operated an electrical-discharge HF laser, it is difficult to prevent the laser medium from superfluorescing.[28,29] Ultee[30] has reported optical gains of 40%/cm from a 0.5-cm path length discharge cell operating with $SF_6/H_2/He$ mixtures. This phenomenon of superfluorescence in HF has been extensively studied by Goldhar et al.[31]

### 1.3. Other Initiation Techniques

There have been three other experimental techniques used for the initiation of non-chain-reaction HF lasers; they are overdriven detonation waves,[32] flashlamp detonation waves,[33] and electron-beam bombardment.[9,34,148] To date, the highest output energy reported from a non-chain-reaction HF laser has been obtained by the means of electron-beam bombardment.[148] An output energy of 2.5 kJ was obtained from a reactant mixture of $SF_6/C_3H_8$. Electron-beam initiation appears to offer many advantages over the other initiation schemes because of its fast rise time, high-energy deposition, and capability to tailor the electron energies to the reactant-gas mixtures.

## 2. CHAIN-REACTION HF LASERS

The maximum overall efficiency that can be derived from a photolysis-initiated non-chain-reaction HF laser is 10% if all the input photons are effective in dissociating $UF_6$. Under actual experimental conditions, however, the coupling between the flashlamp and fluorine-atom source is extremely weak and the actual electrical efficiency is considerably less. As previously mentioned, electrically initiated non-chain-reaction HF lasers have electrical efficiencies of 4%.

A chain reaction has an advantage over a non-chain-reaction system for high-power laser applications because the output of the laser is not directly proportional to the strength of initiation. A good example of a chain-reaction chemical laser is the first hydrogen-halide laser HCl:

$$Cl + H_2 \rightarrow HCl + H, \qquad \Delta H = 1.2 \text{ kcal/mol} \qquad (2)$$

$$H + Cl_2 \rightarrow HCl(v) + Cl, \qquad \Delta H = -45.2 \text{ kcal/mol} \qquad (3)$$

For every chlorine atom produced by photolysis, two HCl molecules are produced and the chlorine atom regenerated. Unfortunately, for the case of the HCl molecule, Reaction (2) is essentially thermoneutral; therefore, vibrationally excited HCl molecules are produced only by Reaction (3). An equivalent reaction between $H_2$ and $F_2$ is shown in Fig. 5. The situation here is somewhat different from that of the HCl system as

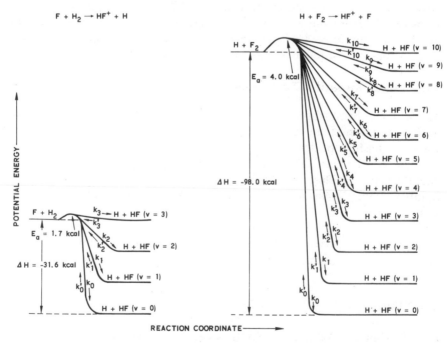

**Fig. 5.** Reaction coordinate diagrams for $F/H_2$ and $H/F_2$ reactions.

vibrationally excited molecules are produced by both reactions in the chain.

The first demonstration of laser action from an $H_2/F_2$ chain reaction was reported in 1969.[4,5] The chain reaction in a mixture of $H_2$ and $F_2$ gases was initiated by electrical discharges or by flash photolysis of the $F_2$ molecules, with laser energy extracted from the vibrationally excited $HF(v)$. At that time, the exact vibrational-level distribution of the $HF(v)$ produced by the reaction

$$H + F_2 \rightarrow HF(v) + F, \qquad \Delta H = -97.9 \text{ kcal/mol} \qquad (4)$$

was not known. However, a spectroscopic analysis of the laser output showed a large number of transitions terminating at the fourth vibrational level of the HF molecule.[5] This was thought to result from the depletion of this level by the chain-branching process

$$HF(v = 4) + F_2 \rightarrow HF(v = 0) + 2F \qquad (5)$$

A chain-branching process, as postulated here, would offer advantages in system efficiency over a straight-chain reaction because it would be necessary to produce far fewer fluorine atoms to start the chain. Later,[62,63] the rate of the chain-branching reaction (5) was measured to be a factor of $10^6$ times slower than the rate of the straight-line reaction (1).

After Batovskii et al.[4] and Basov et al.[5] reported their results, many other researchers joined the investigation of the $H_2/F_2$ chain-reaction laser.[11,13,14,64-80] The initiation techniques used were electrical discharge,[4,5,20,64,66,67,70,71,89] flash photolysis,[5,13,14,65,68,70,72-80] and laser photolysis.[69] The obtained results were almost as varied as the experimental apparatus and gas-mixing procedures that were used.

Research during the 1960's indicated that a mixture of $H_2$ and $F_2$ gases reacted, on the laboratory time scale, by a branched-chain mechanism.[61,62,81-83]

The stability of an $H_2/F_2$ mixture of fixed composition is shown in Fig. 6 in pressure and temperature coordinates. As can be seen, at a fixed temperature, a static mixture of $H_2/F_2$ diluted by an inert gas is stable below the pressure $P_1$. Between pressures $P_1$ and $P_2$, the mixture is unstable. Above pressure $P_2$, the mixture is again stable. This type of explosion-limit behavior has been shown to be indicative of a branched-chain reaction.[84]

Since $H_2/F_2$ mixtures prereact before initiation, techniques have been developed whereby the actual initial conditions of the laser reactants are determined before initiation so that measured results can be compared directly with computer-generated predictions.[85-87] Prereaction of the

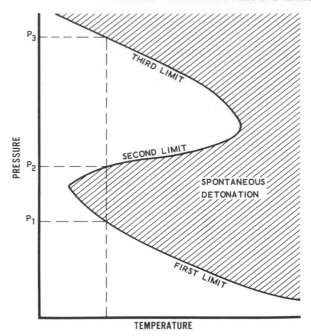

**Fig. 6.** Explosion-limit diagram for chemical system that reacts with chain-branching mechanism. $P_1$ is first-explosion limit, $P_2$ second-explosion limit, and $P_3$ third-explosion limit.

mixture can be inhibited by the addition of small quantities of oxygen.[76,87] Oxygen is believed to inhibit prereaction in this system by scavenging chemically active sites such as hydrogen or fluorine atoms. Experimental results on the effect of $O_2$ on laser performance indicate little or no degradation of laser output power for $O_2/H_2$ ratios less than 0.05.[76]

## 2.1.  Initiation Techniques

**Flash-Photolysis Initiation.**   The two basic types of initiation used for an $H_2/F_2$ laser are transverse electrical discharge and flash photolysis. These techniques are shown schematically in Figs. 1 and 3. The absorption spectra of both $H_2$[88] and $F_2$[89] have been measured and indicate that only the $F_2$ molecules can be dissociated by the flash. Fluorine has a maximum absorption of $3.5 \times 10^{-4}$ (cm-Torr)$^{-1}$ at 2845 Å[89] and consequently will be optically thin to the photolysis light at partial pressures below several hundred torr in 1-cm diameter tubes. Therefore, photolysis initiation will produce a spacially uniform reaction. Experimental techniques have been developed for measuring the fraction of fluorine photodissociated.[87,90] Such measurements indicate that for moderate photolysis energies, that is, 500 to 1000 J, between 0.75 and 1.5% of the fluorine is decomposed into fluorine atoms, which then initiate the chain reaction.

**Electrical-Discharge Initiation.** Exactly which chemically active species are produced by electrical-discharge initiation is still open to question. The possible electron–molecule interactions are

$$H_2 + e \rightarrow 2H + e \qquad \Delta H = 8.9 \text{ eV} \qquad (6a)$$

$$\rightarrow H_2^- \qquad \Delta H = 3.9 \text{ eV} \qquad (6b)$$

$$F_2 + e \rightarrow F + F^- \qquad \Delta H = -1.9 \text{ eV} \qquad (7a)$$

$$\rightarrow 2F + e \qquad \Delta H = 1.5 \text{ eV} \qquad (7b)$$

$$He + e \rightarrow He^* + e \qquad \Delta H = 19.7 \text{ eV} \qquad (8)$$

$$He^* + F_2 \rightarrow He + 2F \qquad \Delta H = -18.2 \text{ eV} \qquad (9)$$

A self-sustaining discharge has a high-energy electron tail extending from 14 to 20 eV with a maximum at 2 to 5 eV, and a non-self-sustaining discharge has a maximum at from 1 to 5 eV with no high-energy tail. Therefore, the actual species produced in the discharge is in question. An additional complication of the self-sustaining discharge initiated $H_2/F_2$ laser is that the initiation is not necessarily spatially uniform. During initiation, discrete discharges are noted between electrode pairs. Consequently, even though a large percentage of the $F_2$ molecules can be dissociated, the dissociation is not spatially uniform, and the laser results cannot be compared directly with computer-predicted results.

**Laser-Photolysis Initiation.** Another technique for initiation is to irradiate the $H_2/F_2$ mixture with a frequency-doubled ruby-laser pulse at 3470 Å.[69] The ruby-laser pulse, which is directed longitudinally along the HF laser axis, photolyzes a small percentage of the $F_2$ to initiate the reaction. Results obtained by this technique are sparse; however, their more interesting aspects will be discussed in the section on efficiently initiated lasers.

## 2.2. Time-Resolved Laser Spectral Output

The method of "arrested-relaxation" infrared chemiluminescence has been used to measure the relative vibrational-level distributions for Reactions (1)[8,91,92] and (4).[93–95] The relative populations of the HF molecules were measured[91] to be $HF(v = 1) = 0.31$, $HF(v = 2) = 1.0$, and $HF(v = 3) = 0.47$ for Reaction (1) and $HF(v = 1) = 0.12$, $HF(v = 2) = 0.13$, $HF(v = 3) = 0.25$, $HF(v = 4) = 0.35$, $HF(v = 5) = 0.78$, and $HF(v = 6) = 1.0$ for Reaction (4).[94] The populations of $HF(v = 2)$ for Reaction (1) and $HF(v = 6)$ for Reaction (4) have been arbitrarily set equal to 1.00. In addition, the overall rates, in conjunction with the relative rates for the production of HF molecules into specific vibrational levels, would lead one to believe that lasing should be expected in the $H_2/F_2$ laser system from $v = 6 \rightarrow 5$ to $v = 1 \rightarrow 0$ transitions if the reacting systems is in a high-$Q$ or low-loss optical cavity.

The first-published spectral output of an $H_2/F_2$ laser confirmed that, indeed, laser energy could be extracted from the HF vibrational levels populated exclusively by Reaction (4). The initial findings were not reproduced again, however, until recently.[76,80] During the interim, published results reported no laser action from $v = 6 \rightarrow 5$,[69] and, in several cases, no laser action from $v = 5 \rightarrow 4$.[11,78] The differences in the observed spectral content of laser pulses from different apparatus can be explained in light of the most recent time-resolved spectral measurements[80] and the chemical laser computer code RESALE.[86] Because of vibrational-cascading effects and the total and relative rates of $HF(v)$ population in Reaction (4), $v = 6 \rightarrow 5$ and $v = 5 \rightarrow 4$ transitions require a high-$Q$, low-loss optical cavity to reach threshold. An $H_2/F_2$ laser for which the $Q$ of the optical cavity can be varied will operate on only one vibrational transition for some minimum $Q$. If the $Q$ is increased, more transitions will be observed until finally all six vibrational levels will contribute to the spectral output. The vibrational-level distributions of Reactions (1) and (4) indicate that $v = 2 \rightarrow 1$ transitions are the only transitions observed in a low-$Q$ cavity, transitions from $v = 4 \rightarrow 3$ to $v = 1 \rightarrow 0$ in a moderate-$Q$ cavity, and transitions from $v = 6 \rightarrow 5$ to $v = 1 \rightarrow 0$ in a high-$Q$ cavity.[86]

With the dependence of the laser output on the condition of the optical cavity in mind, the time-resolved spectral behavior of the laser and its implications for the processes taking place in the chemical reaction are examined. The computer code RESALE[86] predicts for a pulsed $H_2/F_2$ laser that: (1) the first laser transition to reach laser threshold is $P_2(3)$, (2) the time sequencing for laser action from different rotational levels associated with a given vibrational level will increase in a linear manner, (3) only one transition from a given vibrational level can lase at any given time, and (4) and no laser action is seen for transitions of $P_v(J)$, where $J < 2$. These predictions are for a reacting system with infinitely fast rotational-equilibration rates placed inside a Fabry–Perot cavity. The gain of a V–R transition is assumed to increase with the population of vibrationally excited molecules until threshold gain is reached after which

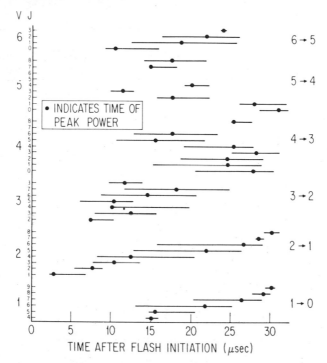

**Fig. 7.** Time-resolved spectroscopy of the observed laser transitions of a $H_2/F_2/He = 0.5/1/40$ mixture. Total pressure = 50 Torr; output coupling = 10%.

it remains constant with all the additional vibrationally excited molecules assumed to either produce laser photons or be deactivated.

Experimental results agree with the computer predictions only to a limited extent.[80] The main source of disagreement appears to be in the assumption of infinitely fast rotational equilibration times. The experimental apparatus used for these studies is shown schematically in Fig. 1. The output coupling mirror had 10% transmission, affording a high-$Q$ cavity. The time-resolved behavior of the observed transitions is shown in Fig. 7. This diagram has $v$, $J$ levels as the ordinate and time measured from the onset of the flashlamp as the abscissa. The observed duration for each transition is displayed as a horizontal bar along its upper level, with the time-of-peak intensity marked with a circle. In contrast to earlier time-resolved measurements of a flash-initiated $H_2/F_2$ laser[11,76,79] and the time-resolved spectra of HF lasers driven by Reaction (1), $P_2(2)$ is the first transition to reach threshold. Wavelengths and peak powers for the observed laser lines along with calculated wavelengths for the transitions are given in Table 2. If the pulse shapes of the individual transitions are

**Table 2**

| MEASURED WAVELENGTH[a] ($\mu$m) | IDENTIFICATION | | CALCULATED WAVELENGTH ($\mu$m) | PEAK POWER (relative units) |
| | VIBRATIONAL BAND | TRANSITION (J) | | |
|---|---|---|---|---|
| 2.6721 | 1-0 | 5 | 2.6720 | 1 |
| 2.7070 | | 6 | 2.7068 | 2190 |
| 2.7432 | | 7 | 2.7434 | 6910 |
| 2.7817 | | 8 | 2.7820 | 4390 |
| 2.8224 | | 9 | 2.8224 | 781 |
| 2.8650 | | 10 | 2.8650 | 89 |
| 2.6955 | 2-1 | 2 | 2.6956 | 872 |
| 2.7267 | | 3 | 2.7267 | 1405 |
| 2.7599 | | 4 | 2.7597 | 2985 |
| 2.7946 | | 5 | 2.7945 | 3850 |
| 2.8313 | | 6 | 2.8311 | 6100 |
| 2.8696 | | 7 | 2.8697 | 4340 |
| 2.9100 | | 8 | 2.9104 | 536 |
| 2.9529 | | 9 | 2.9532 | 449 |
| 2.8531 | 3-2 | 3 | 2.8534 | 622 |
| 2.8881 | | 4 | 2.8881 | 1195 |
| 2.9249 | | 5 | 2.9248 | 1650 |
| 2.9635 | | 6 | 2.9636 | 496 |
| 3.0044 | | 7 | 3.0044 | 1785 |
| 3.0473 | | 8 | 3.0473 | 3750 |
| 3.0926 | | 9 | 3.0926 | 573 |
| 2.9216 | 4-3 | 1 | 2.9212 | 2920 |
| 2.9542 | | 2 | 2.9539 | 3710 |
| 2.9890 | | 3 | 2.9887 | 5980 |
| 3.0257 | | 4 | 3.0254 | 3125 |
| 3.0644 | | 5 | 3.0642 | 2225 |
| 3.1053 | | 6 | 3.1052 | 684 |
| 3.1483 | | 7 | 3.1484 | 1024 |
| 3.2416 | | 9 | 3.2419 | 295 |
| 3.0628 | 5-4 | 1 | 3.0625 | 1120 |
| 3.0973 | | 2 | 3.0972 | 1725 |
| 3.1342 | | 3 | 3.1339 | 500 |
| 3.1733 | | 4 | 3.1728 | 148 |
| 3.2142 | | 5 | 3.2139 | 177 |
| 3.3520 | | 8 | 3.3516 | 377 |
| 3.4031 | | 9 | 3.4026 | 527 |
| 3.2155 | 6-5 | 1 | 3.2151 | 467 |
| 3.2523 | | 2 | 3.2518 | 995 |
| 3.2913 | | 3 | 3.2908 | 458 |
| 3.3325 | | 4 | 3.3320 | 584 |

[a] Estimated accuracy  0.5 cm$^{-1}$
[b] Reference 80

assumed to be similar, the product of the peak-power times the pulse duration can be used as a measure of the pulse energy of individual laser lines. After summing over all rotational lines for each vibrational level, the results show $E(v = 6 \rightarrow 5) = 0.10$, $E(v = 5 \rightarrow 4) = 0.11$, $E(v = 4 \rightarrow 3) = 0.93$, $E(v = 3 \rightarrow 2) = 0.47$, $E(v = 2 \rightarrow 1) = 1.00$, and $E(v = 1 \rightarrow 0) = 0.60$. The energy of the $v = 2 \rightarrow 1$ band was normalized to 1.00.

A set of oscilloscope traces for flashlamp output, total-laser output, and laser output of various lines of the $V = 6 \rightarrow 5$ band is shown in Fig. 8. The observed laser pulse has a duration of 32 $\mu$sec, which is comparable to that reported previously for diluted $H_2/F_2$[11,77] and considerably longer than the results for undiluted $H_2/F_2$.[4,5] The variation in laser-pulse length

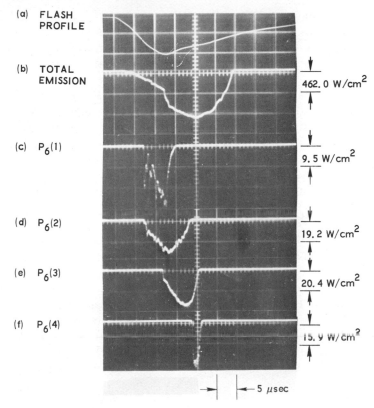

(a) FLASH PROFILE

(b) TOTAL EMISSION — 462. 0 W/cm²

(c) $P_6(1)$ — 9. 5 W/cm²

(d) $P_6(2)$ — 19. 2 W/cm²

(e) $P_6(3)$ — 20. 4 W/cm²

(f) $P_6(4)$ — 15. 9 W/cm²

5 μsec

**Fig. 8.** Selected HF transitions from flash photolysis of $H_2/F_2/He = 0.5/1/40$ mixture. Total sample pressure = 50 Torr; flash energy = 1000 J; output coupling = 10%. (a) Flashlamp profile, (b) total HF laser emission, and (c) through (f) individual HF laser transitions from sixth to fifth vibrational levels. (Adapted from Suchard.[80])

in the different experiments initiated by flash photolysis and electrical discharges is to be expected because of the temperature control afforded by dilution and by the duration of the initiation pulse. The time history of the transitions in the $v = 6 \rightarrow 5$ band, which is pumped exclusively by Reaction (4), indicates partial rotational equilibration. Complete rotational equilibration dictates that only one rotational transition at a time can lase per vibrational band. As the chemical reaction progresses, the laser transitions from a specific vibrational band display rotational equilibrium in that the time of peak power for the specific V–R transition increases in a linear manner as expected. For example, in the $v = 6 \rightarrow 5$ band, typical times at which the transitions reach peak power are $P_6(1)$ at 12.2 μsec, $P_6(2)$ at 16.3 μsec, $P_6(3)$ at 19.5 μsec, and $P_6(4)$ at 21.2 μsec.

The lasing system cannot be said to be in complete rotational equilibrium because the transition of lower $J$ does not terminate when the next transition in that band reaches laser threshold. Transitions in the $v = 5 \rightarrow 4$, $v = 4 \rightarrow 3$, and $v = 3 \rightarrow 2$ bands do not appear to be rotationally equilibrated.

One unusual feature of the data is the lasing on $P_v(1)$ transitions such as $P_4(1)$, $P_5(1)$, and $P_6(1)$. Observations of Kompa et al.[97] for the $WF_6/H_2$ and $WF_6/CH_4$ systems demonstrate the presence of $P_v(1)$ transitions for an HF laser driven by Reaction (1). These observations cannot, however, be explained in terms of the Einstein coefficient of spontaneous emission or the initial rotational distribution of the vibrationally excited HF molecules produced by Reaction (4), as was thought previously.[11]

### 2.3. Experimental–Theoretical Comparison

Several experimental–theoretical comparisons have been reported for the chain reaction HF laser;[13,14,73,77] however, for a complete description of a parameter study of a pulsed $H_2/F_2$ laser, the work of Suchard et al.[77] should be examined. They compare computer predictions and experimental results for the $H_2/F_2$ laser for various $H_2$ pressures. The experimental results compared include the time from flashlamp initiation to laser threshold, total-laser pulse duration and shape, and peak-output intensity. In addition to the $H_2$ variation the dilution ($F_2/He$ ratio) and output coupling was also varied. Their results are given in Figs. 9 through 12.

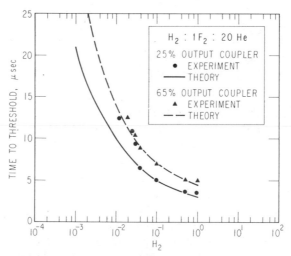

**Fig. 9.** Observed and predicted time to achieve laser threshold as function of $H_2$ for initial composition ratio $H_2/F_2/He = H_2/1/20$ and for 25% and 65% transmitting output mirrors. (Adapted from Suchard et al.[76])

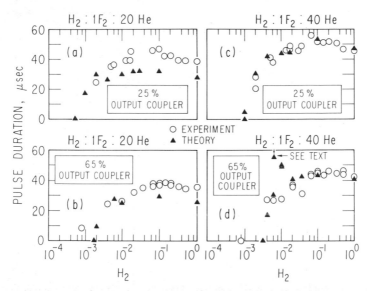

**Fig. 10.** Observed and predicted pulse duration versus parts $H_2$. (Adapted from Suchard et al.[76])

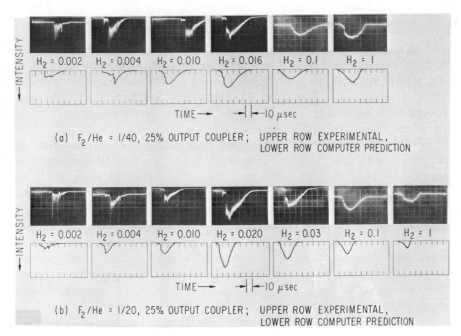

(a) $F_2/He = 1/40$, 25% OUTPUT COUPLER; UPPER ROW EXPERIMENTAL, LOWER ROW COMPUTER PREDICTION

(b) $F_2/He = 1/20$, 25% OUTPUT COUPLER; UPPER ROW EXPERIMENTAL, LOWER ROW COMPUTER PREDICTION

**Fig. 11.** Observed and predicted effect of $H_2$ on intensity and pulse shape for $H_2/F_2$ chain-reaction laser. (Adapted from Suchard et al.[76])

407

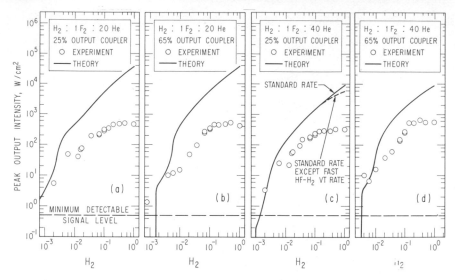

**Fig. 12.** Observed and predicted intensity versus $H_2$. Points are experimental data. Solid curves are predictions based on standard rate coefficient model. Dashed curve in Fig. 10c is peak intensity prediction with an alternate-rate model assumed. (Adapted from Suchard et al.[76])

Observed and predicted values of time to threshold for $F_2/He = 1/20$ and for two different output couplers are in good agreement as seen in Fig. 9. As expected, the time required to reach threshold increases as the initial $H_2$ decreases. For small $H_2$, threshold is not achieved. Similar results have been found for $MoF_6/H_2$ mixtures over a range of mixture pressures.[13]

In Fig. 10, observed and predicted pulse durations are compared as functions of $H_2$ for the four cases studied, and the overall agreement is good. For $H_2/F_2He = H_2/1/40$ and the 25% output mirror, there is excellent agreement. The main trends are the same for the other cases, but there are some deviations between theory and experiment. For example, the high theoretical point in Fig. 10d was calculated for a condition that experimentally produced a double-spiked laser pulse. The small second spike occurred at a time at which the calculated gain on another vibrational transition achieved threshold. The pulse duration of the first spike at this same concentration is in good agreement with experiment.

An upper limit on the laser-pulse duration is imposed by the time required to consume the $H_2$, that is, the reaction duration. The reaction duration is nearly constant for small $H_2$ where the heat release is negligible. For $H_2$ concentrations above $H_2/F_2 = 0.12$, however, the reaction duration decreases with increased $H_2$ because of thermal acceleration of the pumping rates. Experimental and calculated values in all four cases of Fig. 10 show this decreasing trend. The laser pulse is actually shorter than the reaction duration. A finite-time interval of pumping occurs before lasing begins, and deactivation processes quench the laser before the reaction is complete.

Experimental intensity versus time records are found to exhibit three types of pulse shape. Pulses for very low $H_2$ pressures have peak intensity at the beginning and then an irregular decay to zero. Additionally, these records have pronounced transients superposed on the basic pulse shape. Pulses for experiments with somewhat larger $H_2$ pressure rise from threshold-peak intensity in several distinct steps. After peak intensity, there is a smooth decay to zero. Additional transients are less pronounced in these records. Finally, for $H_2$ pressures near stoichiometric, the pulses have small initial steps and a smooth rise to peak intensity before decaying to zero. Records of the observed pulse shapes with the 25% output coupler for $F_2/He = 1/20$ and 1/40 are given in Fig. 11. Also shown in this figure are predicted intensity-versus-time pulse shapes. In order to facilitate comparison, the intensity scale for each computed pulse is selected so that the peak intensity is approximately the same fraction of the ordinate as the observed pulse; the time scales are identical. Except for the transients in the experiment, the main features of these pulse shapes are predicted quite well by the theoretical model. The time location of the observed laser pulses is somewhat arbitrary because flashlamp triggering, in some cases, was delayed relative to oscilloscope triggering.

In Fig. 12, the theoretical curves and experimental data for peak intensity versus $H_2$ pressure are given for the four cases studied. These plots show that below one-tenth part $H_2$, theory and experiment have the same trends. Above this $H_2$ pressure, the agreement vanishes; theoretical peak intensity continues to increase with $H_2$ pressure, and the experiments exhibit an abrupt leveling off. A similar, but more severe, experimental trend is reported by Hess[73] for cases with a composition ratio of $H_2/F_2/He = H_2/1/40$, pressures near 160 Torr, and a 99-cm laser tube. His data show that, when the $H_2/F_2$ ratio is increased from 0.33 to 1.0, the observed peak power decreases by nearly a factor of 3.

In an attempt to explain the discrepancies in the peak-output powers, the rate for V–T deactivation of HF by $H_2$ was varied over a large range of values. In place of the rate coefficient based on experimental data,[98] Suchard et al. hypothesized a "largest plausible" rate, which is used in Fig. 12c. This hypothetical fast deactivation causes a moderate reduction in the predicted peak intensities for the larger values of $H_2$. The fast deactivation effect is not large enough to account for the experimental results. Moreover, it occurs gradually; whereas, the experiments show an abrupt leveling off of the peak intensity. It was known that reasonable variations of the pumping reaction rates, the pumping reaction distribution, or in the HF–HF and HF–$H_2$ V–V rates were incapable of predicting the behavior indicated by the experimental results in Fig. 12.[99] The effect of variations in the HF–F and HF–H V–T rate coefficients was also examined. Neither multiquantum deactivation by these atoms nor increases in their efficiency by a factor of 10 would account for the observed behavior. Explanations for the sharp leveling off of peak intensity may also be sought in the area of undesired optical phenomena. The theoretical predictions are based on some assumptions that are open to question. Spatial nonuniformity of the reaction and concomitant gradients of refractive index may increase cavity losses as well as steer the beam away from the detector. A host of parasitic lasing modes is also conceivable because of the high gain of the medium and because the reaction occurs within a tube whose wall and windows are not perfectly transmitting. Because parasitic oscillations are known to occur in solid-state lasers under high-gain conditions,[100–102] this optical phenomenon was given primary attention.

Parasitic modes are those that reflect off the tube's wall and are regenerative. The discussion was limited to two possible types: circumferential or whispering modes and axial modes. Nonparasitic modes were referred to as fundamental (either longitudinal or transverse) modes. The likelihood of parasitic modes is shown in Fig. 13 where the largest small-signal gain of the medium is compared to the gains required for threshold for several modes of oscillation in the apparatus. Threshold gains for wall-grazing modes with either near-axial or circumferential paths were obtained by use of expressions including Fresnel-law estimates of losses suffered in near-grazing reflections at the laser tube wall (Fig. 13). In making these threshold estimates, gain saturation caused by competing modes was neglected. Also, mode-coupling effects that could cause a lowering of the threshold gain for certain parasitic modes were not accounted for.[103,104] Figure 13 shows that variation of the $H_2$ pressure serves as a convenient way of varying the peak small-signal gain over several orders of magnitude. As expected, peak gains are larger for the less-dilute mixture. As $H_2$ increases, the small-signal gain for either mixture becomes nearly as large as the threshold for circumferential modes and much larger than the threshold values for either fundamental or near-axial grazing modes. The fundamental modes of lowest loss saturate the gain down the center of the laser tube but not near its walls.

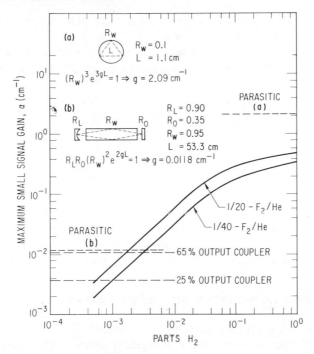

**Fig. 13.** Predicted values of maximum small-signal gain versus $H_2$ for mixtures with initial composition ratio $H_2/F_2/He = H_2/1/20$ and $H_2/1/40$. Also shown are ray paths and threshold gain estimates for some sample parasitic oscillation modes involving grazing reflections at tube wall. (a) Axial mode and (b) circumferential mode. (Adapted from Suchard et al.[76])

In the outer annulus of high gain, the fundamental modes with greater loss, that is, high-order transverse modes, compete with near-axial grazing modes, either of which can degrade the intensity along the laser axis. Grazing modes, once oscillating, are particularly detrimental because they extract energy when passing near the tube center that might otherwise be emitted in a longitudinal mode that contributes to the measured intensity.[100,101]

Measurements of the output intensity of flash photolysis $UF_6/H_2$ and $MoF_6/H_2$ lasers as a function of the total sample pressure by Dolgov-Savel'yev et al.[69] have shown a similar behavior. As the sample pressure was increased, the laser output intensity for the $MoF_6/H_2$ mixture was found first to increase with pressure and then to decrease with further increasing sample pressure. Similar results were found in the $UF_6/H_2$ system. These observations can also be explained in terms of parasitic oscillations because an increase in sample pressure corresponds to an increase in the signal gain of the system if it is assumed that their samples are optically thin to the flashlamp radiation and that Doppler broadening was controlling the gain. Once the system gain has reached a critical value, a further increase in sample pressure will lead to the onset of parasitic oscillations and a decrease in the fundamental-mode laser intensity.

At the present time the state of the art for the HF laser can be summarized as follows: An electrical efficiency of 160% has been obtained in converting electrical energy to laser-output energy (electrical initiation),[79] and an output energy of 2500 J has been demonstrated (electron-beam initiation).[148] The energy density in the electron-beam initiated experiments is not as yet as high as the 80 J/liter reported for photolysis-initiated laser;[77] however, these experiments offer the hope of coupling the efficient-energy conversion of discharge initiation with the large-energy density produced by photolysis initiation. If these experiments prove successful for large-volume initiation, the HF laser will become quite competitive with pulsed $CO_2$ and CO lasers.

## 3. CHAIN AND NON-CHAIN-REACTION DF LASERS

The amount of information that has been published concerning the lasers operating on the DF molecule is extremely sparse. The chemistry and kinetics of the DF molecule would be expected to be quite similar to those of the HF molecule except for the mass effect of the deuterium atom. However, considerable effort has been expended in the measurement of the V–V and V–R–T processes involving DF molecules, and the molecule does not appear to behave as expected.[105–113] Since DF has atmospheric transmission properties superior to those of HF, further study is underway. To date, the primary use of the vibrationally excited DF molecules has been for V–V pumping of $CO_2$ to its laser-active asymmetric stretching mode.[70,113–117]

The first observation of laser action arising from vibrationally excited DF molecules was reported by Deutsch[3] in 1967. After this first demonstration of electrical discharge initiation of the reaction

$$F + D_2 \rightarrow DF(v \leq 4) + D, \qquad \Delta H = -30.63 \text{ kcal/mol} \qquad (10)$$

results were reported of flash-photolysis initiation[2,10,17–19,37,47,118,119] and electrical initiation[30,35,36,59,120] of the same reaction. Experimental results have also been presented for a flash-photolysis initiated $D_2$–$F_2$ chain-reaction laser[70] using Reaction (10) and

$$D + F_2 \rightarrow DF(v \leq 10) + F, \qquad \Delta H = -99.33 \text{ kcal/mol} \qquad (11)$$

Table 3 includes the spectral data obtained from Ref. 70.

**Table 3**

| $\nu$OBS (cm-1) | $\nu$CALC (cm-1) | IDENTIFICATION | | RELATIVE ENERGY OF LINES |
|---|---|---|---|---|
| | | BAND | TRANSITION | |
| 2665.2 | 2665.20 | 1-0 | P(10) | 1 |
| 2727.3 | 2727.28 | 2-1 | P(4) | 0.04 |
| 2704.0 | 2703.97 | | P(5) | 0.26 |
| 2680.1 | 2680.14 | | P(6) | 0.94 |
| 2655.8 | 2655.82 | | P(7) | 1.36 |
| 2631.0 | 2631.02 | | P(8) | 2.37 |
| 2605.8 | 2605.76 | | P(9) | 2.68 |
| 2580.0 | 2580.04 | | P(10) | 4.24 |
| 2553.9 | 2553.90 | | P(11) | 3.46 |
| 2617.3 | 2617.32 | 3-2 | P(5) | 0.41 |
| 2594.1 | 2594.10 | | P(6) | 1.14 |
| 2570.4 | 2570.45 | | P(7) | 1.79 |
| 2546.3 | 2546.30 | | P(8) | 2.74 |
| 2521.7 | 2521.70 | | P(9) | 5.38 |
| 2496.6 | 2496.65 | | P(10) | 3.35 |
| 2471.2 | 2471.18 | | P(11) | 2.26 |
| 2445.3 | 2445.30 | | P(12) | 0.29 |
| 2419.0 | 2419.02 | | P(13)a | |
| 2392.4 | 2392.36 | | P(14)a | |
| 2509.8 | 2509.82 | 4-3 | P(6) | 0.38 |
| 2486.7 | 2486.77 | | P(7) | 0.91 |
| 2463.2 | 2463.25 | | P(8) | 1.08 |
| 2439.3 | 2439.29 | | P(9) | 1.78 |
| 2414.9 | 2414.89 | | P(10) | 0.65 |
| 2390.1 | 2390.07 | | P(11) | 0.19 |
| 2404.6 | 2404.63 | 5-4 | P(7) | 0.016 |
| 2381.7 | 2381.73 | | P(8) | 0.065 |
| 2334.6 | 2334.63 | | P(10)a | |
| 2310.4 | 2310.45 | | P(11)a | |
| 2285.9 | 2285.88 | | P(12)a | |
| 2260.9 | 2260.92 | | P(13)a | |
| 2388.0 | 2388.02 | 6-5 | P(4) | 0.003 |
| 2323.9 | 2323.89 | | P(7) | 0.045 |
| 2301.6 | 2301.60 | | P(8) | 0.09 |
| 2278.9 | 2278.87 | | P(9) | 0.19 |
| 2255.7 | 2255.71 | | P(10) | 0.07 |
| 2232.1 | 2232.15 | | P(11) | |
| 2286.5 | 2286.45 | 7-6 | P(5) | 0.016 |
| 2265.6 | 2265.65 | | P(6) | 0.11 |
| 2244.4 | 2244.38 | | P(7) | 0.06 |
| 2222.7 | 2222.68 | | P(8) | 0.19 |
| 2200.5 | 2200.54 | | P(9)a | |
| 2178.0 | 2177.99 | | P(10)a | |
| 2155.0 | 2155.03 | | P(11)a | |
| 2131.7 | 2131.68 | | P(12)a | |
| 2206.9 | 2206.87 | 8-7 | P(5) | 0.01 |
| 2186.6 | 2186.63 | | P(6) | 0.08 |
| 2165.9 | 2165.93 | | P(7) | 0.18 |
| 2144.8 | 2144.80 | | P(8) | 0.11 |
| 2123.2 | 2123.24 | | P(9) | 0.03 |
| 2101.3 | 2101.27 | | P(10)a | |
| 2056.1 | 2056.14 | | P(12)a | |
| 2033.0 | 2033.01 | | P(13)a | |
| 2108.5 | 2108.48 | 9-8 | P(6) | 0.01 |
| 2088.3 | 2088.34 | | P(7) | 0.06 |
| 2067.8 | 2067.76 | | P(8) | 0.08 |
| 2046.8 | 2046.77 | | P(9) | 0.0008 |
| 2025.4 | 2025.36 | | P(10)a | |
| 2003.6 | 2003.56 | | P(11)a | |
| 1981.4 | 1981.38 | | P(12)a | |

a $E_{discharge}$ = 1.75 J

b Reference 70

The overall rate of Reaction (10) has been calculated[121-123] and measured.[8,17,19,47,124,125] The overall rate and vibrational-level distribution have been calculated for Reaction (11).[126] These results indicate that the rates for Reactions (10) and (11) are very similar to those of Reactions (1) and (4). Experiments measuring the vibrational-level distribution of the DF molecules produced by Reaction (10) have employed infrared

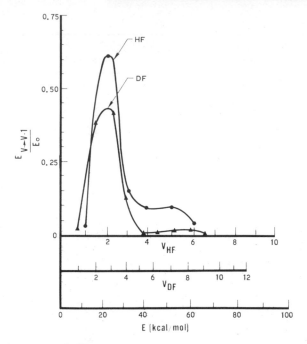

**Fig. 14.** Relative energy of different bands of HF and DF for chemical lasers with mixtures of $F_2/H_2(D_2)$ versus quantum number $v$ of upper vibrational level. (Adapted from Basov et al.[69])

chemiluminescence,[8,125] chemical lasers,[17,19,47] and molecular beams.[124] Both infrared chemiluminescence and chemical-lasers results indicate that $v = 3$ is the level produced with the maximum population. Calculations indicate that the DF molecules produced by Reaction (11) have a peak in their population at $v = 8$.[126] The DF lasers differ from HF lasers in that the energy deposited in the DF product is distributed over more vibrational levels than in the HF.

The only reported results for a $D_2/F_2$ chain-reaction laser are those of Basov et al.[70] Figure 14 shows the relative laser energy emitted from their laser for the different bands of the HF and DF chain-reaction lasers, and Table 3 presents the wavelengths of the observed DF laser lines. Possible explanations for the decreased energy emitted from the DF as opposed to HF include the slower pumping rates for DF and the large number of vibrational levels over which the reaction exothermicity is distributed.

A unique result that has been reported for the DF laser system but not for the HF-laser system is the observation of first-overtone laser emission by Suchard and Pimentel.[47] Using an intracavity filter to lower the gain of the fundamental emission, they observed laser emission on the $P_{3-1}(4)$, $P_{3-1}(5)$, and $P_{3-1}(6)$ transitions of the DF molecule. This is the only reported observation of overtone-laser emission from a hydrogen-halide laser.

## 4. EFFICIENTLY INITIATED HYDROGEN-HALIDE LASERS

The pulsed chemical laser is the only pulsed-laser system from which one could hope to be able to extract more energy from the laser than is necessary to initiate the laser. For example, solid-state lasers have been shown to produce extracted laser energy that is $\leq 5\%$ of the initiation energy; gas lasers have an efficiency of as high as 25% in $CO_2$ with the possibility of 80% in $Xe_2$.[127] Chemical lasers have been operated with 160% efficiency. This does not violate the law of conservation of energy. For all nonchemical (nonhybrid) systems, the initiation energy is used to prepare the laser inversion. If the initiation source does not have unity coupling to the laser medium and if there are any imperfections in the optical apparatus, then the magnitude of the extracted laser energy must necessarily be less than the magnitude of the initiation energy. These conditions can only be met for a two-level system that absorbs the initiation energy with unity efficiency, has no deactivation processes occurring, and is placed in a no-loss optical cavity. This situation cannot be met in practice; consequently, the overall system efficiency is $<100\%$. Nonhybrid systems, in general, have three or more levels and have nonradiative deactivation processes occurring; therefore, their overall efficiencies are $\ll 100\%$.

Pulsed chain-reaction chemical lasers (hybrid systems) offer a way around the overall efficiency problems. The initiation energy is used to produce chemically active sites. In the case of a flash-photolysis $H_2/F_2$ laser, these correspond to fluorine atoms. The sites react with additional reagents ($H_2$) to produce laser-active vibrationally excited molecules (HF) and additional chemically active sites (hydrogen atoms). The chemicals continue to react, producing laser-active molecules, until all the reactants are consumed. The consequence of the chemical reaction is that the energy that can be extracted from this type of laser system is not directly related to the initiation energy and can, in principle, be considerably larger than the initiation energy. This situation has, in fact, been experimentally demonstrated in the chain-reaction $H_2/F_2$ laser system.[79,128]

The first indication that the $H_2/F_2$ laser could be made electrically efficient can be seen in the work of Basov et al.[5] and Dolgov-Savel'yev et al.[69] Basov's work demonstrated that the $H_2/F_2$ reaction was indeed a chain reaction with both cycles of the chain having comparable rates.

Dolgov-Savel'yev's work showed that by the use of a monochromatic initiation source (a frequency-doubled ruby laser) the actual magnitude of the initiation energy absorbed by the sample could be determined and directly compared to the laser output. Their results indicate an efficiency of $\approx 200\%$.

Flash-photolysis initiation of a $H_2/F_2$ laser has failed to produce electrical efficiencies greater than 4%.[77] However, in flash photolysis, electrical energy is first transformed into blackbody radiation that then irradiates the reactants so that losses in the flashlamp make the overall efficiencies low. Only a small portion of the flashlamp light is produced in the wavelength region in which it can be absorbed by the fluorine molecule.[89] Of the energy produced in the proper wavelength region, only a small portion is actually absorbed by the reactants. The efficiency of such a laser can be greater than 100% if it is calculated on the basis of input energy actually absorbed by the reactants and not the electrical input energy. More efficient flashlamps would certainly be valuable for pulsed chemical lasers.

By the proper tuning of the electron energy distribution in an electrical discharge, the coupling of the electrons to the reactants to produce chemically active sites can be efficient.[129] With efficient initiation coupling, a chain-reaction laser system can produce overall efficiencies $>100\%$.[20,128] The efficiency results of Kerber et al.[20] for the $H_2/F_2$ chain reaction laser as a function of the initial reactant pressure are given in Fig. 15. As the pressure is increased, the breakdown voltage of the

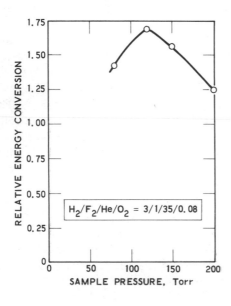

$H_2/F_2/He/O_2 = 3/1/35/0.08$

**Fig. 15.** Relative energy conversion for discharge initiated $H_2/F_2$ laser versus total same pressure for $H_2/F_2/He/O_2 = 3/1/35/0.08$. (Adapted from Kerber et al.[20])

reactant gases increases and the electron energy distribution is raised; as the pressure is lowered, the opposite effect occurs.

A possible solution to the problem of providing efficient initiation for arbitrary reactant pressures is the use of electron-beam initiation. This type of initiation can be used for either non-self-sustaining discharge or direct electron injection. In either case, the discharge parameters are somewhat independent of the composition and pressure of the reactants, and, hence, these systems should be scalable to arbitrary pressures and compositions.

## 5. PULSED HCl LASERS

The pulsed HCl chemical laser is especially important in the development of chemical lasers. It was the first chemical laser to be experimentally demonstrated.[1] A pulsed HCl laser was also the first chemical laser to be operated on a $v(1 \rightarrow 0)$ transition and was the first system for which detailed kinetic models were constructed.[131–133] Although most of the effort has been concerned with HF and DF systems because of their high efficiency, much of our understanding of the phenomenology of chemical lasers originated in studies of HCl systems. Pulsed HCl lasers have been initiated both photolytically and by a pulsed electrical discharge in the reactants. Two distinct types of pulsed HCl lasers exist: Those that have metathetical reaction for the primary inversion mechanism, for example, $H + Cl_2 \rightarrow HCl^\dagger + Cl$; and those that are pumped by photoelimination reactions, for example, $CH_2 = CHCl$ (chloroethylene) $+ h\nu \rightarrow HCl + C_2H_2$.

### 5.1. Reactively Pumped HCl Lasers

**Photolytically Initiated Systems.**   Several reactions have been used as the basis of an HCl laser in addition to the classical reaction

$$H + Cl_2 \rightarrow HCl + Cl, \qquad \Delta H = -45.2 \text{ kcal/mol} \qquad (3)$$

that was used by Kasper and Pimentel.[1] An interesting study of the competitive rates of formation of HCl and DCl in the chain-reaction sequence

$$HD + Cl \rightarrow HCl(DCl) + D(H) \qquad (12)$$

$$H(D) + Cl_2 \rightarrow HCl^\dagger(DCl^\dagger) + Cl \qquad (13)$$

was carried out by Corneil and Pimentel.[134] The halogen-exchange reactions

$$Cl + HI \rightarrow HCl + 1, \qquad \Delta H = -32 \text{ kcal/mol} \qquad (14)$$

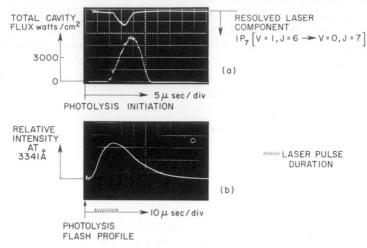

TOTAL CAVITY FLUX watts /cm$^2$

3000

0

5 $\mu$ sec / div

PHOTOLYSIS INITIATION

RESOLVED LASER COMPONENT
$IP_7 \left[ V = 1, J = 6 \rightarrow V = 0, J = 7 \right]$

(a)

RELATIVE INTENSITY AT 3341Å

(b)

⫘⫘⫘ LASER PULSE DURATION

10 $\mu$ sec / div

PHOTOLYSIS FLASH PROFILE

**Fig. 16.** Pulsed HCl laser oscillation from the Cl/HBr reaction. (Adapted from Airey.[131])

and

$$Cl + HBr \rightarrow HCl + Br, \qquad \Delta H = -16 \, kcal/mol \qquad (15)$$

have produced the most efficient HCl lasers. The Cl/HBr laser has been the subject of a detailed analysis[131] in which theory and experiment were compared. Typically, it is easy to produce laser-pulse energies of order 0.1 J/liter of reactive volume at a working pressure of order 10 Torr. The laser pulsewidths are normally in the range of 10 to 30 $\mu$sec at this pressure. Figure 16 is an oscillogram record of the output from a Cl/HBr produced in an apparatus similar to that schematically shown in Fig. 1. Active $Q$-switching experiments have been conducted on the Cl + HBr chemical laser.[135] It was found that very little inversion energy could be stored in the system because of the fast energy-transfer reaction

$$HCl(v = 1) + HBr \rightarrow HCl(v = 0) + HBr \qquad (16)$$

Figure 17 shows the $Q$-switch pulse for a low-loss system (rotating gold-surfaced mirror). The upper trace is an amplification of the lower by a factor of 200. It can be seen that although there are pulses of radiation extending from 8 $\mu$sec, the energy is concentrated in a pulse approximately 0.8 $\mu$sec wide. This pulse corresponds to that point in time when the gain in the medium was at its maximum value. The extent of the wings of pulses could be limited by aperturing the cavity to prevent high-order modes from oscillating.

The Cl + HBr laser oscillated only on the $v(1 \rightarrow 0)$ transition, and the

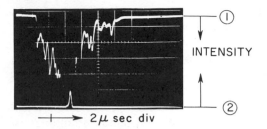

**Fig. 17.** Cl/HBr $Q$-switched laser pulse at different amplifications. Reactants pressures of 6.5 Torr $Cl_2$ and 1.8 Torr HBr. Sensitivity of detector 1 is 200 times that of detector 2. (Adapted from Airey.[131])

number of rotational lines observed ranged from $P_1(4)$ to $P_1(14)$, depending on the loss of the cavity employed. The Cl+HI laser, however, operates on several V–R bands because of the larger exothermicity of the pumping reaction. This laser system has not been studied in as great detail although Polanyi and co-workers[136] have determined accurately the relative rates of production of $HCl(v = 1, 2, 3, 4)$ using the now well-developed techniques for infrared chemiluminescence.

Other reactions that have been employed to pump HCl lasers in flash-photolysis systems are H/NOCl[137] and reactions involving ozone and chlorinated methanes.[138]

**Electrically Initiated Systems.** Deutsch[3,6] was the first to initiate chemical-laser action with a pulsed electrical discharge. Unlike photolysis, dissociation by electron impact may be unspecific and, therefore, such systems have been, in general, harder to quantify. There is also the possibility of direct pumping of the product molecule by electron impact. However, they tend to be easier to operate and have a higher overall electrical efficiency than flash-photolysis systems.

Shortly after the $H_2/Cl_2$ system was operated, Moore[139] initiated the $Cl_2/HI$ system with an electrical discharge. However, for the same reactant pressures, the laser power was significantly less than the photolysis system and oscillated only on the $v = 1-0$ bands.

Electrical initiation by means of transverse excitation, that is, the TEA configuration, has been shown to be quite efficient by a number of research groups.[59,140] Strong HCl and DCl laser emission was observed by Wood and Chang[59] on both chlorine isotopes when reaction was initiated in mixtures of $H_2(D_2)$, $Cl_2$, and inert gas. Typically, pulse lengths were in the 0.3 to 15 $\mu$sec range, and peak-power outputs were in the range of 2 to 50 kW. The HCl emission was also observed when the chlorine was replaced with ICl.

Pure rotational-laser emission on HCl has been observed[141] in discharges in mixtures of water vapor and $BCl_3$.

## 5.2. HCl Photoelimination Lasers

Pulsed HCl laser emission following ultraviolet photolysis of chloroethylene was first observed by Berry and Pimentel.[142] Typically, laser emission is observed on several bands simultaneously. Although this type of laser does not seem to have potential for the efficiency and high-power aspects of laser technology, the information gathered from their study could greatly influence our knowledge of unimolecular chemical kinetics. Berry and Pimentel[142] has shown that the extent of HCl excitation produced differs significantly among the isomers on lasers produced from three isomeric dichloroethylenes. Unlike the exchange reactions discussed in Section 5.1, the pumping rates into the vibrational levels of HCl decrease monotonically with increasing vibrational quantum number.

In summary, pulsed HCl lasers have been instrumental in the development of our understanding of chemical lasers. They do not, however, hold the same promise of high-peak powers and pulse energies that the HF systems presently generate.

## 6. PULSED HBr CHEMICAL LASERS

The HBr chemical lasers are the least well developed of all the hydrogen-halide systems. This, in part, is the result of the lower bond strength of HBr that reduces the exothermicity of many formation reactions below a useful value. Whereas, the reaction $Cl/H_2 \rightarrow HCl/H$ is almost thermoneutral and, therefore, sufficiently rapid to initiate the second step of the chain, the reaction $Br + H_2 \rightarrow HBr + H$ is considerably endothermic and slow. Consequently, most lasers in HBr chemical systems have been initiated with a discharge in which hydrogen atoms are formed by dissociative electron impact.

One reaction that has been employed in flash photolysis induced systems is

$$Br + HI \rightarrow HBr + I, \qquad \Delta H = -15.9 \text{ kcal/mol} \qquad (17)$$

The exothermicity is sufficient to populate the second vibrational level of HBr. Lasing transitions that have been observed[140] are $P_2(5)-P_2(7)$ and $P_1(6)-P_1(9)$, which range in wavelength from 4.078 to 4.263 $\mu$m. A peak-output power of 1.8 kW with a pulse energy of the order of 2 mJ was recorded from flash photolysis of a mixture of $Br_2/HI = 3/1$ at a pressure of 4 Torr in a reactive volume of the order of $100 cm^3$.

Pulsed-discharge initiation of HBr lasers was first observed by Deutsch.[3,6] Subsequently,[58,140] efficient operation by means of transverse-electrical excitation in mixtures of $H_2(D_2)/Br_2$ was attained. Burak et al.[140] recorded emission on the $v(3 \to 2)$, $v(2 \to 1)$, and $v(1 \to 0)$ vibrational bands at kilowatt-power levels from mixtures of 9 Torr $Br_2$ with 108 Torr $H_2$. They, interestingly, observed some degree of mode locking on several transitions. It was surmised that the mode locking was induced by the spatially nonuniform decomposition of the reactants across the pin discharge. In both the flash-photolysis and electrical-discharge induced HBr lasers, it was observed that the $v(2 \to 1)$ transitions preceded the $v(1 \to 0)$ transitions.

In summary, the number of exothermic reactions suitable for pumping chemical HBr lasers is much less than those for HCl or HF. It is unlikely that HBr lasers will ever be scaled to higher powers similar to those observed for HF.

# 7. PULSED CHEMICAL-TRANSFER LASERS

The operation of all chemical-transfer lasers, to date, is based on the selective production of vibrationally excited species in a chemical reaction followed by the fast transfer of energy to a suitable laser species. Several schemes have been used in which the donor–acceptor combinations are: $HCl-CO_2$, $DF(HF)-CO_2$, $HBr-CO_2$, $OD(OH)-CO_2$, and $N_2-CO_2$. In most systems $N_2O$ can replace $CO_2$, but the resulting laser power is lower. By far the most interesting and well-studied system is the $DF-CO_2$ combination in which energy is transferred from vibrationally excited DF, produced in the $D_2-F_2$ chain reaction, to $CO_2$. Transfer lasers have the potential of high chemical efficiency.

The first pulsed chemical-transfer laser[7] was that based on the transfer reaction

$$HCl(v) + CO_2(000) \to HCl(v-1) + CO_2(001) \tag{18}$$

which has a rate constant $\sim(9 \pm 1) \times 10^4$ $(sec\text{-}Torr)^{-1}$ at 300°K. The vibrationally excited HCl was produced by flash-photolysis initiation of Cl/HI. In a laser mixture of 10 Torr $Cl_2$ with 3 Torr HI, 32 mJ of HCl laser energy were produced in the absence of $CO_2$. As $CO_2$ was added incrementally to the system, $CO_2$ laser power at 10.6 $\mu$m gradually overcame the HCl laser action. At a pressure of 1 Torr-added $CO_2$, the HCl laser was completely quenched, and, at 5 Torr, a pulse energy of 59 mJ at 10.6 $\mu$m was recorded. Laser action was also recorded when

$N_2O$ was added to the Cl/HI laser but at a power of only one-third that of the $CO_2$ case. Typical laser pulsewidths ranged from 20 to 60 $\mu$sec. The increased chemical efficiency of the system when operating on $CO_2$ rather than HCl is indicative of the higher stability of the asymmetric stretch mode of $CO_2$ with respect to collisional quenching than is the vibrationally excited HCl.

Almost simultaneously with the demonstration of the HCl–$CO_2$ transfer laser, Gross[114] first operated the DF–$CO_2$ system. The laser was based on the flash photolysis of a mixture of $F_2O$, $D_2$, and $CO_2$. Whereas, in a mixture of $F_2O + D_2$ alone at 40 Torr, the DF pulsewidth was ~15 $\mu$sec, with a 1/1/1 mixture at 40 Torr containing $CO_2$, the pulsewidth was extended to ~30 $\mu$sec and the pulse energy was doubled. Since Gross's initial experiment on the DF–$CO_2$ system there has been considerable research carried out by several groups in both the United States and the Soviet Union on flash-photolysis initiated systems[70,116,117,143] and those initiated by a pulsed electrical discharge.[144]

Suchard et al.[117] performed experiments at a total pressure of 0.5 atm in flash-photolysis apparatus, schematically shown in Fig. 18. The laser tube was quartz, 25-mm i.d., and 55-cm long and was fastened to aluminum fixtures holding KCl windows at the appropriate Brewster's angle. The laser tube was centered within a 1.2-m optical resonator. With a 0.5-atm mixture with mole ratio $D_2/F_2/CO_2/He = 0.33/1/8/10$ and flash energy of 2400 J with ~40 $\mu$sec half width, they observed a laser pulse energy of 2.8 J when using an output-coupling fraction of 35%. The laser

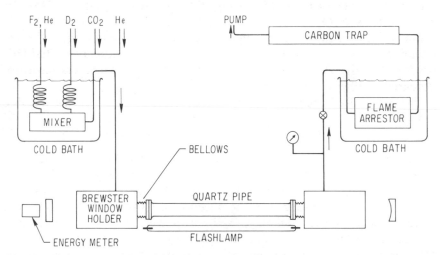

**Fig. 18.** High-pressure flash-photolysis laser apparatus. The laser cooling system and Invar alignment fixtures are not shown. (Adapted from Suchard et al.[117])

pulse width was $20 \mu\text{sec}$. An output energy of 2.8 J corresponded to 5.1% of the available chemical energy in the $290 \text{ cm}^3$ volume of their apparatus. If this volume is corrected to the actual volume usefully employed by the optical cavity modes, the chemical efficiency figure is raised to 15.3%. The latter percentage corresponds to laser-output energy of ~40 J/liter-atm, which is a comparable energy density to electrically pumped atmospheric pressure $CO_2$ lasers.

Poehler and Walker[144] have operated a DF–$CO_2$ laser system using transverse-discharge initiation. They reported laser-pulse energies up to 0.5 J with total-mixture pressures as high as 250 Torr. The laser-pulse duration and intensity was found to be strongly dependent on the gas pressure and the excitation voltage. The laser energy for a typical mixture ($D_2/F_2/CO_2/He = 1/1/3/20$) was found to increase monotonically from 25 to 250 Torr. The pulse duration decreased from $500 \mu\text{sec}$ at 25 Torr to $50 \mu\text{sec}$ at the highest experimental pressure. A maximum chemical efficiency of 2.4% was achieved and an overall electrical efficiency of 20%. The overall electrical efficiency in the case of transverse discharge intiation is orders of magnitude higher than the flash-photolysis initiated systems. However, this is, in part, due to the small size of the apparatus, which is badly matched to the fluorine optical absorption cross-section. It is to be expected that, for large devices the uv-initiated schemes, could compete favorably with electron-impact dissociation.

Pulsed chemical-transfer lasers have been reported[145] on the basis of the flash photolysis of mixtures of $O_3/D_2(H_2)/CO_2$ and $O_3/D_2/N_2O$. Photolysis of ozone yields $O(^1D)$, which then reacts with $D_2$ according to

$$O(^1D) + D_2 \rightarrow OD(v \leq 6) + D \qquad (19)$$

The released deuterium atom then reacts with undissociated ozone, $D + O_3 \rightarrow OD(v \leq 9) + O_2$. In this reaction, the major portion of the exothermicity appears as vibrational energy in the OD radical. The deuteroxyl radical then transfers its energy to $CO_2$ as in the DF–$CO_2$ systems. Only weak and inefficient laser action has been observed for the OD–$CO_2$ system. With 0.85 Torr $O_3$, 50 Torr $D_2$, and 50 Torr $CO_2$, a pulse energy of order 0.7 mJ was observed. Replacing $D_2$ with $H_2$ reduced the laser power by a factor of 2. Replacing $CO_2$ with $N_2O$ reduced the laser power by a factor of 5. The reason for the inefficiency of the OD–$CO_2$ transfer laser is that several deleterious side reactions compete with the pumping and transfer reaction in the laser medium. The OD–$CO_2$ transfer laser has also been reported by a group in the Soviet Union.[146]

Basov and co-workers[147] have also demonstrated a transfer laser based on the explosive decomposition of hydrazoic acid $HN_3$. The decomposition path is a complicated chain reaction in which vibrationally excited nitrogen is formed in one of the fundamental steps

$$NH + N_3H \rightarrow H_2 + 2N_2^\dagger \qquad (20)$$

The vibrationally excited nitrogen $N_2^\dagger$ then transfers its energy to $CO_2$. Although laser action has been reported, it is not known at what power or with what efficiency this system can operate.

Hydrogen bromide has a vibrational energy quantum closely resonant with the asymmetric stretch of $CO_2$. Resonance is not the driving factor in determining the rate of energy exchange between the hydrogen halides and $CO_2$, but it contributes in the case of HBr to a very high cross-section. Burak et al.[140] have studied an HBr–$CO_2$ transfer laser initiated by a transverse discharge in mixtures of $H_2$, $Br_2$, and $CO_2$. Typical mixtures studied were $Br_2/H_2/CO_2 = (1–5)/50/40$ at a total pressure of 100 Torr. Approximately eight times more laser-pulse energy was observed in the transfer laser than in the direct HBr laser.

In summary, it is apparent that the hydrogen-halide transfer lasers are as efficient, if not more efficient, than the hydrogen-halide systems themselves. They have not been scaled to high-pulse energies to date, although there seems no reason why the pulsed DF–$CO_2$ system could not be operated at the several hundred-joule level in a similar manner to the HF system.

## REFERENCES

1. J. V. V. Kasper and G. C. Pimentel, Phys. Rev. Lett. **14,** 352 (1965).
2. K. L. Kompa and G. C. Pimentel, J. Chem. Phys. **47,** 857 (1967).
3. T. F. Deutsch, Appl. Phys. Lett. **10,** 234 (1967).
4. O. M. Batovskii, G. K. Vasil'ev, E. F. Makarov, and V. L. Tal'rose, JETP Lett. **9,** 200 (1969).
5. N. G. Basov, L. V. Kulakov, E. P. Markin, A. I. Nikitin, and A. N. Oraevsky, JETP Lett. **9,** 375 (1969).
6. T. F. Deutsch, IEEE J. Quant. Electron. **QE-3,** 419 (1967).
7. H. L. Chen, J. C. Stephenson, and C. B. Moore, Chem. Phys. Lett. **2,** 593 (1968).
8. J. C. Polanyi and K. B. Woodall, J. Chem. Phys. **57,** 1574 (1972).
9. D. W. Gregg, B. Krawetz, R. K. Pearson, B. R. Schleicher, S. J. Thomas, E. B. Huss, K. J. Pettipiece, J. R. Creighton, R. E. Niver, and Y. L. Pan, Chem. Phys. Lett. **8,** 609 (1971).
10. K. L. Kompa, J. H. Parker and G. C. Pimentel, J. Chem. Phys. **49,** 4257 (1968).
11. S. N. Suchard, R. W. F. Gross, and J. S. Whittier, Appl. Phys. Lett. **19,** 411 (1971).
12. R. W. F. Gross, N. Cohen, and T. A. Jacobs, J. Chem. Phys. **48,** 3821 (1968).

13. A. N. Chester and L. D. Hess, IEEE J. Quant. Electron. **QE-8,** 1 (1972).

14. G. G. Dolgov-Savel'yev, I. A. Polyakov, and G. M. Chumak, Soc. Phys. JETP **31,** 643 (1970).

15. T. D. Padrick and G. C. Pimentel, J. Chem. Phys. **54,** 720 (1971).

16. P. Gensel, K. L. Kompa, and J. Wanner, Chem. Phys. Lett. **7,** 583 (1970).

17. J. H. Parker and G. C. Pimentel, J. Chem. Phys. **51,** 91 (1969).

18. J. H. Parker and G. C. Pimentel, J. Chem. Phys. **55,** 857 (1971).

19. R. D. Coombe and G. C. Pimentel, J. Chem. Phys. **59,** 251 (1973).

20. R. L. Kerber, A. Ching, M. L. Lundquist, and J. S. Whittier, IEEE J. Quant. Electron. **QE-9,** 607 (1973).

21. T. V. Jacobson and G. H. Kimball, Chem. Phys. Lett. **8,** 309 (1971).

22. R. G. Wenzel and G. P. Arnold, IEEE J. Quant. Electron. **QE-8,** 26 (1972).

23. H. Pummer and K. L. Kompa, Appl. Phys. Lett. **20,** 356 (1972).

24. K. J. Pettipiece, Chem. Phys. Lett. **14,** 261 (1972).

25. T. V. Jacobson and G. H. Kimbell, IEEE J. Quant. Electron. **QE-9,** 963 (1973).

26. H. Pummer, W. Breitfeld, H. Welder, G. Klement, and K. L. Kompa, Appl. Phys. Lett. **22,** 319 (1973).

27. T. V. Jacobson, G. H. Kimball, and D. R. Snelling, IEEE J. Quant. Electron. **QE-9,** 496 (1973).

28. T. V. Jacobson and G. H. Kimball, J. Appl. Phys. **42,** 3402 (1971).

29. S. Marcus and R. J. Carbone, IEEE J. Quant. Electron. **QE-7,** 493 (1971).

30. G. J. Ultee, IEEE J. Quant. Electron. **QE-6,** 647 (1970).

31. J. Goldhar, R. M. Osgood, Jr., and A. Javan, Appl. Phys. Lett. **18,** 167 (1971).

32. R. W. F. Gross, R. R. Giedt, and T. A. Jacobs, IEEE J. Quant. Electron. **QE-6,** 168 (1970).

33. R. J. Jensen and W. W. Rice, Chem. Phys. Lett. **8,** 214 (1971).

34. C. P. Robinson, R. J. Jensen, and A. Kolb, IEEE J. Quant. Electron. **QE-9,** 963 (1973).

35. T. F. Deutsch, Appl. Phys. Lett. **11,** 18 (1967).

36. T. F. Deutsch, IEEE J. Quant. Electron. **QE-7,** 174 (1971).

37. J. H. Parker and G. C. Pimentel, J. Chem. Phys. **48,** 5273 (1968).

38. M. J. Berry and G. C. Pimentel, J. Chem. Phys. **49,** 5190 (1968).

39. M. J. Berry and G. C. Pimentel, J. Chem. Phys. **51,** 2274 (1969).

40. K. L. Kompa, P. Gensel, and J. Wanner, Chem. Phys. Lett. **3,** 210 (1969).

41. P. Gensel, K. L. Kompa, and J. Wanner, Chem. Phys. Lett. **5,** 179 (1970).

42. M. C. Lin and W. H. Green, J. Chem. Phys. **53,** 3383 (1970).

43. R. J. Jensen and W. W. Rice, Chem. Phys. Lett. **7,** 627 (1970).

44. C. J. Ultee, Rev. Sci. Instrum. **42,** 1174 (1971).

45. D. P. Akit and J. T. Yardley, IEEE J. Quant. Electron. **QE-6,** 113 (1970).

46. L. E. Brus and M. C. Lin, J. Phys. Chem. **75,** 2546 (1971).

47. S. N. Suchard and G. C. Pimentel, Appl. Phys. Lett. **18,** 530 (1971).

48. M. C. Lin, J. Phys. Chem. **75,** 3642 (1971).

49. M. C. Lin, J. Phys. Chem. **75,** 284 (1971).

50. M. C. Lin and W. H. Green, IEEE J. Quant. Electron. **QE-7**, 98 (1971).

51. M. C. Lin and W. H. Green, J. Chem. Phys. **54**, 3222 (1971).

52. T. V. Jacobson and G. H. Kimbell, Chem. Phys. Lett. **8**, 309 (1971).

52. O. R. Wood, E. G. Burkhard, M. A. Pollack, and T. J. Bridges, Appl. Phys. Lett. **18**, 112 (1971).

54. K. L. Kompa and J. Wanner, Chem. Phys. Lett. **12**, 560 (1972).

55. M. J. Berry, presented at Third Conference on Chemical and Molecular Lasers, St. Louis, Missoui, May 1972.

56. J. L. Roebber and G. C. Pimentel, presented at Third Conference on Chemical and Molecular Lasers, St. Louis, Missouri, May 1972.

57. E. Cuellar-Ferreira and G. C. Pimentel, presented at Third Conference on Chemical and Molecular Lasers, St. Louis, Missouri, May 1972.

58. R. D. Coombe, G. C. Pimentel and M. J. Berry, presented at Third Conference on Chemical and Molecular Lasers, St. Louis, Missouri, May 1972.

59. O. R. Wood and T. Y. Chang, Appl. Phys. Lett. **20**, 77 (1972).

60. G. P. Arnold and R. G. Wenzel, IEEE J. Quant. Electron. **QE-9**, 491 (1973).

61. R. K. Pearson, J. O. Cowles, G. L. Herman, and D. W. Gregg, presented at Third Conference on Chemical and Molecular Lasers, St. Louis, Missouri, May 1971.

62. G. A. Kapralova, E. M. Trofimova, and A. E. Shilov, Kinet. Catal. **6**, 884 (1965).

63. G. A. Kapralova and E. M. Margolina, Kinet. Catal. **10**, 23 (1967).

64. N. G. Basov, E. P. Markin, A. I. Nikitin, and A. N. Oraevsky, IEEE J. Quant. Electron. **QE-6**, 183 (1970).

65. V. S. Burmasov, G. G. Dolgov-Savel'ev, V. A. Polyakov, and G. M. Chumak, JETP Lett. **10**, 28 (1969).

66. N. G. Basov, L. V. Kulakov, E. P. Markin, A. I. Nikitin, and A. N. Oraevsky, "Chemical Laser on Hydrogen Fluoride," presented at International Symposium on Chemical Lasers, Moscow, USSR, Sept. 2–4, 1969.

67. O. M. Batovskii, G. K. Vasil'ev, and V. A. Polyakov, "Lasing in Branched Chain Reactions: $H_2 + F_2$ System," presented at International Symposium on Chemical Lasers, Moscow, USSR, Sept. 2–4, 1969.

68. G. M. Chumak, G. G. Dolgov-Savel'yev, and V. A. Polyakov, "On the Generation Mechanism Acting on Fluorine–Hydrogen Mixtures," presented at International Symposium on Chemical Lasers, Moscow, USSR, Sept. 2–4, 1969.

69. G. G. Dolgov-Savel'yev, V. F. Zharov, Y. S. Neganov, and G. M. Chumak, Zh. Eksp. Teor. Fiz. **61**, 64 (1971); Sov. Phys. JETP **34**, 34 (1972).

70. N. G. Basov, V. T. Galochkin, V. I. Igoshin, L. V. Kulakov, E. P. Markin, A. I. Nikitin, and A. N. Oraevsky, Appl. Opt. **10**, 1814 (1971).

71. N. G. Basov, V. I. Igoshin, E. P. Markin, and A. N. Oraevsky, Kvantovaja Electron. **2**, 3 (1971).

72. L. D. Hess, Appl. Phys. Lett. **19**, 1 (1971).

73. L. D. Hess, J. Chem. Phys. **55**, 2466 (1971).

74. J. Wilson and J. Stephenson, Appl. Phys. Lett. **20**, 64 (1972).

75. L. D. Hess, J. Appl. Phys. **43**, 1157 (1972).

76. N. R. Greiner, IEEE J. Quant. Electron. **QE-8**, 872 (1972).

77. S. N. Suchard, R. L. Kerber, G. Emanuel, and J. S. Whittier, J. Chem. Phys. **57,** 5065 (1972).

78. H. L. Chen and J. Wilson, "On the Atmospheric-Pressure Pulsed HF Chemical Laser," presented at International Electron Device Meeting, Washington, D.C., Dec. 4–6, 1972.

79. J. Wilson, D. Northam, and P. Lewis, IEEE J. Quant. Electron. **QE-9,** 202 (1973).

80. S. N. Suchard, Appl. Phys. Lett. **23,** 68 (1973).

81. J. B. Levy and B. K. W. Copeland, J. Amer. Chem. Soc. **67,** 2156 (1963).

82. J. B. Levy and B. K. W. Copeland, J. Amer. Chem. Soc. **69,** 408 (1965).

83. J. B. Levy and B. K. W. Copeland, J. Amer. Chem. Soc. **72,** 3168 (1968).

84. W. J. Moore, *Physical Chemistry,* third ed. (Prentice-Hall, Englewood Cliffs, New Jersey (1962).

85. S. N. Suchard and Lee D. Bergerson, Rev. Sci. Instrum. **43,** 1717 (1972).

86. G. Emanuel, W. D. Adams, and E. B. Turner, Technical Report TR-0172(2776)-1 (The Aerospace Corp., El Segundo, California, March 1972).

87. S. N. Suchard and D. G. Sutton, IEEE J. Quant. Electron, **QE-10,** 490 (1974).

88. E. J. Bowen, *Chemical Aspects of Light,* second ed. (Clarendon, Oxford, 1946).

89. R. K. Steunenberg and R. C. Vogel, J. Amer. Chem. Soc. **78,** 901 (1956).

90. N. R. Greiner, Chem. Phys. Lett. **16,** 314 (1972).

91. J. C. Polanyi and D. C. Tardy, J. Chem. Phys. **51,** 5717 (1969).

92. N. Jonathan, C. M. Melliar-Smith, and D. H. Slater, Mol. Phys. **20,** 93 (1971).

93. N. Jonathan, C. M. Melliar-Smith, and D. H. Slater, J. Chem. Phys. **53,** 4396 (1972).

94. J. C. Polanyi and J. J. Sloan, J. Chem. Phys. **57,** 4988 (1972).

95. N. Jonathan, S. Okuda, and D. Timlin, Mol. Phys. **24,** 1143 (1972).

96. S. W. Rabideau, H. G. Hecht, and W. B. Lewis, *Proceedings of the Fourth International Symposium on Magnetic Resonance* (Rehovat and Jerusalem, Israel, 1971).

97. K. L. Kompa, J. Wanner, and P. Gensel, presented at Second Conference on Chemical and Molecular Lasers, St. Louis, Missouri, May 1969.

98. N. Cohen, Technical Report TR-0172(2779)-2 (The Aerospace Corp., El Segundo, California, Sept. 1971).

99. R. L. Kerber, G. Emanuel, and J. S. Whittier, Appl. Opt. **11,** 1112 (1972).

100. D. Röss and P. Möckel, Z. Naturforsch. **20a,** 49 (1965).

101. D. Röss, Z. Naturforsch. **20a,** 264 (1965).

102. W. R. Sooy, R. S. Congleton, B. E. Dobratz, and W. K. Ng, *Proceedings of the Third International Congress on Quantum Electronics, Vol. 2,* P. Grivet and N. Bloemboergen, Eds. (Columbia Univ. Press, New York, 1964).

103. R. J. Collins and J. A. Giordmaine, *Proceedings Third International Congress on Quantum Electronics, Vol. 2,* P. Grivet and N. Bloemboergen, Eds. (Columbia Univ. Press, New York, 1964), pp. 1239–1246.

104. C. M. Varma, IEEE J. Quant. Electron. **QE-5,** 78 (1969).

105. J. A. Blauer, W. C. Solomon and T. W. Owens, Int. J. Chem. Kinet. **4,** 293 (1972).

106. R. R. Stephens and T. A. Cool, J. Chem. Phys. **56,** 5863 (1972).

107. J. R. Airey and I. W. M. Smith, J. Chem. Phys. **57,** 1669 (1972).

108. G. C. Berend and R. L. Thommarson, J. Chem. Phys. **58,** 3203 (1973).

109. J. J. Hinchen, J. Chem. Phys. **59,** 233 (1973).

110. J. F. Bott and N. Cohen, J. Chem. Phys. **58,** 934 (1973).

111. J. F. Bott and N. Cohen, *Proceedings of the Ninth International Shock Tube Symposium* (Stanford University, Stanford, California, July 16–19, 1973).

112. R. L. Wilkins, J. Chem. Phys. **59,** 698 (1973).

113. J. F. Bott, J. Chem. Phys. **60,** 427 (1974).

114. R. W. F. Gross, J. Chem. Phys. **50,** 1889 (1969).

115. N. G. Basov, V. T. Galochkin, L. V. Kulakov, E. P. Markin, A. I. Nikitin, and A. N. Oraevsch, Kratk. Sobashch. Fiz. **8,** 10 (1970).

116. T. O. Poehler, M. Shandor, and R. E. Walker, Appl. Phys. Lett. **20,** 497 (1972).

117. S. N. Suchard, A. Ching, and J. S. Whittier, Appl. Phys. Lett. **21,** 274 (1972).

118. J. H. Parker and G. C. Pimentel, IEEE J. Quant. Electron. **QE-6,** 175 (1970).

119. G. G. Dolgov-savel'yev and G. M. Chumak, Soc. J. Quant. Electron. **2,** 383 (1973).

120. N. G. Basov, V. T. Galochkin, L. V. Kulakov, E. P. Martin, A. I. Nikitin, and A. N. Oraevsky, Soc. J. Quant. Electron. **1,** 348 (1972).

121. R. L. Jaffe and J. B. Anderson, J. Chem. Phys. **54,** 2224 (1971).

122. R. L. Wilkins, "Monte Carlo Calculations of Reaction Rates and Energy Distributions Among Reaction Products. $F + D_2 \rightarrow DF + D$," Technical Report TR-0073(3430)-13 (The Aerospace Corp., El Segundo, California, (Feb. 1973).

123. J. T. Muckerman, J. Chem. Phys. **54,** 1155 (1971).

124. T. P. Schafer, P. E. Siska, J. M. Parson, F. P. Tully, Y. C. Wong, and Y. T. Lee, J. Chem. Phys. **53,** 3385 (1970).

125. K. G. Anlauf, P. E. Charters, D. S. Home, R. G. MacDonald, D. H. Maylotle, J. C. Polanyi, W. J. Skiloc, D. C. Tardy, and K. B. Woodall, J. Chem. Phys. **53,** 4091 (1970).

126. R. L. Wilkins, J. Chem. Phys. **58,** 2326 (1973).

127. C. K. Rhodes, "Review of Ultraviolet Lasers," presented at Conference on Laser Engineering and Applications, Washington D.C., May 1973.

128. J. V. Parker and R. R. Stephens, IEEE J. Quant. Electron. **QE-9,** 643 (1973).

129. R. Hofland, M. L. Lundquist, A. Ching, and J. S. Whittier, "Electron-Beam Irradiated Discharge for Initiating High-Pressure Pulsed Chemical Lasers," presented at AIAA Sixth Fluid and Plasma Dynamics Conference, Palm Springs, California, July 16–18, 1973.

130. J. R. Airey, IEEE J. Quant. Electron. **QE-3,** 208 (1967).

131. J. R. Airey, J. Chem. Phys. **52,** 156 (1970).

132. N. Cohen, T. A. Jacobs, G. Emanuel, R. L. Wilkins, Int. J. Chem. Kinet. **1,** 551 (1969).

133. V. I. Igoshin and A. N. Oraevskii, Khimiya Vysikhikh Energii **5,** 397 (1971).

134. P. H. Corneil and G. C. Pimentel, J. Chem. Phys. **49,** 1379 (1968).

135. J. R. Airey, presented at International Meeting on Chemical Lasers, Moscow, Sept. 1969; Report AMP 286 (Avco-Everett Research Laboratories, Everett, Massachusetts, Aug. 1969).

136. J. C. Polanyi, Acct. Chem. Res. **5,** 161 (1972) and references therein.

137. A. Henry, F. Bourein, I. Arditi, R. Charneau, and J. Menard, C.R. Acad. Sci. Ser. B **267,** 616 (1968).

138. M. C. Lin, J. Phys. Chem. **76,** 811 (1972).

139. C. B. Moore, IEEE J. Quant. Electron. **QE-4,** 52, (1968).

140. I. Burak, Y. Noter, A. M. Ronn, and A. Szöke, Chem. Phys. Lett. **13,** 322 (1972).

141. D. P. Akit and J. Yardley, IEEE J. Quant. Electron. **QE-6,** 179 (1970).

142. M. J. Berry and G. C. Pimentel, J. Chem. Phys. **53,** 3453 (1970).

143. J. Wilson and J. C. Stephenson, Appl. Phys. Lett. **20,** 64 (1972).

144. T. O. Poehler and R. E. Walker, Appl. Phys. Lett. **22,** 282 (1973).

145. S. K. Searles and J. R. Airey, Appl. Phys. Lett. **22,** 513 (1973).

146. N. G. Basov, A. S. Bashkin, V. I. Igoshin, A. N. Oraevskii, and N. N. Yuryshev, JETP Lett. **16,** 389 (1972).

147. N. G. Basov, V. V. Gromov, E. L. Koshelev, E. P. Markin, and A. N. Oraevskii, ZhETF **10,** 5 (1969).

148. N. R. Greiner, L. S. Blair, E. L. Patterson, and R. A. Gerber 100 Gigawatt $H_2$–$F_2$ Laser Initiated by an Electron Beam," presented at the Eighth International Conference on Quantum Electronics, San Francisco, California, June 1974.

CHAPTER **7**

# TRANSFER CHEMICAL LASERS

## TERRILL A. COOL

**Cornell University,
Ithaca, New York**

The successful achievement of the proper gas-dynamical conditions for the generation of laser output from chemical reactions, despite the limiting influences of vibrational-energy redistribution and deactivation, has been essential to the extensive recent development of powerful cw chemical-laser sources. The short wavelengths and chain-reaction possibilities of the HF and DF chemical lasers have made them attractive candidates for high-power applications. One practical limitation of these devices has been the need for large vacuum pumps to provide the low static pressure necessary for HF or DF chemical-laser operation. Optimum performance typically occurs at pressures of a few Torr. Even at these low pressures the chemical reaction rates are limited by diffusive mixing, and vibrational populations are controlled by vibrational-energy transfer and deactivation processes. As a result, complete population inversions are not achieved, and the inherent specificity of energy partitioning among reaction products is not as fully utilized as in pulsed chemical lasers.

The transfer chemical laser (TCL), on the other hand, makes efficient use of even substantially relaxed vibrational distributions in the hydrogen-halide molecules for the creation of a complete population inversion in $CO_2$. These devices are "chemical lasers" because a chemical-reaction mechanism resulting in a preferential nonthermal partitioning of product energy into vibration is mandatory for laser operation. This chemically generated vibrational energy is then transferred by collisions to $CO_2$ molecules, which act as the laser medium.

Energy-transfer processes can provide a selective means for coupling the vibrational energy of many V–R states of hydrogen-halide product molecules to the $CO_2(00°1)$ upper laser level. The deactivation processes for the $CO_2(00°1)$ level are sufficiently slow that total inversion between the upper and lower $CO_2$ laser levels can be maintained at relatively high pressures ($\sim100$ Torr). At these pressures, direct population inversion within the product molecules would be virtually impossible because of the influences of deactivation mechanisms. This feature suggests the possibility of an efficient supersonic TCL with a diffuser capable of full atmospheric pressure recovery without vacuum pumps.

The discovery and development of the TCL has had an influence on present concepts of energy transfer between molecules. The now well-known existence of efficient vibration to vibration ($V \rightarrow V$) energy transfer between the hydrogen halides and $CO_2$ could not have been predicted from those theories of energy transfer fashionable before 1969. Only recently have experimental observations been explained in terms of theoretical models that account for the simultaneous role of vibration to rotation ($V \rightarrow R$) transfer in $V \rightarrow V$ exchange processes. Many current refinements in both experimental techniques and theoretical models for vibrational energy transfer have their origins in the need to satisfactorily explain and predict TCL performance. Most of the research connected with the development of the TCL can be classified in four categories: (1) studies of vibrational energy transfer and deactivation mechanisms; (2) investigations of the cw subsonic DF–$CO_2$ chemical laser; (3) development of cw supersonic DF–$CO_2$ devices; and (4) experimental and theoretical studies of pulsed DF–$CO_2$ chemical lasers.

## 1. VIBRATIONAL ENERGY TRANSFER AND DEACTIVATION

The present TCL has evolved from the search for suitable vibrationally excited reaction products that would exhibit a sufficiently strong $V \rightarrow V$ coupling with the upper $CO_2$ laser level to permit operation analogous to that of the $N_2$–$CO_2$ laser. It has been indeed fortunate that the hydrogen- and deuterium-halide products of several simple atom exchange reactions, extensively employed for pulsed chemical lasers,[1-12] exhibit strong $V \rightarrow V$ coupling with the $CO_2(00°1)$ level. The first observations of chemically pumped $CO_2$ laser operation were made by Gross[7] and Chen et al.,[8] for the DF–$CO_2$ and HCl–$CO_2$ systems, respectively. These investigators observed efficient pulsed $CO_2$ laser operation at $10.6 \mu m$ when $CO_2$ was added to pulsed DF and HCl chemical lasers; the addition of large amounts of $CO_2$ caused a quenching of the DF or HCl laser emission, which was replaced by stronger laser action from $CO_2$. The first cw TCL's were reported by Cool et al.[13,14] (HCl–$CO_2$, HF–$CO_2$, DF–$CO_2$), and Cool and Stephens[15] (HBr–$CO_2$). The DF–$CO_2$ and HF–$CO_2$ devices[14,16,17] operated on the chain reaction between hydrogen and fluorine and yielded a higher chemical efficiency than the other systems.

The upper $CO_2$ laser level is pumped by the $V \rightarrow V$, R energy-transfer process:

$$HX(v = n, J = J_1) + CO_2(00°0)$$

$$\rightarrow HX(v = n - 1, J = J_2) + CO_2(00°1) + \Delta E_{int} \quad (1)$$

where X is F, Cl, or Br, and $\Delta E_{int}$ is the internal energy (vibrational and rotational) discrepancy between initial and final states for the HX–$CO_2$ pair. Changes in rotational energy of $CO_2$ are unimportant; however, the rotational-energy spacings in the hydrogen halides are large enough to permit a considerable conversion of the vibrational energy defect of Reaction (1) into increased rotational energy of the hydrogen halide. Only those rotational transitions with quantum-number changes $J_1 \rightarrow J_2$ that cancel most of the large vibrational energy defect associated with Reaction (1) are likely to contribute markedly to the energy transfer. As an example, consider the HF–$CO_2$ case. Table 1 indicates the internal-energy defects associated with various choices of $J_2$ for the values $J_1 = 3$ and $n = 1$. Only the transition $J_1 = 3$ to $J_2 = 9$ has an internal energy defect, $\Delta E_{int}$, of $<100 \text{ cm}^{-1}$. Thus, the probability associated with rotational energy transfer to $J_2 = 9$ ought to greatly exceed that for all other values of $J_2$. This transfer, however, requires the rotational energy of HF to be increased by six quanta. If we postulate an interaction involving the dipole moment of HF, then a description of such an energy-transfer

**Table 1.** Internal Energy Defects for the Reaction:

$HF(v = 1, J_1) + CO_2(00°0)$
$$\rightarrow HF(v = 0, J_2) + CO_2(00°1) + \Delta E_{int}$$

| $J_1$ | $J_2$ | $\Delta E_{int}$, cm$^{-1}$ |
|-------|-------|-----------------------------|
| 3     | 3     | 1612                        |
| 3     | 4     | 1448                        |
| 3     | 5     | 1244                        |
| 3     | 6     | 999                         |
| 3     | 7     | 714                         |
| 3     | 8     | 390                         |
| 3     | 9     | 26                          |
| 3     | 10    | −377                        |

process within the framework of time-dependent perturbation theory would require the consideration of sixth order (and higher) terms.

Before proceeding further, we should inquire whether other possible processes might be invoked to explain the HF–$CO_2$ exchange. One such mechanism is the following[14]:

$$HF(v = 1) + CO_2(00°0) \rightarrow HF(v = 0) + CO_2(10°1) + \Delta E_v = 242 \text{ cm}^{-1} \quad (2)$$

and the subsequent V → V redistribution:

$$CO_2(10°1) + CO_2(00°0) \rightarrow CO_2(10°0) + CO_2(00°1) + \Delta E_v = -21 \text{ cm}^{-1} \quad (3)$$

Such processes involving the intermediate state $CO_2(10°1)$ [or $CO_2(02°1)$] might be highly probable because of the much smaller vibrational energy defects for each step compared with that for the single-step Reaction (1) ($\Delta E_v = 1612$ cm$^{-1}$). On the other hand, if an interaction is present that is strong enough to permit the assumption of several rotational quanta by HF upon collision with $CO_2$, perhaps Reaction (1) could be a possible mechanism for transfer.

Fortunately, the two models predict measurable differences in the laser-excited fluorescences for HF–$CO_2$ mixtures under certain experimental conditions. Stephens and Cool[18] and Hancock and Green[19] have performed experiments that give strong evidence in favor of the direct-coupling Reaction (1). The experiments investigated were based on the prediction that, for a sufficiently small concentration of $CO_2$ in an HF–$CO_2$ mixture, the rate of the second step, Reaction (3), of the intermediate-state coupling mechanism should become small compared to the first step, Reaction (2). The rate of rise of $CO_2(00°1)$ fluorescence

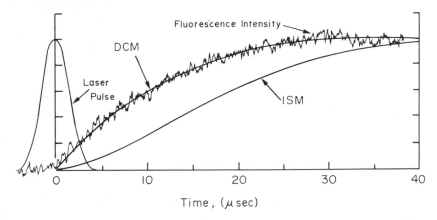

**Fig. 1.** Calculated initial rise in the $CO_2(00°1)$ population (at low $CO_2$ concentration) for two different HF–$CO_2$ coupling mechanisms compared with the observed $CO_2$ fluorescence at 350°K.[18] The calculation for the direct coupling model [Eq. (1)] is labeled DCM; the calculated curve for the intermediate-state processes [Eqs. (2) and (3)] is denoted ISM.

following the laser-induced excitation of the $HF(v = 1)$ level would then be limited by the Reaction (3) under these conditions. Figure 1 shows the result of such an experiment.[18] An HF laser was used to induce the excitation of a small portion of the HF in an HF–$CO_2$ mixture to the $HF(v = 1)$ level. The progress of the subsequent energy transfer to $CO_2$ was followed by monitoring 4.3-$\mu$m fluorescence from $CO_2$. Figure 1 indicates the comparison of the observed rise of 4.3-$\mu$m fluorescence intensity from a mixture with the following composition: HF = 0.63 Torr, $CO_2$ = 0.05 Torr, and Ar = 10.0 Torr at 350°K with approximate theoretical predictions for the two coupling models. The calculated curves show the temporal variation of $CO_2(00°1) \rightarrow CO_2(00°0)$ fluorescence based on reasonable assumptions for the pertinent rate constants. In particular, a rate constant of one-tenth of the gas-kinetic collision rate was assumed for Reaction (3), which is expected to be fast. The close agreement of experiment with calculations based on the direct coupling mechanism of Reaction (1) is convincing evidence in favor of this model, provided that the experimentally observed fluorescence did not include contributions from the 4.3-$\mu$m bands originating from intermediate states.[18] The absence of such contributions was confirmed in experiments by Hancock and Green.[19]

The preceeding example strongly suggests the remarkable conclusion that nearly the entire vibrational energy defect of 1612 cm$^{-1}$ is converted into rotational motion of HF. It would greatly add to the understanding of V–R energy transfer if one could directly ascertain which rotational

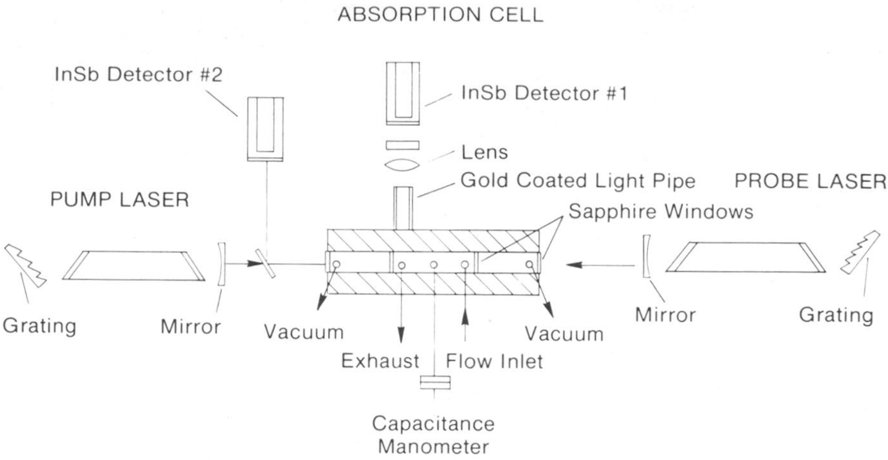

ABSORPTION CELL

**Fig. 2.** Schematic diagram of a laser double-resonance experiment for the study of rotational relaxation in HF. A pump laser is used to excite HF to the first vibrational level at low values of J. A second laser is used as a probe to monitor populations of HF in the ground vibrational state at high vales of J. This may be accomplished by observing the fluorescence from HF($v=1$) at high J produced by the second laser with detector #1; alternatively, the absorption of the high J transitions of the $v=1 \rightarrow v=0$ transitions of the probe laser could be measured with detector #2.

states $J_2$ are favored by Reaction (1). It appears possible to observe the time evolution of the $J_2$ rotational states with a laser double-resonance experiment of the type illustrated in Fig. 2. In such an experiment, a second laser would be employed as a probe to provide an absorption measurement of the $J_2$ populations of the ground vibrational state HF formed in Reaction (1). Experiments of this kind are underway in at least two laboratories.[20] Unfortunately, at present no such direct observations of V → R energy transfer have been accomplished, since all observations of laser-induced fluorescence have been confined to the rotationally unresolved vibrational bands of the HX and $CO_2$ molecules.

## 1.1. Experimental Measurements of V → V, R and V → R, T Transfers in HX–$CO_2$ Systems

A number of laser-induced vibrational-fluorescence measurements have been performed recently over a wide range of temperatures. The various investigators have reported results for the overall (summed over all $J_1$, $J_2$)

processes:

$$HX(v=1)+CO_2(00°0) \underset{k_e'}{\overset{k_e}{\rightleftharpoons}} HX(v=0)+CO_2(00°1) \tag{4}$$

$$HX(v=1)+CO_2(00°0) \xrightarrow{k_{12}} HX(v=0)+CO_2(00°0) \tag{5}$$

$$CO_2(00°1)+HX(v=0) \xrightarrow{k_{21}} CO_2(nm^l0)+HX(v=0) \tag{6}$$

$$HX(v=1)+HX(v=0) \xrightarrow{k_{11}} 2HX(v=0) \tag{7}$$

and

$$CO_2(00°1)+CO_2(00°0) \xrightarrow{k_{22}} CO_2(nm^l0)+CO_2(00°0) \tag{8}$$

The laser-induced fluorescence method provides direct measurements of the quantities $k_e+k_{12}$, $k_e'+k_{21}$, $k_{11}$, and $k_{22}$. Under favorable circumstances, $k_{12}$ and $k_{21}$ can be estimated separately.[21]

Table 2 summarizes the presently available room-temperature measurements of the sum of rate constants $k_e+k_{12}$ for HX–CO$_2$ systems. The data of Table 2 were all obtained with the laser-induced fluorescence method. The sum of probabilities $P_e+P_{12}$ for Reactions (4) and (5) and the vibrational-energy defects $\Delta E_v$ for Reaction (4) are included with the data. In some cases, for example, in the DF–CO$_2$ system, $k_{12}$ is negligible

**Table 2.** Some Experimental Values at 300°K for Rate Constants for Reactions (4) and (5)

| System | Rate constants, $sec^{-1} Torr^{-1}$ | $Z^a$ | $\Delta E_v$, cm$^{-1}$ | Reference |
|---|---|---|---|---|
| HCl–CO$_2$ | $k_e = 8.7 \times 10^4$ | 87 | 537 | 2 |
| DCl–CO$_2$ | $k_e = 2.9 \times 10^4$ | 262 | −258 | 21 |
| HBr–CO$_2$ | $k_e = 2.8 \times 10^5$ | 26 | 209 | 21 |
| DBr–CO$_2$ | $k_e = 7.0 \times 10^2$ | 10,400 | −540 | 21 |
| HI–CO$_2$ | $k_e = 1.3 \times 10^5$ | 57 | −116 | 21 |
| HF–CO$_2$ | $k_e + k_{12} = 7.0 \times 10^4$ | 118 | 1612 | 25 |
| HF–CO$_2$ | $k_e + k_{12} = 3.6 \times 10^4$ | 230 | 1612 | 23 |
| HF–CO$_2$ | $k_e + k_{12} = 5.9 \times 10^4$ | 140 | 1612 | 19 |
| DF–CO$_2$ | $k_e + k_{12} = 2.4 \times 10^5$ | 34 | 558 | 25 |
| DF–CO$_2$ | $k_e + k_{12} = 1.5 \times 10^5$ | 54 | 558 | 24 |

[a] Approximate number of collisions for vibrational-energy transfer.

**Table 3.** Some Experimental Values for the Rate Constants for Reaction (6) at 300°K

| System | Rate constant, $sec^{-1} Torr^{-1}$ | $Z^a$ | Reference |
|---|---|---|---|
| $HCl-CO_2$ | $k_{21} = 4.1 \times 10^3$ | 1900 | 21 |
| $DCl-CO_2$ | $k_{21} < 4.4 \times 10^3$ | >1700 | 21 |
| $HBr-CO_2$ | $k_{21} < 2.4 \times 10^3$ | >3000 | 21 |
| $DBr-CO_2$ | $k_{21} < 1.1 \times 10^3$ | >6500 | 21 |
| $HI-CO_2$ | $k_{21} < 2.3 \times 10^3$ | >3200 | 21 |
| $DI-CO_2$ | $k_{21} < 1.2 \times 10^3$ | >6100 | 21 |
| $HF-CO_2$ | $k_{21} = 5.3 \times 10^4$ | 160 | 19 |
| $HF-CO_2$ | $k_{21} = 4.7 \times 10^4$ | 180 | 25 |
| $(HF-CO_2)^b$ | $(k_{21} = 2.2 \times 10^4)^b$ | $350^b$ | 22 |
| $HF-CO_2$ | $k_{21} = 3.4 \times 10^4$ | 250 | 23 |
| $DF-CO_2$ | $k_{21} = 2.3 \times 10^4$ | 360 | 24 |
| $(DF-CO_2)^b$ | $(k_{21} = 1.9 \times 10^4)^b$ | $440^b$ | 22 |
| $DF-CO_2$ | $k_{21} = 2.6 \times 10^4$ | 320 | 25 |

$^a$ Approximate number of collisions for vibrational deactivation.
$^b$ At $T = 350°K$.

compared to $k_e$. In other cases, for example, in the $HF-CO_2$ case, the relative magnitudes of $k_e$ and $k_{12}$ are difficult to estimate precisely, but useful bounds can sometimes be established.[25] These bounds have been given in Table 2 where possible. Table 2 indicates that in most cases $k_e \geq k_{12}$. The measured rate constants do not show any simple dependence on the vibrational-energy defect $\Delta E_v$; this is not surprising since $\Delta E_v$ for the present systems may be substantially greater than the $\Delta E_{int}$ for Reaction (1), as has been discussed.

Table 3 summarizes the available measurements for the rate constant $k_{21}$ of Reaction (6) obtained with the laser-induced fluorescence technique under room-temperature conditions. The chief difficulty in interpretation of these data is that the $CO_2(nm^l0)$ final states are unknown; however, some theoretical predictions have been given by Shin[26,27] for the $HF-CO_2$ and $DF-CO_2$ cases for chosen final $CO_2$ states.

Another technique developed by Hancock and Smith[28,29] has been recently applied by Airey and Smith[30] to the $HF-CO_2$ and $DF-CO_2$ systems for measurements of the sum of rate constants $k_e + k_{12}$. The experimental technique consists of observations of the quenching of vibrationally excited molecules formed by chemical reaction when a chemically inert species is added to the reacting system. This method provides a unique source of rate information concerning energy transfer from vibrational levels of HF and DF above $v = 1$, that is,

$$HX(v = n) + CO_2(00°0) \xrightarrow{k_e(v=n)} HX(v = n - 1) + CO_2(00°1) \quad (9a)$$

$$HX(v = n) + CO_2(00°0) \xrightarrow{k_{12}(v=n)} HX(v = n - 1) + CO_2(00°0) \quad (9b)$$

The combined probabilities for these processes, $P_e$ and $P_{12}$, are given as a function of vibrational quantum number $n$ in Figs. 3 and 4. The probabilities increase rapidly with $n$ and reach values approaching unity (gas-kinetic rate) for the highest vibrational levels examined.

The temperature dependences of several rates of interest have been established between 300 and 700°K. Many of these measurements have been performed with conventional heated fluorescence cells. However, a shock tube can be used very successfully for gas heating as Bott[31] has shown. This technique combines the advantages of the laser-excited fluorescence method with the high-temperature heating capability of a

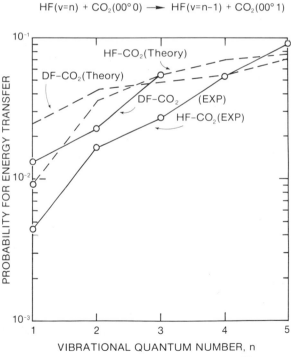

$$HF(v=n) + CO_2(00°0) \longrightarrow HF(v=n-1) + CO_2(00°1)$$

**Fig. 3.** Probabilities for energy transfer between HF(DF) and $CO_2$[30] and the theory of Dillon and Stephenson.[33]

$$HCl(v=n) + CO_2(00°0) \longrightarrow HCl(v=n-1) + CO_2(00°1)$$

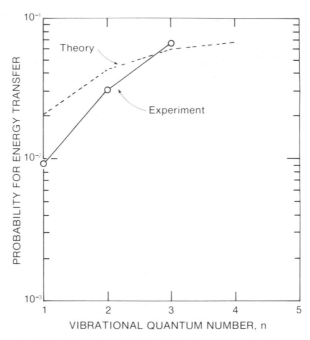

**Fig. 4.** Comparison between measured probabilities for energy transfer between HCl and $CO_2$[30] and the theory of Dillon and Stephenson.[33]

shock tube. Although this technique does not permit repetitive signal averaging, it makes a large temperature range conveniently available without the complications of the heated fluorescence cell.

The temperature dependences of the probabilities for Reactions (4) and (6) for the $HF–CO_2$, $DF–CO_2$, $HCl–CO_2$, and $DCl–CO_2$ systems are given in Figs. 5 and 6. The pronounced increase in energy-transfer probability with decreasing temperature exhibited by the $HF–CO_2$ and $DF–CO_2$ systems in Figs. 5 and 6 is of considerable interest. Proper explanation of this inverse temperature dependence may provide insight concerning the nature of the intermolecular potential between $HF–CO_2$ and $DF–CO_2$ pairs.[25]

Figures 5 and 6 indicate that the $DF–CO_2$ system has a value of $k_e$ which exceeds $k_{21}$ by about an order of magnitude in the temperature regime of interest for TCL operation ($300°K \leq T \leq 400°K$). In contrast, in the $HF–CO_2$ system the deactivation of $CO_2(00°1)$ by HF is comparable to the rate of energy transfer from HF to $CO_2$; this accounts for the

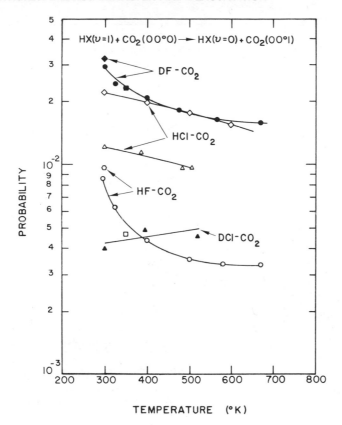

**Fig. 5.** Comparison of the temperature dependences of the V→V,R transfer probabilities for process (4) for the HCl–CO$_2$, DCl–CO$_2$, HF–CO$_2$, and DF–CO$_2$ systems: △, HCl–CO$_2$, Stephenson et al.[2]; ▲, DCl–CO$_2$, Stephenson et al.[2]; ◇, HCl–CO$_2$, Dillon and Stephenson[6]; □, HF–CO$_2$, Stephens and Cool[3]; ○, HF–CO$_2$, present work; , HF–CO$_2$, theory of Dillon and Stephenson[6]; ■, DF–CO$_2$, Stephens and Cool[3]; ●, DF–CO$_2$, present work; ◆, DF–CO$_2$, theory of Dillon and Stephenson.[6]

inferior performance of the HF–CO$_2$ chemical laser compared with that of the DF–CO$_2$ TCL.[32]

Several observations regarding energy transfer and deactivation in the HX–CO$_2$ systems characterize the available data[18,21,25]:

1. Substantial energy transfer from vibration to rotation of the HX molecule occurs. This effect minimizes the internal energy defect and leads to large energy-transfer rates even when the vibrational-energy defects are large.

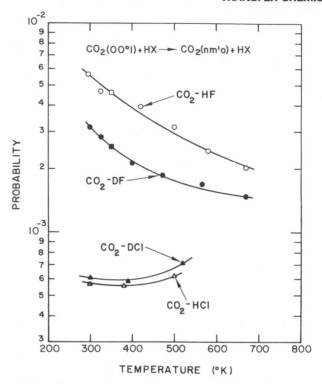

**Fig. 6.** Comparison of the temperature dependences of the $V \rightarrow R,T$ deactivation probabilities of $CO_2(00°1)$ molecules by process (6) for HCl, DCl, HF, and DF: $\triangle$, $CO_2$–HCl, Stephenson et al.[2]; $\blacktriangle$, $CO_2$–DCl, Stephenson et al.[2]; $\blacksquare$, $CO_2$–DF Stephens and Cool[3]; $\bullet$, $CO_2$–DF, present work.

2.  Multiple-quantum rotational energy transfer appears to be important, as exemplified by the HF–$CO_2$ system in particular. Therefore, first- or second-order perturbation theories are not very useful and lead to computed energy-transfer probabilities that are far lower than the experimentally observed rates.

3.  A pronounced inverse temperature dependence of the probabilities for energy transfer and deactivation of the HF–$CO_2$ and DF–$CO_2$ systems exists. This may be the result of an anisotropic short-range intermolecular potential with an associative minimum for the hydrogen-bonded collinear configuration X—H $\cdots$ O=C=O.

## 1.2.  Theory of $V \rightarrow V$, R Energy Transfer in HX–$CO_2$ Systems

In a series of recent papers,[33–35] Dillon and Stephenson have developed a semiclassical theory of $V \rightarrow V$, R energy transfer designed to treat the multiple-quantum rotational energy transfer inherent in the vibrational-energy exchange between HX and $CO_2$ molecules. The theory appears to provide very satisfactory predictions of the probabilities for the V → V, R processes in Reaction (1); theoretical calculations have been performed for the HCl–$CO_2$, HF–$CO_2$, and DF–$CO_2$ systems that are in excellent agreement with experimental results.

The theory of Dillon and Stephenson includes the features:

1.  Curved classical trajectories for the HX–$CO_2$ collision pair were computed with the use of a spherically symmetric potential (Lennard-Jones) having an attractive potential minimum. These classical trajectory calculations were performed for all impact parameters and velocities of interest; the effects of orbiting collisions were specifically included.[35]

2.  The interaction was treated as a combination of V → V and R → R contributions. The V → V transfer was accounted for with the dipole–dipole transition moment interaction; the R → R transfers were described by a permanent moment interaction between the dipole moment of HX and the quadrupole moment of $CO_2$.

3.  The small perturbation approach based on the Born approximation used previously[36,37] was discarded in favor of a description better suited to the relatively strong interactions apparent in the HX–$CO_2$ systems. Recognition was given to the important role played by multi-quantum rotational transitions in the minimization of the overall conversion of internal energy into translational motion.

4.  Short range anisotropy in the intermolecular potential, for example, hydrogen bonding, was neglected. At low temperatures and small impact parameters, such anisotropies could provide a more favorable alignment of the interacting pair leading to an enhancement of both V → V, R and V → R, T energy transfers. Conceivably, the Dillon and Stephenson theory may therefore underestimate the V → V, R energy-transfer probabilities at low temperatures.

Results of the Dillon and Stephenson theory are included with the data of Figs. 3 and 4. The figures indicate that the theory provides values of the rate constant $k_e$ for Reaction (4) that agree with experiment to within a factor of 2. The temperature dependence of the rate constant $k_e$ for the HCl–$CO_2$ transfer is very similar to that calculated;[33] however, the temperature dependence for the HF–$CO_2$ and DF–$CO_2$ systems is more difficult to reproduce theoretically.[34,35] The existence of orbiting collisions[35] and of short-range anisotropy in the intermolecular potential may both provide explanations for the steep inverse dependence of the rate constant $k_e$ observed in experiments.

Much more work is needed, both experimental and theoretical, to determine the mechanism by which vibrational and rotational motions are coupled under the influence of strong attractive interaction potentials. Until now it has not been possible to directly observe the populations of individual rotational states to obtain precise information concerning transfer mechanisms; however, the major experimental difficulties imposed by rotational thermalization may not be unduly restrictive for reactions as rapid as some listed in Table 2.

Laser double-resonance experiments of the type illustrated in Fig. 2 may be useful for studies of $V \rightarrow R$ energy transfer under conditions similar to those encountered in operating lasers. Molecular-beam experiments could provide, in addition, much of the presently lacking detailed knowledge concerning the specific-rate constants of Reaction (1).

## 2.  THE cw SUBSONIC DF–CO₂ CHEMICAL LASER

The subsonic DF–CO$_2$ system was the first example of a chemical laser pumped entirely by chemical means without the use of sustaining external sources for laser excitation.[16] Present subsonic DF–CO$_2$ lasers[38–40] have optical axes aligned transversely to the flow direction, in contrast to the longitudinal optical axis employed in the original cw lasers. Use of the transverse-cavity geometry has also facilitated high-power levels and permitted reliable operation of purely chemical HF and DF lasers at wavelengths of 2.8 and 3.9 $\mu$m, respectively.[32,41]

Figure 7 illustrates the concept of this device, which operates by the simple mixing of bottled gases in a subsonic flow. The gases were mixed in a sequence to maximize the laser output. The important chemical reactions in this laser are:

$$F_2 + NO \xrightarrow{\ k_{10}\ } ONF + F, \qquad \Delta H = -18 \text{ kcal/mol} \quad (10)$$

$$F + NO + M \xrightarrow{\ k_{11}\ } ONF + M, \qquad \Delta H = -57 \text{ kcal/mol} \quad (11)$$

$$F + D_2 \xrightarrow{\ k_{12}(v=n)\ } DF(v=n) + D, \qquad \Delta H = -32 \text{ kcal/mol} \quad (12)$$

and

$$D + F_2 \xrightarrow{\ k_{13}(v=n)\ } DF(v=n) + F, \qquad \Delta H = -98 \text{ kcal/mol} \quad (13)$$

A PURELY CHEMICAL LASER

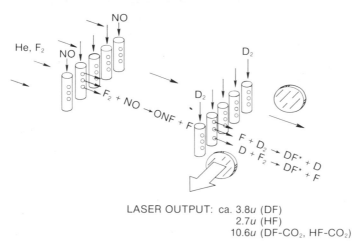

LASER OUTPUT: ca. 3.8$u$ (DF)
2.7$u$ (HF)
10.6$u$ (DF-CO$_2$, HF-CO$_2$)

**Fig. 7.** Reaction steps of the HF, DF, HF–CO$_2$, and DF–CO$_2$ purely chemical lasers.

Optimum performance occurs when the D$_2$ injection is accomplished after an initial F-atom concentration is established with Reactions (10) and (11). The laser can, however, also be operated with all gases injected at the same place in the flow. This operation is analogous to that of the CS$_2$/O$_2$ flame laser.[42,43]

Two measurements of the rate constant of Reaction (10) have been performed with ESR spectroscopy. Mooberry and Wolga[44] reported a value of $k_{10} = 2.6 \pm 0.5 \times 10^9$ cm$^3$/mol-sec at $T = 298°$K; Kim, MacLean, and Valance[45] have measured $k_{10} = 4.75 \pm 0.7 \times 10^9$ cm$^3$/mol-sec at $T = 325 \pm 10°$K. These values are in good agreement and are preferable to a previous measurement of $4.8 \times 10^{10}$ cm$^3$/mol-sec at $T = 298°$K obtained with the diffusion flame technique.[46] A value of $k_{11} = 3 \times 10^{16}$ cm$^6$/mol$^2$-sec has been given[45] for the three-body reaction (11) with M = NO or He. Under conditions of the DF–CO$_2$ subsonic chemical laser, the rates for reactions (10) and (11) are thought to be of comparable magnitude.[47] The rate constants for individual product vibrational states accessible by Reactions (12) and (13) have been extensively studied, and rate-constant data have been published[48–50] and are discussed in the chapter by Cohen and Bott.

Despite the existence of detailed kinetic models for the HF, DF, HF–CO$_2$, and DF–CO$_2$ chemical lasers[48,51–57] and the reasonably adequate state of knowledge concerning the key kinetic rates, to date no detailed computer modeling of the purely chemical subsonic DF–CO$_2$ laser has been reported.

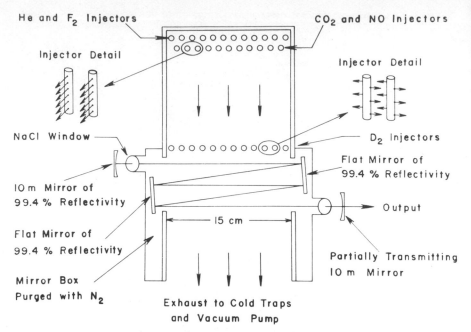

**Fig. 8.** Schematic diagram of a transverse-flow subsonic DF–CO$_2$ chemical laser.

**Table 4.** Operating Characteristics of the DF–CO$_2$ Subsonic Transverse-Flow Laser[a]

| | |
|---|---|
| Optical configuration | 5-path (see Fig. 4) |
| Predominant transverse mode | TEM$_{12}$ |
| Power output (on $P(20)$ transition) | 162 W |
| Chemical efficiency ($F_2 + D_2 \rightarrow 2DF$) | 4.6% |
| Saturation intensity | 143 W/cm$^2$ |
| Maximum local unsaturated gain coefficient | 3.3 m$^{-1}$ |
| Average unsaturated gain coefficient (5-path) | 2.0 m$^{-1}$ |
| Maximum population of CO$_2$(00°1) level | $1.3 \times 10^{16}$ cm$^{-3}$ |
| Optimum mirror transmission | 33% |
| Mean-flow velocity | $2 \times 10^4$ cm/sec |
| Static pressure | 15.4 Torr |
| Rotational temperature in reaction zone | $400 \pm 25$°K |
| Optimum partial-flow rates (mmol/sec) | He: 112 |
| | CO$_2$: 57 |
| | F$_2$: 7.3 |
| | D$_2$: 6.5 |
| | NO: 1.2 |

[a] Ref. 38.

Figure 8 illustrates the design of a transverse flow DF–CO$_2$ laser that was studied in some detail at Cornell[38] to investigate its potential for higher-power operation. At optimum operating conditions, given in Table 4, this laser produced an output of 162 W at 10.6 $\mu$m. Laser operation on HF or DF was also possible if the CO$_2$ was removed and the helium flow rate was substantially increased[32,41]; an output of about 10 W was observed from HF or DF.

Figure 9 shows the relative variation of the population in the upper state of the CO$_2$ laser as a function of distance along the flow direction for the device of Fig. 8. The data of Fig. 9 were obtained by measurement of the CO$_2$(00°1) $\rightarrow$ CO$_2$(00°0) 4.3-$\mu$m fluorescence-emission intensity in the absence of a cavity. An increase in NO flow-rate accelerated the DF chain reaction and caused an increase in peak 4.3-$\mu$m emission intensity and a reduction in the distance between the the injectors and the location of maximum 4.3-$\mu$m emission.[38] This behavior is sketched qualitatively in Fig. 10. When the device was operated as a laser with a given optical cavity of width $W$ and given flow composition, there was a sharply defined optimum NO flow rate for maximum power output.

Subsonic DF–CO$_2$ lasers were subsequently operated at several laboratories. Basov et al.[58] performed longitudinal flow experiments with results similar to those of Cool and Stephens.[17] A transverse flow device was constructed at the Naval Research Laboratory[59] that operated at a power output of 45 W and chemical efficiency of 4%. Brunet and Mabru[40] and Allario[60] have also successfully operated transverse-flow DF–CO$_2$ lasers.

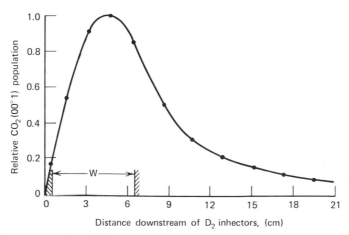

**Fig. 9.** Variation of CO$_2$(00°1) population with time after D$_2$ injection for the device of Fig. 8.

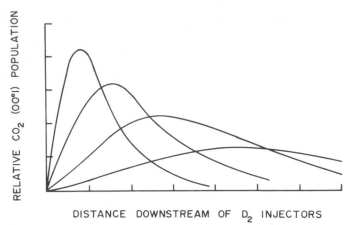

**Fig. 10.** Qualitative representation of the observed influence of NO flow rate on $CO_2(00°1)$ population for the device of Fig. 8.

**Fig. 11.** Transfer chemical laser and gas-handling system operated by Dr. T. J. Falk at an output of 560 W.

Complementing these investigations, an extensive parametric study on DF–CO$_2$ lasers has been recently completed by Falk[39] at Calspan in Buffalo, New York. Falk's laser, Fig. 11, has produced a power output of 560 W. The optical cavity for this device differed from the earlier Cornell laser of Fig. 8. This cavity consisted of a rectangular mirror and a multiple-hole output coupler placed on opposite sides of a flow channel of 2.5×15-cm cross-section. Chemical efficiencies for this device exceed 4%, in good agreement with the results obtained with the Cornell laser. Partial results of the parametric studies of Falk are given in Figs. 12 and 13. Figure 12 shows separate curves of relative power output as a function of the cavity static pressure for three different NO concentrations. The mole fractions of CO$_2$, He, F$_2$, and D$_2$ were the same for all conditions; the cavity pressure was varied by means of a downstream throttle valve. Curves of this type have maxima that define a power envelope. From Fig. 12 it is seen that maximum power output is obtained at a cavity pressure of ~17 Torr. For higher pressures, laser output decreases, and no power was observed above 74 Torr in this device. The lower curve of Fig. 12 indicates that the requisite NO flow rate for maximum power output at a given cavity pressure varies inversely as the square of the cavity pressure. This simple result has not yet been given a direct interpretation in kinetic terms. Now that the key chemical-reaction rates that determine the manner in which F-atom concentration varies with pressure and NO concentration are known,[44,45] it should be possible to model the effects of parametric variations on saturation intensity and power output with some confidence.

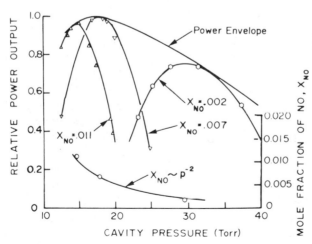

**Fig. 12.** Variation of subsonic DF–CO$_2$ laser output with cavity pressure for differing NO concentrations; dependence of optimum NO concentration on cavity pressure.[39]

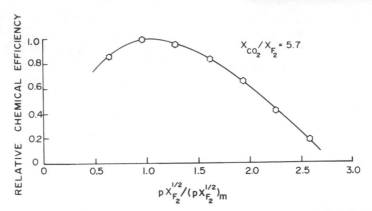

**Fig. 13.** Empirical correlation of laser efficiency with pressure and $F_2$-mole fraction applicable for a wide range of diluent concentrations and cavity pressures. Such a curve applies to a given diluent (AR, $N_2$, He, air) and a given mole ratio of $CO_2$ to $F_2$.[39]

In correlating his experimental data, Falk has introduced an empirical parameter which he calls the "equivalent pressure." It is proportional to the product of cavity pressure and the square root of $F_2$-mole fraction. Given a fixed mole ratio of $CO_2$ to $F_2$, the laser chemical efficiency exhibits a smooth variation as a function of the equivalent pressure, as Fig. 13 shows for a wide variety of conditions of pressure and diluent concentration. The abscissa of Fig. 13 is the equivalent pressure normalized by its optimum value for maximum efficiency. This type of correlation gives rise to families of similar curves for various $CO_2$-to-$F_2$ mole ratios and differing diluents (Ar, $N_2$, He, and air).

Single-pass small-signal gain measurements, for a variety of experimental conditions made at various positions along the flow,[39] have clarified the qualitative picture presented in Fig. 10. Optimization of the NO flow rate has the effect of "tuning" the population inversion to maximize power output for a given flow condition and cavity configuration. A high NO flow rate results in an inversion peak near the upstream end of the cavity; presumably, the large production of DF and ONF causes, in this case, substantial deactivation of $CO_2$[61] before the flow reaches the downstream end of the cavity. Reduction of the NO flow rate below the optimum value causes a decrease in chemical production of F and DF so that inversion extends downstream further than the optimum position for efficient power extraction. Perhaps the most important result of these extensive and careful parametric studies[39] has been the demonstration of significant power output at pressures high enough to encourage the development of supersonic devices with full pressure recovery to an atmospheric exhaust.

Solomon and co-workers[62] at Bell Aerospace have operated a subsonic TCL at power outputs exceeding 15 kW, the highest powers yet reported for the subsonic TCL. Their laser was directly scaled from the smaller lasers already discussed with the use of a detailed laser kinetics computer program. It had an unstable resonator configuration with a 75-cm transverse optical axis and 50% output coupler. The cavity operating pressure was 55 Torr; the specific power was 66 J/g.

These recent results demonstrate the feasibility of the use of a realistic computer model for the design of a subsonic DF–$CO_2$ laser. The inherent simplicity of the transverse subsonic flow configuration should lend itself well to a complete and detailed understanding of this interesting, important, purely chemical laser.

## 3. cw SUPERSONIC DF–CO₂ TCL

The gas-dynamical laser (GDL)[63–65] is an example of a device that utilizes only chemical-energy sources (available as high-pressure bottled gases) and can be operated at a sufficiently high cavity static pressure to permit atmospheric-pressure recovery at the diffuser exhaust. The fact that the population inversion in a GDL is achieved solely by a rapid gas-dynamical expansion, rather than by the nonequilibrium energy release characteristic of chemical lasers, may appear to be an insignificant distinction between the two devices. However, the working fluid of a chemical laser proceeds through a sequence of nonequilibrium conditions that thermodynamically are vastly different from the near-equilibrium path that the gases in a GDL follow. The availability of useful work (laser output) is expected to be much higher for a chemical laser than for a comparable open-cycle GDL.

Figure 14 illustrates the concept of a supersonic TCL that has been proposed by the author. It would permit DF–$CO_2$ chemical-laser opera-

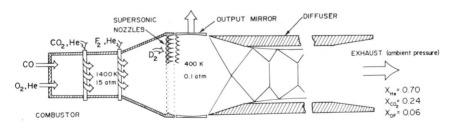

SUPERSONIC TRANSFER CHEMICAL LASER

**Fig. 14.** A chemical laser without a vacuum pump.

tion without the need of a vacuum pump. The flow conditions of this device are quite similar to those found in the present-day GDL.[65] About one-half of the required $CO_2$ is formed by the direct combustion of CO and $O_2$ in the presence of a helium diluent. Premixed $F_2$ and He $(1:9)$ and a mixture of the remaining $CO_2$ and additional He are separately injected into the combustion chamber at a location downstream of the CO combustion zone. Stagnation conditions for the mixed flows should be about 1400°K and 15 atm. Partial thermal dissociation of the $F_2$ occurs before the combined flow is accelerated through an array of supersonic nozzles. Immediately upstream of the nozzle throats, the requisite $D_2$ is injected by means of multiple arrays of small sonic orifices. The flow is then expanded to a Mach number of about 4 as it enters the optical cavity. Mixing is not completed until somewhat after the flow has expanded to the relatively low pressure of the optical cavity. The cavity consists of two rectangular mirrors transverse to a constant-area flow channel. The flow channel is terminated at the downstream end by a supersonic diffuser designed to bring the exhaust gases to ambient pressure without the use of pumps.

The RESALE-laser kinetics model developed for analysis of HF and DF chemical lasers[51] was extended by G. Emanuel[66] to include the best available kinetic rates appropriate to the DF–$CO_2$ system. This modified program was used by Emanuel to calculate the predicted performance, summarized in Table 5 and Figs. 15 through 17, for a supersonic TCL. The calculations correspond to the approximate gas composition given in Fig. 14. A comparison of operating parameters for the subsonic TCL and those predicted for a supersonic TCL is presented in Table 5. The calculated chemical efficiency of 4.7% for the supersonic device is nearly identical to the measured value of 4.6% for the subsonic TCL of Table 4. The specific power of 44 J/g for the supersonic TCL is only slightly below the 55-J/g value measured for the subsonic TCL, despite the fourfold

**Table 5.** DF–$CO_2$ Chemical-Laser Operating Characteristics

|                                           | Subsonic TCL      | Supersonic TCL    |
| ----------------------------------------- | ----------------- | ----------------- |
| Cavity static pressure (Torr)             | 16                | 60                |
| Exhaust pressure (Torr)                    | 16                | 760               |
| Mean-flow velocity (cm/sec)               | $2 \times 10^4$   | $2 \times 10^5$   |
| Specific power (J/g)                      | 55                | 44                |
| Chemical efficiency                        | 4.6               | 4.7               |
| Saturation intensity (W/cm²)              | 143               | 1300              |
| Unsaturated gain (m$^{-1}$)               | 3.3               | 2.0               |
| Temperature in reaction zone (°K)          | 400               | 400               |

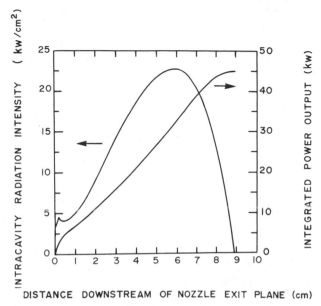

**Fig. 15.** Computed intracavity-radiation flux and integrated laser-power output for the supersonic TLC of Fig. 14.

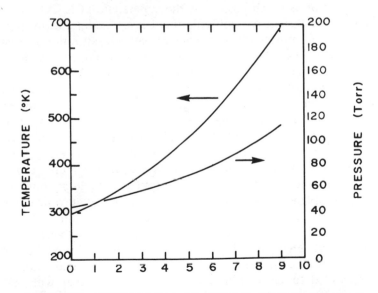

**Fig. 16.** Computed variation of pressure and temperature within the cavity of the device of Fig. 14.

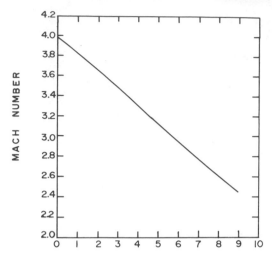

**Fig. 17.**   Theoretical Mach-number variation for the device of Fig. 14.

greater cavity pressure for the supersonic device. Thus the laser outputs per unit mass flow are similar, although the mass-flow rate for a given flow cross-section is much higher for the supersonic TCL. A measure of the useful power output per unit volume of laser medium is given by the product of the unsaturated gain coefficient and the saturation intensity.[67] Thus the supersonic TCL would be expected to have ~5.5 times the power output per unit cavity volume of the subsonic TCL.

Figure 15 shows the calculated variations along the flow direction in the intracavity radiation flux and in integrated power output for the supersonic TCL with a cavity coupling of 15%. The 45-kW total power output figure represents an order-of-magnitude increase in output over the values obtained from GDL's with the same mass flow. Figures 16 and 17 indicate the expected variations in pressure, temperature, and Mach number within the cavity. The laser output ceases 9-cm downstream from the exit plane of the supersonic-nozzle array at which position the gas temperature has climbed to 680°K. The laser flux is quenched at this point because the thermal population of the lower level has reached a value high enough to reduce the gain to zero. To investigate the possibility of improved performance when thermal cutoff of the laser is avoided, a calculation for a divergent channel with an iso-thermal-reaction zone was performed. No improvement in laser performance was found. The reason is that the inversion is reduced to zero by collisional deactivation of the $CO_2(00°1)$ level by ground state DF when the reaction is ~60% complete. Since the thermal cutoff for the constant-area channel occurred at about the same stage of reaction, it is believed that only a small advantage in power output would be realized for an optimally expanded flow (not necessarily isothermal) at a cost of considerable complexity in design. There may be other good reasons for expansion of the flow within the test section, however. Figure 16 indicates a pressure rise in the laser test section caused by chemical-heat release of such a magnitude as to violate the one-dimensional gas-dynamical assumptions made in the computations. It is expected that this thermal choking effect will cause shock formation and boundary-layer separation within the test section; poor optical quality is an expected result.

Another anticipated problem is the possible formation of carbonyl fluoride (COF$_2$) in the combustion chamber. Carbonyl fluoride is known to absorb the 10.6-$\mu$m $P_{20}$ CO$_2$ laser transition to some extent;[68] it is also an efficient deactivator of CO$_2$(00°1) molecules.[61] Although the ultimate formation of COF$_2$ is heavily favored thermodynamically because of its large negative heat of formation, the rate of formation of COF$_2$ may be slow. That is, it is difficult to construct plausible mechanisms for rapid direct attack of CO$_2$ by F and F$_2$. Any excess CO could, however, be converted rapidly to COF$_2$ in the presence of F and F$_2$.[69] The amount of COF$_2$ likely to be present in the proposed device is still an open question; the deleterious effects of COF$_2$ are not, as Fig. 18 illustrates. The experimental data of Fig. 18 show the rapid drop in laser performance when COF$_2$ is deliberately added to the subsonic DF–CO$_2$ TCL.[68]

Another experimental difficulty with the approach is that the CO/O$_2$

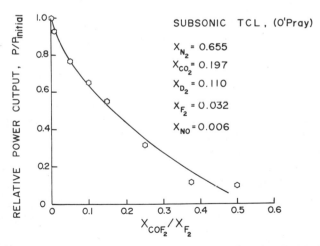

**Fig. 18.**   Influence of COF$_2$ added to the F$_2$ flow of a subsonic TCL.[68]

**Fig. 19.** Supersonic TCL at the Naval Research Laboratory.

combustion process requires a catalyst. The $H_2O$ formed in the combustion of $CH_4$ in a pilot flame is ordinarily used to accomplish this. Unfortunately, both $H_2O$ itself and any HF formed by reaction with $F_2$ can cause significant deactivation of DF and $CO_2$.

Stregack, Watt, and Cool have asembled a supersonic TCL of this type at the Naval Research Laboratory. A photograph of the device is shown in Fig. 19. It was judged more convenient to modify an existing GDL rather than to build a new device completely. In its original GDL configuration, this laser was capable of 4- to 5-kW output at a mass flow of 900 g/sec. Twenty-four supersonic nozzles with 0.8-mm throats expanded the flow into a duct of $3.8 \times 30$-cm cross-section with a transverse cavity. To date, initial experiments with the NRL laser have demonstrated a chemical-laser output of 5 kW; it is not yet clear whether this device will reach its full 45-kW projected performance level.

Other approaches to the design of supersonic DF–$CO_2$ TCL's have been proposed. Thoenes et al.[70] have performed a detailed analysis to predict the performance characteristics of a supersonic TCL of somewhat different configuration from that discussed here. Their analysis is based

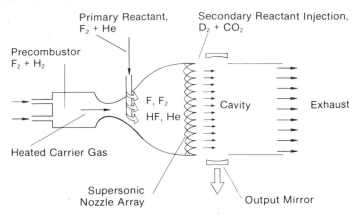

**Fig. 20.** A supersonic TCL with an F$_2$+H$_2$ precombustor.[70]

on the precombustor concept successfully employed in HF and DF chemical lasers,[71] as illustrated in Fig. 20. An H$_2$–F$_2$ precombustor operated at large mole ratios of F$_2$ to H$_2$ produces the requisite thermal dissociation for initiation of the subsequent reaction of F$_2$ and D$_2$. This approach is similar to that of the DF chemical laser; the major difference is that CO$_2$ is added to the injected D$_2$ as illustrated in Fig. 20.

An important feature of the numerical model adopted by Thoenes et al.[70] has been the inclusion of turbulent mixing processes at the relatively high pressures required for supersonic TCL operation with full atmospheric exhaust-pressure recovery. Hofland and Mirels[72] have also analyzed the effects of finite mixing rates on chemical-laser performance. The performance predictions of Thoenes et al.[70] are not as optimistic as those for the NRL laser discussed here. As a result of the finite mixing rates and the HF deactivation inherent in the TCL precombustor concept of Fig. 20, a lower specific power is predicted at cavity pressures (~60 Torr) sufficient for complete atmospheric recovery. The model of Thoenes et al. suggests a specific power of about 11 J/g compared with the value of 49 J/g computed by Emanuel for the NRL device.

The first experiments directed toward the achievement of supersonic TCL operation at high pressures were conducted by Morsell and Shey at the Boeing Company in 1970.[73] These experiments offer much encouragement for supersonic TCL development. Over 500 W of laser output was achieved from a small supersonic TCL at a cavity pressure of 25 Torr. A chemical efficiency of 3% and a specific power of 31 J/g were obtained; ambient exhaust-pressure recovery was not attempted.

Very recently, G. W. Tregay *et al.*[62] of Bell Aerospace Co., Buffalo, N.Y., have successfully operated a supersonic DF–CO$_2$ TCL at power levels exceeding 8 kw and have demonstrated laser operation at pressures to 30 Torr in the supersonic flow within the laser cavity. They have also successfully operated the laser with a $CO-O_2$ precombustor similar to that of the NRL device.

Until more experience is gained with supersonic TCL's the projected figures based on the computer calculations just discussed should be regarded with reservation. If such lasers can be built to perform to theoretical predictions, one can envision the production of large amounts of cw 10.6-$\mu$m laser power in a considerably more efficient manner than has been demonstrated with present GDL technology.[65]

## 4. PULSED DF–CO₂ CHEMICAL LASERS

Since the original observation of pulsed DF–$CO_2$ laser generation by Gross,[7] the emphasis of research on the DF–$CO_2$ system has been focused largely on the cw devices. Recently, however, renewed interest has developed in the pulsed DF–$CO_2$ system.[74–81]

Basov et al.,[74] Wilson and Stephenson,[75] Poehler et al.,[76–78] and Suchard et al.[79] have recently completed detailed studies of the pulsed DF–$CO_2$ TCL. Reaction initiation in mixtures of $D_2$, $F_2$, $CO_2$, and He was accomplished with both flash-photolysis and pulsed-discharge techniques. A schematic diagram of flash-photolysis apparatus similar to those employed in these studies is shown in Fig. 21.

Poehler et al. report a linear increase in laser output energy with increased pressure for fixed-reactant composition and fixed flash lamp-

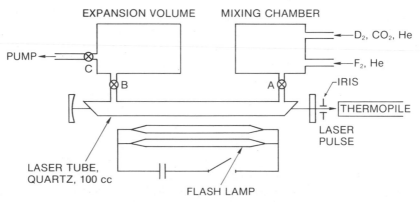

**Fig. 21.** A pulsed flash-photolyzed DF–$CO_2$ chemical laser.

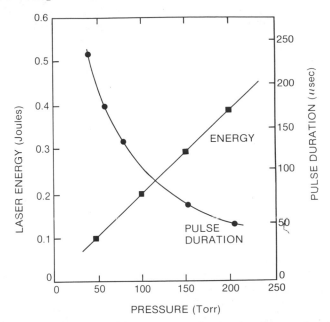

**Fig. 22.** Pulsed DF–CO₂ chemical-laser energy and pulse duration as a function of total gas pressure. $D_2 : F_2 : CO_2 : He :: 1 : 1 : 30 : 20$.[78]

energy as is indicated in Fig. 22. The duration of the laser pulse decreases with increasing pressure. These authors have constructed a theoretical kinetic model that gives results in good agreement with the experiment. It was found that the initial production of F atoms was directly proportional to the flash energy and the initial concentration of $F_2$ molecules. The constant of proportionality for their apparatus was determined experimentally by comparison of the calculated and observed dependence on flash energy of the two experimental parameters: the gain coefficient and the time interval over which positive gain exists. The results of this comparison are given in Fig. 23. The dependence of laser energy and pulse duration on mixture pressure given by the theoretical model are compared with experimental data in Fig. 24. The agreement between experiment and the model predictions is quite satisfactory. The model also gives good qualitative predictions of other experimental features. These include the saturation observed in laser output power for flash energies above 500 J and the existence of an optimum $CO_2$ partial pressure. The data of Figs. 23 and 24 were obtained for the experimental optimum composition for maximum laser output. The laser-output energy per unit volume for this device was 19 J/liter-atm, based upon the total

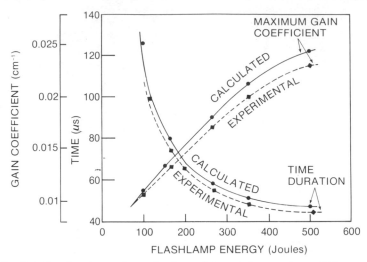

**Fig. 23.** Calculated and experimental values of the maximum-gain coefficient and laser-pulse duration as a function of flashlamp energy at a pressure of 250 Torr for the composition of Fig. 24.[76]

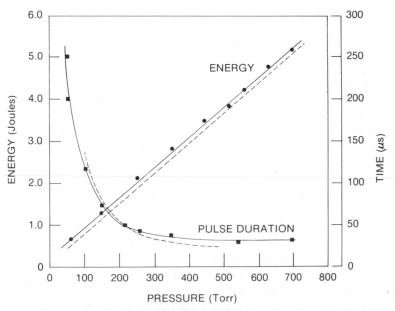

**Fig. 24.** Laser energy and pulse duration as a function of pressure for the mixture: $D_2:F_2:CO_2:He::1:1:6:19$ for a fixed flashlamp energy of 350 J. Solid lines are experimental results; dash lines are computed results.[76]

volume of the reactor. The chemical efficiency, defined as the laser-output energy divided by the chemical-energy release of the total reacting mixture, was 3.2%.

Similar data were obtained by Suchard et al.[79] with a laser that produced 2.8 J at a pressure of 380 Torr for a $D_2$–$F_2$–$CO_2$–He mixture of $0.33:1:8:10$. This energy corresponds to a chemical efficiency of 5.1% and an energy per unit volume of 20 J/liter-atm based on the total reactor volume. The efficiencies of these devices compare favorably with the value of 4.6% found for a carefully optimized device.[38] It should be noted, however, that the chemical efficiency of the pulsed devices is probably much higher because only that portion of the reactor volume coupled to the resonator modes is effective in the production of laser output. If the cavity-mode volume is used as a basis instead of the total reactor volume, the chemical efficiencies of these devices actually exceeded 15%.

When electrical initiation was employed[78] instead of initiation by flash photolysis, it was found that the operating characteristics, including chemical efficiencies and energy densities, were similar to those reported for the flash-initiated TCL. The electrical efficiency, defined as the ratio of total laser-output energy to total electrical-input energy, was 20%. This is in contrast with the overall electrical efficiencies found for initiation by flash photolysis of <1%. Here again, the laser-output energy was observed to vary linearly with mixture pressure to above 200 Torr. However, the system could not be operated reproducibly above 250 Torr because of incipient discharge nonuniformities.

Parker and Stephens[80] have reported electrical efficiencies exceeding 100% (chemical-laser output in excess of electrical-energy input) for an HF chemical laser based upon the $H_2$–$F_2$ chain reaction. A uniform discharge was achieved through use of a spark preionization technique. No data were reported for similar experiments with a DF–$CO_2$ TCL, although operation of the DF–$CO_2$ system at comparable electrical efficiencies seems quite feasible.

In a recent series of papers Kerber et al.[48,52,81] have performed an exhaustive analysis of the kinetics and performance characteristics of the pulsed DF–$CO_2$ TCL. The dominant kinetic processes in characterization of laser performance were found to be:

1. The specific rate constants for chemical excitation of vibrational states in DF:

$$F + D_2 \rightarrow DF(v = n) + D, \qquad n \leq 4$$

$$D + F_2 \rightarrow DF(v = m) + F, \qquad m \leq 9$$

2.  Vibrational deactivation of DF:

$$DF(v = n) + M \rightarrow DF(v = n - 1) + M$$

3.  Vibrational energy transfer from DF to $CO_2$:

$$DF(v = n) + CO_2(00°0) \rightarrow DF(v = n - 1) + CO_2(00°1)$$

4.  Collisional deactivation of $CO_2(00°1)$:

$$CO_2(00°1) + M \rightarrow CO_2(nm^l0) + M$$

The deactivation of the $CO_2(01^10)$ level by helium was found to be sufficiently rapid under typical experimental conditions to ensure thermalization of the $CO_2(10°0)$ population at a value fixed by the translational temperature. The rates for the previously discussed processes are reasonably well known under conditions of laser interest, and it has been possible to model $DF-CO_2$ chemical-laser performance with some accuracy. An experimental difficulty arises in the accurate specification of the initial production of F atoms; as mentioned previously, Poehler et al.[76] were able to specify the initial dissociation by comparison of experimental and theoretical laser gain characteristics, but it would be useful to have a more direct determination. Kerber[52] has calculated the dependence of laser-chemical efficiency as a function of the degree of initial $F_2$ dissociation shown in Fig. 25 for a mixture pressure of 50 Torr in a laser cavity representative of typical experimental apparatus. The overall electrical efficiency of the pulsed $DF-CO_2$ laser is probably optimized at the lower levels of dissociation, despite the fact that the chemical efficiency increases monotonically with the degree of initial dissociation. This is because the degree of initial dissociation depends on the electrical energy input. Chemical efficiencies are shown in Fig. 25 and are in reasonable agreement with the experimental values reported by Poehler et al.[76–78] and Suchard et al.[79]

An important feature of the theoretical models of Kerber et al.[48,52,81] and Poehler et al.[76] is the explanation for the existence of chemical efficiencies somewhat below the theoretical maximum possible value for the $DF-CO_2$ system of $\sim 20\%$. As the chain reaction proceeds the temperature rises, and the population of DF builds up. The temperature rise results in an increase in the population of the lower-laser level with a resultant decrease in gain. The increasing DF population leads to increased deactivation of the upper-laser level. These effects can result in termination of the laser pulse long before reaction is complete. At low levels of initial $F_2$ dissociation ($\leq 1\%$) laser termination occurs typically

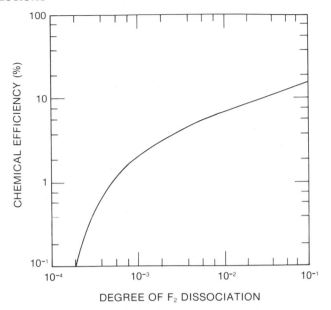

**Fig. 25.** Effect of the degree of $F_2$ dissociation on chemical efficiency for a pulsed DF–$CO_2$ chemical laser.[81]

when the reaction is only half completed. A similar result occurred in calculations of supersonic TCL performance, as was mentioned in Section 3.

## 5. CONCLUSIONS

The TCL has been demonstrated to be an efficient source for the direct conversion of chemical energy into electromagnetic radiation. Most of the development to date has been motivated by the goal of a self-contained cw DF–$CO_2$ laser capable of high power levels at $10.6\mu m$. Model calculations predict that, in its supersonic configuration, the DF–$CO_2$ TCL offers potential advantages that, if achieved, would make such a device a superior alternative to the GDL for large-scale power generation. At present, it appears that cw operation at chemical efficiencies much beyond 5% is an unrealistic goal despite earlier hope for achievement of higher values.[17] Pulsed devices have already been operated at much higher chemical efficiencies;[76–80] values of 15% are possible with proper design. In contrast to the cw devices, electrical energy is required for pulse initiation; nevertheless, overall energy-conversion efficiencies for present pulsed $DF-CO_2$ devices compare quite favorably with electrically pumped $N_2-CO_2$ lasers.

An accurate theoretical understanding of all aspects of these devices is within reach as measurements of the rates for chemical reactions and vibrational transfer and deactivation processes become increasingly available. The kinetic processes controlling the operation of these devices are of consuming interest to many investigators. Fundamental knowledge concerning heretofore poorly understood energy-transfer mechanisms will continue to accrue from laser-related studies.

Computer modeling of the subsonic $DF-CO_2$ TCL with $F_2/NO$ chemistry may yield a better understanding of its potential for high-power operation. Additional experimental investigations of the supersonic TCL are necessary. An accurate assessment of the role of fluid mixing processes and other gas-dynamical influences on supersonic TCL performance may be crucial to further development.

## REFERENCES

1. J. V. V. Kasper and G. C. Pimentel, Phys. Rev. Lett. **14,** 352 (1965).
2. J. R. Airey, IEEE J. Quantum Electron. **QE-3,** 208 (1967).
3. C. B. Moore, IEEE J. Quantum Electron. **QE-4,** 52 (1968).
4. P. H. Corneil and G. C. Pimentel, J. Chem. Phys. **49,** 1379 (1968).
5. K. L. Kompa, J. H. Parker, and G. C. Pimentel, J. Chem. Phys. **49,** 4257 (1968).
6. J. H. Parker and G. C. Pimentel, J. Chem. Phys. **51,** 91 (1969).
7. R. W. F. Gross, J. Chem. Phys. **50,** 1889 (1969).
8. H. L. Chen, J. C. Stephenson, and C. B. Moore, Chem. Phys. Lett. **2,** 593 (1968).
9. M. A. Pollack, Appl. Phys. Lett. **8,** 237 (1966); see also G. Hancock and I. W. M. Smith, Chem. Phys. Lett. **3,** 573 (1969).
10. J. R. Airey, J. Chem. Phys. **52,** 156 (1969).
11. M. C. Lin and S. H. Bauer, Chem. Phys. Lett. **7,** 223 (1970); see also M. C. Lin and L. E. Brus, J. Chem. Phys. **54,** 5423 (1971).
12. L. E. Brus and M. C. Lin, J. Phys. Chem. **75,** 2546 (1971); also J. Phys. Chem. **76,** 1429 (1972).
13. T. A. Cool, R. R. Stephens, and T. J. Falk, J. Chem. Kinet. **1,** 495 (1969).
14. T. A. Cool, T. J. Falk, and R. R. Stephens, Appl. Phys. Lett. **15,** 318 (1969).
15. T. A. Cool and R. R. Stephens, J. Chem. Phys. **52,** 3304 (1970).
16. T. A. Cool and R. R. Stephens, J. Chem. Phys. **51,** 5175 (1969).
17. T. A. Cool and R. R. Stephens, Appl. Phys. Lett. **16,** 55 (1970).
18. R. R. Stephens and T. A. Cool, J. Chem. Phys. **56,** 5863 (1972).
19. J. K. Hancock and W. H. Green, J. Chem. Phys. **56,** 2474 (1972).
20. J. J. Hinchen, private communication.

21. J. C. Stephenson, J. Finzi, and C. B. Moore, J. Chem. Phys. **56,** 5214 (1972).

22. R. S. Chang, R. A. McFarlane and G. J. Wolga, J. Chem. Phys. **56,** 667 (1972).

23. J. F. Bott and N. Cohen, J. Chem. Phys. **58,** 4539 (1973).

24. J. F. Bott and N. Cohen, J. Chem. Phys. **59,** 447 (1973).

25. R. A. Lucht and T. A. Cool, J. Chem. Phys. **60,** 1026 (1974).

26. H. K. Shin, J. Chem. Phys. **57,** 3484 (1972).

27. H. K. Shin, J. Chem. Phys. **60,** 2167 (1974).

28. G. Hancock and I. W. W. Smith, Chem. Phys. Lett. **3,** 573 (1969).

29. G. Hancock and I. W. W. Smith, Appl. Opt. **10,** 1827 (1971).

30. J. R. Airey and I. W. W. Smith, J. Chem. Phys. **57,** 1669 (1972).

31. J. F. Bott, J. Chem. Phys. **57,** 96 (1972).

32. J. A. Shirley, R. N. Sileo, R. R. Stephens, and T. A. Cool, "Purely Chemical Laser Operation in the HF, DF, HF–CO₂, and DF–CO₂ Systems," presented at the AIAA Ninth Aerospace Science Meeting, paper 71-27, Jan. 1971.

33. T. A. Dillon and J. C. Stephenson, J. Chem. Phys. **58,** 2056 (1973).

34. T. A. Dillon and J. C. Stephenson, J. Chem. Phys. **58,** 3849 (1973).

35. T. A. Dillon and J. C. Stephenson, "Energy Transfer During Orbiting Collisions," J. Chem. Phys. **60,** 4286 (1974).

36. R. D. Sharma and C. A. Brau, J. Chem. Phys. **50,** 924 (1969).

37. R. D. Sharma and C. W. Kern, J. Chem. Phys. **55,** 1171 (1971).

38. T. A. Cool, J. A. Shirley, and R. R. Stephens, Appl. Phys. Lett. **17,** 278 (1970).

39. T. J. Falk, "Parametric Studies of the DF–CO₂ Chemical Transfer Laser," CAL RF-2966-A-1, Final Technical Report F29061-70-C-066 (Cornell Aeronautics Laboratory, Buffalo, New York, June 1971).

40. H. Brunet and M. Mabru, "Etude d'un laser chimique DF–CO₂ utilisant un ecoulement gazeux transversal," C. R. Acad. Sci. (Paris) **272,** 232 (1971).

41. T. A. Cool, R. R. Stephens, and J. A. Shirley, J. Appl. Phys. **41,** 4038 (1970) (see first note added in proof).

42. N. Djeu, H. S. Pilloff and S. K. Searles, Appl. Phys. Lett. **18,** 538 (1971).

43. S. K. Searles and N. Djeu, Chem. Phys. Lett. **12,** 53 (1971).

44. E. S. Mooberry and G. J. Wolga "Reaction Kinetics of NO+F₂ → FNO+F and F+NO+M → FNO+M using EPR Detection," May 1972, unpublished.

45. P. Kim, D. I. MacLean and W. B. Valance, "Reaction Kinetics of Nitric Oxide with F and with F₂ by ESR Spectroscopy," unpublished.

46. D. Rapp and H. S. Johnston, J. Chem. Phys. **33,** 695 (1960).

47. J. M. Hoell, Jr., F. Allario, O. Jarrett, Jr., and R. K. Seals, Jr., J. Chem. Phys. **58,** 2896 (1973).

48. R. L. Kerber, N. Cohen, and G. Emanuel, IEEE J. Quantum Electron. **QE-9,** 94 (1973).

49. J. C. Polanyi and K. B. Woodall, J. Chem. Phys. **57,** 1574 (1972).

50. J. C. Polanyi and J. J. Sloan, J. Chem. Phys. **57,** 4988 (1972).

51. G. Emanuel, W. D. Adams, and E. B. Turner, "RESALE-1: A Chemical Laser Computer Program," Report TR-0172(2776)-1 (The Aerospace Corp., El Segundo, California, July 1971); G. Emanuel and W. D. Adams, RESALE-2, CO₂, and

INHALE, Report TR-0172(2776)-5 (The Aerospace Corp., El Segundo, California, April 1972).

52. R. L. Kerber, Appl. Opt. **12**, 1157 (1973).

53. G. Emanuel, J. Quant. Spectrosc. Radiat. Transfer **12**, 913 (1972).

54. G. Emanuel, J. Quant. Spectrosc. Radiat. Transfer **11**, 1481 (1972).

55. R. L. Kerber, G. Emanuel, and J. S. Whittier, Appl. Opt. **11**, 1112 (1972).

56. G. Emanuel and J. S. Whittier, Appl. Opt. **11**, 2047 (1972).

57. S. N. Suchard, R. L. Kerber, G. Emanuel, and J. S. Whittier, J. Chem. Phys. **57**, 5056 (1972).

58. N. G. Basov et al., Sov. Phys. JEPT Lett. **13**, 352 (1971).

59. T. Kan, J. A. Stregack, S. H. Bauer, and G. J. Wolga, Naval Research Laboratory, Sept. 1970, unpublished results.

60. F. Allario, NASA Langley Research Center, private communication (1971).

61. The deactivation rate of $CO_2(00°1)$ molecules by ONF and $COF_2$ have been recently measured, see T. A. Cool and J. R. Airey, Chem. Phys. Lett. **20**, 67 (1973).

62. W. Solomon, private communication (May 1974); G. W. Tregay, M. G. Drexhage, L. M. Wood, and S. J. Andrysiak, IEEE J. Quantum Electron. QE-ii, 672 (1975).

63. V. K. Konykhov and A. M. Prokhorov, Sov. Phys. JEPT Lett. **3**, 286 (1966).

64. I. R. Hurle and A. Hertzberg, Phys. Fluids **8**, 1601 (1965).

65. E. T. Gerry, IEEE Spectrum **7**, 51 (1970); also Bull. Amer. Phys. Soc. **15**, 563 (1970).

66. G. Emanuel, private communications (Jan. 1972).

67. The local gain coefficient for a homogeneously broadened line is given by the expression $\gamma = \gamma_0(1 + I/I_s)^{-1}$, where $\gamma_0$ is the unsaturated gain coefficient, $I$ is the local intracavity radiation intensity, and $I_s$ is the saturation intensity. The power output per unit cavity volume at a given position along the flow can be shown to be proportional to $\gamma_0 I_s$ which is proportional to the quantity $L^{-1} \int_0^L \gamma I \, dX$; the integration is taken along the optical axis of length $L$ [see Eqs. (19) and (23) of T. A. Cool, J. Appl. Phys. **40**, 3563 (1969), for example].

68. J. R. Doughty, J. L. Jack, and J. E. O'Pray, J. Appl. Phys. **44**, 4065 (1973).

69. H. Henrici, M. C. Lin, and S. H. Bauer, J. Chem. Phys. **52**, 5834 (1970).

70. J. Thoenes, A. W. Ratliff, and J. W. Benefield, "Analytical Program for a Continuous Wave Chemical Laser Device," Report No. RK-TR-71-19 (U.S. Army Missile Command, Redstone Arsenal, Alabama, October 1971); A. A. Hayday, "Mixing Phenomena Related to C. W. Chemical Lasers," Report No. RK-TR-71-16, (U.S. Army Missile Command, Redstone Arsenal, Alabama, July 1971); J. Thoenes, A. J. McDanal, A. W. Ratliff, S. D. Smith, L. R. Baker, Jr., J. H. McDermitt, R. R. Mikatarian, L. R. Ring, C. South, D. B. Rensch, J. W. Benefield, and S. C. Kurzius, "Chemical Laser Analysis Development," Technical Report RK-CR-73-2, Vols. 1–5 and Summary Volume (U.S. Army Missile Command, Redstone, Alabama, Oct. 1973).

71. R. A. Meinzer, Int. J. Chem. Kinet. **2**, 335 (1970).

72. R. Hofland and H. Mirels, AIAA J. **10**, 1271 (1972).

73. J. D. McClure, The Boeing Co., private communication (June 1972).

74. N. G. Basov et al., Sb. Kratkie Sobshenia Fis. **1**, 10 (1970).

75. J. Wilson and J. C. Stephenson, Appl. Lett. **20**, 64 (1972).

76. T. O. Poehler, J. C. Pirkle, Jr., and R. E. Walker, IEEE J. Quant. Electron. **QE-9,** 83 (1973).

77. T. O. Poehler, M. Shandor, and R. E. Walker, Appl. Phys. Lett. **20,** 497 (1972).

78. T. O. Poehler and R. E. Walker, Appl. Phys. Lett. **22,** 282 (1973).

79. S. N. Suchard, A. Ching, and J. S. Whittier, Appl. Phys. Lett. **21,** 274 (1972).

80. J. V. Parker and R. R. Stephens, Appl. Phys. Lett. **22,** 450 (1973).

81. R. L. Kerber, "A Parametric Study of a Pulsed DF–$CO_2$ Chemical Transfer Laser." unpublished.

# CHAPTER 8

# NUMERICAL MODELING OF CHEMICAL LASERS

## G. EMANUEL

**TRW Systems Group,
Redondo Beach, California**

The numerical modeling of pulse and cw chemical lasers is reviewed in this article. Of primary interest are the HF, DF, and HCl lasers and the $CO_2$ transfer laser. Our discussion is restricted to the gain-equals-loss (i.e., constant-gain) optical model since physical optics are covered elsewhere. The modeling requires calculations for the coupled processes of chemical reactions and lasing and the gas dynamics in the cw case. The formulations considered provide estimates for laser power and beam intensity, lasing zone length or pulse duration, and spectral content of the beam.

The first attempt at computer simulation of a chemical laser was by Cohen et al.,[1] who modeled the HCl pulse experiments of Corneil and Pimentel.[2] Their object was to explain the complex kinetic processes that cause lasing termination. For this study an existing code[3] was used to investigate a detailed model of the kinetics but without the lasing process. They introduced one feature that has been followed by all subsequent investigators. This was the treatment of individual vibrational levels of HCl as separate species, since a single vibrational temperature does not represent the vibrational population distribution before or during the laser pulse.

Shortly thereafter, Airey[4] modeled his flash-photolysis Cl+HBr pulse experiment. Included in the model was Cl atom production by photolysis, changes in vibrational populations by stimulated emission, and a simple constant-gain lasing condition. With this last condition the radiative intensity adjust instantaneously to the medium's time-dependent state. Figure 1 shows Airey's comparison of the time-resolved lasing transitions (solid lines) versus those predicted by the model (dashed lines). The vertical scale is the quantum number $J$ for the lower vibrational level of the $v = 1 \rightarrow v = 0$ band, the only band that could lase in this experiment. On the whole, agreement is quite good; however, one immediately notes that several adjacent transitions are simultaneously lasing; whereas, the model allows only one transition to lase at a given time.

Much of what has subsequently been done in pulse and cw laser modeling is presaged by these two efforts. This is really not surprising,

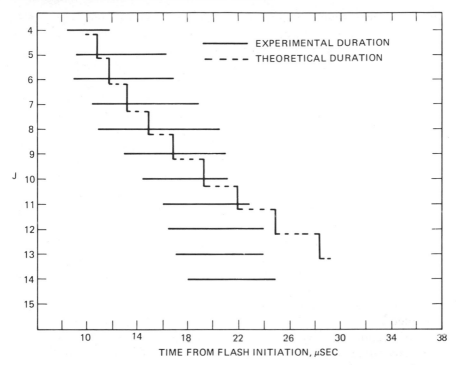

**Fig. 1.** Sequence of the lasing transitions in the $v = 1 \to v = 0$ band of HCl.[4]

since well-developed computer codes and numerical techniques already existed for treating a general nonequilibrium reacting medium (either as a quiescent gas or in a one-dimensional flow) or, with somewhat restricted kinetics, in two-dimensional and axisymmetric flow. (By general, we mean a kinetic model that allows any number of species and reactions to be used subject to computer core storage capacity.) Thus the only major new features introduced are the inclusion of the nonnegligible population changes caused by stimulated emission and a model for the photons. Both aspects are readily treated, providing the gain coefficient can be determined and a simple optical model can be used. The constant-gain model is certainly simple, and, for this reason, has been used in most studies to date. For many observables, such as power and pulse duration, it is adequate. However, it does not provide any information on modes or phase coherence; therefore, the optical quality of the beam cannot be determined.

Because of its importance, we review in Section 1 all parameters and assumptions required for calculating the gain of a V–R transition. We, therefore, cover determination of the rotational, vibrational, and transition energies, rotational-partition function, Einstein coefficients, and Doppler and Lorentz line broadening. Complete results are provided for the HF and DF molecules with references given for the HCl and $CO_2$ molecules.

Section 2 reviews comprehensively the formulation for a premixed cw laser. We treat this case first, because it is the simplest, and other cases are natural extensions of this case. A number of special topics are covered in Section 2. These include the so-called $J$-shift method for laser-line selection, the use of a saturation intensity, and a new technique for combining efficiently a physical optics computation with the constant-gain approach, thereby retaining the strongest features of both methods.

Pulse modeling is reviewed in Section 3. A simple transformation converts the cw premixed formulation in Section 2 to that for a pulse laser. We also discuss flashlamp modeling and the modeling of relaxation oscillations. The section concludes by considering possible reasons why the experimentally observed time sequencing of laser transitions does not agree with current modeling results.

A new formulation for a simple, "leaky stream-tube" mixing model is presented in Section 4. The model differs from the cw premixed treatment of Section 2 primarily through the addition of mixing source terms in the governing equations. A solution is provided for the singularity that occurs at the origin in this model.

The final section reviews the current status of two-dimensional laminar and turbulent cw formulations. Generally, these models utilize the standard boundary-layer equations and the constant-gain optical model. However, we also review work that utilizes the Navier–Stokes equations, which couples a physical-optics code to a two-dimensional flow calculation and considers the presence of transverse pressure gradients in the cavity.

## 1. EQUATION FOR THE GAIN

### 1.1. Gain Coefficient

The gain, or absorption, coefficient $\alpha(\omega)$ was introduced initially as a phenomenological parameter in the radiative transfer equation for the intensity $I$ of a ray of light along path $z$:

$$\frac{1}{I}\frac{dI}{dz} = \alpha(\omega) \tag{1}$$

For a constant $\alpha$ along the path length, Eq. (1) integrates to Beer's law

$$\frac{I}{I_0} = e^{\alpha(\omega)(z-z_0)} \tag{2}$$

where $\alpha$ is positive for gain and negative for absorption.

For a single V–R transition, $\alpha$ can be written in terms of the number densities $N_u$ and $N_l$ of the upper and lower states, the upper and lower degeneracies $g_u$ and $g_l$, the transition wave number $\omega$, a normalized line profile parameter $\phi(\omega - \omega_c)$, where $\omega_c$ is the centerline wave number, and an Einstein absorption coefficient $B_{l\rightarrow u}$. Thus, we have for the gain

$$\alpha(\omega) = \frac{hB_{l\rightarrow u}}{4\pi}\,\omega\left(\frac{g_l}{g_u}N_u - N_l\right) \tag{3}$$

where $h$ is Planck's constant and $4\pi$ is the total solid angle about a point. The quantity $hB_{l\rightarrow u}/4\pi$ is an absorption cross-section per molecule per unit solid angle, and the normalization of $\phi$ is given by

$$\int_{\omega_c-\infty}^{\omega_c+\infty} \phi(\omega)\,d\omega = 1 \tag{4}$$

For convenience as well as consistency, we designate a diatomic line transition by the triplet $(v, J, m)$, where $v$ and $J$ are the vibrational and rotational quantum numbers for the lower state, respectively, irrespective of whether the transition is caused by absorption or emission. The parameter $m$ is $-1$ for $P$-branch transitions and $+1$ for $R$-branch transitions. This definition for $m$ is convenient for computer work, but differs from that in the spectroscopic literature.

To be consistent with the kinetic formulation, we introduce the mol-mass ratio $n(v, J)$ of state $v$, $J$, and $n(v)$ for vibrational level $v$. The number densities in Eq. (3) are then replaced by

$$N_l = N_A\rho n(v, J), \qquad N_u = N_A\rho n(v+1, J+m) \tag{5}$$

where $N_A$ is Avogadro's number and $\rho$ is the gas density. We assume an equilibrium Boltzmann distribution for the rotational populations at the translational temperature $T$:

$$n(v, J) = n(v)\frac{g_l}{Q_{\text{rot}}^{(v)}}\exp\left(-\frac{hc}{k}\frac{E_{v,J}}{T}\right) \tag{6}$$

where $Q_{\text{rot}}^{(v)}$ is the rotational-partition function, $c$ is the speed of light, $k$ is the Boltzmann constant, and $E_{v,J}$ is the rotational energy of state $v, J$. The validity of the Boltzmann assumption is discussed in Section 3.

Inserting Eqs. (5) and (6) into (3) yields our final form for the gain coefficient:

$$\alpha(v, J, m) = \left(\frac{hN_A}{4\pi}\right)\omega(v, J, m)\phi\rho B(v, J, m)(2J+1)$$

$$\times \left[\frac{n(v+1)}{Q_{rot}^{(v+1)}}\exp\left(-\frac{hc}{k}\frac{E_{v+1,J+m}}{T}\right) - \frac{n(v)}{Q_{rot}^{(v)}}\exp\left(-\frac{hc}{k}\frac{E_{v,J}}{T}\right)\right] \quad (7)$$

where we have used $B(v, J, m)$ for $B_{l\to u}$ and $2J+1$ for the degeneracy $g_l$. Later, the gain is evaluated only at line center. Thus, in Section 1.2, formulas, tables, and figures are given for $E_{v,J}$, $\omega_c(v, J, m)$, and $Q_{rot}^{(v)}$. Section 1.3 provides results for $B(v, J, m)$, Section 1.4 for $\phi_c$, and Section 1.5 contains an illustrative numerical example.

## 1.2. Energies and Rotational-Partition Function

We present standard formulas[5] for the vibrational energy $G(v)$, rotational energy $E_{v,J}$, line position $\omega_c(v, J, m)$, and rotational partition function $Q_{rot}^{(v)}$ of a diatomic molecule. All energies are in wave number (cm$^{-1}$) units.

The vibrational energy is given by

$$G(v) = \sum_{l=1} Y_{l0}(v + \tfrac{1}{2})^l \quad (8)$$

where the $Y_{lj}$ are the Dunham coefficients. Zero-point energy is included, and anharmonic corrections are given by the $Y_{20}$, $Y_{30}$, ... terms. The rotational energy is a double expansion in $J(J+1)$ and $(v+\tfrac{1}{2})$

$$E_{v,J} = \sum_{j=1}\sum_{l=0} Y_{lj}(v + \tfrac{1}{2})^l [J(J+1)]^j \quad (9)$$

where the $Y_{01}$ term is the rigid-rotator approximation with other terms providing higher-order corrections. The positions of the V–R lines in the fundamental spectra of a diatomic molecule are then given by

$$\omega_c(v, J, m) = G(v+1) - G(v) + E_{v+1,J+m} - E_{v,J}$$

$$v = 0, 1, \ldots \quad J = \begin{cases} 1, 2, \ldots & m = -1 \\ 0, 1, \ldots & m = 1 \end{cases} \quad (10)$$

Finally, the rotational partition function for vibrational level $v$ is given by

$$Q_{rot}^{(v)} = 1 + \sum_{J=1}(2J+1)\exp\left(-\frac{hc}{k}\frac{E_{v,J}}{T}\right), \quad v = 0, 1, \ldots \quad (11)$$

**Table 1.** Dunham Coefficients ($cm^{-1}$)

| Coefficient | HF (Ref. 6) | DF (Ref. 7) | $H^{35}Cl$ (Ref. 8) | $CO_2(00°1)$ (Refs. 5, 9, and 10) | $CO_2(10°0)$ |
|---|---|---|---|---|---|
| $Y_{10}$ | 4138.73 | 3000.358 | 2990.9463 | — | — |
| $Y_{20}$ | −90.05 | −47.336 | −52.81856 | — | — |
| $Y_{30}$ | 0.932 | 0.373 | 0.22437 | — | — |
| $Y_{40}$ | −1.42(−2) | −6.8(−3) | −1.218(−2) | — | — |
| $Y_{50}$ | −5.9(−4) | — | — | — | — |
| $Y_{01}$ | 20.9555 | 11.0045 | 10.593416 | 0.3783 | 0.3901 |
| $Y_{11}$ | −0.7958 | −0.2936 | −0.307181 | — | — |
| $Y_{21}$ | 1.182(−2) | 1.4(−3) | 1.7724(−3) | — | — |
| $Y_{31}$ | −3.11(−4) | — | −1.201(−4) | — | — |
| $Y_{41}$ | −5.8(−6) | — | — | — | — |
| $Y_{02}$ | −2.153(−3) | −6.5(−4) | −5.31936(−4) | −1.27(−7) | −1.19(−7) |
| $Y_{12}$ | 6.23(−5) | — | 7.51(−6) | — | — |
| $Y_{22}$ | −2.06(−6) | — | −4.0(−7) | — | — |
| $Y_{03}$ | 1.68(−7) | — | 1.74(−8) | — | — |
| $Y_{13}$ | −6.5(−9) | — | −6.34(−10) | — | — |
| $Y_{04}$ | −1.9(−11) | — | −9.93(−13) | — | — |

The summation in Eq. (11) terminates when the last computed term does not significantly alter the sum. This is frequently at a large $J$, since the sum is slowly converging at high temperatures. Some care must be exercised to ensure that the higher-order terms in $E_{v,J}$ do not become dominant. If this occurs before the sum for $Q_{rot}^{(v)}$ has converged, it is necessary to recalculate with the highest-order terms (those $Y_{lj}$ terms with high $j$ subscripts) deleted in $E_{v,J}$.

Table 1 contains the Dunham coefficients for HF, DF, $H^{35}Cl$, and for the upper (00°1) and lower (10°0) vibrational levels of the 10.6-$\mu$m band of $^{12}C^{16}O_2$.[5-10] For the lower level of the $CO_2$ band, $J$ assumes even values 2, 4, 6, . . .; for the upper level, it assumes odd values 1, 3, 5, . . . . The vibrational energy, based on the data in Table 1, is given in Table 2. For the 10.6-$\mu$m band of $CO_2$, the energies are

$$G(10°0) = 3921.7 \ cm^{-1}, \qquad G(00°1) = 4882.8 \ cm^{-1}$$

where a zero-point energy, 2533.4 $cm^{-1}$, is included, and the energy of the 10°0 state contains a 43.0 $cm^{-1}$ correction caused by the 10°0−02°0 Fermi resonance.

**Table 2.** Vibrational Energy (cm$^{-1}$)

| $v$ | HF | DF | H$^{35}$Cl |
|---|---|---|---|
| 0 | 2046.97 | 1488.39 | 1482.30 |
| 1 | 6008.55 | 4395.26 | 4368.27 |
| 2 | 9797.96 | 7210.61 | 7150.28 |
| 3 | 13420.0 | 9936.36 | 9829.08 |
| 4 | 16878.8 | 12574.3 | 12405.1 |
| 5 | 20178.1 | 15125.9 | 14878.6 |
| 6 | 23320.9 | 17592.7 | 17249.4 |
| 7 | 26309.4 | 19975.9 | 19517.2 |
| 8 | 29145.2 | 22276.6 | 21681.1 |
| 9 | 31828.7 | 24495.7 | 23740.3 |

Based on the constants in Table 1, $P$- and $R$-branch line positions, Eq. (10), have been calculated and are shown in Tables 3 and 4 for HF and DF, respectively. A similar table for H$^{35}$Cl can be found in Ref. 11, while the H$^{37}$Cl transitions are approximately 2 cm$^{-1}$ removed from those of H$^{35}$Cl.[12] The $P$-branch values in the tables can be compared with the HF and DF laser wavelength measurements of Deutsch.[13] He measured the positions of 31 HF lines of the three lasing bands with an estimated accuracy of $\pm 0.12$ cm$^{-1}$. Agreement with Table $3a$ is quite good, the worst case, $P_3(3)$, differing by 0.20 cm$^{-1}$. Twenty-five of the 31 wavelengths fall within the estimated experimental accuracy. For DF, the wavelength positions of 30 lines of the four lasing bands were measured[13] with an estimated accuracy of $\pm 0.08$ cm$^{-1}$. Agreement with Table $4a$ is not as good as in the HF case, particularly at high $J$ (12 and above), since only three nonrigid rotator terms are used in constructing the DF table. In the worst case, $P_1(16)$, the difference is 2.24 cm$^{-1}$. At low $J$ values the difference is about 0.25 cm$^{-1}$.

The rotational-partition function, Eq. (11), is shown in Figs. 2 and 3 for HF and DF, respectively. As the temperature increases, the classical limit for $Q_{rot}^{(v)}$ becomes valid and

$$Q_{rot}^{(v)} \cong \frac{kT}{hc\sigma[Y_{01} + Y_{11}(v + \frac{1}{2})]} \tag{12}$$

where the symmetry factor $\sigma$ is unity for HF and DF, and the principal rigid-rotator correction term, $Y_{11}(v + 1/2)$, is retained to approximate the separation of the different $v$-curves in the figures. For CO$_2$, the exact classical limit, where $\sigma = 2$ and $Y_{11} = 0$, is generally used.[14]

**Table 3a.** Line Positions for HF, *P*-Branch.

HF(V,J,-1) , 1/CM , P-BRANCH , (V,J LOWER LEVEL)

| J= 1 | 1 | 2 | 3 | 4 | 5 | 6 | 7 |
|---|---|---|---|---|---|---|---|
| V=0 | 3.92047E+03 | 3.87786E+03 | 3.83381E+03 | 3.78837E+03 | 3.74160E+03 | 3.69354E+03 | 3.64425E+03 |
| 1 | 3.74984E+03 | 3.70882E+03 | 3.66640E+03 | 3.62262E+03 | 3.57754E+03 | 3.53121E+03 | 3.48369E+03 |
| 2 | 3.58394E+03 | 3.54446E+03 | 3.50360E+03 | 3.46143E+03 | 3.41799E+03 | 3.37334E+03 | 3.32751E+03 |
| 3 | 3.42223E+03 | 3.38425E+03 | 3.34493E+03 | 3.30432E+03 | 3.26246E+03 | 3.21941E+03 | 3.17523E+03 |
| 4 | 3.26415E+03 | 3.22763E+03 | 3.18980E+03 | 3.15070E+03 | 3.11037E+03 | 3.06888E+03 | 3.02627E+03 |
| 5 | 3.10904E+03 | 3.07395E+03 | 3.03756E+03 | 2.99997E+03 | 2.96109E+03 | 2.92210E+03 | 2.88001E+03 |
| 6 | 2.95616E+03 | 2.92247E+03 | 2.88751E+03 | 2.85139E+03 | 2.81391E+03 | 2.77583E+03 | 2.73575E+03 |
| 7 | 2.80474E+03 | 2.77243E+03 | 2.73885E+03 | 2.70406E+03 | 2.66805E+03 | 2.63093E+03 | 2.59272E+03 |
| 8 | 2.65389E+03 | 2.62294E+03 | 2.59072E+03 | 2.55772E+03 | 2.52267E+03 | 2.48692E+03 | 2.45008E+03 |

| J= 8 | 8 | 9 | 10 | 11 | 12 | 13 | 14 |
|---|---|---|---|---|---|---|---|
| V=0 | 3.59381E+03 | 3.54225E+03 | 3.48964E+03 | 3.43604E+03 | 3.38150E+03 | 3.32608E+03 | 3.26983E+03 |
| 1 | 3.43504E+03 | 3.38530E+03 | 3.33453E+03 | 3.28280E+03 | 3.23015E+03 | 3.17653E+03 | 3.12233E+03 |
| 2 | 3.28058E+03 | 3.23259E+03 | 3.18359E+03 | 3.13365E+03 | 3.08280E+03 | 3.03112E+03 | 2.97856E+03 |
| 3 | 3.12995E+03 | 3.08364E+03 | 3.03634E+03 | 2.98810E+03 | 2.93899E+03 | 2.88904E+03 | 2.83831E+03 |
| 4 | 2.98259E+03 | 2.93788E+03 | 2.89221E+03 | 2.84561E+03 | 2.79814E+03 | 2.74985E+03 | 2.70079E+03 |
| 5 | 2.83786E+03 | 2.79470E+03 | 2.75058E+03 | 2.70555E+03 | 2.65965E+03 | 2.61294E+03 | 2.56546E+03 |
| 6 | 2.69507E+03 | 2.65340E+03 | 2.61076E+03 | 2.56722E+03 | 2.52282E+03 | 2.47761E+03 | 2.43162E+03 |
| 7 | 2.55345E+03 | 2.51321E+03 | 2.47200E+03 | 2.42989E+03 | 2.38691E+03 | 2.34311E+03 | 2.29854E+03 |
| 8 | 2.41220E+03 | 2.37331E+03 | 2.33346E+03 | 2.29271E+03 | 2.25108E+03 | 2.20863E+03 | 2.16539E+03 |

| J=15 | 15 | 16 | 17 | 18 | 19 | 20 | 21 |
|---|---|---|---|---|---|---|---|
| V=0 | 3.21283E+03 | 3.15512E+03 | 3.09676E+03 | 3.03780E+03 | 2.97831E+03 | 2.91834E+03 | 2.85793E+03 |
| 1 | 3.06726E+03 | 3.01150E+03 | 2.95510E+03 | 2.89832E+03 | 2.84061E+03 | 2.78262E+03 | 2.72423E+03 |
| 2 | 2.92543E+03 | 2.87153E+03 | 2.81700E+03 | 2.76190E+03 | 2.70626E+03 | 2.65015E+03 | 2.59360E+03 |
| 3 | 2.78686E+03 | 2.73473E+03 | 2.68198E+03 | 2.62864E+03 | 2.57478E+03 | 2.52044E+03 | 2.46566E+03 |
| 4 | 2.65101E+03 | 2.60055E+03 | 2.54947E+03 | 2.49781E+03 | 2.44562E+03 | 2.39295E+03 | 2.33983E+03 |
| 5 | 2.51726E+03 | 2.46837E+03 | 2.41888E+03 | 2.36870E+03 | 2.31817E+03 | 2.26706E+03 | 2.21549E+03 |
| 6 | 2.38491E+03 | 2.33753E+03 | 2.28951E+03 | 2.24090E+03 | 2.19175E+03 | 2.14209E+03 | 2.09196E+03 |
| 7 | 2.25324E+03 | 2.20725E+03 | 2.16062E+03 | 2.11339E+03 | 2.06560E+03 | 2.01729E+03 | 1.96849E+03 |
| 8 | 2.12141E+03 | 2.07674E+03 | 2.03140E+03 | 1.98545E+03 | 1.93892E+03 | 1.89185E+03 | 1.84428E+03 |

**Table 3b.** Line Positions for HF, $R$-Branch.

$WC(V,J,J+1)$ , 1/CM , R-BRANCH , (V,J LOWER LEVEL)

| | J=0 | 1 | 2 | 3 | 4 | 5 | 6 |
|---|---|---|---|---|---|---|---|
| V=0 | 4.00115E+03 | 4.03912E+03 | 4.07545E+03 | 4.11008E+03 | 4.14299E+03 | 4.17411E+03 | 4.20342E+03 |
| 1 | 3.92742E+03 | 3.86390E+03 | 3.89889E+03 | 3.93215E+03 | 3.96373E+03 | 3.99357E+03 | 4.02164E+03 |
| 2 | 3.65859E+03 | 3.69367E+03 | 3.72719E+03 | 3.75910E+03 | 3.78937E+03 | 3.81795E+03 | 3.84479E+03 |
| 3 | 3.49399E+03 | 3.52766E+03 | 3.55981E+03 | 3.59049E+03 | 3.61937E+03 | 3.64669E+03 | 3.67231E+03 |
| 4 | 3.33306E+03 | 3.36536E+03 | 3.39616E+03 | 3.42542E+03 | 3.45311E+03 | 3.47917E+03 | 3.50357E+03 |
| 5 | 3.17515E+03 | 3.20509E+03 | 3.23555E+03 | 3.26350E+03 | 3.28990E+03 | 3.31470E+03 | 3.33787E+03 |
| 6 | 3.01953E+03 | 3.04911E+03 | 3.07725E+03 | 3.10388E+03 | 3.12899E+03 | 3.15251E+03 | 3.17443E+03 |
| 7 | 2.86537E+03 | 2.89362E+03 | 2.92042E+03 | 2.94574E+03 | 2.96953E+03 | 2.99177E+03 | 3.01241E+03 |
| 8 | 2.71182E+03 | 2.73872E+03 | 2.76418E+03 | 2.78816E+03 | 2.81064E+03 | 2.83156E+03 | 2.85089E+03 |

| | J=7 | 8 | 9 | 10 | 11 | 12 | 13 |
|---|---|---|---|---|---|---|---|
| V=0 | 4.23087E+03 | 4.25642E+03 | 4.28005E+03 | 4.30171E+03 | 4.32138E+03 | 4.33903E+03 | 4.35643E+03 |
| 1 | 4.04791E+03 | 4.07232E+03 | 4.09486E+03 | 4.11549E+03 | 4.13417E+03 | 4.15088E+03 | 4.16559E+03 |
| 2 | 3.86988E+03 | 3.89315E+03 | 3.91460E+03 | 3.93417E+03 | 3.95185E+03 | 3.96759E+03 | 3.98139E+03 |
| 3 | 3.69621E+03 | 3.71835E+03 | 3.73868E+03 | 3.75710E+03 | 3.77383E+03 | 3.78854E+03 | 3.80142E+03 |
| 4 | 3.52628E+03 | 3.54723E+03 | 3.56645E+03 | 3.58387E+03 | 3.59945E+03 | 3.61317E+03 | 3.62501E+03 |
| 5 | 3.35935E+03 | 3.37918E+03 | 3.39721E+03 | 3.41349E+03 | 3.42797E+03 | 3.44061E+03 | 3.45140E+03 |
| 6 | 3.19469E+03 | 3.21327E+03 | 3.23014E+03 | 3.24525E+03 | 3.25857E+03 | 3.27008E+03 | 3.27975E+03 |
| 7 | 3.03141E+03 | 3.04874E+03 | 3.06437E+03 | 3.07826E+03 | 3.09037E+03 | 3.10068E+03 | 3.10916E+03 |
| 8 | 2.86853E+03 | 2.88465E+03 | 2.80895E+03 | 2.91156E+03 | 2.92240E+03 | 2.93145E+03 | 2.93866E+03 |

| | J=14 | 15 | 16 | 17 | 18 | 19 | 20 |
|---|---|---|---|---|---|---|---|
| V=0 | 4.36817E+03 | 4.37961E+03 | 4.38894E+03 | 4.39614E+03 | 4.40120E+03 | 4.40410E+03 | 4.40483E+03 |
| 1 | 4.17828E+03 | 4.18892E+03 | 4.19750E+03 | 4.20400E+03 | 4.20840E+03 | 4.21069E+03 | 4.21085E+03 |
| 2 | 3.99320E+03 | 4.00301E+03 | 4.01080E+03 | 4.01654E+03 | 4.02023E+03 | 4.02184E+03 | 4.02135E+03 |
| 3 | 3.81231E+03 | 3.82124E+03 | 3.82818E+03 | 3.83311E+03 | 3.83602E+03 | 3.83687E+03 | 3.83567E+03 |
| 4 | 3.63493E+03 | 3.64292E+03 | 3.64899E+03 | 3.65301E+03 | 3.65504E+03 | 3.65507E+03 | 3.65305E+03 |
| 5 | 3.46029E+03 | 3.46723E+03 | 3.47232E+03 | 3.47546E+03 | 3.47660E+03 | 3.47560E+03 | 3.47267E+03 |
| 6 | 3.28755E+03 | 3.29345E+03 | 3.29742E+03 | 3.29945E+03 | 3.29951E+03 | 3.29758E+03 | 3.29363E+03 |
| 7 | 3.11795E+03 | 3.12052E+03 | 3.12333E+03 | 3.12421E+03 | 3.12312E+03 | 3.12004E+03 | 3.11495E+03 |
| 8 | 2.94402E+03 | 2.94745E+03 | 2.94904E+03 | 2.94865E+03 | 2.94629E+03 | 2.94194E+03 | 2.93556E+03 |

**Table 4a.** Line Positions for DF, P-Branch.

4G(V,J,J-1) , 1/CM , P-BRANCH , (V,J LOWER LEVEL)

| V | J= 1 | 2 | 3 | 4 | 5 | 6 | 7 |
|---|---|---|---|---|---|---|---|
| 0 | 2.88515E+03 | 2.86287E+03 | 2.84004E+03 | 2.81668E+03 | 2.79279E+03 | 2.76841E+03 | 2.74353E+03 |
| 1 | 2.79422E+03 | 2.77253E+03 | 2.75029E+03 | 2.72752E+03 | 2.70424E+03 | 2.68047E+03 | 2.65621E+03 |
| 2 | 2.70520E+03 | 2.68408E+03 | 2.66243E+03 | 2.64026E+03 | 2.61758E+03 | 2.59441E+03 | 2.57076E+03 |
| 3 | 2.61791E+03 | 2.59738E+03 | 2.57631E+03 | 2.55472E+03 | 2.53264E+03 | 2.51006E+03 | 2.48701E+03 |
| 4 | 2.53221E+03 | 2.51225E+03 | 2.49175E+03 | 2.47075E+03 | 2.44925E+03 | 2.42727E+03 | 2.40482E+03 |
| 5 | 2.44793E+03 | 2.42853E+03 | 2.40860E+03 | 2.38818E+03 | 2.36726E+03 | 2.34586E+03 | 2.32400E+03 |
| 6 | 2.36489E+03 | 2.34605E+03 | 2.32670E+03 | 2.30687E+03 | 2.28649E+03 | 2.26558E+03 | 2.24441E+03 |
| 7 | 2.28295E+03 | 2.26466E+03 | 2.24587E+03 | 2.22687E+03 | 2.20680E+03 | 2.18556E+03 | 2.16587E+03 |
| 8 | 2.20193E+03 | 2.18420E+03 | 2.16595E+03 | 2.14722E+03 | 2.12801E+03 | 2.10834E+03 | 2.08823E+03 |
| 9 | 2.12168E+03 | 2.10449E+03 | 2.08679E+03 | 2.06861E+03 | 2.04996E+03 | 2.03086E+03 | 2.01132E+03 |

| V | J= 8 | 9 | 10 | 11 | 12 | 13 | 14 |
|---|---|---|---|---|---|---|---|
| 0 | 2.71818E+03 | 2.69238E+03 | 2.66613E+03 | 2.63946E+03 | 2.61238E+03 | 2.58490E+03 | 2.55705E+03 |
| 1 | 2.63148E+03 | 2.60630E+03 | 2.58059E+03 | 2.55465E+03 | 2.52821E+03 | 2.50139E+03 | 2.47419E+03 |
| 2 | 2.54664E+03 | 2.52209E+03 | 2.49710E+03 | 2.47170E+03 | 2.44590E+03 | 2.41971E+03 | 2.39316E+03 |
| 3 | 2.46351E+03 | 2.43957E+03 | 2.41520E+03 | 2.39043E+03 | 2.36526E+03 | 2.33971E+03 | 2.31380E+03 |
| 4 | 2.38192E+03 | 2.35859E+03 | 2.33484E+03 | 2.31068E+03 | 2.28614E+03 | 2.26122E+03 | 2.23599E+03 |
| 5 | 2.30170E+03 | 2.27898E+03 | 2.25583E+03 | 2.23224E+03 | 2.20837E+03 | 2.18402E+03 | 2.15945E+03 |
| 6 | 2.22270E+03 | 2.20057E+03 | 2.17804E+03 | 2.15511E+03 | 2.13180E+03 | 2.10813E+03 | 2.08412E+03 |
| 7 | 2.14475E+03 | 2.12322E+03 | 2.10128E+03 | 2.07995E+03 | 2.05625E+03 | 2.03320E+03 | 2.00981E+03 |
| 8 | 2.06769E+03 | 2.04674E+03 | 2.02540E+03 | 2.00367E+03 | 1.98158E+03 | 1.95919E+03 | 1.93636E+03 |
| 9 | 1.99135E+03 | 1.97099E+03 | 1.95023E+03 | 1.92900E+03 | 1.90760E+03 | 1.88576E+03 | 1.86359E+03 |

| V | J=15 | 16 | 17 | 18 | 19 | 20 | 21 |
|---|---|---|---|---|---|---|---|
| 0 | 2.52883E+03 | 2.50026E+03 | 2.47137E+03 | 2.44215E+03 | 2.41264E+03 | 2.38284E+03 | 2.35277E+03 |
| 1 | 2.44663E+03 | 2.41873E+03 | 2.39050E+03 | 2.36197E+03 | 2.33313E+03 | 2.30402E+03 | 2.27465E+03 |
| 2 | 2.36626E+03 | 2.33902E+03 | 2.31146E+03 | 2.28359E+03 | 2.25543E+03 | 2.22701E+03 | 2.19832E+03 |
| 3 | 2.28755E+03 | 2.26096E+03 | 2.23405E+03 | 2.20686E+03 | 2.17938E+03 | 2.15163E+03 | 2.12362E+03 |
| 4 | 2.21034E+03 | 2.18441E+03 | 2.15816E+03 | 2.13162E+03 | 2.10480E+03 | 2.07772E+03 | 2.05039E+03 |
| 5 | 2.13447E+03 | 2.10918E+03 | 2.08358E+03 | 2.05770E+03 | 2.03154E+03 | 2.00512E+03 | 1.97846E+03 |
| 6 | 2.05978E+03 | 2.03512E+03 | 2.01017E+03 | 1.98493E+03 | 1.95943E+03 | 1.93367E+03 | 1.90768E+03 |
| 7 | 1.98610E+03 | 1.96207E+03 | 1.93776E+03 | 1.91316E+03 | 1.88831E+03 | 1.86320E+03 | 1.83787E+03 |
| 8 | 1.91326E+03 | 1.88986E+03 | 1.86618E+03 | 1.84222E+03 | 1.81801E+03 | 1.79356E+03 | 1.76888E+03 |
| 9 | 1.84111E+03 | 1.81934E+03 | 1.79628E+03 | 1.77195E+03 | 1.74836E+03 | 1.72457E+03 | 1.70054E+03 |

**Table 4b.** Line Positions for DF, R-Branch.

WC(V,J,+1) , 1/CM , R-BRANCH , (V,J LOWER LEVEL)

**J= 0**

| V | 0 | 1 | 2 | 3 | 4 | 5 | 6 |
|---|---|---|---|---|---|---|---|
| 0 | 2.92800E+03 | 2.94553E+03 | 2.96845E+03 | 2.98775E+03 | 3.00640E+03 | 3.02439E+03 | 3.04170E+03 |
| 1 | 2.83591E+03 | 2.85587E+03 | 2.87523E+03 | 2.89396E+03 | 2.91206E+03 | 2.92959E+03 | 2.94627E+03 |
| 2 | 2.74574E+03 | 2.76514E+03 | 2.78393E+03 | 2.80212E+03 | 2.81966E+03 | 2.83656E+03 | 2.85280E+03 |
| 3 | 2.65732E+03 | 2.67516E+03 | 2.69440E+03 | 2.71204E+03 | 2.72904E+03 | 2.74541E+03 | 2.76111E+03 |
| 4 | 2.57049E+03 | 2.58778E+03 | 2.60548E+03 | 2.62357E+03 | 2.64004E+03 | 2.65587E+03 | 2.67105E+03 |
| 5 | 2.48510E+03 | 2.50283E+03 | 2.51939E+03 | 2.53654E+03 | 2.55248E+03 | 2.56778E+03 | 2.58244E+03 |
| 6 | 2.40095E+03 | 2.41816E+03 | 2.43477E+03 | 2.45080E+03 | 2.46621E+03 | 2.48099E+03 | 2.49514E+03 |
| 7 | 2.31793E+03 | 2.33459E+03 | 2.35067E+03 | 2.36617E+03 | 2.38105E+03 | 2.39533E+03 | 2.40897E+03 |
| 8 | 2.23583E+03 | 2.25196E+03 | 2.26752E+03 | 2.28250E+03 | 2.29688E+03 | 2.31064E+03 | 2.32377E+03 |
| 9 | 2.15451E+03 | 2.17012E+03 | 2.18516E+03 | 2.19960E+03 | 2.21349E+03 | 2.22675E+03 | 2.23939E+03 |

**J= 7**

| V | 7 | 8 | 9 | 10 | 11 | 12 | 13 |
|---|---|---|---|---|---|---|---|
| 0 | 3.05832E+03 | 3.07424E+03 | 3.08944E+03 | 3.10390E+03 | 3.11760E+03 | 3.13054E+03 | 3.14269E+03 |
| 1 | 2.96236E+03 | 2.97775E+03 | 2.99242E+03 | 3.00635E+03 | 3.01955E+03 | 3.03197E+03 | 3.04362E+03 |
| 2 | 2.86835E+03 | 2.88321E+03 | 2.89735E+03 | 2.91079E+03 | 2.92347E+03 | 2.93539E+03 | 2.94654E+03 |
| 3 | 2.77614E+03 | 2.79048E+03 | 2.80412E+03 | 2.81703E+03 | 2.82921E+03 | 2.84064E+03 | 2.85129E+03 |
| 4 | 2.68556E+03 | 2.69389E+03 | 2.71251E+03 | 2.72492E+03 | 2.73660E+03 | 2.74754E+03 | 2.75771E+03 |
| 5 | 2.59644E+03 | 2.60976E+03 | 2.62238E+03 | 2.63429E+03 | 2.64549E+03 | 2.65593E+03 | 2.66562E+03 |
| 6 | 2.50853E+03 | 2.52144E+03 | 2.53357E+03 | 2.54499E+03 | 2.55569E+03 | 2.56566E+03 | 2.57487E+03 |
| 7 | 2.42195E+03 | 2.43427E+03 | 2.44590E+03 | 2.45584E+03 | 2.46706E+03 | 2.47655E+03 | 2.48530E+03 |
| 8 | 2.33626E+03 | 2.34908E+03 | 2.35923E+03 | 2.36968E+03 | 2.37943E+03 | 2.38845E+03 | 2.39674E+03 |
| 9 | 2.25138E+03 | 2.26271E+03 | 2.27338E+03 | 2.28336E+03 | 2.29264E+03 | 2.30119E+03 | 2.30902E+03 |

**J=14**

| V | 14 | 15 | 16 | 17 | 18 | 19 | 20 |
|---|---|---|---|---|---|---|---|
| 0 | 3.15404E+03 | 3.16457E+03 | 3.17428E+03 | 3.18314E+03 | 3.19113E+03 | 3.19825E+03 | 3.20447E+03 |
| 1 | 3.05447E+03 | 3.06452E+03 | 3.07374E+03 | 3.08211E+03 | 3.08963E+03 | 3.09628E+03 | 3.10204E+03 |
| 2 | 2.95691E+03 | 2.96646E+03 | 2.97520E+03 | 2.98310E+03 | 2.99015E+03 | 2.99634E+03 | 3.00164E+03 |
| 3 | 2.86117E+03 | 2.87025E+03 | 2.87851E+03 | 2.88594E+03 | 2.89253E+03 | 2.89825E+03 | 2.90310E+03 |
| 4 | 2.76710E+03 | 2.77570E+03 | 2.78350E+03 | 2.79047E+03 | 2.79659E+03 | 2.80187E+03 | 2.80627E+03 |
| 5 | 2.67454E+03 | 2.68267E+03 | 2.69000E+03 | 2.69651E+03 | 2.70219E+03 | 2.70701E+03 | 2.71097E+03 |
| 6 | 2.58332E+03 | 2.59099E+03 | 2.59785E+03 | 2.60392E+03 | 2.60915E+03 | 2.61333E+03 | 2.61705E+03 |
| 7 | 2.49328E+03 | 2.50049E+03 | 2.50691E+03 | 2.51252E+03 | 2.51731E+03 | 2.52126E+03 | 2.52435E+03 |
| 8 | 2.40426E+03 | 2.41102E+03 | 2.41699E+03 | 2.42216E+03 | 2.42651E+03 | 2.43003E+03 | 2.43269E+03 |
| 9 | 2.31609E+03 | 2.32240E+03 | 2.32793E+03 | 2.33266E+03 | 2.33658E+03 | 2.33968E+03 | 2.34193E+03 |

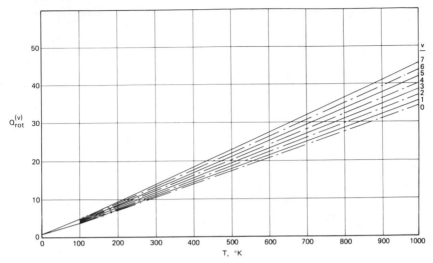

**Fig. 2.** $Q_{rot}^{(v)}$ versus temperature for HF($v$), $v = 0, \ldots, 7$.

### 1.3. Einstein Coefficients

The probability of transition $(v+1, J+m) \rightarrow (v, J)$ is given by the square of the matrix element $M_l^u$. The quantum-mechanical calculation is described in Refs. 15 through 17, with a detailed error discussion in Refs. 16 and 17. The Einstein coefficient in Eq. (7) is related to the matrix element by[16]

$$B(v, J, m) = \frac{16\pi^4 \times 10^{-7}}{3h^2 c} \left( \frac{2J+1+m}{2J+1} \right) |M_l^u|^2 \tag{13}$$

where $10^{-7}$ converts ergs to joules, and $B$ is defined in terms of an isotropic intensity, not an energy density. Accurate values of the hydrogen-halide matrix elements for the fundamental and overtone transitions have recently been produced by Meredith and Smith[18] and Herbelin and Emanuel.[19] Both sources yield the same values for the fundamental transitions. Tables 5 and 6 contain the $P$- and $R$-branch Einstein $B$ coefficients for HF and DF, respectively, based on these matrix elements. A similar table for $H^{35}Cl$ can be found in Ref. 11.

Matrix elements for the individual transitions in the 10.6-$\mu$m band of $CO_2$ have not been reported. Reference 20, however, provides a value for the Einstein $A$ coefficient for the $P(20)$ transition (the one most commonly encountered when lasing) of $1/5.38 \text{ sec}^{-1}$. This is converted to the

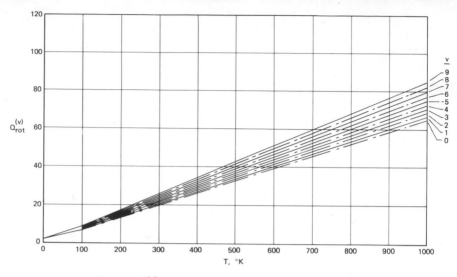

**Fig. 3.** $Q_{rot}^{(v)}$ versus temperature for DF($v$), $v = 0, \ldots, 9$.

square of the matrix element as follows:[16]

$$|M_{100,20}^{001,19}|^2 = \frac{3 \times 10^7 h}{64\pi^4 \omega^3} \frac{2J-1}{J} A$$

$$= \frac{3 \times 10^7 \times 6.6256 \times 10^{-35} \times 39}{64\pi^4 \times 944.44^3 \times 20 \times 5.38}$$

$$= 1.372 \times 10^{-39} \text{ erg-cm}^3$$

To a first approximation, $|M_l^u|^2$ is independent of $J$.[21,22] The $J$-dependent Einstein $B$ coefficients for all $P$- and $R$-branch transitions are then found using Eq. (13) with the foregoing value for $|M_l^u|^2$.

Other values for the $P(20)$ $A$ coefficient are found in the literature. For example, a value of $1/5.65 \text{ sec}^{-1}$ can be deduced from Ref. 23; whereas, Ref. 24 gives $1/5.2 \text{ sec}^{-1}$. The value of $1/4.7 \text{ sec}^{-1}$ given in Ref. 25 is reported to be too large.[26]

### 1.4. Line-Profile Function

**Assumptions.** Only collisional and Doppler broadening is considered in developing expressions for the line profile function $\phi$; all other broadening processes are negligible in comparison. At low pressures ($\sim 5$ Torr),

**Table 5a.** Einstein Fundamental *P*-Branch Coefficients for HF.

B(V,J,-1) , CM**2/MOLECULE-JOULE-SEC , P-BRANCH , (V,J LOWER LEVEL)

J= 1

| V | 1 | 2 | 3 | 4 | 5 | 6 | 7 |
|---|---|---|---|---|---|---|---|
| V=0 | 2.68951E+13 | 3.40062E+13 | 3.83456E+13 | 4.18040E+13 | 4.48978E+13 | 4.78260E+13 | 5.06819E+13 |
| 1 | 5.24404E+13 | 6.64453E+13 | 7.50689E+13 | 8.19852E+13 | 8.81967E+13 | 9.40900E+13 | 9.98459E+13 |
| 2 | 7.56363E+13 | 9.60814E+13 | 1.08806E+14 | 1.19085E+14 | 1.28359E+14 | 1.37181E+14 | 1.45809E+14 |
| 3 | 9.52909E+13 | 1.21433E+14 | 1.37912E+14 | 1.51337E+14 | 1.63510E+14 | 1.75125E+14 | 1.86503E+14 |
| 4 | 1.10055E+14 | 1.40809E+14 | 1.60496E+14 | 1.76696E+14 | 1.91472E+14 | 2.05617E+14 | 2.19498E+14 |
| 5 | 1.18508E+14 | 1.52410E+14 | 1.74528E+14 | 1.92945E+14 | 2.09863E+14 | 2.26119E+14 | 2.42102E+14 |
| 6 | 1.19330E+14 | 1.54527E+14 | 1.78043E+14 | 1.97913E+14 | 2.16317E+14 | 2.34082E+14 | 2.51586E+14 |
| 7 | 1.11575E+14 | 1.45880E+14 | 1.69510E+14 | 1.89843E+14 | 2.08870E+14 | 2.27340E+14 | 2.45584E+14 |
| 8 | 9.51033E+13 | 1.26117E+14 | 1.48358E+14 | 1.67968E+14 | 1.85551E+14 | 2.04720E+14 | 2.22730E+14 |

J= 8

| V | 8 | 9 | 10 | 11 | 12 | 13 | 14 |
|---|---|---|---|---|---|---|---|
| V=0 | 5.35155E+13 | 5.63558E+13 | 5.92714E+13 | 6.21249E+13 | 6.50750E+13 | 6.80784E+13 | 7.11403E+13 |
| 1 | 1.05561E+14 | 1.11291E+14 | 1.17072E+14 | 1.22927E+14 | 1.28874E+14 | 1.34924E+14 | 1.41088E+14 |
| 2 | 1.54382E+14 | 1.62980E+14 | 1.71651E+14 | 1.80430E+14 | 1.89339E+14 | 1.98394E+14 | 2.07606E+14 |
| 3 | 1.97816E+14 | 2.09160E+14 | 2.20598E+14 | 2.32161E+14 | 2.43893E+14 | 2.55794E+14 | 2.67882E+14 |
| 4 | 2.33307E+14 | 2.47152E+14 | 2.61100E+14 | 2.75191E+14 | 2.89449E+14 | 3.03891E+14 | 3.18526E+14 |
| 5 | 2.58010E+14 | 2.73954E+14 | 2.89998E+14 | 3.06577E+14 | 3.22512E+14 | 3.39011E+14 | 3.55676E+14 |
| 6 | 2.69016E+14 | 2.86475E+14 | 3.04014E+14 | 3.21660E+14 | 3.39421E+14 | 3.57293E+14 | 3.75263E+14 |
| 7 | 2.63764E+14 | 2.81956E+14 | 3.00194E+14 | 3.18485E+14 | 3.36818E+14 | 3.55173E+14 | 3.73516E+14 |
| 8 | 2.40691E+14 | 2.58646E+14 | 2.76599E+14 | 2.94529E+14 | 3.12403E+14 | 3.30174E+14 | 3.47788E+14 |

J=15

| V | 15 | 16 | 17 | 18 | 19 | 20 | 21 |
|---|---|---|---|---|---|---|---|
| V=0 | 7.42652E+13 | 7.74566E+13 | 8.07181E+13 | 8.40529E+13 | 8.74638E+13 | 9.09540E+13 | 9.45263E+13 |
| 1 | 1.47372E+14 | 1.53783E+14 | 1.60326E+14 | 1.67008E+14 | 1.73832E+14 | 1.80804E+14 | 1.87927E+14 |
| 2 | 2.16985E+14 | 2.26541E+14 | 2.36277E+14 | 2.46198E+14 | 2.56311E+14 | 2.66617E+14 | 2.77121E+14 |
| 3 | 2.80166E+14 | 2.92651E+14 | 3.05341E+14 | 3.18239E+14 | 3.31344E+14 | 3.44657E+14 | 3.58177E+14 |
| 4 | 3.33356E+14 | 3.48384E+14 | 3.63606E+14 | 3.79018E+14 | 3.94612E+14 | 4.10381E+14 | 4.26311E+14 |
| 5 | 3.72501E+14 | 3.89477E+14 | 4.06591E+14 | 4.23826E+14 | 4.41162E+14 | 4.58573E+14 | 4.76035E+14 |
| 6 | 3.93314E+14 | 4.11420E+14 | 4.29550E+14 | 4.47672E+14 | 4.65744E+14 | 4.83722E+14 | 5.01555E+14 |
| 7 | 3.91811E+14 | 4.10012E+14 | 4.28066E+14 | 4.45915E+14 | 4.63494E+14 | 4.80730E+14 | 4.97544E+14 |
| 8 | 3.65181E+14 | 3.82282E+14 | 3.99013E+14 | 4.15288E+14 | 4.31012E+14 | 4.46084E+14 | 4.60330E+14 |

**Table 5b.** Einstein Fundamental R-Branch Coefficients for HF.

B(V,J,+1) , CM**2/MOLECULE-JOULE-SEC , R-BRANCH , (V,J LOWER LEVEL)

**J = 0**

| V | 0 | 1 | 2 | 3 | 4 | 5 | 6 |
|---|---|---|---|---|---|---|---|
| 0 | 7.24005E+13 | 4.56286E+13 | 3.87640E+13 | 3.47940E+13 | 7.18289E+13 | 2.93512E+13 | 2.71623E+13 |
| 1 | 1.40496E+14 | 8.83056E+13 | 7.48003E+13 | 6.69250E+13 | 6.10084E+13 | 5.60451E+13 | 5.16494E+13 |
| 2 | 2.01459E+14 | 1.26200E+14 | 1.06510E+14 | 9.49168E+13 | 8.61458E+13 | 7.87565E+13 | 7.21954E+13 |
| 3 | 2.51964E+14 | 1.57176E+14 | 1.32041E+14 | 1.17071E+14 | 1.05557E+14 | 9.59961E+13 | 8.73949E+13 |
| 4 | 2.88308E+14 | 1.78830E+14 | 1.49377E+14 | 1.31566E+14 | 1.17867E+14 | 1.06213E+14 | 9.58117E+13 |
| 5 | 3.06705E+14 | 1.88941E+14 | 1.56527E+14 | 1.36641E+14 | 1.21196E+14 | 1.07991E+14 | 9.61853E+13 |
| 6 | 3.03801E+14 | 1.85333E+14 | 1.51857E+14 | 1.30923E+14 | 1.14494E+14 | 1.00387E+14 | 8.77722E+13 |
| 7 | 2.77541E+14 | 1.66950E+14 | 1.34616E+14 | 1.13940E+14 | 9.75430E+13 | 8.34330E+13 | 7.08602E+13 |
| 8 | 2.28426E+14 | 1.34454E+14 | 1.05705E+14 | 8.68498E+13 | 7.17779E+13 | 5.85543E+13 | 4.74790E+13 |

**J = 7**

| V | 7 | 8 | 9 | 10 | 11 | 12 | 13 |
|---|---|---|---|---|---|---|---|
| 0 | 2.51689E+13 | 2.33211E+13 | 2.15900E+13 | 1.99570E+13 | 1.84100E+13 | 1.69405E+13 | 1.55423E+13 |
| 1 | 4.76401E+13 | 4.39211E+13 | 4.04358E+13 | 3.71491E+13 | 3.40374E+13 | 3.10847E+13 | 2.82794E+13 |
| 2 | 6.52026E+13 | 6.06406E+13 | 5.54289E+13 | 5.05176E+13 | 4.58740E+13 | 4.14759E+13 | 3.73077E+13 |
| 3 | 7.95294E+13 | 7.22784E+13 | 6.53928E+13 | 5.89518E+13 | 5.28962E+13 | 4.71700E+13 | 4.17662E+13 |
| 4 | 8.62945E+13 | 7.74675E+13 | 6.92200E+13 | 6.14854E+13 | 5.42223E+13 | 4.74049E+13 | 4.10177E+13 |
| 5 | 8.53891E+13 | 7.54008E+13 | 6.61080E+13 | 5.74453E+13 | 4.93750E+13 | 4.18767E+13 | 3.49420E+13 |
| 6 | 7.62716E+13 | 6.56931E+13 | 5.59343E+13 | 4.69407E+13 | 3.86360E+13 | 3.11628E+13 | 2.43766E+13 |
| 7 | 5.94910E+13 | 4.91641E+13 | 3.98019E+13 | 3.13721E+13 | 2.38705E+13 | 1.73116E+13 | 1.17249E+13 |
| 8 | 3.73986E+13 | 2.85050E+13 | 2.07615E+13 | 1.41715E+13 | 8.76449E+12 | 4.58930E+12 | 1.71048E+12 |

**J = 14**

| V | 14 | 15 | 16 | 17 | 18 | 19 | 20 |
|---|---|---|---|---|---|---|---|
| 0 | 1.42108E+13 | 1.29425E+13 | 1.17345E+13 | 1.05840E+13 | 9.49191E+12 | 8.45437E+12 | 7.47142E+12 |
| 1 | 2.56122E+13 | 2.30791E+13 | 2.06733E+13 | 1.83927E+13 | 1.62331E+13 | 1.41951E+13 | 1.22776E+13 |
| 2 | 3.33584E+13 | 2.96205E+13 | 2.60888E+13 | 2.32117E+13 | 1.96338E+13 | 1.70997E+13 | 1.39902E+13 |
| 3 | 3.66734E+13 | 3.18850E+13 | 2.73978E+13 | 2.33978E+13 | 1.93296E+13 | 1.57569E+13 | 1.25020E+13 |
| 4 | 3.50528E+13 | 2.95080E+13 | 2.43650E+13 | 1.96938E+13 | 1.54428E+13 | 1.16485E+13 | 8.33116E+12 |
| 5 | 2.85712E+13 | 2.27722E+13 | 1.75590E+13 | 1.29522E+13 | 8.97853E+12 | 5.67140E+12 | 3.07162E+12 |
| 6 | 1.83437E+13 | 1.30891E+13 | 8.44730E+12 | 5.06050E+12 | 2.38136E+12 | 6.71994E+11 | 6.00674E+09 |
| 7 | 7.15187E+12 | 3.64589E+12 | 1.27197E+12 | 1.07374E+11 | 2.42909E+11 | 1.78434E+12 | 4.85434E+12 |
| 8 | 2.07729E+11 | 1.74031E+11 | 1.72928E+12 | 4.97271E+12 | 1.00767E+12 | 1.71989E+13 | 2.65295E+13 |

484

**Table 6a.** Einstein Fundamental *P*-Branch Coefficients for DF.

B(V,J-1) , CM**2/MOLECULE-JOULE-SEC , P-BRANCH = (V,J LOWER LEVEL)

| | J= 1 | 2 | 3 | 4 | 5 | 6 | 7 |
|---|---|---|---|---|---|---|---|
| V=0 | 1.92508E+13 | 2.40070E+13 | 2.67059E+13 | 2.87432E+13 | 3.04345E+13 | 3.20879E+13 | 3.36114E+13 |
| 1 | 3.84575E+13 | 4.79477E+13 | 5.34327E+13 | 5.75579E+13 | 6.10056E+13 | 6.43454E+13 | 6.74526E+13 |
| 2 | 5.75751E+13 | 7.19216E+13 | 8.01467E+13 | 8.64196E+13 | 9.17874E+13 | 9.67505E+13 | 1.01501E+14 |
| 3 | 7.66002E+13 | 9.57551E+13 | 1.06809E+14 | 1.15236E+14 | 1.22537E+14 | 1.29280E+14 | 1.35718E+14 |
| 4 | 9.54492E+13 | 1.19456E+14 | 1.33390E+14 | 1.44037E+14 | 1.53287E+14 | 1.61852E+14 | 1.70072E+14 |
| 5 | 1.14108E+14 | 1.42960E+14 | 1.59785E+14 | 1.72701E+14 | 1.83986E+14 | 1.94424E+14 | 2.04499E+14 |
| 6 | 1.32513E+14 | 1.66199E+14 | 1.85956E+14 | 2.01210E+14 | 2.14559E+14 | 2.26976E+14 | 2.38923E+14 |
| 7 | 1.50567E+14 | 1.89072E+14 | 2.11796E+14 | 2.29438E+14 | 2.44891E+14 | 2.59368E+14 | 2.73282E+14 |
| 8 | 1.68260E+14 | 2.11547E+14 | 2.37266E+14 | 2.57333E+14 | 2.74988E+14 | 2.91541E+14 | 3.07532E+14 |
| 9 | 1.85423E+14 | 2.33478E+14 | 2.62194E+14 | 2.84747E+14 | 3.04670E+14 | 3.23415E+14 | 3.41518E+14 |

| | J= 8 | 9 | 10 | 11 | 12 | 13 | 14 |
|---|---|---|---|---|---|---|---|
| V=0 | 3.51006E+13 | 3.65640E+13 | 3.80251E+13 | 3.94878E+13 | 4.09544E+13 | 4.24382E+13 | 4.39386E+13 |
| 1 | 7.04812E+13 | 7.34748E+13 | 7.64533E+13 | 7.94540E+13 | 8.24594E+13 | 8.55010E+13 | 8.85761E+13 |
| 2 | 1.06146E+14 | 1.10731E+14 | 1.15300E+14 | 1.19888E+14 | 1.24515E+14 | 1.29184E+14 | 1.33921E+14 |
| 3 | 1.42009E+14 | 1.48266E+14 | 1.54520E+14 | 1.60776E+14 | 1.67098E+14 | 1.73493E+14 | 1.79956E+14 |
| 4 | 1.78115E+14 | 1.86085E+14 | 1.94072E+14 | 2.02094E+14 | 2.10193E+14 | 2.18393E+14 | 2.26673E+14 |
| 5 | 2.14351E+14 | 2.24142E+14 | 2.33940E+14 | 2.43803E+14 | 2.53759E+14 | 2.63862E+14 | 2.74104E+14 |
| 6 | 2.50691E+14 | 2.62364E+14 | 2.74065E+14 | 2.85857E+14 | 2.97780E+14 | 3.09860E+14 | 3.22139E+14 |
| 7 | 2.86980E+14 | 3.00659E+14 | 3.14354E+14 | 3.28179E+14 | 3.42179E+14 | 3.56351E+14 | 3.70784E+14 |
| 8 | 3.23293E+14 | 3.39036E+14 | 3.54814E+14 | 3.70757E+14 | 3.86931E+14 | 4.03304E+14 | 4.20031E+14 |
| 9 | 3.59305E+14 | 3.77292E+14 | 3.95278E+14 | 4.13478E+14 | 4.31929E+14 | 4.50669E+14 | 4.69751E+14 |

| | J=15 | 16 | 17 | 18 | 19 | 20 | 21 |
|---|---|---|---|---|---|---|---|
| V=0 | 4.54663E+13 | 4.70093E+13 | 4.85777E+13 | 5.01556E+13 | 5.17717E+13 | 5.34179E+13 | 5.50936E+13 |
| 1 | 9.16931E+13 | 9.48658E+13 | 9.80822E+13 | 1.01339E+14 | 1.04663E+14 | 1.08038E+14 | 1.11488E+14 |
| 2 | 1.38717E+14 | 1.43582E+14 | 1.48552E+14 | 1.53606E+14 | 1.58716E+14 | 1.63941E+14 | 1.69256E+14 |
| 3 | 1.86561E+14 | 1.93233E+14 | 1.99993E+14 | 2.06920E+14 | 2.13969E+14 | 2.21148E+14 | 2.28442E+14 |
| 4 | 2.35104E+14 | 2.43722E+14 | 2.52455E+14 | 2.61335E+14 | 2.70396E+14 | 2.79645E+14 | 2.89123E+14 |
| 5 | 2.84499E+14 | 2.95104E+14 | 3.05878E+14 | 3.16893E+14 | 3.28103E+14 | 3.39511E+14 | 3.51230E+14 |
| 6 | 3.34628E+14 | 3.47770E+14 | 3.60335E+14 | 3.73536E+14 | 3.87045E+14 | 4.00798E+14 | 4.14933E+14 |
| 7 | 3.84770E+14 | 4.00447E+14 | 4.15689E+14 | 4.31310E+14 | 4.47279E+14 | 4.63516E+14 | 4.80110E+14 |
| 8 | 4.37021E+14 | 4.54402E+14 | 4.72071E+14 | 4.90149E+14 | 5.08685E+14 | 5.27650E+14 | 5.47027E+14 |
| 9 | 4.89239E+14 | 5.09126E+14 | 5.29447E+14 | 5.50200E+14 | 5.71457E+14 | 5.93234E+14 | 6.15536E+14 |

**Table 6b.** Einstein Fundamental *R*-Branch Coefficients for DF.

B(V,J,J+1) , CM**2/MOLECULE-JOULE-SEC , R-BRANCH , (V,J LOWER LEVEL)

| V | J= 0 | 1 | 2 | 3 | 4 | 5 | 6 |
|---|---|---|---|---|---|---|---|
| V=0 | 5.33878E+13 | 3.41799E+13 | 2.95232E+13 | 2.69552E+13 | 2.51092E+13 | 2.36004E+13 | 2.22858E+13 |
| 1 | 1.06442E+14 | 6.80837E+13 | 5.87533E+13 | 5.35964E+13 | 4.98703E+13 | 4.68229E+13 | 4.41573E+13 |
| 2 | 1.59080E+14 | 1.01658E+14 | 8.76523E+13 | 7.98511E+13 | 7.42210E+13 | 6.96064E+13 | 6.55745E+13 |
| 3 | 2.11221E+14 | 1.34831E+14 | 1.16078E+14 | 1.05655E+14 | 9.81116E+13 | 9.18896E+13 | 8.64297E+13 |
| 4 | 2.62648E+14 | 1.67484E+14 | 1.44034E+14 | 1.30970E+14 | 1.21428E+14 | 1.13565E+14 | 1.06681E+14 |
| 5 | 3.13313E+14 | 1.99538E+14 | 1.71372E+14 | 1.55625E+14 | 1.44070E+14 | 1.34540E+14 | 1.26226E+14 |
| 6 | 3.62962E+14 | 2.30845E+14 | 1.98020E+14 | 1.79548E+14 | 1.66002E+14 | 1.54805E+14 | 1.44962E+14 |
| 7 | 4.11324E+14 | 2.61274E+14 | 2.23793E+14 | 2.02592E+14 | 1.87035E+14 | 1.74102E+14 | 1.62736E+14 |
| 8 | 4.58366E+14 | 2.90710E+14 | 2.48609E+14 | 2.24706E+14 | 2.07071E+14 | 1.92389E+14 | 1.79493E+14 |
| 9 | 5.03630E+14 | 3.18895E+14 | 2.72272E+14 | 2.45642E+14 | 2.25948E+14 | 2.09512E+14 | 1.95035E+14 |

| V | J= 7 | 8 | 9 | 10 | 11 | 12 | 13 |
|---|---|---|---|---|---|---|---|
| V=0 | 2.10012E+13 | 1.99893E+13 | 1.89481E+13 | 1.79612E+13 | 1.70222E+13 | 1.61214E+13 | 1.52615E+13 |
| 1 | 4.17432E+13 | 3.95085E+13 | 3.74024E+13 | 3.54057E+13 | 3.35063E+13 | 3.16876E+13 | 2.99412E+13 |
| 2 | 6.19103E+13 | 5.85029E+13 | 5.53040E+13 | 5.22797E+13 | 4.93944E+13 | 4.66371E+13 | 4.39854E+13 |
| 3 | 8.14795E+13 | 7.59001E+13 | 7.25476E+13 | 6.85084E+13 | 6.46734E+13 | 6.08942E+13 | 5.73355E+13 |
| 4 | 1.00433E+14 | 9.46284E+13 | 8.91740E+13 | 8.40241E+13 | 7.91015E+13 | 7.43867E+13 | 6.98837E+13 |
| 5 | 1.18635E+14 | 1.17976E+14 | 1.10947E+14 | 9.87131E+13 | 9.27435E+13 | 8.70430E+13 | 8.15415E+13 |
| 6 | 1.36012E+14 | 1.27702E+14 | 1.19883E+14 | 1.12500E+14 | 1.05466E+14 | 9.87473E+13 | 9.23192E+13 |
| 7 | 1.52397E+14 | 1.42792E+14 | 1.33788E+14 | 1.25265E+14 | 1.17139E+14 | 1.09403E+14 | 1.01983E+14 |
| 8 | 1.67760E+14 | 1.58334E+14 | 1.46600E+14 | 1.36910E+14 | 1.27675E+14 | 1.18907E+14 | 1.10504E+14 |
| 9 | 1.81851E+14 | 1.69508E+14 | 1.58144E+14 | 1.47271E+14 | 1.36950E+14 | 1.27119E+14 | 1.17721E+14 |

| V | J=14 | 15 | 16 | 17 | 18 | 19 | 20 |
|---|---|---|---|---|---|---|---|
| V=0 | 1.44315E+13 | 1.36309E+13 | 1.28509E+13 | 1.21056E+13 | 1.13840E+13 | 1.06882E+13 | 1.00161E+13 |
| 1 | 2.82626E+13 | 2.66388E+13 | 2.50727E+13 | 2.35690E+13 | 2.21121E+13 | 2.07117E+13 | 1.93617E+13 |
| 2 | 4.14293E+13 | 3.89843E+13 | 3.66247E+13 | 3.43368E+13 | 3.21423E+13 | 3.00265E+13 | 2.79911E+13 |
| 3 | 5.39082E+13 | 5.05951E+13 | 4.74062E+13 | 4.43503E+13 | 4.14109E+13 | 3.85644E+13 | 3.58339E+13 |
| 4 | 6.55512E+13 | 6.13801E+13 | 5.73854E+13 | 5.35327E+13 | 4.98185E+13 | 4.62706E+13 | 4.28331E+13 |
| 5 | 7.63451E+13 | 7.12931E+13 | 6.64638E+13 | 6.18081E+13 | 5.73739E+13 | 5.30482E+13 | 4.89041E+13 |
| 6 | 8.61566E+13 | 8.02269E+13 | 7.45526E+13 | 6.90906E+13 | 6.38515E+13 | 5.88436E+13 | 5.40289E+13 |
| 7 | 9.49005E+13 | 8.80654E+13 | 8.15627E+13 | 7.53140E+13 | 6.93077E+13 | 6.35524E+13 | 5.80654E+13 |
| 8 | 1.02467E+14 | 9.47474E+13 | 8.73910E+13 | 8.03290E+13 | 7.35707E+13 | 6.71323E+13 | 6.09907E+13 |
| 9 | 1.08737E+14 | 1.00155E+14 | 9.19479E+13 | 8.41043E+13 | 7.66191E+13 | 6.94777E+13 | 6.26665E+13 |

where cw lasers operate, Doppler broadening dominates, while at high pressure (above 70 Torr), where pulse lasers frequently operate, collisional broadening dominates. By introducing the Voigt function, a unified treatment is provided encompassing both processes. As will be evident from the ensuing discussion, Doppler broadening is conceptually straightforward and is treated identically for all molecular radiators. On the other hand, collisional broadening is complex and may require different treatment for different gas mixtures. Since it is still an area of active research, our treatment of collisional broadening is thus approximate and subject to later revision.

To evaluate the possibility of line overlap, we first consider overlap between lines within a V–R band and then overlap between lines from different bands. Adjacent lines of a band of an anharmonic molecule, such as HF, do not overlap appreciably until the pressure is as high as 5 atm.[27] There is overlap at the $R$-branch bandhead, but the $R$-branch transitions have lower gain, [28] and consequently lasing does not occur at the bandhead. No overlap is thus assumed between adjacent lines of the same band.

We also assume that lasing occurs only in a narrow spectral region at the center of a transition since this is where the gain is highest. We, therefore, neglect overlap between transitions of different bands. This assumption may be questioned for some transitions at high gas pressure where the width of the collisionally broadened lines is large. In particular, if a pulse laser is operating on the HF or DF chain at high pressure, with many bands participating in the lasing, then overlap on some transitions from different bands may occur. (We return to this topic at the end of the pulse section.) So far, the question of overlap has not been addressed in any analytical or numerical study. We thus adhere to the nonoverlap assumption, realizing that under special conditions it may prove invalid part of the time on certain transitions.

Our line-profile treatment assumes homogeneous broadening where the photon flux interacts with the full populations in the upper and lower rotational states. This assumption is strictly valid when the transitions are primarily collisional broadened as in most pulse lasers. Continuous-wave lasers, however, operate at low pressures where Doppler broadening dominates. This implies that not all of the population in the frequency distribution of the line participate, and that "hole burning" should be present. However, gas lasers deplete the lines homogeneously, *even at low pressure*, under several conditions[29]: (1) when the half-width of the hole burned in the gain profile is comparable to that of the transition itself; (2) when the collision frequency is fast enough to replenish rapidly the depleted population producing the hole; and (3) when the number of modes that can oscillate per transition are sufficient to deplete the entire gain profile.

To date, the only published work that investigates hole burning in a chemical laser is that of Glaze.[30] He measured the dip in the $P_2(4)$ transition with a small cw, single-mode, HF laser using HE or $N_2$ diluent. The depth of the hole, with either diluent, was quite small (~17%) at the lowest pressures investigated (0.25 Torr for $N_2$, 0.5 Torr for He). Glaze thus concludes that velocity diffusion into the region of the hole is effective in replenishing the depleted population of the Doppler profile. At ~0.67 Torr with $N_2$ and 3 Torr with He, no hole is present. The second mechanism thus appears to be effective in preventing hole burning in low-pressure (~3 Torr) HF lasers.

Many chemical-laser experiments use a stable oscillator cavity with no mode-limiting aperatures. As a result of the high gain, typical of chemical lasers, this arrangement results in multimode operation. The gain profile thus saturates uniformly to the extent that outcoupling and losses allow. In this type of experiment, mechanism (3) is also satisfied, and the line is depleted homogeneously. It thus appears that the homogeneous broadening assumption is often valid at low pressure despite the fact that the line profile is Doppler broadened.

To recapitulate, we utilize three assumptions to simplify the line profile formulation:

1.  Lasing occurs at the centerline of the profile.

2.  Transitions do not overlap.

3.  Transitions are homogeneously broadened.

**Voigt Function.** The simultaneous broadening of a transition by the Doppler and collisional mechanisms is mathematically described by the Voigt function. Here we assume that the collisional broadening, by itself, results in a pure Lorentz profile. This profile is accurate in the vicinity of the transition centerline, but may deviate somewhat from the observed profile in the wings.[31] Since we restrict our interest to the centerline of the transition, this deviation can be neglected.

The Voigt function is a convolution integral of the profiles, and is defined by[32]

$$K(x, y) = \frac{y}{\pi} \int_{-\infty}^{\infty} \frac{e^{-t^2} \, dt}{y^2 + (x - t)^2} \tag{14}$$

Armstrong[32] has reviewed the extensive literature on this function, while

Ref. 33 gives an approximate algebraic equation for $K$. We do not need this result, however, since lasing is assumed to occur at the center of the line, and an exact formula is then available. The line profile at an arbitrary wave number $\omega$ is given by

$$\phi(\omega; v, J, m) = \left(\frac{\ln 2}{\pi}\right)^{1/2} \frac{1}{\alpha_{DP}(v, J, m)} K(x, y) \tag{15}$$

where $\phi$ satisfies Eq. (4). The quantities $x$ and $y$ are given by

$$x = (\ln 2)^{1/2} \frac{|\omega - \omega_c(v, J, m)|}{\alpha_{DP}(v, J, m)} \tag{16a}$$

and

$$y = y(v, J, m) = (\ln 2)^{1/2} \frac{\alpha_{LR}(v, J)}{\alpha_{DP}(v, J, m)} \tag{16b}$$

where $\alpha_{DP}$ and $\alpha_{LR}$ are the Doppler and Lorentz half widths at half maximum height, respectively. The computation of these half widths is described shortly.

At line center, where $x = 0$, the Voigt function becomes[32]

$$K(0, y) = [1 - \text{erf}(y)] \exp(y^2) \tag{17}$$

In the limit of pure Doppler broadening, $y = 0$ and $K(0, 0) = 1$. In the Lorentz broadening limit, $y \to \infty$ and $K(0, y) \sim 1/(\pi^{1/2} y)$. We thus have

$$\phi_{DP}(\omega_c) = \frac{(\ln 2/\pi)^{1/2}}{\alpha_{DP}} \quad \text{and} \quad \phi_{LR}(\omega_c) = \frac{1}{\pi \alpha_{LR}} \tag{18}$$

for the two limits. Note that, by substituting either limiting value into Eq. (7), we have maximum gain when the half width is a minimum.

The error function in Eq. (17) can be computed by means of Eq. (7.1.28) in Ref. 34 for $y \leq 2.4$. In the range $0 \leq y \leq 2.4$, this formula results in an accurate value for $K(0, y)$. For $y$ larger than 2.4, however, it can produce a large error in $K(0, y)$, since erf$(y) \sim 1$ as $y \to \infty$. Hence, for $y > 2.4$, we use the truncated asymptotic expression for $\phi_c$:

$$\phi_c = \frac{1}{\pi \alpha_{LR}} [1 - (2y^2)^{-1} + 3(2y^2)^{-2} - 15(2y^2)^{-3} + 105(2y^2)^{-4}] \tag{19}$$

which is based on Eq. (7.1.23) in Ref. 34. A numerical comparison in the vicinity of $y = 2.4$ shows a very smooth transition between the two methods. Since the error in the asymptotic expression is largest for small $y$, we conclude that the error in $\phi_c$ is also quite small for all positive $y$.

**Doppler Half Width.** The doppler half width is given by the well-known expression[35]

$$\alpha_{DP}(v, J, m) = \left(\frac{2 \times 10^7 \, N_A k \ln 2}{c^2}\right)^{1/2} \omega_c(v, J, m)\left(\frac{T}{W^*}\right)^{1/2} \quad (20)$$

where $W^*$ is the molecular weight of the lasing molecule.

**Lorentz Half Width.** For billiard-ball, or hard-sphere, collisions between molecule $a$ and perturber molecules $i$, the Lorentz half width at half amplitude can be written as[36]

$$\alpha_{LR} = \frac{1}{4\pi c} \sum_i N_i D_{ai}^2 \left[2\pi kT\left(\frac{1}{m_a} + \frac{1}{m_i}\right)\right]^{1/2} \quad (21a)$$

where $N_i$ is the number density of species $i$, $D_{ai}$ is the collision diameter, and $m_a$ and $m_i$ are the masses of the colliding molecules. We introduce into Eq. (21a) the mole-mass ratio $n_i$, molecular weight of species $i$ $W_i$, and the collision cross-section $\sigma_{a-i}$ to obtain

$$\alpha_{LR} = \left(\frac{2kN_A^3}{\pi^3 c^2}\right)^{1/2} \rho T^{1/2} \sum_i \sigma_{a-i}\left(\frac{1}{W_a} + \frac{1}{W_i}\right)^{1/2} n_i \quad (21b)$$

where

$$\sigma_{a-i} = \frac{\pi}{4} D_{ai}^2$$

$$N_i = N_A \rho n_i \quad (22)$$

and

$$W_i = N_A m_i$$

The equation of state

$$p = \frac{\mathscr{R}\rho T}{W} \quad (23)$$

is now used, where $p$ is pressure, $\mathscr{R}$ is the universal gas constant (82.059 cm³-atm/mol-°K), and $W$ is the average molecular weight of the gas. We thus obtain

$$\alpha_{LR} = \frac{pW}{T^{1/2}} \sum_i a_i n_i \quad (21c)$$

where the broadening coefficient $a_i$ is defined by

$$a_i = \left(\frac{2kN_A^3}{\pi^3 \mathscr{R}^2 c^2}\right)^{1/2}\left(\frac{1}{W_a} + \frac{1}{W_i}\right)^{1/2} \sigma_{a-i} \quad (24)$$

In addition to billiard-ball collisions, molecules with electric-dipole or higher-multipole moments, such as HF, have enchanced resonance broadening. This type of broadening was treated by the billiard-ball, resonant-dipole theory of Anderson[37,38] and was used in the initial modeling[39] of chemical lasers to account for the enhanced broadening of HF($v$)—HF($v'$) collisions. In this approach, resonance broadening is considered the result of collisions with perturber molecules only in the upper or lower vibrational levels of the transition. A probability function $F(T, J)$ was used to account for the strong $J$ and temperature dependence of resonance broadening.[40,41] This function is defined by

$$F(T, J) = \frac{\sum_{v} \sum_{|m^*|=-2}^{|m^*|=+1} (2J+1) \exp\left\{ -\frac{hc}{k}[(E_{v,J} + G(v))/T] \right\}}{\sum_{v} \sum_{J} (2J+1) \exp\left\{ -\frac{hc}{k}[(E_{v,J} + G(v))/T] \right\}} \tag{25}$$

where $m^* = -J$ for the $P$-branch, $m^* = J+1$ for the $R$-branch, and $(2J+1)$ is the rotational degeneracy. Note that the $v$-summations cancel if we assume the molecule is a rigid-rotator harmonic-oscillator. Figure 4

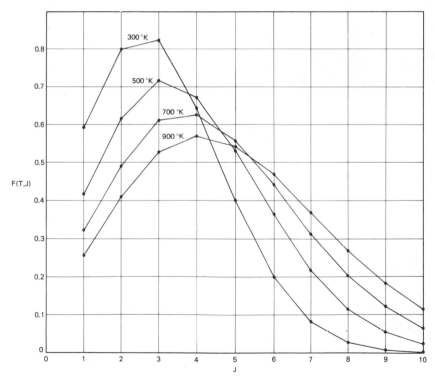

**Fig. 4.** Resonance dipole function for HF according to Eq. (25). (Courtesy R. E. Meredith.)

shows $F(T, J)$ for HF. It is evident from the figure that resonance broadening decreases with increasing temperature, and that the broadening is most pronounced for intermediate $J$ values of 3 and 4 with a sharp falloff at higher and lower $J$ values. The validity of this $J$ dependence has been verified experimentally for HCl[40] and for HF[42].

The expression used for $\alpha_{LR}$ in the early modeling[39] was

$$\alpha_{LR}(v, J) = \frac{pW}{T^{1/2}} \left\{ \sum_i a_i n_i + [n(v) + n(v+1)] F(T, J)(b^* T^{-1/2} - a^*) \right\} \quad (21d)$$

where $v$, $J$ designate the $P$- or $R$-branch transition under consideration, and $a^*$ and $b^*$ are empirical constants. For HF these constants are estimated[39] as $a^* = 1.74(°K)^{1/2}$/cm-atm and $b^* = 252°K$/cm-atm, and consequently the $a^*$ term is negligible compared to the $b^*$ term.

The validity of the approximation in Eq. (21d) that only HF molecules in levels $v$ and $v+1$ lead to resonance broadening can be assessed by the calculations of Smith et al.[43] shown in Fig. 5. These are based on the Anderson theory including higher-order multipole moments.[42] At large $J$,

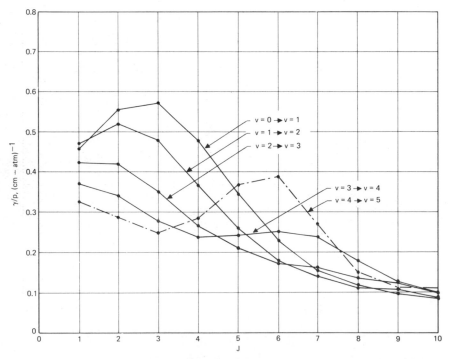

**Fig. 5.** Calculated half width for HF($v$) → HF($v+1$) transitions broadened by HF(0) at 373.1°K.[43]

not shown in the figure, the line widths achieve an asymptotic value of about 0.06 $(cm\text{-}atm)^{-1}$. This is the billiard-ball value for HF–HF collisions at 373.1°K. With this in mind, it is evident that the approximation is only qualitatively correct since appreciable resonance broadening occurs on the transitions in the upper bands due to collisions with HF(0). (Transitions $v \rightarrow v+1$ and $v+1 \rightarrow v$ have identical half widths.)

Half width at half-height line values are given in the literature as $\gamma/p$ $((cm\text{-}atm)^{-1})$. For foreign gas broadening, this converts to $a_i$ by means of the formula

$$a_i = \frac{T^{1/2}\gamma}{p} \qquad (26)$$

where the molecule being perturbed is in an infinite bath of the perturber molecules at temperature $T$. Typical room-temperature hard-sphere values for $\gamma/p$ are 0.05 $(cm\text{-}atm)^{-1}$. Measurements of self-broadening are generally done in a pure gas with the perturber in the ground state; hence, $W[n(v)+n(v+1)] \cong 1$. Ignoring the smaller $a^*$ and billiard-ball terms, we have for $b^*$

$$b^* \cong \frac{T}{F(T, J)} \cdot \frac{\gamma}{p} \qquad (27)$$

Examination of Fig. 5 gives $\gamma/p = 0.57$ $(cm\text{-}atm)^{-1}$ for the $P_1(3)$ transition at 373°K. Using $F(373, 3) \cong 0.775$ from Fig. 4, we have $b^* = 274$°K/cm-atm, which is in fair agreement with the earlier estimates of 252.

An appreciable number of papers have appeared pertaining to line broadening of the hydrogen halides. (Optical-collision cross-sectional data for $CO_2$ may be found in Refs. 20 and 36.) In addition to Benedict et al.,[40] Refs. 44 through 48 treat foreign and self-broadening of HCl. Line broadening of HF or DF is treated in Refs. 27, 41 through 43, 47, and 49 through 53. These studies show that the variation in line width with $J$ is often considerable and complex. For example, the line widths of the transitions in the $1 \rightarrow 0$ band of HCl vary by a factor of 6 when broadened by Ar, a factor of 2 when broadened by Ne, and almost no variation when broadened by He.[46] Figures 6 and 7 compare measurements of the HF $2 \rightarrow 0$ band broadened by $N_2$ and $H_2$, respectively, versus the Van-Kranendonk theory and the Anderson theory.[53] (Line widths of the first overtone band are somewhat broader than are the $1 \rightarrow 0$ linewidths.[45,46,52]) The interaction here is dipole–quadrapole, and there is no falloff at low $J$. The experimental variation with $J$ in the $P$-branch is considerable and should tend to favor lasing on high $J$ transitions due to their narrower line widths. Figure 8 shows computations for the $1 \rightarrow 0$ band broadening of DF by DF, DF by HF, and HF by

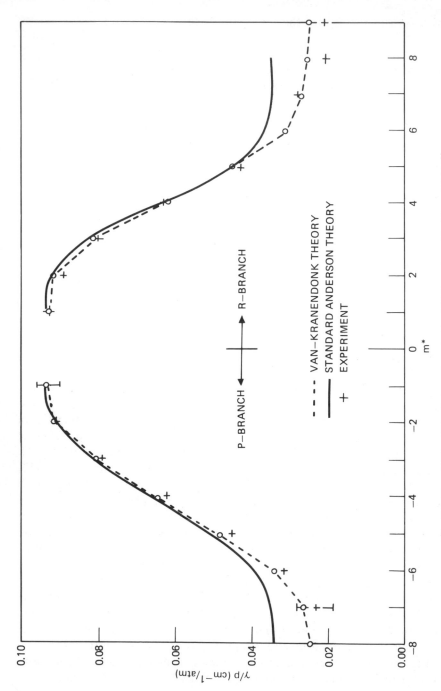

**Fig. 6.** Line widths of the first overtone band of HF broadened by $N_2$ at $373°K$.[53]

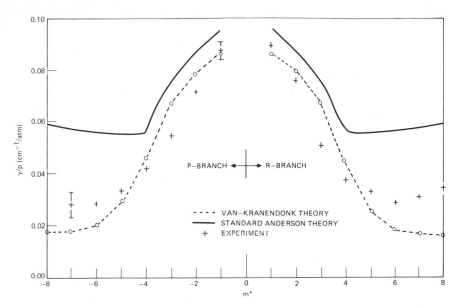

**Fig. 7.** Line widths of the first overtone band of HF broadened by $H_2$ at 373°K.[53]

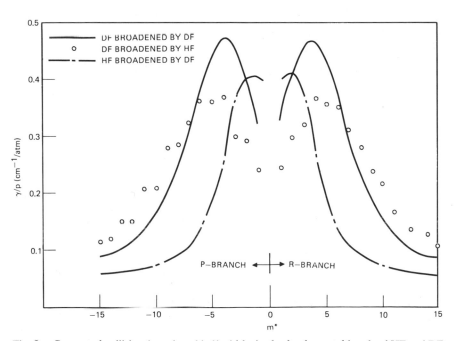

**Fig. 8.** Computed collision-broadened half widths in the fundamental bands of HF and DF at 400°K.[53]

DF using the Anderson theory.[53] As expected for dipole–dipole interaction, the half widths decrease at small $J$. For DF broadened by HF the decrease in the $P$-branch is stepwise, a consequence of the difference of 2 in the DF and HF rotational constant $Y_{01}$ (Table 1).

The line-width formulation encompassed by Eq. (21d) allows a $J$ dependence only for self-broadening. It is clear from the foregoing discussion that this is inadequate for pulse operation when collisional broadening dominates, except when He is the diluent. Equation (21d) should thus be viewed as a first estimate for the collisional broadening process.

## 1.5. Numerical Example

We calculate the peak, or centerline, gain of the $P_1(5)$ transition for a 400°K gas consisting of 19-Torr $N_2$ and 1-Torr HF. For simplicity, we assume half of the HF is in the ground state and the other half is in the first vibrational level. A number of readily determined quantities are shown, along with the table, equation, or figure that provides the value:

$$\omega_c(0, 5, -1) = 3741.6 \text{ cm}^{-1} \qquad \text{Table } 3a$$

$$B(0, 5, -1) = 4.49 \times 10^{13} \text{ cm}^2/\text{J-sec} \qquad \text{Table } 4a$$

$$\left.\begin{aligned} E_{0,5} &= 614.9 \text{ cm}^{-1} \\ E_{1,4} &= 394.9 \text{ cm}^{-1} \end{aligned}\right\} \qquad \text{Eq. (9) and Table 1}$$

$$Q_{\text{rot}}^{(0)}(400) = 13.90 \qquad \text{(Fig. 2)}$$

$$Q_{\text{rot}}^{(1)}(400) = 14.43 \qquad \text{(Fig. 2)}$$

$$\frac{hN_A}{4\pi} = 3.175 \times 10^{-11} \text{ J-sec/mol}$$

$$\frac{hc}{k} = 1.4388 \text{ cm-°K}$$

The vibrational concentrations in Eq. (7) are given by the equation of state

$$\rho n(1) = \rho n(0) = \frac{p_{\text{HF}(v)}}{\mathscr{R}T} = \frac{1}{2 \times 760 \times 82.06 \times 400} = 2.00 \times 10^{-8} \text{ mol/cm}^3$$

Equation (20) provides the Doppler half width as

$$\alpha_{\text{DP}} = 3.581 \times 10^{-7} \times 3741.6 \left(\frac{400}{20}\right)^{1/2} = 6.00 \times 10^{-3} \text{ cm}^{-1}$$

where $W^* = 20$ g/mol for HF.

We neglect the small $a^*$ term and introduce partial pressures into Eq. (21$d$) to obtain the Lorentz half width:

$$\alpha_{LR} = T^{-1/2}(a_{N_2}p_{N_2} + a_{HF(0)}p_{HF(0)} + a_{HF(1)}p_{HF(1)})$$
$$+ b^*(p_{HF(0)} + p_{HF(1)})T^{-1}F(T, J)$$

From Fig. 6 with $J = 5$, we estimate

$$a_{N_2} = (373)^{1/2} \times 0.045 = 0.867(°K)^{1/2}/\text{cm-atm}$$

The billiard-ball value for HF–HF broadening leads to

$$a_{HF(0)} = a_{HF(1)} \cong (373)^{1/2} \times 0.06 = 1.16(°K)^{1/2}/\text{cm-atm}$$

while $b^* = 274°K/\text{cm-atm}$ and $F(400, 5) \cong 0.475$ from Fig. 4. The Lorentz half width, thus, is

$$\alpha_{LR} = \frac{1}{20}\left(0.867 \times \frac{19}{760} + 1.16 \times \frac{1}{760}\right) + \frac{274 \times 0.475}{760 \times 400} = 1.78 \times 10^{-3} \text{ cm}^{-1}$$

A second method for determining $\alpha_{LR}$ would use Fig. 5 directly for both the resonant and billiard-ball HF–HF broadening as follows:

$$a_{HF(0)} = (373)^{1/2} \times 0.3448$$
$$a_{HF(1)} \cong a_{HF(0)} = 6.65(°K)^{1/2}/\text{cm-atm}$$

and

$$\alpha_{LR} = \frac{1}{20}\left(0.867 \times \frac{19}{760} + 6.65 \times \frac{1}{760}\right) = 1.72 \times 10^{-3} \text{ cm}^{-1}$$

which agrees well with the first estimate for $\alpha_{LR}$.

Using the second $\alpha_{LR}$ estimate, we have for $y$ [Eq. (16$b$)], $K(0, y)$ [Eq. (17)], and $\phi_c$ [Eq. (15)]:

$$y = 0.8326\frac{1.72 \times 10^{-3}}{6.00 \times 10^{-3}} = 0.238$$

$$K(0, y) = [1 - \text{erf}(0.238)]\exp(0.238^2) = 0.779$$

$$\phi_c - 0.4697 \times \frac{0.779}{6.00 \times 10^{-3}} - 61.0 \text{ cm}$$

We thus have for the gain [Eq. (7)]

$$\alpha(0, 5, -1) = 3.175 \times 10^{-11} \times 3741.6 \times 61.0 \times 4.49 \times 10^{13}$$

$$\times 11 \times 2 \times 10^{-8}\left\{\frac{1}{14.43}\exp\left(-\frac{1.4388 \times 394.9}{400}\right)\right.$$

$$\left. -\frac{1}{13.90}\exp\left(-\frac{1.4388 \times 614.9}{400}\right)\right\} = 0.635 \text{ cm}^{-1}$$

This large-gain value is a direct result of the assumption of equal concentrations for HF(0) and HF(1). Normally, lasers operate on partial inversions, where HF(0) > HF(1), and the first exponential term in $\alpha$ is substantially reduced relative to the second term.

## 2. PREMIXED CW MODEL

### 2.1. Introductory Remarks

In general in the cold-reaction HF or DF laser, the rate of pumping by the $F + H_2(D_2)$ reaction is rapid, and the slower rate of mixing is the controlling factor. However, mixing is faster, or achieved earlier, than the kinetics and is no longer the rate-controlling process under the following conditions:

1. When the reactant gases are premixed and the kinetics is triggered by some technique, such as a discharge, $E$-beam, or shock wave.

2. Under HF or DF chain operation where both hot- and cold-reactions pump, but where the overall rate of excited-state production is controlled by the small concentrations of the H and F chain carriers, so that considerable mixing can occur before the reaction is completed.

3. Under transfer-laser operation where the deactivation of $CO_2$ relative to HF or DF is slow.

4. In cold-reaction operation where the initial temperature and pressure are low enough to allow rapid mixing relative to the pumping reaction.

In addition to being appropriate for these cases, a premixed calculation provides a useful upper bound for the power of a cw cold reaction mixing laser, discussed later in Section 4.

The premixed formulation, of course, is simpler than that with mixing, and therefore is presented first. Furthermore, the modifications needed to extend it to the important pulse and mixing cases are straightforward. In fact, most of the formalism developed in this section is directly applied to these cases.

Along with the premixed, one-dimensional flow assumption, we adopt a standard chemical-rate equation approach for the kinetics. Our constant-gain optical treatment is consistent with this approach, that is, only the local composition and local radiative intensity affect the inversion. In addition to the assumptions already introduced, such as rotational equilibrium and homogeneous broadening, we neglect spontaneous emission, which is quite weak for the V–R transitions of diatomics. Oscillator operation is generally assumed, although the modifications needed for amplifier modeling are also presented.

The formulation of an optical model in Section 2.2 is based on the analysis in Ref. 29. The next three sections then present the entire system of equations in a manner suitable for efficient solution by a high-speed computer. The underlying theory for this approach was first given in Ref. 28, while Refs. 39 and 54 first presented the numerical formulation. Extensive parametric analysis of the premixed HF and HCl chain lasers are found in Refs. 55 and 11, respectively.

The last three sections consider topics of related interest. In Section 2.6, we discuss the analysis of the computed solutions beyond that of power and efficiency. In Section 2.7, we discuss the saturation intensity parameter, generally not used for chemical lasers but frequently used for other gas lasers. In the last section, we present a new, simple method for coupling the rate equation approach for a premixed (or mixing) laser to a physical optics code.

## 2.2. Optical Model

**Gain-Equals-Loss Condition.** The coordinate system used is shown in Fig. 9, where $x$ is the direction of flow with $z$ parallel to the optical axis. The mirrors are located on either side of the jet, which is of width $L$, and provides a regenerative cavity. Each mirror is characterized by constant values for reflectivity $r$, absorptivity $a$, and transmissivity $t$, where these satisfy

$$r_i + a_i + t_i = 1, \qquad i = 1, 2 \tag{28}$$

The outcoupled beams consist of the intensities $I_i$, which are related via the transmissivities to the intracavity fluxes (designated by a superscript $+$ or $-$ according to the beam's direction)

$$I_1 = t_1 I_1^-, \qquad I_2 = t_2 I_2^+ \tag{29}$$

From Eq. (2), and assuming the gain $\alpha$ is independent of $z$, we have for $I_1^-$ and $I_2^+$

$$I_1^- = I_2^- e^{\alpha L}, \qquad I_2^+ = I_1^+ e^{\alpha L} \tag{30}$$

The intensities $I_2^-$ and $I_1^+$, however, are given by the reflectivity conditions

$$I_2^- = r_2 I_2^+, \qquad I_1^+ = r_1 I_1^- \tag{31}$$

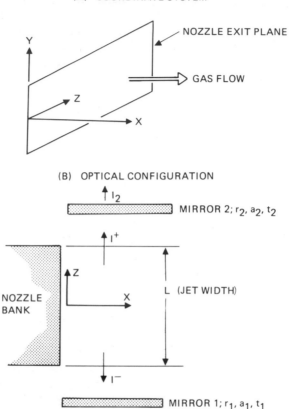

**Fig. 9.** Schematic of coordinate system and optical configuration.

From Eqs. (30) and (31) we readily obtain the gain-equals-loss (or laser-threshold) condition

$$r_1 r_2 e^{2L\alpha} = 1 \qquad (32a)$$

and the relation

$$\frac{I_1^+}{I_2^-} = \left(\frac{r_1}{r_2}\right)^{1/2} \qquad (32b)$$

According to Eq. (32a), the increase in intensity $e^{2L\alpha}$ over one round trip is balanced by absorption and transmission losses at each mirror. Later on we designate the gain, when lasing, provided by Eq. (32a) as $\alpha_{th}$. This gain has a constant value, which is why the approach is called a constant-gain model.

**Ratio of Outcoupled Power to Total Power.** It is convenient for the later analysis to determine the ratio $f$ of outcoupled intensity $I_1 + I_2$ (or power) to the outcoupled intensity plus the mirror absorbed intensity $(I_2 + I_2)_{loss}$. With the foregoing relations, this ratio is given by

$$f = \frac{I_1 + I_2}{I_1 + I_2 + (I_1 + I_2)_{loss}} = \frac{t_1(r_2)^{1/2} + t_2(r_1)^{1/2}}{(a_1 + t_1)(r_2)^{1/2} + (a_2 + t_2)(r_1)^{1/2}}$$

$$= \frac{t_1(r_2)^{1/2} + t_2(r_1)^{1/2}}{[(r_1)^{1/2} + (r_2)^{1/2}][1 - (r_1 r_2)^{1/2}]} \tag{33}$$

Let us consider the case of a nontransmitting back mirror, mirror 2, and a nonabsorbing dielectric transmitting mirror. We thus have

$$a_1 = t_2 = 0$$

$$f = \frac{1}{1 + (a_2/t_1)(r_1/r_2)^{1/2}}$$

If the transmissivity $t_1$ equals $a_2$, then $f \approx \frac{1}{2}$, and one-half the power is lost in mirror absorption. (A typical absorptivity at HF or DF wavelengths for a gold-coated mirror is 0.02.)

**Closed-Cavity Model.** To model closed-cavity operation with identical mirrors, set $t_1 = t_2 = a$, $a_1 = a_2 = 0$, and $r_1 = r_2 = r$. In this case, $f = 1$ since all the power is coupled into the mirrors.

**Zero-Power Model.** To obtain a zero-power solution, often referred to as a small-signal gain solution, we need only set $r_1 r_2 = 0$. From Eq. (32a) we then observe that threshold cannot be achieved.

**Ratio of Minimum to Maximum Cavity Intensity.** The earlier assumption that $\alpha$ is independent of $z$ is most nearly fulfilled when the intensity in both directions, $I^+ + I^-$, is also independent of $z$. The most stringent condition occurs when $r_2 = 1$ and $r_1 = r$, with mirror 1 having a maximum intensity and mirror 2 a minimum intensity. The ratio of the intensities at the two mirrors equals[29]

$$\frac{I_2^+ + I_2^-}{I_1^+ + I_1^-} = \frac{2(r)^{1/2}}{1 + r} \tag{34}$$

This ratio is between 0.75 and 1 when $1 \geq r \geq 0.2$. The assumption is thus approximately valid even at a fairly large outcoupling fraction, that is, low values of $r$. Examination of Eq. (1) also shows that an average gain, $(1/L) \int_0^L \alpha\, dz$, can be used in Eq. (30) if necessary.[56]

**Amplifier Model.**   For simulation of an amplifier we dispense with Eq. (32a). In its place an arbitrary intensity distribution $I(x)$ is used that is independent of $z$.[57] In an actual amplifier, the intensity is unidirectional with generally a large difference between the input and output values. Consequently, the assumption that $\alpha$ is independent of $z$, or can be averaged properly with $z$, is less tenable than for an oscillator cavity.

**Selection of Lasing Transitions.**   We now discuss which transitions within a band can lase. Consider a thin slice of gas in the cavity at axial location $x$ with a known density, temperature, and composition, including the vibrational levels $n(v)$ of the excited species. Hence, the gain can be determined as a function of $J$, and it can be shown that the peak gain occurs in the $P$-branch.[28] Let us denote the rotational quantum number for the $P$-branch transition with maximum gain as $J_m$. It is, therefore, natural to expect that if any transition in the band is to lase it would be this one. However, if transition $J_m$ is lasing, then Eq. (32a) is also satisfied, and $\alpha$ can be eliminated by combining this equation with Eq. (7). Thus, as long as this transition is lasing, there is a linear relation between $n(v)$ and $n(v+1)$. Furthermore, at this axial location only the $J_m$ transition in the band can lase since all other transitions have lower gain and, hence, do not satisfy Eq. (32a).

As we move downstream, transition $J_m$ will continue to lase until either the pumping can no longer maintain the gain at the value required by Eq. (32a), or the gain on a different transition in the band grows to where it equals $\alpha(v, J_m, m)$. This second transition now satisfies Eq. (32a), and it will also lase. If both transitions were allowed to lase simultaneously, we would have two different linear relations connecting $n(v)$ and $n(v+1)$. These two equations would fully determine $n(v)$ and $n(v+1)$, but now the problem is overdetermined since we also have rate equations, given later, for the vibrational populations. To avoid this overdeterminancy, we require that only one transition in a band can lase at any axial location in the cavity. As a consequence, when the gain of a nonlasing transition overtakes the lasing one, lasing abruptly shifts to this new transition. As might be anticipated, the gain on the terminated transition immediately begins to fall.[28] The process we are describing is termed $J$-shifting and is analyzed theoretically in detail in Ref. 28.

The requirement that only one transition in a band can lase at a given axial location is a consequence of one of two major assumptions. The most evident of these is rotational equilibrium used in the derivation of

Eq. (7). Deleting this assumption simply means deriving a gain equation without the use of the Boltzmann relation, Eq. (6). Lasing on transition $(v, J, m)$ then provides a linear relation between $n(v, J)$ and $n(v+1, J+m)$. Multiline lasing per band now poses no overdeterminancy with the vibration populations given by

$$n(v) = \sum_J n(v, J), \qquad v = 0, 1, \ldots \tag{35}$$

So far, the evidence for rotational nonequilibrium stems from pulse experiments. We, therefore, discuss this evidence and the physical justification for rotational nonequilibrium at the end of Section 3.

   The second major assumption is the use of an optics model that results in a local lasing condition, Eq. (32a). If we retain the rotational-equilibrium assumption, but replace Eq. (32a) by a mode approach, then multiline lasing per band is again feasible. The underlying reason for this is that the coherent modes, either of the stable or unstable variety, are volumetric entities. Equation (32a) is then replaced by an integral equation that balances round-trip gain versus losses for the volume encompassed by the mode. The highly restrictive linear relation between $n(v)$ and $n(v+1)$ no longer applies. Each transition within a band with adequate gain can now lase providing a mode can be established for that transition where the net volumetric gain balances the mode's outcoupling and losses. The modes, of course, interact with each other via the Boltzmann equation. As a consequence of this relation, lasing on various transitions within the band alter the inversion between $n(v+1)$ and $n(v)$ so that adequate volumetric gain sufficient for lasing on other transitions may not occur.

## 2.3.  Formulation of Equations

**Flow Conservation Equations.**   The equations of state, momentum, and continuity have their usual form for a one-dimensional flow

$$p = \mathscr{R}\rho T\left(\sum_i n_i + \sum_v n(v)\right) \tag{36}$$

$$u\frac{du}{dx} + 1.01325 \times 10^6 \frac{1}{\rho}\frac{dp}{dx} = 0 \tag{37}$$

and

$$\dot{m} = \rho u A \tag{38}$$

where $u$ is velocity, $\dot{m}$ is the mass-flow rate, and $A$ is the cross-sectional area of the flow. In Eqs. (36) through (38) pressure is in atmospheric units, and all other quantities are in cgs units, hence, the conversion factor in Eq. (37). Conservation of energy is given by

$$h + \frac{u^2}{2 \times 4.184 \times 10^{10}} + \frac{P}{4.184 \times 10^3 \, f\dot{m}} = h_{in} + \frac{u_{in}^2}{4.184 \times 10^{10}} \tag{39}$$

where $h$ (in kcal/g) is the specific enthalpy

$$h = \sum_i H_i n_i \tag{40}$$

$H_i(T)$ is the molar enthalpy of species $i$, and $P$ is the outcoupled (useful) power in watts. Several conversion factors appear, and the subscript "in" denotes initial conditions. The ratio $P/f$ is the total power, outcoupled plus losses, produced by the medium. The summation in Eq. (40) is over all constituents, including the vibrational levels of the excited species. (On occasion, it may be desirable to also treat the vibrational levels of a nonlasing molecule, such as $H_2$, as separate species. This is done in order to model accurately $HF-H_2$ V–V exchange. The summations, of course, include these vibrational levels.)

As is always the case with one-dimensional flows, it is necessary to specify one of the variables $A$, $p$, $T$, ... as a function of $x$.

**Rate Equations.** All reactions are represented by

$$\sum_i \alpha_{ri} X_i \underset{k_{br}}{\overset{k_{fr}}{\rightleftharpoons}} \sum_i \beta_{ri} X_i$$

where $X_i$ stands for the species whose mole-mass ratio is $n_i$ or $n(v)$; $\alpha_{ri}$ and $\beta_{ri}$ are stoichiometric coefficients; $k_{fr}$ and $k_{br}$ are rate coefficients; and the summation is over all species including any catalytic ones. Those species not involved in reaction $r$ have $\alpha_{ri} = \beta_{ri} = 0$, while any catalytic species has $\alpha_{ri} = \beta_{ri} = 1$. A catalytic species may be a single species or it may be a sum of species. The sum form is merely a simple way of combining a number of reactions, all of which differ only in the choice of catalyst, into a single reaction. This is permissible when all the summed reactions have the same rate coefficients.

The rate equations for nonlasing species are given by

$$\rho u \frac{dn_i}{dx} = \chi_{ch}(i) = \sum_r (\beta_{ri} - \alpha_{ri}) L_r \tag{41}$$

where $L_r$ is

$$L_r = k_{fr} \prod_j (\rho n_j)^{\alpha_{rj}} - k_{br} \prod_j (\rho n_j)^{\beta_{rj}} \tag{42}$$

and the summation in Eq. (41) is over all reactions that alter $n_i$. For excited species $n(v)$, the rate equations are

$$\rho u \frac{dn(v)}{dx} = X_{ch}(v) + X_{rad}(v) - X_{rad}(v-1), \qquad v = 0, 1, \ldots \qquad (43)$$

where $X_{rad}(-1) = 0$, and $X_{ch}$ has already been defined. The quantity $X_{rad}(v)$ accounts for the increase in $n(v)$ per cubic centimeter per second caused by lasing on transition $(v+1, J_m + m) \to (v, J_m)$, where $J_m$ denotes the transition in the $(v+1) \to v$ band with maximum gain. Similarly, $X_{rad}(v-1)$ accounts for the decrease in $n(v)$ caused by lasing from $v$ to $v-1$.

Before threshold, all the $X_{rad}$ are zero, and there is no formal difference between Eqs. (41) and (43). At the first lasing threshold, one of the $X_{rad}$ becomes positive and now appears with opposite signs in two consecutive rate equations. Gradually, other $X_{rad}$ terms turn on as lasing becomes multiband. Since $X_{rad}(v)$ represents the effect of stimulated emission plus absorption, it cannot be negative.[28] Downstream, when $X_{rad}(v)$ first becomes zero, lasing terminates on the $v+1 \to v$ band. After lasing has fully terminated, Eqs. (41) and (43) are again indistinguishable.

**Power and Intensity.**  We designate the average, two-way intensity $I^+ + I^-$ inside the medium by $I$ in Watts per square centimeter. It can be shown that $I$ and $X_{rad}$ are related by[28]

$$X_{rad}(v) = \frac{\alpha(v, J_m, m)I(v)}{hcN_A\omega_c(v, J_m, m)} \qquad (44)$$

where $\alpha$ must satisfy both Eqs. (7) and (32a).

To evaluate $I$ when $\alpha$ is independent of $z$, note that

$$I^+(z) = I_1^+ e^{\alpha z}$$

$$\frac{1}{L}\int_0^L I^+(z)\,dz = \frac{e^{\alpha L}-1}{\alpha L} I_1^+$$

With similar formulas for $I^-(z)$ and its average, we obtain

$$I = \frac{e^{\alpha L}-1}{\alpha L}(I_1^+ + I_2^-) = \frac{e^{\alpha L}-1}{\alpha L}\frac{(r_1)^{1/2}+(r_2)^{1/2}}{(r_2)^{1/2}} I_2^- \qquad (45)$$

where Eq. (32b) is used to obtain the right-hand expression.

We next evaluate, with the assistance of Eqs. (29), (30), and (32b), the outcoupled power $P$ in terms of the transmitted intensities to obtain

$$
\begin{aligned}
dP &= (I_1 + I_2)y\,dx \\
&= e^{\alpha L}(t_1 I_2^- + t_2 I_1^+)y\,dx \\
&= \frac{e^{\alpha L}}{L}\frac{t_1(r_2)^{1/2} + t_2(r_1)^{1/2}}{(r_2)^{1/2}}I_2^- A\,dx
\end{aligned}
\qquad (46a)
$$

where $y\,dx$ is the mirror area for power $dP$ and $A = Ly$. By means of Eqs. (32$a$), (33), (45), and (46$a$), the power can be further written as

$$
\frac{dP}{dx} = fA\alpha I \qquad (46b)
$$

Combining Eqs. (44) and (46$b$) produces our final version

$$
\frac{dP(x;v)}{dx} = hcN_A fA(x)\omega_c(v, J_m, m)X_{\text{rad}}(v) \qquad (46c)
$$

where $P(x; v)$ is the outcoupled power from threshold to $x$ for the $v+1 \to v$ band. The total outcoupled power to $x$ is simply given by

$$
P(x) = \sum_v P(x; v) \qquad (47)
$$

Once the $X_{\text{rad}}(v)$ are known, we determine $I(v)$ by Eq. (44) and the band power $P(x; v)$ by integration of Eq. (46$c$). Thus $X_{\text{rad}}(v)$, which also appears in the rate equations, is a logical choice for the independent radiative variable.

## 2.4. Numerical Formulation

**Preliminary Remarks.** A number of important considerations govern the numerical treatment of the equations. These are summarized as follows:

1. We would like to evolve a linear system of equations in order to take advantage of efficient matrix inversion algorithms.

2. The formulation should encompass solutions for stream tubes with prescribed pressure or area for cw modeling. Conversion to a pulse code should involve a minimum of changes.

3. A general approach should be used that allows an arbitrary number of reactions, nonlasing species, and vibrational levels of the excited species.

4. Within the bounds of the previously mentioned constraints, the numerical procedure should be as simple as possible for ease in coding. Advantage should be taken, when lasing is not occurring, of simpler and computationally faster approaches.

Since the momentum and rate equations are first-order, ordinary, differential equations, we develop a linear system where the unknown variables are

$$\frac{1}{T}\frac{dT}{dx}, \quad \frac{1}{A}\frac{dA}{dx}, \quad \frac{1}{\rho}\frac{d\rho}{dx}, \quad \frac{1}{p}\frac{dp}{dx}, \quad \frac{dn(v)}{dx}, \qquad v = 0, 1, \dots \qquad (48)$$

This system is solved by straightforward matrix inversion for a stream tube with a given temperature, area, density, or pressure distribution. These differentials, along with the rate equations, are then numerically integrated. When lasing is not occurring, the same procedure is used, except that the rate equations for $n(v)$ are now treated in the same manner as the other rate equations.

**Rate Equations.** Equation (43) cannot be integrated directly as is done for the nonlasing species since the $X_{rad}(v)$ are unknown. To overcome this difficulty, it is necessary to reformulate these equations. We expedite this reformulation by the introduction of two vectors $\delta(v)$ and $v^{(k)}$ such that

$$\delta(v) = \begin{cases} 0, & \text{if } X_{rad}(v) = 0 \\ 1, & \text{if } X_{rad}(v) > 0 \end{cases} \qquad (49)$$

where $v = 0, 1, \dots, v_{fin} - 1$. ($v_{fin}$ is the highest vibrational level being considered.) For convenience, we adjoin to this vector the values $\delta(-1) = \delta(v_{fin}) = 0$. The $v^{(k)}$ vector is defined as

$$0 \le v^{(1)} < v^{(2)} < \cdots < v^{(2N)} \le v_{fin} \qquad (50a)$$

where

$$\begin{aligned} v^{(1)} &= \text{smallest } v \text{ such that } \delta[v^{(1)}] = 1 \\ v^{(2)} &= \text{next larger } v \text{ such that } \delta[v^{(2)}] = 0 \\ v^{(3)} &= \text{next larger } v \text{ such that } \delta[v^{(3)}] = 1 \\ v^{(4)} &= \text{next larger } v \text{ such that } \delta[v^{(4)}] = 0, \text{ and so on.} \end{aligned} \qquad (50b)$$

The $v^{(k)}$ vector contains $2N$ entries, where $N$ is defined by Eq. (50a). By means of the $\delta(v)$, we keep track of those bands that are lasing, while the $v^{(k)}$ keeps track of a string of adjacent bands that are simultaneously lasing.

One can readily show by writing out Eq. (43) with some of the $X_{rad}(v)$ equal to zero that the following equations apply:

$$\sum_{v=v^{(1)}}^{v^{(2)}} \frac{dn(v)}{dx} = \frac{1}{\rho u} \sum_{v=v^{(1)}}^{v^{(2)}} X_{ch}(v)$$

$$\sum_{v=v^{(3)}}^{v^{(4)}} \frac{dn(v)}{dx} = \frac{1}{\rho u} \sum_{v=v^{(3)}}^{v^{(4)}} X_{ch}(v) \qquad (51)$$

$$\vdots$$

There are $N$ such equations.

The nonlasing rate equations are written as

$$\frac{dn_i}{dx} = \frac{1}{\rho u} X_{ch}(i) \qquad (52)$$

for purposes of numerical integration. When lasing is not occurring between $v+1$ and $v$ or between $v$ and $v-1$, that is, $\delta(v)+\delta(v-1)=0$, the rate equation for $n(v)$ has the form of Eq. (52).

The $dn(v)/dx$ are finally determined by simultaneous solution of a set of equations consisting of Eq. (51), Eq. (52) when applicable to $n(v)$, flow equations, and gain-equals-loss equations that are discussed later. Once values for the $dn(v)/dx$ and $X_{ch}(v)$ are known, we find the $X_{rad}(v)$ by writing Eq. (43) as

$$X_{rad}(0) = -X_{ch}(0) + \rho u \frac{dn(0)}{dx}$$

$$\vdots \qquad (53)$$

$$X_{rad}(v) = -X_{ch}(v) + \rho u \frac{dn(v)}{dx} + X_{rad}(v-1)$$

The intensity and power are then calculated from Eqs. (44) and (46c), respectively.

**Flow Equations.**  Differentiation of Eqs. (36) and (38), when combined with momentum, results in

$$\frac{1}{T}\frac{dT}{dx} + \frac{1}{\rho}\frac{d\rho}{dx} - \frac{1}{p}\frac{dp}{dx} = \Lambda_2 \qquad (54)$$

and

$$\frac{1}{A}\frac{dA}{dx} + \frac{1}{\rho}\frac{d\rho}{dx} - \Gamma_2\frac{1}{p}\frac{dp}{dx} = 0 \qquad (55)$$

where $u$ is eliminated by means of the continuity equation, and where $\Gamma_2$ and $\Lambda_2$ are

$$\Gamma_2 = \frac{1.01325 \times 10^6 \mathcal{R}T}{Wu^2} \tag{56}$$

$$\Lambda_2 = -W\left(\sum_i \frac{dn_i}{dx} + \sum_v \frac{dn(v)}{dx}\right)$$

$$= -\frac{W}{\rho u}\left(\sum_i \chi_{ch}(i) + \sum_{v=0}^{v_{fin}} \chi_{ch}(v)\right) \tag{57}$$

and

$$W = \left(\sum_i n_i + \sum_v n(v)\right)^{-1} \tag{58}$$

Differentiating Eq. (39) for the energy yields

$$\Gamma_1 \frac{1}{T}\frac{dT}{dx} - \frac{1}{p}\frac{dp}{dx} = \Lambda_1 - \frac{10}{1.01325\,puAf} \sum_{v=0}^{v_{fin}-1} \frac{dP(x;v)}{dx} \tag{59a}$$

In this equation $\Gamma_1$ and $\Lambda_1$ are given by

$$\Gamma_1 = \frac{W}{R}\left(\sum_i n_i C_{pi} + \sum_v n(v)C_{pv}\right) \tag{60}$$

$$C_{pi} = 10^3 \frac{dH_i}{dT}, \qquad C_{pv} = 10^3 \frac{dH(v)}{dT} \tag{61}$$

$$\Lambda_1 = \Lambda_1' - \frac{10^3\,W}{RT} \sum_{v=0}^{v_{fin}} H(v)\frac{dn(v)}{dx} \tag{62a}$$

and

$$\Lambda_1' = -\frac{10^3\,W}{RT} \sum_i H_i \frac{dn_i}{dx} = -\frac{10^3\,W}{RT\rho u} \sum_i H_i \chi_{ch}(i) \tag{62b}$$

where $R$ $(= 1.98725$ cal/mol-°K$)$ is the universal gas constant, and $C_{pi}$ and $C_{pv}$ are molar-heat capacities at constant pressure. It is important to note that when lasing is not occurring the summation term in Eq. (62a) becomes part of the summation term in $\Lambda_1'$ and $\Lambda_1 = \Lambda_1'$. Introducing Eqs. (46c) and (62) and $\delta(v)$ into Eq. (59a) results in

$$\Gamma_1 \frac{1}{T}\frac{dT}{dx} - \frac{1}{p}\frac{dp}{dx} + \frac{10^3\,W}{RT} \sum_{v=0}^{v_{fin}} H(v)\frac{dn(v)}{dx}$$

$$= \Lambda_1' - \frac{10hcN_A}{1.01325\,pu} \sum_{v=0}^{v_{fin}-1} \delta(v)\,\omega_c(v)\,\chi_{rad}(v) \tag{59b}$$

Equation (53) is next used to finally obtain

$$\Gamma_1 \frac{1}{T}\frac{dT}{dx} - \frac{1}{p}\frac{dp}{dx} + \frac{10^3 W}{RT} \sum_{v=0}^{v_{\text{fin}}} \frac{dn(v)}{dx}\left[ H(v) + \frac{hcN_A}{4.184\times10^3}\delta(v) \sum_{j=v}^{v^{(k+1)}-1} \omega_c(j)\right]$$

$$= \Lambda_1' + \frac{10hcN_A}{1.01325\,pu}\sum_{v=0}^{v_{\text{fin}}}\delta(v)\,\omega_c(v)\sum_{j=v^{(k)}}^{v}\chi_{ch}(j) \quad (59c)$$

The two $j$-sums need to be computed only when $\delta(v)=1$. In these sums, therefore, $k$ is an odd integer and $v^{(k)}\le v < v^{(k+1)}$. With this in mind, we see that the double sum on the left-hand side stems from the relation

$$\sum_{v=0}^{v_{\text{fin}}}\delta(v)\,\omega_c(v)\sum_{j=v^{(k)}}^{v}\frac{dn(j)}{dx} = \sum_{v=0}^{v_{\text{fin}}}\frac{dn(v)}{dx}\delta(v)\sum_{j=v}^{v^{(k+1)}-1}\omega_c(j) \quad (63)$$

Note that the left-hand sides of Eqs. (54), (55), and (59c) are linear in the variables of array (48). The coefficients of these quantities and the right-hand sides of the equations are known algebraic functions of $\rho$, $T$, $p$, $A$, $n_i$, and $n(v)$.

**Flow-Model Equation.** The flow model is controlled by the identity:

$$a_1\frac{1}{T}\frac{dT}{dx} + a_2\frac{1}{A}\frac{dA}{dx} + a_3\frac{1}{\rho}\frac{d\rho}{dx} + a_4\frac{1}{p}\frac{dp}{dx} = b \quad (64)$$

Its use is best described by an example. If the flow is defined by specification of density as a function of distance, then $a_1 = a_2 = a_4 = 0$, $a_3 = 1$, and $b = (1/\rho)(d\rho/dx)$. Including Eq. (64) in the set of simultaneous equations to be solved avoids the need to define a separate set of equations for each flow model.

In general, $b$ is a function of $x$, such as $(1/A)\,dA/dx$. It is then necessary to specify $A(x)$ (e.g., by a curve-fitted table) and to obtain its derivative.

**Gain-Equals-Lose Equation.** To be consistent with the foregoing formulation it is necessary to differentiate Eq. (7). A considerable algebraic simplification occurs if we first set $a^* = b^* = 0$ in Eq. (21d). These terms are generally negligible, even at high pressure where collision broadening dominates. Only at high-pressure and at low-diluent conditions will the $b^*$ term make a significant contribution. With this approximation, we find from Eqs. (16b), (20), and (21d) that $y$ is proportional to the quantity

$$\rho\left[\sum_i a_i n_i + \sum_v a_v n(v)\right]$$

and, hence

$$\frac{1}{y}\frac{dy}{dx} = \frac{1}{\rho}\frac{d\rho}{dx} + \frac{\sum_i a_i \dfrac{dn_i}{dx} + \sum_v a_v \dfrac{dn(v)}{dx}}{\sum_i a_i n_i + \sum_v a_v n(v)} \tag{65}$$

From Eq. (20) the derivative of $\alpha_{DP}$ is

$$\frac{1}{\alpha_{DP}}\frac{d\alpha_{DP}}{dx} = \frac{1}{2}\frac{1}{T}\frac{dT}{dx} \tag{66}$$

The derivative of the line-profile function, Eqs. (15) and (17), is given by

$$\frac{1}{\phi_c}\frac{d\phi_c}{dx} = \left[2y - \frac{2}{\pi^{1/2}}\frac{e^{-y^2}}{1-\operatorname{erf}(y)}\right]\frac{dy}{dx} - \frac{1}{\alpha_{DP}}\frac{d\alpha_{DP}}{dx} \tag{67a}$$

Combining the foregoing equations produces:

$$\frac{1}{\phi_c}\frac{d\phi_c}{dx} = -\frac{1}{2}\frac{1}{T}\frac{dT}{dx} - F(y)\left[\frac{1}{\rho}\frac{d\rho}{dx} + \frac{\sum_i a_i \dfrac{dn_i}{dx} + \sum_v a_v \dfrac{dn(v)}{dx}}{\sum_i a_i n_i + \sum_v a_v n(v)}\right] \tag{67b}$$

where

$$F(y) = -2y^2 + \frac{2}{\pi^{1/2}}\frac{ye^{-y^2}}{1-\operatorname{erf}(y)} \tag{68a}$$

Because $1-\operatorname{erf}(y) \to 0$ as $y \to \infty$, we use the asymptotic form

$$F(y) = \frac{2y^2 - 3}{2y^2 - 1} \tag{68b}$$

when $y \geq 8$.

We next define two functions as follows:

$$A_1(v) = \frac{\delta(v)n(v)}{Q_{rot}^{(v)}}\exp\left(-\frac{hc}{k}\frac{E_{v,J_m}}{T}\right) \tag{69}$$

and

$$A_2(v) = \frac{\delta(v)n(v+1)}{Q_{rot}^{(v+1)}}\exp\left(-\frac{hc}{k}\frac{E_{v+1,J_m+m}}{T}\right) \tag{70}$$

where the $J_m$ subscript in Eq. (70) is associated with $v$ not $v+1$. The gain, Eq. (7), thus has the form

$$\alpha(v, J_m, m) = \frac{hN_A}{4\pi}\omega_c B(2J_m+1)\phi_c\rho(A_2(v) - A_1(v)) \tag{71}$$

We note from Eq. (32a) that during lasing $(d\alpha/dx) = 0$. Differentiating Eq. (71) with the derivative equal to zero and using Eqs. (67b), (69), and (70) yields the final form for the gain-equals-loss condition

$$
\begin{aligned}
\left\{\left[\frac{hc}{k}\frac{E_{v,J_m}}{T} - C(v, T) - \tfrac{1}{2}\right]A_1(v)\right. \\
- \left[\frac{hc}{k}\frac{E_{v+1,J_m+m}}{T} - C(v+1, T) - \tfrac{1}{2}\right]A_2(v)\left.\right\}\frac{1}{T}\frac{dT}{dx} \\
+ [A_1(v) - A_2(v)][1 - F(y)]\frac{1}{\rho}\frac{d\rho}{dx} + \frac{A_1(v)}{n(v)}\frac{dn(v)}{dx} \\
- \frac{A_2(v)}{n(v+1)}\frac{dn(v+1)}{dx} + \frac{[A_2(v) - A_1(v)]F(y)}{\sum_i a_i n_i + \sum_v a_v n(v)}\sum_{v=0}^{v_{\text{fin}}} a_v \frac{dn(v)}{dx} \\
= [A_1(v) - A_2(v)]F(y)\frac{\sum_i a_i(dn_i/dx)}{\sum_i a_i n_i + \sum_v a_v n(v)}
\end{aligned}
\tag{72}
$$

where

$$
C(v, T) = \frac{d \ln Q_{\text{rot}}^{(v)}}{d \ln T} \cong 1
\tag{73}
$$

The unity value for $C(v, T)$ is based on the classical limit, Eq. (12). This limit is accurate to within 3% for all vibrational levels of HF below $v = 10$ when the temperature exceeds 300°K.

**Structure-of-the-Matrix Equation.** There are $\sum_v \delta(v)$ nontrivial Eqs. (72), since $A_1$ and $A_2$ are proportional to $\delta(v)$. Equations (51), (52), and (72) total $v_{\text{fin}} + 1$ in number for the $dn(v)/dx$, $v = 0, 1, \ldots, v_{\text{fin}}$.

When lasing is not occurring, we have a $4 \times 4$-matrix equation that encompasses Eqs. (54), (55), (59c), and (64). During lasing a $(v_{\text{fin}} + 5) \times (v_{\text{fin}} + 5)$-matrix equation is used that encompasses these four equations plus those for the $dn(v)/dx$. The solution of this matrix equation provides the variables, array (48), which are then numerically integrated. All of the pertinent equations are written so that their right-hand sides constitute the right-hand side of the matrix equation. While the order of the matrix does not change during lasing, the elements do change. Thus any time a threshold, $J$-shift, or band termination occurs, the elements must be reevaluated.

## 2.5. Numerical Procedure

The matrix equation and Eq. (52) for the $n_i$ are solved by stepwise forward integration. As is well known, the chemical-rate equations have a small time constant, that is, they are "stiff." Efficient solution, consequently, requires the use of one of the special numerical methods developed for this type of differential equation.[58] Two methods in common use are those of Gear[59] and an easily coded, modified Runge–Kutta procedure.[3]

Power is determined by integration of Eq. (46c). This equation, however, is auxiliary to the matrix and rate equations and, thus, can be computed by any quadrature method.

We next discuss the conditions for threshold, $J$-shift, and lasing termination or cutoff. Threshold on band $v + 1 \rightarrow v$ occurs when $\alpha(v, J_m, m)$ first satisfies Eq. (32a). A $J$-shift occurs on a lasing band when the gain $\alpha(v, J_m \pm 1, m)$ first equals that on the lasing transition $\alpha(v, J_m, m)$. Finally, cutoff on band $v + 1 \rightarrow v$ occurs when $\chi_{\text{rad}}(v) = 0$, which is equivalent to the intensity $I(v)$ going to zero. All three processes involve discontinuities. Thus for threshold or a $J$-shift on band $v + 1 \rightarrow v$, $\chi_{\text{rad}}(v)$, $dn(v)/dx$, and $dn(v+1)/dx$ all change discontinuously. At cutoff, $d\chi_{\text{rad}}(v)/dx$ changes discontinuously.

Accurate location of these discontinuities is achieved by an under/over relaxation procedure.[54] We illustrate this procedure by discussing how threshold is found for the nonlasing band, $v + 1 \rightarrow v$, that has the largest gain. The current value for $\alpha(v, J_m, m)$ and that at the previous integration step are used to determine, via linear extrapolation to a value based on Eq. (32a), an estimated threshold location $x_{\text{th}}$. A possible step size of $0.5 (x_{\text{th}} - x)$ is then considered, where $x$ is the current axial location. We refer to the step size as only possible because numerical stability or a $J$-shift on some other band may dictate a still smaller step size. If the threshold-step size is controlling, then the next step uses $0.7(x_{\text{th}} - x)$, where $x_{\text{th}}$ here is an updated estimate. The next step used $0.9(x_{\text{th}} - x)$, with all steps thereafter using $1.001(x_{\text{th}} - x)$ until threshold is actually achieved. This method has been found to produce very accurate threshold locations with reasonable efficiency, that is, the first considered 1.001 step. During this searching process, it is necessary to keep track of the current and preceding values of $\alpha(v, J_m \pm 1, m)$ since these gains occasionally achieve threshold slightly ahead of $\alpha(v, J_m, m)$. The reason for this is that the slope of $\alpha(v, J_m \pm 1, m)$ can be slightly greater than $\alpha(v, J_m, m)$ with a crossover in gain occurring near the threshold value.

For $J$-shifts, the same procedure is used to track the adjacent gains, $\alpha(v, J_m \pm 1, m)$, with 0.8 and 1.001 used for the extrapolation coefficients.

The procedure is applied to $X_{rad}(v)$ to determine the location of cutoff with extrapolation coefficients 0.9 and 1.02.

It is necessary to reduce the step size from its nominal value after a discontinuity has occurred since the slopes of the nonlasing gains $\alpha(v, J_m \pm 1, m)$ can discontinuously increase. After the discontinuity two steps are required to reestablish an accurate threshold extrapolation for these gains. Consequently, the first step after a discontinuity is reduced to one quarter, the next step to half, and the third step to three quarters of their nominal values.

The foregoing methods for treating the discontinuities require locating $J_m$ for each $v$ at the start of an integration step. After the first few steps are completed, this time-consuming search procedure can be sharply reduced by checking $\alpha(v, J, m)$ only in the neighborhood of the $J_m$ from the previous step.

Accurate location between integration steps of the threshold, $J$-shift, and cutoff points requires frequent modification of the step size. In addition, a number of derivatives change discontinuously at these points. Multistep integration methods,[58] where the step size can only be halved or doubled, are thus at a severe disadvantage for this computation. The step size is easily altered an arbitrary amount with Runge–Kutta,[3] however, and, for this reason, it is utilized in the numerical formulation of Ref. 54. Note also that Runge–Kutta, being a single-step method, requires no special start or discontinuity provisions.

It is worth mentioning one special technique that has proved to be of great value in expediting the debugging of the complex system of equations in Section 2.4. For computational purposes, we use a system of equations linear in variables (48). For debugging purposes, however, after an integration step is completed, it is only necessary to check the vastly simpler set of equations given at the start of the derivation. As part of the coding process, it is desirable to construct a subroutine that directly checks Eqs. (36) through (39), (41), and (43). Formulation or programming errors are quickly detected from the print out of these calculations.


## 2.6.  Analysis of Solution

The detailed information that can be obtained from a computed solution can be grouped into four general categories:

1.  *Gas Dynamic*   Here we include the pressure, temperature, velocity, Mach number, jet cross-sectional area, and so on, as a function of

distance. Since the ratio of specific heats is readily computed, one can also estimate normal-shock pressure recovery downstream of the cavity.

2. *Composition* This includes the $n_i$ and $n(v)$ [or concentrations $\rho n_i$ and $\rho n(v)$] versus distance.

3. *Laser Parameter* We include intensity and power distribution versus distance. For example, one can obtain the power emitted per line $P(v, J)$ from the solution. This is done by printing $P(x; v)$ each time a discontinuity occurs. For a zero-power solution the small-signal gain is in this category.

4. *Laser Mechanism* This is most easily discussed by considering vibrational level $v$. Let us first decompose $X_{ch}(v)$ into

$$X_{ch}(v) = X_p(v) - X_d(v)$$

where $X_p$ represents the chemical pumping by the cold and hot reactions into vibrational level $v$. The net effect on level $v$ of all other reactions is treated as deactivation, although some of these produce level $v$ while others deplete it. Thus, $X_{rad}(v)$ and $X_p(v)$ produce level $v$, while $X_{rad}(v-1)$ and $X_d(v)$ deplete it. The term $X_d$ can be further decomposed into the specific V–V and V–T processes that alter level $v$. In this manner, one can determine the relative importance of the various deactivation reactions, as well as that of deactivation relative to pumping, and finally cascading effects.

## 2.7. Saturation Intensity

Gordon et al.[60] provide a general description of a four-level laser where the excited transition is simultaneously Doppler broadened and collisionally broadened. For the steady-state case, $dn(v)/dt = 0$, they derive an equation for the intensity

$$\alpha(z) = \frac{1}{I}\frac{dI}{dz} = \frac{\alpha_0}{(1 + I/I_s)^{1/2}} K(0, (1 + I/I_s)^{1/2} y) \tag{74}$$

In Eq. (74) $\alpha_0$ is the Doppler-broadened small signal gain, $I_s$ is the saturation intensity whose value is presumed constant, and $y$ and $K$ are defined by Eqs. (16$b$) and (17), respectively. In the limits of pure Doppler broadening, $y = 0$, or pure collisional broadening, $y \to \infty$, we obtain the familiar equations for inhomogeneous and homogeneous broadened gain, respectively.[60] These simple equations, however, are inappropriate for chemical lasers. First, $u\,dn(v)/dx = dn(v)/dt$ is generally not zero even as a rough approximation. [See, for example, the solution for HF($v$) in Ref. 55.] Second, the derivation in Ref. 60 provides the variation in intensity only in the direction of the optical axis, not in the flow direction.

Another major problem in attempting to utilize Eq. (74) for a chemical laser is that $\alpha_0$ and $I_s$ both vary in a complex fashion in the flow direction.[56] In addition the values for $\alpha_0$ and $I_s$ are line dependent. For instance, Ref. 61 gives $I_s = 160$ to $180 \, \text{W/cm}^2$ for $P_1(6)$ and $I_s = 60$ to $100 \, \text{W/cm}^2$ for $P_2(5)$. For these reasons we avoid the use of a saturation intensity.

## 2.8.  Coupling to Physical Optics

**Preliminary Remarks.**  Examination of the zero-power and power-on solutions in Ref. 55 shows little difference between them, except for the temperature and the $n(v)$. Thus $\rho$, $u$, and $n_i$ are little affected by the lasing process. This result is anticipated since the lasing directly affects the energy equation among the conservation and state equations, and only the rate equations for $n(v)$. For the same reason, a physical optics solution would differ negligibly from a constant-gain solution for $\rho$, $u$, and $n_i$. Furthermore, if the outcoupling is approximately the same for the two solutions, then the temperature also will not differ significantly.

We take advantage of the foregoing result to develop an efficient method for coupling a constant-gain solution to one that includes physical optics. The basic idea is to first perform a constant-gain calculation saving on tape certain quantities such as $\rho$ and $T$. A physical optics solution, say, using a Fresnel integral for $I^2$, is then performed for the same outcoupling. In addition to the Fresnel integral, this latter calculation involves only greatly simplified rate equations for the $n(v)$. This calculation does not require the conservation and state equations, or the general-rate equations for $n_i$ and $n(v)$. We thus have a two-step procedure wherein the first step provides a solution for the fluid mechanic variables and the $n_i$. In the second step, a physical-optics solution is obtained for the intensities and $n(v)$.

For simplicity, we retain the assumption of a Boltzmann distribution for the rotational populations at the translational temperature. As discussed in Section 2.2, simultaneous multiline lasing per band is possible with a physical optics approach. We include this feature.

**Formulation.** Equations (7), (15), (17), and (20) are combined to rewrite the gain as

$$\alpha(v, J, m) = \frac{hcN_A}{(32\pi^3 10^7 kN_A)^{1/2}} \left(\frac{W^*}{T}\right)^{1/2} \rho(2J+1)B(v, J, m)$$

$$\times [1 - \mathrm{erf}\,(y)]e^{y^2} \left[\frac{n(v+1)}{Q_{\mathrm{rot}}^{(v+1)}} \exp\left(-\frac{hc}{k}\frac{E_{v+1,J+m}}{T}\right)\right.$$

$$\left. -\frac{n(v)}{Q_{\mathrm{rot}}^{(v)}} \exp\left(-\frac{hc}{k}\frac{E_{v,J}}{T}\right)\right] \quad (75)$$

If we ignore the $J$ dependence of $F(T, J)$ in Eq. (21$d$), we have the simple relation used later:

$$y(v, J, m) = \frac{\omega_c(v, J', m)}{\omega_c(v, J, m)} y(v, J', m) \quad (76)$$

that connects the $y$ values for different transitions of the same band.

We now consider $\rho$, $T$, $u$, and $n_i$ as known functions of $x$ provided by the constant-gain solution. The $X_{ch}(v)$ can then be written as

$$X_{ch}(v) = A_v + \sum_{v'=0}^{v_{\mathrm{fin}}} A_{v,v'} n(v') + \sum_{v''=0}^{v_{\mathrm{fin}}} \sum_{v'=0}^{v_{\mathrm{fin}}} A_{v,v',v''} n(v')n(v'') \quad (77)$$

where $v'' \geq v'$ and the $A$ coefficients depend on $T$, $\rho$, and $n_i$, and, therefore, are known functions of $x$. Pumping is accounted for by the $A_v$ term, while the last two terms account for V–T and V–V deactivation. Explicit equations in terms of the rate coefficients $k_{fr}$ and $k_{br}$ and the $\rho n_i$ can be written for the $A$ coefficients once the reaction mechanisms are specified. The rate equations for $n(v)$ are now of the form

$$\rho u \frac{dn(v)}{dx} = X_{ch}(v) + \frac{1}{hcN_A} \sum_J \frac{\alpha(v, J, m)I(v, J)}{\omega_c(v, J, m)}$$

$$-\frac{1}{hcN_A} \sum_J \frac{\alpha(v-1, J, m)I(v-1, J)}{\omega_c(v-1, J, m)} \quad (78)$$

where $\alpha$ and $X_{ch}$ are given by Eqs. (75) and (77), respectively. Each term inside the summations represents one mode, and we allow at most one mode per transition. Of course, for the mode to have positive intensity the round-trip gain must balance all losses.

The connection between the constant-gain and physical-optics solutions are given by the following:

1.  A constant-gain calculation is performed. At various $x_i$, we save on
    tape

    $$\rho, u, T, A_v, A_{v,v'}, A_{v,v',v''} \tag{79}$$

2.  Similarly, for some fixed $J'$, we save on tape $y(v, J', m)$, $v = 0, 1, \ldots, v_{fin} - 1$, at the same $x_i$.

If the physical-optics code requires any of the foregoing quantities at axial
locations different from $x_i$, they may be obtained by interpolation. Alternatively, the quantities in array (79) and the $y$'s may be curve fit. For a
given $v$, the various $y$-values required for $\alpha$ in Eq. (78) are obtained from
the constant-gain $y(v, J', m)$ and Eq. (76).

The system of equations now consists of the Fresnel equations, one for
each mode, and rate equations, one for each $n(v)$. These equations are
then solved iteratively, as in Ref. 62.

## 3.  PULSE MODEL

### 3.1.  Introductory Remarks

Pulse modeling is a straightforward extension of the cw premixed case
with the same assumptions, except that now the gas is motionless. Two
optical models are considered; the first uses Eq. (32a), which assumes
that the radiative flux adjusts instantaneously to the uniform gas condition, that is, the speed of light is infinite. In the second optical model a
time-dependent, rate equation approach is used for the intensity.

Early attempts at modeling a pulse chemical laser are by Cohen et al.[1]
and by Airey.[4] Airey's results are particularly impressive in that he
obtained good agreement between theory and experiment for the choice
of lasing transitions, the sequence in which they lase, and the cavity flux
(intensity). Simultaneous lasing, however, is observed on many adjacent
transitions of the one lasing band, with the theoretically predicted transition centered among these (Fig. 1). With time, the $J$ values for the lasing
transitions shift upward in both experiment and theory. The fair agreement for cavity flux makes it evident that the theoretical intensity, on one
line, is approximately the sum of the intensities of the simultaneous lasing
transitions.

The most extensive comparison between model and experiment for an
HF laser is given in Ref. 63. Four items are compared for an $H_2 + F_2$

chain laser initiated by flash photolysis. For three of the four items—time to threshold, pulse duration, and pulse shape—the agreement is good, although, for low gains, considerable oscillation occurs in the experimental intensity. The fourth item compared is the peak output intensity. Agreement here is barely adequate at low gains. At high gains, the computation overpredicts the intensity by 1 to 2 orders of magnitude.

Reference 64 also contains a comparison for an HF laser of theory and experiment with good agreement for time to threshold as affected by pressure, flashlamp intensity, and optical-cavity loss. Other important pulse-modeling papers are contained in Refs. 65 through 69. Most of these deal with HF, including the electric discharge study of Lyman.[65] The one exception is Ref. 66, which treats the $DF/CO_2$ transfer laser. Reference 69 is the only chemical-laser modeling paper to treat relaxation oscillations. Purely analytical papers relevant to pulse laser modeling are Refs. 70 through 72.

In the next section, the pulse-modeling formulation is obtained from the cw premixed case by a simple transformation. We also discuss the flash-lamp formulation and relaxation oscillations. The last section discusses time sequencing of the lasing transitions as affected by a number of phenomena, including rotational nonequilibrium.

## 3.2. Formulation of Equations

**Premixed cw to Pulse Transformation.** Our premixed cw model is readily converted to the pulse case by means of the transformation[39] in Table 7. Referring to this table, we observe that $x$ becomes time (in seconds). Velocity has the value of unity, not zero, in order that $x = t$ and

**Table 7.** Premixed cw Pulse Transformation

| Item | Premixed cw value | Pulse value |
|------|-------------------|-------------|
| Independent variable | $x$ | $t$ |
| Velocity | $u$ | 1 |
| Area | $A$ | 1 |
| Density | $\rho$ | Constant |
| Momentum equation | Eq. (55) | Delete |
| $P$ | Cumulative power | Cumulative energy density |
| $dP/dx$ | Eq. (46c) | Instantaneous output power |
| $I$ | Average intracavity intensity | Average intracavity intensity |

$ud(\ )/dx = d(\ )/dt$. For pulse operation, density is constant in Eq. (64), and Eq. (55) for momentum is deleted. Thus, the matrix equation has one less row, and, since the flow area is also deleted, one less column. Note that the constant-pressure molar-heat capacity $C_{pi}$ is not replaced by the constant-volume molar heat capacity.

The quantities $X_{rad}$ and $I$ have the same meaning as before; however, $P$ is now the energy density (J/cm$^3$). In other words, $P$ is the cumulative laser energy from 1 cm$^3$ of cavity gas up to time $t$, as is evident from Eq. (46$b$) with $x \rightarrow t$ and $A = 1$. The output power $P'$ is $dP/dt$ and is given by the right-hand side of Eq. (46$c$), again with $A = 1$. Pulse energy $E(t, v)$ to time $t$ is then given by

$$E(t, v) = \int_0^t P'(t', v) \, dt' = hcN_A f \int_0^t \sum_{J_m} \omega_c(v, J_m, m) X_{rad}(v) \, dt' \quad (80)$$

where the total pulse energy is $\sum_v E(t, v)$. The model is formulated to describe lasing from 1 cm$^3$ of active medium with length $L$. It is only necessary to multiply the output power or energy by the overall volume to obtain the actual output power or energy.

**Flashlamp.** Many pulse laser experiments use a flashlamp for photolytic initiation of the kinetics. All modeling efforts[3,4,64] assume a gas that is optically thin to the flashlamp light, whose absorbing species mole-mass ratio $n_a$ is governed by

$$\left( \frac{dn_a}{dt} \right)_f = -I_f(t) n_a \quad (81)$$

where $I_f$ is a normalized intensity such that $I_f = 1$ corresponds to a $1/e$ decrease in $n_a$ sec$^{-1}$. Of course, $I_f(t)$ is a required input to any calculation. [The time dependency of $I_f(t)$ is easily measured by monitoring directly the flashlamp's output; hence $I_f(t) = \lambda_f I_f(\text{measured})$. The constant $\lambda_f$ is then determined by measuring in a nonreacting gas the total change in $n_a$ caused by the flashlamp's pulse.] If the photolytic decomposition of $n_a$ produces species $n_j$ then

$$\left( \frac{dn_j}{dt} \right)_f = c_j I_f(t) \, n_a \quad (82)$$

where $c_j$ is the number of $n_j$ molecules produced by the decomposition of one $n_a$ molecule. In practice, the right-hand sides of Eqs. (81) and (82) are actually added to the right-hand side of rate Eq. (52).

Energy is also absorbed by the gas. Let $\omega_a$ be the average wave number of the absorbed energy. Energy absorption then requires adding to the right-hand side of Eq. (59c) the term[3]

$$\frac{hcN_A}{1.01325 \times 10^6} \frac{\omega_a}{p} \rho n_a I_f(t) \tag{83}$$

**Relaxation Oscillations.** A simple analysis of relaxation oscillations is given in Ref. 73. When the deviations from the steady state are small, the analysis provides an approximate condition required for oscillation

$$\frac{\tau_{\text{deact}}}{\tau_{\text{cav}}} > \frac{1}{4} \frac{R^2}{R-1} \tag{84}$$

where $\tau_{\text{deact}}$ is the collisional-deactivation time of the upper state, and $\tau_{\text{cav}}$ is the cavity lifetime, which is a function of the round-trip time $2L_{\text{cav}}/c$ ($L_{\text{cav}}$ is the mirror-separation distance) and the outcoupling fraction. $R$ is the ratio of the pumping rate to that required for threshold and is roughly the small-signal gain $\alpha_o$ divided by the threshold gain $\alpha_{\text{th}}$, given by Eq. (32a). Relaxation oscillations, therefore, are a function of both the gain medium length $L$ and the mirror separation $L_{\text{cav}}$. Inequality (84) predicts no relaxation oscillations at very small gain ($R \cong 1$) or very high gain ($R \gg 1$). The high-gain result appears to be in accord with the experimental results of Ref. 63, where relaxation oscillations are observed only for low and medium gains. It is not surprising, however, that inequality (84) is not applicable to the low-gain regime, since the relaxation oscillations in this case are severest just after threshold where the small-perturbation assumption is not valid.

Relaxation oscillation after threshold is a consequence of the gain exceeding its nominal threshold value while the photon flux is still in the process of increasing. Because of this time-lag, the photon flux subsequently also exceeds its nominal threshold value, thereby producing a high peak-power spike similar to that from a Q-switched laser. The rate of flux buildup depends on both $\alpha_o L$ and the cavity lifetime. Therefore, as in Eq. (84), both lengths should appear in any relaxation model.

In Ref. 69 a phenomenological rate equation is used for the intensity

$$\frac{1}{c} \frac{dI(v, J)}{dt} = (\alpha(v, J, m) - \alpha_{\text{th}}) I(v, J) \tag{85}$$

that is based on the radiative-transfer equation plus a loss term. The laser gas is assumed to be uniform, and the only given length in the formulation is $L$. Nevertheless, a pronounced gain spike occurs each time a new transition reaches threshold. For the conditions examined, the gain then quickly decays to its threshold value after the intensity has achieved its threshold value.[69] Simultaneous multiline lasing per band is included in the model. Nevertheless, the predicted trends of $J$-shifting are quite close to that discussed in Section 2.2, with transition $(v, J)$ rapidly terminating just after transition $(v, J+1)$ turns on. In one other important aspect is the constant-gain model verified. Total pulse energy and pulse energy per band are in good agreement for the two models.[69]

Reference 74 provides an interesting derivation of a rate equation for the intensity

$$\frac{1}{c}\frac{dI}{dt} = \frac{1}{2L_{\text{cav}}}(r_1 r_2 e^{2\alpha L} - 1)I \tag{86}$$

that depends on both $L$ and $L_{\text{cav}}$. The derivation for Eq. (86) is similar to that in Section 2.2, except the flux at mirror 1 is not in balance after one round trip through the cavity. In the model, outcoupling can occur at either mirror. The derived relation between $X_{\text{rad}}$ and $I$ is, however, asymmetric in the mirror reflectivities $r_1$ and $r_2$.[74] This peculiar circumstance is a consequence of the arbitrary selection of mirror 1 for the location of the imbalance in intensity.

Both Eqs. (85) and (86) predict a very small intensity when $\alpha < \alpha_{\text{th}}$ or $r_1 r_2 \exp(2\alpha L) < 1$. The photon flux, therefore, can increase only after a threshold gain is achieved. Since neither equation includes spontaneous emission, a small, but arbitrary value is used for the initial intensities. Fortunately, the solution is quite insensitive to this value.[69]

### 3.3. Time Sequence of Lasing Transitions

**Preliminary Remarks.** A number of pulse experiments[4,74–76] have investigated the time sequencing of the lasing transitions by time-resolved spectroscopy. In a number of these studies,[75,76] the results are interpreted as requiring a nonequilibrium rotational population distribution. All of the experiments utilized a stable oscillator with flash-photolysis for initiation except Ref. 74, where a pin-discharge is used.

In the experiments, three different phenomena are observed that are not predicted by the constant-gain model. In the first, $J$-shifting is correctly predicted, but many lines are simultaneously lasing (see Fig. 1). This is the case for the $2 \rightarrow 1$ and $6 \rightarrow 5$ bands of the HF chain laser experiment of Ref. 75, but not the case for the other bands. Second, lasing is observed on lower $J$ transitions than predicted by the model. For example, in Ref. 75, lasing is observed on $P_4(1)$, $P_5(1)$, and $P_6(1)$, none of which is predicted by the model. Third, the time sequence of the lasing transitions within a band is random following no obvious pattern.[74,75]

Besides rotational nonequilibrium, there are other mechanisms that could, in part, explain the above observations. Our main objective here is to assess these mechanisms, which include: (1) rotational nonequilibrium, (2) collisional broadening, (3) line overlap, and (4) multimode operation.

**Rotational Nonequilibrium.** It is important to note that without lasing the rotational populations are in very good agreement with a Boltzmann distribution, as shown by the cw low-pressure (3.5 Torr) small-signal gain measurements of Ref. 77. Thus if rotational nonequilibrium is to occur, it must be due to perturbations caused by stimulated emission, rather than by the kinetics. In his HCl pulse laser experiment, Airey[4] estimated the minimum stimulated emission time two different ways, each producing $10^{-5}$ sec. At 8 Torr this is adequate for 400 to 500 collisions, which should be more than enough to rotationally equilibrate HCl. Note that simultaneous lasing is observed from upper $J = 3$, 4, and 5 rotational populations (see Fig. 1), where rotational relaxation should be rapid. Furthermore, when the experiment was repeated with added Ar diluent at twice the pressure, the same simultaneous lasing was observed.[4]

In the DF–CO$_2$ pulse experiment of Ref. 76 performed at 55 Torr, simultaneous lasing is observed on transitions $P(14)$ through $P(20)$. The rotational equilibration time is given as $2 \times 10^{-9}$ sec; however, the data are insufficient for estimating an emission time, although the author implies that it must be considerably larger than the rotational-relaxation time.

A stimulated emission time $\tau_{em}$ for transition $(v+1, J+m) \rightarrow (v, J)$ is defined by

$$\frac{dn(v, J)}{dt} \approx \frac{n(v, J)}{\tau_{em}}$$

From Eqs. (43) and (44) we, thus, have

$$\tau_{em} = \frac{\rho n(v, J)}{X_{rad}(v)} = \frac{hcN_A \omega_c \rho n(v, J)}{\alpha(v, J, m) I(v, J)} \tag{87}$$

where Eq. (32a) provides an estimate of the gain.

Let us estimate $\tau_{em}$ for the $P_6(3)$ HF transition whose intracavity intensity is approximately 1 kW/cm$^2$ in the experiment in Ref. 75. The gas initially is at 300°K and 50 Torr with about 0.5 Torr of H$_2$ and 1 Torr of F$_2$. We roughly estimate the average HF(6) partial pressure to be 1.0% of the initial H$_2$ pressure, to obtain

$$\rho n_{HF(6)} = \frac{0.01}{2 \times 760 \times 82 \times 300} = 2.67 \times 10^{-10} \text{ mol/cm}^3$$

For a Boltzmann distribution the HF(6, 2) upper-state concentration is then

$$\rho n_{HF(6,2)} = \rho n_{HF(6)} \frac{5}{13.25} \exp\left(-\frac{1.4388 \times 97}{300}\right) = 6.35 \times 10^{-11} \text{ mol/cm}^3$$

The active medium length is 43.5 cm and the outcoupling is 10%, which provides, via Eq. (32a), $\alpha = 1.15 \times 10^{-3} \text{ cm}^{-1}$. From Eq. (87), we, thus, have

$$\tau_{em} = \frac{11.96 \times 3.038 \times 10^3 \times 6.35 \times 10^{-11}}{1.15 \times 10^{-3} \times 10^3} = 1.4 \times 10^{-6} \text{ sec}$$

Since a collision time at 50 Torr is $\approx 10^{-8}$ sec, we conclude that the $P_6(3)$ transition cannot be affected by rotational nonequilibrium.

Based on this estimate and the work of Airey,[4] we conclude that rotational nonequilibrium is not likely the cause of the experimental observations. This is not necessarily the case in cw experiments, where pressures on the order of 3 Torr or less are encountered. To date, there is no unambiguous verification of rotational nonequilibrium in a low-pressure pulse or cw experiment. It should be noted that hole burning and rotational nonequilibrium occur under the same physical conditions of low pressure and high intensity. If the two phenomena are simultaneous, interpretation of experimental results would be difficult.

Rotational nonequilibrium may also occur at somewhat higher pressures as a result of the pumping reaction. For instance, Ref. 78 shows that excited HCl from the $H + Cl_2$ reaction is initially formed in high-rotational states near $J = 13$. Collisional relaxation from these high-rotational states is relatively slow because of the large-energy decrement at high $J$.[79] Our discussion is also not applicable to very short pulses whose duration is equal to or smaller than a collision time, and where rotational nonequilibrium would occur.

Inclusion of rotational nonequilibrium in the model is straightforward. The dependence of the gain on the rotational mole-mass ratios must be maintained. Second, Eq. (35) for $n(v)$ is used, and finally, rate equations, similar to Eq. (43), are used for the $n(v, J)$. Modeling of rotational nonequilibrium is well within the core capacity and computational speed of current computers. Instead, the biggest single obstacle is our lack of knowledge of rotational relaxation rates, although this situation is currently improving.[79,80]

**Collisional Broadening.** As discussed in Section 1.4, collisional broadening at high pressures can have a pronounced effect on the relative gains of different transitions. Nevertheless, this mechanism cannot explain simultaneous multiline lasing per band; nor is broadening a likely factor in experiments heavily diluted with He, such as in Ref. 75. The decrease in line width, however, at small $J$, as shown in Fig. 8, could be responsible for the lasing on $P_4(1)$, $P_5(1)$, and $P_6(1)$.[75] The only alteration needed to properly account for the effects of collision broadening is a more realistic $J$ dependence for $\alpha_{LR}$ than that provided by Eq. (21$d$).

**Line Overlap.** As mentioned earlier, Ref. 75 observed HF lasing on $P_4(1)$, $P_5(1)$, and $P_6(1)$ at a total pressure of 50 Torr. A scan of Tables 3$a$ and 3$b$ shows that two of the three transitions have rather close neighbors, as shown in Table 8. Recall from the discussion in Section 1.2 that

**Table 8.** Nearest Neighbors for $P_4(1)$, $P_5(1)$, and $P_6(1)$ of HF

| | Transition | | Nearest neighbor |
| Designation | Wave number | Designation | Wave number |
|---|---|---|---|
| $P_4(1)$ | 3422.23 | $P_3(5)$ | 3417.99 |
| | | $R_5(3)$ | 3425.42 |
| $P_5(1)$ | 3264.15 | $P_4(5)$ | 3262.46 |
| | | $R_6(3)$ | 3263.50 |
| $P_6(1)$ | 3109.04 | $P_5(5)$ | 3110.37 |

the accuracy of the line positions in Tables 3$a$ and 3$b$ is about $\pm 0.12$ cm$^{-1}$. Thus, any gain present in the wings of the nearest neighbor transitions would enhance the gain of the transitions under discussion. (An alternative possibility is line misidentification in interpreting the experimental results.) All the nearest-neighbor transitions in Table 8 are likely to have some gain. It, thus, appears possible that line overlap may account for some of the lasing on the low $J$ transitions and, perhaps, for some of the chaotic sequencing of the lasing transitions observed in Ref. 75.

**Multimode Operation.** We already mentioned that all of the time-resolved pulse experiments utilized a stable oscillator. With this type of cavity, a high-gain medium, and spatial nonuniformities in the medium, multimode operation is a certainty.[81] It is evident from the foregoing discussions and Section 2.2 that multimode operation is, by far, the most tenable explanation for the observation of simultaneous lasing of different transitions within a band. Spatial nonuniformities that result in multimode operation may also be the cause of the occasionally chaotic time sequencing of the lasing transitions.

Modeling of multimode lasing for a spatially uniform but time-dependent medium is dealt with in Section 2.8. It would seem desirable, however, to repeat some of the time-resolved spectroscopy experiments with a single-mode oscillator. This is most readily done with an unstable oscillator configuration. Although this type of oscillator is frequently used for cw experiments, it is rarely used for pulse work. Aperturing can also be used with a stable oscillator to obtain single-mode operation. The suppressed modes, however, may still lase parasitically, thereby compromising the experiment. In contrast, the large-mode volume obtainable with an unstable oscillator assists in parasitic mode suppression. The suggested single-mode, time-resolved pulse experiment would be of great importance in resolving the uncertainties in the modeling and in providing direction for future development.

## 4. SIMPLE MIXING MODEL

### 4.1. Introductory Remarks

The simplest cw mixing model one can conceive of would involve three stream tubes, one each for the oxidizer and fuel streams with a third stream tube between these two. The mixing and reaction occur in the third stream tube.[82] This approach is realtively simple and is computationally much faster than multidimensional codes, with a speed comparable to the premixed cw code. Furthermore, its formulation is a natural extension of the premixed case in Section 2. The most serious problem with this so-called "leaky stream-tube" approach is that the rates of mixing must be specified. One must utimately resort to experiment or a multidimensional-mixing calculation for this information. Another limitation is that shock waves and a transverse-pressure gradient are not included. The approach, however, does allow for either a specified axial pressure gradient or a specified cross-sectional flow area. The only previously published leaky stream tube formulation is in Ref. 83, where both the momentum and chemical-rate equations are incorrect. Furthermore, no consideration is given to the singularity at the origin.

### 4.2. Formulation of Equations

**Preliminary Remarks.** Figure 10 shows a schematic plane view of the nozzle bank and the two-dimensional mixing regions. The $z$-coordinate is along the optical axis with mirrors located parallel to the flow as in Fig. 9. $G$ and $J$ subscripts are used to designate the flows that emanate from the

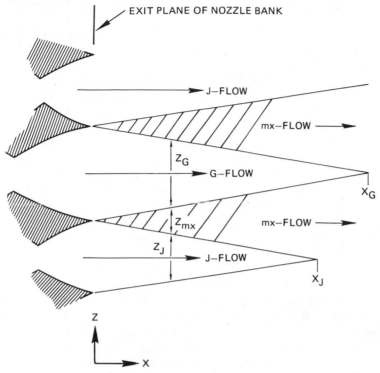

**Fig. 10.**  Schematic of coordinate system and flow regions for leaky stream-tube model.

$N_{BG}$ and $N_{BJ}$ individual nozzles. There are $N_{mx}$ separate mixing regions

$$N_{mx} = N_{BG} + N_{BJ} - 1 \tag{88}$$

where an $mx$ subscript designates this region.

We use the following key assumptions:

1.  There are no gaps between adjacent parallel flows.
2.  At any axial location the static pressure is uniform across all stream tubes.
3.  At any axial location flow conditions are uniform within a stream tube.

These assumptions enable us to use the one-dimensional conservation equations. For purposes of simplicity and brevity, the following secondary assumptions are also introduced.

4.  The $G$ and $J$ flows are isentropic with constant values for their ratios of specific heats, $\gamma_G$ and $\gamma_J$.

5.  The $G$ and $J$ flows contain no absorbing gas.

6.  The nozzle trailing edges are sharp, that is, no base relief or wake regions occur.

7.  Threshold occurs downstream of $x = 0$.

Assumptions 4 through 7 are not essential. For example, absorbing gas, such as ground-state HF from a combustor, is readily accounted for in the gain-equals-loss equation. Should base relief be present with the initial streams at different static pressures, then a preliminary calculation can be performed that would allow the two flows to isentropically expand (or compress) to the simpler initial conditions utilized here.

We again use the matrix equation approach of Section 2. A set of equations linear in the variables

$$\frac{1}{T}\frac{dT}{dx}, \quad \frac{1}{y_B}\frac{dy_B}{dx}, \quad \frac{1}{\rho}\frac{d\rho}{dx}, \quad \frac{1}{p}\frac{dp}{dx}, \quad \frac{1}{u}\frac{du}{dx}, \quad \frac{dn(v)}{dx}, \quad v = 0, 1, \ldots, v_{\text{fin}}$$

(89)

are obtained, where $y_B(x)$, the height of the jet, replaces the area $A$ used in the premixed case. Our choice of variables is similar to the premixed case, except that it is convenient to add to array (89) the velocity $u$. It is important to note that all quantities in this array, except for $p$ and $y_B$, are defined only for the gas in the mixing region.

A self-consistent set of equations is obtained for these variables. At any axial location, we thus satisfy continuity, momentum, and energy for the individual flows as well as for the conglomerate. As in the premixed case, initial conditions are required. These normally include specification of $T$, $z$, $u$, and $n_i$ at $x = 0$ for both the $G$ and $J$ flows. (We assume that $n(v)$ and $z_{mx}$ are zero at $x = 0$.) Additionally, the distribution with $x$ of either $p$ or $y_B$ is needed. The rate of mixing, which also must be specified, can be accounted for in a number of ways. We specify the mass-entrainment rate, since it is of primary concern and results in the simplest formulation.

**Solution for the G and J Flows.**   As a result of assumption 4, we can immediately write for the $G$ flow.

$$T_G = T_G(0)\left(\frac{p}{p(0)}\right)^{(\gamma_G - 1)/\gamma_G}$$

(90a)

$$\rho_G = \rho_G(0)\left(\frac{p}{p(0)}\right)^{1/\gamma_G}$$

(90b)

$$u_G = (2 \times 4.184 \times 10^7 R)^{1/2}\left[\frac{\gamma_G}{\gamma_G - 1}\frac{1}{W_G}(T_{Gst} - T_G)\right]^{1/2}$$

(90c)

and

$$n_{iG} = n_{iG}(0) \tag{90d}$$

where $T_{Gst}$ is the (constant) stagnation temperature of the $G$ flow. A truncated polynomial is used for the mass-flow rate

$$\dot{m}_G(x) = \dot{m}_G(0)\left[1 + g_1\left(\frac{x}{x_G}\right) - (1 + g_1 - g_2)\left(\frac{x}{x_G}\right)^2\right], \qquad 0 \le x \le x_G$$

$$= \dot{m}_G(0)g_2, \qquad x_G < x \tag{91a}$$

where $g_1$, $g_2$, and $x_G$ are input constants. A linear rate of entrainment is given by $g_2 = 1 + g_1$, and $x_G$ is the mixing length for the $G$ stream. Similarly, for the $J$ flow, we have

$$\dot{m}_J(x) = \dot{m}_J(0)\left[1 + j_1\left(\frac{x}{x_J}\right) - (1 + j_1 - j_2)\left(\frac{x}{x_J}\right)^2\right], \qquad 0 \le x \le x_J$$

$$= \dot{m}_J(0)j_2, \qquad x_J < x \tag{91b}$$

The width of the $G$ flow is provided by continuity as

$$z_G(x) = \frac{\dot{m}_G(x)}{N_{BG}y_B\rho_G u_G} \tag{92}$$

Exact counterparts of Eqs. (90) and (92) are used for the $J$ flow.

**Mixed-flow Equations.**   Figure 11 shows the fluxes into a control volume for conservation of momentum, energy, and species. The cross-sectional area $A_{mx}$ of a single mixing layer is $z_{mx}y_B$; the total cross-sectional area for the mixed flow is, therefore, $(NA)_{mx}$. By balancing the fluxes shown in Fig. 11a, we readily obtain for the mixed region momentum equation

$$\frac{du}{dx} + \frac{1.01325 \times 10^6}{\rho u}\frac{dp}{dx} = Q_{mom} \tag{93}$$

where the momentum source term is

$$Q_{mom} = \frac{1}{\dot{m}}\left[(u - u_G)\frac{d\dot{m}_G}{dx} + (u - u_J)\frac{d\dot{m}_J}{dx}\right] \tag{94}$$

A mixing term appears on the right-hand side of the equations as a driving or source term. In Eq. (94) $d\dot{m}_G/dx$ and $d\dot{m}_J/dx$ are obtained from Eq. (91), and $\dot{m}$ is given by

$$\dot{m}(x) = \dot{m}_{mx}(x) = \dot{m}_G(0) - \dot{m}_G(x) + \dot{m}_J(0) - \dot{m}_J(x) \tag{95}$$

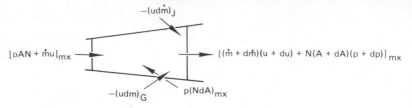

(A)    MOMENTUM BALANCE

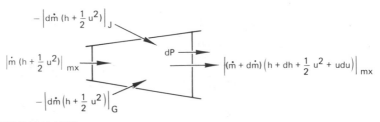

(B)    ENERGY BALANCE

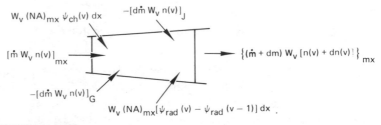

(C)    SPECIES BALANCE

**Fig. 11.**  Mixing-region fluxes into a control volume for momentum, energy, and species conservation. (Conversion factors not shown).

Similarly, the energy equation has the form (Fig. 11$b$)

$$h + \frac{u^2}{2 \times 4.184 \times 10^{10}} + \frac{P}{4.184 \times 10^3 \, f\dot{m}}$$

$$= \frac{1}{\dot{m}} [(\dot{m}_G(0) - \dot{m}_G) h_{Gst} + (\dot{m}_J(0) - \dot{m}_J) h_{Jst}] \quad (96)$$

where $f$ is given by Eq. (33), and the outcoupled power $P$ is

$$\frac{dP(x; v)}{dx} = hcN_a f(Nz)_{mx} y_B \omega_c(v, J, m) X_{rad}(v) \quad (97)$$

The intensity $I(v)$ is still given by Eq. (44) with $\alpha = \alpha_{mx}(v, J_m, m)$. The equation of state is given by Eq. (36), and continuity can be written as

$$z_{mx} = \left(\frac{\dot{m}}{N\rho u}\right)_{mx} \frac{1}{y_B} \tag{98}$$

We also assume a constant jet width, which results in

$$N_{BG}z_G + N_{BJ}z_J + N_{mx}z_{mx} = N_{BG}z_G(0) + N_{BJ}z_J(0) \tag{99}$$

where Eqs. (92) and (98) provide the $z$'s on the left-hand side.

From Fig. 11c, we obtain the rate equation

$$\frac{dn(v)}{dx} = \frac{1}{\rho u}[\chi_{ch}(v) + \chi_{rad}(v) - \chi_{rad}(v-1)] + Q_{sp}(v) \tag{100}$$

where the species source term is

$$Q_{sp}(v) = \frac{1}{\dot{m}}\left[(n(v) - n_G(v))\frac{d\dot{m}_G}{dx} + (n(v) - n_J(v))\frac{d\dot{m}_J}{dx}\right] \tag{101}$$

These equations also hold for the $n_i$, except that the $\chi_{rad}$ terms are zero. (Note that $n_G(v) = n_J(v) = 0$, in accord with assumption 5. They are included in order to cover the nonlasing species case where they become $n_{Gi}$ and $n_{Ji}$.) Equations (51) and (52) still apply, except that $\chi_{ch}$ is replaced by $\chi_{ch} + \rho u Q_{sp}$.

Finally, the gain-equals-loss equation has the form

$$r_1 r_2 e^{2(Nz)_{mx}\alpha_{mx}} = 1 \tag{102}$$

and thus $\alpha_{mx}(v, J_m, m)$ is no longer constant with $x$.

The foregoing equations, along with those for the $G$ and $J$ flows, determine the solution in the mixing region. In Appendix A the equations are reorganized to provide a linear system in variables (89), that closely resembles the premixed system in Section 2.4. Two types of solutions are encompassed: one where a pressure distribution is given, the other where a distribution for $y_B(x)$ is specified. As previously discussed, initial conditions for the $G$ and $J$ flows are required. Initial conditions for $u$, $T$, $n_i$, and so on, in the mixed region are not specified, but are obtained by systematically expanding all dependent variables in powers of $x$. The results for this expansion are shown in Appendix B. We observe that $u(0)$, $T(0)$, and $n_i(0)$ depend only on the linear part of the entrainment rates, that is, the $g_1$ and $j_1$ terms, and not at all on the kinetics. On the other hand, the source terms $Q_{sp}(0)$, $Q_{en}(0)$, and $Q_{mom}(0)$ depend on both the linear and quadratic terms in the entrainment rates. Only $Q_{sp}(0)$ depends on the kinetics in the mixing region at $x = 0$.

Since $\dot{m}(0) = 0$, a number of source terms are singular at the origin. The same expansion procedure also resolves this difficulty. Appendix B, therefore, provides formulas for the source terms evaluated at the origin. The equations in the appendixes have been coded and solved with the modified Runge–Kutta procedure discussed in Section 2.5. At the origin, the solution in Appendix B is used, and no difficulty is encountered in the subsequent integration.

Frequently, the solution extends beyond $x_G$ (or $x_J$). Since the derivative $d\dot{m}_G/dx(d\dot{m}_J/dx)$ is discontinuous at $x_G(x_J)$, it is advisable to treat this point as an additional discontinuity (see Section 2.5).

As in the premixed case, the flow can choke when $y_B(x)$ is specified. The choking condition is found by setting the determinant of the governing matrix equal to zero. After some manipulation, we obtain

$$\left(NA\frac{M^2-1}{\gamma M^2}\right)_{mx} + \left(N_B A\frac{M^2-1}{\gamma M^2}\right)_{G} + \left(N_B A\frac{M^2-1}{\gamma M^2}\right)_{J} = 0$$

where $M$ is the Mach number, and both $\gamma_{mx}$ and $M_{mx}$ are given by their chemically "frozen" definitions. This equation is a generalization of the well-known $M = 1$ choking condition.

## 4.3. Mixing-model Results

The importance of mixing for the cold reaction HF laser is readily demonstrated with this model. Table 9 summarizes the more important input conditions for a series of constant pressure calculations where only

**Table 9.** Conditions Used in Mixing-model Analysis

|  | Nozzle exit plane conditions | |
| --- | --- | --- |
|  | Fluorine | Hydrogen |
| $z(0)$, cm | 0.19 | 0.1 |
| $u(0)$, cm/sec | $2.2 \times 10^5$ | $1.8 \times 10^5$ |
| $T(0)$, °K | 201 | 195 |
| $p(0)$, Torr | 9.2 | 9.2 |

Molar ratios: $H_2/F = 5$; diluent/F $= 10.9$.
Mixing conditions: $x_G = x_J =$ variable; constant pressure; $g_1 = -1$, $g_2 = 0$, $j_1 = -0.2$, $j_2 = 0.8$.
HF kinetics: provided by N. Cohen of The Aerospace Corporation.

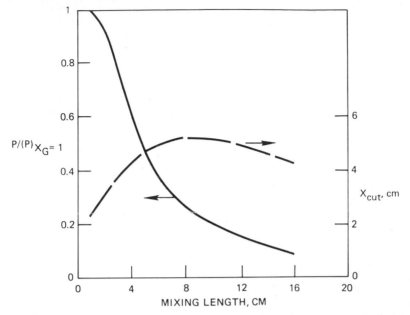

**Fig. 12.** Normalized power and cutoff distance $x_{cut}$ versus mixing length. See Table 9 for conditions used in these calculations.

the mixing length, $x_G = x_J$, is varied. Linear mixing is used with the $g_i$ and $j_i$ chosen to provide an equal molar rate of entrainment for the $H_2$ and F, in accord with the expected flame-sheet behavior.

Figure 12 shows the power, normalized by its value when $x_G = 1$, and the cutoff distance versus the mixing length. The power changes by an order of magnitude as the mixing length changes from 1 to 16 cm. A limiting value is achieved, corresponding to the premixed case, as the mixing length shrinks to zero. The distance to cutoff $x_{cut}$ has a maximum value at a mixing length of $\approx 8$ cm. At this and larger mixing lengths, much of the fluorine is unmixed and, therefore, unreacted at cutoff, with a consequently low chemical efficiency. At mixing lengths less than 4.5 cm, cutoff occurs after all the fluorine is entrained, with an obvious improvement in efficiency. The variation of power with mixing length is in good accord with the analysis in Ref. 82. The analysis predicts a hyperbolic variation of power when the mixing length exceeds 4.5 cm (the high-pressure regime), and a linear variation terminating with the premixed power for shorter mixing lengths. Furthermore, at the crossover point where the mixing length equals $x_{cut} = 4.5$ cm, the analysis predicts the power to be one-half of the premixed value.

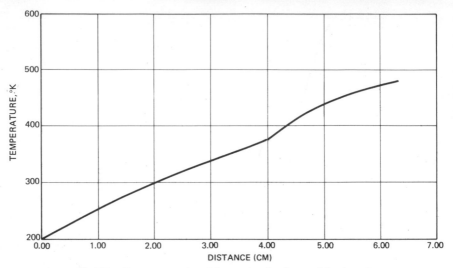

**Fig. 13.**   Temperature in mixing region for 4-cm mixing length.

Figures 13 through 15 show the temperature, $n(v)$, and laser mechanisms for the 4-cm mixing-length case. The velocity is not shown, since it is constant, as can be verified directly with the formulas in the appendices (with $p$ constant and linear entrainment rates). The kink in the temperature profile, Fig. 13, occurs at $x_G = x_J = 4$ cm, where further entrainment abruptly ends. In the calculations, $v_{fin} = 7$, although pumping stops with $v = 3$. The higher vibrational levels, however, are populated by HF–HF V–V exchange. The topmost curve in Fig. 14 shows the total HF mole-mass ratio in the mixing region. The rapid increase in $n(v)$ at small $x$ is somewhat misleading since the $z$-averaged $n(v)$ increases at a slower rate. The rather severe kinks in the $n(2)$ and $n(3)$ curves at small $x$ are caused by the onset of lasing, which prevents these vibrational levels from increasing at their previous rate. As expected, the model predicts that the laser operates on partial inversions with a total inversion occurring only at very small $x$.

In Fig. 15 we show the $\chi_{rad}(v)$, $\chi_{rad}(v-1)$, $\chi_p(v)$, and $\chi_d(v)$ (the last two variables are the pumping and deactivation parameters defined in Section 2.6) for the first four vibrational levels. Since the HF kinetics used in the calculations provide no pumping into $v = 0$, all HF(0) is produced by $1 \rightarrow 0$ lasing and deactivation from upper levels, with both processes important in causing cutoff. The jagged nature of all the $\chi_{rad}$ curves is caused by frequent $J$-shifting. Figure 15$b$ shows that cascading is the dominant HF(1) mechanism for the first 2 cm. In other words, lasing from

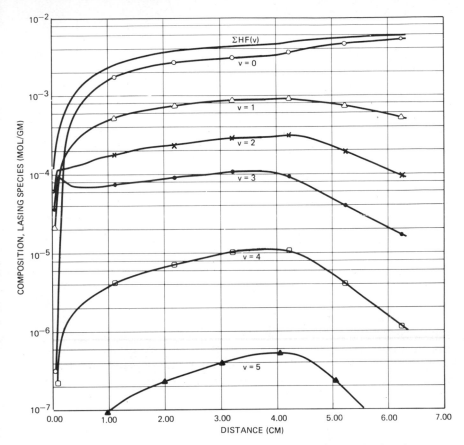

**Fig. 14.** HF($v$) mole-mass ratios in mixing region for 4-cm mixing length.

$2 \rightarrow 1$ is approximately counterbalanced by that from $1 \rightarrow 0$. Pumping and deactivation, which also roughly counterbalance each other, only become significant close to cutoff. The situation for HF(2) and HF(3), Figs. 15c and d, are clearly different with pumping now a major factor. Despite the linear entrainment rates $X_p(2)$ and $X_p(3)$ decay exponentially: this result is a consequence of the pumping rate exceeding the $H_2$ and F entrainment rates. Thus the $H_2$ and F concentrations in the mixing region steadily decline from their initial values at the origin, which are based only on the entrainment rates.

In summary, we observe that rapid-mixing produces optimum efficiency, but also a very short lasing zone length. Detailed examination of a particular case shows that, besides the obvious importance of pumping,

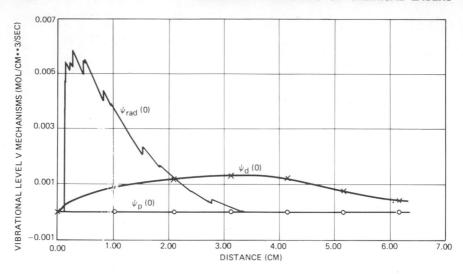

**Fig. 15a.**    HF(0) mechanisms for 4-cm mixing length.

both deactivation and cascading are significant processes. The approximate exponential decay of the $X_{rad}(v)$, often seen experimentally in burns with plastic targets, stems from the pumping rate depleting the F and $H_2$ concentrations faster than mixing replenishes them.

## 5.  COMPLEX MIXING MODELS

Some of the material in this section has already been reviewed from the gas-dynamical and laser-physics point of view. Consequently, we concentrate here on the modeling aspects with primary emphasis on assumptions and numerical procedures. Generally, the standard compressible, two-dimensional, boundary-layer equations in von Mises' coordinates are utilized for a reacting, rotationally equilibrated, perfect gas mixture. These equations represent the starting point of most efforts to date, and, since the equations are well known, we do not reproduce them.

The first multidimensional analysis to appear is that in Ref. 61, where the HF cold-reaction laser is analyzed with a simplified kinetic formulation involving 11 reactions. The two-dimensional flow field consists of a single semichannel, that is, from the centerline of the hydrogen nozzle to the centerline of the adjacent fluorine nozzle. An amplifier-type optical model is used in which an intensity uniform in $x$ is transmitted across the

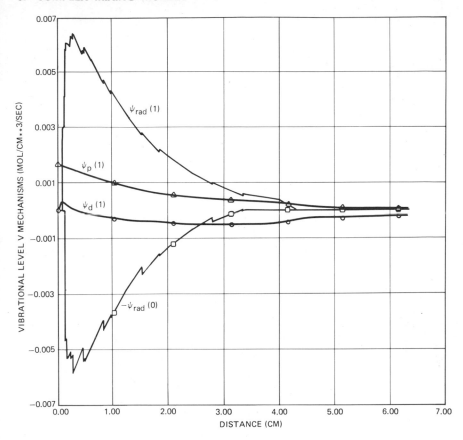

**Fig. 15b.** HF(1) mechanisms for 4-cm mixing length.

semichannel. The radiation field consists of two transitions, $P_1(6)$ and $P_2(5)$. In several calculations with increasing values for the initial intensity, the effect of gain saturation is shown. At a high initial intensity, the calculation should approximate a well-saturated oscillator cavity.

The model adheres to the assumptions of the first paragraph. The generated solutions are for a constant-pressure flow field, uniform initial profiles across each nozzle for velocity, temperature, and composition, and laminar mixing with all Lewis numbers unity and the Prandtl number equal to 0.72. Computation time with a CDC 6600 is approximately 30 min per solution.

References 84 and 85 contain a more general model kinetically (i.e., more reactions) and optically. Equation (102) is used for the gain, except

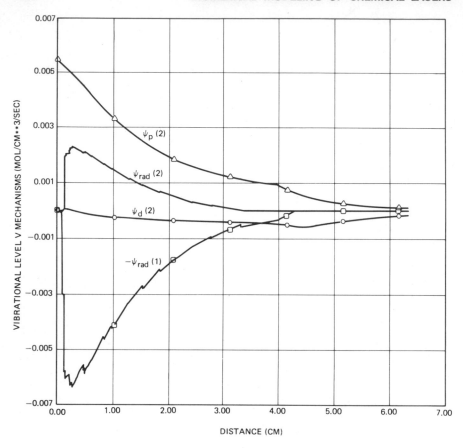

**Fig. 15c.**   HF(2) mechanisms for 4-cm mixing length.

that $\alpha_{mx}$ is averaged across the mixing layer. The usual boundary-layer equations for either laminar or turbulent mixing are used. In the latter case, a variety of eddy viscosity models are available. All Lewis and Prandtl numbers for both laminar and turbulent flow are assumed equal to unity. Nonuniform initial profiles are permitted, and an axial gradient for either the pressure or area can be specified.

A succinct presentation of the equations can be found in Ref. 84. Nevertheless, the sequential order of the numerical procedure requires clarification on a number of points. First, the kinetic and stimulated emission terms are decoupled from the boundary-layer equations. The boundary-layer equations are then solved in the next integration step

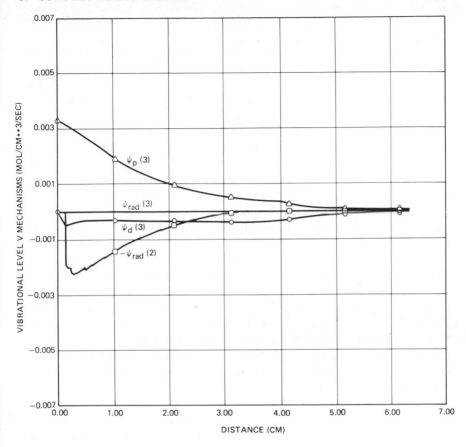

**Fig. 15d.** HF(3) mechanisms for 4-cm mixing length.

without these terms. The result is then used to calculate chemical-production rates that are then used to update the concentrations. The accuracy of this procedure is uncertain, and whether iteration is required is not addressed. An analogous procedure is used for the gain. A preliminary estimate of the gain distribution at the next step is obtained by evaluating the unsaturated gain starting with the saturated gain from the current step. This new distribution is then scaled to satisfy, on average, the gain-equals-loss relation. Again, accuracy is uncertain, and the actual extrapolation procedure is unclear. For example, the first procedure for extending the flow and concentrations is also sufficient to establish the gain. Finally, no mention is made of how or what transitions are selected for lasing. The computation time per solution is $\approx 45$ min on a CDC 6600.

Another large-scale computer model, with capabilities similar to the one above, is described in Ref. 86. Constant Prandtl and Lewis numbers are used for both laminar and turbulent mixing, where the latter uses one of several eddy viscosity models or a turbulent kinetic-energy transport equation. A differential form for the gain is developed, similar to that in Appendix A, except that the $dn(v)/dx$ in this equation is replaced by its rate equation. A linear system of equations in the intensities is, thus, developed that is solved in conjunction with the conservation equations. A description is provided of the explicit variable grid, finite difference integration procedure. Threshold, $J$-shifting, and cutoff are treated in a manner analogous to that in Section 2. Computation time per case on a CDC 6600 ranges from 5 min to 1 hr, depending on the complexity of the case.

The foregoing code[86] has recently been combined with a physical optics code[87,88] as described in Ref. 89. The code is limited to a slit resonator, that is, one with no variations in $y$, a Fresnel number less than 60, and geometric optics inside the gain medium. Effects caused by shock waves cannot be treated. The full set of equations governing the flow and kinetics is a subroutine of the Fresnel integral code and generally is called several times per case, since the Fresnel integral equation is solved iteratively. Computational time for cases without boundary layers is $\approx 25$ min on a CDC 6600.

The Lockheed group has also developed a code[90] that treats transverse pressure gradients using a method initially described in Ref. 91. The code is limited to a single-lasing transition that satisfies the gain-equals-loss condition. The method assumes that the equations can be split into a set of inviscid equations, solved by the method of characteristics, and a set of viscous equations. The latter represent a perturbation to the inviscid equations. Iteration at each solution point in the flow is required in order to couple the equations. In this treatment, shock waves appear in a smeared fashion. Run time is $\approx 25$ min per case on a CDC 6600.

Not all approaches rely on the boundary-layer equations. At Los Alamos Scientific Laboratory, a code is under development that calculates the full two-dimensional Navier–Stokes equations including species transport, chemical kinetics, and the effect of energy release from reactions and laser-power extraction. An implicit Eulerian scheme, valid at all flow speeds, is used for the conservation equations.[92] Both laminar and turbulent flows can be calculated with a transport equation used for the turbulent kinetic energy. The method for numerically treating transport coefficients is given in Ref. 93.

An interesting approach to turbulent mixing in a chemical laser is contained in Ref. 94. The starting point of the analysis is two Fokker–Planck equations that govern the distribution functions of the fluid elements and the chemical species.[95]. The formalism is applied to the HF cold-reaction laser, where turbulence is found to affect performance primarily through its effect on the pumping rates, which are governed by the turbulent dissipation rate. In this approach, the equations are hyperbolic, and the method of characteristics is employed for their solution. The characteristic directions and compatibility relations are well defined and make this approach numerically straightforward.

## APPENDIX A.   LEAKY STREAM-TUBE MODEL

Except where a $G$ or $J$ subscript appears, all quantities refer to the mixing layer. Of course, $y_B$ and $p$ refer to all stream tubes. The double $i, v$ subscript means that the summation is over both lasing and nonlasing species. The first summation term on the right-hand side of the energy equation is over both $i$ and $v$ in the nonlasing case, and the $H(v)$ term on the left-hand side is then deleted. In the nonlasing case, the matrix is reduced to the first five equations. (New parameters are defined after the equations are given.)

### Energy Equation

$$T \sum_{i,v} C_{pi} n_i \left(\frac{1}{T}\frac{dT}{dx}\right) + \frac{u^2}{4.184 \times 10^7}\left(\frac{1}{u}\frac{du}{dx}\right)$$

$$+ \sum_{v=0}^{v_{fin}} \left[10^3 H(v) + \frac{hcN_A}{4.184}\delta(v)\sum_{j=v}^{v^{(k+1)}-1}\omega_c(j)\right]\frac{dn(v)}{dx}$$

$$= -\sum_i 10^3 H_i \frac{dn_i}{dx} + \frac{hcN_A}{4.184\rho u}\sum_{v=0}^{v_{fin}}\delta(v)\omega_c(v)\sum_{j=v^{(k)}}^{v}\tilde{X}_{ch}(j) + 10^3 Q_{en}$$

### State Equation

$$\frac{1}{T}\frac{dT}{dx} + \frac{1}{\rho}\frac{d\rho}{dx} - \frac{1}{p}\frac{dp}{dx} = -\frac{W}{\rho u}\sum_{i,v}\tilde{X}_{ch}(i)$$

### Momentum Equation

$$\frac{1.01325 \times 10^6 p}{\rho u}\left(\frac{1}{p}\frac{dp}{dx}\right) + u\left(\frac{1}{u}\frac{du}{dx}\right) = Q_{mom}$$

**Solution Type**

$$a_y \frac{1}{y_B} \frac{dy_B}{dx} + a_p \frac{1}{p} \frac{dp}{dx} = b$$

**Jet Width**

$$[N_{BG} z_G(0) + N_{BJ} z_J(0)] y_B \left( \frac{1}{y_B} \frac{dy_B}{dx} \right) + \frac{\dot{m}}{\rho u} \left( \frac{1}{\rho} \frac{d\rho}{dx} \right)$$

$$+ \left[ \left( \frac{\dot{m}}{\rho u} K \right)_G + \left( \frac{\dot{m}}{\rho u} K \right)_J \right] \frac{1}{p} \frac{dp}{dx} + \frac{\dot{m}}{\rho u} \left( \frac{1}{u} \frac{du}{dx} \right) = Q_{\text{jet}}$$

**Nonlasing n(v)**

$$\frac{dn(v)}{dx} = \frac{1}{\rho u} \tilde{X}_{\text{ch}}(v), \qquad \delta(v) + \delta(v-1) = 0$$

**Sequence of Lasing $n(v)$**

$$\sum_{v=v^{(k)}}^{v^{(k+1)}} \frac{dn(v)}{dx} = \frac{1}{\rho u} \sum_{v=v^{(k)}}^{v^{(k+1)}} \tilde{X}_{\text{ch}}(v), \qquad k = 1, 3, \ldots$$

**Gain Equation**

$$\frac{1}{A_{2-1}} \left[ A_2 \left( \frac{hc}{k} \frac{E_{v+1,J+m}}{T} - C(v+1, T) - \tfrac{1}{2} \right) - A_1 \left( \frac{hc}{k} \frac{E_{v,J}}{T} - C(v, J) - \tfrac{1}{2} \right) \right] \left( \frac{1}{T} \frac{dT}{dx} \right)$$

$$- \frac{1}{y_B} \frac{dy_B}{dx} - F(y) \frac{1}{\rho} \frac{d\rho}{dx} - \frac{1}{u} \frac{du}{dx} - \frac{F(y)}{\sum_{v,i} a_i n_i} \sum_{v=0}^{v_{\text{tip}}} a_v \frac{dn(v)}{dx} - \frac{A_1}{n(v) A_{2-1}} \frac{dn(v)}{dx}$$

$$+ \frac{A_2}{n(v+1) A_{2-1}} \frac{dn(v+1)}{dx}$$

$$= \frac{F(y)}{\rho u \sum_{v,i} a_i n_i} \sum_i a_i \tilde{X}_{\text{ch}}(i) + Q_{\alpha(v)}, \qquad \delta(v) = 1$$

The following definitions are used in the foregoing equations (all $G$ definitions have a $J$ counterpart):

$$\tilde{X}_{\text{ch}}(i) = X_{\text{ch}}(i) + \rho u Q_{\text{sp}}(i)$$

$$Q_{\text{sp}}(i) = \frac{1}{\dot{m}} \left[ (n_i - n_{iG}) \frac{d\dot{m}_G}{dx} + (n_i - n_{iJ}) \frac{d\dot{m}_J}{dx} \right]$$

(same equations applying for lasing species)

$$Q_{en} = \frac{1}{\dot{m}}\left\{\left[h + \frac{1}{2}\frac{u^2}{4.184 \times 10^{10}} - h_{Gst}\right]\frac{d\dot{m}_G}{dx}\right.$$
$$\left. + \left[h + \frac{1}{2}\frac{u^2}{4.184 \times 10^{10}} - h_{Jst}\right]\frac{d\dot{m}_J}{dx}\right\}$$

$$h = \sum_{i,v} n_i H_i$$

$$h_{Gst} = \sum_i n_{iG} H_i(T_G(0)) + \frac{u_G^2(0)}{2 \times 4.184 \times 10^{10}}$$

$$Q_{mom} = \frac{1}{\dot{m}}\left[(u - u_G)\frac{d\dot{m}_G}{dx} + (u - u_J)\frac{d\dot{m}_J}{dx}\right]$$

$$b = \frac{1}{y_B}\frac{dy_B}{dx} \quad \text{or} \quad \frac{1}{p}\frac{dp}{dx} \quad \text{(See Section 2.4)}$$

$$K_G = \left[\frac{1}{\gamma}\frac{T_{st} - (\gamma + 1)T(x)/2}{T_{st} - T(x)}\right]_G$$

$$T_{Gst} = T_G(0) + \left[\frac{\gamma - 1}{\gamma}\frac{Wu^2(0)}{2 \times 4.184 \times 10^7 \times R}\right]_G$$

$$Q_{jet} = \left[\frac{1}{(\rho u)_G} - \frac{1}{\rho u}\right]\frac{d\dot{m}_G}{dx} + \left[\frac{1}{(\rho u)_J} - \frac{1}{\rho u}\right]\frac{d\dot{m}_J}{dx}$$

$A_1$ is given by Eq. (69). $A_2$ is given by Eq. (70). $A_{2-1} = A_2 - A_1$. $C(v, t)$ is given by Eq. (73). $F(y)$ is given by Eq. (68a).

$$Q_{\alpha(v)} = \frac{1}{\dot{m}}\left(\frac{d\dot{m}_G}{dx} + \frac{d\dot{m}_J}{dx}\right)$$

## APPENDIX B.   SOLUTION AT INITIAL SINGULARITY

For convenience, we define

$$a_G = \frac{\dot{m}_G(0)g_1}{x_G}, \qquad a_J = \frac{\dot{m}_J(0)j_1}{x_J}, \qquad \theta = a_G + a_J$$

The initial velocity and mole-mass ratios are then given by

$$u(0) = \frac{1}{\theta}(a_G u_G(0) + a_J u_J(0))$$

and

$$n_i(0) = \frac{1}{\theta}(a_G n_{iG}(0) + a_J n_{iJ}(0))$$

The specific enthalpy and temperature are obtained from

$$h(0) = \frac{1}{\theta}(a_G h_{Gst} + a_J h_{Jst}) - \frac{u^2(0)}{2 \times 4.184 \times 10^{10}}$$

and

$$h(0) = \sum_i n_i(0) H_i(T(0))$$

From the foregoing, we readily determine $W$, $\gamma$, $\rho$, and the Mach number at the origin in the mixing layer. The initial value of $Q_{jet}$ is not singular, but $Q_{sp}$, $Q_{en}$, and $Q_{mom}$ are singular. ($Q_{\alpha(v)}$ is not evaluated at $x = 0$ because lasing starts downstream of $x = 0$.) To find their values at $x = 0$, we first write

$$u = u(0) + \bar{u}x + \cdots$$
$$u_G = u_G(0) + \bar{u}_G x + \cdots$$
$$u_J = u_J(0) + \bar{u}_J x + \cdots$$
$$p = p(0) + \bar{p}x + \cdots$$
$$y_B = y_B(0) + \bar{y}x + \cdots$$
$$h = h(0) + \bar{h}x + \cdots$$

and

$$A_{mx} = \bar{a}x + \cdots$$

Substituting the above into the appropriate equations yields at $x = 0$:

$$Q_{sp}(i) = -\frac{\chi_{ch}(i)}{2\rho u} + \frac{1}{\theta}[b_G(n_i(0) - n_{iG}) + b_J(n_i(0) - n_{iJ})]$$

where

$$b_G = \frac{\dot{m}_G(0)}{x_G^2}(1 + g_1 - g_2)$$

$$b_J = \frac{\dot{m}_J(0)}{x_J^2}(1 + j_1 - j_2)$$

$$Q_{jet} = a_G\left(\frac{1}{(\rho u)_G} - \frac{1}{\rho u}\right) + a_J\left(\frac{1}{(\rho u)_J} - \frac{1}{\rho u}\right)$$

$$K_{jet} = \frac{y_B}{p}[(N_{Bz}K)_G + (N_{Bz}K)_J]$$

(see Appendix A for the definition of $K$)

$$[N_{BG}z_G(0) + N_{BJ}z_J(0)]\bar{y} + K_{\text{jet}}\bar{p} = Q_{\text{jet}}$$

(this equation is solved for either $\bar{y}$ or $\bar{p}$, whichever is unknown)

$$\bar{a} = -\frac{\theta}{\rho u N_{\text{mx}}}$$

$$\bar{u}_G = -\frac{1}{2}\frac{\bar{p}}{p}\left(2 \times 4.184 \times 10^7 \frac{\gamma - 1}{\gamma} \frac{R}{W} \frac{T^2}{T_{\text{st}} - T}\right)^{1/2}_G$$

(there is a similar equation for $\bar{u}_J$),

$$\bar{u} = \frac{1}{2\theta}[1.01325 \times 10^6 \,\bar{p}\bar{a}N_{\text{mx}} + 2b_G(u - u_G) + 2b_J(u - u_J) + a_G\bar{u}_G + a_J\bar{u}_J]$$

$$\bar{h} = -\frac{u\bar{u}}{4.184 \times 10^{10}} + \frac{1}{\theta}\left\{b_G\left[h + \frac{u^2}{2 \times 4.184 \times 10^{10}} - h_{G\text{st}}\right]\right.$$

$$\left. + b_J\left[h + \frac{u^2}{2 \times 4.184 \times 10^{10}} - h_{J\text{st}}\right]\right\}$$

$$Q_{en} = \frac{1}{\theta}\left\{b_G\left[h + \frac{u^2}{2 \times 4.184 \times 10^{10}} - h_{G\text{st}}\right]\right.$$

$$\left. + b_J\left[h + \frac{u^2}{2 \times 4.184 \times 10^{10}} - h_{J\text{st}}\right]\right\}$$

$$Q_{\text{mom}} = \bar{u} + \frac{1.01325 \times 10^6 \,\bar{p}}{\rho u}$$

## REFERENCES

1. N. Cohen, T. A. Jacobs, G. Emanuel, and R. L. Wilkins, Int. J. Chem. Kinet. **I,** 551 (1969); also **II,** 339 (1970).

2. P. H. Corncil and G. C. Pimentel, J. Chem. Phys. **49,** 1379 (1968).

3. E. B. Turner, G. Emanuel, and R. L. Wilkins, "The NEST Chemistry Computer Program," Technical Report TR-0059 (6240-20)-1 (The Aerospace Corp., Los Angeles, California, 1970).

4. J. R. Airey, J. Chem. Phys. **52,** 156, (1970).

5. G. Herzberg, *Molecular Spectra and Molecular Structure*, 2nd ed. (Van Nostrand, New York, 1950).

6. D. E. Mann, B. A. Thrush, D. R. Lide, Jr., J. J. Ball, and N. Acquista, J. Chem. Phys. **34,** 420 (1961).

7. R. M. Talley, H. M. Kaylor, A. H. Nielsen, Phys. Rev. **77,** 529 (1950).

8. D. H. Rank, B. S. Rao, and T. A. Wiggins, J. Mol. Spectrosc. **17,** 122 (1965).

9. K. Rossmann, W. L. France, K. N. Rao, and H. H. Nielsen, J. Chem. Phys. **24,** 1007 (1956).

10. G. Emanuel and W. D. Adams, "RESALE-2, $CO_2$, and INHALE," Technical Report TR-0073 (3430)-4 (The Aerospace Corp., Los Angeles, California, 1972).

11. G. Emanuel and N. Cohen, J.Q.S.R.T. **14,** 613 (1974).

12. T. F. Deutsch, IEEE K. Quant. Electron. **QE-3,** 419 (1967).

13. T. F. Deutsch, Appl. Phys. Lett. **10,** 234 (1967).

14. P. V. Avizonis, D. R. Dean, and R. Grotbeck, Appl. Phys. Lett. **23,** 375 (1973).

15. W. Benesch, J. Chem. Phys. **40,** 422 (1964).

16. R. E. Meredith and F. G. Smith, J.Q.S.R.T. **13,** 89 (1973).

17. F. G. Smith, J.Q.S.R.T. **13,** 717 (1973).

18. R. E. Meredith and F. G. Smith, "Investigation of Fundamental Laser Processes. Vol. II: Computation of Electric Dipole Matrix Elements for Hydrogen Fluoride and Deuterium Fluoride," Report 84130-39-T(II) (Willow-Run Laboratory, Ann Arbor, Michigan, 1971).

19. J. M. Herbelin and G. Emanuel, J. Chem. Phys. **60,** 689 (1974).

20. A. D. Wood, "Program for High Power Laser Techniques, Vol. II," Report AFAL-TR-68-361 (Avco Everett Research Laboratory, Everett, Massachusetts, 1968).

21. H. Statz, C. L. Tang, and G. F. Koster, J. Appl. Phys. **37,** 4278 (1966).

22. M. Rotenberg, R. Bivens, N. Metropolis, and J. K. Wooten, Jr., *The* 3-*j and* 6-*j Symbols* (The Technical Press, MIT, Cambridge, Massachusetts 1959).

23. T. K. McCubbin, Jr., and T. R. Mooney, J.Q.S.R.T. **8,** 1255 (1968).

24. C. P. Christensen, C. Freed, and H. A. Haus, IEEE J. Quant. Electron. **QE-5,** 276 (1969).

25. E. T. Gerry and D. A. Leonard, Appl. Phys. Lett. **8,** 227 (1969).

26. N. Djeu, T. Kan, and G. J. Wolga, IEEE J. Quant. Electron. **QE-4,** 256 (1968).

27. G. A. Kuipers, J. Mol. Spectrosc. **2,** 75 (1958).

28. G. Emanuel, J.Q.S.R.T. **11,** 1481 (1971).

29. W. W. Rigrod, J. Appl. Phys. **36,** 2487 (1965).

30. J. A. Glaze, Appl. Phys. Lett. **23,** 300 (1973).

31. W. F. Herget, W. E. Deeds, N. M. Gailar, R. J. Lovell, and A. H. Neilsen, J. Opt. Soc. Amer. **52,** 1113 (1962).

32. B. H. Armstrong, J.Q.S.R.T. **7,** 61 (1967).

33. E. E. Whiting, J.Q.S.R.T. **8,** 1379 (1968).

34. M. Abramowitz and I. A. Stegun, ed., *Handbook of Mathematical Functions* (National Bureau of Standards, Applied Math. Ser. 55, 1965).

35. S. S. Penner, *Quantitative Molecular Spectroscopy and Gas Emissivities* (Addison-Wesley, Reading, Massachusetts, 1959).

36. R. R. Patty, E. R. Manring, and J. A. Gardner, Appl. Opt. **7,** 2241 (1968).

37. P. W. Anderson, Phys. Rev. **76,** 647 (1949).

38. C. J. Tsao and B. Curnutte, J.Q.S.R.T. **2,** 41 (1962).

39. G. Emanuel, W. D. Adams, and E. B. Turner, "RESALE-1: A Chemical Laser Computer Program," Technical Report TR-0172 (2776)-1 (The Aerospace Corp., Los Angeles, California, 1972).

40. W. S. Benedict, R. Herman, G. E. Moore, and S. Silverman, Can. J. Phys. **34,** 850 (1956).

41. F. S. Simmons, "Infrared Spectroscopic Study of Hydrogen-Fluorine Flames," Report 4613-123T (Willow Run Laboratory, Ann Arbor, Michigan, 1966).

42. R. E. Meredith, J.Q.S.R.T. **12,** 485 (1972).

43. G. H. Lindquist, R. E. Meredith, C. B. Arnold, F. G. Smith, and R. L. Spellicy, "Investigation of Chemical Laser Processes," Report 191300-1-P Environmental Research Institute of Michigan (Ann Arbor, Michigan, 1973). The results shown in Fig. 5 are corrected values, courtesy of F. G. Smith and R. E. Meredith.

44. H. Babrov, G. Ameer, and W. Benesch, J. Chem. Phys. **33,** 145 (1960).

45. D. H. Rank, D. P. Eastman, B. S. Rao, and T. A. Wiggins, J. Mol. Spectrosc. **10,** 34 (1963).

46. R. H. Tipping and R. M. Herman, J.Q.S.R.T. **10,** 881 (1970).

47. P. Varanasi, S. K. Sarangi, and G. D. T. Tejwani, J.Q.S.R.T. **12,** 857 (1972).

48. C. Boulet, P. Isnard, and A. Levy, J.Q.S.R.T. **13,** 897 (1973).

49. D. F. Smith, "Molecular Properties of Hydrogen Fluoride," Proc. Second U.N. Geneva Conf. **28,** 130 (1958).

50. D. F. Smith, Spectrochim. Acta **12,** 224 (1958).

51. B. M. Shaw and R. J. Lovell, J. Opt. Soc. Amer. **59,** 1598 (1969).

51. T. A. Wiggins, N. C. Griffen, E. M. Arlin, and D. L. Kerstetter, J. Mol. Spectrosc. **36,** 77 (1970).

53. R. E. Meredith, T. S. Chang, F. G. Smith, and D. R. Woods, "Investigations in Support of High Energy Laser Technology. Vol. I," Report SAI-73-004-AA(I), (Science Applications, Inc., Ann Arbor, Michigan, 1973). [See also R. E. Meredith and F. G. Smith, J. Chem. Phys. **60,** 3388 (1974); F. G. Smith and R. E. Meredith, J.Q.S.R.T. **14,** 385 (1974).]

54. E. B. Turner, W. D. Adams, and G. Emanuel, J. Comp. Phys. **11,** 15 (1973).

55. G. Emanuel, T. A. Jacobs, and N. Cohen, J.Q.S.R.T. **13,** 1365 (1973).

56. T. A. Cool, J. Appl. Phys. **40,** 3563 (1969).

57. J. G. Skifstad, Comb. Sci. Tech. **6,** 287 (1973).

58. J. H. Steinfeld, L. Lapidus, and M. Hwang, I & EC Fund. **9,** 266 (1970).

59. C. W. Gear, Commun. Assoc. Comp. Mach. **14,** 176 (1971).

60. E. I. Gordon, A. D. White, and J. D. Rigden, *Proceedings of the Symposium on Optical Lasers,* Vol. 13 (Polytechnic Press, Brooklyn, New York, 1963), p. 309.

61. W. S. King and H. Mirels, AIAA J. **10,** 1647 (1972).

62. D. B. Rensch and A. N. Chester, Appl. Opt. **12,** 997 (1973).

63. S. N. Suchard, R. L. Kerber, G. Emanuel, and J. S. Whittier, J. Chem. Phys. **57,** 5065 (1972).

64. A. N. Chester and L. D. Hess, IEEE J. Quant. Electron. **QE-8,** 1 (1972).

65. J. L. Lyman, Appl. Opt. **12,** 2736 (1973).

66. R. L. Kerber, N. Cohen, and G. Emanuel, IEEE J. Quant. Electron. **QE-9,** Part II, 94 (1973).

67. L. D. Hess, J. Chem. Phys. **55,** 2466 (1971).

68. R. L. Kerber, G. Emanuel, and J. S. Whittier, Appl. Opt. **11,** 1112 (1972).

69.  J. J. T. Hough and R. L. Kerber, "Effect of Relaxation Oscillations in Pulsed $H_2 + F_2$ Chemical Lasers," unpublished results.

70.  R. L. Kerber, Appl. Opt. **12,** 1157 (1973).

71.  G. Emanuel and J. S. Whittier, Appl. Opt. **11,** 2047 (1972).

72.  B. P. Curry and R. E. Kidder, "A Multi-Level Pulsed Chemical Laser Model," (Lawrence Livermore Laboratory, Livermore, California, Report UCRL-74024 1973).

73.  A. E. Siegmann, *An Introduction to Lasers and Masers* (McGraw-Hill, New York, 1971).

74.  R. K. Pearson, J. O. Cowles, G. L. Hermann, D. W. Gregg, and J. R. Creighton, IEEE J. Quant. Electron. **QE-9,** 879 (1973).

75.  S. N. Suchard, Appl. Phys. Lett. **23,** 68 (1973).

76.  S. N. Suchard, IEEE J. Quant. Electron. **QE-10,** 87 (1974).

77.  R. A. Chodzko, D. J. Spencer, and H. Mirels, IEEE J. Quant. Electron. **QE-9,** 550 (1973).

78.  K. G. Anlauf, D. S. Horne, R. G. Macdonald, J. C. Polanyi, and K. B. Woodall, J. Chem. Phys. **57,** 1561 (1972).

79.  J. C. Polanyi and K. B. Woodall, J. Chem. Phys. **56,** 1563 (1972).

80.  R. R. Jacobs, K. J. Pettipiece, and S. J. Thomas, Appl. Phys. Lett. **24,** 375 (1974).

81.  C. L. Tang, H. Statz, and G. deMars, J. Appl. Phys. **34,** 2289 (1963).

82.  J. E. Broadwell, Appl. Opt. **13,** 962 (1974).

83.  R. R. Mikatarian and L. R. Ring, "Lockheed One-Dimensional Laser and Mixing Program Guide," Vol. III, Technical Report RK-CR-73-2 (Lockheed Missiles and Space Co., Huntsville, Alabana, 1973).

84.  R. Tripodi, L. J. Coulter, B. R. Bronfin, and L. S. Cohen, "A Coupled Two-Dimensional Computer Analysis of CW Chemical Mixing Lasers" AIAA Paper No. 74-224, presented at AIAA Twelfth Aerospace Sciences Meetings, Washington, D.C., 1974).

85.  B. R. Bronfin, L. S. Cohen, L. J. Coulter, H. McDonald, S. Shamroth, and R. Tripodi, "Development of Chemical Laser Computer Models," Vol. 1, Technical Report AFWL-TR-73-48 (United Aircraft Research Laboratory, East Hartford, Connecticut, 1973).

86.  J. Thoenes, A. J. McDanal, A. W. Ratliff, and S. D. Smith, "Laser and Mixing Program Theory and User's Guide," Vol. I, Technical Report RK-CR-73-2 (Lockheed Missiles and Space Co., Huntsville, Alabama, 1973).

87.  D. B. Rensch and A. N. Chester, "Chemical Laser Mode Control Program," Final Technical Report, Contr. DAAH01-70-C-1082, (Hughes Research Laboratory, Malibu, California, 1971).

88.  C. South and D. B. Rensch, "User's Manual, Two-Dimensional Fresnel Integral Computer Program," Final Technical Report, Contr. DAAH01-73-C-0298 (Hughes Research Laboratory, Malibu, California, 1973).

89.  J. Thoenes, C. South, and D. B. Rensch, "Two-Dimensional Fresnel Integral Laser and Mixing Program User's Guide," Vol. II, Technical Report RK-CR-73-2 (Lockheed Missiles and Space Co., Huntsville, Alabama, 1973).

90.  A. W. Ratliff, J. Thoenes, and S. D. Smith, "Method-of-Characteristics Laser and Mixing Program Theory and User's Guide," Vol. IV, Technical Report RK-CR-73-2 (Lockheed Missiles and Space Co., Huntsville, Alabama, 1973).

91. G. Moretti, "Analysis of Two-Dimensional Problems of Supersonic Combustion Controlled by Mixing," AIAA Preprint 64–69 (1964).

92. F. H. Harlow and A. A. Amsden, J. Comp. Phys. **8,** 197 (1971).

93. W. C. Rivard, O. A. Farmer, T. D. Butler and P. J. O'Rourke, Report LA-5426-MS (Los Alamos Scientific Laboratory, Los Alamos, New Mexico, 1973).

94. P. M. Chung and H. T. Shu, "HF Chemical Laser Amplification Properties of a Uniform, Turbulent Mixing Layer," *Proceedings of the ACTA Astronautica Conference on Explosive and Reactive Gases* (1974), to be published.

95. P. M. Chung, Phys. Fluids **15,** 1735 (1972).

# CHAPTER 9

# CLASSICAL DYNAMICS OF BIMOLECULAR REACTIONS

## ROGER L. WILKINS

**The Aerospace Corporation**
**El Segundo, California**

A chemical laser is a device whose radiant energy is due to a nonequilibrium distribution of the chemical energy among the reaction products in the active medium. The vibrational degrees of molecular freedom are the prime accumulator of this chemical energy. It is for this reason that many experimental and theoretical studies have been made of the energy distributions among reaction products in several exothermic chemical reactions. Experimental investigations have employed infrared chemiluminescence,[1-8] chemical lasers,[9-10] and crossed molecular beams.[11-12] We do not review these experimental methods, but rather confine our discussion to the quasiclassical trajectory method[13-27] of calculating the collision dynamics of three-atom systems on semiempirical potential-energy hypersurfaces. We present a brief discussion of how the semiempirical potential-energy surfaces are constructed and how trajectory studies use them to simulate the actual collision processes. We discuss how certain features on the potential energy surface can be used to predict (without detailed trajectory calculations) the reaction dynamics across the surface, including the product-energy distributions.

## 1.  POTENTIAL-ENERGY SURFACES

The calculation of the potential energy of the polyatomic system is the necessary prerequisite for a discussion of molecular structure of chemical reaction. The potential-energy surface depends parametrically on the nuclear coordinates. The actual calculations of the surface is a quantum-mechanical problem[28,29] requiring the solution of the electronic Schroedinger equation with the full Hamiltonian of the system. For collision energies of chemical interest, the nuclear velocities are sufficiently small relative to those of the electrons, and, thus, the Born–Oppenheimer approximation can be applied to separate the nuclear and electronic coordinates. We shall assume for the chemical processes discussed in this chapter that the motion is adiabatic, which means that the system is restricted to a single electronic state so that a single electronic wave function can be used to represent the state of the electrons throughout the collision encounter.

In principle, the problem of solving the Schroedinger equation is well defined. In practice, it is a very difficult problem to calculate a surface to such an accuracy that it can be regarded as known. There are many computational difficulties. The chemical reactions discussed in this book require solution of a multicenter problem. This requires evaluation of many complicated and time-consuming multicenter integrals. To calculate cross-sections for chemical reactions, it is necessary to have knowledge of the complete potential energy surface. For a three-atom system, about 100 points are required to determine the surface, and even more points are required as the number of atoms increase. For many chemical reaction problems, a very high accuracy is essential. For example, a change of 1.5 kcal/mol in the barrier height can change the rate constant at room temperature by one order of magnitude. It is necessary to calculate total energy to an accuracy of 0.1% or better for problems where potential-energy barriers are important.

Another problem is that there is very little direct information available on potential-energy surfaces required for chemical reactions. This lack of experimental information makes the determination of potential surfaces an important part of chemical kinetics. Karplus[30] has discussed three ways of handling this problem. They are: (1) purely theoretical, (2) semitheoretical, and (3) purely empirical. We will deal only with the last one. The purely empirical methods interest us for several reasons. Even if the potential surface were known exactly, it would have to be expressed in a way useful for numerical calculations. Explicit derivatives are required for easy computational manipulation. It is our opinion that a parameterized function can best serve as our working definition of the potential energy. The semiempirical method developed by London–Eyring–Polanyi–Sato (LEPS) for the construction of a potential-energy surface is very valuable for use in trajectory calculations. The reaction attributes calculated with quasiclassical trajectories on an optimized LEPS potential-energy surface have been found to be in quantitative agreement with available experimental data. We shall describe in Section 1.1 the empirical method used to develop the three-atom LEPS potential-energy surface.

## 1.1. Computation Procedure

In the LEPS technique,[31] the empirical potential-energy surface used for a three-atom system is given by

$$U(r_{12}, r_{23}, r_{13}) = \frac{1}{(1+K)} [Q_{12} + Q_{23} + Q_{13}$$
$$- (\tfrac{1}{2}(J_{12} - J_{23})^2 + \tfrac{1}{2}(J_{12} - J_{13})^2 + \tfrac{1}{2}(J_{23} - J_{13})^2)^{1/2}] \quad (1)$$

where $K$ is a single adjustable parameter, and $Q_{ij}$ is the Coulomb integral, and $J_{ij}$ is the exchange integral for each atom pair. The coordinates $r_{12}$, $r_{23}$, and $r_{13}$ refer to the internuclear separation between each atom pair. The calculation of $Q_{ij}$ and $J_{ij}$ requires information on the parameters of the interaction potentials for the ground electronic state and the first repulsive state of each atom pair in the system. The parameters of the Morse interaction potential for the bonding states $E_{ij}^b(r_{ij})$ are obtained from spectroscopic data. Information on the shape of the first repulsive state is usually not available. To obtain an analytic expression for the first repulsive state, Sato[31] modified the Morse equation by changing the sign between the two exponential terms from minus to plus and dividing by two. He found, that for $H_2$, the resulting anti-Morse interaction potential gave fairly good agreement with the shape of the $^3\Sigma_u^+$ curve for $H_2$. The potential function used by Sato for the antibonding state of each atom pair $(i, j)$ becomes:

$$E_{ij}^a(r_{ij}) = \tfrac{1}{2} D_{ij} \{ \exp[-2\beta_{ij}(r_{ij} - r_{ij}^0)] + 2 \exp[-\beta_{ij}(r_{ij} - r_{ij}^0)] \} \tag{2}$$

where $D_{ij}$ is the dissociation energy, $\beta_{ij}$ is the Morse parameter, and $r_{ij}^0$ is the equilibrium internuclear distance of the ground electronic state. The parameter $K$ in Eq. (1) is used to adjust the value of the energy $E_{ij}^a(r_{ij})$ of the antibonding state of each atom pair. This can be understood by examining the expressions for the Coulomb integral $Q_{ij}$ and the exchange integral $J_{ij}$:

$$\frac{Q_{ij}}{1+K} = \frac{1}{2} \left[ E_{ij}^b(r) + \left( \frac{1-K}{1+K} \right) E_{ij}^a(r) \right] \tag{3}$$

and

$$\frac{J_{ij}}{1+K} = \frac{1}{2} \left[ E_{ij}^b(r) - \left( \frac{1-K}{1+K} \right) E_{ij}^a(r) \right] \tag{4}$$

The ratio $(1-K)/(1+K)$ adjusts the energy $E_{ij}^a(r_{ij})$ for the antibonding state of each atom pair. The value of $K$ can be positive, zero, or negative to compensate for errors in the energies of the antibonding states. The optimum potential-energy surface is one which gives: (1) an activation energy in agreement with the known experimental value, and (2) a vibrational-energy distribution of the product molecule in agreement with experiment. It is difficult to construct a surface that fulfills both requirements. Table 1 lists the parameters used in constructing the LEPS surfaces for the $F+H_2$ and $H+HF$ reactions and their isotopic analogs.

When the LEPS method was applied to the $H+F_2$ reaction, it was not possible to construct a potential-energy surface with Eq. (2) that agreed

**Table 1.**    Potential Parameters[a] Used in LEPS Surface for $FH_aH_b$

| | | Bond | |
| Parameter | F–$H_a$ | $H_a$–$H_b$ | $H_b$–F |
| --- | --- | --- | --- |
| $\beta$ $(A^{-1})$ | 2.232 | 1.942 | 2.232 |
| $D_e$ (kcal/mol) | 141.1 | 109.2 | 141.1 |
| $r_0$ $(A)$ | 0.9170 | 0.7417 | 0.9170 |
| $K$ | 0.160 | 0.160 | 0.160 |

[a] F. D. Rossini et al., Natl. Bur. Std. (U.S.) Circ. 500 (1952).

**Table 2.** Parameters Used in Constructing the LEPS Surface for the Reaction $H+F_aF_b$

| | | Bond | |
|---|---|---|---|
| Parameter | $H+F_a$ | $F_a-F_b$ | $F_b-H$ |
| $\beta$ (A$^{-1}$) | 2.232 | 2.902 | 2.232 |
| $D_e$ (kcal/mol) | 141.1[a] | 38.02[b] | 141.1[a] |
| $r_0$ (A) | 0.9171 | 1.4170 | 0.9170 |
| $K$ | 0.05 | 0.05 | 0.05 |
| $A$ (kcal/mol) | 75.0[c] | 46.0[c] | 75.0[c] |

[a] R. F. Barrow and J. W. C. Johns, Proc. Roy. Soc. London Ser. A **251**, 504 (1959).
[b] G. Das and A. C. Wahl, Phys. Rev. Lett. **24**, 440 (1970).
[c] Trajectory calculations.

with the experimental-activation energy[32] of 2.4 kcal/mol without using a negative value for $K$. Analysis of the results of trajectories on this surface show that the peak of the vibrational distribution was found to be in the eighth vibrational level of HF rather than the experimentally observed fifth[6] or sixth[7,8] vibrational level. To alleviate this situation, the anti Morse potential given by Eq. (2) was modified by substituting $A_{ij}$ for $D_{ij}/2$. The values of $A_{ij}$ were adjusted through trajectory calculations to give HF populations in most of the vibrational ($v'=2$ through $v'=9$) levels. This was possible in only a small range of values of $A_{ij}$ for which the surface lacked a potential-energy barrier. In this parametric study, to obtain values for $A_{ij}$, $K$ was initially assigned a value of zero. The resulting interaction potentials are given in a recent paper by Wilkins.[20] The molecular parameters given in Table 2 were used to calculate the modified LEPS potential-energy surface for the $H+F_2$ reaction. Once the values of $A_{ij}$ were chosen, a value of $K$ was found that, when used in Eq. (1), gave a potential-energy surface with an activation energy in agreement with the known experimental-activation energy. The optimum value of $K$ was found to be 0.05, which when used in Eq. (1) gave a potential-energy surface with a barrier height of 2.3 kcal/mol.

The properties of several linearly activated complexes for three-atom systems containing the atoms hydrogen, deuterium, and fluorine are listed in Table 3. The data in this table can be used to make absolute rate calculations and to make preliminary decisions on the expected dynamics across the surface. The highly exothermic $F+H_2$ and $H+F_2$ reactions and their isotopic analogs have low-energy barriers. For a surface with a low-energy barrier, it is relatively easy for the representative mass point

**Table 3.** Parameters of Linear Activated Complexes

| Parameter | Complex | | | | | | | | | | | |
|---|---|---|---|---|---|---|---|---|---|---|---|---|
| | F-HH | F-HD | F-DH | F-DD | H-FH | H-F-D | D-F-H | D-FD | H-FF | D-FF | F-H-F | F-DF |
| $r_{12}$ (Å)[a] | 1.54 | 1.54 | 1.54 | 1.54 | 1.055 | 1.055 | 1.055 | 1.055 | 2.050 | 2.050 | 1.109 | 1.109 |
| $r_{23}$ (Å)[a] | 0.765 | 0.765 | 0.765 | 0.765 | 1.055 | 1.055 | 1.055 | 1.055 | 1.445 | 1.445 | 1.109 | 1.109 |
| $\omega_1$ (cm$^{-1}$)[b] | 3785. | 3201. | 3350. | 2678. | 2725. | 2359. | 2359. | 1927. | 851. | 846. | 618. | 618. |
| $\omega_2$ (cm$^{-1}$)[c] | 461. | 431. | 379. | 332. | 375. | 330. | 330. | 227. | 42. | 34. | 470. | 337. |
| $\omega_3$ (cm$^{-1}$)[d] | 448i | 383i | 366i | 331i | 236i | 198i | 198i | 175i | 336i | 242i | 2506i | 1794i |
| $F_{11}$ (mdyn/Å) | 4.8 | 4.8 | 4.8 | 4.8 | 2.19 | 2.19 | 2.19 | 2.19 | 4.2 | 4.2 | 1.23 | 1.23 |
| $F_{22}$ (mdyn/Å) | -0.14 | -0.14 | -0.14 | -0.14 | 2.19 | 2.19 | 2.19 | 2.19 | -0.054 | -0.054 | 1.23 | 1.23 |
| $F_{12}$ (mdyn/Å) | 0.52 | 0.52 | 0.52 | 0.52 | 2.22 | 2.22 | 2.22 | 2.22 | 0.20 | 0.20 | 3.05 | 3.05 |
| $F_\varphi/r_{12}r_{23}$ (mdyn/A)[e] | 0.019 | 0.019 | 0.019 | 0.019 | 0.038 | 0.038 | 0.038 | 0.038 | 0.0011 | 0.0011 | 0.032 | 0.032 |
| $E_c$ (kcal/mol)[e] | 1.01 | 1.01 | 1.01 | 1.01 | 1.43 | 1.43 | 1.43 | 1.43 | 2.28 | 2.28 | 21.66 | 21.66 |
| $E_0$ (kcal/mol)[f] | 1.47 | 1.36 | 1.43 | 1.33 | 0.48 | 1.46 | -0.17 | 0.70 | 2.40 | 2.38 | 17.98 | 19.22 |

[a] Saddle-point position.
[b] Nondissociative stretch of transition state.
[c] Bending frequency of transition state.
[d] Dissociative stretch of transition state.
[e] Saddle-point height.
[f] Classical activation energy at absolute zero of temperature.

to cross its barrier. We would expect the exchange reaction H+ FH($v = 1$) → HF($v' = 0$)+H to take place. The vibrational energy in HF($v = 1$) is more than sufficient for the representative mass point to cross the low-energy barrier. We would not expect the exchange reaction F+HF($v = 1$) → FH($v' = 0$)+F to occur. In this latter case, the vibrational energy in HF($v = 1$) is not sufficient for the representative mass point to cross the very high-energy barrier for this reaction. Thus, in the former reaction, we would expect both reactive and nonreactive collision processes to contribute to the deactivation mechanism; whereas, in the latter reaction, only nonreactive collisions are important in the relaxation process. Trajectory calculations[21–27] substantiate these conclusions. In the former reaction, chemical exchange provides an important mechanism for the efficient relaxation of vibrationally excited HF by H atoms. The rate constants calculated from the trajectory calculations[21–25] indicate that the rates of atom-transfer reactions of H with HF are very fast, and, therefore, represent a large loss of potential power in HF chemical lasers. The results of a recent flow-tube experiment[25] indicate also that the transfer rate of H with HF is very fast. In the F+HF reaction, chemical exchange becomes an important mechanism for the relaxation of HF($v$) by F atoms, for HF in vibrational states higher than $v = 2$. The rate constants derived from the trajectory calculations[26,27] indicate that the rate of atom-transfer reactions of F with HF or DF is very slow and does not represent a large loss of potential power for most operational conditions in HF or DF chemical lasers.

Parr and Truhlar[33] have discussed the presence of wells in some LEPS surfaces for some atom-transfer reactions involving hydrogen with HCl, HI, and HBr. The potential-energy surfaces for the reactions F+H$_2$ and H+F$_2$ are shown in Figs. 1 and 2, respectively, and do not have wells. The saddle point for the F+H$_2$ reaction occurs early in the entrance valley where the H$_2$ bond distance is only slightly perturbed (0.765 Å instead of 0.742 Å); and the H–F distance is about 1.6 times its equilibrium separation. Likewise, for the H+F$_2$ reaction, the saddle point occurs early where the F$_2$ bond distance is only slightly perturbed (1.445 Å instead of 1.417 Å) and the H–F distance is more than twice its equilibrium separation.

The potential-energy surface should be independent of the direction of the reaction. Therefore, the surface used for the reaction F+H$_2$ should be adequate for the reaction between H and HF. The potential-energy surface for the reaction between H and HF must satisfactorily represent the energies of both the H–HF and H–FH configurations. The surface used for the reaction H+F$_2$ should be adequate for the reaction between F and HF. The surface for the reaction between F and HF must

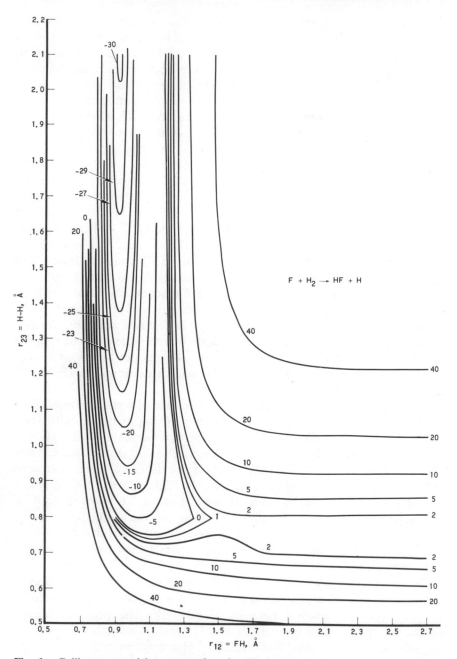

**Fig. 1.** Collinear potential-energy surface for the exothermic reaction $F + H_2 \rightarrow HF + H$, $\Delta H_r = -31.6$ kcal/mol. Barrier height: $E_b = 1.01$ kcal/mol. The saddle point occurs at the configuration $r_{12}^{\ddagger} = 1.540$ A, $r_{23}^{\ddagger} = 0.765$ A.

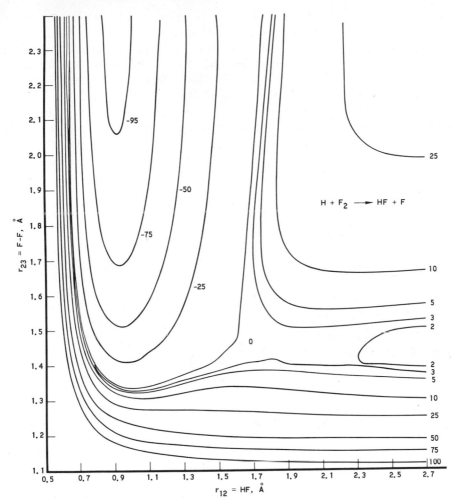

**Fig. 2.** Collinear potential-energy surface for the exothermic reaction $H+F_2 \rightarrow HF+F$, $\Delta H_r = -98$ kcal/mol. Barrier energy: $E_b = 2.3$ kcal/mol. The saddle point occurs at the configuration $r_{12}^{\ddagger} = 2.05$ A, $r_{23}^{\ddagger} = 1.455$ A.

satisfactorily represent the energies of both the F–HF and F–FH configurations. Parr and Truhlar[33] discussed this problem for atom-transfer reactions involving atomic hydrogen with several hydrogen halides. They discussed the possibility of using different Sato parameters for each saddle point of the energy surface and how a failure of the LEPS procedure would be fatal to classical and quantum-mechanical calculations, in which the full-energy surface is used. In this chapter, it is assumed that the constant Sato parameter used for the reaction $F+H_2$ can be used to describe the full energy surface for the configurations F–HH, H–HF, and H–FH. Also, it is assumed that the constant Sato parameter used for the reaction $H+F_2$ can be used to describe the full energy surface for the configurations H–FF, F–HF, and F–FH. The effect of this assumption on the surfaces for the reactions $F+H_2$ and $H+F_2$ can be investigated only when additional experimental data become available. The potential-energy surfaces are shown in Figs. 1 through 4 for the reactions $F+H_2$, $H+F_2$, $H+HF$, and $F+HF$, respectively. In Section 1.2, we describe the effect of significant features of the energy surface on the reaction dynamics across the surface.

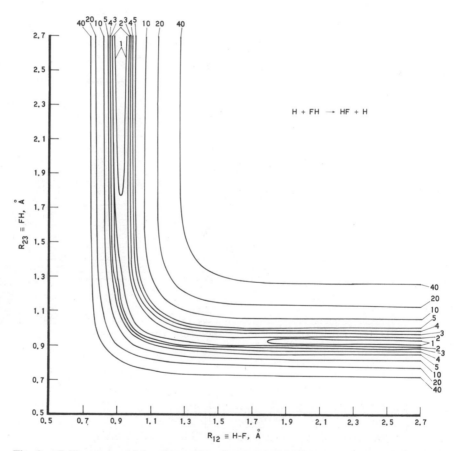

**Fig. 3.** Collinear potential-energy surface for the thermoneutral reaction $H+FH \to HF+H$, $\Delta H_r = 0$. Barrier height: $E_b = 1.4\,\text{kcal/mol}$. The saddle point occurs at the configuration $r_{12}^{\ddagger} = r_{23}^{\ddagger} = 1.055\,\text{A}$.

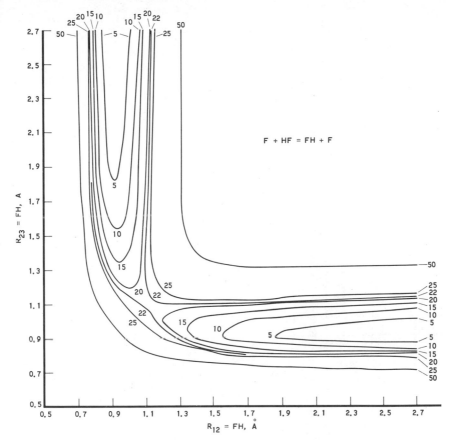

**Fig. 4.** Collinear potential-energy surface for the thermoneutral reaction $F+HF\rightarrow FH+F$, $\Delta H_r = 0$. Barrier height: $E_b = 21.7$ kcal/mol. The saddle point occurs at the configuration $r_{12}^{\ddagger} = r_{23}^{\ddagger} = 1.109$ A.

## 1.2. Significant Features of Energy Surfaces

In this section we shall discuss the type of information that can be obtained from a potential-energy surface. We will show in Section 1.3 how the potential-energy surface can be used to provide preliminary information on the fraction of the energy of an exothermic reaction that will appear as vibrational excitation in the new bond.

Potential-energy surfaces are categorized according to the components of energy released in going from reactants to products.[34] These components are: (1) attractive, the energy is released as the reagents approach one another; (2) mixed, the energy is released as the products separate

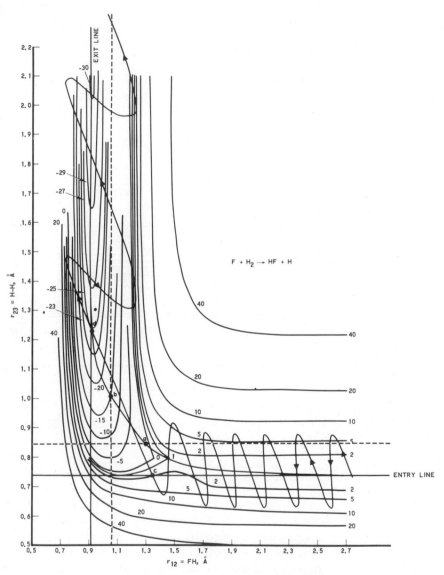

**Fig. 5.** Rectilinear, minimum-path, and collinear-trajectory categorizations of the energy release on the $F + H_2 \rightarrow HF + H$ potential-energy surface.

562

but the reactants are still attracting each other; and (3) repulsive, the energy is released as the products separate. Each of these three types of energy release can be identified with a qualitative feature of the potential-energy surface: attractive energy release with a drop in potential energy along the entrance valley, mixed energy release with a drop in potential along the curved path that cuts the corner of the surface, and repulsive energy release with a drop in the potential along the exit valley. Kuntz et al.[35] have provided three methods for calculating the percentages of attractive, mixed, and repulsive energy release ($\% \mathscr{A}$, $\% \mathscr{M}$, $\% \mathscr{R}$) in terms of collinear potential-energy surfaces. They have shown that one can correlate the character of the potential-energy surface with the fraction of reaction energy that becomes vibrational energy in the product molecule. The three methods introduced by Kuntz et al.[35] for calculating the percentages of attractive, mixed, and repulsive energy release on collinear surfaces are as follows (see Figs. 5 and 6).

**Rectilinear-Path Method.** The system enters the entrance valley at $r_{12} = \infty$, $r_{23} = r_{23}^0$, and leaves the exit valley at $r_{12} = r_{12}^0$, $r_{23} = \infty$. The two solid lines of this rectilinear path (see Figs. 5 and 6) define the *entry line* along $r_{12}$ with $r_{23} = r_{23}^0$ and the *exit line* along $r_{23}$ with $r_{12} = r_{12}^0$. Kuntz et al.[35] point out that this definition is too crude to be generally useful. This definition does not take into account that the reacting system in some cases is not able to pass through the configuration $r_{12} = r_{12}^0$, $r_{23} = r_{23}^0$, since this configuration can be at a high energy compared with that of the reactants. A more precise definition provided by Kuntz et al.[35] is as follows: The system proceeds along a linear path along the *entry line* until core repulsion sets in and thereafter takes any route to the final state. Core repulsion is $A \cdot BC$ repulsion, which cannot be released by closer approach of $A$ to $BC$, that is to say, by further decrease in $r_{12}$ with $r_{23} = r_{23}^0$. The attractive energy release $\mathscr{A}_\perp$ is the drop in potential along the entry line until core repulsion sets in. The repulsive energy release $\mathscr{R}_\perp$ is defined as $\Delta H_c - \mathscr{A}_\perp$, where $\Delta H_c$ is the energy of the reagents minus the energy of the products without inclusion of the zero-point, energy.

**Minimum-Path Method.** To use this method, we must first calculate the outer classical turning point for the reactant molecule in its most probable $(v, J)$ state at 300°K (dashed line parallel to $r_{12}$ axis) and for the product molecule in its most probable $(v', J')$ state at 300°K (dashed line parallel to $r_{23}$ axis). The attractive energy release $\mathscr{A}_M$ in the minimum-path method is the drop-in potential along the reaction path up to the outer classical turning point of the reactant molecule. The mixed-energy release $\mathscr{M}_M$ begins when the reaction path crosses the outer classical turning point

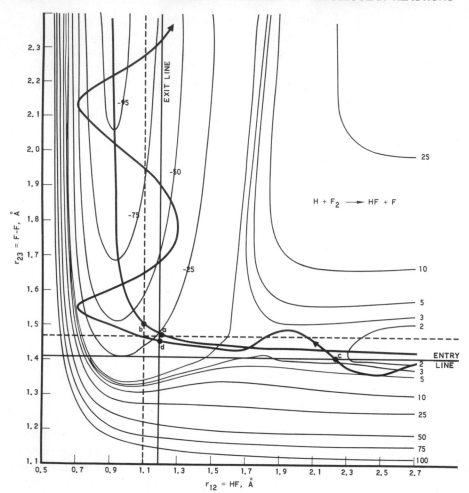

**Fig. 6.** Rectilinear, minimum-path, and collinear-trajectory categorizations of the energy release on the $H+F_2 \rightarrow HF+F$ potential-energy surface.

of the reactant molecule and terminates when the reaction path crosses the outer classical turning point of the product molecule. The repulsive energy release $\mathscr{R}_M$ is defined as $\Delta H_c - (\mathscr{A}_M + \mathscr{M}_M)$.

**Collinear-Trajectory Method.** The attractive energy release $\mathscr{A}_T$ is the energy released at the point on the potential surface where the collinear trajectory last crossed the equilibrium internuclear separation of the reactant molecule. The collinear trajectory is calculated at the collision energy $E_R = V^{\pm}$, where $V^{\pm}$ is the barrier height of the potential-energy surface. The mixed-energy release $\mathcal{M}_T$ is the potential energy released in going from the point where the collision trajectories last crossed the equilibrium internuclear separation of the reactant molecule to the point where it first crosses that of the product molecule. The repulsive energy release is given by $\mathcal{R}_T = \Delta H_c - (\mathcal{A}_T + \mathcal{M}_T)$. The $\%\mathcal{A}$ is defined as the ratio $\mathcal{A} \times 100/(-\Delta H_c)$. The others: $\%\mathcal{M}$ and $\%\mathcal{B}$, are similarly defined.

There is a definite relationship between the type of energy released on the potential-energy surface and the fraction of the energy of reaction that appears as vibrational excitation in the new bond. It has been found that upward to 90% of the energy of reaction becomes internal energy in $AB$ for reactions that occur on the attractive type of potential-energy surface. On attractive surfaces, the product molecule is highly vibrationally excited. Evans and Polanyi[36] showed that alkali-metal plus halogen reactions have attractive surfaces. Trajectory calculations show that attractive-energy surfaces tend to channel the reaction energy quite efficiently into product vibration. Repulsive-energy surfaces tend to channel the reaction energy into product translation. Polanyi[34] noted a special feature of the repulsive surface. There was a dramatic lowering of product-vibrational excitation when the incident atom mass was much lighter than the mass of the molecule under attack. Polanyi called this effect the "light-atom anomaly." What happens to the percentage of product-vibrational excitation in going from an attractive surface to a repulsive surface? Polanyi[34] found only a moderate decrease in product-vibrational excitation in going from attractive surface to a repulsive surface. The reason for this moderate decrease in product-vibrational excitation is as follows. For most mass combinations, the typical trajectory cuts the corner of the energy surface and drops into the exit valley at a point where the valley is quite deep. Thus, even on a repulsive surface, by cutting the corner of the surface (known as mixed-energy release) the representative mass point is able to move to and fro across the exit valley. The repulsive energy released is converted quite efficiently into product vibration. When the mass of the incident atom is much lighter than the mass of the molecule under attack, the mixed-energy release is a minimum, and, therefore, vibrational excitation is much lower.

The potential-energy surface can be used to obtain information on product-angular distribution in molecular-beam experiments. Trajectory calculations indicate that repulsive surfaces give backward scattering of the product molecule (back along the direction of the incident atom), and attractive surfaces give forward scattering. If the energy release is mixed on the repulsive surface, there can be forward scattering even on repulsive surfaces. The foregoing correlations for angular distribution hold for cases where the collision energy is modest and the encounters are direct.

Polanyi and Wong[37] have discussed the effect of barrier location on the dynamics of chemical reactions. They found that, if the crest of the barrier location of a thermoneutral reaction is shifted slightly into the entry valley, then translational energy in the reagent is vastly more effective than reagent vibrational energy in promoting reaction. If the crest of the barrier is displaced slightly into the exit valley of the surface, they found that vibration in the bond under attack is more effective than reagent translational energy in promoting reaction. For exothermic reactions, the crest of the barrier is located in the entry valley regardless of the type of potential-energy surface (see Figs. 1 and 2). This means that the crest of the barrier is located in the exit valley for endothermic reactions. The particles attempting to surmount the barrier in the endothermic direction see a surface whose crest is located in the exit valley; therefore, for successful reaction, the endothermicity must be present as vibrational energy in the bond under attack. This has been verified with both experimental and theoretical studies. There are other interesting features of potential-energy surfaces that we have not discussed in this chapter because of space limitations. For those who would like to read more on this subject we would recommend a recent review article by Polanyi and Schreiber.[38]

### 1.3. Illustrative Examples

For the reactions $F + H_2$ and $H + F_2$, Table 4 lists $\% \mathscr{A}$, $\% \mathscr{M}$, and $\% \mathscr{R}$, determined by the methods discussed in Section 1.2. For each system, it can be seen that $(\mathscr{A} + \mathscr{M})_T$ correlates quite well with the fraction of reaction energy that becomes product-vibrational energy. These results are in agreement with the qualitative rule proposed by Kuntz et al.[35] that the fraction of the reaction energy that enters into product vibration should be approximately equal to the $\% (\mathscr{A} + \mathscr{M})$ for any surface. Our results are in qualitative agreement with those of Muckerman.[14] Muckerman found that, for the reaction $F + H_2$ and its isotopic analogs, the collinear trajectory method gives values for $(\mathscr{A} + \mathscr{M})_T$ in much closer agreement with the relation suggested by Kuntz et al. than does the minimum-path method. It should be stated, however, that our value of $(\mathscr{A} + \mathscr{M})_T$ is not a collinear trajectory, but is for a random selection of initial-state parameters for a three-dimensional trajectory calculation.

Figures 5 and 6 show typical reactive collisions on the $F + H_2$ and $H + F_2$ energy surfaces. The two perpendicular solid lines represent the entry and exit lines, respectively, of the rectilinear method. The two dashed lines parallel to the entry and exit lines of the rectilinear method correspond to the outer classical turning points of reactant (horizontal line)

**Table 4.** Categorization of Energy Release on $F+H_2$ and $H+F_2$ Surfaces

|  | $F+H_2$ | $H+F_2$ |
|---|---|---|
| % $\mathscr{A}_\perp$ | 3 | 33 |
| % $\mathscr{R}_\perp$ | 97 | 67 |
| % $\mathscr{A}_M$ | 10 | 30 |
| % $\mathscr{M}_M$ | 44 | 13 |
| % $\mathscr{R}_M$ | 46 | 57 |
| % $\mathscr{A}_T$ | 17 | 2 |
| % $\mathscr{M}_T$ | 55 | 52 |
| % $\mathscr{R}_T$ | 28 | 46 |
| $(\mathscr{A}+\mathscr{M})_M$ | 54 | 43 |
| $(\mathscr{A}+\mathscr{M})_T$ | 72 | 54 |
| % $\bar{E}_{vib}$ [a] | 71[b] | 54[c] |
| % $\bar{E}_{vib}$ [d] | 66[e] | 54[f] |

[a] Trajectory calculations.
[b] R. L. Wilkins, J. Chem. Phys. **57**, 912 (1972).
[c] R. L. Wilkins, J. Chem. Phys. **58**, 2326 (1973).
[d] Infrared chemiluminescence experiments.
[e] J. C. Polanyi and D. C. Tardy, J. Chem. Phys. **51**, 5617 (1969).
[f] J. C. Polanyi and J. J. Sloan, J. Chem. Phys. **57**, 4988 (1972).

and product (vertical line) molecules at 300°K. The mixed-energy release $\mathscr{M}_M$ along the minimum path commences at point $a$ and terminates at point $b$. The attractive-energy release $\mathscr{A}_M$ is the energy released along the minimum path up to point $a$. The attractive-energy release $\mathscr{A}_T$ for the collinear-trajectory method is the energy released along the collision path up to point $c$. The mixed-energy release $\mathscr{M}_T$ is the energy released along the trajectory starting at point $c$ and terminating at point $d$. The surface (Fig. 6) for the reaction $H+F_2$ illustrates the *light-atom anomaly effect*. A representative mass point on this repulsive surface cuts the corner of the surface at a H–FF configuration at which the exit valley of the surface is quite deep. This produces oscillations of the representative point across the exit valley; thus, the energy released is converted into product vibration. Vibrational excitation in the newly formed HF molecule is, therefore, much lower in the reaction $H+F_2$ than in the reaction $F+H_2$. It can be seen in Table 4 and in Figs. 5 and 6 that the mixed-energy release is smaller in the $H+F_2$ reaction than in the $F+H_2$ reaction.

## 2.  TRAJECTORY STUDIES

The quasiclassical-trajectory method has been used to calculate the reaction dynamics of several important reactions in the active medium of both HF and DF chemical lasers.[13-27] The important reaction attributes that have been calculated include overall rate constants, total reaction cross-sections, specific rate constants for formation of product molecules in specified V–J states, angular distributions of reaction products, and energy distributions among the reaction products. The quasiclassical-trajectory method has been extremely valuable in the prediction of actual experimental results, such as those found in chemical lasers, infrared chemiluminescence, and crossed molecular-beam experiments. It is our opinion, based on theoretical results of recent trajectory calculations for the pumping and the deactivation reactions of HF and DF, that the quasiclassical method provides a proved procedure for obtaining accurate and realistic dynamical information for collision systems of interest.

In this section, we shall briefly describe: (1) the procedure generally used to solve the classical equations of motion, (2) how the initial values of the dynamical variables are chosen, (3) how the final classical internal energy is partitioned into product vibrational and rotational energies, (4) how the results of individual trajectories are placed on a statistical basis in order to calculate reaction cross-sections and rate constants, and (5) how the mechanism of the reaction affects the details of the collision.

### 2.1.  Computation Procedure

The quasiclassical procedure described by Karplus, Porter, and Sharma[39] can be used to examine the collision dynamics of an atom–diatomic molecule reaction. The Hamiltonian that describes the three-particle system is written in generalized coordinates, and the resulting set of 12 Hamilton's equations is solved by numerical integration. The calculation starts with the attacking atom $A$ sufficiently far from the molecule $BC$ so that their mutual interaction is small. The calculation ends when either atom $A$ is sufficiently far from $BC$, or atom $B$ is sufficiently far from $AC$, or when atom $C$ is sufficiently far from $AB$. When one of these three conditions occurs, we consider the collision to be over. The first condition corresponds to a nonreactive elastic or inelastic collision depending on what has happened to the initial state of the reagent molecule. The second and third conditions correspond to rearrangement collisions. The rearrangement collisions are referred to as exchange reactions in this paper. The exchange reactions can be classified as reactive elastic collisions or reactive inelastic collisions.

The twelve differential equations can be integrated with four steps of a Runge–Kutta–Gill integration routine followed by a modified Adam–Moulton predictor–corrector method. The required procedure has been adequately described by Smith and Wood.[40] Bunker[41] has described in great detail the types of numerical errors that can occur in the numerical integration of the classical equations of motion. One verification of numerical accuracy is to check at each point along the trajectory for conservation of total energy and total angular momentum. If the total energy is not conserved, then the error is usually due to the step size. The accuracy of the integration should also be tested by making changes in the step size and by integrating the classical equations of motion backward through the use of the values of the 12 dynamical variables at the end of a complete trajectory as starting conditions. To find the best value for the step size, preliminary calculations are carried out with different values for the step size. For the system $F + H_2$, it was found that with a step size of $1.5 \times 10^{-16}$ sec, the total energy was conserved to better than 0.1% in the worst case. Too large a step size affects the final-energy distribution among the various degrees of freedom of the reaction products. The appropriate step size is quite sensitive to the potential surface and other parameters· of the reaction; if it is too small, then the computer time required to integrate the equations of motion becomes prohibitive. To minimize computer time, one must readjust the step size for each chemical system until its optimum value is obtained. The efficiency of the computation depends greatly on the selection of step size. This last point cannot be over emphasized. We found that the computation time for a single trajectory when optimized was dependent on initial parameters, but was on the average about 1 sec per trajectory on the CDC 7600 computer.

## 2.2. Initial Values of Dynamical Variables

The integration of the 12 classical equations of motion requires values of 12 dynamical variables at the start of each trajectory calculation. Since it is not possible to calculate all trajectories for a particular reaction, it is necessary to use a representative sample of trajectories. The Monte Carlo technique[42] provides a procedure by which the initial position and momentum coordinates are chosen so as to simulate a representative sample of points in the phase space corresponding to their statistical distribution. Karplus, Porter, and Sharma[39] have presented the actual equations used to select the initial dynamical variables. We will not reproduce them, but instead will give a brief description of what is involved in their selection and some pointers on things that can be done to speed up the computation of individual trajectories.

The initial values of the coordinates in each collision are determined by a fixed initial relative velocity $V_R$ of the reagent molecules, by a chosen initial relative separation, $\rho$ of the atom from the molecule, by selected values of the vibrational $v$ and rotational $J$ quantum numbers of the reagent molecule, and by randomly selected values of the impact parameter $b$, the three orientation angles, and the vibrational phase of the reagent molecule. The coordinate system that defines the position of $A$ relative to $BC$ can be oriented in such a way that $A$ and the center of the mass of $BC$ lie in the $x$–$y$-plane with $V_R$, parallel to the z-axis. This orientation of the coordinate system allows an initial choice of the other three dynamical variables required to start the integration. At each relative velocity, it is necessary to select the maximum value of the impact parameter $b_{max}$ such that for $b \geq b_{max}$, there should be no reactions. Before the trajectory calculations are started for a given reaction at a fixed velocity of reagents, it is necessary to compute about 50 random trajectories at different values of $b$ until a value of $b$ is found for which 50 randomly selected trajectories produce no reactions. By this procedure, a different value of $b_{max}$ is chosen at each initial relative velocity of reagents. A value of 5 Å is usually assumed for the initial relative separation of atom and molecule. The time required to complete individual trajectories on a computer is quite sensitive to the choice of the initial value of this variable. It is best to choose a value of the relative atom–molecule separation $\rho$ as small as possible but such that the interaction is negligible between the incident atom and the molecule under attack. Proper choice of the initial value of $\rho$ can give large reductions in computer time. At each relative velocity, approximately 200 trajectory calculations are usually sufficient for each initial set of parameters $(V_R, v, J)$. The actual number of calculations depends on the amount of accuracy demanded by the researcher.

## 2.3. Final-State Analysis

A collision is over for the reaction $A + BC \rightarrow AB + C$ when the distance of the atom $C$ from the center of mass of the molecule $AB$ is greater than or equal to the initial separation of the incident atom $A$ from the molecule $BC$ under attack. This definition can easily be extended to the other two possible reactions that must be considered in these calculations, namely, $A + BC \rightarrow A + BC$ and $A + BC \rightarrow AC + B$. We will describe briefly how the classical energy is partitioned into rotational and vibrational energies of the product molecule $AB$ for the reaction $A + BC \rightarrow AB + C$. The rotational angular momentum of the product molecule is used to calculate the rotational quantum number $J'$ of $AB$. The value of

$J'$ is rounded to the nearest whole integer. The rotational energy of $AB$ is calculated with this value of $J'$ with the equation for the rotational energy of a rigid rotator. The vibrational energy of $AB$ is calculated as the difference of the internal energy of $AB$ minus the rotational energy of $AB$. The vibrational quantum number $v'$ of $AB$ is obtained by solving the energy expression for an anharmonic oscillator with the vibrational energy of $AB$. The value of $v'$ is rounded to the nearest whole number. The actual equations have been given in a recent paper by Muckerman.[14] The molecular product of each trajectory was assigned a $(v', J')$ state to calculate the required cross-sections for chemical- and energy-transfer reactions. Comparison between the trajectory and experimental results for product-energy distributions and specific rate constants indicate that this method of estimating $(v', J')$ is indeed a valid approach. The procedure for separating the final internal energy of the product molecule into vibrational and rotational energies, and assigning the product molecule to $(v', J')$ quantum states will be the same for the reactions $A + BC \rightarrow A + BC$ or $A + BC \rightarrow AC + B$ as for the reaction $A + BC \rightarrow AB + C$.

## 2.4. Analysis of Trajectories

To compute the reaction attributes that can be compared with experimental data, it is necessary to perform a large number of trajectory calculations. Three matrices: $R^{AB}(v', J')$, $R^{AC}(v', J')$, and $NR^{AB}(v', J')$, are used to facilitate calculations of cross-sections $\sigma(v, J, v', J', E_R)$ for both reactive and nonreactive collisions. For example, the element $(v', J')$ of the matrix $R^{AB}$ is advanced by one each time the reaction $A + BC(v, J) \rightarrow AB(v', J') + C$ occurs. The cross-section $\sigma(v, J, v', J', E_R)$ is defined as

$$\sigma(v, J, v', J', E_R) = \pi b_{max}^2 (E_R) \frac{N_r(v, J, v', J', E_R)}{N(v, J, E_R)} \tag{5}$$

where $N_r$ is the number of collisions that occur at energy $E_R$ with the reagent molecule in the initial $(v, J)$ state and the product molecule in the final $(v', J')$ state; $N(v, J, E_R)$ is the total number of such collisions. The equation used to calculate the rate constants from the cross-sectional data is

$$k_{v, J, v', J'}(T) = \left(\frac{8k_B T}{\pi \mu}\right)^{1/2} \frac{1}{k_B^2 T^2} \int_0^\infty \sigma(v, J, v', J', E_R) E_R \exp\left(\frac{-E_R}{k_B T}\right) dE_R \tag{6}$$

where $k_B$ is the Boltzmann constant and $\mu$ is the reduced mass of the colliding species. To obtain the rate constants $k_{v', J'}(T)$, the rate constants

$k_{v, J, v', J'}(T)$ are averaged over a Boltzmann distribution of $(v, J)$ states of the initial reagent molecule. The rate constants $k_{v'}(T)$ are calculated from the rate constants $k_{v', J'}(T)$ by summation of $k_{v', J'}(T)$ over all $J'$ states for a fixed value of $v'$.

The angular distribution, or the differential polar cross-section, of the product molecule is defined by

$$q_r = \frac{\pi b_{\max}^2 \, N_r(v, J, E_R, \theta)}{N(v, J, E_R) \, \Delta\theta} \tag{7}$$

where $N_r(v, J, E_R)$ is the number of reactive collisions in which the product molecule scatters into the angular interval $\theta$ to $\theta + \Delta\theta$.

## 2.5. Details of the Collision

In this section, we will describe how bond and force plots are used to obtain information on the mechanism of chemical- and energy-transfer reactions. A bond plot for a three-atom system is a plot of the three-particle distances $r_{12}$, $r_{23}$, and $r_{13}$ as a function of time. Bond plots are given in Figs. 7 through 9 for the reaction $F + H_2$. Figure 7 illustrates the

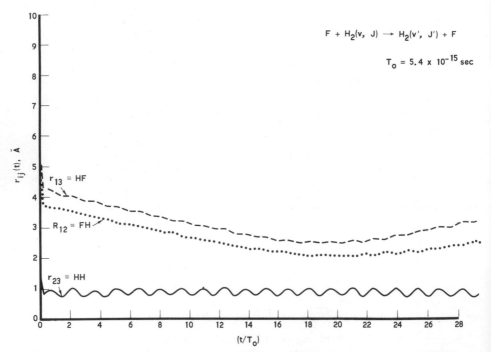

**Fig. 7.** Bond plot of a typical nonreactive collision between an F atom and an $H_2$ molecule.

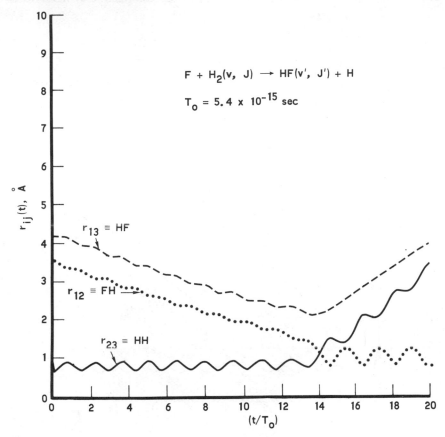

**Fig. 8.** Bond plot of a typical reactive collision between an F atom and an $H_2$ molecule. This reaction occurs by a direct interaction mechanism.

typical collision where a reaction does not occur. The function $r_{23}(t)$ shows the initial and final vibrational oscillations of $H_2$. This plot illustrates a nonreactive elastic collision. Figure 8 shows a plot of the typical reactive collision. The weaker bond $H_2$ is broken, and the stronger bond HF is formed in a vibrationally excited state. The interaction time is short; the F atom interacts strongly with the $H_2$ molecule for approximately $10^{-14}$ sec. The mechanism of this collision is called "direct interaction," since a collision complex was not formed. Approximately 90% of the reactive collisions that occur on the $F + H_2$ energy surface correspond to the direction interaction mechanism. An example of a complex formation is shown in Fig. 9. This collision is an example of an atypical trajectory in that the F atom and the $H_2$ molecule stay together for a

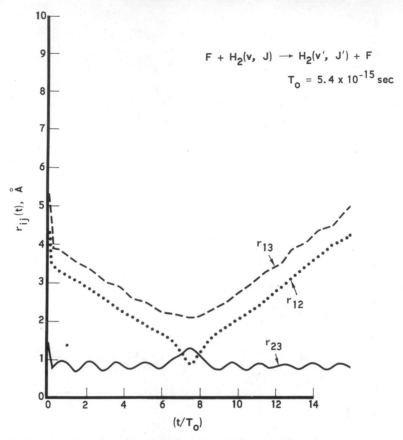

**Fig. 9.** Bond plot of an atypical reactive collision between an F atom and an $H_2$ molecule. A collision complex is formed for a brief period of time. This reaction corresponds to nonreactive inelastic collision.

longer period of time than is usually the case. A complex is formed for about $1.1 \times 10^{-14}$ sec, which is much shorter than the rotational period of $H_2$. We have concluded, therefore, that the mechanism for the reaction between F atoms and $H_2$ molecules is the simple direct type and not the complex-formation type. The trajectory displayed in Fig. 9 corresponds to a nonreactive inelastic collision, since the frequency of oscillation of $H_2$ in the final state is less than the frequency of oscillation of $H_2$ in its initial state.

The force plot shows the forces operating between the three atoms in the case of a three-atom system. Figure 10 shows a force plot for the $F + H_2$ reaction. The procedure for producing such a force plot has been

discussed by Polanyi and Wong.[37] The operating forces during an encounter are separated into three components: $F_{12}$, $F_{23}$, and $F_{13}$. These
three force components as a function of time are shown in Fig. 10. A
positive value of the force component corresponds to a repulsive force
and a negative value to an attractive force. The force $F_{13}$ between the end
atoms is low throughout and is indicated by the dashed line. The force of
attraction between the incident F atom and one of the H atoms in $H_2$
starts at the point designated 1* and becomes even more strongly
attractive at the point 2*. The arrow in Fig. 10 indicates the "switch-
over" point, that is, a point where the force of attraction in the newly
formed bond (dotted curve) exceeds the force of attraction in the old

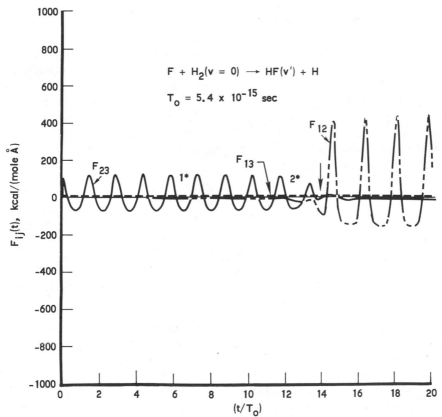

**Fig. 10.** Force plot for the reactive collision between an F atom and an $H_2$ molecule. The
force $F_{13}$ between the end atoms is quite small and corresponds to the dotted curve along
the x-axis; $F_{23}$ is the force between $H_B H_C$, and $F_{12}$ is the force between $FH_B$. The arrow
near $t/T_0 = 14$ indicates the switchover point.

bond (solid curve). The large regular oscillations beyond the switch-over point indicate a large amount of vibrational excitation in the new bond.

The forces plots can also be used to examine the collision for "secondary encounters." A secondary encounter[34] has occurred when the force between the products, having begun to decrease in absolute magnitude, exhibits a secondary peak. If the secondary peak in the force is positive, then the secondary encounter is said to be "clouting." If the secondary peak in the force is negative, then the secondary encounter is said to be "clutching." Clouting involves a repulsive encounter, and clutching, an attractive encounter. The effect of secondary encounters is to reduce vibrational excitation on highly attractive surfaces. Secondary encounters are rare on repulsive surfaces. The reason for this is that the reaction products on a repulsive surface are thrown apart before secondary interaction can occur. However, on surfaces that correspond to reaction with a large amount of mixed-energy release, it is possible to observe secondary encounters. On a surface that exhibits mixed-energy release, the repulsive energy is channeled quite efficiently into internal excitation of the products. Therefore, the products separate slowly; hence, secondary encounters are possible. Secondary encounters can have a marked effect on the vibrational-excitation energy of the newly formed molecule.[34] The net effect of a large number of secondary encounters is likely to be a broadening of the energy and angular distributions.

## REFERENCES

For recent infrared chemiluminescence studies on the pumping reactions $F + H_2$, $F + D_2$, see:

1. J. C. Polanyi and D. C. Tardy, J. Chem. Phys. **51,** 5717 (1969).
2. K. G. Anlauf, P. E. Charters, D. S. Horne, R. G. MacDonald, D. H. Maylotte, J. C. Polanyi, W. J. Skrlac, D. C. Tardy, and K. B. Woodall, J. Chem. Phys. **53,** 4091 (1970).
3. N. Jonathan, C. M. Melliar-Smith, and D. H. Slater, J. Mol. Phys. **20,** 93 (1971).
4. N. Jonathan, C. M. Melliar-Smith, D. Timlin, and D. H. Slater, Appl. Opt. **10,** 1827 (1971).
5. J. C. Polanyi and K. B. Woodall, J. Chem. Phys. **57,** 1574 (1972).

For recent infrared chemiluminescence studies on the pumping reactions $H + F_2$, see:

6. N. Jonathan, C. M. Melliar-Smith, and D. H. Slater, J. Chem. Phys. **53,** 4396 (1970).
7. N. Jonathan, S. Okuda, and D. Timlin, J. Mol. Phys. **24,** 1143 (1972).
8. J. C. Polanyi and J. J. Sloan, J. Chem. Phys. **57,** 4988 (1972).

For recent chemical laser studies on the pumping reactions $F + H_2$, $F + D_2$, see:

9. J. H. Parker and G. C. Pimentel, J. Chem. Phys. **51,** 91 (1969).

10. R. D. Coombe and G. C. Pimentel, J. Chem. Phys. **59,** 251 (1973).

For recent crossed-molecular beam studies on the pumping reaction $F+H_2$, $F+D_2$, see:

11. T. P. Shafer, P. E. Siska, J. M. Parson, F. P. Tully, Y. C. Wong, and Y.-T. Lee, J. Chem. Phys. **53,** 3385 (1970).

12. Y. T. Lee, *Physics of Electronics and Atomic Collisions, Proceedings of the Seventh International Conference of Physics of Electronics and Collisions,* Govers and F. J. deHeer, Eds. (North-Holland, Amsterdam, 1972), p. 357.

For recent trajectory calculations on the pumping reactions $F+H_2$, $F+D_2$, and $F+HD$, see:

13. R. L. Jaffe and J. B. Anderson, J. Chem. Phys. **54,** 2224 (1971); **56,** 682 (1972).

14. J. T. Muckerman, J. Chem. Phys. **54,** 1155 (1971); **56,** 2997 (1972); **57,** 3388 (1972).

15. R. L. Wilkins, J. Chem. Phys. **57,** 912 (1972).

16. R. L. Wilkins, J. Chem. Phys. **77,** 3081 (1973).

17. R. L. Wilkins, J. Mol. Phys. **28,** 21 (1974).

18. N. C. Blais and D. G. Truhlar, J. Chem. Phys. **58,** 1090 (1973).

For recent trajectory calculations on the pumping reactions $H+F_2$, $D+F_2$, see:

19. N. Jonathan, S. Okuda, and D. Timlin, J. Mol. Phys. **24,** 1143 (1972).

20. R. L. Wilkins, J. Chem. Phys. **58,** 2326 (1973).

For recent trajectory calculations on the deactivation reactions $H+HF(v)$, $H+DF(v)$, $D+HF(v)$, and $D+DF(v)$, see:

21. D. L. Thompson, J. Chem. Phys. **57,** 4170 (1972).

22. R. L. Wilkins, J. Chem. Phys. **58,** 3038 (1973).

23. R. L. Wilkins, J. Mol. Phys. **29,** 555 (1975).

24. J. B. Anderson, J. Chem. Phys. **52,** 3849 (1970).

25. M. A. Kwok and R. L. Wilkins, J. Chem. Phys. **60,** 2189 (1974).

For recent trajectory calculations on the deactivation reactions $F+HF(v)$ and $F+DF(v)$, see:

26. D. L. Thompson, J. Chem. Phys. **57,** 4164 (1972); **60,** 2200 (1974).

27. R. L. Wilkins, J. Chem. Phys. **59,** 698 (1973); **60,** 2201 (1974).

For a recent quantum-mechanical calculation of the potential-energy surface for the $F+H_2$ reactions, see:

28. C. F. Bender, P. K. Pearson, S. V. O'Neil, and H. F. Schaefer, J. Chem. Phys. **56,** 4626 (1972); Science **176,** 1412 (1972).

For a recent quantum-mechanical calculation of the potential-energy surface for the $H+F_2$ reaction, see:

29. S. V. O'Neil, P. K. Pearson, and H. F. Schaefer, J. Chem. Phys. **58,** 1126 (1974).

30. M. Karplus, "Potential-Energy Surfaces," *Proceedings of the International School of Physics "Enrico Fermi" Course XLIV, Molecular Beams and Reaction Kinetics* (Academic, New York, 1970).

31. S. Sato, J. Chem. Phys. **23,** 592, 2465 (1955).

32. R. G. Albright, A. F. Dononov, G. K. Lavrovskaya, J. J. Morosov, and V. L. Tal'rose, J. Chem. Phys. **50,** 3632 (1970).

33. C. A. Parr and D. G. Truhlar, J. Phys. Chem. **75,** 1844 (1971).

34. J. C. Polanyi, "Some Significant Features of Potential Energy Surfaces," *Proceedings of the Conference on Potential Energy Surfaces in Chemistry (Aug. 10–13, 1970), University of California, Santa Cruz*, W. A. Lester, Jr., Ed.

35. P. J. Kuntz, E. M. Nemeth, J. C. Polanyi, S. D. Rosner, and C. E. Young, J. Chem. Phys. **44,** 1168 (1966).

36. M. G. Evans and M. Polanyi, Trans. Faraday Soc. **35,** 178 (1939).

37. J. C. Polanyi and W. H. Wong, J. Chem. Phys. **51,** 1439, 1451 (1969).

38. J. C. Polanyi and J. L. Schreiber, "The Dynamics of Bimolecular Reactions," *Physical Chemistry—An Advanced Treatise, Volume VI, Kinetics of Gas Reactions*, H. Eyring, W. Jost, and D. Henderson, Eds. (Academic Press, New York, 1975), Chapter 9.

39. M. Karplus, R. N. Porter, and R. D. Sharma, J. Chem. Phys. **43,** 3259 (1965).

40. I. W. M. Smith and P. M. Wood, J. Mol. Phys. **25,** 441 (1973).

41. D. L. Bunker, J. Chem. Phys. **51** (1969).

42. J. M. Hammersley and D. C. Handscomb, *Monte Carlo Methods* (Wiley, New York, 1964).

# CHAPTER 10

# STATISTICAL AND DYNAMICAL MODELS OF POPULATION INVERSION

## AVINOAM BEN-SHAUL[*]

## G. LUDWIG HOFACKER

**Technische Universitat München, München, Germany**

[*] Now at the Department of Physical Chemistry, The Hebrew University, Jerusalem, Israel.

The theory of inelastic and reactive collisions has developed rapidly in recent years. Nevertheless, it has not yet reached the stage where quantitative predictions can be made starting from *ab initio* calculated potential-energy surfaces, not even for three-atom systems.[1] An additional challenge to the theory of molecular collisions is, therefore, to provide schemes and models to analyze experimental findings and interpret their physical meaning. Some models designed for this purpose and their applications in chemical-laser problems will be described. Particular attention will be paid to a recent statistical–dynamical approach that is based on some fundamental concepts and relations from information theory. There is hope that in the near future increased experimental insight into the relevant dynamical processes, on one hand, and improved quantal, semiclassical, and classical molecular dynamics methods (using rather accurate potential-energy surfaces), on the other, will close the gap between theory and experiment in the area of small reactive systems. Until then, relatively simple but physically meaningful models can, besides being of value in their own right, contribute much towards that goal.

    The most detailed and reliable information on small reactive systems comes from infrared chemiluminescence,[2] molecular beam,[3] and chemical-laser studies.[4] In some sense, the classical-trajectory calculations[5] that have now become a routine tool can be considered as another source of quasiexperimental-kinetic information. Their reliability is, however, inescapably dependent upon the accuracy of the potential-energy surfaces they use. Product-energy distributions determined in these studies are of primary importance for the development of efficient chemical lasers. A major problem arising in the experiments is the interpretation of the data produced in frightening amounts by modern equipment. Since their *ab initio* reproduction is not yet feasible, the immediate task for the theory must be to provide means of analyzing and condensing the mounds of experimental results and relate them to dynamical features of the reaction. These constitute the prime motivation for our article. We will show that elegant mathematical schemes to arrange and analyze very complex fields of experimental data are available (Section 1), and physical models can be constructed in the same conceptual frame as used for data interpretation (Section 3).

The formalism of information theory was first invoked by Bernstein and Levine[6] in the analysis of kinetic data. They introduced the "informational entropy" and the "surprisal"[7-12] to provide a measure for what one might call, the statistical nature of a reaction. Since then this approach has been further developed; many reactions have been analyzed, and new concepts were found to be useful in characterizing the product energy distributions in chemical reactions[13-25] and inelastic collisions.[26] For example, it was found that product vibrational distributions in many exothermic reactions can be characterized by closed-form expressions containing only one parameter which, for reasons to be clarified later, was called the "vibrational temperature parameter." The value of such expressions in chemical-laser studies is obvious. Information theory also provides means for the physical interpretation[10-12] of these expressions and allows new insights into the processes of chemical-reaction dynamics.

Hofacker and Levine designed a model for population inversion[27-30] that could be cast into the concepts of information theory and evaluated within this formalism.[14] They showed that the information-theoretical approach (although in physical terms it does not go beyond the realm of statistical mechanics) allowed them to handle certain problems in a more lucid and straightforward manner. In that way the dynamical nature of the temperature parameters could be elucidated. Of course, evaluation according to information-theoretical principles is not restricted to any specific model. The essential requirement for such a model is that it accounts for the most relevant dynamical features of the reaction without completely specifying the mechanics of the system.

## 1. CHARACTERIZATION OF PRODUCT-STATE DISTRIBUTIONS

### 1.1 Population Inversion

The degree of population inversion between two V–R levels of a diatomic molecule is usually defined as

$$\Delta N_{vJ}^{v'J'} = N_{v'J'} - \frac{2J'+1}{2J+1} N_{vJ} \tag{1}$$

Here $N_{vJ}$ is the population of the lower level characterized by the vibrational quantum number $v$ and the $(2J+1)$-fold degenerate rotational level $J$. $N_{v'J'}$ and $2J'+1$ are the corresponding quantities for the higher energy level.

The first requirement for lasing transitions to be possible is a positive net gain. Equivalently, the difference between the gain of photons per unit time or unit length in the laser cavity because of stimulated emission and the corresponding loss of photons through output coupling, diffraction, absorption, or any other mechanism must be positive. Since the stimulated emission gain is proportional to $\Delta N$, the condition

$$\Delta N_{vJ}^{v'J'} > 0 \tag{2}$$

is a necessary one for lasing action between levels $v'$, $J'$ and $v$, $J$.[4,31]

Typical chemical-laser pumping reactions are characterized by large exothermicities and highly inverted product populations. If the populations of the excited molecules are known, the "inversion condition," Eq. (2), provides a selection criterion for transitions that can possibly lase. Under some restrictive assumptions this condition can be expressed in a conveniently simple form. Later on, we shall show how Eqs. (1) and (2), by the introduction of information-theoretical concepts, can be formulated in a compact and physically clear form under less restrictive assumptions and for different conditions.

We start out with the conventional approach, thereby concentrating on exothermic reactions of the kind

$$A + BC \rightarrow AB^* + C \tag{3}$$

where $AB^*$ indicates the possible appearance of internally excited product molecules. For the energy levels of the diatomics, we shall mostly employ the "vibrating rotor anharmonic oscillator" (VRAO) level scheme,[32] that is,

$$E_v = \omega_e v - \omega_e x_e v (v+1) \tag{4}$$

$$E_J(v) = B_v J(J+1); \qquad B_v = B_e - \alpha_e(v + \tfrac{1}{2}) \tag{5}$$

where the energies are measured from the vibrational ground state; $\omega_e$, $B_v$, and so on, are the familiar spectroscopic constants (in Herzberg's notation).

A criterion for the inversion condition, Eq. (2), rests on the simultaneous fulfillment of two basic assumptions:

1. The relaxation of the $AB^*$ molecules occurs in two stages: (1) fast vibrational ($V \rightarrow V$) energy transfer and fast R–T ($R \rightarrow R$, $T \rightarrow T$, and $R \leftrightarrow T$) energy transfer and (2) relatively slow $V \leftrightarrow RT$ energy exchange.

2. The main laser action takes place after the first relaxation stage.

These assumptions imply that the V–R levels of the lasing $AB$ molecules are populated according to the partial equilibrium distribution that characterizes the system after the first relaxation step, namely,

$$P(v, J) = \frac{(2J+1)}{q} \exp\left[ -\frac{E_v}{(kT_V)} - \frac{E_J(v)}{(kT_R)} \right] \qquad (6)$$

Here $T_V$ is a vibrational-temperature characteristic for thermal equilibrium in a system of interacting oscillators. $T_R = T$ is the R–T temperature, equal to the heat-bath temperature. The normalization constant $q = q(T_V, T_R)$ is a mixed V–R partition function. If the coupling between the vibration and the rotation is weak, that is, $E_J(v)$ is practically independent of $v$, $q$ can be factored into rotational and vibrational contributions $q(T_V, T_R) = q_V(T_V)q_R(T_R)$.

The populations $N_{vJ}$ are proportional to $P(v, J)$. Equations (6) and (7), therefore, yield[4,31]

$$\frac{T_R}{T_V} < -\frac{E_{J'}(v') - E_J(v)}{E_{v'} - E_v} \qquad (7)$$

To be consistent with the assumptions leading to this inequality, both $T_V$ and $T_R = T$, being quantities characterizing equilibrium conditions, must be positive. [The intermediate equilibrium described by Eq. (6) can be thought of as a complete thermal equilibrium in a system of diatomic molecules where V↔R, T transfer is forbidden. $T_V$ and $T_R$ are thus determined by the averages of the energies originally available for the oscillators and the "moving-rotors," respectively; see also Section 2.] Since $T_V > 0$ implies $N_{v'} < N_v$ for $v' > v$, the assumptions cannot account for the very common situation in chemical lasers where the inversion is complete, that is, $N_{v'} > N_v$. The restricted validity of the criterion Eq. 7 is apparent when considering, for instance, $R$-branch transitions ($v' = v + 1$, $J' = J + 1$). Lasing $R$-branch transitions are frequently observed in chemical lasers.[4] If, as argued previously, $T_V$, $T_R > 0$, then the right-hand side of Eq. (7) is negative, and the inequality cannot be satisfied. Nevertheless, Eqs. (6) and (7) are still often used to describe even highly inverted populations where $T_V$ must take negative values. In this case, however, $T_V$ must be given different values for different vibrational levels; its physical meaning is then rather obscure.

Although we are only indirectly concerned with relaxation phenomena, some qualitative remarks regarding the assumption leading to Eqs. (6) and (7) will be made to bring things into better perspective and to set the basis for the discussion in Section 2. Before doing so, it should be mentioned that in exceptional cases Eq. (6) may adequately describe the real molecular-distribution function, if the relaxation mechanism is not as in assumption (1). An initial Boltzmann distribution of harmonic oscillators, characterized by temperature $T_V$ and contained in a heat bath of temperature $T$, will relax by V–T and V–V transfer through a continuous sequence of Boltzmann distributions with well-defined vibrational temperatures until the heat bath temperature $T_V = T$ is reached.[34] However, anharmonicity effects may drastically influence this behavior,[35] and, more importantly, initial product-state distributions in typical exothermic pumping reactions are usually "completely non-Boltzmann."

Inspection of the temporal evolution and spectral distribution of the outputs of some chemical lasers operating under reduced pressures[4,35] indicate that lasing may occur from nonequilibrium rotational populations [in contrast to Eq. (6)]. Moreover, vibrational relaxation is not necessarily negligible. In other words, the two-step relaxation mechanism described above is not always correct. Let us consider this point in a little more detail.

Because of its near resonant character vibrational-energy exchange in collisions between like molecules is usually a very effective process.[36–38] However, as a result of vibrational anharmonicities, V–V exchange is always accompanied by some V $\rightarrow$ R, T energy transfer. Thus, the distinction between these two processes cannot be strict. Another factor that may reduce the effective rate of V–V equilibration in chemical lasers is the dilution of the excited molecules by other molecular species in the laser cavity. These are mostly reactant molecules and other reaction products. R–T energy-transfer rates are large even in collisions between unlike molecules. Yet, the ir chemiluminescence experiments of Polanyi and co-workers[39–43] indicate that at least for halogen halides, the R $\rightarrow$ T rates are rapidly decreasing functions of $J$. This is interpreted as a result of the increased rotational spacing: $\Delta E = 2BJ$ for $\Delta J = 1$, and so on. On the other hand, V $\rightarrow$ R, T processes are not necessarily slow. An extreme example is provided by their rates in HF–HF collisions.[44] These processes are believed to proceed, mainly at low temperatures, a V $\rightarrow$ R mechanism.[44–48] They are especially fast at low temperatures because of the strong dipole attraction forces, and at high temperatures because of the large rotational and translational velocities. Considering, in addition, the high deactivation rates of HF by H, F, and $H_2$, one may conclude that V $\rightarrow$ R, T rates are comparable to V $\rightarrow$ V rates in the HF chemical laser. (For a list of recommended HF deactivation rates see Ref. 49. More recent experiments indicate higher V–V rates.[44,50,51]) In view of the relatively low R–T rates at high $J$'s and of some preliminary investigations[48,52] indicating increased V $\rightarrow$ R, T rates at such $J$'s, the suggestion[52] of a possible competition between these two processes does not seem unlikely.

## 1.2. Entropy and Surprisal

The method used here to characterize product-energy distributions is based on concepts from information theory and statistical mechanics. We first state some of the fundamental definitions.

Considering a molecular or any other system, with $N$ available states and with probabilities $P_n$ of finding the system in state $n$ by some measurement, a generalized entropy function $S$ can be defined by

$$S = - \sum_{n=1}^{N} P_n \log P_n \qquad (8)$$

The logarithm can be taken to any basis. Unless stated otherwise, however, we will use the natural logarithm. $S$ is the fundamental function of information theory,[7-12] where it is known as "uncertainty" or "information content." In statistical mechanics,[9-12] a function of the same type is known as "entropy." This is why $S$, given by Eq. (8), is often called "information entropy." The importance of $S$ lies in the fact that it is the only function satisfying the intuitive a priori requirements for an "uncertainty," for instance, its additivity property. For a general discussion of $S$, the reader is referred to the literature.[7-12] Here we will only cite some of its properties relevant to the forthcoming discussion.

We first consider the special case where all states have zero probability except one, say $n = n^*$. Then, we have $P_{n^*} = 1$, and there is no uncertainty regarding the outcome of any measurement on the system. Thus, the uncertainty is minimal and $S = 0$. In the other extreme, we may assume $P_n = 1/N$ for all $n$. It can be shown that $S$ now reaches its maximum value $S = \log N$. In equilibrium statistical mechanics, a distribution characterized by equal probabilities for all states is known as "microcanonical." Other equilibrium distributions, for example, the canonical distribution, also correspond to maximal entropies. In these cases, however, the maxima are conditional, that is, the equilibrium distribution maximizes $S$ subject to some additional constraints (e.g., $\langle E \rangle = \sum P_n E_n = \text{constant}$ for the canonical ensemble, where $E$ is the energy). The maximization procedure, also known as Jaynes' principle, provides the basis for the information theoretic approach to statistical mechanics[9-12] (cf., Section 3).

The entropy can be considered as the distribution average of the quantity

$$I_n = -\log P_n \qquad (9)$$

Thus

$$S = \sum P_n I_n = \langle I \rangle \qquad (10)$$

If $S$ is considered as the average, or integral, uncertainty associated with the outcomes of the observations on the system, then $I_n$ can be thought of as a detailed measure for the uncertainty associated with a specific state $n$. In the context of chemical reactions the term "surprisal" for $I$ became more widely used.[6] For the sake of consistency with the literature, we kept this nomenclature also. There is some variance in the literature regarding the semantics of "information," "entropy," "uncertainty," and "surprisal," depending on the objectives of the treatment, which are, however, of little concern to us. (Compare, for instance the definitions of the uncertainty measure in Refs. 8 and 12, see also Ref. 53.) As to the surprisal, we leave interpretations to a later stage where the importance of this concept will be apparent.

The definitions of entropy and surprisal can be extended to cases where $n$ stands for a set of numbers, for example, quantum states; some of them can also be continuous variables. These extended definitions will often be used throughout the subsequent sections.

## 1.3.  Product-State Distributions

When the irrelevant center of mass motion is disregarded and Reaction (3) is assumed to be electronically adiabatic, six quantum numbers are required to specify the quantum state of the triatomic-product system $AB + C$. One possible set of numbers is $\mathbf{k}$, $v$, $J$, and $m_J$, where $\mathbf{k}$ is the momentum vector of the relative translational motion and $v$, $J$, and $m_J$ are the internal quantum numbers of $AB$. In molecular beam experiments, the measured variables are usually the components of $\mathbf{k}$, that is, the angular and the translational-energy distributions. The internal quantum numbers are only rarely resolved. In chemiluminescence experiments and in chemical-laser work, the focus is on the internal-energy distribution, that is, $v$ and $J$ distributions are measured.

For many exoergic reactions, it can be assumed that the total energy available for the reaction products is confined to a small range $E$, $E + \delta E$,[40–43] (a more general description can be given[2,22]). $E$ is usually estimated as $E \approx E_a + (\frac{5}{2})RT + \Delta E_0$ where $E_a$ is the activation energy and $\Delta E_0$ the exoergicity.[40–43] The width $\delta E \sim RT$ is primarily determined by the reactant translational-energy distribution. For highly exoergic reactions, for example, typical chemical-laser reactions, it is $\delta E \ll E \sim \Delta E_0$.

The total-product energy is distributed among the translational, vibrational, and rotational degrees of freedom:

$$E = E_t + E_v + E_J \tag{11}$$

If, as above, $E$ is well defined, it is actually more convenient to introduce the reduced (dimensionless) quantities $f_x = E_x/E$, ($x = t$, $v$, $J$) satisfying

$$f_t + f_v + f_J = 1 \tag{12}$$

If $E$ is given, the complete product-energy distribution is determined by the V–R distribution. This is a conditional distribution denoted by $P(v, J | E)$ or, since $E$ is given, simply by $P(v, J)$. $P(v, J)$ is proportional to the detailed rate constant $k(v, J)$ into the product state $v$, $J$. (For thermal-reactant distributions, $k(v, J)$ is approximately the rate from the most populated reactant level $\hat{v}'\hat{J}'$ into the product level $v$, $J$.[54]) Thus,

$$P(v, J) = \frac{k(v, J)}{\displaystyle\sum_{v,J} k(v, J)} \tag{13}$$

where the sum extends over all $v$, $J$ compatible with $E_v + E_J \leq E$.

There is a large number of translational quantum states compatible with any given $v$, $J$ and total energy between $E$ and $E + \delta E$. The number of these states is $\rho_t(E_t)\,\delta E$, where $\rho_t(E_t) \propto E_t^{1/2} = (E - E_v - E_J)^{1/2}$ is the density of translational states.[15,55] If all quantum states of the $AB + C$ product system compatible with $v$, $J$ and $E$, $E + \delta E$ are equally probable, the V–R distribution will be microcanonical:

$$P_0(v, J) = Q_0^{-1}(2J+1)(E - E_v - E_J)^{1/2} \propto (2J+1)\rho_t(E_t) \tag{14}$$

where $Q_0$ is the normalization factor or the partition function. (This is a microcanonical partition function of the three-atom center of mass system $AB + C$, whose total energy is $E$.) When it is remembered that $P_0$ is proportional to the number of quantum states consistent with $v$; $J$; $E$, $E + \delta E$, then $P(v, J)/P_0(v, J)$ is proportional to the probability that the reaction will end up in one of these states. Therefore, up to an additive constant,

$$I(v, J) = -\log\left[\frac{P(v, J)}{P_0(v, J)}\right] \tag{15}$$

is the surprisal associated with this probability.[13–26] The average of this surprisal corresponds to the joint V–R entropy[15]:

$$S[v, J] = -\sum_{v,J} P(v, J) \log\left[\frac{P(v, J)}{P_0(v, J)}\right] \tag{16}$$

The square brackets on the left-hand side are used to denote the summation variables, not function arguments. Note that Eqs. (8) and (16) define $S$ as a dimensionless quantity, and that the thermodynamic dimensions are obtained by multiplying with the Boltzmann constant $k$. The connection between product state entropies and thermodynamic reaction entropies is an integral part of the general approach[21-23] but will not be discussed in detail here. Because of the normalization of the density of states through Eq. (14), $S[v, J]$ as distinguished from the $S$ in Eqs. (8) and (10) is not positive. It can be shown[6,15] that $S[v, J] \leq 0$ and that only in the case of a microcanonical distribution, that is, $P = P_0$, do we have $S[v, J] = 0$. The difference between the maximum of the entropy (i.e., the microcanonical entropy) and the entropy of the experimentally observed distribution is $\Delta S = S_0 - S = -S$. This is a positive quantity that was called the "entropy deficiency."[6,15,18,21-23] There are two possible interpretations of $\Delta S$. First, in the lack of any information regarding the dynamics of the reaction, our first guess will be that all product quantum states are equally probable. By performing the experiment and recognizing the real distribution, we gain information and $\Delta S$ is a measure for the gained information. For a second interpretation, let us suppose the dynamics of the reaction is such that all final states are equally likely. This is called statistical behavior (see Sections 3 and 4). Thus $\Delta S$ is an integral measure of the deviation from statistical behavior. Highly inverted populations, for instance, will have large $\Delta S$ values.

We shall next list some of the most useful relations for the entropy analysis of product energy distributions. The joint V–R distribution can be decomposed as

$$P(v, J) = P(v)P(J/v) \tag{17}$$

with

$$P(v) = \sum_{J \leq J^*(v)} P(v, J) \tag{18}$$

and

$$\sum_{J \leq J^*(v)} P(J/v) = \sum_{v \leq v^*} P(v) = \sum_{v,J} P(v, J) = 1 \tag{19}$$

where $J^*(v)$ is the largest $J$ compatible with $E_J \leq E - E_v$, and $v^*$ is the largest $v$ satisfying $E_v \leq E$; $P(v)$ is the vibrational distribution, and $P(J/v)$ is the conditional rotational distribution, that is, the rotational distribution in the vibrational manifold $v$. Applying Eqs. (16) through (19) to $P$ and $P_0$, we find[15] for $\Delta S = -S$,

$$\Delta S[v, J] = \Delta S[v] + \Delta S[J/v]$$
$$= \sum_v P(v) \log \left[ \frac{P(v)}{P_0(v)} \right] + \sum_v P(v) \sum_J P(J/v) \log \left[ \frac{P(J/v)}{P_0(J/v)} \right] \tag{20}$$

Explicit expressions for the statistical distributions $P_0$ are given in the forthcoming sections. The first term in Eq. (20) measures the deviations of the vibrational distribution from statistical behavior, while the second represents the average of the corresponding rotational deviations over all vibrational levels.

In the following sections, we will focus attention on the analysis of experimental results. Theoretical interpretations are left to Sections 3 and 4.

## 1.4.  Vibrational Distributions

Experimental results[25,40-43] and classical-trajectory studies[43,56] for isotopic reactions (e.g., $Cl + HI \rightarrow HCl^* + I$ and $Cl + DI \rightarrow DCl^* + I$ or $F + H_2 \rightarrow HF^* + H$ and $F + D_2 \rightarrow DF^* + D$) show that when $f_v$, Eq. (12), is used rather than $v$ as the distribution variable, $P(f_v)$ is practically invariant with respect to isotopic substitution. This is a reason for considering $f_v$ as a more significant variable than the corresponding quantum number.

The vibrational surprisal, expressed now in terms of $f_v$, can be expanded as

$$I(f_v) = -\log\left[\frac{P(f_v)}{P_0(f_v)}\right] = \lambda_0 + \lambda_v f_v + \cdots \qquad (21)$$

where $\lambda_0$, $\lambda_v$, and so on, are $v$-independent coefficients. (Notice the difference between $v$ and $v$; $v$ is used throughout as a subscript for "vibration," while $v$ denotes the vibrational state.) Examination of experimental data has shown that for many exoergic, population-inverting reactions the first two terms of Eq. (21) are sufficient for a description, accurate to within the experimental error bounds of the observed distributions.[13-25] Among the reactions following this rule are $Cl + HI \rightarrow HCl^* + I$, $F + H_2 \rightarrow HF^* + H$, $O + CS \rightarrow CO^* + S$ and $F + HBr \rightarrow HF^* + Br$ (see also Fig. 1). For this class of reactions we then have

$$P(f_v) = Q_V^{-1} P_0(f_v) \exp\left(-\lambda_v f_v\right) \qquad (22)$$

where $Q_V = \exp(\lambda_0)$ is the partition function, that is, the normalization constant. The statistical–microcanonical distribution is determined from Eqs. (4), (5), (14), (17) through (19). When $J$ is considered as a continuous variable $P_0(f_v)$ reduces[15,55] to

$$P_0(f_v) = \frac{(1-f_v)^{3/2}}{\sum_{v \leq v^*} (1-f_v)^{3/2}} \approx (\tfrac{2}{5})(1-f_v)^{3/2} \qquad (23)$$

The approximation expressed by the second equality corresponds to the classical rigid-rotor harmonic oscillator (RRHO) model where $x_e$ and $\alpha_e$

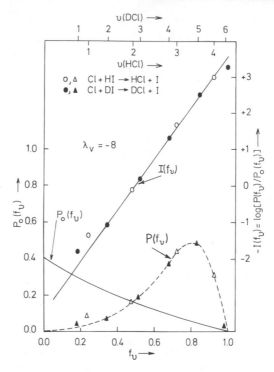

**Fig. 1.** Distribution of final vibrational states in the reactions Cl+HI→HCl*+I and Cl+DI→DCl*+I. Experimental data from Ref. 40. Calculation of $I(f_v)$ and $P_0(f_v)$ from Ref. 15. The scales correspond to the Cl+HI reaction where $P(f_v)$ is normalized according to $\sum P(f_v) = 1$. The results for Cl+DI were renormalized so that the equally sloped $I(f_v)$ plots corresponding to the two reactions will also overlap.

in Eqs. (4, 5) are zero and where $J$ and $v$ (or $f_v$) are considered as continuous variables. The reaction of $H + Cl_2 \rightarrow HCl^* + Cl$, which was investigated, for instance, by the arrested relaxation method,[42] provides an example for a nonlinear $I(f_v)$ behavior. However, this reaction belongs to the class of "angular-moment limited" reactions,[15–18] which are characterized by large differences between the average orbital-angular momentum of reactants and products, due to the large differences in the reduced masses. The $P_0$ of Eq. (23), which does not explicitly consider the conservation of total angular momentum in the reaction must then be modified (see Section 4.2). The resulting surprisal plot may be quite different from the one obtained with Eq. (23).

Equation (22) has the form of the Boltzmann canonical distribution in statistical mechanics, where $P_0$ is proportional to the density of states or the degeneracy factor. Since $f_v = E_v/E$, then $E/k\lambda_v$ is playing the role of a temperature and $\lambda_v$ is called the (reciprocal) vibrational-temperature parameter. It should be borne in mind, however, that the system considered is the nonequilibrium microscopic $AB^* + C$ center-of-mass system (see also the chapter by Wilkins).

Together with the definitions of entropy, the formal analogy of Eq. (22) to statistical mechanics is complete. Theoretical considerations based on the information-theoretical approach to elucidate this analogy will be presented in Sections 3 and 4.

Physically, the temperature parameter is a direct measure of the extent of population inversion. For typical chemical laser reactions $\lambda_v$ was found[13-25] to be negative, ranging between $-5$ and $-10$. A measure of the inversion $\langle f_v \rangle$ is a decreasing function of $\lambda_v$. The proof[20] follows along lines from statistical mechanics. Using Eq. (23), one finds

$$\frac{d\langle f_v \rangle}{d\lambda_v} = \langle f_v \rangle^2 - \langle f_v^2 \rangle \leq 0 \tag{24}$$

This means that if $\lambda_v < 0$ then $|\lambda_v|$ is a direct measure of the degree of inversion. It should be noted here that the degree of inversion is not a well-defined concept. The simple definition provided by Eq. (1), for instance, has the disadvantage of giving different $\Delta N$ for each pair of levels. On the other hand, $\langle f_v \rangle$ is too general a function, since it does not specify the distribution. Therefore, for reactions characterized by a well-defined $\lambda_v$, the vibrational-temperature parameter determines both $\Delta N$ and $\langle f_v \rangle$ (see the following equations) provides a unique measure of the degree of vibrational inversion.

In the spirit of thermodynamics, an expression for the vibrational entropy follows from Eqs. (20) and (22):

$$S[v] = S[f_v] = \log Q_V + \lambda_v \langle f_v \rangle \tag{25}$$

Other important expressions in the same spirit are

$$\langle f_v \rangle = -\frac{\partial \log Q_V}{\partial \lambda_v} \tag{26}$$

and

$$\lambda_v = -\frac{\partial S}{\partial \langle f_v \rangle} \tag{27}$$

Note that all quantities in Eqs. (25) and (27) are dimensionless and that the thermodynamic units are obtained if the right-hand side of Eqs. (16), (20), and (25) are multiplied by $k$.

## 1.5.  Rotational Distributions

The role of the rotational distributions within the various vibrational manifolds is no less important in determining the output of a chemical laser than that of the vibrational distributions. Figure 2 presents a specific example: The product V–R distribution in the reaction $Cl + HI \rightarrow HCl^* + I$. At room temperature, the total reaction energy is sufficient to populate the levels $v = 0, \ldots, 4$ of the HCl molecule. The full curves represent the initial, completely nonrelaxed V–R distributions, extrapolated to zero-rotational relaxation as determined by the arrested-relaxation method.[40] The height of the curve at $v, J$ is proportional to $P(v, J)$; the area under each curve is $P(v)$ and when these areas are normalized to 1 each curve is $P(J/v)$. Broken lines represent $P(v, J)$ after R–T thermalization at room temperature. The vibrational distribution is assumed to remain completely unrelaxed, reflected by the areas under each of the curves, which

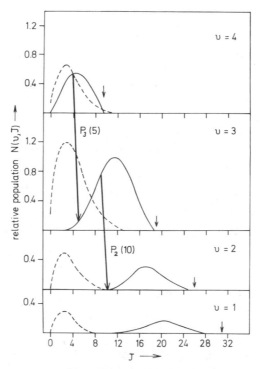

**Fig. 2.**  V–R distributions of HCl. Solid lines correspond to the original distributions in the $Cl + HI \rightarrow HCl + I$ reaction, measured by the arrested-relaxation method.[40]

are the same as in the completely nonrelaxed case. The $P(f_v)$ of this reaction fits Eq. (22) with $\lambda_v = -8$.[13-15]

The influence of the rotational distribution on $\Delta N_{vJ}^{v'J'}$ and, therefore, on the laser output, is apparent. If, as proposed for some pulsed chemical lasers,[4,33] the rotational relaxation is partially arrested, laser transitions such as the $P_2(10)$, $P_3(5)$ lines are expected to be intense. On the other hand, if the rotational distributions are thermal, they have almost the same shape and maxima. Because of this overlap, laser transitions are expected only at low $J$'s and with relatively small intensity. We will now turn to the characterization of the rotational distributions using the tools prepared in the previous sections.

Transforming Eqs. (14) and (17) through (19) to the $f_x$ variables, the microcanonical-conditional rotational distribution is given by[16-18]

$$P_0(f_J/f_v) = \frac{(3/2)[1 - (f_J/(1 - f_v))]^{1/2}}{(1 - f_v)} \tag{28}$$

Note that $P(f_J/f_v) = P(f_J/v)$ and that due to $P(f_J/f_v) = P(J/v) \, dJ/df_J$ we have $P(f_J/v) = P(J/v)E/(2J + 1)B_v$. Similarly, as in Eq. (21), we can expand[20]

$$I(f_J/f_v) = -\log\left[\frac{P(f_J/f_v)}{P_0(f_J/f_v)}\right] = \theta_0^{(v)} + \theta_R^{(v)}\left(\frac{f_J}{1 - f_v}\right) + \cdots \tag{29}$$

The expansion variable $f_J/(1 - f_v)$, ranging for a given $v$ between 0 and 1, represents the fraction of the nonvibrational energy in the rotation. The notation in Eq. (29) accounts for the possibility of R–V coupling, that is, that the shape of the rotational distribution determined by the $J$-independent $\theta$ coefficients may change with $v$.

By analysis of experimental and classical-trajectory data, it was found[16,17] that, for a number of exoergic reactions of the type $X + LY \rightarrow LX^* + Y$ where X and Y are halogens and L is an hydrogen isotope, all the $\theta$ coefficients are approximately zero. The remarkable result is that, within the experimental error bounds, one may assume for these reactions that

$$P(f_J/f_v) = P_0(f_J/f_v) \tag{30}$$

The reactions for which Eq. (30) was found to hold were all characterized by vibrational distributions of the type of Eq. (22). The reaction $F + H_2 \rightarrow HF^* + H$ was extensively investigated by various methods. Its $P(f_v)$ is given by Eq. (22) with $\lambda_v = 6.5$,[13,15,25] but there is still some controversy concerning its $P(f_J/f_v)$; see Refs. 17, 19, 41, and 43. The chemilumines-cence results,[41] for example, can be represented by Eq. (29) with $v$-

independent, nonzero $\theta_0$ and $\theta_R$, and with no higher-order coefficients.[20] $\theta_R$ was then called the rotational-temperature parameter (cf., Section 3).

Keeping in mind that Eq. (29) cannot be considered to be generally valid, let us discuss here briefly some of its implications. Using Eqs. (28) and (29), one can show[17]

$$\langle f_J \rangle_\upsilon = \tfrac{2}{3} \langle f_t \rangle_\upsilon = \tfrac{2}{5}(1 - f_\upsilon) \tag{31}$$

and

$$\langle f_J \rangle = \tfrac{2}{3} \langle f_t \rangle = \tfrac{2}{5}(1 - \langle f_\upsilon \rangle) \tag{32}$$

Here $\langle f_J \rangle_\upsilon$ is the average fraction of rotational energy for products in vibrational state $\upsilon$; $\langle f_J \rangle$ is the average of this quantity over all $\upsilon$'s. The partitioning of the nonvibrational energy between rotation and translation with a ratio of $2:3$ is according to the equipartition principle. This is obvious in the light of Eq. (28), which represents a microcanonical distribution, where all the R–T quantum states compatible with given $\upsilon$ and $E$ are equally probable.

Equations (31) and (32) provide a quick check for the R–T microcanonical-distribution assumption of Eq. (28), by comparing these averages with experimental data.[17,19] Another diagnostic means is provided by the predicted maxima of the rotational distributions. Transforming back from $f_J$ to $J$ we find that Eq. (29) reads:

$$P_0(J/\upsilon) = a(2J+1)[1 - f_\upsilon - b_\upsilon J(J+1)]^{1/2} \tag{33}$$

where the normalization constant $a = (\tfrac{2}{3})b_\upsilon(1 - f_\upsilon)^{3/2}$ and where $b_\upsilon \equiv B_\upsilon/E$. From Eq. (33), we find that the predicted most-probable rotational level is

$$\hat{J}(\upsilon) \approx \left[ \frac{(1 - f_\upsilon)}{2b_\upsilon} \right]^{1/2} - \frac{1}{2} \tag{34}$$

In the next section we shall consider the gain coefficients and the laser-rate equations. Equations (30) through (34) will serve there as zeroth-order approximations to the actual rotational distributions.

## 2. APPLICATION TO CHEMICAL LASERS

The development of $\Delta N_{\upsilon J}^{\upsilon' J'}$ in time and the output intensity in the laser are determined by the solution of the laser-rate equations. This requires detailed kinetic, optical, and mechanical information. The kinetic data involve the detailed reactive-rate constants, which determine the pumping rate of the excited molecular states and, furthermore, the rate constants of inelastic collisions for the various deactivation processes. At present, a wealth of data has been collected concerning product-state distributions (detailed reactive-rate constants). On the other hand, our knowledge about relaxation processes, especially rotational deactivation, is much more limited. To the same degree, our ability to perform highly accurate studies of laser systems is restricted. Some clues as to what intensities to expect from the various laser transitions can sometimes be obtained even without solving the rate equations. Estimates of this kind are possible if reasonable assumptions regarding the energy distribution among the lasing molecules during the main emission period can be made. In view of the discussion in Section 1.1, three special cases are analyzed in the following three sections.

## 2.1. Nonrelaxed Populations

As the first case, we consider the situation in very fast pulsed chemical lasers where a considerable amount of the output radiation may take place before rotational and vibrational deactivation is completed.[4,19] In such cases the condition for lasing, Eq. (2), can be approximated by $N_{vJ} \propto P(v, J)$, where $P(v, J)$ is the original product-state distribution. In accordance with experimental data and classical-trajectory calculations on a large number of chemical-laser reactions, we assume that $P(v)$ is given by Eq. (22). Recognizing the restricted generality of this choice, we take for $P(J/v)$ the microcanonical form of Eqs. (28) and (30). Equation (17) then yields

$$P(v, J) = Q^{-1}(2J+1)[1 - f_v - f_J]^{1/2} \exp(-\lambda_v f_v) \tag{35}$$

where $Q$ is a normalization constant determined by Eq. (19). We then define the dimensionless quantities

$$q_v = \left(\frac{b_v}{b_{v+1}}\right)^{1/2} \exp(2\lambda_v \, \Delta f_v) \tag{36}$$

$$s_v = \frac{(1 - f_v)}{b_v} = J^*(v)[J^*(v) + 1] \tag{37}$$

and

$$\Delta f_v = f_{v+1} - f_v \tag{38}$$

As above $b_v = B_v/E$, $J^*(v)$ is the maximal $J$ allowed by energy conservation. The inversion condition for the $P$-branch then reads[19]

$$0 < \Delta N_{v,J}^{v+1,J-1} = C(2J+1)\{q_v^{-1/2}[S_{v+1} - J(J-1)]^{1/2} - [S_v - J(J+1)]^{1/2}\} \exp(-\lambda_v f_v) \tag{39}$$

where $C$ is a proportionality constant to be determined from $P(v, J) \propto N_{vJ}$. For $R$-branch transitions the terms $(2J-1)$ and $J(J-1)$ should be replaced by $(2J+3)$ and $(J+1)(J+2)$, respectively. Using Eq. (39), one may derive, for example, the maximal $J$ for which $\Delta N > 0$ is still fulfilled. This is given by

$$J_m(v) = (2r_v)^{-1}\left\{1+\left[\frac{4r_v^2(S_{v+1}-q_vS_v)}{(1-q_v)}\right]^{1/2}\right\}$$

$$\approx (2r_v)^{-1}+\left[\frac{(S_{v+1}-q_vS_v)}{(1-q_v)}\right]^{1/2} \tag{40}$$

where $r_v = (1-q_v)/(1+q_v)$. For reactions like $Cl+HI \rightarrow HCl^*+I$, $F+HCl \rightarrow HF^*+Cl$ or $F+H_2 \rightarrow HF^*+H$ and their isotopic analogs ($\Delta f_v \approx 0.15$ through 0.35 and $-\lambda_v \approx 5$ through 8), we find $q_v \ll 1$ and $r_v \approx 1$. Thus,

$$J_m(v) \approx s_{v+1}^{1/2}+\tfrac{1}{2}=J^*(v+1) \tag{41}$$

Equation (41) means that the inversion is positive for most of the levels populated by the reaction. If the vibrational spacing $\Delta f_v$ is small, as in the $O+CS \rightarrow CO^*+S$ reaction where $\Delta f_v \approx 0.05$, $q_v$ is no longer negligible and $J_m(v) < J^*(v+1)$, subject in this case to the unverified assumption of Eq. (30); Eq. (22) is valid here, with $\lambda_v \approx -7.7$.[17]

It should be remembered that Eqs. (39) through (41) were derived for the initial nonrelaxed-produce distribution. Therefore, they only indicate which transitions are expected to lase if collisional deactivation and radiational losses are small. As already mentioned, a situation like this may prevail in some short pulse lasers. If saturation effects[57,58] are small, the relative intensities of the various possible transitions can be estimated from the (zeroth-order) gain coefficient[4,31]:

$$\alpha_{vJ}^{v'J'} = \sigma_{vJ}^{v'J'}\,\Delta N_{vJ}^{v'J'} \tag{42}$$

Here $\sigma_{vJ}^{v'J'} = aB_{vJ}^{v'J'}$ is the cross-section for stimulated emission, $B_{vJ}^{v'J'}$ is Einstein's absorption coefficient, and $\alpha$ is a constant depending on line broadening. A calculation of the $\alpha$'s in the HF laser, based on spectroscopically known $\sigma$ values and $\Delta N$ from Eq. (39), was presented elsewhere.[19] In the same investigation, an attempt was made to resolve the combined effects of pumping and cascading. To this purpose, relaxation effects have been ignored (see following remarks), and the laser-rate equations, (see the chapter by G. Emanuel), Eqs. (43), were solved:

$$\frac{dN_{vJ}}{dt} = k_{vJ}N_1N_2+\sigma_{vJ}Q_{vJ}c\,\Delta N_{vJ}^{v'J'}-\sigma_{v-1,J+1}Q_{v-1,J+1}c\,\Delta N_{v-1,J+1}$$

$$-A_{v-1,J+1}N_{vJ}+A_{vJ}N_{v+1,J-1} \tag{43a}$$

and

$$\frac{dQ_{vJ}}{dt} = A_{vJ}N_{v+1,J-1}+\sigma_{vJ}Q_{vJ}c\,\Delta N_{vJ}-\frac{Q_{vJ}}{\tau_P} \tag{43b}$$

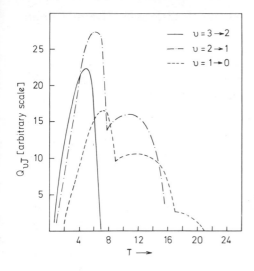

**Fig. 3.** Integral-output intensities from a model solution of the rate equations, Eqs. (42) and (43), for the $F + H_2 \rightarrow HF^* + H$ chemical laser, taken from Ref. 19. The $k_{vJ} \propto P(v, J)$ are assumed to be given by Eq. (35) with $\lambda_v = -6.5$.[15]

The notation accounts for the fact that only $P$-branch transitions were considered, that is, $\Delta N_{vJ} \equiv \Delta N_{vJ}^{v+1,J-1}$, and so on. The reactive-rate constant $k_{vJ}$ into $v$, $J$ was calculated in accordance with Eqs. (13) and (35); $N_1$ and $N_2$ are the reactant concentrations (in the example, F and $H_2$); $Q_{vJ}$ is the density of photons with $h\nu = E_{v+1,J-1} - E_{v,J}$; $\tau_P$ is the average photon life time in the laser cavity, and $c$ is the velocity of light. The first term in Eq. (42) represents the rate of pumping into level $v$, $J$ by the reaction. The next two terms account for the stimulated transitions that feed and deplete $v$, $J$ and the last two terms represent the spontaneous processes, which are important only during the early stages of laser action. The last term in Eq. (43) accounts for cavity losses. The results of the calculations for the integrated output, shown in Fig. 3, give us some idea about the output to expect from the laser. Although these calculations account for some preliminary measurements in a pulsed HF laser,[19,33] there are two reasons for not taking them literally. First, even though the pulse of the laser is short and the pressure is low, relaxation effects, mainly rotational, still influence the final results. The second reason is the uncertainty regarding the applicability of Eqs. (28) and (30) in the HF case, as mentioned previously. It is hard to correct for these two factors because of lack of accurate data. Yet, an investigation that utilizes the most recent available relaxation data, on one hand, and that bridges the informational gaps by reasonable models (Refs. 39, 46, and 48), on the other, is now in progress[59] and will hopefully shed some more light on these problems (see 'note added in proof').

## 2.2. The Vibrationally Nonrelaxed but Rotationally Thermal Populations

In this section we deal with the case where most of the laser emission occurs after R–T equilibration; Equation (22) can be used again for $P(v)$, but $P(J/v)$ is the Boltzmann rotational distribution

$$P(J/v) = \frac{(2J+1)\exp[-B_v J(J+1)/kT]}{q_R(v)} \tag{44}$$

$P(J/v)$ and the rotational partition function $q_R(v)$ are very slowly varying functions of $v$ (entering through $B_v$). $T = T_R$ is the T–R, or simply, the heat-bath temperature. Using Eqs. (17), (22), and (44) and the dimensionless quantities $b_v = B_v/E$ and $\beta = E/kT$, we have

$$N_{vJ} \propto (1-f_v)^{3/2}(2J+1)\exp[-\lambda_v f_v - \beta b_v J(J+1)] \tag{45}$$

The shape and maxima $\hat{J} \simeq (2\beta b_v)^{-1/2} - \frac{1}{2}$ of the rotational distributions are practically $v$-independent (see Fig. 2). The inversion condition, Eq. (2), may now be fulfilled for some $P$-branch transitions, even in the case of partial inversion, that is, $N_{v+1,J-1} > N_{v,J}$ and $N_{v+1} < N_v$. It is obvious, however, that $\Delta N$ will be much larger in the case of complete inversion, that is, when $N_{v+1} > N_v$. From Eqs. (22), (23), and (45), it can be seen that complete inversion between certain neighboring vibrational levels is only possible if $\lambda_v < 0$ (see Fig. 1). A quick estimate of the highest $v$ for which $N_{v+1} > N_v$ is satisfied can be obtained by treating $f_v$ in Eq. (22) as a continuous variable. This shows that for $\lambda_v < 0$, $P(f_v)$ has a maximum at $f_{\hat{v}} \simeq 1 + (3/2)\lambda_{\hat{v}}$ and complete inversion exists for all $v < \hat{v}$.

Substituting Eq. (45) into Eq. (2), we find that the necessary condition for lasing on $P$-branch transitions is

$$-\lambda v\, \Delta f_v + \beta J\left[2b_v + \alpha(J-1) - \frac{3}{2}\log\left(1 - \frac{\Delta f_v}{1-f_v}\right)\right] > 0 \tag{46}$$

where $\alpha \equiv \alpha_e/E$. The $\alpha$-dependent term can be neglected for molecules such as HF ($\hat{J} \simeq 2$; $\Delta f_v \simeq 0.3$) but not for molecules with small rotational- and vibrational-energy spacing such as CO ($\hat{J} \simeq 25$; $\Delta f_v \simeq 0.05$).

## 2.3. Partially Equilibrated Populations

The assumptions that lead to Eq. (7) can be summarized as: (1) the main-laser action takes place after the rotational and translational modes are already in thermal equilibrium at the heat-bath temperature $T_R = T$, and (2) the vibration is separately equilibrated at a temperature $T_V$. With the limitations mentioned in Section 1.1 in mind, we are now looking for a relation between $T_V$ and $\lambda_v$, so that Eq. (7) can be expressed in terms of $T_R$ and $\lambda_v$.

It was argued in Section 1.1 that vibrational equilibration is most likely to be reached through resonant $V \rightarrow V$ transfer taking place in collisions between the excited $AB$ molecules. We, therefore, assume that only the $AB$ molecules exchange vibrational energy among themselves. For the sake of simplicity, we take these molecules as harmonic oscillators. Our assumptions imply that the vibrational energy can be redistributed among the $AB$ oscillators but cannot be exchanged with other modes, or with unlike molecules. Consequently, the average vibrational energy per molecule $\langle E_v \rangle$ does not change in the course of the relaxation process.[60] Before relaxation $\langle E_v \rangle$ can be expressed as a function of $\lambda_v$ by the use of Eq. (27). In the classical RRHO approximation, it can be shown[22] that $\langle f_v \rangle = \langle E_v \rangle / E$ is given by

$$\langle f_v \rangle = 1 + \tfrac{5}{2}\lambda + \lambda^{3/2} \frac{\exp(-\lambda)}{\gamma(\tfrac{5}{2}; \lambda)} \qquad (47)$$

where $\lambda \equiv -\lambda_v$ (for $\lambda_v < 0$) and $\gamma$ is the incomplete gamma function. When equilibration is completed,

$$P(v) = q_V^{-1} \exp\left(-\frac{E_v}{kT}\right) \qquad (48)$$

holds, and

$$\langle E_v \rangle = \hbar\omega \left[\exp\left(\frac{\hbar\omega}{kT_V}\right) - 1\right]^{-1} \qquad (49)$$

where $\omega$ is the oscillator frequency, and the vibrational energy is measured from the $v = 0$ level. Comparison of Eqs. (47) and (49) leads through some harmless approximations to[22]

$$T_V \simeq \left(\frac{E}{k}\right)(1 - \tfrac{5}{2}\lambda) \qquad (50)$$

Here only the relevant case of inverted populations, that is, $\lambda_v = -\lambda < 0$, is considered. Substituting Eq. (50) into Eq. (7), we obtain the lasing condition in terms of $\lambda_v$.

It is known[35,36] that anharmonicity effects can invalidate the existence of a well-defined $T_V$. Therefore, the discussion in this section is limited to cases where the V–V transfer is very fast, very nearly resonant, and leading to a well-defined $T_V$. Equations (47) through (50) have a physical meaning if in addition to these requirements $\lambda_v$ is also well defined. When all these conditions are fulfilled, we may say that, since $\langle f_v \rangle$ and $\lambda_v$ are quantities of dynamical origin (see Sections 3 and 4), we have found a dynamical interpretation for $T_V$.

## 3. CONSTRAINTS AND TEMPERATURE PARAMETERS

### 3.1. The Maximum-Entropy Principle

Entropy, surprisal, and temperature-like distribution parameters have proved to be useful tools for analyzing product-state distributions. Entropy deficiency and surprisal were interpreted as overall and detailed measures for the deviation from microcanonical equilibrium. Because of their appearance in distributions of thermodynamical type the parameters $\lambda_v$ and $\theta_R$ were called temperature parameters. In this chapter, we will lay down a broader basis for the statistical aspects of reaction dynamics and explore the dynamical significance of the temperature parameters.

The particular advantage of the information-theoretical formalism,[10–12] for our purpose, lies in its primary aspect, namely, to make the best predictions on a system, compatible with a given set of *a priori* data. We do not expect more detailed results by information-theoretical arguments than by statistical mechanics. Information theory, however, provides an analytical tool in the form of the maximum entropy principle, which seems perfectly tailored to the field of phenomena we are investigating.

The central concept of this approach is the entropy $S$ defined by Eqs. (8) and (16). It enters the predictive scheme of the maximum entropy principle, or Jaynes' principle, stated as follows: "Given some limited information on a physical system, the best prediction of its real probability distribution function is provided by the distribution that maximizes $S$ under the constraints of the given information." The probability-distribution function is the set of numbers determining the probabilities of finding the system in its various states.

Jaynes' principle can also be exploited by reversing the argument, that is, given the observed distribution and when the entropy corresponding to this distribution is assumed to be maximal, then the constraints can be identified. As a specific example, let us first consider the vibrational distribution of Eq. (22). It can be shown[14,17,20] that $P(f_v)$ is the distribution that maximizes the vibrational entropy $S[f_v]$, defined through Eq. (20), but subject to the constraint

$$\langle f_v \rangle = \sum_v f_v P(f_v) = \mathsf{A} \tag{51}$$

where $\mathsf{A}$ is an *a priori* given value. Specifically, from all distributions satisfying Eq. (51), only the one given by Eq. (22) maximizes $S[f_v]$. The entropy is then given by Eq. (25). Mathematically, the maximization of $S$ can be done by the method of Lagrange multipliers; $\lambda_v$ is then the Lagrangian multiplier of the constraint on the vibrational energy, Eq. (51), because of some peculiar mechanism of energy distribution in the reacting system (c.f., discussion of this point in Section 4.1). The analogy of $1/\lambda_v$ to the thermodynamic temperature, which is the (reciprocal) Lagrangian multiplier of the average energy of the system under consideration, is now obvious. We, therefore, call all distribution parameters arising from dynamic constraints "temperature" in a generalized sense.

The foregoing procedure allows us to detect dynamical constraints but does not exhibit their nature. Equation (22) can be thought of as the most random distribution compatible with the constraint of Eq. (51). Specifically, dynamical features of the reaction, due to the masses, the initial state distribution, and the potential-energy surface, only restrict the value of $\langle f_v \rangle$ to be A. Apart from this, the product-state distribution is random (in the sense of maximum entropy), and no further dynamical information can be extracted from the observed distribution. We may say that reactions characterized by product distributions, such as in Eq. (22), exhibit statistical behavior in the canonical ensemble sense. The introduction of general dynamical constraints may, thus, be regarded as a nontrivial extension of some established statistical–microcanonical theories of chemical reactions (c.f., Section 4.2). If the number of constraints is small, then we are able to determine the behavior of complex systems on the basis of very little dynamical information.

The finding that Eq. (22) reflects most gross features of many highly exoergic reactions suggests that simple dynamical models that take into account the most relevant dynamical factors may be helpful in investigating the dynamical character of the constraints. The first attempt in this direction was based on a one-dimensional nonadiabatic model for population inversion,[14–27] which yields, in the high inversion limit,

$$\langle f_v \rangle = \frac{\frac{1}{2}g^2\hbar\omega}{E} \tag{52}$$

Here $g$ is the V–T coupling constant, which depends on the curvature of the reaction path, the local kinetic energy, and the vibrational frequency at the region of highest curvature. The model is discussed in more detail in the next section; let us only mention here that the mass dependence of $g$ is such that $\langle f_v \rangle$ is nearly invariant under isotopic substitution, in accordance with experimental findings.

## 3.2. Rotational Temperature Parameters

There are no *a priori* restrictions on the number or the character of the dynamical constraints of a reactive system. The larger the number of constraints, the smaller the degree of indeterminacy (i.e., the statistical nature) of the reaction. If the number of constraints is not small compared to the number of accessible final states, statistical models cease to provide an adequate means to predict the outcomes of reactive processes, and one has to resort to the much more complicated classical or quantum-mechanical description of the reaction.

A systematic general treatment of constraints based on moment analysis of the distributions was given by Levine and Bernstein.[20] Here we confine ourselves to the situation where the number of constraints is two. To this end we consider the joint V–R distribution $P(f_v, f_J)$.

By analysis of chemiluminescence results from the reaction of $F + H_2 \rightarrow HF^* + H$,[41] it was found[20] that the observed distribution can be represented by the formula

$$P(f_v, f_J) = Q_V^{-1} P_0(f_v) \exp\left(-\lambda_v f_v\right) Q_R^{-1} P_0(f_J/f_v) \exp\left(-\theta_R \frac{f_J}{1-f_v}\right) \quad (53)$$

The various quantities appearing here are the same as defined by Eqs. (21) through (23), (28), and (29). It is not difficult to verify that Eq. (53) is exactly the distribution that maximizes $S[f_v, f_J]$ of Eqs. (16) and (20), subject to the two constraints

$$\langle f_v \rangle = \sum_v f_v P(f_v) = \mathbf{A} \quad (54)$$

and

$$\left\langle \frac{f_J}{1-f_v} \right\rangle = \sum_{J < J^*(v)} \left(\frac{f_J}{1-f_v}\right) P(f_J/f_v) = \mathbf{B} \quad (55)$$

where $\mathbf{A}$ and $\mathbf{B}$ are a priori given constants. Moreover, $\mathbf{B}$ is independent of $v$, since otherwise $\theta_R$ would depend on $v$. The number of constraints will then be $v^* + 1$ where $v^*$ is the number of accessible vibrational states. Equations (54) and (55) now represent a hierarchy of constraints. First, Eq. (54) determines the overall vibrational distribution exactly as in the previous section. The second constraint, Eq. (55), has no influence on the vibrational distribution and determines only the conditional rotational distribution $P(f_J/f_v)$. Equation (55), thus, determines how the nonvibrational energy is partitioned between the rotational and translational degrees of freedom. If $B = \frac{2}{5}$ as in Eqs. (31) and (32), then $\theta_R = 0$ and the special case of an R–T microcanonical distribution, Eqs. (28), (30), and (35) are obtained.[17] The same result would be obtained if the constraint,

Eq. (55), was not imposed. At present, simple dynamical estimates of B, of the type of Eq. (52) for A, are not available.

## 4. DYNAMICAL AND STATISTICAL MODELS

It was pointed out previously that there is a need to devise simple models whose mechanistic features can be transposed into dynamical constraints by a suitable analysis. Considering, in particular, the constraints given by Eq. (51), we would like to find a model that simulates all the dynamical characteristics relevant for the vibrational distribution of products and, at the same time, allows an evaluation of $\langle f_v \rangle$ in closed form.

There is a growing amount of reliable kinetic information (for reviews see Refs. 1 and 61) from classical, semiclassical and quantum-mechanical calculations that may provide guidelines for the construction of tractable models of the desired kind. Many of these studies set out to explore the role of various dynamical factors determining the detailed reactive rate constants $k(v'J' \to vJ)$.[61-79] Generally speaking, these factors can be divided into two classes: (1) effects of the potential-energy surface, for example, its curvature, the location of barriers or the change in the vibrational frequency along the reaction path; and (2) kinetic factors, for example, the translational and internal energies of the reactants and the mass combinations. Most helpful for understanding the influence of the different factors are the qualitative arguments of Polanyi and co-workers.[78,79] Thus far, however, only very few models[27-30,62-67] provide closed-form expressions in terms of dynamically significant variables for quantities such as $\langle E_v \rangle$ or $\langle E_J \rangle$. In other words, only very few models exist that allow one to relate the temperature parameters to real product distributions and their dynamic constraints.

A nonadiabatic model[27-30] to serve this purpose will subsequently be described. To date, it is the only one that could be fitted completely into the conceptual framework of information theory. This model leads to an extension of the common transition-state theory to the diabatic-transition state concept,[14] which allows a prediction of inverted product distributions (see 'note added in proof').

### 4.1. The Nonadiabatic Model of Population Inversion

If one chooses internal coordinates $\xi$, $\eta$ of the three-atom reaction system such that $\eta$ represents small distances perpendicular to the reaction path, (i.e., the line of minimum curvature leading from the valley of reactants into the valley of products) and $\xi$ represents distances along the reaction path, the potential-energy surface can be represented by an harmonic approximation:

$$\mathcal{U} = \mathcal{U}_0(\xi) + \tfrac{1}{2}K(\xi)\eta^2 \tag{56}$$

where $\mathcal{U}_0(\xi)$ is the potential along the reaction path and $K(\xi)$ is the force constant for vibration perpendicular to it. The mass-weighted coordinate system can be chosen to be orthogonal, eliminating cross terms in the kinetic energy operator which takes the form[74,80–83]

$$\mathcal{T} = \mathcal{T}^\xi(\xi, \eta) + \mathcal{T}^\eta(\xi) \tag{57}$$

The first term in Eq. (57) is the kinetic energy of translation, and the second is the vibrational kinetic energy. To first order in $\eta$, the deviations from the reaction path, $\mathcal{T}^\xi$ is given by

$$\mathcal{T}^\xi(\xi, \eta) = \mathcal{T}^\xi(\xi, 0) + \eta \frac{\partial \mathcal{T}^\xi}{\partial \eta}\bigg]_{\eta=0} \tag{58}$$

The first term represents the kinetic energy along the reaction path; whereas, the second represents the change in kinetic energy as the system deviates from it. If the potential-energy valley is shallow along $\eta$, that is, $K(\xi)$ is small, the wave function may be large at large $\eta$ values. In this case, higher-order terms in $\eta$ have to be included in Eq. (58). The classical analog of the second term in Eq. (58) is an internal centrifugal force (the derivative of the kinetic energy along the reaction path towards the radius of curvature). This force was introduced by Fischer, Hofacker, and Seiler[82] as the main coupling mechanism between the vibrational and translational motions.

The total Hamiltonian can now be written as

$$\mathcal{H} = \mathcal{T} + \mathcal{U} = \mathcal{H}_0 + \mathcal{H}_1 \tag{59}$$

where the unperturbed Hamiltonian,

$$\mathcal{H}_0 = \mathcal{H}^\xi + \mathcal{H}^v(\xi) \tag{60}$$

with

$$\mathcal{H}^\xi = \mathcal{T}^\xi(\xi, 0) + \mathcal{U}_0(\xi) \tag{61a}$$

and

$$\mathcal{H}^v = \mathcal{T}^\eta(\xi) + \tfrac{1}{2}K(\xi)\eta^2 \tag{61b}$$

This allows for adiabatic decoupling between translational and vibrational motion over the basis

$$\{|\alpha\rangle |v; \xi\rangle\} \tag{62}$$

where $|\alpha\rangle$ and $|v; \xi\rangle$ are defined as the eigenfunctions

$$\mathcal{H}^\xi |\alpha\rangle = E_\alpha |\alpha\rangle$$

and

$$\mathcal{H}^v |v; \xi\rangle = \hbar\omega(\xi)(v + \tfrac{1}{2}) |v; \xi\rangle \qquad (63)$$

The T–V coupling term constitutes $\mathcal{H}_1$, which to first order in $\eta$ is equal to the centrifugal force

$$\mathcal{H}_1 = \eta \frac{\partial \mathcal{T}^\xi}{\partial \eta}\bigg]_{\eta=0} \qquad (64)$$

We could now proceed with a perturbation treatment, with $\mathcal{H}_1$ as perturbation. However, if $\mathcal{H}_1$ in this model will be capable of producing transitions over several vibrational quanta, low-order perturbation theory cannot be a suitable means to describe such processes. Later on we shall outline an elegant way to overcome this difficulty. Let us first, through classical considerations, estimate $\mathcal{H}_1$ by order of magnitude. For a classical particle being carried up the walls of the potential energy valley by the centrifugal force, $\partial T^\xi/\partial \eta$ should be about equal to the harmonic restoring force $K(\xi)\eta$, that is,

$$K(\xi)\bar{\eta} = \frac{\partial T^\xi}{\partial \eta} = 2\kappa(\xi)T^\xi \qquad (65)$$

where both the classical local kinetic energy $T^\xi$ and its derivative are to be evaluated at $\eta = 0$; $\kappa$ is the curvature and $\bar{\eta}$ the mean deviation from the reaction path.

The work done by the centrifugal force divided by $\hbar\omega$ is the maximal number $v^*$ of vibrational quanta that can be transferred to the product molecule by T–V coupling;

$$v^* = \frac{K\bar{\eta}^2}{2\hbar\omega} = \left(\frac{\bar{\eta}}{\eta_0}\right)^2 \qquad (66)$$

where $(1/2)K\eta_0^2 \equiv \hbar\omega$. At the point of maximum curvature of an early downhill potential-energy surface $T^\xi$ is large and may be several times $\hbar\omega$. If $K(\xi)$ does not change drastically with $\xi$ we see from Eqs. (65) and (66) that $\bar{\eta}$ and $v^*$ attain their maximum values at the point of maximum curvature of the reaction path. Estimates[29,30] based on some reasonable potential-energy surfaces show that at his point

$$v^* = \frac{4\kappa^2(T^\xi)^2}{K^2(2\hbar\omega/K)} > 1 \qquad (67)$$

In the case of $F + H_2 \rightarrow HF^* + H$, where an *ab initio* potential-energy surface is available[84] it was found[30] that $v^* \approx 3$. On the basis of the notion that in some instances $\eta_0 \approx 1/\kappa$, Eq. (67) may yield $v^* \approx (T^\xi/\hbar\omega) \gg 1$.

The problem of how to handle the centrifugal force in a perturbation scheme was treated by Hofacker and Levine[27] and by Fischer and Ratner.[83] $\mathcal{H}_1$ is split into diagonal and nondiagonal parts with respect to the eigenstates $|\alpha\rangle$ of $\mathcal{H}^\xi$. The diagonal term is then absorbed into $\mathcal{H}_0$, and the remaining nondiagonal term is small and can, therefore, be handled by low-order perturbation theory. The modified $\mathcal{H}_0$ has T–V eigenstates implying an harmonic oscillator motion shifted away from the reaction path by an amount $\bar{\eta} = 2\kappa T^\xi/K$, exactly as in the classical estimate, Eq. (69). The eigenfunctions of the modified $\mathcal{H}_0$ now constitute a new adiabatic basis given by[27,83]

$$\{|\alpha\rangle e^{-\eta \partial/\partial\eta} |v; \xi\rangle\} \tag{68}$$

The original and the shifted oscillators are shown in Fig. 4. It can be seen that the shift away from the reaction path raises the reference energy of the oscillator. The quantum-mechanical treatment also shows that the transition probabilities for vibrational excitation are simply related to the overlap of the wave functions of the original and the shifted oscillators. The total energy of the classical reactive wave packet travelling along $\xi$ is given by

$$E = T^\xi + \tfrac{1}{2}K\bar{\eta}^2 + \hbar\omega(v + \tfrac{1}{2}) \tag{69}$$

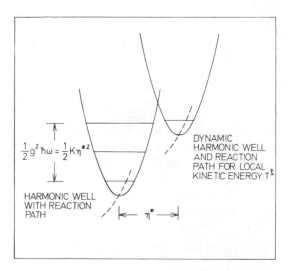

**Fig. 4.**   Harmonic oscillators shifted by centrifugal force.

where the middle term on the right-hand side represents the internal centrifugal potential.

The propensity rules for the T–V transitions are given by this treatment exactly as one would expect them from classical considerations. If there is just one curved section in the reaction path and the potential does not change appreciably along $\xi$ in the region of curvature, the transitions from a vibrationally cold ($v = 0$) reactant into a hot-product state ($v \geq 1$) center around a vibrational energy $K\eta^{*2}/2$, where

$$\eta^* = \max \bar{\eta}(\xi) \tag{70}$$

In other words, the point of maximum curvature along the reaction path and the local kinetic energy determine the T–V energy transfer. Hofacker and Levine,[14] therefore, assumed that even in three-dimensional exoergic reactions with an early downhill potential, the average vibrational energy transferred from vibrationally cold reactants to hot products is approximately

$$\langle E_v \rangle = \tfrac{1}{2} K\eta^{*2} \equiv \tfrac{1}{2} g^2 \hbar\omega \tag{71}$$

where $g$ is the coupling constant mentioned in Eq. (52) ($g > 1$ in the case of strong-population inversion). Combining Eqs. (65), (66), and (70) we find $g = 2\kappa T^\xi/(K\hbar\omega)^{1/2}$. Thus

$$\langle f_v \rangle = \frac{g^2 \hbar\omega}{2E} = \frac{2(\kappa T^\xi)^2}{KE} \tag{72}$$

For isotopic reactions (e.g., $F + H_2 \rightarrow HF + H$ and $F + D_2 \rightarrow DF + D$), the total energy release $E$ and the force constant $K$ do not depend on the kind of the isotope. Thus, if the curvature is insensitive to isotopic substitution (as in the foregoing example), the isotopic invariance of $\langle f_v \rangle$, found experimentally,[40] is explained.[14,30]

In the previous section, it was demonstrated, in general terms, how constraints of the type of Eqs. (51) and (52) can determine, through Jaynes' principle, the form of the distribution functions. For the sake of completeness, we will repeat here the procedure and emphasize the special consequences of the model just described. Equation (71) has the form of a dynamical constraint on $E_v$, which can be exploited in the maximum-entropy algorithm to predict or, more precisely, to resolve the product-state distribution. Since no further information is available, this is the only dynamical constraint. When the quantum-statistical terminology is used,[10–12,14] where $\rho$ denotes the density matrix corresponding to the three-atom system $AB + C$, the maximization procedure can be summarized as follows:

$$S = -\mathrm{Tr}\,(\rho \log \rho) \rightarrow \max \tag{73}$$

subject to

$$1 = \mathrm{Tr}\,(\rho) \tag{74}$$

$$E = \mathrm{Tr}\,(\rho \mathcal{H}) \tag{75}$$

and

$$\langle E_v \rangle = \mathrm{Tr}\,(\rho \mathcal{H}^v) = \tfrac{1}{2} g^2 \hbar \omega \tag{76}$$

The trace has to be taken over all the states of $AB + C$. Equation (74) is the standard normalization condition, Eq. (75) restricts the results to a specific total reaction energy $E$, and Eq. (76) is the dynamical constraint. If the restriction, Eq. (75), is taken as an initial condition, that is, $\rho$ is *a priori* restricted to the given total energy $E$, then Eq. (75) is redundant, Eq. (73) will be equivalent to Eq. (16), and Eq. (76) to Eqs. (51) and (52).[10-12] The density matrix that maximizes $S$ is given, for a microcanonical ensemble of reactive systems,[14] by

$$\rho_\lambda = Q^{-1}\,\delta(E - \mathcal{H})\exp{(-\lambda_v' \mathcal{H}')} \tag{77}$$

where

$$Q = \mathrm{Tr}\,\{\delta(E - \mathcal{H})\exp{(-\lambda_v' \mathcal{H}^v)}\} \tag{78}$$

and $\lambda_v'$ is evaluated through

$$\langle E_v \rangle = \frac{-(\partial \log Q)}{\partial \lambda_v'} \tag{79}$$

Since Eq. (71) is the only constraint and the trace extends over all product states, including the translational and rotational states, Eq. (77) is equivalent to Eq. (35) with $\lambda_v' = \lambda_v E$ (see the discussion in Section 3.2), and the vibrational distribution is, thus, given by Eq. (22). In the RRHO approximation, Eq. (79) leads to Eq. (47). Comparing this with Eq. (72), we obtain the following relation between $\lambda_v$ and $g$:

$$\frac{g^2}{v^*} = 1 + \tfrac{5}{2}\lambda + \frac{\lambda^{3/2}\exp{(-\lambda)}}{\gamma(\tfrac{5}{2};\lambda)} \tag{80}$$

where $v^* = E/\hbar\omega$ and $\lambda = -\lambda_v$ (for $\lambda_v < 0$). It should be noted that, for $\lambda \to 0$, $\rho_\lambda$ in Eq. (77) represents the unrestricted microcanonical ensemble, and we obtain the prediction of common transition-state theory, where

$$\langle f_v \rangle = \frac{\mathrm{Tr}\,(\rho_0 \mathcal{H}^v)}{E} \tag{81}$$

would be the same for the products as for the reactants, save for minor frequency and density-of-state effects.

The role of the temperature parameter in the sense of statistical mechanics is apparent. Formally, the result (77) is the same as that for the probability distribution of the subsystem of a larger system where the total energy of the system and the average energy of the subsystem are specified. In statistical mechanics, the temperature concept is reserved for systems so large that the energy content of the subsystem is small compared to the total energy [thus the $\delta$ function in Eq. (77) is irrelevant]. For three-atom reactive systems the energy content of the vibration perpendicular to the reaction path is by no means small compared to the total energy $E$. Therefore, the temperature-like distribution of vibrational energy is truncated by $E$. What we called the "temperature parameter" $\lambda$ can at best be called an extension of the temperature concept of statistical mechanics.

The results given previously suggest the following interpretation. The interaction of the translational and rotational degrees of freedom with the oscillator is not statistical in the equipartition sense. Instead, the strong T–V interaction shifts the center of the vibrational distribution away from the statistical average, by $g^2 \hbar \omega / 2 - \langle E_v \rangle_0$, $(\langle E_v \rangle_0 = 2E/7)$. The weak V–T interaction, caused by the nondiagonal perturbation over the basis set of shifted oscillator wave functions, Eq. (68), has an analogy in the heat-bath, subsystem interaction. The large width of the vibrational distribution, witnessed by Fig. 1, is due to the small number of degrees of freedom participating in the reactive energy exchange.

To date, the range of application of the strong-coupling model described previously, is limited to cases where only one dynamical constraint dominates the product-state distribution. Refinements are necessary towards a better estimate of Eq. (71), on one hand (to draw more precise conclusions from the results of the information theoretical analysis of experiments), and, on the other hand, simple rules to predict which reactions lead to population inversion from structural and spectroscopic information.[64] Furthermore, better distributions for the a priori estimates of $P_0$ (see Section 4.2) must be used in more sophisticated applications of the model.

## 4.2. Statistical Theories

A number of statistical models, most of them older than the one described in the previous section, use statistical mechanics to make up for lacking dynamical information. In the literature on chemical-reaction dynamics, the term "statistical theory" is commonly used for these models where assumptions on equal distribution in phase space play the key role in determining the probabilities of product states for given reactant states.[85,86] For an extensive discussion on statistical theories and related topics (such as the statistical theory of rotational excitation, unimolecular reactions, transition-state theory, and references to statistical models of nuclear reactions), see Ref. 85.

Of the various statistical models of chemical reactions, the one proposed by Light and co-workers[86-92] has become the most widely known (see, e.g., Refs. 86-97). Their line of approach resembles that of Keck's earlier theory of recombination reactions.[98,99] Pechukas and Light[88] and Nikitin[100] simultaneously derived equivalent expressions for the cross-sections $\sigma(v', J' \to v, J)$, corresponding to the reaction $A + BC(v', J') \to AB(v, J) + C$, at a given total reaction energy $E$. Their results are quite general in the sense that they can be derived with various approaches. Light's treatment will be outlined later. Nikitin's derivation is based on a variation of transition-state theory in a microcanonical ensemble constrained by energy and total angular-momentum conservation in the collision. Similar results can be obtained with formal scattering theory,[85] where the statistical assumptions are imposed directly on the $S$ matrix, in analogy with the statistical theory of rotational excitation.[85,101,102] Different theories designed for the same purpose were formulated by Eu and Ross[103] and Marcus.[104]

In this section we want to briefly outline the differences in physical content and range of validity between the previous statistical theories and the information-theoretical treatment. To this end, we sketch the main features of Light's model for bimolecular exchange reactions without activation barriers. This model leads to expressions for the detailed cross-section $\sigma(v', J' \to v, J)$, and the theory can be extended to determine product-angular distributions as well.[75-92] It is assumed that every reactive collision proceeds through a "strong-coupling complex," whose mode of decomposition into products is independent of its mode of formation. The complex is defined by a region in configuration space where the interaction between the colliding molecules is strong. The life time and the structure of the complex are considered irrelevant. The boundaries of the strong-coupling region are determined by the location of the orbital angular-momentum barrier, provided it is smaller than the hard-sphere collision diameter. These considerations are then used in the calculation of the cross-section for complex formation for any possible total angular momentum $K$. The crucial statistical assumption is that the breakdown probabilities of the complex into the various available final channels are proportional to the fluxes into the corresponding phase-space volumes. (The flux is the product of the relative translational velocity and the volume of the phase-space element.) For instance, the decomposition probability into products $AB(v, J) + C$ is taken proportional to the flux into the phase-space volume defined by $v$, $J$, and $K$, and by the small-energy range $E$, $E + \delta E$, subject to the dynamical conditions on the strong-coupling region. The final result for the detailed cross-section is

$$\sigma(v', J' \to v, J) = \frac{\pi\hbar^2}{2\mu' E_t'(2J'+1)} \sum_K (2K+1) \frac{\mathcal{N}^K(v', J')\mathcal{N}^K(v, J)}{N(K)} \quad (82)$$

where $\mathcal{N}^K(v, J)$ is the number of possible $L$, $J$ combinations ($L$ is the product orbital angular momentum) compatible with $v$, $J$, $K$, $E$, and the dynamical constraints; $\mathcal{N}^K(v', J')$ is the corresponding quantity for the reactants, and $N(K)$ is the sum of $\mathcal{N}^K(v, J)$ on all open channels, including that of the reactants.

The predictions of Eq. (82) were compared with some experimental data (see, e.g., Refs. 86 and 96). It was found then, and is even more evident now, that experimental vibrational distributions in exoergic population-inverting reactions deviate extremely from the predicted behavior. This finding does not invalidate these expressions, if they are taken as a diagnostic tool to characterize actual product-state distributions.[86]

It can be shown[16,81,89] that in many reactions $E$ conservation is much more restrictive than $K$ conservation in determining the value of $\sigma(v', J' \to v, J)$. This is mainly true for reactions where there is no appreciable difference between the reduced masses, and, therefore, between the orbital angular momenta of reactants and products. An example is the reaction of $Cl + HI \to HCl + I$, a counterexample: $H + Cl_2 \to HCl + Cl$. Here the reduced mass of the products is about 35 times that of the reactants. In the derivation[15] of Eqs. (23) and (28) for the microcanonical distributions $P_0(v)$ and $P_0(v, J)$, angular-momentum conservation is not explicitly considered. It can be shown[54] that for many largely exothermic reactions taking place at room temperature $\langle\sigma(v', J' \to v, J)\rangle \approx \sigma(\hat{v}', \hat{J}' \to v, J)$ where $\hat{v}'$, $\hat{J}'$ is the most probable V–R reactant level at this temperature and the averaging is over the thermal reactant distribution. Therefore, according to Eq. (13), $P(v, J)$ is approximately proportional to $\sigma(\hat{v}', \hat{J}' \to v, J)$. Hence, for those cases, where $E$ conservation is the decisive restriction, we should expect that our $P_0(v, J)$ will be proportional to the right-hand side of Eq. (82). In the same fashion, one could, in principle, expect[16] from Eq. (82) to provide the appropriate $P_0(v, J)$ distribution when $K$ conservation is also important. However, it should be mentioned that the relation between the expressions provided by the "pure" statistical theories and the $P_0$ distributions are not as simple as one may think. Discussion of these fine points is beyond the scope of this work: They are considered in detail elsewhere.[105]

The foregoing arguments imply that the phase-space models[86-104] can, at best, equip us with the accurate zeroth-order microcanonical distribution functions $P^0(v)$ or $P^0(v, J)$ used in the surprisal-entropy analysis. Clearly, they cannot account for any inverted product distributions as shown in Fig. 1. In conclusion, we should like to emphasize that these statistical models may provide zeroth-order estimates to the real distributions, which, by the maximum entropy principle and implementation of the relevant constraints, allow us to determine more accurately the higher-order contributions that contain the desired dynamical information.

Since 1973, when this article was written, the information theory approach has made further significant progress and has been successfully applied to a variety of problems (Reviews[106,107] by Levine and Bernstein cover the developments until 1975). Considerable progress has also been achieved with respect to other topics considered in the article. It is, therefore, quite obvious that some statements, remarks, and points of emphasis seem now a little anachronistic. Yet, since this article is not intended to be comprehensive, a few brief comments referring to some selected points may suffice to bring matters into the right perspective.

Entropy-surprisal analyses of the kind described for product state distributions in exoergic chemical reactions were carried out, among other, to characterize: selective energy consumption by the reactants in chemical reactions[108–109], state-to-state rate constants in vibrational[110] and rotational[111] energy transfer collisions; determination of branching ratios[112]; vibrational distributions in excited states[113,114]; and rate constants for electronic transitions.[115,116] In addition, the method has been applied to describe the development of entropy along the reaction path[117] and the changes in entropy and energy during the passage from nascent (completely nonrelaxed) reaction products to completely thermal products $(T_v = T_R = T)$.[23,118] This enables one, at least in principle, to set bounds on the efficiencies of chemical lasers.[118] A development of particular interest is the possibility of "surprisal or rate constant synthesis."[106–111] The synthesis (prediction) requires the identification of constraints and their implementation in the framework of the maximum entropy principle (Section 3). The constraints employed so far fall in two classes. First, dynamical constraints of the kind described previously, Eqs. (52), (71), and (72), on the basis of the nonadiabatic model (Section 4.1). Similar dynamical constraints can be evaluated from other simple models.[119–122] Predictions of rate constants based on "momentum transfer constraints" were proven to be often adequate.[121,122] The second kind of constraints, mainly applied in energy transfer problems, stem from information on the bulk behavior of macroscopic systems,[106,107,110,123] for example, the frequent observation that the average vibrational energy decays exponentially can be taken as a macroscopic constraint. Detailed vibrational relaxation rates can then be synthesized[110] with the aid of appropriate sum rules.[124]

The rapid accumulation of detailed experimental and theoretical information on relaxation processes pertinent to chemical laser systems[44,125] facilitates the possibility of performing reasonably accurate modellings of chemical lasers. A number of recent studies[35,59,125-129] confirm the notion, mentioned in Sections 1 and 2, that effects of rotational nonequilibrium can play a significant role in determining the output characteristics of chemical lasers. These, and previous studies (see, e.g., the chapter by G. Emanuel[129] and Refs. 130, 131) also indicate that considerable portions of the laser energy may be extracted from partially inverted populations. Moreover, fast rotational relaxation (achieved, e.g., by the addition of an inert gas) may enhance the intensity. Hence it should be stressed that the lasing criteria in Section 2 are referred primarily to the first lasing stages (soon after threshold).

## REFERENCES

1. R. D. Levine, MTP Int. Rev. Sci. *Theoretical Chemistry* Phys. Chem. Ser. I, Vol. 1, W. Byers Brown, Ed. (Butterworths, London, 1972), Chap. 7.

2. J. C. Polanyi and T. Carrington, MTP Int. Rev. Sci. *Chemical Kinetics*, Phys. Chem. Ser. I, Vol. 9, J. C. Polanyi, Ed. (Butterworths, London, 1972), Chap. 5.

3. J. L. Kinsey, MTP Int. Rev. Sci. *Reaction Kinetics*, Phys. Chem. Ser. I, Vol. 9, J. C. Polanyi, Ed. (Butterworths, London, 1972), Chap. 6.

4. K. L. Kompa, *Topics in Current Chemistry, Vol. 37, Chemical Lasers* (Springer, Berlin, 1973).

5. See the chapter by R. Wilkins.

6. R. B. Bernstein and R. D. Levine, J. Chem. Phys. **57,** 434 (1972).

7. C. E. Shannon, Bell Syst. Tech. J. **27,** 379, 623 (1948); reprinted in C. E. Shannon and W. Weaver, *The Mathematical Theory of Communication* (Univ. of Illinois Press, Urbana, Illinois, 1949).

8. R. Ash, *Information Theory* (Interscience, New York, 1965).

9. A. J. Khinchin, *Mathematical Foundations of Information Theory* (Dover, New York, 1957).

10. E. T. Jaynes, in *Statistical Physics, 1962 Brandeis Lectures, Vol. 3* (Benjamin, New York, 1963), p. 81.

11. A. Katz, *Principles of Statistical Mechanics* (Freeman, San Francisco, 1967).

12. A. Hobson, *Concepts in Statistical Mechanics* (Gordon and Breach, New York, 1971).

13. A. Ben-Shaul, R. D. Levine, and R. B. Bernstein, Chem. Phys. Lett. **15,** 160 (1972).

14. G. L. Hofacker and R. D. Levine, Chem. Phys. Lett. **15,** 165 (1972).

15. A. Ben-Shaul, R. D. Levine, and R. B. Bernstein, J. Chem. Phys. **57,** 5427 (1972).

16. R. D. Levine, B. R. Johnson, and R. B. Bernstein, Chem. Phys. Lett. **19,** 1 (1973).

17. A. Ben-Shaul, Chem. Phys. **1,** 244 (1973).

18. R. D. Levine and R. B. Bernstein, Disc. Faraday Soc.: Molecular Beam Scattering **55,** 100 (1973).

19. A. Ben-Shaul, G. L. Hofacker, and K. L. Kompa, J. Chem. Phys. **59,** 4664 (1973).

20. R. D. Levine and R. B. Bernstein, Chem. Phys. Lett. **22,** 217 (1973).

21. R. D. Levine, Univ. of Wisconsin, Theoretical Chemistry Institute Report WIS-TCI-495 (1973).

22. A. Ben-Shaul, Mol. Phys. **27** 1585 (1974).

23. A. Ben-Shaul, O. Kafri and R. D. Levine Chem. Phys. **10,** 367 (1975).

24. R. D. Levine, Ber. Bunsenges. Phys. Chem. **78,** 111 (1974).

25. M. J. Berry, J. Chem. Phys. **59,** 6220 (1973).

26. M. Rubinson and J. I. Steinfeld Chem. Phys. **4,** 467 (1974).

27. G. L. Hofacker and R. D. Levine, Chem. Phys. Lett. **9,** 617 (1971).

28. R. D. Levine, Chem. Phys. Lett. **10,** 510 (1971).

29. G. L. Hofacker and K. W. Michel, Ber. Bunsenges. Phys. Chem. **78,** 174 (1974).

30. G. L. Hofacker and N. Rösch, Ber. Bunsenges. Phys. Chem. **77,** 661 (1973).

31. J. C. Polanyi, Appl. Opt. Suppl. 2, Chemical Lasers 109, (1965).

32. G. Herzberg. *Spectra of Diatomic Molecules*, Vol. 1, 2nd ed. (Van-Nostrand, Princeton, 1950).

33  E. W. Montroll and K. E. Shuler, J. Chem. Phys. **26,** 454 (1957).

33a. T. D. Padrick and M. A. Gusinow, Chem. Phys. Lett. **24,** 270 (1974).

34  C. E. Treanor, J. W. Rich, and R. G. Rehm, J. Chem. Phys. **48,** 1798 (1968).

35  Pummer and Kompa, App. Phys. Lett. **20,** 356 (1972).

36. C. B. Moore, Adv. Chem. Phys. **23,** 41 (1973).

37. J. I. Steinfeld in MTP Int. Rev. Sci. *Chemical Kinetics*, Phys. Chem. Ser. I. Vol. 9, J. C. Polanyi, Ed. (Butterworths, London, 1973), Chap. 8.

38. See the chapter by Cohen, Herbelin, and Benson.

39. J. C. Polanyi and K. B. Woodall, J. Chem. Phys. **56,** 1563 (1972).

40. D. H. Maylotte, J. C. Polanyi, and K. B. Woodall, J. Chem. Phys. **57,** 1547 (1972).

41. J. C. Polanyi and K. B. Woodall, J. Chem. Phys. **57,** 1574 (1972).

42. K. G. Anlauf, D. S. Horne, R. G. McDonald, J. C. Polanyi, and K. B. Woodall, J. Chem. Phys. **57,** 1561 (1972).

43. A. M. G. Ding, J. L. Kirsch, D. S. Perry, J. C. Polanyi, and J. L. Schreiber, Disc. Faraday Soc., Molecular Beam Scattering **55,** 252 (1973).

44. See the chapter by Bott and Cohen.

45. J. F. Bott and N. Cohen, J. Chem. Phys. **55,** 3698 (1971).

46. H. K. Shin, Chem. Phys. Lett. **10,** 81 (1971).

47. L. H. Sentman, Chem. Phys. Lett. **18,** 493 (1973).

48. G. C. Berend and R. L. Thommarson, J. Chem. Phys. **58,** 3203 (1973).

49. S. N. Suchard, R. L. Kerber, G. Emanuel, and J. S. Whittier, J. Chem. Phys. **57,** 5065 (1972).

50. J. F. Bott, J. Chem. Phys. **57,** 96 (1972).

51. J. F. Bott, Chem. Phys. Lett. **23,** 335 (1973).

52. J. K. Hancock and W. H. Green, J. Chem. Phys. **56,** 2474 (1972).

53. A. Hobson and B. K. Cheng, J. Statist. Phys. **7,** 301 (1973).

54. K. G. Anlauf, D. H. Maylotte, J. C. Polanyi, and R. B. Bernstein, J. Chem. Phys. **51,** 5716 (1969).

55. J. L. Kinsey, J. Chem. Phys. **54,** 1206 (1971).

56. C. A. Parr, J. C. Polanyi, and W. H. Wong, J. Chem. Phys. **58,** 5 (1973).

57. A. E. Siegmann, *An Introduction to Lasers and Masers* (McGraw-Hill, New York, 1971).

58. A. Maitland and M. H. Dunn, *Laser Physics* (North-Holland, Amsterdam, 1969).

59. A. Ben-Shaul, K. L. Kompa and U. S. Schmailzl, J. Chem. Phys. (in press).

60. K. E. Shuler, J. Chem. Phys. **32,** 1692 (1960).

61. See for example, Faraday Discussions of the Chemical Society **55** (1973).

62. F. T. Smith, J. Chem. Phys. **31,** 1352 (1959).

63. M. V. Basilevski, Mol. Phys. **26,** 765 (1973).

64. M. D. Pattengill and J. C. Polanyi Chem. Phys. **3,** 1 (1974).

65. P. J. Kuntz, Trans. Faraday Soc. **66,** 2980 (1970).

66. P. J. Kuntz, Mol. Phys. **23,** 1035 (1972).

67. M. T. Marron, J. Chem. Phys. **58,** 153 (1973).

68. J. C. Light, Adv. Chem. Phys. **19,** 1 (1971).

69. J. C. Light, in *Methods of Computational Physics,* Vol. 10, (Academic, New York, 1971), p. 111, and references therein.

70. D. J. Diestler, J. Chem. Phys. **56,** 2092 (1972).

71. S.-F. Wu, B. R. Johnson, and R. D. Levine, Mol. Phys. **25,** 839 (1973).

72. N. H. Hijazi and K. J. Laidler, J. Chem. Phys. **58,** 349 (1973).

73. E. A. McCullough and R. E. Wyatt, J. Chem. Phys. **54,** 3578, 3592 (1971).

74. R. A. Marcus, J. Chem. Phys. **45,** 4493, 4500 (1966).

75. K. T. Tang, B. Kleinman, and M. Karplus, J. Chem. Phys. **50,** 1119 (1969).

76. D. G. Truhlar and A. Kuppermann, J. Chem. Phys. **56,** 2232 (1972).

77. J. Manz, Mol. Phys. **28,** 399; **30,** 899 (1974).

78. J. C. Polanyi, Acct. Chem. Res. **5,** 161 (1972), and references therein.

79. J. C. Polanyi, Disc. Faraday Soc., Molecular Beam Scattering **55,** (1973) and references therein.

80. G. L. Hofacker, Z. Naturforsch. **18a,** 607 (1963).

81. R. A. Marcus, J. Chem. Phys. **41,** 603, 610 (1965).

82. S. F. Fischer, G. L. Hofacker, and R. Seiler, J. Chem. Phys. **51,** 395 (1969).

83. S. F. Fischer and M. Ratner, J. Chem. Phys. **57,** 2769 (1972).

84. H. F. Schaefer, *The Electronic Structure of Atoms and Molecules* (Addison Wesley, London, 1972).

85. R. D. Levine, *Quantum Mechanics of Molecular Rate Processes* (Clarendon, Oxford, 1968).

86. J. C. Light, Disc. Faraday Soc. **44,** 14 (1967).

87. J. C. Light, J. Chem. Phys. **40,** 3221 (1964).

88.   P. Pechukas and J. C. Light, J. Chem. Phys. **42,** 3281 (1965).

89.   P. Pechukas, J. C. Light, and C. Rankin, J. Chem. Phys. **44,** 794 (1966).

90.   J. C. Light and J. Lin, J. Chem. Phys. **43,** 3209 (1965).

91.   J. Lin and J. C. Light, J. Chem. Phys. **45,** 2545 (1966).

92.   R. A. White and J. C. Light, J. Chem. Phys. **55,** 379 (1971).

93.   D. G. Truhlar and K. Kuppermann, J. Phys. Chem. **73,** 1722 (1969).

94.   R. D. Levine, F. A. Wolf, and J. A. Maus, Chem. Phys. Lett. **10,** 2 (1971).

95.   D. C. Fullerton and T. F. Moran, Chem. Phys. Lett. **10,** 626 (1971).

96.   W. H. Wong, Can. J. Chem. **50,** 633 (1972).

97.   D. G. Truhlar, J. Chem. Phys. **51,** 4617 (1969); **56,** 1481 (1972); **57,** 4063 (1972).

98.   J. Keck, J. Chem. Phys. **29,** 410 (1958).

99.   J. Keck, Adv. Chem. Phys. **13,** 85 (1967).

100.   E. E. Nikitin, Theoret. Exp. Chem. **1,** 135, 428 (1965).

101.   R. B. Bernstein, A. Dalgarno, H. Massey, and I. C. Percival, Proc. Roy. Soc. London Ser. **A-274,** 427 (1963).

102.   W. A. Lester and R. B. Bernstein, J. Chem. Phys. **53,** 11 (1970).

103.   B. C. Eu and J. Ross, J. Chem. Phys. **44,** 2467 (1966).

104.   R. A. Marcus, J. Chem. Phys. **45,** 2630 (1966).

105.   A. Ben-Shaul, R. D. Levine and R. B. Bernstein J. Chem. Phys. **61,** 4937 (1974). A. Ben-Shaul (to be published).

106.   R. D. Levine and R. B. Bernstein in *Modern Theoretical Chemistry, Vol. III Dynamics of Molecular Collisions,* W. H. Miller, Ed. (Plenum, New York, 1975).

107.   R. B. Bernstein and R. D. Levine, *Adv. in Atom. and Mol. Phys. Vol.* 11, 215, D. R. Bates, Ed. (Academic, New York, 1975).

108.   R. D. Levine and J. Manz, J. Chem. Phys. **63,** 4280 (1975).

109.   H. Kaplan, R. D. Levine, and J. Manz, J. Chem. Phys. **12,** 447 (1976).

110.   I. Procaccia and R. D. Levine, J. Chem. Phys. **63,** 4261 (1975).

111.   R. B. Bernstein, Int. J. Quan. Chem. (to be published).

112.   R. D. Levine and R. Kosloff, Chem. Phys. Lett. **28,** 300 (1974).

113.   D. G. Sutton and S. W. Suchard, Appl. Opt. **14,** 1898 (1975).

114.   D. M. Manos and J. M. Parson, J. Chem. Phys. **63,** 3575 (1975).

115.   U. Dinur, R. Kosloff, R. D. Levine, and M. J. Berry, Chem. Phys. Lett. **34,** 199 (1975).

116.   M. J. Berry, Chem. Phys. Lett. **29,** 323, 1974.

117.   G. L. Hofacker and R. D. Levine, Chem. Phys. Lett. **33,** 404 (1975).

118.   R. D. Levine and O. Kafri, Chem. Phys. Lett. **27,** 175 (1974).

119.   M. J. Berry, Chem. Phys. Lett. **27,** 73 (1974).

120.   U. Halavee and M. Shapiro, J. Chem. Phys. (in press).

121.   A. Kafri, E. Pollak, R. Kosloff, and R. D. Levine, Chem. Phys. Lett. **33,** 201 (1975).

122.   H. Kaplan and R. D. Levine, J. Chem. Phys. **63,** 5064 (1975).

123.   I. Procaccia, Y. Shimoni, and R. D. Levine, J. Chem. Phys. (in press).

124.   I. Oppenheim, K. E. Shuler, and G. H. Weiss, Adv. Mol. Relaxation Processes **1,** 13 (1967).

125. S. Ormonde, Rev. Mod. Phys. **47,** 193 (1975).

126. H. L. Chen, R. L. Taylor, J. Wilson, P. Lewis, and W. Fyfe, J. Chem. Phys. **61,** 306 (1974).

127. L. H. Sentman, J. Chem. Phys. **62,** 3523 (1975).

128. E. Keren, R. B. Gerber, and A. Ben-Shaul, W. Chem. Phys. (to be published).

129. G. Emanuel, this volume.

130. R. W. Chester, J. Chem. Phys. **53,** 3595 (1970).

131. J. R. Airey, J. Chem. Phys. **52,** 156 (1970).

# CHAPTER 11

# THE CO CHEMICAL LASER

**BARRY R. BRONFIN**[*]

United Technologies Research Center
East Hartford, Connecticut

**WILLIAM Q. JEFFERS**[**]

McDonnell Douglas Research Laboratories
St. Louis, Missouri

[*] *Present address:* Scientific Leasing Inc., Founders Plaza, E. Hartford, Conn. 06108.
[**] *Present address:* Helios Incorporated, P.O. Box 2190, Boulder, CO 80302.

The carbon-monoxide laser constitutes a major chemical-laser class; others (hydrogen halides, hydroxyl) are discussed in other chapters. In the well-studied $O$–$CS_2$ system, the primary CO laser pumping reaction mechanism is now accepted to be $O + CS \rightarrow CO(v) + S$.

The CO chemical-laser system has lent itself to a wide variety of experimental configurations: premixed $O_2$–$N_2O$–$CS_2$ flames, $O$–$O_2$–$CS_2$ continuous-flow mixing lasers, and various premixed pulsed lasers. The mixing-laser devices have been studied in greatest detail; best-device performance has yielded cw output powers of the order of 35 W and separately specific laser energies of 31.3 J/g of total gas flow. Based only upon reactant flow (exclusive of inert gas diluents), the specific energy rises to 293 J/g. The observed emission spectra in these chemical-laser devices generally include the $13 \rightarrow 12$ to $2 \rightarrow 1$ vibrational bands of CO, consistent with the separately measured vibrational distribution yielded from the $O + CS$ pumping reaction. These transitions correspond to the 4.85 to 5.7 $\mu$m infrared wavelength band. One of the major problems hindering the development of large-scale CO chemical lasers is the requirement for continuous generation of oxygen atoms to initiate the reaction sequence. A second problem is that of atmospheric absorption of CO laser radiation. Acceptable transmission coefficients for CO lines exist only on particular rotational transitions of the lower vibrational bands.

Modeling codes for calculating the theoretical behavior of CO chemical lasers have been developed, and results are within a factor of 2 of predicting many experimental data. The slow vibrational relaxation of CO with most collision partners leads to predictions of very favorable laser performance with increased cavity pressures. For the same reason, the laser performance is predicted to be generally insensitive to mixing rates of the O-atom and fuel flows. Finally, slow V–T relaxation is also favorable to high-chemical efficiency; this prediction has been borne out by experiment.

## 1. THE CHEMISTRY OF THE CO LASER MEDIUM

In the most widely studied method of production of vibrationally excited CO, oxygen atoms are reacted with carbon disulfide, in the gas phase, to produce the CS radical [Reaction (1), Table 1]. The labile CS species then react with additional O atoms to produce vibrationally excited CO [Reaction (2), Table 1]. This reaction pair serves to pump the CO chemical laser. A group of other side reactions also can proceed in the O–CS$_2$

**Table 1.** Elementary Reaction and Kinetic Data Pertinent to the O–CS$_2$ Chemical-laser System (after Ref. 82)

| Reaction | Heat of reaction, $\Delta H°$,$^a$ kcal/mol | Reaction rate constant, $k_1$,$^b$ cm$^3$/mol-sec | References |
|---|---|---|---|
| 1. $O + CS_2 \rightarrow CS + SO$ | −22.0 | $5.00 \times 10^{13} \exp(-1.9/RT)$ | 4 |
| 2. $O + CS \rightarrow CO^* + S$ | −85.0 | $2.4 \times 10^{14} \exp(-2/RT)$ | 3, 5, 8 |
| 3. $S + O_2 \rightarrow SO + O$ | −5.6 | $1.4 \times 10^{12}$ | 6, 7 |
| 4. $SO + O_2 \rightarrow SO_2 + O$ | −12.8 | $3.5 \times 10^{11} \exp(-6.5/RT)$ | 3 |
| 5. $O + CS_2 \rightarrow COS + S$ | −31.1 | $1.0 \times 10^{14} \exp(-8/RT)$ | 3, 8 |
| 6. $O + COS \rightarrow CO + SO$ | −51.4 | $1.9 \times 10^{13} \exp(-4.5/RT)$ | 4 |
| 7. $O + S_2 \rightarrow SO + S$ | −22.1 | $4.0 \times 10^{12}$ | 3 |
| 8. $S + CS_2 \rightarrow CS + S_2$ | +1.6 | $1.0 \times 10^{14} \exp(-4/RT)$ | 3 |
| 9. $O_2 + CS \rightarrow CO + SO$ | −90.0 | $5.5 \times 10^{10} \exp(-2/RT)$ | 9 |
| 10. $O_2 + CS \rightarrow CPS + O$ | −39.0 | $1.0 \times 10^{13} \exp(-12/RT)$ | 68 |
| 11. $O_2 + CS_2 \rightarrow CS + SO_2$ | 24.0 | $1.0 \times 10^{12} \exp(-43/RT)$ | 87 |
| 12. $S + COS \rightarrow CO + S_2$ | −22.0 | $1.9 \times 10^{13} \exp(-4.5/RT)$ | 4 |
| 13. $CS + SO \rightarrow CO + S_2$ | −62.0 | $1.0 \times 10^9$ | 87 |
| 14. $O_3 + O \rightarrow 2O_2$ | −93.7 | $3.1 \times 10^{13} \exp(-5.7/RT)$ | 88 |
| 15. $O_3 + SO \rightarrow O_2 + SO_2$ | −48.8 | $1.5 \times 10^{12} \exp(-2.1/RT)$ | 89 |
| 16. $SO + O \rightarrow SO_2 + h\nu$ | | $4.5 \times 10^{10}/T$ | 87 |
| | | $k_M$ (cm$^6$/mol$^2$-sec) | |
| 17. $O + O + M \rightarrow O_2 + M$ | −119.1 | $(4.6 \times 10^{17}/T) \exp(-0.172/T)$ | 12 |
| 18. $O + S + M \rightarrow SO + M$ | −124.7 | $K_{18} = K_{17}$ | —$^c$ |
| 19. $S + S + M \rightarrow S_2 + M$ | −101.7 | $k_{19} = k_{17}$ | —$^c$ |
| 20. $CO + O + M \rightarrow CO_2 + M$ | −127.2 | $5.9 \times 10^{15} \exp(-4.1/RT)$ | 90 |
| 21. $CS + S + M \rightarrow CS_2 + M$ | −103.3 | $k_{21} = k_{20}$ | —$^c$ |
| 22. $CO + S + M \rightarrow COS + M$ | −73.0 | $k_{22} = k_{20}$ | —$^c$ |
| 23. $O_2 + O + M \rightarrow O_3 + M$ | −25.5 | $1.7 \times 10^{13} \exp(2.1/RT)$ | 12, 91 |
| 24. $O + SO + M \rightarrow SO_2 + M$ | −131.9 | $9.6 \times 10^{19}/T$ | 4 |

$^a$ Data obtained from JANAF Tables of Thermochemical Data,$^{95}$ with the exception that Okabe's$^{17}$ value for $\Delta H_f^o$ for the CS was assumed.
$^b$ Activation energy $E_A$ (kcal/mol); the gas constant $R$ (kcal/mol-°K).
$^c$ Assumed values in absence of kinetic data.

medium as presented in Table 1, reproduced from Howgate and Barr[82] (see also Refs. 1 and 2). The kinetic data in Table 1 are based mainly upon the measurements of Homann, Krome and Wagner[3]; Westenberg and deHaas[4]; Hancock, Morley, and Smith[5]; Fair and Thrush[6]; Sheen[9]; Halstead and Thrush[10]; Arnold and Kimbell[68]; and Donovan and Little.[86]

The first three reactions listed dominate the kinetics of many CO chemical-laser media of interest, hence, they have been used in a simplified computer analysis of the O–CS$_2$ chemical laser, as discussed in Section 6.

As discussed in the following section, Reaction (2) (O + CS → CO* + S) provides the process by which a large fraction of the reaction exothermicity (~66%) is converted into nonequilibrium-vibrational excitation of the CO molecule, leading to laser emission from that species. That reaction is a member of a class of bimolecular exchange reactions, examples of which are discussed in other chapters, which yield nonequilibrium-vibrational excitation in the newly formed diatomic product of an atom–molecule reaction.[13] Reaction (9), also highly exothermic, was suggested as an alternative pumping step, first by Arnold and Kimbell[68] and later by Rosenwaks and Yatsiv.[14] Hancock and Smith,[8] however, conclusively demonstrated that Reaction (1) indeed was exclusively responsible for the pumping of CO($v$) in this system. Earlier, Callear[79] determined that the CS fragment has a very long lifetime in mixtures with O$_2$, thus precluding Reaction (9) as an important pumping reaction. A body of experimental data collected by Suart, Dawson, and Kimbell[1] and compared with computer simulations performed by that same group also confirms that Reaction (2) is the probable pumping step. Rate constants for Reaction 2, which are presented in Table 1, have been obtained by extrapolating the room-temperature measurements of Hancock and Smith[8] to the 1100°K data obtained by Homann, Krome, and Wagner.[3]

Reaction (1) serves to begin the O-atom reactions in this system and supplies the CS species. As observed by Smith,[15] some of the exothermicity of Reaction (1) is deposited in the vibrational states of both the CS and SO diatomic products. This vibrational excitation could accelerate the reaction rates of these diatomic species participating in subsequent steps. The rate constants for Reaction (1) that are presented in Table 1 are those determined by Westenberg and deHaas[4] with the temperature dependence of Homann, Krome, and Wagner.[3]

Reaction (3) is particularly interesting in that the S + O$_2$ → SO + O process regenerates O-atoms and, thereby, can contribute to chain propagation. The rate coefficients for Reaction (3) that are presented in Table 1 are those recently measured by Davis, Klemm, and Pilling,[7] which are in close agreement with the measurements of Fair and Thrush.[6] These values are recommended in favor of the earlier measurements by Homann, Krome, and Wagner,[3] which originally appeared in the Howgate–Barr tabulation.[82] Other possible contributors to chain propagation are: Reaction (4), which could be accelerated by vibrationally excited SO produced through Reaction (1); and Reaction (10), which would propagate the chain in conjunction with Reaction (1), particularly at elevated temperatures. Some of the likely gas-phase termination reactions, (17) through (24), which are generally slow, are also presented in Table 1.

A central requirement for the continuous pumping of CO laser emission by the elementary reactions of Table 1 is the continuous generation of oxygen atoms. As discussed in Section 6, various, rather conventional approaches to O-atom production have been utilized by different research groups, for example, (1) electric discharge dissociation (dc, rf, and microwave) of $O_2$, (2) thermal arc dissociation of $O_2$. No convenient, separate, purely chemical method of O-atom generation for this chemical laser system has yet been found. In the $CS_2$–$O_2$ flame laser, discussed in Section 7, however, oxygen atoms are produced by the flame chemistry.

## 2. VIBRATIONAL DISTRIBUTION OF THE O+CS REACTION

The reaction between oxygen atoms and CS molecules is well established as the elementary reaction that produces vibrationally excited CO in $O_2$–$CS_2$ chemical lasers.[8] In cw flow-mixing lasers, the reactants are in their ground-electronic and vibrational states, so the pumping reaction is

$$O(^3P) + CS(^1\Sigma) \rightarrow CO(v)(^1\Sigma) + S(^3P) - 85 \text{ kcal/mol} \qquad (2)$$

[see Reaction (2), Table 1]. With the present uncertainty in the heat of formation of CS,[17,18] the reaction exothermicity probably should be considered to be $85 \pm 5$ kcal/mol.

This energy is sufficient to populate the level $v = 15$ of CO if all of the available energy went into CO vibration. In general a distribution of the excited states of CO with very little population in $v = 15$ and above is expected. There is no a priori reason to expect an inverted distribution suitable for laser action on CO vibrational transitions; in this respect, O+CS is a remarkable reaction.[13]

The distribution of $CO(v)$ from Reaction (2) has been studied experimentally by several investigators using differing techniques. Hancock, Morley, and Smith[5] examined the $CO(\Delta v = 2)$ infrared chemiluminescence of a flowing $O + CS_2$ reaction very dilute in Ar. The observation time for molecules in their apparatus was long compared to radiative

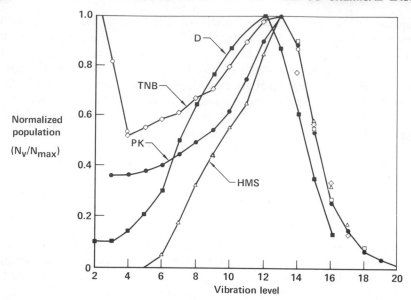

**Fig. 1.** Chemical distribution of CO normalized to $N_{13} = 1.0$ versus vibrational level.

lifetimes of the excited states, and collisional relaxation was also significant, leading to an observed distribution of $CO(v)$ highly relaxed from its initial state. The relative rate of formation $k_v$ was deduced by Hancock, Morley, and Smith[5] through a steady-state analysis of the infrared overtone intensities accounting for radiative, collisional, pumping, and wall deactivation processes. This distribution is shown in Fig. 1, labeled *HMS*.

The same investigators also performed a pulsed photolysis of $O_2$–$CS_2$–Ar mixtures, monitoring the gain or loss on the $5 \rightarrow 4$ through $20 \rightarrow 19$ CO bands with a cw CO probe laser. Their reactant concentrations were sufficiently high that vibrational relaxation caused some redistribution of the excited molecules as early as 50 $\mu$sec after the photolysis flash. They determined vibrational populations from the observed optical gain coefficients at 50, 100, and 150 $\mu$sec, and then extrapolated back to $t = 0$ to derive an initial distribution. Hancock, Morley, and Smith[5] concluded that the results of the steady-state chemiluminescence and the pulsed flash-photolysis experiments were in satisfactory agreement.

Tsuchiya, Nielsen, and Bauer[19] used a pulsed discharge through $O_2$–$CS_2$–He mixtures and observed time delays to oscillation in a wavelength-selective laser oscillator. They interpreted the delay times in terms of the initial distribution of $CO(v)$. They also measured the pulsed chemiluminescence from their device at various times following the discharge pulse, and the measured spectral distributions were used to estimate vibrational populations. Their results from the pulsed chemiluminescence experiment are plotted in Fig. 1 as TNB.

The most recent experiments on the CO($v$) distribution have been performed by Powell and Kelley.[20] They used a pulsed discharge through a flowing-gas mixture with $O_2 : CS_2$ of about $25:1$. A cw probe laser was used to monitor gain or absorption on the $3 \rightarrow 2$ through $20 \rightarrow 19$ vibrational transitions. The reactant pressures in the pulsed discharge were $O_2 : CS_2 : O = 750 : 5.5 : 20$ mTorr, where the O-atom pressure was measured by $NO_2$ titration.[21] Following the 1-$\mu$sec discharge pulse, the gain or loss on all of the bands was found to increase linearly with time up to 50 $\mu$sec. This behavior implies a constant rate of CO($v$) formation with no collisional redistribution for a time long (50 $\mu$sec) compared to the discharge initiation pulse (1 $\mu$sec). The observed gain and absorption data are shown in Fig. 2. The 50-$\mu$sec data were analyzed to obtain the initial distribution labeled as PK in Fig. 1.

The intracavity laser technique has been used by Djeu[84] to measure gain or loss in a continuously flowing reaction cell fueled with O-atoms and CS-molecules. Very small (<0.01%) loss or gain due to the reaction products forming in the cell can be measured with this technique, essentially by measuring the change in oscillation frequency range of a stable CO laser occupying a common active optical path with the reaction cell within a single resonator. Djeu's results, labeled D, are shown in Fig. 1, and lie intermediate between PK and HMS for $v \leq 6$.

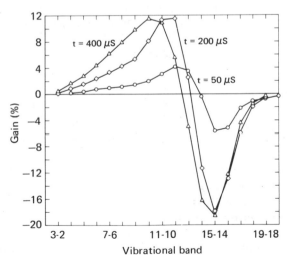

**Fig. 2.** Gain of the CO vibrational bands for various times after discharge pulse.

If one assumes that the *PK* distribution of Fig. 1 can be extrapolated smoothly to $v = 1$ and $v = 0$, the fraction of the 85 kcal/mol of reaction energy going into CO vibration is 0.66. The distribution is very favorable for optical gain in the CO transitions since total inversion is predicted from $13 \to 12$ to $4 \to 3$, with inversion ratios of 1.15 at $12 \to 11$ down to $1.00 \pm 0.05$ at $4 \to 3$.

## 3. MOLECULAR ENERGY TRANSFER IN CO CHEMICAL-LASER MEDIA

In many chemical-laser devices, the vibrationally excited CO is generally not formed very rapidly compared to V–V collisional processes, which redistribute the excited populations. Thus, the actual distribution of excited states depends to a great deal on the species and densities of collision partners. In discussing collisional transfer in the CO chemical laser, a distinction may be made between species that are present as reactants or products and those added intentionally as nonreactive gas additives that can enhance CO inversion ratios and, hence, lasing intensity. In the first category of reactants and products, the major species are: $O_2$, $CS_2$, $SO$, $SO_2$, and $OCS$. Vibrationally cold CO and $N_2O$ have been shown to be favorable nonreactive gas additives. With very intense stimulated emission, the CO vibrational-state distribution also can be affected by lasing. Spontaneous emission is generally slow enough as to be unimportant in CO laser devices.[82]

The simplest collisional event for the excited CO molecules to undergo is a $V \to T$ energy transfer (see Refs. 22, 23, and 82 for a general discussion). The single quantum $V \to T$ process can be expressed as

$$CO(v) + M \xrightarrow{k_v^M} CO(v-1) + M - \Delta E(v)$$

where M denotes any collision partner and $v \geq 1$. The entire energy of the CO transition $\Delta E(v)$ is released to translation of $CO(v-1)$ and M subsequent to the collision. For most species, the rate $k_v^M$ for deactivation of CO is slow,[27] particularly compared with similar rates for hydrogen halide molecules.[96]

Table 2 is a collection of experimental $V \to T$ rates for CO. Helium is a commonly used diluent in CO chemical lasers, however the CO–He $V \to T$ rate is orders of magnitude smaller than $V \to V$ transfer rates. The $V \to T$ deactivation by free O-atoms is efficient[31] and can represent an important channel for vibrational-energy decay. Although $V \to T$ processes are possible with diatomic and triatomic molecular collision partners, there are generally more rapid $V \to V$ processes that will dominate.

**Table 2.** Experimental $V \to T$ Rates for CO at 300°K

| $CO(v)+M \xrightarrow{k_v^M} CO(v-1)+M$ | | | |
|---|---|---|---|
| M | $v$ | $k_v^M (\sec^{-1} \text{Torr}^{-1})$ | References[a] |
| He | 1 | $0.64 \pm 0.07$ | 24–27 |
| | 9 | 5.30 | 28 |
| | 10 | 9.19 | 28 |
| | 11 | 12.0 | 28 |
| | 12 | 18.7 | 28 |
| | 13 | 21.9 | 28 |
| Ar | 1 | $0.02 \pm 0.002$ | 27–29 |
| $H_2$ | 1 | $14.4 \pm 0.6$ | 27–29 |
| $D_2$ | 1 | $0.45 \pm 0.15$ | 24 |
| CO | 1 | 0.01 | 27, 30 |
| O | 1 | $610.^b$ | 31 |

[a] Rates listed are taken from the first reference.

[b] Extrapolated to 300°K.

The $V \to V$ energy transfer collisions (see Refs. 22, 23, 80, and 81 for general discussion) are possible with the diatomic and triatomic molecular species that exist in CO chemical lasers. Such collisions may be written as

$$CO(v) + AB \xrightarrow{k_v^{AB}} CO(v-1) + AB^* - \Delta E$$

or

$$CO(v) + ABC \xrightarrow{k_v^{ABC}} CO(v-1) + ABC^* - \Delta E$$

where $AB$, $ABC$ is initially unexcited vibrationally and $AB^*$, $ABC^*$ is left excited after the collision. The energy $\Delta E$ released to translation in the collision is the difference between the CO transition $(E_v - E_{v-1})$ and the $AB$, $ABC$ transition $E_{AB^*} - E_{AB}(E_{ABC^*} - E_{ABC})$. Such collisions can be either exo- or endothermic; when $\Delta E$ is small compared to $kT$, the collision is termed near resonant. The $V \to V$ energy-transfer rates are generally large when one or more of the following criteria are met;[32]

1. the collision is nearly resonant,

2. the molecular transitions of both partners are allowed according to electric-dipole selection rules, or

3. the intermediate complex in the collision is quasibound.

**Table 3.** Experimental $V \to V$ Rates for CO-Diatomic Molecule Collisions at 300°K

| | | $CO(v) + AB \xrightarrow{k_v^{AB}} CO(v-1) + AB^*$ | |
|---|---|---|---|
| $AB$ | $v$ | $k_v^{AB}(sec^{-1} Torr^{-1})$ | References[a] |
| CO | 2 | $6.2 \pm 0.4 \times 10^4$ | 33, 34 |
| | 3 | $7.3 \pm 0.4 \times 10^4$ | 33 |
| | 4 | $7.0 \pm 0.4 \times 10^4$ | 33, 35 |
| | 5 | $5.0 \pm 0.4 \times 10^4$ | 33, 35 |
| | 6 | $2.8 \pm 0.3 \times 10^4$ | 33, 35 |
| | 7 | $1.25 \pm 0.15 \times 10^4$ | 33, 35 |
| | 8 | $6.2 \pm 1.0 \times 10^3$ | 33, 35 |
| | 9 | $3.8 \pm 1.0 \times 10^3$ | 33, 35 |
| | 10 | $2.4 \pm 0.75 \times 10^3$ | 33, 35 |
| | 11 | $1.5 \pm 0.6 \times 10^3$ | 33 |
| $O_2$ | 1 | $2.67 \pm 0.46$ | 24 |
| | 12 | 742 | 28 |
| | 13 | 1556 | 28 |
| $N_2$ | 1 | $179 \pm 26$ | 24, 34, 36 |
| | 4 | 84.8 | 28 |
| | 5 | 60.1 | 28 |
| | 6 | 38.5 | 28 |
| | 7 | 30.0 | 25 |
| | 8 | 25.5 | 28 |
| | 9 | 17.3 | 28 |
| | 10 | 14.1 | 28 |
| | 11 | 7.8 | 28 |

[a] Rates are taken from the first listed reference.

In Table 3 the $V \to V$ rates for the diatomic species CO, $O_2$, and $N_2$ are listed. The CO–CO rates are for the process

$$CO(v) + CO(0) \longrightarrow CO(v-1) + CO(1)$$

which is the dominant CO–CO $V \to V$ transfer process when vibrationally cold CO is intentionally added to the medium. The CO–$O_2$ rates become larger with increasing $v$, since CO is nearly resonant with the $0 \to 1$ $O_2$ transition at $v = 24$. Even though the CO–$O_2$ rates are only of order $10^3$/sec/Torr at $v = 13$, the large amount of $O_2$ usually present in CO chemical lasers effectively prevents upward $V \to V$ pumping of CO into the levels $v \gtrsim 13$. In the lower levels where the CO chemical laser operates, the $CO-O_2$ rates are presumably too slow to remove appreciable vibrational energy from CO. Similarly, the $CO-N_2$ rates listed in Table 3 imply that there would be very little energy transfer from CO to $N_2$ unless very large excesses of $N_2$ were present.

Of the triatomic species present as reactants or products, the $CS_2$ fuel is an unavoidable and efficient deactivator of CO (see Table 4). The $CO-CS_2$ $V \rightarrow V$ rates for $v > 1$ have not been measured; however, they are also likely to be large. OCS is a minor product in $O-O_2-CS_2$ reactions and is a very rapid deactivator of CO. OCS is an example that satisfies the first two criteria stated above for rapid $V \rightarrow V$ transfer: the $00°0 \rightarrow 00°1$ transition of OCS is strongly allowed, and is nearly resonant with the $4 \rightarrow 3$ transition of CO.

Some species have been shown to enhance lasing in CO chemical lasers due to favorable $V \rightarrow V$ transfer effects.[40,63] When the $V \rightarrow V$ transfer rates between CO and a gas additive decrease with increasing $v$ in the CO levels, the effect of the additive is to increase the inversion ratio on the CO bands at the cost of removing quanta from CO. Vibrationally cold CO and $N_2O$ are both examples of such additive effects.[41] $N_2O$ is the most favorable since the $V \rightarrow V$ transfer rates of Table 4 decrease monotonically from $v = 1$ to $v = 8$, thus, increasing the inversions on the $8 \rightarrow 7$ through $2 \rightarrow 1$ bands.

## 4. OPTICAL DATA

Pertinent to the CO chemical laser are the values for the V–R radiative–transition matrix elements $R$ of the CO molecule. Young and Eachus[42] have reported the matrix element for the fundamental band, $R_{10} = 1.04 \times 10^{-19}$ esu-cm. For higher lying transitions, the simple harmonic oscillator scaling law[43] applies, that is, $R_{v,v-1} = v^{1/2} R_{10}$. The optical-broadening cross-sections for some colliding species have been determined from absorption measurements by Williams et al.,[44] including their variation with the rotational quantum number $J$. The pertinent collision cross-sectional data are presented in Table 5.

## 5. MODELING OF CW MIXING LASERS

### 5.1. Fluid Flow and Mixing

In this section a model of the subsonic transverse flow-mixing laser will be developed to elucidate the basic effects in such devices. The geometry of the model, shown in Fig. 3, is chosen on the basis of experimental

**Table 4.** Experimental $V \rightarrow V$ Rates for CO-Triatomic Molecule Collisions at 300 K

$$CO(v) + ABC \xrightarrow{k_v^{ABC}} CO(v-1) + ABC^*$$

| $ABC$ | $v$ | $k_v^{ABC}$ $(\text{sec}^{-1}\,\text{Torr}^{-1})$ | References |
|-------|-----|-----|-----|
| $CS_2$ | 1 | $1.49 \pm 0.17 \times 10^4$ | 24, 37, 38 |
| $SO_2$ | 1 | $11.6 \pm 0.2$ | 24, 37 |
| OCS | 1 | $2.63 \pm 0.48 \times 10^5$ | 24, 37 |
| | 4 | $1.59 \times 10^6$ | 28 |
| | 5 | $2.12 \times 10^6$ | 28 |
| | 6 | $1.99 \times 10^6$ | 28 |
| | 7 | $1.27 \times 10^6$ | 28 |
| | 8 | $4.91 \times 10^5$ | 28 |
| | 9 | $2.51 \times 10^5$ | 28 |
| | 10 | $1.38 \times 10^5$ | 28 |
| | 11 | $1.09 \times 10^5$ | 28 |
| | 12 | $6.82 \times 10^4$ | 28 |
| | 13 | $3.50 \times 10^4$ | 28 |
| $N_2O$ | 1 | $1.51 \pm 0.13 \times 10^5$ | 24 |
| | 2 | $5.8 \times 10^4$ | 39 |
| | 3 | $3.0 \times 10^4$ | 39 |
| | 4 | $1.8 \times 10^4$ | 39 |
| | 5 | $7.8 \times 10^3$ | 39 |
| | 6 | $5.3 \times 10^3$ | 28 |
| | 7 | $3.46 \times 10^3$ | 28 |
| | 8 | $3.04 \times 10^3$ | 28 |
| | 9 | $3.04 \times 10^3$ | 28 |
| | 10 | $3.50 \times 10^3$ | 28 |
| | 11 | $3.82 \times 10^3$ | 28 |
| | 12 | $3.29 \times 10^3$ | 28 |
| | 13 | $3.29 \times 10^3$ | 28 |
| $CO_2$ | 1 | $2.0 \pm 0.1 \times 10^3$ | 24 |
| | 4 | 813 | 25 |
| | 5 | 601 | 28 |
| | 6 | 566 | 28 |
| | 7 | 537 | 28 |
| | 8 | 572 | 28 |
| | 9 | 742 | 28 |
| | 10 | $1.17 \times 10^3$ | 28 |
| | 11 | $1.87 \times 10^3$ | 28 |
| | 12 | $3.04 \times 10^3$ | 28 |
| | 13 | $4.24 \times 10^3$ | 28 |

**Table 5.** Collision Cross-sections for Broadening of CO V–R Optical Transitions by Various Gases

| CO collision partner | Cross-section[a] $(cm^2 \times 10^{-14})$ |
| --- | --- |
| CO | 1.10 |
| He | 0.30 |
| Ar | 0.96 |
| $H_2$ | 0.35 |
| $N_2$ | 1.04 |

[a] values listed for first $[P(1)]$ rotational line; see Ref. 44 for variation with rotational quantum number.

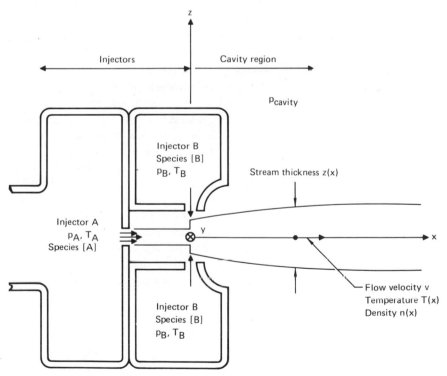

**Fig. 3.** Cross-sectional schematic drawing of a transverse continuous flow-mixing laser.

devices. Some initial set of species $[A]$ is contained in the main injector at pressure $p_A$ and temperature $T_A$. This gas is pumped through a slot of width $w_A$ and length $L$ into the cavity region, which is assumed to be at pressure $p_{cav}$. No pressure gradients are allowed within the cavity region. The initial stream is assumed to be nonreactive. A specification of the flow rates $F_A$ of all the species in $[A]$ is given, and all species are assumed to behave as ideal gases. The expansion of species $[A]$ into the cavity region is assumed to be adiabatic and, given as initial stream thickness $z$ (cm), temperature $T$ (°K), flow velocity $v$ (cm/sec), and number density $n$ (particles/cm³), are computed from the following gas dynamic and perfect gas-law equations:

$$v = -\frac{\bar{C}_p z L N_0 p_{cav}}{F_A \bar{m}_A} + \left[\frac{2\bar{C}_p N_0 k T_A}{\bar{m}_A} + \left(\frac{\bar{C}_p z L N_0 p_{cav}}{F_A \bar{m}_A}\right)^2\right]^{1/2} \tag{25}$$

$$T = \frac{T_A p_{cav} L z v}{F_A k} \tag{26}$$

and

$$n = \frac{p_{cav}}{kT} \tag{27}$$

where $p_{cav}$ is the cavity pressure (cgs-units), $\bar{C}_p$ is the average heat capacity (excluding vibration) in $[A]$ at constant pressure (cal/g), $\bar{m}_A$ the average mass per particle in $[A]$ (g), $z$ is the thickness of the initial stream (cm), $L$ is the length of the initial stream (cm), $F_A$ is the total flow rate of initial stream (cm$^{-3}$ sec$^{-1}$), $k$ is Boltzmann's constant (cgs units), $N_0$ is Avogadro's number, and $T_A$ is the injector temperature (°K).

All of the stream parameters remain constant until the transverse injection point at $x = 0$ is reached. The flow of reactants $[B]$ from the transverse injector pair is treated in the same manner as the expansion from the main injector $A$ with the use of the parameters $[B]$, $p_B$, $T_B$, $w_B$, $L$, and $F_B$, The flow from injectors $B$ is assumed to be in the plus and minus $z$-directions, and the stream quantities are recomputed under the assumption that the flows $[A]$ and $[B]$ combine instantly.

In order to compute the stream quantities at $x = 0$, the following conservation laws are used:
conservation of particles,

$$F = F_A + F_B \tag{28}$$

conservation of mass,

$$F\bar{m} = F_A \bar{m}_A - F_B \bar{m}_B \tag{29}$$

conservation of number of degrees of freedom,

$$F\bar{g} = F_A \bar{g}_A + F_B \bar{g}_B \tag{30}$$

conservation of $x$-momentum,

$$Fmv|_{x=0} = F_A \bar{m}_A v|_{x>0} \tag{31}$$

conservation of energy (at constant pressure),

$$(F_A + F_B)\left(\tfrac{1}{2}mv^2 + gkT + p\frac{1}{n}\right)$$

$$= F_A\left(\tfrac{1}{2}\bar{m}_A v^2 + \bar{g}_A kT + p\frac{1}{n_A}\right) + F_B\left(\tfrac{1}{2}\bar{m}_B v_B^2 + \bar{g}_B kT_B + p\frac{1}{n_B}\right) \tag{32}$$

where $\bar{g}_A$ is one-half the average number of degrees of freedom per particle in the mixture $[A]$ (excluding vibration) of an ideal gas at constant volume. Thus, at $x = 0$, the gases have "mixed" (for fluid-flow purposes), and the new chemical species set $[A + B]$ is known, along with the new values $g$, $m$, $T$, $n$, $v$, and $z$.

To model the real situation of finite mixing times, the concentrations of the reaction set $[B]$ are allowed to react with $[A]$ as

$$\frac{dy_i}{dx} = \frac{n_i - y_i}{l_{\text{mix}}} - \frac{y_i dz}{z \, dx} \tag{33}$$

where $n_i$ is the density of the $i$th species in $[B]$, and $y_i$ is that portion of $n_i$ that has been allowed to react with set $[A]$. The second term arises from stream expansion, $z$ being the stream thickness. The mixing length $l_{\text{mix}}$ must be furnished as an input parameter. The quantity $l_{\text{mix}}$ might be calculated on the basis of some model of the flow and mixing situation (such as diffusive mixing), or its value might be based on an experimental estimate.

## 5.2. Chemical Kinetics, Stream Heating, and Expansion

The detailed chemical kinetics has been limited to three reactions, since all other collisional effects on CO are to be allowed with all reactant and product species. The reactions used to approximate the chemistry of the system for this model,[94] as condensed from Table I, are

$$O + CS_2 \xrightarrow{k_1} CS + SO - 22 \text{ kcal} \tag{1}$$

$$O + CS \xrightarrow{k_2} CO^* + S - 85 \text{ kcal} \tag{2}$$

and

$$S + O_2 \xrightarrow{k_3} SO + O - 5.6 \text{ kcal} \tag{3}$$

To calculate stream properties at $x > 0$, the chemical kinetics must be included since the chemical species are changing and the reactions release heat. As before, the stream is allowed to expand or contract so that its internal pressure $p_{cav}$ is constant. Let the $l$th reaction be described by a rate $k_l$ per unit mass, energy release $\epsilon_l$, and $r_{il}$, the number of particles of species $i$ created or destroyed in each elementary step of the reaction. Then the differential equation for the temperature of the stream is found by energy balance:

$$dT = \frac{\bar{m}}{\bar{g}k}\left[\frac{1}{v}\sum_l \epsilon_l k_l \, dx - d\tilde{Q}_{rad} - \frac{kT}{\bar{m}z}\,dz - d\tilde{U}_v - \frac{kT}{v}\sum_i g_i \sum_l r_{il}k_l \, dx\right] \quad (34)$$

where $\tilde{Q}_{rad}$ is the net energy loss through radiative processes per unit mass, and $\tilde{U}_v$ is the energy stored in vibration per unit mass (only CO).

The stream thickness $z$ is calculated from the differential equation related to the perfect gas law:

$$dz = \frac{\bar{m}z}{v(\bar{g}+1)kT}\left[\sum_l \epsilon_l k_l \, dx - v\,d\tilde{Q}_{rad} - v\,d\tilde{U}_v - kT\sum_i (g_1 - \bar{g})\sum_l r_{il}k_l \, dx\right]$$
$$(35)$$

## 5.3.  CO Formation and Relaxation

After the first injection point $x = 0$, Reaction (2) begins to form CO*. The probability of formation by Reaction (2) into each excited state $v$ is taken to be the normalized initial distribution of $PK$ in Fig. 1. The distribution of CO* is calculated by integrating the master equations for vibrational relaxation. Both V $\rightarrow$ T and V $\rightarrow$ V collisional energy exchange processes are allowed for each chemical species existing in the set of Reactions (1) through (3).

For the V $\rightarrow$ V processes, the following collisions are considered (see also Section 3):

$$CO(v) + CO(v') \rightarrow CO(v \pm 1) + CO(v' \mp 1) \quad (36)$$

$$CO(v) + CS_2(00^\circ 0) \rightarrow CO(v - 1) + CS_2(00^\circ 1) \quad (37)$$

$$CO(v) + CS(v = 0) \rightarrow CO(v - 1) + CS(v = 1) \quad (38)$$

$$CO(v) + SO(v = 0) \rightarrow CO(v - 1) + SO(v = 1) \quad (39)$$

$$CO(v) + O_2(v = 0) \rightarrow CO(v - 1) + O_2(v = 1) \quad (40)$$

All of the species except CO are assumed to exist entirely in their ground vibrational states, so that the reverse processes of Reactions (36) through (40) are not allowed.

Probabilities for both $V \rightarrow V$ and $V \rightarrow T$ energy transfer in CO–M collisions (M = CO, He, $O_2$) are calculated by previously developed techniques.[46,47] The $V \rightarrow V$ calculations for CO–CO exchanges,

$$CO(v) + CO(v') \rightarrow CO(v \pm 1) + CO(v' \mp 1) \tag{41}$$

are performed for both long-range and short-range interactions, and for $(v, v') \leq 50$. The exchange probabilities for short-range interactions are limited to values of 0.125 for the CO molecule by the theory[47] used. The long-range interactions are treated by perturbation theory, and an upper limit is set on the calculated long-range probabilities $P$ by using a pseudoprobability $P'$ defined by

$$P' = \frac{P}{1 + eP} \qquad (e = 2.71828 \ldots) \tag{42}$$

The $V \rightarrow T$ exchange with He is treated with the standard SSH theory[48] of $V \rightarrow T$ transfer adjusted to experimental data[24] by a scale factor. The exchange with $O_2$ also was treated in the relaxation equations as a $V \rightarrow T$ collision, but the probability for this event,

$$CO(v) + O_2(0) \rightarrow CO(v - 1) + O_2(1) \tag{43}$$

was calculated from the $V \rightarrow V$ theory for short-range interactions.[47] Again, the computed results were adjusted to experimental data[24] by a scale factor. The CO–$CS_2$, CO–CS, and CO–SO collisions were treated with only short-range forces, as was done for CO–$O_2$. The $V \rightarrow V$ and $V \rightarrow T$ probabilities are allowed to vary with temperature:

$$P(T) = \exp(a + bT^{-1/3} + cT^{-1}) \tag{44}$$

where the parameters $a$, $b$, and $c$ are chosen to fit the computed probabilities at 100, 300 and 700°K.

## 5.4. Stimulated Emission

The model chosen to treat stimulated emission in the chemical-laser theory is that of an amplifier. An input optical field is assumed to exist at one end of the active medium of length $L$. The frequency spectrum of the field is comprised of up to 50 of the $P$- or $R$-branch CO transition center frequencies. The input field intensity is allowed to vary only along the $x$-axis (the flow axis), and it may assume such shapes as Gaussian, Lorentzian, or a step function in $x$. For Gaussian fields, the parameter $x_w$ is the half width at half intensity.

The peak intensity and $x$-variation details are input parameters for each of the 50 optical input fields. At each position in the flow, beginning at $x = 0$, the program calculates the optical-gain coefficient for every transition with an input optical field. In the incremental distance $dx$ along the stream, the power transfer to or from the optical fields is given by a differential form of Beer's law:

$$dP_{v,J} = I^0_{v,J} f_{v,J}(x)(e^{\alpha_{v,J}L} - 1)z(x)\, dx \tag{45}$$

where $I^0_{v,J}$ is the peak field intensity along $x$ for the transition described by the upper-state $v$ and lower-state $J$ quantum numbers $f_{v,J}(x)$ is the $x$-variation of input optical-field intensity, $\alpha_{v,J}$ is the optical-gain coefficient, $L$ is the length of the stream in the $y$-direction, and $z(x)$ is the stream thickness in the $z$-direction.

Note that if the gain $\alpha_{v,J}$ for any transition is negative, the power transfer from the medium to that field is negative, that is, the medium is absorbing energy from the input optical field. Thus, the differential equation for power delivered to the optical field is

$$\frac{dP_{v,J}}{dx} = I^0_{v,J} f_{v,J}(x)(e^{\alpha_{v,J}L} - 1)z(x) \tag{46}$$

and the total power delivered to the optical fields from $x = 0$ to $x$ is the simple summation

$$P_{\text{rad}}(x) = \sum_{v,J} \int_0^x \frac{dP_{v,J}}{d\xi}\, d\xi \tag{47}$$

The saturating effects of the fields on the medium is accounted for through the expression[49]

$$\frac{dn_v}{dx} = -\frac{1}{v_{\text{flow}}} \sum_J \frac{\alpha_{u,J} \bar{I}_{v,J}}{h\nu_{v,J}} = -\frac{dn_{v-1}}{dx} \tag{48}$$

where $v_{\text{flow}}$ is the stream-flow velocity, and $\bar{I}_{v,J}$ is the average flux along the $y$-direction

$$\bar{I}_{v,J} = \frac{1}{v} \int_0^L I^0_{v,J} f_{v,J}(x) e^{\alpha_{v,J}y}\, dy$$

$$= \frac{I^0_{v,J} f_{v,J}(x)}{\alpha_{v,J}L}(e^{\alpha_{v,J}L} - 1) \tag{49}$$

The remaining quantities computed are the vibrational power in the medium and the chemical efficiency of power extraction by the optical fields. The vibrational power is computed as

$$\frac{dP_{\text{vib}}}{dx} = \bar{E} \frac{dn_{\text{CO}}}{dx}\Big|_{\text{chem}} z(x) L v_{\text{flow}} \tag{50}$$

where $\bar{E}$ is the average vibrational energy per CO molecule formed by the O+CS reaction (*PK* distribution of Fig. 1), and $dn_{CO}/dx$ |$_{chem}$ is the derivative of the CO density due only to chemical formation. Then the total vibrational power delivered to the medium from $x = 0$ to $x$ is found through the simple integration

$$P_{vib}(x) = \int_0^x \frac{dP_{vib}}{d\xi} \, d\xi \tag{51}$$

Finally, the chemical efficiency for these calculations is defined as

$$\eta'_{chem}(x) = \frac{P_{rad}(x)}{P_{vib}(x)} \tag{52}$$

## 5.5. Modeling Results

For the sample calculation to be discussed in this section, the geometry, cavity pressure, and reactant and diluent flow rates have been chosen to match an experimental transverse flow-laser device. Table 6 lists the numerical values for these input parameters; all quantities were measured experimentally except for the mixing length which was estimated. The calculations were performed assuming first the absence and then the presence of input optical fields.

**Table 6.** Input Parameters for Modeling Calculations

| | | |
|---|---|---|
| Injector A (slotted type, $W_A = 0.05$ cm, $L = 50.0$ cm) | | |
| Flow rates (mmol/sec) | He | 21.5 |
| | $O_2$ | 12.45 |
| | O | 0.3 |
| Pressure (Torr) | | 11.0 |
| Temperature (°K) | | 315 |
| $z$ (cm) | | 0.5 |
| Injector B (slotted type, $W_B = 0.10$ cm, $L = 50.0$ cm) | | |
| Flow rates (mmol/sec) | Ar | 3.5 |
| | $CS_2$ | 0.5 |
| Mixing length (cm) | | 0.5 |
| Pressure (Torr) | | 10.0 |
| Temperature (°K) | | 300 |
| $z$ (cm) | | 0.1 |
| Cavity conditions | | |
| Pressure (Torr) | | 5.0 |
| Temperature (°K) | | 300 |

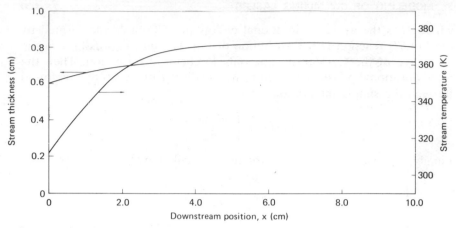

**Fig. 4.** Computed stream quantities $z(x)$ (stream thickness) and $T(x)$ (stream temperature).

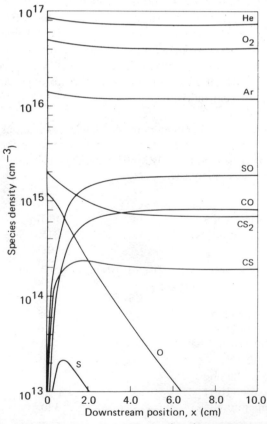

**Fig. 5.** Chemical species densities as a function of stream position.

The calculated stream thickness $z(x)$ and temperature $T(x)$ are shown in Fig. 4. The stream expands slightly due to the temperature rise of about 60°K. This increase in stream thickness is reflected in the diluent densities as shown in Fig. 5. As shown in Fig. 5, the reactions are nearly complete at $x = 3.0$ cm. The final CO concentration, $8.2 \times 10^{14}$ cm$^{-3}$, corresponds to a partial pressure of 0.032 Torr compared to the total stream pressure of 5.0 Torr.

The vibrational-population distributions of excited CO are shown in Fig. 6, plotted at $x = 0.5$, 1.0, 3.0, and 10.0 cm. The *PK* distribution of Fig. 1 is used in these calculations and is quite evident at $x = 0.5$ cm. The distribution evolves by continued formation and by rearrangement through V → V collisions. The major effects are a shift in the peak to

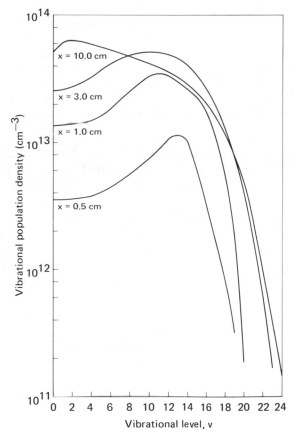

**Fig. 6.** The CO vibrational-state population distributions in absence of external radiation fields.

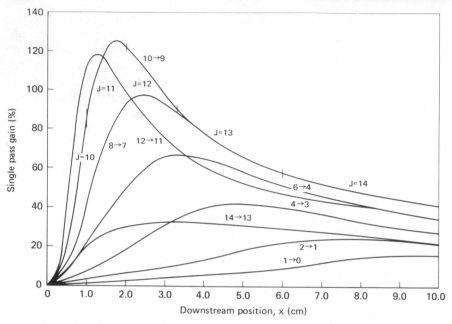

**Fig. 7.** Maximum small-signal single-pass gains for a 50-cm long active medium.

lower $v$ and a broadening of the initial peak at $v = 13$. At very large downstream distances, the predicted distribution begins to approach the $V \to V$ relaxed distribution characteristics of electric discharge CO lasers. The small-signal single-pass gains for a 50 cm long active medium are shown in Fig. 7. The gains plotted constitute an envelope of the maximum gains that exist on the particular vibrational band at each position $x$. The $J$ value for maximum gain shifts as the inversion and temperature change. The particular $J$ values for maximum gain are shown for the $10 \to 9$ band. For any $v \to v - 1$ band, the gain versus $x$ on any single $P$-branch transition would be peaked more sharply. Finally, the positive slopes at low $v$ of the distributions of Fig. 6 imply gain in the $R$-branch of CO. On any band, the $R$-branch gains are always smaller than $P$-branch gains, so $R$-branch emission can be expected only with wavelength-selective optical cavities, and cw $R$-branch emission, in fact, has been observed from CO chemical lasers.[85]

The previous calculation was repeated using the same input parameters with the addition of input optical fields for extraction of laser power. One rotational line on each vibrational band was selected for the bands $13 \to 12$ down to $2 \to 1$ (see Fig. 8). The choice of rotational transitions was made on the basis of maximum atmospheric transmission. The

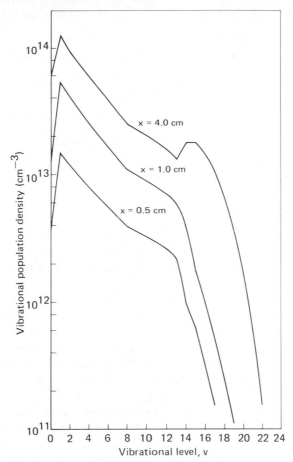

**Fig. 8.** The CO vibrational-state population distributions on the $13 \rightarrow 12$ through $1 \rightarrow 0$ bands at various downstream positions $x$ in presence of external radiation fields.

imposed field intensities are taken to be $100\,\text{W/cm}^2$ for each of the selected transitions and are uniform from $x = 0$ to $x = 4.0\,\text{cm}$. These relatively high input intensities are used to ensure saturation of the medium and thus predict an upper limit on output-power extraction. Figure 8 shows the CO vibrational distributions at $x = 0.5$, 1.0, and $4.0\,\text{cm}$ under the influence of these intense input optical fields. By comparison with the distributions without imposed optical fields of Fig. 6, the power extraction case shows that the medium is held near saturation as the reactions continue to produce excited CO. In Fig. 9, the calculated total output power and chemical efficiency are plotted. The total power delivered to the optical fields is 22.3 W at $x = 4.0\,\text{cm}$, which corresponds

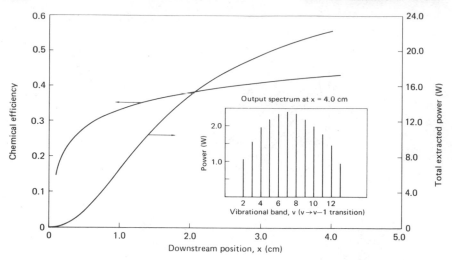

**Fig. 9.** Computed chemical efficiency and output power.

to 43% of the vibrational power delivered to CO by the reactions. From the shape of the chemical efficiency curve a limiting value of about 50% might be estimated. The spectrum of the extracted power at $x = 4.0$ cm is shown as an inset in Fig. 9. Note that the spectrum does not mirror the shape of the reaction distribution curve (*PK* distribution) seen approximately at $x = 0.5$ cm in Fig. 6.

The model and results discussed in this section are intended to illustrate the main features of transverse-flow cw CO chemical lasers. Many additional effects could be included to describe the fluid flow, mixing, and chemical kinetics more accurately at the cost of greatly increasing the complexity of modeling programs.

## 6. CW CO CHEMICAL-LASER EXPERIMENTS

Following the discovery by Hancock and Smith[50] that the O–CS$_2$ reactions yielded vibrational-state population distributions in the CO product molecules that were inverted, the demonstration of continuous-laser emission on V–R transitions between the states pumped by those reactions was quick to follow.[52] Pulsed laser emission, pumped by O–CS$_2$ and similar reactions, had been observed first, much earlier, by Pollack.[51] The pulsed CO chemical-laser experiments are discussed separately in Section 8.

## 6.1. Longitudinal-Flow cw Lasers

The first demonstration of cw CO chemical-laser emission was reported by Wittig, Hassler, and Coleman[52] with a longitudinal-flow laser apparatus diagrammed in Fig. 10. In this device $O_2$–He mixtures were passed through a simple, continuous, 60-Hz discharge (12 kV, 60 mA) to induce partial dissociation of the $O_2$ to O-atoms. This atom-rich flow was sent into one end of a 15-cm long, 1.0-cm i.d. Pyrex tube connected at the opposite end to a 500 liter/min vacuum pump. $CS_2$ vapor was introduced separately through a simple peripheral ring of gas inlet holes located near the O–$O_2$–He inlet, as shown. O–$CS_2$ reactions proceeded as the reagents mixed and flowed down the length of the reaction tube. The reacting medium was maintained at pressures typically between 10 and 15 Torr with the gas feed rates listed in Table 7. The design of the apparatus afforded an unobstructed 15-cm active optical path along the length of the reaction tube, which proved to be sufficient to sustain cw oscillation within the near confocal cavity shown in the figure. Laser output was detected through a flat 0.5-mm i.d. hole-coupling mirror.

As discussed in Section 3, the lifetime of the $CO(v \geq 1)$ states is

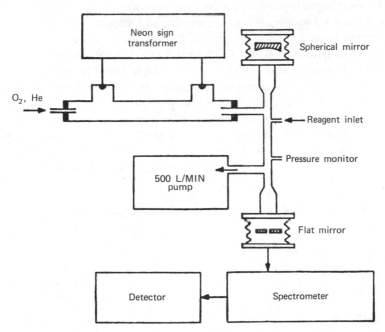

**Fig. 10.** Early longitudinal flow cw CO chemical laser apparatus, after Wittig, Hassler, and Coleman.[52]

**Table 7.** Gas Feed Rates in Longitudinal-flow cw CO Chemical Laser of Wittig, Hassler, and Coleman[52]

| Reagent | Feed rate[a] (STP liter/min) |
|---------|------------------------------|
| He      | 34.0                         |
| $O_2$   | 7.5                          |
| $CS_2$  | 0.3                          |

[a] Laser output: 10.0 mW.

sufficiently long to expect that the region of positive gain for many transitions extends for the full length of the reaction tube. An inherent consequence of the longitudinal design is that the intracavity laser flux propagates through a nonuniform reacting medium, which may exhibit high gain in the vicinity of the $O-CS_2$ mixing zone and low gain at the other end of the Pyrex tube.

The laser initially produced a cw output power of 1.0 mW, emitted over a band from 5.2 to 5.6 $\mu$m, corresponding to a progression of 28 $P$-branch transitions from $v = 12$ to $v = 7$.

Similar longitudinal flow CO chemical lasers were assembled by Suart, Kimbell, Dawson, and Arnold[1,40] in which the $O_2$ dissociation was accomplished by a microwave discharge. In the experiments of Suart, Dawson, and Kimbell,[1] a 2.3-W cw output was achieved from their 1.0-m cavity longitudinal-flow laser with an efficiency of conversion of chemical-reaction exothermicity to laser power of 0.92%. This efficiency level was achieved with the addition of cold CO to the reacting medium as indicated in Table 8 and with water cooling of the reaction-tube walls. A

**Table 8.** Flow Conditions in longitudinal CO Laser of Suart, Dawson, and Kimbell[1]

| $O_2$ | Flow He | (STP l/min) $CS_2$ | CO | Pressure (Torr) | flow velocity (m/sec) | Notes |
|-------|---------|--------------------|------|-----------------|------------------------|-------|
| 5.7 | 14.0 | 0.32 | 0 | 3.0 | 237 | Temperature: no CO $370 \pm 20°$K, |
| 5.7 | 14.0 | 0.32 | 1.9 | 3.1 | 237 | with CO $360 \pm 20°$K; Power 0.3 W with CO |
| 6.8 | 67 | 1.1 | 3.4 | 9 | 292 | Maximum power obtained 2.3 W |
| 3.1 | 14.4 | 0.35 | 1 | 4.0 | 188 | Sidelight fluorescence measured |
| 3.1 | 14.4 | 0.35 | 0.96 | 4.2 | 188 | Power 0.3 W with CO |

discussion of the role of added $CO(v=0)$ is presented later in the section.

Hirose, Hassler, and Coleman[53] adapted their earlier longitudinal-flow laser (see Fig. 10) so a continuous flow of "active nitrogen" (formed by continuous glow discharge through $N_2$) could be admixed to a separate, unexcited $O_2$–$CS_2$ feed within the laser cavity. Continuous laser emission on V–R bands of CO was also achieved through this initiation technique.

Barry, Boney, and Brandelik[54] have studied the development of cw laser emission on CO V–R transitions produced from continuous dc discharges maintained in various low-pressure hydrocarbon-oxidizer mixtures buffered with helium, for example, $CH_4$–air, $C_3H_8$–air, $C_2H_2$–air, $C_2N_2$–NO, and HCHO–air. The continuous discharge maintained in an 80-cm long double-walled glass tube. The inner tube was fed continuously with reactive gas mixtures highly diluted with helium to a total gas pressure of ~4 Torr; the outer tube jacket was fed continuously with liquid $N_2$ to maintain temperature control in the inner medium. Multiline laser-power levels of ~3 W were obtained over CO bands from $v = 15 \rightarrow 14$ to $v = 3 \rightarrow 2$, corresponding to electrical-to-optical conversion efficiency levels of ~4%. Since electrical excitation was maintained continuously throughout the reacting laser medium the proportion of electrical and chemical pumping is difficult to assign.

The longitudinal flow CO chemical laser is a device of convenience and simplicity. However, data acquired from a transverse-flow geometry, discussed in the following section, offers considerably more insight into the reaction mechanisms in the CO chemical-laser medium.

## 6.2. Transverse-Flow cw Lasers

A second cw CO chemical-laser geometry has received considerable attention: the transverse flow laser. In this device diagrammed in Fig. 3, the laser cavity is oriented transverse to the flow direction so that the optical path includes a more uniform reaction zone than a longitudinal optical path of a cavity that would be oriented parallel to the flow axis. Therefore, the transverse cavity geometry may offer higher laser efficiency and greater understanding of the laser medium. Experiments have been reported with both fast subsonic and supersonic flow velocities that use this transverse-cavity geometry.

## 6.3. Subsonic-Flow Experiments

The most widely investigated CO chemical-laser devices have used subsonic-flow transverse to the optical-cavity axis. The combination of mixing rates and reaction rates under typical operating conditions is such as to build up the optical gain to a maximum at a few centimeters downstream of the fuel injectors. Gain depletion processes such as vibrational relaxation then cause the gain to decay on a similar distance scale. With stable optical resonators, the optical fields may occupy a distance of 1 to 2 cm along the flow direction. Thus the distances for creating of gain, power extraction, and removal of the saturated medium from the optical fields are comparable. This combination of conditions has led to the best device performance with cw lasers.

Transverse-flow mixing lasers have been reported in some detail by Foster,[55] by Ultee and Bonczyk,[56] and by Jeffers and Wiswall.[57] The major characteristics of these devices are summarized in Table 9. The operating conditions for the three devices are remarkably comparable. The device reported by Jeffers and Wiswall[57] produced 6.0 W when Ar was used as a diluent in the $CS_2$ stream. This value is quite comparable to the Ultee and Bonczyk[56] value of 4.5 W. The similarity of laser operation in these different devices illustrates the dominance of $V \rightarrow V$ transfer effects in establishing output power.

Many gas additives have been investigated for power enhancement in CO chemical lasers. Only "cold" CO and $N_2O$ are firmly established as capable of increasing the total output power. The explanation for the beneficial effects of CO and $N_2O$ was first suggested by Suart, Arnold, and Kimbell;[1] they pointed out that the $V \rightarrow V$ transfer between the chemically formed CO and the gas additive could increase the inversion

**Table 9.** Performance of Typical Subsonic Transverse Flow cw CO Chemical Lasers

| Device | Foster[a] | Ultee and Bonczyk[b] | Jeffers and Wiswall[c] |
|---|---|---|---|
| Reactant flow rates (mmol/sec) | | | |
| $He:O_2:O$ | $14.6:9.9:(nm)^d$ | $32.2:17.4:(nm)^d$ | $21.5:12.4:0.32$ |
| $CS_2:N_2O$ | $0.30:0$ | $0.53:0$ | $0.50:4.0$ |
| $m_{total}$ (g/sec) | 0.40 | 0.72 | 0.71 |
| $P_{out}$ (W)[e] | 1.0 | 4.5 | 22.0 |
| $p_{cav}$ (Torr) | 4.3 | 5.4 | 5.0 |
| $v_{flow}$ (m/sec) | 61 | 40 | 50 |
| $T_{gas}$ (°K) | 366 | 295 | 490 |
| Flow area (cm²) | 22.8 | 39.5 | 50 |
| $\sigma$ (J/g) | 2.50 | 6.25 | 31.0 |
| Band spectra | $13 \rightarrow 12$ to $5 \rightarrow 4$ | $14 \rightarrow 13$ to $2 \rightarrow 1$ | $12 \rightarrow 11$ to $2 \rightarrow 1$ |

[a] Ref. 55.
[b] Ref. 56.
[c] Ref. 57.
[d] (nm), not measured.
[e] cw.

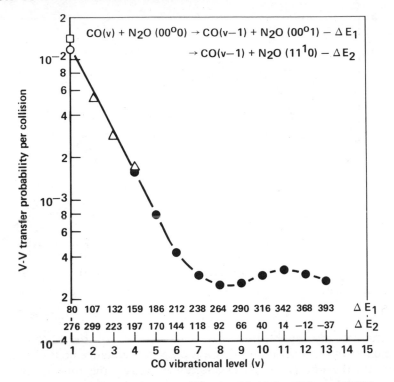

**Fig. 11.** The V–V transfer probabilities for CO–$N_2O$ collisions at 300°K.

ratios on the lasing vibrational bands. As an illustration of this process, consider the case of $N_2O$, as a gas additive for a CO chemical lasers. The probabilities for CO–$N_2O$ V → V transfer[58] are shown in Fig. 11. For the CO levels $1 \leq v \leq 8$, the transfer probability to $N_2O$ increases with decreasing $v$. Thus the added $N_2O$ will increase the inversion ratio on the 2→1 through 8→7 bands. A similar effect occurs for cold CO addition, except that the transfer probabilities decrease with increasing $v$ for $v \geq 4$ at $T - 300°K$.

These collisional-transfer effects are seen quite clearly in the emission spectra of certain CO chemical-laser devices. Figure 12 shows the total band intensities produced by the Jeffers and Wiswall[57] device under the conditions of Table 9. The operating conditions for the three sets of spectra were identical except for the use of Ar, CO, or $N_2O$ as a diluent in the $CS_2$ fuel. The total output powers were 6.0, 9.4, and 16.0 W with Ar, CO, or $N_2O$ diluent flows of 4.0 mmol/sec. Since Ar can only participate as a slow V → T collision partner with excited CO, the band

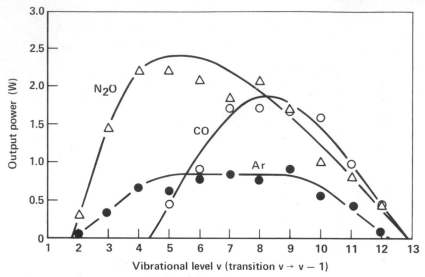

**Fig. 12.** Output spectra of a cw CO chemical laser with Ar, CO, and $N_2O$ diluents.

spectra with ·Ar might be called characteristic of the chemical-laser medium. Cold CO addition gives a total power enhancement of 56% but quenches the low bands. $N_2O$ addition increases the total output by a factor of 2.7, with fairly large increases in low band emission.

## 6.4.  Supersonic-Flow Experiments

Following the demonstration of cw CO laser emission from $CS_2$–O reactions in media flowing at subsonic velocities, experiments were undertaken to develop high-power cw CO laser from $CS_2$–O reactions in supersonic flows. Boedeker, Shirely, and Bronfin[59] employed a small-scale continuous thermal-arc apparatus, as diagrammed in Fig. 13, for these experiments. Molecular oxygen was introduced into the inert-gas arc jet, and the gases were allowed to mix and stabilize at temperatures sufficient to ensure considerable oxygen dissociation (>3000°K) in a plenum. Plenum pressures were held at levels from 0.2 to 1.0 atm. The resultant O–Ar mixtures were exhausted through a simple two-dimensional nozzle that (1) freezes the O-recombination kinetics,[60] thereby preserving the high O-atom concentrations developed in the plenum, and (2) generates a low translational-temperature supersonic flow downstream. $CS_2$ vapor, carried in an inert-gas stream, was added to the supersonic O-atom stream through a pair of opposing slot injectors. This

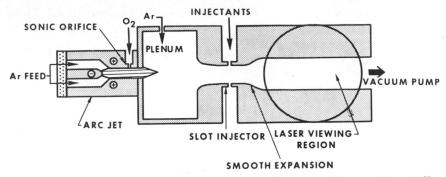

**Fig. 13.** Arc-excited cw CO chemical laser, after Boedeker, Shirley, and Bronfin.[59]

sequential thermal-arc expansion-mixing configuration previously had been employed to generate efficient cw HF laser emission (see the chapter by Gross and Spencer). A near-confocal optical cavity was oriented transverse to the flowing reacting CO medium so as to include a 5 cm active optical path. Over 35-W cw was observed[61] on 48 individual CO V–R bands, a sample spectrum of which is presented in Fig. 14. By on-line mass spectrometric analysis of the gas composition of the cavity

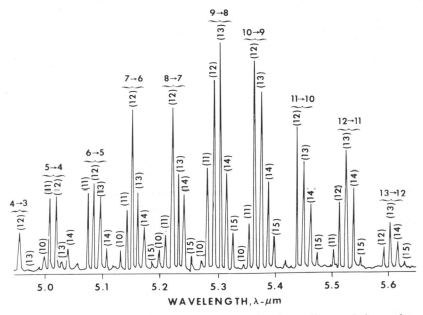

**Fig. 14.** Typical spectrum of laser emission through a 1-mm diameter hole coupler.

flow, the continuous molar flow rate of CO could be determined. From that datum, the efficiency of the conversion of reaction exothermicity to laser emission was established as ~6%.

## 7.  THE CW CO FLAME LASER

One of the more interesting aspects of CO chemical-laser research has been the observation that cw CO laser emission can be achieved from a simple free-burning flame fed with carbon disulfide and oxygen. The development of laser emission directly from a free-burning flame appears to be a unique feature of the $O_2$–$CS_2$ chemistry. Foster and Kimbell[62] first reported the observation of a complete population inversion of certain CO vibrational levels in the low-temperature, low-pressure flame formed from a slot burner fed with $CS_2$, $O_2$, and $N_2$. The inversion among CO vibrational levels from approximately $v = 9$ to $v = 5$ and $v = 3$ to $v = 2$ was determined from detailed spontaneous infrared emission spectra in the 2.6 $\mu$m overtone band observed along an axis positioned ~2.5 cm above and parallel to the burner channel. These same flame spectra, observed without additive (feed $CS_2 : O_2 = 1 : 4$, molar ratio), did not exhibit the inversion apparent with additive (feed $CS_2 : O_2 : N_2 = 1 : 4 : 2$ molar ratio); see Fig. 15. The enhancement of population inversion in the flame system via favorable $V \rightarrow V$ exchange between CO and $N_2$ is suggested by Foster and Kimbell (see Section 3). However, since $N_2$–CO $V \rightarrow V$ exchange proceeds relatively slowly (see Table 3), it is difficult to estimate the degree of influence by that process. Similar relative vibrational-state population-density determinations have been made by Linevsky and Carabetta[63] in which $N_2O$ was found to be an effective additive to induce a population inversion within the $CS_2$–$O_2$ flame. The detailed role of $N_2O$ in this system is not well understood.

Piloff, Searles, and Djeu[64] reported the first cw flame laser that uses the same $O_2$–$CS_2$ combustion system. Used was a multiple tube and screen burner that generated a uniform self-sustaining flame with a 60-cm active optical path. When the burner was fed with separate $CS_2$ and $O_2$ streams, without the benefit of any additive gases, cw-laser emission was stabilized within a low-loss optical cavity to include the 60-cm long flame. Low-power (~1 mW) laser oscillations were observed on three CO V–R transitions:

$$v = 8 \rightarrow v = 7, \quad P(11) \text{ at } 5.216 \ \mu\text{m}$$

$$v = 9 \rightarrow v = 8, \quad P(12) \text{ at } 5.297 \ \mu\text{m}$$

$$v = 11 \rightarrow v = 10, \quad P(10) \text{ at } 5.421 \ \mu\text{m}$$

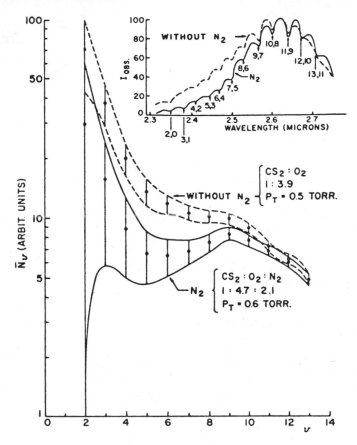

**Fig. 15.** Vibrational-state distributions observed in free-burning $O_2 + CS_2$ flame.

An extensive examination of the $O_2$–$CS_2$ flame laser has been conducted by Linevsky and Carabetta[63] using a 60-cm long honeycomb-faced burner fed through multiple-hole injector tubes supplied from gas manifolds, as shown in Fig. 16. Maximum CO-laser output through a 2% transmitting mirror, ~35-W cw, was achieved in supplying the flame with a feed of $CS_2 : N_2O : O_2 = 1.0 : 216 : 18$ (molar ratio) at a total pressure of ~30 Torr. This laser output represents a chemical-to-optical conversion efficiency of ~2.5% at these conditions. CO laser lines were observed on a cascade of V–R transitions from $v = 12 \rightarrow v = 11$ through $v = 5 \rightarrow v = 4$. An end view of the actual free-burning flame, along the 60-cm long optical path, is presented in Fig. 17, in which the region of maximum gain is indicated. Significant laser output was maintained at total gas pressures

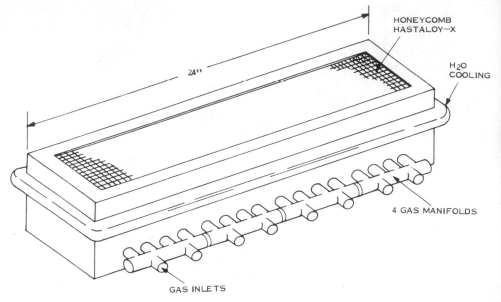

**Fig. 16.** Honeycomb burner used for CO flame laser studies by Linevsky and Carabetta.[63]

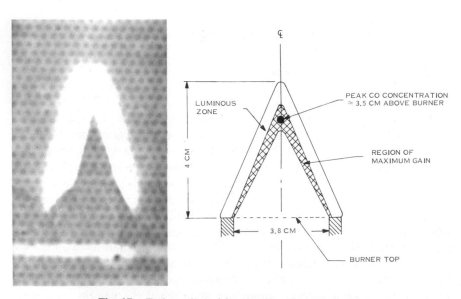

**Fig. 17.** End-on view of free-burning $O_2 + CS_2$ flame.

up to 90 Torr in the flame zone. In general, laser output was observed to scale directly with burner length or gas throughput.

In addition, the influence of selected gas additives was studied, as tabulated in Table 10. As expected, the addition of vibrationally cold molecules (e.g., $N_2O$, CO, and $CO_2$) yielded significant increases in laser output. The mechanism suggested for the observed improvement is that of favorable $V \to V$ transfer (see Section 5). The addition of $NO_2$ to the flame caused the termination of laser output, probably through the rapid consumption of O-atoms[3] by the process

$$O + NO_2 \xrightarrow{k'} NO + O_2$$

where $k' = 1.0 \times 10^{13} \exp(-580/RT)$.

Also investigated by Linevsky and Carabetta[63] was the interesting question of whether alternative carbon-containing fuels could yield vibrational populations in CO within free-burning flames. A listing of the alternative fuels that were studied is presented in Table 11. As noted, no laser activity was achieved for any of the substitute fuels while burning the flames over a wide range of pressures and stoichiometry. Nevertheless, several of these alternative fuels did produce laser output in pulsed electric discharge lasers (see Section 8).

**Table 10.** Measured Influence of Various Molecular Additives on $O_2$–$CS_2$ Flame–Laser Output Power, Determined by Linevsky and Carabetta[63]

| Additive | Effect of additive on $CS_2$–$O_2$ flames | Effect of additive on $CS_2$–$N_2O$–$O_2$ flames |
|---|---|---|
| CO | Doubled power in optimal quantity | Reduced power |
| $SF_6$ | Little or no effect | Little or no effect |
| $C_2N_2$ | Reduced power | Reduced power |
| OCS | Tripled power in optimal quantity | Reduced power |
| NO | Inhibited flame, caused blow-off | Inhibited flame, caused blow-off |
| $C_2H_2$ and other hydrocarbons | Terminated lasing when added in small quantities | Terminated lasing when added in small quantities |
| $CO_2$ | Increased power by factor of 3 or 4 | No effect |
| $SOCl_2$ | Reduced power | Reduced power |
| $H_2S$ | Terminated lasing | Terminated lasing |
| $SO_2$ | Little or no effect | Little or no effect |

**Table 11.** Listing of Fuels Explored by Linevsky and Carabetta[63] as CO Flame-laser Feeds Used in Conjunction with $O_2$ and $O_2$–$N_2O$ Oxidizer Feeds

| Fuel | Maximum laser output, W | Remarks |
|------|--------------------------|---------|
| $CS_2$ | 35 | $O_2$–$N_2O$ oxidizer |
| $C_2N_2$ | none | $T_v \approx 5000°K$ |
| $CH_4$ | none | |
| $C_2H_4$ | none | Strong $H_2O$ absorption |
| $CH_3CCH$ | none | in CO laser-emission |
| $CH_3$–$O$–$CH_3$ | none | region |
| $CHOCH$ | none | |
| $CH_3OH$ | none | |

## 8. PULSED CO CHEMICAL LASERS

Laser oscillation on V–R bands of CO has been pumped by several different chemical reactions developed within numerous different reagent mixtures, as exemplified by Table 12. The first observation of laser emission from chemically pumped CO was reported in 1966 by Pollack[51] who flash-photolyzed gaseous $CS_2$–$O_2$–He mixtures positioned directly within an optical cavity. A few microjoules of output were observed over V–R transitions in CO from $v = 14 \rightarrow 13$ through $v = 6 \rightarrow 5$. The same approach was utilized later by Gregg and Thomas[66] to generate pulsed laser output from various mixtures of $CS_2$–$O_2$ and $CS_2$–air. By inserting a grating in the cavity, 270 separate lasing lines were identified and assigned in tuning through the dispersion of the grating. Interestingly, pulsed laser emission on both $P$- and $R$-branches were observed (see Section 6) on vibrational bands from $v = 16 \rightarrow 15$ through $v = 1 \rightarrow 0$. From detailed, time-resolved spectral measurements, Gregg and Thomas inferred the selectivity of the chemical excitation and anticipated the cw CO flame laser.

Rosenwaks and Yatsiv[14] obtained pulsed laser emission on the V–R transitions of CO formed from the flash photolysis of $CS_2$–$NO_2$–He mixtures. Subsequently, Rosenwaks and Smith[67] developed a pulsed laser by flash-photolyzing $CS_2$–$SO_2$–Ar and $CSe_2$–$SO_2$–Ar mixtures in addition to the familiar $CS_2$–$O_2$ system. When sufficient inert-gas buffer was added to the reactive mixtures to minimize the effects of gas heating, approximately equal pulsed-laser outputs were observed from the three chemically different systems.

Laser emission on CO V–R transitions also has been achieved by pulsed electric-discharge initiation of low-pressure $CS_2$–$O_2$ mixtures. Laser action could be generated by spark initiation through either static mixtures or continuously flowing gases. The flowing systems also allowed repetitive laser-pulse operation.[68,69] Jacobson and Kimbell[69] obtained peak CO laser powers exceeding 1 kW by initiating $CS_2$–$O_2$ inert-gas mixtures at ~30 Torr in a TEA laser apparatus,[70] an example of which is shown in Fig. 18.

Ahlborn, Gensel, and Kompa[73] also reported pulsed CO laser emission generated by transverse multispark discharges through transverse flows of $CS_2$ and $O_2$. The transverse flow allowed ~100 Hz repetition rates yielding 6-W multiline average power along a 1.2-m active path at ~70-Torr cavity pressure. Typical laser energy per pulse was observed to be 0.1 mJ, distributed over about 50 spectral lines from 4.91 to 5.6 $\mu$m. Bashkin and Yuryshev[74] investigated the initiation of low pressure (typically 14 Torr) mixtures by longitudinal electric-discharge pulse along an 80-cm cavity. Maximum laser energy per pulse was observed to be 16 mJ over 4 $\mu$sec, obtained from a $CS_2 : O_2 : He = 1 : 12 : 15$ mixture, initiated by a 4-J electric pulse. In general, these photolytically initiated and spark-initiated pulsed CO chemical lasers exhibit rather low electric input conversion to optical output efficiencies, for example, 0.4%,[74] in comparison to the high efficiencies, for example, >50%,[75] observed in purely electrically excited CO lasers.[76] However, more sophisticated $e$-beam controlled-pulsed discharge techniques have not been applied to the initiation of CO chemical lasers. Such excitation techniques have yielded very high electrical-to-optical conversion efficiencies in pulsed HF chemical lasers.

As shown in Table 12, pumping of CO-laser emission has been achieved in several pulse-initiated chemical systems besides the $O_2$–$CS_2$ reaction chemistry. Lin and Bauer[71] obtained CO laser emission in a band from 5.2 to 5.6 $\mu$m from electrically initiated mixtures of $C_3O_2$–$O_2$–He. In that system, the CO laser emission is thought to be pumped by the reaction

$$C_2O + O_2 \rightarrow 2CO(v) + O, \qquad \Delta H° = -85 \text{ kcal/mol}$$

Further experiments showed that the same chemical system could be employed to achieve cw laser output from a continuous high-voltage dc glow discharge through the gas mixtures, although laser emission then was observed from both CO and $CO_2$ species in the same medium. In separate experiments flash-photolyzing $O_3$–$C_3O_2$ mixtures, Lin and Brus[72] hypothesize that the observed CO laser emission is pumped directly by the electronically excited O-atom

$$O(^1D) + C_3O_2 \rightarrow 3CO(v), \qquad \Delta H° = -160 \text{ kcal/mol}$$

**Table 12.** Reported Pulsed CO Chemical Lasers

| Reagents and diluents | Initiation | Primary initiation and pumping reactions[a] | Exothermicity of pumping reaction $\Delta H°$ [kcal/mol] | Laser transitions[b,c] | Peak laser output[c] | Electrical-to-optical efficiency[c] | Notes | Comments | Ref. |
|---|---|---|---|---|---|---|---|---|---|
| 1. $O_2$–$CS_2$–He | Flash photolysis | $O+CS_2 \rightarrow CS+SO$<br>$O+CS \rightarrow CO^*+S$ | −85 | $P_{14}(12) \cdots P_6(13)$<br>$5.676 \cdots 5.089$ μm | 0.5 W | n.m. | d | First observation of CO chemical laser | 51 |
| 2. $O_2$–$CS_2$–He | Flash photolysis | $O+CS_2 \rightarrow CS+SO$<br>$O+CS \rightarrow CO^*+S$ | −85 | $P_{16}(9) \cdots P_1(9)$<br>$5.805 \cdots 4.745$ μm<br>$R_{15}(28) \cdots R_9(6)$<br>$5.374 \cdots 5.109$ μm | n.m. | n.m. | d | 270 laser lines observed. Intracavity grating used. | 66 |
| 3. $O_2$–$CS_2$–He | Longitudinal spark | $O+CS_2 \rightarrow CS+SO$<br>$O+CS \rightarrow CO^*+S$ | −85 | $P_{13}(13) \cdots P_6(8)$<br>$5.608 \cdots 5.047$ μm | n.m. | n.m. | d | Continuous reagent flow. | 68 |
| 4. $O_2$–$CS_2$–He | Transverse spark[e] | $O+CS_2 \rightarrow CS+SO$<br>$O+CS \rightarrow CO^*+S$ | −85 | $P_{13}(14) \cdots P_2(9)$<br>$5.620 \cdots 4.805$ μm | >1.0 kW | n.m. | d | Continuous reagent flow. Pulse repetition rate ~2 Hz. | 69 |
| 5. $O_2$–$CS_2$ | Transverse spark | $O+CS_2 \rightarrow CS+SO$<br>$O+CS \rightarrow CO^*+S$ | −85 | $P_{12}(19) \cdots P_3(14)$<br>$5.640 \cdots 4.914$ μm | 6 W (average) | 0.05% | | Transverse gas flow. Pulse repetition rate 100 Hz. | 73 |
| 6. $O_2$–$CS_2$–He | Longitudinal spark | $O+CS_2 \rightarrow CO^*+S$<br>$O+CS \rightarrow CO^*+S$ | −85 | n.m. | >4.0 kW | 0.04% | | 4 μsec output pulse width. | 74 |
| 7. $NO_2$–$CS_2$–He | Flash photolysis | $NO_2 \xrightarrow{h\nu} NO+O$<br>$O+CS_2 \rightarrow CS+SO$<br>$O+CS \rightarrow CO^*+S$ | −85 | n.m. | n.m. | n.m. | | Output less than $O_2$–$CS_2$ systems. | 14 |
| 8. $O_2$–$C_3O_2$ | Longitudinal pulsed discharge | $O+C_3O_2 \rightarrow CO+C_2O$<br>$C_2O+O_2 \rightarrow 2CO^*+O$ | −85 | $P_{11} \cdots P_7$ | n.m. | n.m. | | Also reports cw emission. | 71 |
| 9. $O_3$–$C_3O_2$–Ar | Flash photolysis | $O_3 \xrightarrow{h\nu} O(^1D)+O_2$<br>$O(^1D)+C_3O_2 \rightarrow 3CO^*$ | −160 | $P_{13}(13) \cdots P_5(12)$<br>$5.608 \cdots 5.022$ μm | n.m. | n.m. | | | 72 |

| No. | System | Method | Reaction[a] | ΔH (kcal) | Laser transition[b] | | | Notes | Ref. |
|---|---|---|---|---|---|---|---|---|---|
| 10. | $O_3$–XCN–$SF_6$ X≡Cl, Br, I, and CN | Flash photolysis | $O_3 \xrightarrow{h\nu} O(^1D) + O_2$<br>$XCN \rightarrow X + CN$<br>$O(^1D) + CN \rightarrow CO^* + N$ | −164 | $P_{13}(13) \cdots P_5(12)$ 5.633···5.075 μm | n.m. | n.m. | $SF_6$ added to control gas heating. | 72 |
| 11. | $SO_2$–XCN; X≡Cl, Br, and CN | Flash photolysis | $SO_2 \xrightarrow{h\nu} SO + O$<br>$O + CN \rightarrow CO^* + N$ | −74 | | n.m. | n.m. | SF added to control gas heating. | 72 |
| 12. | $SO_2$–$C_2H_2$ | Flash photolysis | $SO_2 \xrightarrow{h\nu} SO + O$<br>$O + C_2H_2 \rightarrow CO^* + CH_2$<br>$O + CH_2 \rightarrow CO^* + 2H$ | −47<br>−75 | | n.m. | n.m. | λ≥1650 Å $SF_6$ added to control gas heating. | 72 |
| 13. | $SO_2$–$CH_2Br_2$ | Flash photolysis | $SO_2 \xrightarrow{h\nu} SO + O$<br>$CH_2Br_2 \rightarrow CH_2 + 2Br$<br>$O + CH_2 \rightarrow CO^* + 2H$ | −75 | | n.m. | n.m. | λ≥1650 Å $SF_6$ added to control gas heating. | 72 |
| 14. | $SO_2$–$CHBr_3$ | Flash photolysis | $SO_2 \xrightarrow{h\nu} SO + O$<br>$CHBr_3 \rightarrow CH + 3Br$<br>$O + CH \rightarrow CO^* + H$ | −176 | $P_{18}(13) \cdots P_2(13)$ 6.031···4.8·4 μm | r.m. | n.m. | λ≥1650 Å $SF_6$ added to control gas heating. | 72 |
| 15. | $SO_2$–$CS_2$–He | Flash photolysis | $SO_2 \xrightarrow{h\nu} SO + O$<br>$O + CS_2 \rightarrow CS + SO$<br>$O + CS \rightarrow CO^* + S$ | −85 | $P_{14} \cdots P_3$ | r.m. | n.m. | Ar a suitable diluent also. | 92 |
| 16. | $SO_2$–$CSe_2$–He | Flash photolysis | $SO_2 \xrightarrow{h\nu} SO + O$<br>$O + CSe_2 \rightarrow CSe + SeO$<br>$O + CSe \rightarrow CO^* + Se$ | −118 | $P_{20} \cdots P_9$ | r.m. | n.m. | | 92 |
| 17. | $O_2$–$CSe_2$–He | Longitudinal pulsed discharge | $O + CSe \rightarrow CO^* + Se$ | −118 | $P_{18} \cdots P_{17}$ | n.m. | n.m. | | 93 |

[a] Species are assumed to exist in ground electronic states unless otherwise identified with parenthetic notation; vibrational excitation is identified by *.

[b] The symbol $P_v(J)$ indicates laser transitions: $(v) \rightarrow (v-1)$, $(J-1) \rightarrow (J)$; $R_v(J)$, $(v) \rightarrow (v-1)$, $(J+1) \rightarrow (J)$; where $v$ is the vibrational quantum number and $J$ is the rotational quantum number of the lower state.

[c] n.m. indicates no reported measurement.

[d] $N_2$ also satisfactory as an inert buffer gas.

[e] "TEA" laser apparatus, after Beaulieu.[70]

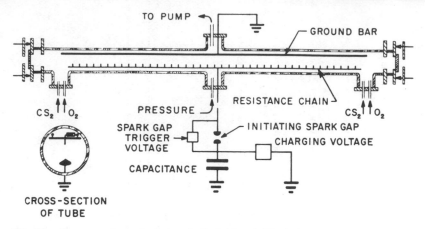

**Fig. 18.** Transversely excited electrically initiated (TEA) CO chemical-pulse laser.

Additional experiments by Lin and Brus[72] efficiently generated $O(^1D)$ and $O(^3P)$ atoms by uv flash photolysis of $O_3$ and $SO_2$. In various reagent mixtures, several new pumping reactions were demonstrated:

$$O + CN \rightarrow CO(v) + N, \qquad \Delta H° = -74 \text{ kcal/mol}$$

$$O + CH_2 \rightarrow CO(v) + 2H, \qquad \Delta H° = -75 \text{ kcal/mol}$$

$$O + CH \rightarrow CO(v) + H, \qquad \Delta H° = -176 \text{ kcal/mol}$$

The latter reaction is sufficiently exothermic to yield CO laser transitions from $v = 18 \rightarrow v = 17$ through $v = 2 \rightarrow v = 1$. Wittig and Smith[92] also followed the flash-photolytic dissociation of $SO_2$ route to produce O-atoms in $SO_2$–$CS_2$ mixtures. This system achieved CO laser transitions similar to those seen in $O_2$–$CS_2$ flash-photolysis pulse lasers. In additional experiments by Wittig and Smith[92] and Rosenwaks and Smith,[93] another new pumping process was demonstrated in pulse-initiated $O_2$–$CSe_2$ and $SO_2$–$CSe_2$ mixtures:

$$O + CSe \rightarrow CO(v) + Se, \qquad \Delta H° = -118 \text{ kcal/mol}$$

This reaction is sufficiently exothermic to yield CO laser transitions from $v = 20 \rightarrow v = 19$ through $v = 9 \rightarrow c = 8$.

The various pulse-initiated systems are interesting as potentially efficient ir lasers and as interesting media in which to study complex exothermic gas-phase reaction kinetics. Probably other new chemical-pumping reactions will be identified in similar pulse-initiated experiments.

## 9. COMPARISON OF CO AND HF–DF LASER SYSTEMS

A comparison of the well-studied cw CO and HF–DF chemical-mixing lasers has been summarized in Table 13. Both of these laser media are pumped by atom–molecule exchange reactions, as shown in line 1 of the table; however, the reaction leading to the production of $CO(v)$ has an exothermicity, $\Delta H_p$, greater than 2.5 times the comparable energy for the reaction producing $HF(v)$, as listed in line 2. While similarly large fractions $f_v$ of the reaction exothermicity are converted into vibration, as shown in line 3, the larger $\Delta H$ associated with the CO pumping reaction

**Table 13.** Comparison of CO and HF–DF Chemical-Laser Systems

|  | CO laser | HF laser |
|---|---|---|
| 1. Pumping reaction | $O + CS \rightarrow CO(v) + S$ | $F + H_2 \rightarrow HF(v) + H$ |
| 2. Exothermicity, $\Delta H_p$ (kcal/mol) | $-85$ | $-31.7$ |
| 3. Fraction of exothermicity converted to vibration, $f_v$ | 0.66 | 0.75 |
| 4. Highest accessible vibrational level, $v_{max}$ | 15 | 3 |
| 5. Rate constant (for Reaction line 1), $k_p$ (cm$^3$/molecule-sec) | $1.4 \times 10^{-11} \,(300°K)$ | $1.77 \times 10^{-11} \,(300°K)$ |
| 6. Dissociation reaction | $O_2 \rightarrow 2O$ | $F_2 \rightarrow 2F$ |
| 7. Dissociation enthalpy, $\Delta H_d$ (kcal/mol) | $+119.1$ | $+37$ |
| 8. Observed laser spectra, $\lambda$ ($\mu$m) | 4.87–5.60 | 2.64–2.87 |
| 9. Operating laser cavity pressure, $p_{cav}$ (Torr) | 5.0–50.0[e] | 5.0[c] |
| 10. Laser efficiency, $\eta$[a] | 0.37 | 0.13[c] |
| 11. Specific-laser energy (total mixture), $\sigma$ (J/g) | 31.0[d] | 86.0[c] |
| 12. Specific-laser energy (diluent free), $\sigma'$ (J/g) | 293[d] | 289[c] |
| 13. $V \rightarrow T$ decay rate,[b] $k_{V \rightarrow T}$ (sec$^{-1}$ Torr$^{-1}$) | 0.7 | $8.7 \times 10^4$ |
| 14. $V \rightarrow V$ exchange rate,[b] $k_{V \rightarrow V}$ (sec$^{-1}$ Torr$^{-1}$) | $6.2 \times 10^4$ | $6.6 \times 10^5$ |

[a] Laser efficiency $\eta$ is defined as the ratio of output energy to energy into vibrational modes of the laser molecule, and, thus, has a maximum of 1.00 when all of the vibrational energy is extracted as optical output.

[b] The rates $k_{V \rightarrow T}$ and $k_{V \rightarrow V}$ are given for self deactivation at 300°K, that is, CO–CO or HF–HF[97] collisions. The $V \rightarrow T$ rates apply to relaxation of the first level, $v = 1$. The $V \rightarrow V$ rates are for the $(2, 0) \rightleftarrows (1, 1)$ exchange, referring respectively to the initial and final vibrational states in the $V \rightarrow V$ exchange collision.

[c] Ref. 77.

[d] Ref. 57.

[e] Refs. 59 and 61.

allows population of higher vibrational states, up to $v_{max}$, as listed in line 4. While CO vibrational population up to the fifteenth level is accessible, significant population in the nascent products exists in levels less than the 10th. The rates of the pumping reactions, at 300°K, are quite similar, as listed on line 5.

The pumping reactions, line 1, are initiated by the free atoms O and F, respectively; see line 6. The heat of dissociation, $\Delta H_d$, for these molecules is given on line 7, which shows that $O_2$ dissociation requires more than three times the enthalpy of $F_2$ dissociation.

The spectral range observed from the CO laser is deeper into the infrared than from the HF laser; see line 8. Typical gas pressures in cw-laser cavities are listed in line 9.

The laser efficiency $\eta$ observed in cw lasers is presented in line 10. Here $\eta$ is defined as the ratio of laser-energy output to the vibrational energy produced by the pumping reactions; thus, $\eta$ has a maximum of 1.00 when all vibration is extracted as laser output. These efficiency levels are reflected in the specific laser energy observed for these media; the laser output achieved divided by the mass of gas mixture involved including reactants and diluents $\sigma$ and, on another basis, divided by the mass of reactants involved free of diluents $\sigma'$. These values are listed on lines 11 and 12.

Typical rates for key molecular energy transfer processes are presented also: line 13, self-deactivation of vibrational quanta into translation at 300°K; and, line 14, the exchange of vibrational quanta at 300°K.

## 10.  CURRENT DIRECTIONS IN CO CHEMICAL-LASER RESEARCH

### 10.1.  Improvements in Laser Performance

Improvements in the performance of the cw CO chemical laser are desirable in: (1) the chemical-to-optical conversion efficiency $\eta$; (2) the specific power $\sigma$, which is the ratio of laser output power to total mass flow rate; and (3) the elevation of the pressure in the laser cavity $p_{cav}$.

Jeffers et al.[78] have shown that significant performance improvements can be found in substituting the feed of semistable CS molecules for $CS_2$ in a cw CO chemical laser. CS flow was prepared from a continuous low-pressure gas discharge through $CS_2$ vapor; the O-atom flow also was produced in a conventional gas discharge. The by-passing of the $O + CS_2 \rightarrow CS + SO$ reaction (see Table 1) via this approach limits the waste-heat addition to the medium, which results in low translational and rotational temperatures in the medium, thereby enhancing laser output.

Several researchers[1,41,62] have demonstrated that higher CO-laser efficiencies can be achieved through the addition to the active medium of molecule species chosen for their favorable $V \rightarrow V$ exchange properties. In varying degrees, CO, $N_2$, and $N_2O$ have been shown to be effective additives.

Improvement in the specific power $\sigma$ of the cw CO chemical laser may be available by reducing the amount of inert diluent added to the medium. (Specific power is defined as the ratio of laser-output power to mass flow rate.) The inert-gas feed, typically helium, appears beneficial in two ways: (1) The diluent absorbs waste heat thereby controlling gas temperature; and (2) the diluent stabilizes the plasma in regimes favorable to the dissociation of $O_2$. This latter effect is of importance in CO chemical-laser systems employing low-pressure gas-discharge initiation. The substitution of aerodynamic cooling[59] for inert-gas buffering appears to be a useful approach to temperature control in the medium. For diffusion-controlled systems, cryogenic cooling of the walls of the laser tube[54] gives sufficient temperature control. Little work, however, has been directed toward the optimization of the production of O-atoms in various dc, rf, or microwave discharges.

## 10.2. Practical Considerations

The absence of an efficient and convenient method for the generation of large quantities of O-atoms remains a major obstacle to the utilization of CO chemical lasers in high-power cw applications. The convenient flame-lasers (see Section 8) generate O directly within the flame; however, the CO flame laser has been demonstrated at efficiencies of only a few percent. A consideration of the thermochemistry of various fuel-oxidizer systems suggested that systems such as $M-OF_2$ (where M is an alkali metal) could yield high concentrations of O at thermochemical equilibrium temperatures between 3000 to 4500°K; whether such systems could be developed practically is an open question. Highly efficient thermal O-atom generation may be attractive for high-power cw applications. Thermal arcs have been demonstrated already[59] in cw CO chemical-laser operation. In any of the purely chemical, thermal, or discharge methods of O-atom generation, higher atom yields may be possible by the use of reagents with weaker O-bonding than $O_2$, for example, $N_2O$ or $O_3$.

One of the inherent advantages of the CO chemical laser is that the vibrationally excited CO-laser species exhibits an exceptionally long lifetime with respect to collisional decay.[25–30] Exploration of cw chemical-laser operation at high cavity pressures >100 Torr has not begun. Such studies, as discussed previously, are somewhat hindered in that high-pressure O-atom sources are not available. However, the thermal-arc technique was employed by Boedeker, Shirely, and Bronfin[59] to achieve cw CO-laser output at $p_{cav} \leq 50$ Torr.

In the sea-level atmosphere, water vapor, and to a lesser extent CO and $CO_2$, induce severe absorption losses in the spectral band from approximately 4.8 to 8.0 $\mu$m. For applications involving transmission of CO-laser radiation over long distances through the atmosphere, shifting the CO chemical-laser transitions to wavelengths of less than 5 $\mu$m is necessary. Such downshifts have been demonstrated in CO chemical lasers by (1) addition of favorable $V \rightarrow V$ exchange molecules,[57] (2) cooling of the medium,[54,59] and (3) single-line selection by use of a grating within the cavity.[57,66]

## 10.3.  Basic Understanding

The chemical kinetics and reaction mechanisms pertinent to the CO chemical laser are not fully understood (as discussed in Section 1). Considerable additional information is needed in the reaction kinetics of the $O_2$–$CS_2$ system over wider temperature ranges, with low temperatures ($<300°$K) being of particular interest. Higher gain and shorter-wavelength emission are favored at low temperatures. While the pumping process in the $O + CS$ reaction appear to be well described (see Section 2), other pumping routes to vibrationally inverted CO have been demonstrated, yet are less clearly understood. Many questions exist as to the possible generality of the population inversions seen in some exothermic reactions yielding CO.

Many of the molecular-energy transfer processes pertinent to the CO laser medium are also not fully understood (as discussed in Section 3). The temperature dependence of many of the important $V \rightarrow V$ and $V \rightarrow T$ processes are not known, particularly in the low-temperature regime. Little data have been collected relating to $V \rightarrow V$ exchange among dissimilar molecular collision partners. Such processes appear to be effective in enhancing CO laser output and downshifting CO laser transitions (as discussed in Section 3).

Interestingly, the CO-laser medium is being researched actively both as a chemically pumped laser and as a purely electrically pumped laser. Data from these different experimental areas are often mutually useful, leading to continuing progress in the understanding of both these systems. The CO chemical laser is an inherently efficient laser device with interesting and advantageous characteristics.

## ACKNOWLEDGMENTS

The authors acknowledge the interesting discussions and helpful suggestions of Dr. D. P. Ames, Dr. J. D. Kelley, Dr. T. J. Menne, and Dr. H. T. Powell of the McDonnell Douglas Research Laboratories; Dr. R. Tripodi and Dr. C. J. Ultee of the United Technologies Research Center; Dr. M. J. Linevsky of the General Electric Co., Space Sciences Laboratory; Professor C. Wittig of the Electrical Engineering Department at the University of Southern California; and Dr. N. Djeu of the U.S. Naval Research Laboratory.

## REFERENCES

1. R. D. Suart, P. H. Dawson, and G. H. Kimbell, J. Appl. Phys. **43,** 1022 (1972).

2. K. Schofield, J. Phys. Chem. Ref. Data **2,** 25 (1973).

3. K. H. Homann, G. Krome, and H. Gg. Wagner, Ber. Bunsenges. Phys. Chem. **72,** 998 (1968); **74,** 654 (1970).

4. A. Westenberg and N. deHaas, J. Chem. Phys. **50,** 707 (1969).

5. G. Hancock, C. Morley, and I. W. M. Smith, Chem. Phys. Lett. **12,** 193 (1971)

6. R. W. Fair and B. A. Thrush, Trans. Faraday Soc. **65,** 1557 (1969).

7. D. D. Davis, R. B. Klemm, and M. Pilling, Int. J. Chem. Kinet. **4,** 367 (1972).

8. G. Hancock and I. W. M. Smith, Trans. Faraday Soc. **67,** 2586 (1971).

9. D. B. Sheen, J. Chem. Phys. **52,** 648 (1970).

10. C. J. Halstead and B. A. Thrush, Proc. Roy. Soc. London Ser. A **295,** 363 (1966).

11. N. Cohen and R. W. F. Gross, "Temperature Dependence of Chemiluminescent Reactions V3: The $SO_2$ Afterglow," TR-100 (2240-20)-6, Vol. III (The Aerospace Corp., El Segundo, California, 1968).

12. H. S. Johnston, "Gas Phase Reaction Kinetics of Neutral Oxygen Species," Nat. Stand. Ref. Data Ser. *NSRDS-NBS* **20,** (Sept. 1968).

13. J. C. Polanyi, Appl. Optics, Suppl. **2,** Chemical Lasers, 109 (1965).

14. S. Rosenwaks and S. Yatsiv, Chem. Phys. Lett. **9,** 266 (1971).

15. I. W. M. Smith, Disc. Faraday Soc. **44,** 194 (1967).

16. I. W. M. Smith, Trans. Faraday Soc. **64,** 378 (1968).

17. H. Okabe, J. Chem. Phys. **56,** 4381 (1972).

18. D. L. Hildenbrand, Chem. Phys. Lett. **15,** 379 (1972).

19. S. Tsuchiya, N. Nielsen, and S. H. Bauer, J. Phys. Chem. **77,** 2455 (1973).

20. H. T. Powell and J. D. Kelley, J. Chem. Phys. **60,** 2191 (1974).

21. F. Kaufman, *Progress in Reaction Kinetics* (Pergamon, New York 1961).

22. T. L. Cottrell and J. C. McCoubrey, *Molecular Energy Transfer in Gases* (Butterworths, London, 1961).

23. P. Borrell, Adv. Molecular Relaxation Processes **1,** 69 (1967–68).

24. W. H. Green and J. K. Hancock, J. Chem. Phys. **59,** 4326 (1973).

25. R. C. Millikan, J. Chem. Phys. **40,** 2594 (1964).

26. J. T. Yardley, J. Chem. Phys. **52,** 3983 (1970).

27. R. C. Millikan, J. Chem. Phys. **38,** 2855 (1963).

28. G. Hancock and I. W. M. Smith, Appl. Opt. **10,** 1827 (1971).

29. W. J. Hooker and R. C. Millikan, J. Chem. Phys. **38,** 214 (1963).

30. M. G. Ferguson and A. W. Reed, Trans. Faraday Soc. **61,** 1559 (1965).

31. R. E. Center, J. Chem. Phys. **58,** 5230 (1973).

32. C. B. Moore, Acct. Chem. Res. **2,** 104 (1969).

33. H. T. Powell, J. Chem. Phys. **59,** 4937 (1973).

34. J. C. Stephenson, Appl. Phys. Lett. **22,** 576 (1973).

35. I. W. M. Smith and C. Wittig, Trans. Faraday Soc. **69,** 939 (1973).

36. P. F. Zittel and C. B. Moore, Appl. Phys. Lett. **21,** 81 (1972).

37. M. A. Kovacs, J. Chem. Phys. **58,** 4704 (1973).

38. D. D. Edey and J. E. Mayer, *International Encyclopedia of Physical Chemistry and Chemical Physics* (Pergamon, London 1967).

39. H. T. Powell, private communication.

40. R. D. Suart, S. J. Arnold, and G. H. Kimbell, Chem. Phys. Lett. **7,** 337 (1970).

41. W. Q. Jeffers and C. E. Wiswall, Appl. Phys. Lett. **23,** 626 (1973).

42. L. A. Young and W. J. Eachus, J. Chem. Phys. **44,** 4195 (1966).

43. G. Herzberg, *Molecular Spectra and Molecular Structures,* 2nd ed. (Van Nostrand, Princeton, New Jersey, 1950).

44. D. Williams et al., Mol. Phys. **20,** 769 (1971).

45. L. S. Bender, R. Roback and B. R. Bronfin, Report No. M911176 (United Aircraft Research Laboratories, East Hartford, Connecticut, March 1973).

46. W. Q. Jeffers and J. Daniel Kelley, J. Chem. Phys. **55,** 4433 (1971).

47. J. D. Kelley, J. Chem. Phys. **56,** 6108 (1972).

48. K. J. Herzfeld and T. A. Litovitz, *Absorption and Dispersion of Ultrasonic Waves* (Academic, New York, 1959).

49. G. Emanuel, J. Quant. Spectrosc. Radiat. Transfer **11,** 1481 (1971).

50. G. Hancock and I. W. M. Smith, Chem. Phys. Lett. **3,** 573 (1969).

51. M. A. Pollack, Appl. Phys. Lett. **8,** 257 (1966).

52. C. Wittig, J. C. Hassler, and P. D. Coleman, Appl. Phys. Lett. **16,** 117 (1970); Nature **226,** 845 (1970).

53. Y. Hirose, J. C. Hassler, and P. D. Coleman, IEEE J. Quantum Electron. **QE-9,** 114 (1973).

54. J. D. Barry, W. E. Boney, and J. E. Brandelik, Appl. Phys. Lett. **18,** 15 (1971); IEEE J. Quantum Electron. **QE-7,** 208 (1971); Appl. Phys. Lett. **19,** 141 (1971); J. D. Barry et al., Appl. Phys. Lett. **20,** 243 (1972).

55. K. D. Foster, J. Chem. Phys. **57,** 2451 (1972).

56. C. J. Ultee and P. A. Bonczyk, IEEE J. Quantum Electron. **QE-10,** 105 (1974).

57. W. Q. Jeffers and C. E. Wiswall, Appl. Phys. Lett. **23,** 626 (1973).

58. W. Q. Jeffers and J. D. Kelly, J. Chem. Phys. **55,** 4433 (1971).

59. L. R. Boedeker, J. A. Shirley, and B. R. Bronfin, Appl. Phys. Lett. **21,** 247 (1972).

60. K. L. Wray, *10th Symposium (International) on Combustion* (Combustion Institute, Pittsburgh, Pennsylvania, 1965), pp. 523–37.

61. L. R. Boedeker and A. E. Mensing, Report No. M911176 (United Aircraft Research Laboratories, East Hartford, Connecticut, March 1973).

62. K. D. Foster and G. H. Kimbell, J. Chem. Phys. **53,** 2539 (1970).

63. M. J. Linevsky and R. A. Carabetta, Report No. 73SD4223 and 73SD4278 (Space Sciences Laboratory, General Electric Co., Philadelphia, Pennsylvania, 1973); Appl. Phys. Lett. **22,** 288 (1973).

64. H. S. Pilloff, S. K. Searles, and N. Djeu, Appl. Phys. Lett. **19,** 9 (1971).

65. S. K. Searles and N. Djeu, Chem. Phys. Lett. **12,** 53 (1971).

66. D. W. Gregg and S. J. Thomas, J. Appl. Phys. **39,** 4399 (1968).

67. S. Rosenwaks and I. W. M. Smith, Trans. Faraday Soc. **69,** 1416 (1973).

68. S. J. Arnold and G. H. Kimbell, Appl. Phys. Lett. **15,** 351 (1969).

69. T. V. Jacobson and G. H. Kimbell, J. Appl. Phys. **41,** 5210 (1970).

70. J. A. Beaulieu, Appl. Phys. Lett. **16,** 504 (1970).

71. M. C. Lin and S. H. Bauer, Chem. Phys. Lett. **7,** 223 (1970).

72. M. C. Lin and L. E. Brus, J. Chem. Phys. **54,** 5423 (1971); J. Phys. Chem. **75,** 1425 (1972); Int. J. Chem. Kinet. **5,** 173 (1973); **6,** 1 (1974).

73. B. Ahlborn, P. Gensel, and K. L. Kompa, J. Appl. Phys. **43,** 2487 (1972); Report No. IV/13 (Max-Planck-Institute für Plasmaphysik, Garching, Germany, March 1971).

74. A. S. Bashkin and N. N. Yuryshev, Sov. J. Quant. Electron. **2,** 499 (1973), English transl.

75. M. M. Mann, M. L. Bhaumik, and W. B. Lacina, Appl. Phys. Lett. **16,** 430 (1970); M. L. Bhaumik, Appl. Phys. Lett. **17,** 188 (1970); M. M. Mann, D. K. Rice, and R. G. Eguchi, "Experimental Investigation of High Energy CO Lasers," 8th International Quantum Electronics Conference, San Francisco, California, June 1974.

76. N. N. Sobolev and V. V. Sokovikov, Sov. J. Quant. Electron. **2,** 305 (1973), English transl.

77. H. Mirels and D. J. Spencer, IEEE J. Quant. Electron. **QE-7,** 501 (1971).

78. W. Q. Jeffers et al., Appl. Phys. Lett. **22,** 587 (1973).

79. A. B. Callear, Proc. Roy. Soc. London Ser. A **276,** 401 (1963); A. B. Callear, "Energy Transfer in Molecular Collisions," *Photochemistry and Reaction Kinetics*, P. G. Ashmore, F. S. Dainton and T. M. Sugden, Eds. (University Press, Cambridge, England, 1967), Chap. 7; A. B. Callear and I. W. M. Smith, Nature **213,** 381 (1967).

80. C. E. Treanor, J. W. Rich, and R. G. Rehm, J. Chem. Phys. **48,** 1798 (1968).

81. B. F. Gordiets, A. I. Osipov, E. V. Stapochenko, and I. A. Shelepin, Sov. Phys. Usp. **15,** 759 (1973).

82. Reviewed by: D. W. Howgate and T. A. Barr, Jr., J. Chem. Phys. **59,** 2815 (1973).

83. H. Gueguen, I. Arditi, M. Margottin-Maclou, and L. Henry, Comp. Rend. Acad. Sci. Paris, **272,** 1139 (1971).

84. N. Djeu, to be published.

85. W. Q. Jeffers and H. T. Powell, "Carbon Monosulfide and Probe Laser Research," Report No. MDC Q0498 (McDonnell Douglas Research Laboratories, St. Louis, Missouri, July, 1973).

86. R. J. Donovan and D. J. Little, Chem. Phys. Lett. **13,** 488 (1972).

87. Estimated by S. H. Bauer, as reported in Ref. 8.

88. J. L. McCrumb and F. Kaufman, J. Chem. Phys. **57,** 1270 (1972).

89. C. J. Halstead and B. A. Thrush, Photochem. Photobiol. **4,** 1007 (1965); Proc. Roy. Soc. London Ser. A **295,** 363 (1966).

90. Recent measurements recommended in preference to those presented in Ref. 82: R. Simonaitis and J. Heicklen, J. Chem. Phys. **56,** 2004 (1972).

91. S. W. Benson, *The Foundation of Chemical Kinetics* (McGraw-Hill, New York, 1960), p. 402. Presented as $k_{-24}$.

92. C. Wittig and I. W. M. Smith, Appl. Phys. Lett. **21,** 536 (1972).

93. S. Rosenwaks and I. W. M. Smith, J. Chem. Soc.—Faraday II **69,** 1416 (1973).

94. In these calculations, the activation energies for both $k_1$ and $k_2$ have been set at 1.9 kcal/mol; otherwise, the data of Tables 1 through 5 were used.

95. D. R. Stull and H. Prophet, JANNAF Tables of Thermochemical Data, Natl. Stand. Ref. Data Sec. **37** (1972).

96. J. K. Hancock and W. H. Green, J. Chem. Phys. **57,** 4515 (1972).

97. W. Q. Jeffers, Appl. Phys. Lett. **21,** 267 (1972).

# CHAPTER 12

# THE PHOTOCHEMICAL IODINE LASER

**KRISTIAN HOHLA**

**KARL L. KOMPA**
**Max-Planck-Institut für Plasmaphysik, Euratom Association,**
**Garching, Germany**

In 1938 Porret and Goodeve[1] analyzed the uv absorption of alkyl iodides and concluded that photolysis of $CH_3I$ should give mainly $^2P_{1/2}$ iodine atoms. In 1964 Kasper and Pimentel[2] observed stimulated emission signals in the flash photolysis of $CF_3I$ and $CH_3I$, which they interpreted as laser action caused by the following scheme:

$$CX_3I + h\nu \rightarrow CX_3 + I(^2P_{1/2}), \qquad X = H, F \tag{1}$$

$$I(^2P_{1/2}) \rightarrow I(^2P_{3/2}) + h\nu_0 \qquad (\lambda_0 = 1.315 \ \mu m) \tag{2}$$

The emission was not discovered in an ordinary-laser cavity but in a Raman-type multiple reflection cell. The laser exhibited a very high gain even in these first experiments, and an amplification of 3.4 over only a 5-cm active length was measured. In 1965, the study of this laser was extended to several other alkyl iodides.[3] The most intense emission was found with the fluorinated compounds, and no laser action was obtained in the photolysis of $i$-$C_3H_7I$, HI, or $I_2$. Not only did isopropyl iodide and molecular iodine not produce laser emission, but these compounds even acted as quenchers when added to other photolysis mixtures. The reason for this has still not become clear, as will be seen later. In 1966 the pressure dependence of the laser output was investigated, and a first attempt was made to provide a model interpretation of the data.[4] Also in 1966 Donovan and Husain[5] published the first of a series of papers studying, by kinetic spectroscopy in the vacuum uv, the production and decay of $I^2(P_{1/2})$ in $CF_3I$ photolysis. This work provided many of the collisional deactivation-rate constants. In the same year it was shown by DeMaria and Ultee[6] that this laser could be scaled to energies as high as 65 J.

Starting in 1968 a series of Russian papers began to appear. Basov et al.[7] found that the final amount of $I_2$ formed in $C_3F_7I$ photolysis when no cavity mirrors were present was 2.4 times larger than that measured inside a cavity with two 100% reflecting mirrors. They concluded that the $I_2$ formation was related to reactions of excited iodine atoms and suggested as a possible explanation the reaction

$$I^* + C_3F_7I \rightarrow C_3F_7 + I_2 \tag{3}$$

where $I^*$ represents the excited iodine $^2P_{1/2}$ state. Andreeva et al.[8] considered the possibility of obtaining excited iodine atoms as a result of chemical reactions of the type

$$CF_3 + CF_3I \rightarrow C_2F_6 + I^* \tag{4}$$

They found that the extent of photolytic decomposition varied inversely with inert-gas pressure and concluded that only hot $CF_3$ radicals play a role in Reaction (4). The photolysis of $CF_3I$ can even generate fluorine atoms: This was shown by using $CF_3I$ as the fluorine source to pump a hydrogen-fluoride chemical laser.[9] The pyrolytic decomposition of $CF_3I$ during flash photolysis was examined by Zalesskii and Moskalev[10] by kinetic-absorption spectroscopy of $CF_3I$. Pyrolysis of $CF_3I$ appears to be the reason for a sudden increase in its decomposition rate at some point during the photolysis. The increase, however, disappeared if a large excess of argon was added or if an iodide with a larger specific heat like $C_3F_7I$ were used. If excessive heating is avoided, the main reason for termination of the laser pulse is collisional quenching of the excited iodine atoms, mainly by $I_2$. It became apparent, as more experimental laser studies were done, that the reaction mechanism is indeed more complicated than Reactions (1) and (2) suggest. Small-signal gain measurements during and after the photolysis[11] showed that the maximum amplification is reached some time after the light input has terminated. This effect cannot be described as resulting exclusively from the photolytic reaction (1). Two explanations have been proposed, including additional chemical pumping[12] and diffusion of the excited iodine.[13] Furthermore, under some conditions, the dissociation of certain iodides seems to be largely reversible.[14] In this way, the same photolysis mixture can be used many times. These remarks indicate that our knowledge of the kinetic processes in this laser is still rather incomplete. Kinetic modeling studies have been published by Hohla and Kompa,[12] O'Brien and Bowen,[15] Turner and Rapagnani,[16] Zalesskii,[17] Andreeva et al.,[18] and Franklin.[19] None of these studies, however, could account for all of the observed features already mentioned. Nevertheless, as will be shown in this paper, in spite of these insufficiencies, conditions have been found under which efficient laser operation is possible.

Interest in the technical development of this laser only appeared rather late. The first $Q$-switching and mode-locking experiment was done by Ferrar.[20] Splitting and shifting of the laser line by Zeeman effect were investigated by several authors with the aim to increase the energy storage of iodine laser amplifiers,[21,22] It is now believed, however, that the necessary line broadening can be more easily and simply achieved by adding a foreign gas. The total transition probability was determined in several studies, and the hyperfine splitting was analyzed.[23,45] This will be discussed in more detail in the next section. Starting in 1970 a systematic study of the high-power potential of this laser was conducted by Hohla, Gensel, and Kompa.[24,25] This work has led to the scaling of the laser to the 100-J range for pulse durations of 1 nsec. Somewhat later, similar studies were also begun, particularly in Russia.[26] The growing interest in this laser has been motivated largely by the need of very high laser powers for experiments directed toward controlled nuclear fusion.

## 1. CHEMICAL KINETICS OF THE IODINE-LASER SYSTEM

In a first approach, one might look at the primary and secondary processes in this laser, as are shown in Fig. 1. On the basis of this scheme, one would expect the initial I* concentration to be determined by the photolysis light input and to be controlled at later times by collisional deactivation. The competing role of stimulated emission and collisional quenching is apparent, especially for iodine-laser amplifiers where energy storage is necessary for some time before the oscillator signal passes through the gain medium. The simple scheme of Fig. 1, however, is not capable of explaining many of the finer details of the laser operation. A more specific discussion is thus intended in this section, which will be organized in such a way that the spectroscopy of the parent iodides, problems of quantum yield and branching ratio of I*/I, and the time development of the gain are treated first. The next paragraph deals with problems of the self-regeneration of the laser medium. Finally, a set of recommended rate constants for the various collisional processes is discussed, and the results of numerical modeling studies are mentioned.

### 1.1. Primary Photolysis

Iodine-laser pumping is initiated through the absorption of photons by the parent iodides in the near-uv range of the spectrum. Absorption bands and cross-sections for three relevant perfluoro-alkyliodides are shown in Fig. 2.[34] The excitation involves the promotion of a nonbonding electron to an antibonding orbital of the iodine-carbon bond. No potential-energy level diagram has been constructed directly for these compounds that allows one to relate the primary absorption to any

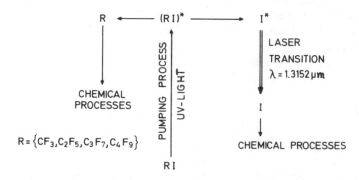

**Fig. 1.** Principal mechanisms in the iodine laser.

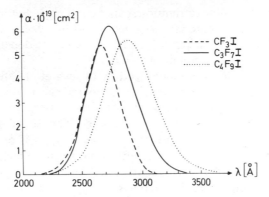

**Fig. 2.** Absorption bands of perfluoro-alkyliodides.

subsequent photochemical processes. However, some information can be obtained by analogy from the corresponding diagram for HI that has been compiled by Mulliken (Fig. 3).[27] It is seen from Fig. 3 that the excitation in this wavelength range can go into several states, all totally repulsive in nature. If this diagram also applies to the iodides used in the laser, dissociation is expected to occur with a primary quantum yield of unity under all conditions in the gas-phase photolysis. The apparent overall

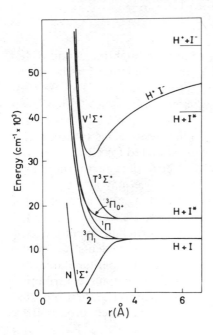

**Fig. 3.** Molecular-energy levels of hydrogen-iodide (after Mulliken[27]).

quantum yield of iodine formation can be much lower as a result of recombination of the photolysis fragments and has been determined, for instance, by Dacey[28] in cw photolysis to be (pressure-dependent) 0.02 to 0.13. For flash-photolysis conditions values of $\Phi = 0.43$ to 0.68 depending on mixture composition were measured by Ogawa et al.[29] for the yield of formation of $CF_3$ in $CF_3I$ photolysis. All this suggests that regenerative processes for the parent molecules can effectively operate in this photolysis system at least under some conditions. The scheme of potential curves of Fig. 3 for HI is not directly applicable to $CF_3I$ or $C_3F_7I$. Figure 3 shows that either $^2P_{1/2}$ or $^2P_{3/2}$ iodine could result, which is indeed the case for HI. On the other hand, for the alkyl iodides, there is fairly strong experimental evidence that the higher state is formed predominantly.[30,35] It is conceivable that the respective curves for $CF_3I$ and similar compounds are not to be drawn parallel but should cross at some point. This would mean that, as the excited molecules move on the potential curve in the direction of separation, they cross over to the path that leads to I* formation. There has been some argument in the literature concerning the branching ratio of I*/I formation and the possibility of additional chemical-pumping contributions in the iodine laser.[8,11-13] Such effects are conceivable directly through Reaction (5) or indirectly through Reaction (6), which removes ground-state iodine:

$$CF_3^{\ddagger} + CF_3I \rightarrow C_2F_6 + I^* \quad (?) \tag{5}$$

$$CF_3 + I \rightarrow CF_3I \tag{6}$$

A contribution to I* formation by Reaction (5) through the action of hot $CF_3$ radicals ($CF_3^{\ddagger}$) has been suggested as the explanation for the difference in laser output with and without thermalization of $CF_3$.[8] Besides, an apparent quantum efficiency slightly above unity, $\Phi_{I^*} > 1$, has been found by various authors.[22,31] If the value of $\Phi_{CF_3} = 0.68$ of Ogawa et al.[29] is accepted, the I* production in $i$-$C_3F_7I$ photolysis seems to occur with $\Phi_{I^*} \leq 1.2$.[31] A similarly high figure has been estimated for $CF_3I$ photolysis from output measurements based on radiometric data of the flashlamp input to the laser medium.[22] Experimental evidence may also be seen in (1) the appearance of an emission signal at times long after completion of photolysis in lasers that are operated very near threshold, (2) in the delayed-energy maximum in time-dependent $Q$-switching experiments,[22] and (3) in time-dependent small-signal gain measurements.[12] On the other hand, the precision of these rather inconsistent experimental results does not warrant a clear conclusion on a chemical-pumping contribution to this laser. For the late-time gain maximum, for instance, first reported by Hohla et al.,[11,22] an alternative explanation has been proposed by

Gusinow et al.[13] ascribing the effect of diffusion. Again the experimental conditions are not strictly comparable, and it appears that this point has to await further experimental clarification.

The uv absorption of $CF_3I$ involves photon energies between 122 kcal/mol (230 nm) and 95 kcal/mol (300 nm).[32] This gives an average excitation energy of 107 kcal/mol. Of this absorbed energy, 54 kcal/mol are consumed in breaking the $CF_3$–I bond,[33] and 22 kcal/mol are carried away by the electronically excited I*; the reaction is, therefore, 31 kcal/mol exothermic. If this exothermicity were deposited in translational energy of the products it would be distributed according to momentum conservation: ~20 kcal to $CF_3$ and ~11 kcal to I*. This ratio changes if part of the excess energy is taken up in internal modes of the $CF_3$ radicals. Secondary reactions involving hot radicals are, therefore, probable, unless effective thermalization occurs. Mass-spectrometric analysis of $CF_3I$ photodissociation with no diluent or other means of thermalization has been reported to give $CF_4$ (70%), $C_2F_6$ (8–8.5%) and $C_2F_4$ (5%).[17] Such a situation is reminiscent of the gamma-radiolysis of $CF_3I$ where only $CF_4$, $I_2$, and $C_2F_2I_2$ (no $C_2F_6$) have been identified as final products.[36] However, even if hot-radical reactions can be excluded, the kinetic mechanisms describing the laser behavior are quite complex. The temperature rise in $CF_3I$ photolysis may be estimated on the basis of the known specific heat of this molecule, $\bar{C}_v^{300-1000°K} = 17$ cal/deg-mol, $C_v^{600°K} = 20$ cal/deg-mol.[37] A rough calculation shows that the temperature can rise to >1000°K because of the heat released from the dissociation even for a moderate light input, if no inert-gas diluent is added. The thermal explosion of $CF_3I$ seems to start at ~1200°K.[32] The exothermic chain decomposition of $CF_3I$ occurs through the reaction of $CF_3$ radicals and iodine atoms with $CF_3I$, and the overall process is described by

$$2CF_3I \rightarrow C_2F_6 + I_2, \qquad \Delta H = -21 \text{ kcal/mol} \qquad (7)$$

In analyzing the onset of pyrolysis, Parker[32] found that reaction

$$I + CF_3I \rightarrow CF_3 + I_2 \qquad (8)$$

which is exothermic by about 3 kcal/mol, must contribute to the decomposition even at relatively low temperatures. The half-time for pyrolysis of $CF_3I$ according to Eq. (7) is 1 sec at 900°K, 600 $\mu$sec at 1200°K, and 2.4 $\mu$sec at 1600°K.[32]

The linewidth of the laser transition must be broadened to permit sufficient energy storage in iodine high-power amplifiers. This can be accomplished most conveniently by the addition of inert diluents at or above atmospheric pressure. However, this poses two kinetic problems: One is the danger of increased collisional deactivation of I*, and the other

**Table 1.** Broadening Coefficients $\beta_M$, Rate Coefficients $k_M$ for Collisional Deactivation of I*, Tolerable Pressures at Which the Deactivation Reaches 10% in 10 $\mu$sec, and Cross-sections $\sigma$

| M | $\beta_M/10^{15}$, cm$^{-2}$ Torr$^{-1}$ | Reference | $k_M/10^{-16}$, cm$^3$/sec | Reference | $P_{10}$ Torr | $\sigma/10^{-19}$ cm$^2$ for $P_{10}$ | $\sigma/10^{-19}$ cm$^2$ for 700 Torr |
|---|---|---|---|---|---|---|---|
| He | 3.1 | 41 | 0.02 | 5 | 170,000 | | 4.6 |
| Ar | 3.8 | 41 | | | | | 3.8 |
| | 3.6 | 44 | | | | | 4.0 |
| | 4.1 | 45 | | | | | 3.5 |
| N$_2$ | 3.55 | 38 | 0.02 | 5 | 170,000 | | 4.0 |
| CO | 5.1 | 38 | 2 | 5 | 1,700 | 1.15 | 2.8 |
| CO$_2$ | 6.3 | 38 | 12 | 5 | 280 | 5.7 | 2.0 |
| | 7.0 | 38 | 1.5 | | 2,200 | 0.65 | 1.95 |
| | 7.3 | 38 | 4.6 | | 700 | | |
| SF$_6$ | 5.0 | 38 | 0.24 | 5 | 14,000 | | 2.9 |
| CF$_2$Cl$_2$ | 11.3 | 38 | 25 | 38 | 135 | 6.6 | |
| CF$_3$I | 8.7 | 38 | 2.2 | 43 | 1,500 | 0.77 | 1.64 |
| | | | 0.65 | 43 | | | |
| i-C$_3$F$_7$I | 15.5 | 38 | 8.0 | | 420 | 1.5 | |
| CF$_3$Br | 5.4 | 38 | —[a] | | (600)[a] | 3.1 | |
| (CF$_3$)$_2$CO | 14.4 | 38 | —[a] | | (350)[a] | 2.0 | |
| I,I* | 12.5 | 38 | | | | | |

[a] Nonexponential decay is found with CF$_3$I (see Ref. 38).

problem is the increased rate of three-body recombination of I*. However, effective pressure broadening can be accomplished by a suitable compromise between broadening and collisional quenching. Rather comprehensive data on I* collisional deactivation have been collected by Donovan and Husain.[5] Fuss[38] has reexamined and partly completed these data with respect to high-power laser operation (Table 1). Under high-pressure conditions, the quenching by impurities like $O_2$ may be more severe than quenching by the broadening additive.

## 1.2. Self-Regeneration of the Laser Medium

The primary process in the photolysis of perfluoro-alkyliodides is the splitting of the carbon–iodine bond, other modes of dissociation (e.g., carbon–fluorine bond cleavage) being negligible. Among the various alkyl iodides, $i$–$C_3F_7I$ is now preferred primarily because of its broad uv absorption, which yields a good efficiency of conversion of the pump light into laser radiation. If reproducibility of the laser output is of concern, the laser gas is exchanged after each shot. Under typical conditions, roughly 10% of the gas is photolyzed each time. In 1972, Hohla and Kompa discovered that nearly total regeneration of the laser gas can occur under certain conditions. This means that the same filling can be used so often that each iodine atom contributes more than one time to the laser operation. The effect has been investigated in detail by Fuss et al.[14] While laser operation in this fashion has long been employed in oscillators, it can be used in amplifiers only with some sacrifice in reproducibility, since the amplification depends critically on the stored energy.

An explanation for this effect can be found, if regeneration of the parent iodide because of the reaction of a radical R with molecular iodine formed in the preceding experiment

$$R + I_2 \rightarrow RI + I \tag{9}$$

is assumed.[14] Consequently, more RI should be regenerated if additional radicals are formed in the reaction volume. This has been shown by the addition of hexafluoro-acetone as a photolytic $CF_3$ source[39]:

$$(CF_3)_2CO \overset{h\nu}{\rightarrow} 2\,CF_3 + CO \tag{10}$$

The effect of this additive is particularly noticeable for $CF_3I$, which normally exhibits only a weak tendency for self-regeneration (Fig. 4).

If the molecular iodine is removed after each shot by circulating the laser gas through a cold trap, the regenerating Reaction (9) is eliminated. Surprisingly, under these conditions, some self-regeneration still exists for

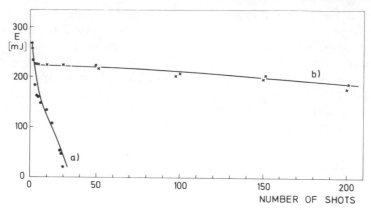

**Fig. 4.** Self-regeneration in a $CF_3I$ photodissociation laser: (*a*) without and (*b*) with $(CF_3)_2CO$ as an additional $CF_3$ radical source. The decrease in energy in the first shots is caused by the accumulation of $I_2$.

$C_3F_7I$ (Fig. 5). To explain this observation, it can be assumed[39] that $C_3F_7$, unlike smaller radicals, dimerizes only slowly, thus existing in the mixture long enough to react finally with ground-state iodine atoms in a reaction analog to Reaction (6). This assumption is also supported by the finding that small radicals regenerate the parent iodide according to Reaction (6) if ground-state iodine is produced rapidly by stimulated emission. Figure 5 shows that this process may also contribute under conditions where the

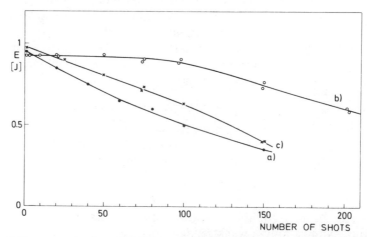

**Fig. 5.** Self-regeneration in a $C_3F_7I$ photodissociation laser with removal of $I_2$ after each shot: (*a*) with added $SF_6$ as a heat bath to avoid further pyrolytic decomposition, (*b*) without added $SF_6$ and with laser energy extraction, (c) without added $SF_6$ and without laser energy extraction.

primary $C_3F_7$ radicals decompose pyrolytically. The yield of iodides that are capable of I* formation (not necessarily $C_3F_7I$ alone) in subsequent photolysis reactions depends then critically on the rapid formation of ground-state iodine atoms by laser emission. Through this effect, the earlier observation of Basov et al.[7] that the yield of $I_2$ is definitely different with and without stimulated emission (see previous discussion) may be partly explainable.

## 1.3. Kinetic Modeling Studies

Kinetic analyses have been reported by O'Brien and Bowen,[15] Zalesskii and Venediktov,[40] Zalesskii,[17] Andreeva et al.,[18] Hohla and Kompa,[12] and Franklin.[19]. The most recent and comprehensive study is that of Turner and Rapagnani,[16] and we will base the following discussion on the work of these authors. This study, like the majority of the other experimental and computational papers, is motivated by the interest in the high-power development of this laser, particularly for laser-fusion experiments. This poses some boundary conditions on calculations with respect to the pressure range of interest. Since this application calls for very high energies in very short pulses, three requirements have to be met: (1) The gain must be reduced to permit sufficient energy storage; (2) to obtain short pulses, a correspondingly large band width of the laser transition is necessary; and (3) for good energy extraction, it is desirable to have sufficient overlap of the frequency profiles of the various hyperfine components (see the following discussions). Thus a high foreign-gas pressure has to be used for pressure broadening, which in turn raises questions of collisional deactivation, not only by the broadening gas but also by impurities unavoidable under these conditions and of termolecular reactions not encountered in low-pressure operation. The goal of this study was to investigate the pressure scaling and the influence of three-body reactions, and to predict the energy storage and inversion history at high foreign-gas pressures. The results were compared to experimental data obtained by Hohla and Kompa[12] and by Aldridge.[41] The rate constants were taken from the literature or, where not directly available, from a fit to the experimental results mentioned.

The particles considered in the modeling are I, I*, $CF_3$ or $C_3F_7(R)$, $CF_3I$, or $C_3F_7I(RI)$, $C_2F_6$ or $C_6F_{14}(R_2)$, $I_2$, and He. The types of reactions that occur between these particles can then be grouped as follows:

1.  *Collisional deactivation of I\**

    $I^* + M \rightarrow I + M$, with M being any of the species present in the mixture. Of particular interest is deactivation by radicals, by $O_2$ and by $I_2$.

2. *Radical recombinations*

   $I + R \rightarrow RI$, $2R \rightarrow R_2$

3. *Radical-molecule reactions*

   $R + RI \rightarrow R_2 + I$

   $R + I_2 \rightarrow RI + I$

   The reactions mentioned under 2 and 3 are important for the regeneration of the parent material.

4. *Three-body recombination*

   $I + I^{(*)} + M \rightarrow I_2 + M$      $(M = RI, He, I_2)$

   This refers to the recombination of either two ground-state atoms or of one excited and one unexcited iodine atom. The case of two excited iodine atoms is unimportant. This conclusion is based on experimental experience as well as on theoretical arguments. As Fig. 6 shows, the recombination of two excited iodine atoms leads to a repulsive state of the iodine molecule, while in the other cases, stable $I_2$ states can be formed.

The modeling calculations of Turner and Rapagnani[16] assumed an energy density that is typical of any iodine high-power amplifier and that

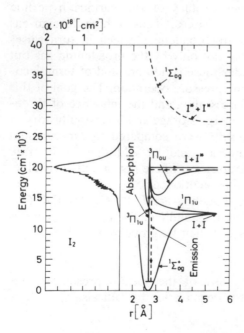

**Fig. 6.** Absorption bands and energy level scheme of $I_2$ (after Wilson[35]).

**Table 2.** Chemical Reactions in a RI Photodissociation Laser[a]

| Reactants | Products | $K\left(\dfrac{cm^3}{sec}\right)$ $R = CF_3$ | Reference | $R = C_3F_7$ |
|---|---|---|---|---|
| $I^* + RI$ | $I + RI$ | $5.4E-17$ | (5, 46) | $8.0E-18$ |
| $I^* + R$ | $I + R$ | $3.7E-18$ | 18 | |
| $I^* + R_2$ | $I + R_2$ | $4.7E-16$ | 46 | |
| $I^* + O_2$ | $I + O_2$ | $8.6E-12$ | 47, 48 | |
| $I^* + I_2$ | $I + I_2$ | $1.3E-14e^{1650/T}$ | | |
| | | $3.2E-12$ | 49 | |
| | | | | |
| $R + R$ | $R_2$ | $1.5E-11$ | 29, 50, 51 | $1.0E-11$ |
| $I + R$ | $RI$ | $5.0E-11$ | 29 | |
| $I^* + R$ | $RI$ | $3.0E-12$ | 18 | |
| | | | | |
| $R + RI$ | $R_2 + I$ | $3.0E-16$ | 18 | |
| $I^* + RI$ | $I_2 + R$ | $2.5E-19$ | 18, 32 | |
| $R + I_2$ | $RI + I$ | $4.0E-12$ | 52 | |
| $I + RI$ | $I_2 + R$ | $1.6E-23$ | 52, 53 | |
| | | | | |
| $2I + M$ | $I_2 + M$ | $4.3E-34$ | b | |
| $2I + RI$ | $I_2 + RI$ | $7.0E-34e^{1600/T}$ | b | $2.1E-33e^{1600/T}$ |
| | | $1.5E-31$ | b | $4.5E-31$ |
| $2I + He$ | $I_2 + He$ | $8.27E-29T^{-1.7}$ | b | |
| | | $4.6E-33$ | b | |
| $2I + I_2$ | $2I_2$ | $1.1E-15T^{-5.9}$ | b | |
| | | $2.9E-30$ | b | |
| $2I + O_2$ | $I_2 + O_2$ | $3.7E-32$ | b | |
| $2I + R_2$ | $I_2 + R_2$ | $5.8E-32$ | b | |
| | | | | |
| $I^* + I + M$ | $I_2 + M$ | $4.3E-34$ | b | |
| $I^* + I + RI$ | $I_2 + RI$ | $4.3E-34$ | b | $1.6E-33$ |
| $I^* + I + He$ | $I_2 + He$ | $9.4E-34$ | b | $2.2E-33$ |
| $I^* + I + I_2$ | $2I_2$ | $4.3E-32$ | b | |

[a] Unless otherwise mentioned, rate constants are given at 300°K. The data are taken from Ref. 16, other references as noted.

[b] A detailed discussion of this group of reactions is given in Ref. 16. The extensive data on iodine recombination, in the literature, are, therefore, not fully quoted here.

applies approximately to the 1-kJ system under construction at Garch-ing.[42] The degree of photodissociation was chosen to be roughly 10% for $\approx 10$-Torr RI and is generated by a 10-$\mu$sec flash. No precision is claimed for times of the same order as the photolysis flash, and emphasis is placed primarily on the inversion history at longer times.

The catalog of reactions to be considered together with their rate coefficients are shown in Table 2. It is beyond the scope of this survey to justify the choice of rate data in detail, but some general justification may be seen in the excellent agreement of the computation with both the experimental time dependence of the small-signal gain as given by Hohla and Kompa[11,12] and of the half-time measurements by Aldridge[41] as indicated in Fig. 7. The conclusions that may be drawn at this point can be summarized as follows:

1.   Despite some doubts, the rate model that can be constructed in this way is capable of explaining and predicting the energy storage and lifetime of the inversion under high-power amplifier conditions.

2.   In agreement with experimental observations for the high-pressure conditions chosen, no collisional losses for times shorter than $10^{-4}$ sec are expected. The lifetime of the inversion is determined principally by the formation of $I_2$ and subsequent deactivation of $I^*$.

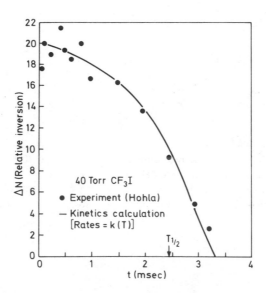

**Fig. 7.**   Experimental- and theoretical-inversion history for a $CF_3I$ photodissociation laser (data from Refs. 46 and 16).

## 2. SPECTROSCOPY OF THE LASER TRANSITION

The photolysis produces iodine atoms in the lowest-lying electronically excited state $(5^2P_{1/2})$. The transition to the ground state $(5^2P_{3/2})$ is a magnetic-dipole transition. The radiative lifetime of the excited state has recently been redetermined by Derwent and Thrush to be 170 msec.[54] This is in satisfactory agreement with calculations and experimental results of Zuev et al.,[45] who calculated 130 msec as compared to his experimental value of 180 msec. In contrast to these consistent data, various spectroscopic measurements gave lifetimes between 0.017 and 0.045 msec.[47,55] These very short lifetimes may be due to collisional deactivation not fully accounted for in the analysis.

### 2.1. Hyperfine Structure of the Transition

A detailed spectroscopic analysis shows hyperfine splitting of both the $^2P_{1/2}$ and $^2P_{3/2}$ levels.[45] The degeneracy of the upper and lower level is canceled by the magnetic-dipole and the electric-quadrupole moment of the nucleus of the $^{127}I$ atom (natural occurrence 100%, nuclear spin 5/2). The angular momentum of the upper level can be parallel or antiparallel to the nuclear spin. The total angular momentum can then be characterized by the values $F = 5/2 + 1/2 = 3$ and $F = 5/2 - 1/2 = 2$. In the same way, the lower level is split into four different hyperfine levels with the $F$ numbers, $F = 1, 2, 3, 4$. In Fig. 8, the different levels are depicted, including the $F$ numbers. The single hyperfine levels are again degenerated, because the nuclear spin with its value of 5/2 can be oriented in $[2 \times 5/2 + 1] = 6$ different ways with respect to the direction of an external magnetic field. This means the total degeneracy g of the $^2P_{1/2}$-state is 12, while the degeneracy of the $F = 2$ level is 5 and the degeneracy of the $F = 3$ level is 7. These values are listed on the right-hand side of the sublevels in Fig. 8. The g factors are of great importance because they determine the relation of the inversion in the two upper states. Therefore, 5/12 of the iodine atoms are produced in the $F = 2$ and 7/12 are in the $F = 3$ level.

The possible transitions between the various hyperfine levels are indicated in Fig. 8. One can distinguish that two groups are drawn with the relative intensities in the lower part of Fig. 8: a group at shorter wavelengths starting from the $F = 3$ level, and a longer wavelength group starting from the $F = 2$ level. Each of these groups consists of three lines, whose distances in wave numbers are given in Fig. 8. The theoretical values given here are in good agreement with the experimental results of Jaccarino[56] and Verges.[57]

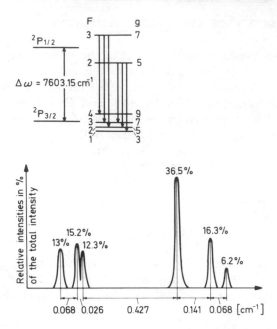

**Fig. 8.** Hyperfine splitting of the iodine-laser transition (after Zuev et al.[45]).

## 2.2. The Cross-Section for Stimulated Emission

A most important parameter of a laser is the cross-section for stimulated emission $\sigma$. It governs the small signal amplification $V$ of the laser medium by means of the relation:

$$V = \exp(\sigma \, \Delta N l)$$

where $\Delta N$ represents the inversion density and $l$ the length of the laser medium. Often the laser is characterized by the gain coefficient $g_0$, which is equal to the product of $\sigma$ and $\Delta N$:

$$g_0 = \sigma \, \Delta N$$

The cross-section $\sigma$ itself is a function of the Einstein coefficient for spontaneous emission $A$, the frequency of the transition $\nu$, and the line shape. For a Lorentzian shape, which results from pressure broadening, the following equation holds:

$$\sigma = \frac{A c^2}{4 \pi \nu^2} \frac{f(\nu - \nu_0)}{\Delta \nu}$$

where $\Delta\nu$ is the full width at half maximum, and

$$f(\nu - \nu_0) = \left[1 + \left(\frac{2(\nu - \nu_0)}{\Delta\nu}\right)^2\right]^{-1}$$

The maximum value of $\sigma$ occurs at line center, where

$$\sigma_0 = \frac{Ac^2}{4\pi^2\nu^2\,\Delta\nu}$$

For the iodine-laser transition there exist, corresponding to the six hyperfine transitions, six cross-sections $\sigma_{F_u \to F_e}$. Assuming that all the single lines have the same line width $\Delta\nu$ the following relation is valid:

$$\frac{\sigma_i}{\sigma_j} = \frac{A_i}{A_j}$$

In the following table the theoretical $A_{i-j}$ values for the various transitions are listed:[45]

| | | | | | | |
|---|---|---|---|---|---|---|
| Spontaneous emission probability, $A_{l-j}$ | 5.0 | 2.1 | 0.6 | 2.4 | 3.0 | 2.3 |
| Transition $F_u \to F_l =$ | 3–4 | 3–3 | 3–2 | 2–3 | 2–2 | 2–1 |

One finds the total $A$ by summing up all the intensities of the various lines, taking into account the various degeneracies $g$ of the upper levels:

$$NA = N\frac{g_3}{g_3 + g_4}(A_{3-4} + A_{3-3} + A_{3-2}) + N\frac{g_4}{g_3 + g_4}(A_{2-3} + A_{2-2} + A_{2-1})$$

where $N$ is the total number of particles in the state $^2P_{1/2}$. In this way, one finds a value of 7.7 sec$^{-1}$ for $A$.

It is common to define the cross-section for stimulated emission in a similar way by referring it to the total number of particles in the upper level.

$$\sigma_{F_u \to F_l} = \sigma_{F_u \to F_l}|_{\text{real}} \frac{\Delta N_{F_u \to F_l}}{\Delta N_{\text{total}}}$$

To calculate the total $\sigma$ at a frequency $\nu$, one has only to sum up all the single $\sigma_{F_u \to F_l}$ at this frequency.

## 2.3.  Pressure Broadening

One of the most striking features of the iodine laser is the possibility to change the $\sigma$ coefficient of the laser medium. From the equation for $\sigma_0$, it follows that $\sigma_0$ is inversely proportional to the line width $\Delta\nu$. On the other hand, $\Delta\nu$ is directly related to the pressure $p$:

$$\Delta\nu \sim pa \text{ and therefore } \sigma_0 \sim \frac{1}{p}$$

$\sigma_0$ represents the maximum value of one transition. By superimposing all the lines, the relation is modified at pressures where a considerable overlap of the lines occurs. For a more detailed discussion, see, for example Ref. 23. The cross-section for stimulated emission was measured under various conditions by a number of authors. Hohla[46] combined large- and small-signal amplification measurements to calculate $\sigma$. These experiments were performed with $CF_3I$ as the iodine source; the duration of the probe signal was 100 nsec. In a similar way, Zuev et al.[45] have measured the cross-section for $C_3F_7I$ in the pure gas and in gas mixtures of argon with $C_3F_7I$; the pulse duration in these experiments was in the microsecond range. Aldridge[41] calculated the cross-section from threshold measurements in mixtures of argon and helium with $C_3F_7I$. Fuss and Hohla[38] have repeated the experiments[41,45] but with 1-nsec pulses, in $CO_2$ and argon mixtures. The only line-resolved measurements were performed by Zuev et al.[45] From experiments of Alekseev et al.,[58] it can be deduced that the other workers measured the cross-section of the $F = 3 \rightarrow F = 4$ transition.

In the following table, a list of the maximum values of $\sigma$ for the various transitions are given according to Zuev et al.[45] The pressure of $C_3F_7I$ was 20 Torr.

| Transition $F_u \rightarrow F_l =$ | 3–4 | 3–3 | 3–2 | 2–3 | 2–2 | 2–1 |
|---|---|---|---|---|---|---|
| $\sigma \times 10^{18}$ cm$^2$ | 6.0 | 2.4 | 0.66 | 2.67 | 3.3 | 2.55 |

With the assumption that $A$ is given by 7.7 sec$^{-1}$ as determined from the previously tabulated values of the various $A_{i-j}$, it is possible to calculate an effective line width for the composite transition at high pressures. Such calculations by Fuss[38] are presented in Fig. 9, assuming that the oscillator was operating on the $3-4$ transition.

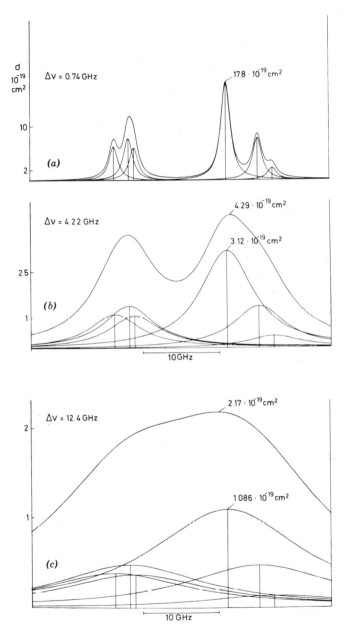

**Fig. 9.** Calculated spectra of the iodine laser transition for various line widths.[38] (a) 20 Torr of $C_3F_7I$, (b) 50 Torr of $C_3F_7I + 150$ Torr of $CO_2$, and (c) 50 Torr of $C_3F_7I + 700$ Torr of $CO_2$.

Figure 9$a$ shows the calculated line structure at a pressure of 60 Torr. Experimentally, a maximum value of $\sigma = 1.78 \times 10^{-18}$ cm$^2$ was measured. The corresponding line width is $\Delta \nu = 0.74$ GHz. It can be seen that only the 2–1 and 2–2 lines overlap and that the 3–4 line is relatively uneffected by its neighbors. In Fig. 9$b$, the total pressure is 350 Torr (10 Torr of $C_3F_7I$ plus 340 Torr of $CO_2$) and the maximum measured $\sigma = 4.29 \times 10^{-19}$ cm$^2$. This corresponds to an individual line width of 4.22 GHz, and the two groups of hyperfine lines are completely overlapping. At a pressure of 750 Torr, the entire hyperfine line structure is smeared into a single-gain profile. The pressure broadened widths of the individual lines is now 12.4 GHz, roughly equal to the separation between the two line groups of the hyperfine structure. The total feature has an effective full-width at half-height of almost 40 GHz. Since the line width corresponds to the ultimate homogeneous band width of the medium, amplification of pulses of 10 to 20-psec duration should be possible.

The stimulated emission cross-section $\sigma$ controls two of the most important laser amplifier parameters, the threshold amplification $V_{PM}$ for the onset of parasitic oscillations, and the energy density $E_{st}$ storable in a high-gain amplifier. The threshold amplification and $\sigma$ define the maximum storable inversion density

$$\Delta N_{PM} = \frac{\ln V_{PM}}{\sigma_{max}}$$

where $\Delta N_{PM}$ is the inversion density per unit area of the amplifier at which parasitic oscillations begin. The storable energy $E_{st}$ will always be smaller than $\Delta N_{PM} h\nu$ and, therefore,

$$E_{st} \le \frac{h\nu \ln V_{PM}}{\sigma_{max}}$$

Since $\sigma_{max}$ is inversely proportional to the line width, and for pressure-broadened lines inversely proportional to the pressure, the storable energy can be increased by the addition of foreign gases to the active medium. Obviously only those gases are candidates for this purpose that do not appreciably quench the excited iodine atoms during the pumping time. Figure 10 shows the dependence of $\sigma_{max}^{-1}$ on the total pressure of $C_3F_7I$ plus that of various added foreign gases. The experimental data are from three different sources[38,41,45] and were obtained with two different techniques; Despite this fact, they are in good agreement. The broadening effect of $C_3F_7I$ is five times, and that of $CO_2$ is twice that of argon; since the quenching effect of $CO_2$ on I* is small, $CO_2$ is the preferred broadening additive in our high-power lasers. It can be seen that it is

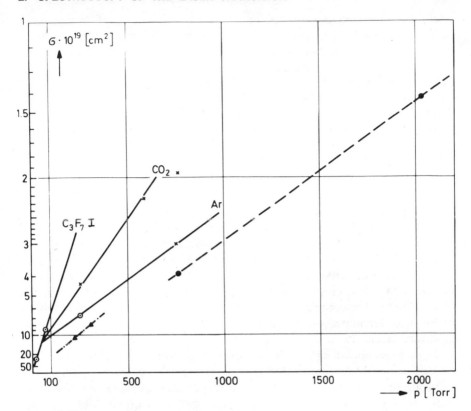

**Fig 10.**  Stimulated emission cross-section $\sigma$ as a function of Ar, $CO_2$, and $C_3F_7I$ pressure. ● absolute measurements, × relative measurements, both performed with nanosecond pulses that contained only frequencies of the $F = 3 \rightarrow F = 4$ transition,[38] —— absolute measurements by Aldridge with Ar,[41] ·—·—· absolute measurements with Ar by Zuev et al.[45]

possible to reduce $\sigma_{max}$ to values as small as $2 \times 10^{-19}$ cm$^2$. At a pressure of 50 Torr of $C_3F_7I$ and 650 Torr of $CO_2$, it is possible to store 6 J/cm$^2$, if a critical threshold amplification of $V_{PM} \sim 10^4$ can be achieved.

## 2.4.  Zeeman Splitting

It is also possible to increase the effective line width of the iodine laser and, therefore, to reduce the stimulated emission cross-section by Zeeman splitting of the hyperfine structure. It was first pointed out by Gregg et al.[21] that the degeneracies of the iodine hyperfine transitions are canceled if an external magnetic field is superimposed on the active medium. The high degeneracy factor of the various levels results in a splitting of the $^2P_{1/2}$ upper level into 12 and that of the $^2P_{3/2}$ lower level into 24 sublevels. Between these levels, all transitions are allowed that are accompanied by a change of the magnetic quantum number $m = \pm1$, 0. The shift of the lines is proportional to the external magnetic field strength $H$ (Gauss). The proportionality constant is different for each hyperfine component. If a magnetic field is applied that varies in its strength along the direction of propagation of the laser beam, usually the axis of the laser tube, the lines will be shifted by different amounts at different positions along the laser tube. To photons traveling along the axis of the laser tube, the medium will therefore, "appear" to have a much increased effective line width consisting of the variously shifted, overlapping hyperfine transitions. Hohla[46] has measured an average value for the broadening effect of a magnetic field with a nearly linear axial gradient; he finds that the effective stimulated emission cross-section $\sigma_{\text{eff}} = 2.2 \times 10^{-15} \Delta H^{-1}$ cm$^2$ where $\Delta H$ (Gauss) is the difference in field strength between its maximum and minimum value. It is, therefore, possible to reach cross-sections of $10^{-19}$ cm$^2$ with reasonable magnetic fields. Due to the complexity and high installation costs for such magnetic fields this technique is, however, of little practical interest for large aperture lasers.

Inhomogeneous magnetic fields could, however, be used to advantage in the mode locked oscillator of a high-power amplifier system. The splitting of the lines would increase the spectral content of the oscillator output pulses needed to fully saturate the subsequent high-pressure amplifiers. Zeeman splitting of the lines of the oscillator medium could also be used to control the effective band width of the oscillator and to decrease the duration of the mode-locked pulses.

Finally, it has to be borne in mind that the high currents of the flashlamps produce strong magnetic fields in the laser medium unless a coaxial lamp geometry is used. Measurements of the influence of the magnetic fields of the flashlamps on the laser parameters have been performed by Hwang and Kasper,[23] who also show how to correct for these fields in noncoaxial geometries.

## 2.5. Sublevel Relaxation

Relaxation times between the two sublevels of the $^2P_{1/2}$ upper state and the four sublevels of the $^2P_{3/2}$ lower state are finite and have to be given special consideration if one wants to produce or amplify short pulses. The subject has been investigated theoretically by Yukov[64] and experimentally by Alekseev et al.[58] These investigations show that the relaxation mechanisms between the sublevels of the upper and those of the lower states are significantly different. Only particles with an electronic angular moment can effectively relax the hyperfine levels of the upper state; whereas, the relaxation of the ground state is due to van der Waal's collisions. Relaxation of the upper state levels is, therefore, mainly due to iodine atoms and $CF_3$ or $C_3F_7$ radicals by resonance excitation transfer with an estimated rate of $1.7 \times 10^{-10}$ cm$^3$/sec. The lower levels by contrast relax in collisions with all particles of the active medium, and the mixing rate has been estimated to be of the order of $2.7 \times 10^{-9}$ cm$^3$/sec. The concentration of iodine atoms and radicals under normal amplifier conditions is of the order of $10^{17}$ cm$^{-3}$ and the relaxation time of the upper levels should, hence, be of the order of $10^{-8}$ to $10^{-7}$ sec. This implies that, during laser-pulse times in the nanosecond regime, the fine structure levels of the upper iodine state cannot relax. This is the reason why the pulse output of an oscillator driving a short-pulse high-power amplifier chain has to contain frequencies of both hyperfine line groups in order to be able to saturate the entire spectral line of the amplifier medium. As indicated previously, Zeeman splitting of the oscillator medium may be the simplest way to assure this. Mixing of the levels of the ground state can be achieved easily by the addition of a foreign gas. At a pressure of 1 atm, relaxation of the lower levels would be accomplished in less than $10^{-9}$ sec. It should also be pointed out that, in any modeling of the iodine laser, the finite relaxation time between the hyperfine levels has to be taken into account, and a complete set of equations for all six transitions is necessary for the correct description of this laser.

## 3. IODINE SHORT-PULSE OSCILLATORS

The term "short pulse" can refer to different time regimes, depending on the application. In our context, pulse durations between $10^{-10}$ and $10^{-7}$ sec are discussed. Depending on the application, the energy in the laser pulse varies between several milli-Joules and several thousand Joules. While the duration of the pulses is determined by an oscillator, which creates low-energy pulses, the total pulse energy is provided by the amplifier stages.

Depending on the time duration there are a number of oscillator versions to chose from. In the following sections, the relevant devices for the iodine laser will be discussed.

### 3.1. Pulse Durations of $5 \times 10^{-8}$ to $5 \times 10^{-7}$ sec

This time domain is covered by so-called gain-switch devices. By the term "gain switch," the following mechanism is understood. The gain in the

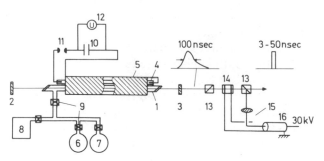

**Fig. 11.** Iodine-laser oscillator with pulse cutting system (1—laser tube; 2, 3—resonator mirrors; 4—flashlamps; 5—optical reflector; 6, 7—bulbs for storage of $C_3F_7I$ and waste material; 8—pumps; 9—valves; 10, 11, 12—high-voltage pulse generator with capacitor, switching spark gap and power supply; 13—glan-prism polarizers; 14—Pockels cell; 15, 16—laser-triggered cable discharge).

laser medium rises so quickly that it reaches and exceeds the oscillator threshold in a time interval short compared to the built up time of oscillations.

The "dump spike" at the beginning of the emission signal contains the major part of the oscillator energy. Flashlamp pumping times of $1 \times 10^{-6}$ to $2 \times 10^{-6}$ sec and rise times of $\sim 100$ nsec are possible with low inductance flashlamp circuits. In iodine, the gain-switch effect is complicated by the interaction of the magnetic field of the flashlamps with the gain medium. The duration of the dump spike is determined by the length of the resonator, the length of the active medium, and the rate of pumping. Typical dump pulse durations of 50 to 100 nsec have been measured with flashlamp durations of $\sim 2$ $\mu$sec. Figure 11 shows a typical oscillator that is capable of delivering pulses of 1 to 10 J in 100 nsec. A quartz laser tube (1) of 2-cm i.d. and 100-cm length is connected to a vacuum system (9). The laser tube is placed inside a resonator cavity of 2-m length.

After the laser tube is pumped down with a mechanical pump to $\sim 10^{-3}$ Torr, it is filled with $C_3F_7I$. After several laser shots, the waste material is condensed into a bulb (7), and the tube can be refilled. Next to the laser tube, one or more Xe flashlamps are mounted that are connected to a high-voltage pulse capacitor (10) and are discharged in parallel by the triggered spark gap (11). Flashlamps and laser tube are coupled by means of an optical reflector, which also serves as the return lead for the flashlamp current. With 1-kJ of capacitor energy, about 5 J of laser energy are obtained, about 50% of which are contained in the first emission pulse.

## 3.2. Pulse Durations of 3 to 50 nsec

Further pulse shortening is achieved either with switching elements in the oscillator, or with an external optical-pulse cutting system. In addition to the usual $Q$-switching, the gain in the iodine laser can be controlled by rapid switching of an external magnetic field, for instance with oscillating capacitor discharges. In this way, the stimulated emission cross-section can be switched in times of about 100 nsec. This technique can also be called gain-switching because it is the gain $g = \sigma \Delta N$ that changes its value in a short time. For normal $Q$-switching, the same optical elements used with neodymium lasers can be employed, as for instance Kerr or KD*P Pockels cells and rotating mirrors. To date no passive $Q$-switching elements such as dyes, are available for 1.315 $\mu$m. There is hope, however, that passive nonlinear absorbers may be found for this laser, too.

In general, external-switching techniques in which a pulse is cut out of the output signal are simple and flexible. The technique of pulse cutting is illustrated in Fig. 11. When there is no voltage on the Pockels cell, the incoming pulse, which has been polarized with the aid of a Glan prism, passes through the Pockels cell and is deflected by a second Glan prism. This part of the pulse is used to trigger a cable discharge which applies a voltage to the Pockels cell. The time during which the Pockels cell remains activated is determined by the traveling time of the high-voltage pulse in the cable. During this time, the Pockels cell rotates the orientation of the laser polarization by 90°, permitting the beam to pass the second Glan prism. The advantages of this type of pulse cutting are twofold: First, the duration can be varied over a wide range without any changes in the optical elements, and, second, the beam passes through the switch only once. This technique results in less distortion of the optical quality of the laser beam than when intracavity switches are used.

## 3.3. Pulse Durations of 0.1 to 3 nsec

Pulse durations around 1 nsec are of interest in experiments directed towards laser-induced controlled nuclear fusion. This pulse-time regime is accessible by mode-locking techniques. Active mode locking has first been applied to the iodine laser by Ferrar.[20] A quartz tube within the resonator is excited to acoustic oscillations producing acousto-optic mode-locking. In this way, a modulation is introduced in the laser the period of which can be matched to the traveling time of the photons in the cavity. Thus, those photons that pass through the modulator at certain times suffer a lower loss than others. The transmitted photon packet is amplified as it travels back and forth in the cavity and contracts in time. This is a result of the nonlinear gain characteristics of the oscillator medium. Since on each transversal part of the signal is coupled out, a train of short pulses is obtained that are separated by one cavity round-trip time. The duration of the individual pulses depends primarily on the line width of the transition. In principle, it should be possible to achieve band width limited pulses[59] by mode-locking as has been demonstrated, for instance, for ruby lasers. This means that the shortest possible pulse duration is obtained with the largest line width. The line width of the iodine laser depends on the gas pressure as has been discussed in Section 2. For instance, with a gas mixture of 50 Torr of $C_3F_7I + 550$ Torr of $CO_2$, the effective line width (full width at half height) is of the order of 40 GHz, and pulses of 10 to 20 psec duration should be possible.

In our iodine oscillator, we have coupled the modes by using a conventional acousto-optic quartz modulator. We then selected a single pulse from the resulting pulse train by a pulse-cutting system. The contrast ratio of the signal to the remaining background was found to be larger than $3 \times 10^4$. The pulses were recorded with a Valvo XA-1003 photocell connected to a Tektronix 519 oscilloscope. The pressure of $C_3F_7I$ was varied between 20 and 300 Torr. The resulting pulse duration as a function of pressure is shown in Fig. 12. It is seen that, at 20 Torr, 2.7-nsec pulses are produced and that the duration is reduced with

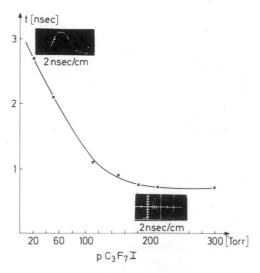

**Fig. 12.**   Oscillator-pulse duration as function of pressure of $i$-$C_3F_7I$.

increasing pressure. At 180 Torr, a pulse length of 0.7 nsec is reached. A further shortening of the pulse length could not be observed because of the finite resolution time of our diagnostic system.

## 4. IODINE HIGH-POWER AMPLIFIERS

The energy of the oscillator depends on the pulse duration and the laser volume. The lasing volume in an oscillator is determined by the mode structure. In order to get a Gaussian beam, one has to work in the lowest transverse $TEM_{00}$ mode, which results in small beam diameters. For very short pulses, which are created by mode-locking devices, the pulse energy is usually restricted to several milli-Joules. For pulse durations of 10 to 100 nsec, the energy is increased to several hundred milli-Joules. In many applications, primarily in laser-plasma experiments, very large laser-pulse energies are required. A mode-locked pulse from the oscillator might have an energy of 1 to 5 mJ, while $10^3$ to $10^4$ J are desired for such experiments. Extreme pulse amplification by factors of $10^6$ to $10^7$ has, therefore, to be considered. The design of the required amplifiers may be divided into three parts: (1) The generation of the inversion, (2) the storage of the inversion, and (3) the release of the energy by the oscillator signal.

### 4.1  Generation of the Inversion

The inversion in the iodine laser is caused by a photolytic process. There are two problems that have to be considered: the efficiency of the laser system and the homogeneity of the laser medium. The efficiency of converting capacitor energy into laser output depends on the spectral width of the uv absorption band of the parent iodides and on the energy output of the flashlamps into this absorption band. In the experiments at the Max-Planck-Institut für Plasmaphysik, Xe lamps are used that are operated at a pressure of 20 to 70-Torr of Xe. The largest efficiency was found to be 0.8% when $i\text{-}C_3F_7I$ was used. This value was found in oscillator measurements: The amplifier was put into a resonator, and the laser energy was compared with the stored electrical energy. As discussed in Section 1, this energy can also be extracted by an oscillator signal. Any further improvement of the efficiency rests with better flashlamps. Lamps spectrally tailored to the $C_3F_7I$ absorption band by adding, for example, mercury to the Xe discharge are the principal prospect. The final limitation on the efficiency will be the 21% quantum efficiency of the conversion of uv light at 270 nm to the laser emission of 1315 nm.

Besides flashlamps, exploding wires can be used as light sources for pumping the iodine laser as was demonstrated by Basov et al.[26] They reported an efficiency of 0.4%. The advantages of exploding wires seem to be their cheapness and their reliability.

The homogeneity of the laser medium depends on two parameters: the duration of the flashlamps and the geometrical cross-section of the laser medium. The first dependence arises from shock waves that start from the laser-tube wall initiated by the flashlight.[10,60,61] As was pointed out by Golubev et al.[61] gases adsorbed at the walls are flash-evaporated and give rise to shock waves that travel into the laser medium with sonic speed. The shock waves produce density gradients that distort the wave fronts of the laser beam and destroy the beam quality. To overcome this problem, fast flashlamps are used. The uv light is deposited in the laser gas before the shock wave has traveled into the center of the tube. Assuming a shock velocity of $3 \times 10^4$ cm/sec in the laser medium, a flash duration of $5 \times 10^{-6}$ sec is sufficient for laser tubes with diameters of more than 2 cm. In this example, the shock wave has traveled only 1.5 mm from the wall, nearly the whole gas volume is still undistributed, and a good beam quality is obtainable. With increasing diameter, longer flash durations are tolerable without loss of usable aperture.

The size of the geometrical cross-section of the amplifier that can be uniformly pumped is of particular importance, since it determines the maximum energy obtainable in one beam. A systematic experimental study of this question has not been made so far. A qualitative estimate may be obtained, however, by considering the penetration depth of the uv radiation. The penetration depth $l$ is related to the absorption cross-section $\alpha$ (Fig. 2) and $n$, the number density of the alkyliodide by $l = 1/\alpha n$. For a cylindrical laser tube with a diameter of 7 cm, good homogeneity of illumination has been found at 25-Torr of $C_3F_7I$.[63] The energy density in these experiments was 30 J/liter. The energy density was measured in an oscillator device and is again extractable with nanosecond pulses (see Section 1). With various laser tubes with diameters $D$ of 2.5 and 20 cm, the radial homogeneity of the energy distribution was measured by means of small-signal amplification measurements.[62] At $C_3F_7I$, pressures that obey the experimentally determined relationship[62]

$$p = 170 D^{-1} \text{ Torr}$$

radial gain variations were found to be less than 10%. The degree of photodissociation amounts to 0.2 to 0.3. With this formula for $p$, it is easy to see how an iodine-laser amplifier can be scaled in volume and, thus, in energy: The total number of particles is proportional to $pD^2$ and, therefore, proportional to the diameter. Assuming the degree of photodissociation constant, the energy that can be pumped per unit length of the amplifier tube is proportional to the tube's diameter. This is because the number of flashlamps is proportional to the circumference of the tube and, therefore, proportional to the diameter.

## 4.2. Storage of the Energy in the Amplifier

As was pointed out already in Section 2, an essential problem of any amplifier system is saturation by self oscillations. At a given overall small-signal amplification, the small feedback resulting from laboratory walls or scattering particles suffices to start lasing in parasitic modes. Beyond this threshold, the amplifier becomes an oscillator, and further input of pump energy is directly converted into radiation. This situation is schematically illustrated in Fig. 13. From the threshold condition

$$V^2 R_1 R_2 T^2 = 1$$

where $V$ is the small signal amplification, and $R_1$ and $R_2$ are the reflectivities of any walls on both ends of the amplifier. $T$ refers to the transmission of the active medium. Together with the expression for the small-signal amplification (see Section 1), an expression for the storable energy may be written as follows:

$$E_{st} = \frac{h\nu \ln (R_1 R_2 T^2)^{-1/2}}{\sigma_{max}}$$

While the influence of $\sigma$ is quite strong, as was discussed earlier, the influence of $R_1$ and $R_2$ is only weak because of the logarithmic dependence. With black surfaces several meters away from the amplifier, values of $10^{-6}$ to $10^{-8}$ for $R_1$ or $R_2$ are easily obtained. Improving $R_1$ or $R_2$ from $10^{-6}$ to $10^{-8}$ changes the storable energy only by a factor of 1.33. In

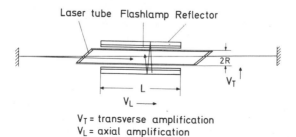

$V_T$ = transverse amplification
$V_L$ = axial amplification

**Fig. 13.** Illustration of the threshold of transverse- versus longitudinal-laser oscillation.

many cases, paths inside the amplifier can have a high amplification and, therefore, a comparatively low threshold. The reflectors of the flashlamps act as reflectors for the ir-laser light. Therefore, oscillations at an angle to the laser axis are possible. Additionally, the smooth surface of the quartz tube allows ring modes with a low threshold. Along so-called zig-zag paths, the amplification can become so high that a considerable part of the stored energy can be released in such parasitic modes.

In the experiments of Hohla et al.,[63] an aluminum spiral was fitted into the quartz tube to prevent the zig-zag parasitic modes. Small-signal amplification measurements proved this technique to be very successful; the maximum amplification increased from $10^3$ to $10^5$. But, even at $V = 10^5$, the reflectivities outside the amplifier device were not the limitations. Putting a second amplifier in series with the first, the overall amplification could be increased to $10^7$. This was achieved by separating the two amplifiers by 40 m. Therefore, the main problem in attaining high amplifications appears not to be the "outside" reflectivities but internal light paths with low threshold.

In Fig. 14, the storable energy is given as a function of the cross-section for stimulated emission and as a function of the highest sustainable small-signal amplification. On the right abscissa, the total pressure of the

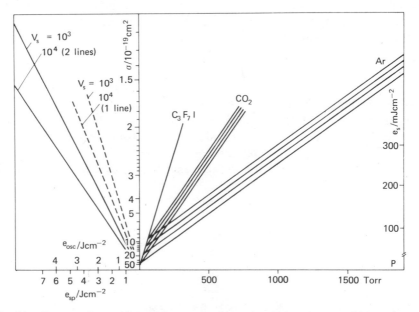

**Fig. 14.** Cross-sections and storable energies for various mixture composition and pressures. Parameter is the small-signal amplification, the solid lines refers to the full frequency spectrum while the dotted lines correspond to frequencies of the $F = 3 \rightarrow F = 4$ transition.

laser medium is plotted, while the ordinate gives the corresponding cross-section. The left abscissa gives the storable energy density $e_{st}$ (in $J/cm^2$). The straight lines represent the relation between $\sigma$ and $e_{st}$ with the amplification $V_s$ as the parameter. To read the figure correctly, one has to pick the chosen pressure, which gives the $\sigma$ value on the right side. Crossing over horizontally to the left-hand side of the figure, the line with the corresponding amplification is met. The abscissa value then represents the storable energy.

If more than $10^7$ to $10^8$ overall small-signal amplification is needed, several amplifiers have to be installed. But now they must be optically isolated against each other. For smaller beam diameters ($d \leq 3$ cm), Pockels or Kerr cells can be used. They have the advantage to work in both directions, since their transmission in both directions, $T_1$ and $T_2$, are equal and very low, for example, $10^{-2}$ to $10^{-3}$. At the moment, the oscillator pulse passes the amplifier, they are opened for a short time interval. For larger diameters, only Faraday rotators are available. These act only as one-way valves: The path from the oscillator to the target is open, while the reverse path is closed, that is, $T_1 \gg T_2$. Because only the mean square root of the transmission, $(T_1 T_2)^{1/2}$, enters into the expression of the storable energy, the effect of Faraday rotators on the storable energy is small, and their main use is to protect the oscillator from light reflected from the target. The materials for Pockels cells, Kerr cells, and Faraday rotators are the same as those used in Nd-glass laser systems. Bleachable absorbers for 1.315 $\mu$m radiation are as yet unavailable.

Even though the dependence of the storable energy on the amplification $V$ is only small, $V$ is a very important factor in determining the needed input energy to extract the energy from an amplifier.

## 4.3. Energy Extraction

The energy stored in the amplifier is extracted by the traveling pulse generated in the oscillator. For a theoretical description the following assumptions are made: (1) The pulse duration is long compared to the transverse relaxation time, that is, the inverse band width; (2) the incoming pulse contains frequencies of both line groups (see Section 1); and (3) the pressure broadening is so high that both line groups are smeared together. Condition (3) is normally fulfilled in amplifiers with a total pressure of more than 1 atm $CO_2$ or Ar. Condition (1) restricts the argument to pulse durations of more than 100 psec. Condition (2) has to be demonstrated spectroscopically. If only one line group is oscillating in the oscillator, the degeneracy factor in the following formula has to be changed.

The amplification for an arbitrary incoming pulse can be described by the formula[46]

$$V = (b\sigma e_0)^{-1} \ln \{1 + [\exp(b\sigma e_0) - 1] \exp \sigma \, \Delta N l\}$$

where $e_0$ stands for the total incoming energy density in photons per square centimeter, $b$ for the degeneracy factor 1.5, $\sigma$ for the cross-section for stimulated emission, $\Delta N$ for the inversion density (atoms/cm$^3$), and $l$ for the active amplifier length. In the limit of very small $e_0$, the equation reduces to the familiar expression for the small-signal amplification $V_s = \exp(\sigma \, \Delta N l)$, while for large values of $e_0$ it gives the formula for large-signal amplification: $V_L = 1 + \Delta N / b e_0$, that is, the entire stored energy is extracted by the pulse. Normally $(b\sigma)^{-1}$ is called the local saturation energy density $e_s$; the inversion at a certain point in an amplifier is reduced to $1/e$ if $e_s$ has passed this point. To be distinguished from the local saturation density is the overall saturation energy density $e_{os}$, which represents the input energy density that releases the fraction $1 - e^{-1} = 0.63$ of the stored energy. This value depends on $V_s = \exp(\sigma \, \Delta N l)$ and is normally considerably smaller than $e_s$.

A very informative parameter is the so-called extraction efficiency of an amplifier, which is defined as:

$$\eta = \frac{E_{out} - E_{in}}{E_{stored}}$$

where $E_{out}$ and $E_{in}$ represent the total output and input energies, and $E_{stored}$ is the total stored energy of an amplifier. From the foregoing relation for the total amplification $V$, $\eta$ can be calculated. The corresponding curves are plotted in Fig. 15. The maximum $\eta$ for an iodine laser is 0.66 because of the degeneracy of the various levels. The small-signal amplification $V_s$ is the parameter that varies from $10^3$ to $10^5$ and $10^7$ in the three diagrams. In each diagram three curves are plotted, representing three possible $\sigma$ values. With $V_s$ and $\sigma$, the storable energy is given as the value in brackets. As one can see, for example, from the second diagram, $\sim 10^{-3}$ J/cm$^2$ is enough energy to extract 30% of the stored 5.75 J/cm$^2$, if the total amplification is $10^5$.

The consequence of this discussion is that in the high-gain medium large energies can be extracted with moderate input energies, and, therefore, it is possible to work with only one large energy amplifier that is activated by small input pulses. As a result, iodine lasers consist normally of an oscillator, one or two small amplifiers, and one large energy amplifier.

The highest reported energies to date are 50 J in 5 nsec measured by

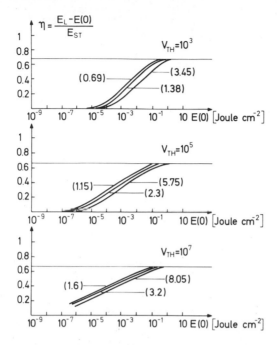

**Fig. 15.** Efficiency $\eta$ of iodine laser amplifiers (the numbers in parentheses are the stored energies in joules per square centimeter).

Basov and co-workers[26] and 60 J in times below 1 nsec measured by Hohla et al.[63]

Finally, a short remark on nonlinear effects in the iodine laser medium: Up to now, no detailed theoretical or experimental study of self-focusing and self-phase modulation existed. Rough estimations show that, with argon as the pressure broadener, these effects should occur at powers that are at least an order of magnitude higher than in solid-state laser devices.

## REFERENCES

1. D. Porret and C. F. Goodeve, Proc. Roy. Soc. London Ser. A **165,** 31 (1938).

2. J. V. V. Kasper and G. C. Pimentel, Appl. Phys. Lett. **5,** 231 (1964).

3. J. V. V. Kasper, J. H. Parker, and G. C. Pimentel, Phys. Rev. Lett. **14,** 352 (1965); J. Chem. Phys. **43,** 1827 (1965).

4. M. A. Pollack, Appl. Phys. Lett. **8,** 36 (1966).

5. For a review see R. J. Donovan and D. Husain, Ann. Rept. Chem. Soc. London Ser. A **68** (1971).

6. A. J. DeMaria and C. J. Ultee, Appl. Phys. Lett. **9,** 67 (1966).

7. N. G. Basov, D. K. Gavrilina, Yu. S. Leonov, and V. A. Sautkin, JETP Lett. **8,** 106 (1968).

8. T. L. Andreeva, V. I. Malyshev, A. I. Maslov, I. I. Sobel'man, and V. N. Sorokin, JETP Lett. **10,** 271 (1969).

9. M. J. Berry, Chem. Phys. Lett. **5,** 269 (1972).

10. V. Yu. Zalesskii and E. I. Moskalev, Sov. Phys. JETP **30,** 1019 (1970).

11. K. Hohla and K. L. Kompa, Chem. Phys. Lett. **14,** 445 (1972).

12. K. Hohla and K. L. Kompa, Z. Naturforsch. **27a,** 938 (1972).

13. M. A. Gusinow, J. K. Rice, and T. D. Padrick, Chem. Phys. Lett. **21,** 197 (1973).

14. W. Fuss, K. Hohla, and K. L. Kompa, unpublished results.

15. D. E. O'Brien and J. R. Bowen, J. Appl. Phys. **40,** 4767 (1969); D. E. O'Brien and J. R. Bowen, J. Appl. Phys. **42,** 1010 (1971).

16. C. Turner and N. L. Rapagnani, Laser Fusion Program—Semiannual Report, UCRL-50021-73-1, Lawrence Livermore Laboratory, Livermore, California (Jan.–June 1973), UCID–16935.

17. V. Yu. Zalesskii, Sov. Phys. JETP **34,** 474 (1972).

18. T. L. Andreeva, S. V. Kuznetsova, A. I. Maslov, I. I. Sobel'man, and V. N. Sorokin, JETP Lett. **13,** 449 (1971).

19. R. D. Franklin, Technical Report ARL 73-0071 (Wright Patterson Air Force Base, Dayton, Ohio April 1973).

20. C. M. Ferrar, Appl. Phys. Lett. **12,** 381 (1968).

21. D. W. Gregg, R. E. Kidder, and C. V. Dobler, Appl. Phys. Lett. **13,** 297 (1968).

22. P. Gensel, K. Hohla, and K. L. Kompa, Appl. Phys. Lett. **18,** 48 (1971).

23. W. C. Hwang and J. V. V. Kasper, Chem. Phys. Lett. **13,** 511 (1972).

24. K. Hohla, P. Gensel, and K. L. Kompa, Proc. Second *Workshop on Laser Interaction and Related Plasma Phenomena,* J. Schwarz, and H. Hora, Eds. (Plenum, New York, 1972).

25. K. Hohla, Third *Workshop on Laser Interaction and Related Plasma Phenomena,* J. Schwarz and H. Hora, Eds. (Plenum, New York, 1974).

26. N. G. Basov, L. E. Golubev, V. S. Zuev, V. A. Katulin, V. N. Netemin, V. Y. Nosach, O. Y. Nosach, and A. L. Petrov, Kvant. Elektr. **6,** 116 (1973).

27. R. S. Mulliken, Phys. Rev. **51,** 310 (1937).

28. J. R. Dacey, Disc. Faraday Soc. **14,** 84 (1953); J. R. Majer and J. P. Simons, Adv. Photochem. **2,** 137 (1964).

29. T. Ogawa, G. A. Carlson, and J. C. Pimentel, J. Phys. Chem. **74,** 2090 (1970).

30. R. J. Donovan and D. Husain, Trans. Faraday Soc. **62,** 11 (1966).

31. D. W. Gregg, Semiannual Report, UCRL-50021-73-1 Lawrence Livermore Laboratory, Laser Fusion Program, Livermore, California (Jan.–June 1973), p. 130.

32. J. H. Parker, Thesis, University of California, Berkeley, California (1968).

33. R. K. Boyd, G. W. Downs, J. S. Gow, and C. Horrex, J. Phys. Chem. **67,** 719 (1963).

34. W. Fuss, unpublished data.

35. K. R. Wilson, *Excited State Chemistry,* J. N. Pitts, Jr., Ed. (Gordon and Breach, New York, 1970), p. 44.

36. I. McAlpine and H. Sutcliffe, J. Phys. Chem. **73**, 3215 (1969).

37. W. Fuss and K. Hohla, IEEE J. Quant. Electron. **QE-9**, 765 (1974).

38. W. Fuss and K. Hohla, Report IPP IV/67 (Max-Planck-Institut für Plasmaphysik, Garching, Germany, 1974).

39. W. Fuss, unpublished data.

40. V. Yu. Zalesskii and A. A. Venediktov, Sov. Phys. JETP **28**, 1104 (1969).

41. F. T. Aldridge, Appl. Phys. Lett. **22**, 180 (1973).

42. G. Brederlow, K. Hohla, and K. J. Witte, Report IPP IV/74 (Max-Planck-Institut für Plasmaphysik, Garching, Germany, 1974).

43. V. Y. Zalesskii and T. I. Krupenikova, Opt. Spectrosc. **30**, 439 (1973).

44. F. T. Aldridge, IEEE J. Quant. Electron. **11**, 215 (1975).

45. V. S. Zuev, V. A. Katulin, V. Yu. Nosach, and O. Yu. Nosach, Sov. Phys. JETP **35**, 870 (1972).

46. K. Hohla, Report IPP IV/33 (Max-Planck-Institut für Plasmaphysik, Garching, Germany, 1971).

47. R. J. Donovan and D. Husain, Trans. Faraday Soc. **62**, 11, 2023 (1966); E. W. Abrahamson, L. J. Andrews, D. Husain, and J. R. Wiesenfeld, J. C. S. Faraday II, **68**, 48 (1972).

48. J. J. Deakin, D. Husain, and J. R. Wiesenfeld, Chem. Phys. Lett. **10**, 146 (1971).

49. R. J. Donovan and D. Husain, Nature **206**, 171 (1965).

50. P. B. Ayscough, J. Chem. Phys. **24**, 944 (1956).

51. N. Basco and F. G. M. Hathorn, Chem. Phys. Lett. **8**, 291 (1971).

52. J. C. Amphlett and E. Whittle, Trans. Faraday Soc. **63**, 2695 (1967).

53. G. S. Lawrence, Trans. Faraday Soc. **63**, 1155 (1967).

54. R. G. Derwent and B. A. Thrush, Chem. Phys. Lett. **9**, 591 (1971).

55. J. J. Deakin, D. Husain, and J. R. Wiesenfeld, Chem. Phys. Lett. **10**, 146 (1971); Nature **213**, 227 (1967); Trans. Faraday Soc. **63**, 1349 (1967).

56. V. Jaccarino, Phys. Rev. **94**, 1798 (1954).

57. J. Verges, Spectrochim. Acta **24B**, 177 (1969).

58. V. A. Alekseev et al., Sov. Phys. JETP **36**, 238 (1973).

59. B. C. Johnson, Report UCRL-50021-73-2 (Lawrence Livermore Laboratory, Livermore, California, 1973).

60. I. M. Belotsova, C. B. Danilov, I. A. Sinitsina, and V. V. Spiridonov, Sov. Phys. JETP **31**, 791 (1970).

61. L. E. Golubev, V. S. Zuev, V. A. Katulin, V. Y. Nosach, and O. Y. Nosach, Kvant. Electr. **6**, 23 (1973).

62. K. Hohla, R. Volk, K.-J. Witte, IEEE J. Quant. Electron. **QE-9**, 764 (1974).

63. K. Hohla, G. Brederlow, W. Fuss, K. L. Kompa, J. Raeder, R. Volk, S. Witkowski, and K. J. Witte, J. Appl. Phys., in press.

64. E. A. Yukov, Sov. J. Quant. Electron. **3**, 117 (1973).

# CHAPTER 13

# METAL-ATOM OXIDATION LASERS*

**REED J. JENSEN**

**Los Alamos Scientific Laboratory,**

**Los Alamos, New Mexico**

It has been known since the thirties that alkali metal atoms, when abstracting halogen atoms, deposit much of the energy of reaction in vibration of the product molecules.[2–6] When molecular beam techniques were developed, results quickly corroborated this fact.[7–10] The $K + I_2$ system has been studied in some detail by this technique.[11] Most of the energy of reaction appears as internal energy of the KI product, and a reactive cross-section of $170 \pm 50$ Å$^2$ has been measured for the reaction. Similar results have been obtained for all the alkali metals in reactions with each diatomic halogen. For the case of $Cs + SF_6$, however, only about

* Work performed under the auspices of the U.S. Atomic Energy Commission.[1]

6% of the reaction energy appears as internal energy of the CsF product.[12] The majority appears in the $SF_5$ fragment and translational energy.

Trajectory calculations designed to provide insight into beam-scattering data indicate for many reactions that most of the energy of reaction appears as vibrational rather than rotational or electronic energy.[13,14] Recently, ab initio calculations of potential energy surfaces for $Li + F_2 \rightarrow LiF + F$ have shown the potential surfaces to be the early downhill type,[15,16] which tends to deposit energy in vibration of the product molecule. These developments all indicate that vibrational levels of a metal-halide product molecule should be inverted initially and capable of laser action as a result of an atom–diatom exchange chemical reaction.

Cross-beam chemiluminescence studies with group IIa metal fluorides have shown inversions in the vibrational states of electronically excited product fluorides when the metal vapor is burned in $F_2$ gas. For example, when calcium metal is burned in $F_2$ gas many of the CaF molecules are born in the $B^2\Sigma$ state, and chemiluminescence data[17] indicate that the vibrational population of the $B$ state is sharply peaked around $v = 26$. Similar data have been obtained for the reaction $Ca + ClF \rightarrow CaF + Cl$ where the vibrational population of the $B$ state is sharply peaked around $v = 9$. This observed vibrational distribution is consistent with data from molecular-beam work that indicate that most of the exothermicity of the reaction appears in internal energy of the reaction product.

Prior to the operation of the first chemical laser, Polanyi[18] suggested that reactions of the type

$$M + XY \rightarrow MX^* + Y$$

show particular promise of producing chemical-laser action where M is an alkali metal atom, XY is a covalent halogen or halide, and MX is an ionic metal halide. After Kasper and Pimentel[19] announced the successful operation of the HCl chemical laser, attention was turned to reactions involving hydrogen. Chemiluminescence methods were used to obtain the initial distribution over product states for a number of reactions including: $H + Cl_2 \rightarrow HCl + Cl,$[20] $Cl + HI \rightarrow HCl + I,$[20] $F + H_2 \rightarrow HF + H,$[21] $D + Cl_2 \rightarrow DCl + Cl,$[22] $H + Br_2 \rightarrow HBr + Br,$[22] $H + ClI \rightarrow HCl + H,$[22] $Cl + DI \rightarrow DCl + I,$[22] $F + H_2 \rightarrow HF + H,$[22] $F + D_2 \rightarrow DF + D,$[22] that produce inverted populations of product vibrational states. All of these results indicate the likely success of a metal-atom oxidation laser where the inversion results from an exothermic atom–diatom exchange, chemical reaction.

Table 1 displays 27 combinations of metals and oxidizers that have produced laser emission in early work at Los Alamos[23,24] along with the heat of dissociation, the enthalpy of reaction, and the harmonic

**Table 1.** Thermochemical and Spectral Data for Laser Molecules

| Metal M | Oxidizer XY | $D^{\circ}_{298}$ of MX[a] kcal/mol | $\Delta H^{\circ}_{298}$ of reaction (1)[a] kcal/mol | $10^3$ cm$^{-1}$ | $\omega_e$ of MX[a] cm$^{-1}$ |
|---|---|---|---|---|---|
| Li | $F_2$ | 139 | −101 | 35.3 | 914.33 |
| C | $O_2$ | 257 | −138 | 48.2 | 2169.52 |
| C | $F_2$ | 129 | −91 | 31.8 | 1308.1 |
| Mg | $F_2$ | 107 | −70 | 24.3 | 717.6 |
| Al | $F_2$ | 160 | −123 | 42.9 | 801.95 |
| Ti | $O_2$ | 169 | −50 | 17.4 | 1008.4 |
| Ti | $F_2$ | 148 | −110 | 38.5 | 593[b] |
| Ti | $NF_3$ | 148 | −87 | 30.6 | 593[b] |
| V | $O_2$ | 147[c] | −28[c] | 9.8[c] | 1012.7[d] |
| V | $F_2$ | 128[c] | −90[e] | 31.4[e] | — |
| Fe | $F_2$ | 107 | −69 | 24.2 | 630.0 |
| Ni | $F_2$ | 104[e] | −66[e] | 23.2[e] | 740[f] |
| Cu | $F_2$ | 88 | −50 | 17.5 | 621.89 |
| Zn | $F_2$ | 87[e] | −49[e] | 17.3[e] | 630[g] |
| Zr | $O_2$ | 194 | −75 | 26.1 | 978.0 |
| Zr | $F_2$ | 147 | −110 | 38.3 | 628 |
| Mo | $O_2$ | 124 | −5 | 1.8 | 955 |
| Mo | $F_2$ | 120[e] | −82[e] | 28.8[e] | — |
| Ag | $F_2$ | 83[e] | −46[e] | 16.0[e] | 513.4[h] |
| Ta | $O_2$ | 197[i] | −78[i] | 27.2[i] | 1028.69[j] |
| Ta | $F_2$ | 145[e] | −107[e] | 37.4[e] | — |
| W | $O_2$ | 161 | −42 | 14.8 | 1055 |
| W | $F_2$ | 130 | −92 | 32.3 | 726.5 |
| Pt | $F_2$ | — | — | — | — |
| Au | $F_2$ | 73[e] | −35[e] | 12.3[e] | — |
| U | $O_2$ | 179[k] | −60[k] | 21.1[k] | 920[k] |
| U | $F_2$ | 156[l] | −118[l] | 41.4[l] | 483[m] |

[a] Except as noted below, all values were calculated from Ref. 25.
[b] Ref. 26.
[c] VO data from Ref. 27; V(g) data from Ref. 28.
[d] Ref. 29.
[e] Calculated from data in Ref. 28.
[f] Ref. 30.
[g] Ref. 31.
[h] Ref. 32.
[i] Calculated from $D^{\circ}_0$ instead of $D^{\circ}_{298}$ from Ref. 33.
[j] Ref. 34.
[k] Ref. 35.
[l] Ref. 36.
[m] Calculated from U–F stretching force constant given in Ref. 37.

vibrational-frequency constant for each of the lasing molecules. In this work, it is assumed that the lasing species is the neutral diatomic monofluoride, although the possibility of lasing from a charged or polyatomic species must be considered.

Tables 2 and 3 display the known electronic states of product molecules with energy less than, or within $kT$, the exothermicity of the reactions of some metal atoms with $F_2$ and $O_2$, respectively. For many of these molecules, data are scanty or nonexistent, but it can be seen that, for molecules such as UO, TaO, and TiO, a very large number of accessible electronic states exist. Except for one letter by Zuev et al.,[45] laser

**Table 2.**   $M + F_2 \rightarrow MF + F$ Based on $F_2$ Bond Energy of $13{,}200 \text{ cm}^{-1}$

| Metal atom | Ground state of metal atom | Exothermicity, $\text{cm}^{-1}$ | Product MF states with energy less than exothermicity[a] | Reference |
|---|---|---|---|---|
| Li | $^2S_{1/2}$ | 35,300 | $^1\Sigma^+$ | b |
| C | $^3P_0$ | 31,800 | $X^2\pi$ | c |
| Mg | $^1S_0$ | 24,300 | $X^2\Sigma^+$ | |
| Al | $^2P_{1/2}$ | 42,900 | $^1\Sigma^+$; $A^1\pi$ 43,947.3 | |
| Ca | $^1S_0$ | 31,000 | $^2\Sigma^+$; $A^2\pi_{1/2}$ 16,483; | |
|  | $F_2$ energy |  | $A^2\pi_{3/2}$ 16,558; $B^2\Sigma^+$ | |
|  |  |  | 18,857.51; $C^2\pi$ 30.285.2; | d |
| Ti | $^3F_2$ | 38,500 | $^4\Sigma^-$; $^4\pi \sim 25{,}000$ | |
| V | $^4F_{3/2}$ | 31,400 | | |
| Fe | $^5D_4$ | 24,200 | | |
| Ni | $^3F_4$ | 23,200 | | |
| Cu | $^2S_{1/2}$ | 17,500 | $^1\Sigma$; $^1\pi$ 17,546.8 | |
| Zn | $^1S_0$ | 17,100 | $^2\Sigma$ | |
| Zr | $^3F_2$ | 42,500 | | |
| Mo | $^7S_3$ | 28,700 | | |
| Ag | $^2S_{1/2}$ | 16,100 | $X^1\Sigma^+$ | |
| Ta | $^4F_{3/2}$ | 37,400 | | |
| W | $^5D_0$ | 32,200 | | |
| Pt | $^3D_3^1$ | | | |
| Au | $^2S_{1/2}^0$ | 12,200 | | |
| U | $^5L_6$ | 41,400 | | |

[a] The first symbol is the metal monofluoride ground state followed by the excited states and their energies ($\text{cm}^{-1}$).
[b] Data in this table taken from Ref. 33.
[c] See Ref. 38.
[d] See Ref. 26.

**Table 3.** $M + O_2 \rightarrow MO + O$

| Metal atom | Ground state of M | Exothermicity $\Delta H^\circ_{298}$, cm$^{-1}$ | Product MO states with energy less than $\Delta H^\circ_{298}$ | Reference |
|---|---|---|---|---|
| C | $^3P_0$ | 48,300 | $^1\Sigma^+$; $A\,^3\pi$ 48,473.97 | b |
| Ti | $^3F_2$ | 17,400 | $^3\Delta_{1,2,3}$, $^3\phi_2$ 14,019.74; $^3\phi_3$ 14,095.85; | c |
| | | | $^1\Delta$ 580; $^1\Sigma^+$ 2,798; $^1\pi$ 11,852 | |
| V | $^4F_{3/2}$ | 9,800 | $^4\Sigma^-$ | |
| Zr | $^3F_2$ | 26,100 | $^3\pi\,^3\Sigma$ 16,088; $B$ 18,053–18,134 | d |
| Mo | $^7S_3$ | $-700 \pm 5200$ | $^3\Sigma$ | e |
| Ta | $^4F_{3/2}$ | 27,200 | $X^2\,\Delta_{5/2}$ 3505.43; $A''\,\Delta_{3/2}$ 10,860.95; $A'\,\pi_{1/2}$ 11,062; $B\Phi_{5/2}$ 12,852; $C\Delta_{3/2}$ 13,569; $E\Phi_{5/2}$ 15,880.6; $K'\Phi_{7/2}$ 22,918.75; $L\pi_{1/2}$ 23,341.74; $M\phi_{5/2}$ 24,058.42; $N\pi_{3/2}$ 25,593.13; $O\phi_{7/2}$ 26,121.50; $P\Delta_{3/2}$ 26.673.04; $Q'\Delta_{5/2}$ 27,290.63 | |
| W | $^5D_0$ | 14,800 | $X$, $A \sim 14,160$ | |
| U | $^5L_6$ | 21,100 | $X^7\pi$; $^7\pi$ 2420; $^7\pi$ 5647; $^7\pi$ 11,698; $^7\Delta$ 806; $^7\Delta$ 3227; $^7\Delta$ 4437; $^7\Delta$ 9680; $^5\pi$ 4840; $^5\pi$ 6212; $^5\pi$ 7019; $^5\pi$ 8632; $^5\pi$ 10,892; $^7\Sigma^+$ 4437; $^7\Sigma^+$ 4500; $^7\Sigma^+$ 8068; $^7\Sigma^-$ 9682; $^7\Sigma^-$ 13,715; $^5\Delta$ 2097; $^5\Delta$ 4034; $^5\Delta$ 6857; $^5\Delta$ 7261; $^2\Sigma^+$ 3630; $^2\Sigma^-$ | f |

[a] The first symbol is the metal-monofluoride ground state followed by the excited states and their exothermicities.
[b] References in this table taken from Ref. 33.
[c] See Ref. 39 through 42.
[d] See Ref. 43.
[e] See Ref. 25.
[f] Values read from a graph two significant figures of accuracy in Ref. 44.

oscillation as a result of direct chemical pumping of electronic transitions of product molecules has not yet been reported. In view of the widespread evidence that reaction energy is deposited in internal energy of products, the possibility of laser oscillation on an electronic transition pumped by a chemical reaction should be explored thoroughly. Laser oscillation could be invaluable in providing basic spectroscopic constants and initial distributions over electronic states of product molecules. Refractory metal halides and oxides have many low-lying electronic states and large reaction exothermicities. Population inversions of electronic states of these product molecules seem likely.

Large inversion-energy storage may be possible for a class of metal-atom oxidation lasers that operate on spin forbidden transitions. Angular momentum must be conserved in chemical reactions, and, under certain conditions, electron-spin and electron-orbital angular momentum do not couple or mix with molecular-rotational angular momentum. In these cases, spin and orbital angular momentum must be conserved separately. This conservation can force a product atom or molecule to be borne in a state other than the ground state and can channel a large fraction of the reaction energy into laser energy. Such a scheme could be used on either electron-orbital or electron-spin angular momentum, although the electron-spin scheme is more general, since there are no requirements for a linearly activated complex.

Consider the atom–diatom reaction

$$A + BC \rightarrow AB + C$$

Because the spins of the reactants can couple together to give a resultant spin $S_R$ that can have any value between $|S_A - S_{BC}|$ and $S_A + S_{BC}$, products can only be forced to particular excited states if one of the reactants has spin 0. In the most obvious example, if the reactants are chosen so that the atom has spin or orbital angular momentum of one unit, while the reactant molecule has no electronic angular momentum, and the products both have zero electronic angular momentum in their ground states, the product molecule or atom has to be borne in an excited electronic state with one unit of angular momentum. The principle of conservation of electron spin is known as the Wigner spin-conservation rule. However, it should be noted that, since rotation of the product molecule can exchange quanta of angular momentum with electrons, the spin-conservation rule does not always hold. The degree of isolation of electronic angular momentum varies from case to case, but generally holds best for light species. An example of a reaction utilizing spin conservation is the following:

$$C^3P + N_2O^1\Sigma \rightarrow N_2{}^1\Sigma + CO^3\pi$$

The ground state of CO, of course, is $^1\Sigma$ so that the reaction may force the CO to the $^3\pi$ excited state initially.

Each of the metals listed in Tables 1, 2, and 3 require high temperature for evaporation and, therefore, some new techniques for preparation of metal vapor in sufficient quantities for lasing. A number of techniques such as nozzles, shocked-metal plates, and exploding wires might meet the requirements. Laser action has indeed been achieved with exploding wires and shocked-metal plates to produce metal atoms. Each of these techniques will be discussed in their respective sections of this chapter.

## 1. EXPLODING-WIRE METAL-ATOM OXIDATION LASERS

The innovation necessary for the development of a metal-atom oxidation laser must provide and mix metal atoms with the oxidizer gas in a time comparable to, or shorter than, the lifetime of the upper laser level of the lasing species. Radiative lifetimes of vibrational states of polar molecules are in the range of one to several milliseconds. Vibrational states are deactivated by V–V energy transfer in a few dozen collisions and by V–T energy transfer in a few hundred collisions. Since V–V relaxation of a product molecule with itself does not quench laser action from the $v = 1 \rightarrow v = 0$ transition, it is possible to achieve lasing on that transition, if the pumping rate exceeds the rate of spontaneous emission and the rate of V–T energy deactivation. For example, in the HF or HCl lasers operating at 10 Torr, the time available for pumping is at least a few hundred collision periods or a few microseconds. The time available for atomization and mixing of metals with oxidizers in metal-atom oxidation lasers was presumed to be similar.

Electrical explosions of small-diameter metal wires are known[46–49] to produce irregular jets of vapor or colloidal material with velocities near $10^6$ cm/sec that would seem to fill the requirements for atomization and mixing of a sufficient amount of material in the appropriate time for laser action on vibrational transitions. The mixing would be accomplished by diffusion of the metal vapor out of the small-diameter jets and into the oxidizer gas. Exploding wires have already been used in the synthesis of some binary inorganic compounds,[50] and diatomic product emission has been observed in exploding-wire experiments.[51] For these reasons, an exploding-wire, metal-atom oxidation laser was designed and built.

The first successful laser tube,[23] illustrated in Fig. 1, was constructed of stainless steel and glass. It could be evacuated to sub-Torr pressures, was noncorrodable, and was easy to disassemble and clean. The glass tubes supporting the wires were 23.8-cm long with either a 22-mm or 17.5-mm i.d. An outer glass tube provided the vacuum seal. Tubes plated with thin metallic films and tubes containing graphite smears were also used in this setup. Figure 2 shows this laser tube incorporated in the experimental arrangement. The laser tube was centered between two gold-surfaced mirrors of 10-m radii of curvature that were 140 to 160 cm apart. One mirror has a 2-mm hole through its center for output coupling. The optical cavity between the KBr windows and the laser mirrors was purged with

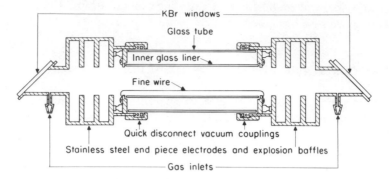

**Fig. 1.** Glass and stainless-steel laser tube (reproduced with permission of *Applied Physics Letters*).

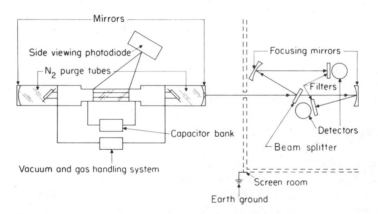

**Fig. 2.** Exploding-wire laser set-up with instrumentation.

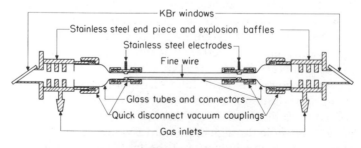

**Fig. 3.**

710

nitrogen to remove and exclude moisture and gases that might attenuate the laser radiation.

The exploding-wire setup was later modified (Fig. 3) to give a shorter wire-reloading time between shots. This modification doubled the repetition rate of the laser experiments. The 2.54 to 5.08-cm glass adapters at each end of the tube can slide within the end piece quick-disconnects, thereby allowing the glass liner containing the wire(s) to be quickly changed without disturbing the alignment of the KBr windows. The laser

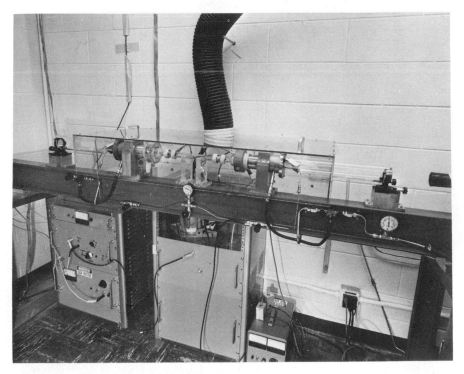

**Fig. 4.**

mirrors and laser tube were mounted on a massive steel I-beam to give mechanical stability to the optical cavity (Fig. 4). In several experiments, a 1-mm coupling-hole output mirror was used to give high feedback. Later experiments have employed 2-m radii of curvature mirrors with a near confocal separation of 186 cm.

Before the fine wires were installed in the laser tube, the laser cavity was aligned by operating the laser as a $CO_2$ laser; that is, by flowing a $CO_2$ laser-gas mixture through the tube and providing a pulsed electric

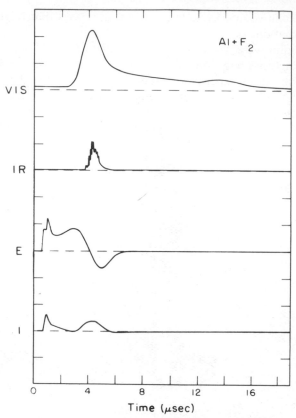

**Fig. 5.** AlF Laser characteristics: VIS is visible luminescence; IR is infrared laser oscillogram; E is exploding-wire voltage; and I is exploding-wire current.

discharge between the end-pieces. The laser mirrors were then carefully adjusted to optimize the $CO_2$ laser signal.

Some early results obtained with the apparatus (Fig. 1) are shown in Fig. 5. The top oscillogram in the figure is the signal from a photodiode mounted at the side of the laser tube. It monitors the visible luminescence from the wire burst and from the burning of the metal in the fluorine. The second trace is an oscillogram of the infrared-laser oscillation. The infrared laser was much more intense than the visible fluorescence, and the oscillogram had to be recorded at low amplification. Notice the sharp threshold, termination, and characteristic spiking of the laser oscillation. That is to be contrasted with the visible luminescence, which continues at a low level for over 18 $\mu$sec. The voltage trace $E$ and the current trace $I$ are characteristic of exploding wires and will be discussed in more detail later. It is important to note that the lasing occurs after the burst of the wire and during the expansion of the vapor and droplets into the oxidizing gas.

Lithium wire was prepared by extruding Li through a 0.200-mm die. The corrosion of Li metal under various gases was examined by placing small chunks of the metal in evacuable tubes under dry argon, evacuating the tubes, and backfilling with 100 Torr of oxidizing gas. None of the gases—$F_2$, $NF_3$, $SF_6$, $Cl_2$, or $O_3$—reacted violently with the shiny metal or darkened the surface rapidly. In fact, the gases seemed to form a tough protective coating on the lithium. A tube containing 120 Torr of $F_2$ and Li did not react violently even when heated to incandescence! Lithium wires treated with $F_2$ or $CO_2$ can be left in air several minutes without tarnishing. Preparation and extrusion of the Li wires was performed in a drybox under a $CO_2$ atmosphere to minimize corrosion.

To implode metallic films into oxidizing gases, several of the glass liners (Figs. 1 and 3) were plated on the inside with fine films of Cu, Al, and Mg. The Cu and Al films were prepared under vacuum by vaporizing the metals off filaments positioned at the center of the tube. The Mg films were prepared by exploding two 0.127-mm Mg wires onto the inner surface of Pyrex tubes at $<10^{-5}$ Torr. The films were electrically connected to the stainless-steel end pieces with spring clips. Electrically conducting carbon smears were prepared by rubbing graphite onto 3-mm grit-roughened glass rods. These were used in place of the wires for CO and CF laser experiments.

When a bus bar was installed in place of the wire in the laser, an underdamped sine wave was obtained from the current-viewing resistor. Analysis of this waveform yielded the following circuit characteristics for the 3.0-$\mu$F 20-kV power supply: peak current $I_{max} = 45$ kA; ringing period $P_R = 7.78$ $\mu$sec (frequency, $f = 129$ kHz); inductance $L = 0.512$ $\mu$H; and resistance $R = 0.0373$ $\Omega$.

Current and voltage traces typical of the 22.9-cm long wires fired in the laser are shown in Fig. 6. (It should be noted that the bridgewire current can never exceed $I_{max}$ of the bus-bar test.) The burst phase constitutes only a small part of a typical exploding-wire time history. It is frequently followed by a phase known as dark pause, where the current falls to a constant low value, and the light output of the wire is very small. The last phase, restrike, occurs when electrical avalanche begins, and a ringing discharge completes the capacitor discharge. From Table 4, it is evident that the restrike current $I_{max}$ is often an order of magnitude greater than $I_p$, the largest current observed in the burst phase. The ringing period $P_R$ of the restrike current approximates that observed with the bus bar. Light intensity during the restrike is much greater than that of the burst phase.

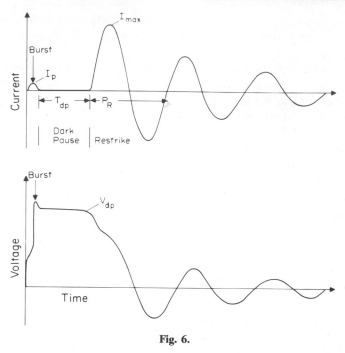

Fig. 6.

Numerous and sometimes contradictory theories of exploding bridge-wire (EBW) phenomenon can be found in the literature, but a model[52] that agrees with these data states that the wire is reduced to a high-density colloid of liquid-metal droplets and vapor by the end of the burst phase. Since no continuous conduction path exists and the vapor density results in an electron mean-free path less than that required to accelerate the thermionic electrons to avalanching velocities, a high resistance exists, and current flow is carrier-limited. The vapor expands to lower densities until avalanching and, hence, restrike occurs.[52] This model suggests that the duration of dark pause $t_{dp}$, should correlate with the ionization potential of the EBW material, high pressures should retard restrike, and the phase change is facilitated when glass liners are inserted inside the laser tube. All three of these effects are observed in the data.

Since restrike produces a great deal of ionization, perhaps as high as 90% for lithium,[53] it is probably desirable to deposit more energy during the burst phase and less in the restrike phase to minimize ionization. To reduce the energy in excess of that required to burst the fine wires, a $0.48$-$\mu$F capacitor was used for many of the laser shots. A bus-bar waveform indicated the following circuit characteristics for the smaller

**Table 4.** Exploding-Wire Function Tests

| Material[a] | Initial conditions | | | Burst | Dark pause | | Restrike | |
|---|---|---|---|---|---|---|---|---|
| | $V_0$,[b] kV | $P_0$, Torr | Inner liner | $I_F$, kA | $t_{dp}$, $\mu$sec | $V_{dp}$, kV | $I_{max}$, kA | $P_R$, $\mu$sec |
| Li | 20 | 91 air | yes | 3.83 | 1.0 | 23.0 | 33.4 | 8.4 |
| Al | 20 | 88 air | yes | 7.61 | 2.2 | 22.9 | 28.2 | 8.0 |
| Mg | 20 | 87 air | yes | 4.32 | 3.6 | 23.2 | 30.4 | 8.2 |
| Cu | 20 | 87 air | yes | 5.76 | 4.0 | 23.0 | 26.7 | 8.4 |
| Fe | 20 | 86 air | yes | 3.29 | 5.0 | 23.0 | 24.0 | 8.0 |
| Fe | 20 | 87 air | no | 2.88 | 7.4 | 22.2 | 26.7 | 8.0 |
| Cu (4) | 20 | 88 air | yes | 14.40 | 0 | none | 24.7 | 7.4 |
| Cu | 15 | 84 air | no | 4.32 | 9.8 | 16.8 | 26.3 | 7.5 |
| Cu | 20 | 578 air | no | 4.94 | 9.6 | 23.0 | 30.4 | 8.1 |
| Cu | 20 | 112 air | no | 4.74 | 8.0 | 23.0 | 34.9 | 7.6 |
| Cu | 20 | 6 air | no | 4.93 | 0 | none | 37.0 | 7.6 |
| Cu | 20 | 578 SF$_6$ | yes | 4.73 | 6.2 | 23.0 | 30.3 | 6.2 |
| Cu | 20 | 586 SF$_6$ | no | | 87.0 | 23.0 | 28.0 | |
| Li | 20 | 586 SF$_6$ | no | | 3.0 | 22.2 | 22.6 | |

[a] Single-wire shots except where indicated.
[b] Not calibrated; ±2.5%.

**Table 5.** Representative Metal-Atom Oxidation-Laser Results

| Metal[a] M | Oxidizer XY | Oxidizer pressure Torr | Time to laser onset, $\mu$sec | Laser-pulse duration, $\mu$sec | Relative laser intensity | Laser wavelength, $\mu$m |
|---|---|---|---|---|---|---|
| Li | $F_2$ | 25.6 | 2.0 | 2.6 | 5.0 | $16 > \lambda > 13$ |
| C[b] | $O_2$ | 71.8 | 8.4 | 2.0 | 3.1 | $24 > \lambda > 5$ |
| C[b] | $F_2$ | 41.0 | 8.0 | 2.0 | 7.1 | $24 > \lambda > 10.5$ |
| Mg | $F_2$ | 28.2 | 4.2 | 1.5 | 4.8 | $13.5 > \lambda > 12.8$ |
| Mg[c] | $F_2$ | 30.1 | 1.3 | 2.8 | 6.6 | $14 > \lambda > 12.8$ |
| Al | $F_2$ | 24.6 | 3.0 | 1.7 | 6.5 | $13.5 > \lambda > 12.5$ |
| Al[d] | $F_2$ | 31.7 | 1.5 | 1.4 | $\gg 10.0$ | $14 > \lambda > 8.8$ |
| Ti | $O_2$ | 65.5 | 5.7 | 1.4 | 3.4 | $24 > \lambda > 10.5$ |
| Ti | $F_2$ | 24.8 | 3.8 | 1.6 | $\gg 1.4$ | $24 > \lambda > 11.1$ |
| Ti | $NF_3$ | 26.2 | 4.9 | 1.7 | 4.3 | $24 > \lambda > 5$ |
| V | $O_2$ | 55.8 | 5.3 | 1.0 | 1.3 | $14 > \lambda > 8.8$ |
| V | $F_2$ | 59.0 | 4.0 | 1.7 | 3.1 | $14 > \lambda > 8.8$ |
| Fe | $F_2$ | 28.4 | 4.2 | 2.2 | 6.2 | $24 > \lambda > 11.1$ |
| Ni | $F_2$ | 25.8 | 3.4 | 2.2 | 7.0 | $24 > \lambda > 10.5$ |
| Cu | $F_2$ | 26.7 | 3.5 | 2.1 | 6.6 | $24 > \lambda > 11.1$ |
| Cu[e] | $F_2$ | 35.6 | 4.3 | 0.4 | 4.6 | $14 > \lambda > 8.8$ |
| Zn | $F_2$ | 34.2 | 6.3 | 0.9 | 2.2 | $14 > \lambda > 8.8$ |
| Zr | $O_2$ | 58.9 | 4.8 | 1.3 | 7.5 | $14 > \lambda > 8.8$ |
| Zr | $F_2$ | 65.7 | 2.3 | 2.4 | $\gg 2.1$ | $14 > \lambda > 8.8$ |
| Mo | $O_2$ | 61.5 | 4.3 | 1.2 | 1.9 | $14 > \lambda > 8.8$ |
| Mo | $F_2$ | 61.9 | 3.4 | 2.3 | $\gg 2.9$ | $16 > \lambda > 8.8$ |
| Ag | $F_2$ | 79.8 | 2.4 | 1.0 | $\gg 3.4$ | $14 > \lambda > 8.8$ |
| Ta | $O_2$ | 60.6 | 4.4 | 1.0 | 3.4 | $14 > \lambda > 8.8$ |
| Ta | $F_2$ | 31.8 | 4.5 | 1.8 | $\gg 1.7$ | $14 > \lambda > 8.8$ |
| W | $O_2$ | 60.1 | 4.3 | 1.3 | 5.9 | $14 > \lambda > 8.8$ |
| W | $F_2$ | 37.2 | 3.1 | 2.6 | $\gg 1.7$ | $14 > \lambda > 8.8$ |
| Pt | $F_2$ | 28.1 | 3.6 | 2.4 | 6.6 | $24 > \lambda > 11.1$ |
| Au | $F_2$ | 28.6 | 10.4 | 1.3 | 5.5 | $24 > \lambda > 10.5$ |
| U | $O_2$ | 51.7 | 3.6 | 1.9 | 5.8 | $16 > \lambda > 8.8$ |
| U | $F_2$ | 30.5 | 3.7 | 3.0 | 7.3 | $24 > \lambda > 10.5$ |

[a] All metals are in fine-wire form except as noted.
[b] Graphite smears on glass rods, $>0.22$ m mol C.
[c] Mg film on glass liner $<0.43$ mmol Mg.
[d] Al film on glass liner $<0.15$ mmol Al.
[e] Cu film on glass liner $<0.24$ mmol Cu.

capacitor: $I_{max} = 19.0\,\text{kA}$; $P_R = 2.96\,\mu\text{sec}$ ($f = 338\,\text{kHz}$); $L = 0.463\,\mu\text{H}$; and $R = 0.0916\,\Omega$.

A preliminary energy calculation has revealed that about 10% of the energy required to totally vaporize the ion wire has been deposited by the end of dark pause. Thus only part of the wire is vaporized up to the inception of restrike. Many workers have observed segmentation of long bridgewires as vaporization occurs. There are observed striations in the metal deposition patterns on the glass liners.[1] The burst pattern is indicative of both wire segmentation and vapor dispersion as it expands across the laser cavity. The striations are typically 1 to 2 mm apart in agreement with published observations.[54] The iron-wire burst is particularly nonuniform, and small metal beads indicate that part of the wire does not vaporize. The fine semitransparent films deposited on the walls of the liners are evidence of metallic-vapor deposition. Rapid mixing of metal vapor and oxidizer gas is achieved because of segmentation and nonuniform burst; however, segmentation itself may be undesirable because the metal vapor is generated over a prolonged period of time.

Using the experimental arrangements of Figs. 1 through 4, infrared laser emission was obtained from the molecules listed in Table 1. Intense pulses of infrared radiation of 0.4- to 3.0-$\mu$sec duration were observed 1.3 to 10.4 $\mu$sec after triggering the capacitor discharge. Table 5 lists representative results and experimental conditions for various combinations of metals and oxidizers. Oscillograph tracings for some of these lasers are shown in Figs. 7 through 9. Figure 7 shows characteristic laser,

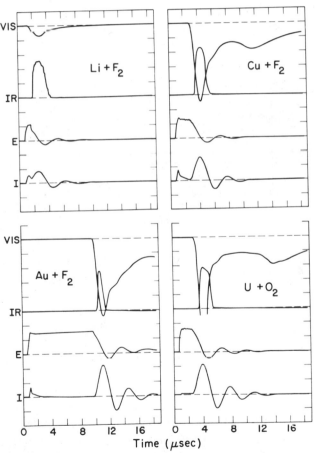

**Fig. 7.** Oscillograms of exploding wires in oxidizing gases showing visible fluorescence (VIS); infrared laser laser pulse (IR); voltage (E), 15 kV/div; and current (I), 5 kA/div (except Li at 10 kV/div). (Figures used with permission of *Chemical Physics Letters*.)

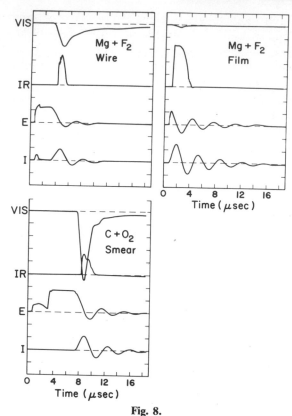

**Fig. 8.**

visible fluorescence, current, and voltage traces for metals with widely differing properties. Results from the three electrical-evaporation methods used in atomizing conducting solids, that is, exploding wire, cylindrically imploding metallic film, and exploding graphite smear, are shown in Fig. 8. With a cylindrically imploding metal film, the visible fluorescence is blocked by the film itself. Figure 9 contrasts the visible fluorescence and laser pulse for exploding Ti wires in three different oxidizing gases.

In experiments with $F_2$ as the oxidizer, decrease in pressure indicated consumption of $F_2$ by a chemical reaction producing a nonvolatile fluoride. The amount of $F_2$ consumed corresponds to oxidation of varying amounts of metal from 10% for Mg to 100% for Ti, Ni, and Mo. In all experiments, except those using the smallest diameter Ti, Ni, and Mo wires, a striated film of metal was deposited on the glass tube. The exploding-wire shots in $O_2$ either caused an increase or a decrease in

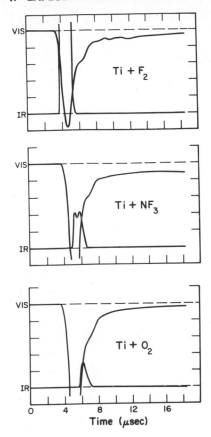

**Fig. 9.**

pressure depending on whether the final reaction products were volatile, as in $C + O_2 \rightarrow CO_2$, or not, as in $Ti + O_2 \rightarrow TiO_2$.

The current and voltage traces for the exploding wires exhibit burst phase, dark pause during vapor expansion, and sinusoidal restrike, as discussed earlier and shown in Fig. 6. Figures 5, 7, and 8 show wide variations in burst current and dark pause with different materials and techniques. For the Mg film, no dark pause was observed, and the burst current was followed immediately by restrike. In the graphite smear, the burst phase was not detected on the current trace, but its presence was indicated by the first dip in the voltage curve. The slow burst might be attributed to the high resistance of the graphite smear.

A comparison of the electrical characteristics of the Al wire in a 54.4-Torr $F_2$ laser shot (Fig. 5) and the Al exploding-wire shot of Table 5 is instructive. The 0.48-$\mu$F capacitor used in the laser shot had been substituted for the 3.0-$\mu$F capacitor used in the exploding-wire tabulation. This resulted in a slight drop in the peak burst current $I_p$ from 7.61 to 6.26 kA; however, the peak restrike current $I_{max}$ dropped from 28.4 kA to only 4.12 kA. A detailed calculation of the electrical behavior of the Al+$F_2$ shot shown in Fig. 5 was made by computer analysis. The analysis showed that wire burst, defined as the time when $dI/dt$ is at its first minimum, occurred at 0.60 $\mu$sec. Peak burst power of 92.3 MW and peak resistance of 4.12 $\Omega$ occur at or near this time. (Initial resistance of this 0.127-mm diameter wire was 0.53 $\Omega$.) Of the 96 J originally stored in the capacitor, 21.1 J are deposited by the time of wire burst. The 8.14-mg mass of the Al wire requires 8.7 J for complete liquification and an additional 100 J for complete vaporization. Therefore, no more than 12% of the wire could be vaporized at burst time. Appreciable energy deposition during the dark pause results from the high resistance of the wire (as much as 36.7 $\Omega$), even though the dark-pause current falls from 6 kA to 235 A. By the start of restrike, 49.6 J are deposited, but no more than 41% of the wire could be vaporized. Restrike behavior is that of a negative resistance arc in which no further electrical energy is deposited. In fact, energy is extracted from the plasma, and cooling of the hot vapor is probably occurring. The previously mentioned vapor-fraction calculations are upper limits, because energy lost to kinetic energy, radiation, ionization, and other nonvaporization processes has been neglected.

When wires were strung diagonally across the tube so that they were confined only inertially, they failed to produce laser oscillation. This was probably due to reduced metal-vapor production resulting from the lack of interaction with the walls of the glass liner.[50]

The Al+$F_2$ laser operated over a wide range of pressures from 12.7 to 251 Torr of $F_2$. Pulse duration and laser onset time both appeared to increase with increasing pressure. The effect of pressure on the TiF laser output was studied. Single 0.041-mm diameter Ti wires were fired in $F_2$ gas over a pressure range of 25 to 230 Torr. The relative intensities, pulse durations, and onset times were measured from oscillograph traces of a PbSnTe ir detector with a 8.8-$\mu$m long-pass filter. The results are shown in Fig. 10. All of the graphs indicate that the most significant changes in laser characteristics with respect to pressure occur between 25 and 75 Torr. From 75 to 230 Torr, the changes are more gradual. Laser onset time would be proportional to the reciprocal of the fluorine pressure if it were controlled solely by bimolecular chemical kinetics. However, the increase of laser onset time with increasing pressure probably is due to the effect of gas pressure on the wire-burst phenomenon.

The following four observations support the conclusion that the laser oscillation is driven by chemical reaction and not by electrical pumping.

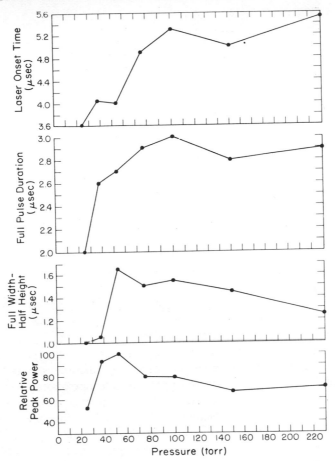

**Fig. 10.**

(1) In all experiments the onset of visible fluorescence coincided with the restrike current, but the onset of the ir laser pulse occurs anytime after the visible fluorescence begins and varies with the conditions of the experiment; (2) when wires were exploded into an evacuated laser tube, negligible infrared emission was observed, but with oxidizing gases in the tube, the exploding wires produced ir fluorescence for tens of microseconds, with short pulses of ir laser radiation 2 to 3 orders of magnitude more intense than the fluorescence; (3) the lasers demonstrated oxidizer-pressure threshold effects as illustrated in Fig. 11 for the $U + O_2$ system; and (4) laser action from shocked metal-plate vaporization lasers (discussed in the following section) from the $Mg + F_2$ reaction

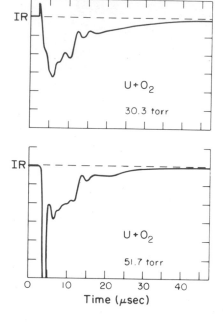

**Fig. 11.**

was achieved without electrical discharge. Laser characteristics observed from this shocked metal-plate laser system are indistinguishable from the results obtained in exploding-wire apparatus and corroborate the results. In all the work performed to date, the lasing species are assumed to be diatomic molecules or ions because of the difficulties involved in producing inverted populations of polyatomic molecules by chemical reaction.[55] Further oxidation of the metal to polyatomic species is believed to produce a nearly thermal distribution of product molecules.

The MgF, ZnF, ZrF, AgF, WF, and UF lasers exhibit shorter wavelengths than would be expected from calculated $P$-branch transitions. The wavelength ranges for LiF and CF are at longer wavelengths than their respective $\omega_e$ values might indicate. This is probably due to a high degree of vibrational excitation that results in transitions between high vibrational levels. No $\omega_e$ values were found in the literature for VF, MoF, TaF, PtF, or AuF. Much infrared spectroscopic work is needed to assign laser species, elucidate important reaction mechanisms, and obtain spectroscopic data for little-known molecules.

The exploding-wire laser experiments can already remove some of the uncertainty in published thermodynamic data. In the conversion of chemical energy into laser-pumping energy, the upper limit on the energy available to the products may be expressed as the sum of the activation energy and enthalpy of reaction $E_A - \Delta H°$,[56] or, more conservatively, as the difference in ground-state dissociation energy of the oxidizer and diatomic produce $D_0°(YX) - D_0°(MX)$.[5] For the reaction $Mo + O_2 \rightarrow MoO + O$, one calculates $\Delta H°_{298} = -5 \pm 15$ kcal/mol.[25] Since the reaction must populate at least the first excited vibrational state in order to achieve laser action, the $\Delta H°_{298}$ of reaction must be at least a negative 2.7 kcal/mol, leading to a maximum positive uncertainty limit of $+2.3$ kcal/mol, instead of $+15$ kcal/mol. In the years ahead, detailed spectroscopy will be used to determine how far up the vibrational ladder laser action is originating, and what kinetic and thermochemical data can be obtained from exploding-wire metal-atom oxidation lasers.

## 2. SHOCKED METAL-PLATE VAPORIZATION LASERS

Shock heating of compressible materials is an alternative to exploding wires for production of metal vapor. The jets that emanate from grooved metal plates when strongly shocked can provide the vapor required for metal-atom oxidation lasers. A flash X-radiograph of a uranium plate and the jets produced by the shock wave is shown in Fig. 12. The upper picture shows a 2.54-cm uranium plate with 90° $V$-grooves and a 10-cm wide Composition B (60% RDX, 40% TNT) explosive charge before it is fired. The lower picture is a stop-action photograph of the same assembly taken 25 $\mu$sec after the shock wave (generated by the explosive) arrived at the free surface of the uranium.[58] The jets from the $V$-grooves had moved 49.7 mm from the original free surface in that time, at a velocity of 1.99 mm/$\mu$sec. The vapor fraction of these jets is unknown. In general, the amount of vapor produced in this type of event depends on the metal used. Kilogram quantities of barium vapor with velocities $>1$ cm/$\mu$sec have been produced from metal-lined conical-shaped charges.[59,60]

The groove shape has a considerable effect on the nature and velocity of the resulting jet.[1,61] It is clear from this photograph and from theoretical calculations[61] that shocked plates can yield appreciable metal vapor, and this technique has been employed successfully to produce metal vapor for the $Mg + F_2$ metal-atom oxidation laser. The experiments are now described.

To minimize damage from the 450-g charge of explosive, the laser assembly was mounted in a protective blast shield inside an optical cavity similar to those used for the exploding-wire laser. The laser is aligned by demounting the explosive, running the device as a $CO_2$ laser, optimizing all mirror adjustments, and then reinstalling the charge on the laser tube.

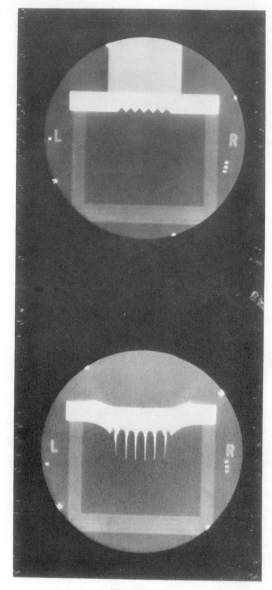

Fig. 12.

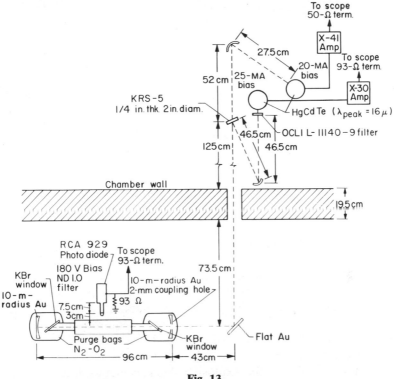

**Fig. 13.**

Figure 13 shows the layout of the laser and optical diagnostics for the shocked-plate vaporization laser.

In the first series of shots with uranium plates and oxygen gas, only weak ir fluorescence was produced along the laser axis, but very bright visible luminescence was incident on a photodiode situated a few centimeters from the side window of the laser tube. This luminescence was probably chemiluminescence from the oxidation of uranium vapor or particles by the oxygen gas.

In the second series of experiments, $5.08 \times 30.5$-cm square grooved magnesium plates were used with fluorine gas. Intense ir laser action with a duration of $0.3$-$\mu$sec FWHM and a wavelength of $>11$ $\mu$m was observed. This laser pulse was very similar in power, duration, and wavelength to the MgF laser pulses described in the previous section. Oscillograms of the laser pulse are shown in Fig. 14. The abrupt termination of the visible-side luminescence signal is caused by the shock-wave destruction of the photodiode.

Perhaps the most important result of this experiment is that metal atom

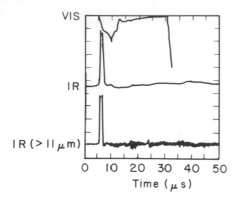

**Fig. 14.** Shocked plate Mg+F metal-atom oxidation laser. Upper trace is from side-looking photodiode; center trace marked IR is on-axis unfiltered oscillogram of HgCdTe detector filtered with an 11-$\mu$m long-pass filter. (Figure used with permission of *Journal of Applied Physics.*)

oxidation lasing has been observed in a system in which no electrical discharge is present. This verifies that the pumping mechanism is purely chemical and should remove doubt concerning the exploding-wire metal-atom oxidation-laser results. The explosives technique is too time-consuming to use as a primary tool for the study of metal-atom oxidation-laser properties, because damage to the medium containment box is severe and much set-up and preparation is required for every shot. Continuous-flow and exploding-wire methods are much more suitable for most data gathering.

The ultimate utility of the explosive shocked metal-plate technique is unknown. It will depend upon the development and performance of metal-atom oxidation lasers—their efficiency, energy storage capacity, and so on. The explosives themselves represent a compact source of energy, ~2 MJ for the charge used in this experiment, and even very small efficiencies would make such lasers practical for certain applications.

## 3. FLOW-MIXING, cw METAL-ATOM OXIDATION LASERS

The continuous, high-velocity stream of metal-vapor atoms necessary for a cw metal-atom oxidation laser may be produced by such methods as the injection of metallic powders into high-enthalpy gaseous flows (plasma jet or flame spray), the use of consumable electrode arcs, and the heating of the liquid metal to temperatures where its vapor pressure is sufficiently high. Of these general methods, the latter is the most direct and the simplest, and it will be the only one discussed in this section.

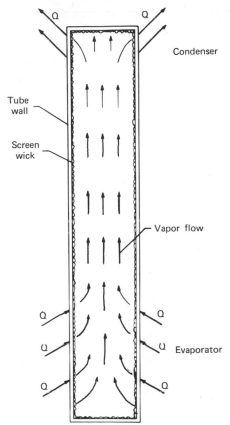

Condenser

Tube
wall

Screen
wick

Vapor flow

Evaporator

**Fig. 15.** Heat pipe.

A device that can be used as a compact, large surface area, essentially isothermal, metal vaporizer is the heat pipe.[62] The basic heat pipe is shown schematically in Fig. 15. The pipe is an evacuated, closed-off tube with a tightly fitting fine-mesh screen structure along all its interior surfaces. Heat added at the evaporator end vaporizes the working fluid from the screen surface. The vapor flows to the condenser end of the pipe where it recondenses, giving up its heat of vaporization. The condensed liquid then recycles to the vaporizer section through the screen structure by capillary action. Since the liquid and vapor phases of the working fluid are in equilibrium along the pipe, the entire pipe is essentially isothermal. This basic heat pipe operation has been extended to an open-ended heat-pipe design for the continuous production of lithium vapor for a continuous hypersonic lithium jet.[63,64]

An open-ended heat-pipe metal-vapor injection system is simple to construct and relatively easy to operate. In addition, by making the metal-vapor nozzles an integral part of the heat pipe, recondensation of the metal vapor on the nozzle walls can be prevented.[64] The mass-flow rate of metal vapor is easily calculated from the temperature of the heat pipe. The pressure of metal vapor in the nozzle plenum is the vapor pressure of the metal at the heat-pipe temperature, and, thus, the mass flow may be determined from the isentropic flow relation:

$$\dot{m} = A^* \left[ \frac{\gamma}{R} \left( \frac{2}{\gamma+1} \right)^{(\gamma+1)/(\gamma-1)} \right]^{1/2} P_0 (T_0)^{-1/2}$$

where $\dot{m}$ is the mass flow, $A^*$ is the nozzle-throat area, $\gamma$ is the ratio of specific heats, $R$ is the gas constant, $T_0$ is the plenum-gas temperature taken to be the heat-pipe temperature, and $p_0$ is the plenum pressure taken to be the vapor pressure at temperature $T_o$.

Results from the exploding-wire and shocked metal-plate vaporization lasers should be a guide in defining the operating parameters of a transverse-flow laser utilizing a heat-pipe vaporizer. These results indicate that an optical gain length of 20 to 30 cm is sufficient to obtain laser oscillation. Consideration of probable minimum vibrational-state lifetimes and formation rates indicate that metal-vapor velocities of $10^5$ to $10^6$ cm/sec and cavity temperatures of 200 to 400°K are appropriate for successful operation of a continuous-flow metal-atom oxidation laser. Fresnel-number considerations indicate that a beam diameter near 1 cm is needed for lasers in the 10- to 20-$\mu$m spectral region with a 30-cm length.

A computer program was used to calculate nozzle-exit conditions of a two-dimensional slit nozzle for a range of inlet temperatures, exit Mach numbers, and beam widths. Flows of Mg, Li, Na, and Zn were investigated. The flow was assumed to be an ideal isentropic flow of a monatomic vapor, as is known to be the case for hypersonic lithium-vapor flow.[64] Computed quantities were: nozzle plenum pressure $p_0$; nozzle-exit height; nozzle-exit width; nozzle-throat height; metal-vapor mass flow, metal-vapor Mach number, and velocity; and metal-vapor exit temperature, pressure, and density. Typical results, computed for the MgF laser, are listed in Table 6. While these calculations were carried out for a single nozzle, the results would be the same for an array of smaller nozzles having the same total throat area. In practice, an array of small nozzles will be required so that fluorine injection ports may be placed within the beam to provide more uniform mixing of the reactants than would be provided by purely peripheral fluorine injection. The exact contour design of the metal vapor and fluorine nozzles were not considered in detail;

**Table 6.** Operating Parameters of a Transverse-Flow Laser

| | |
|---|---|
| Pumping reaction | $Mg(g) + F_2 \rightarrow MgF^* + F$ |
| $-\Delta H^\circ_{298}$ of reaction | 69.5 kcal/mol |
| $\omega_e$ | 717.6 cm$^{-1}$ |
| $\lambda$ | 13.9 $\mu$m |
| Operating temperature | 1000°K |
| Vapor pressure of Mg | 7.5 Torr |
| Nozzle dimensions | 31-cm wide, 1-cm high |
| Nozzle-throat height | 0.33 cm |
| Power input for vaporization | 7 kW |
| Velocity of Mg atoms | Mach 3, $v = 1.13 \times 10^5$ cm/sec |
| Nozzle pressure of Mg | 0.235 Torr |
| Mass flow rate of Mg | 0.053 mol/sec |
| Maximum pumping speed[a] | 990 liter/sec |
| Mass flow rate of $F_2$[b] | 0.053 mol/sec |

[a] Assuming from the stoichiometry of the pumping reaction that 0.026 moles $F_2$/sec will be the maximum flow.

[b] Assuming a stoichiometric reaction between Mg and $F_2$.

neither was the flow-mixing process downstream of the injectors. Nevertheless, the results show that a transverse-flow laser can be constructed at reasonable values of vaporizer temperature, vaporizer power, and pumping speeds.

The construction of a flowing metal-atom oxidation laser, operating on the reaction between fluorine and metal vapor, presents a formidable problem in materials compatibility because of the coincidence of high temperature, corrosive metal, and fluorine. Nevertheless, successful operation of a heat pipe-supplied Na + $F_2$ flame has been achieved.[65] Infrared chemiluminescence from the Na + $F_2$ flame was observed in the experiment. Continuous-flow metal-atom oxidation lasers will likely be reported in the near future, and they should be useful in providing basic-rate data on the oxidation of metals atoms and basic spectroscopic data on lasing molecules.

# REFERENCES

1. W. W. Rice, W. H. Beattie, J. G. DeKoker, D. B. Fradkin, P. F. Bird, and R. J. Jensen, "Metal Atom Oxidation Lasers," Report LA-5452 (Los Alamos Scientific Laboratory, Los Alamos, New Mexico, Sept. 1973).

2. M. G. Evans and M. Polanyi, Trans. Faraday Soc. **35**, 173 (1938).

3. M. Polanyi, *Atomic Reactions* (Williams and Northgate, London, 1932).

4. H. Beutler and M. Polanyi, Z. Phys. Chem. **B1,** 3 (1928).

5. S. V. Bogdandy and M. Polanyi, Z. Phys. Chem. **B1,** 21 (1928).

6. M. Polanyi and G. Schay, Z. Phys. Chem. **B1,** 30 (1928).

7. D. R. Herschbach, Appl. Opt. Suppl. 2, 128 (1965), and references therein.

8. R. Grice and P. B. Empedocles, J. Chem. Phys. **48,** 5352 (1967).

9. D. D. Parrish and R. R. Herm, J. Chem. Phys. **51,** 5467 (1969).

10. T. T. Warnock, R. B. Bernstein, and A. E. Grosser, J. Chem. Phys. **46,** 1685 (1967).

11. J. Troe and H. G. Wagner, Ann. Rev. Phys. Chem. **23,** 311 (1972), and references therein.

12. S. J. Riley, "Velocity Analysis of Reactive Scattering," Ph.D. Thesis, Harvard Univ., Cambridge, Massachusetts (1970).

13. N. Blais, J. Chem. Phys. **49,** 9 (1968).

14. P. J. Kuntz, M. H. Mok, and J. C. Polanyi, J. Chem. Phys. **50,** 4623 (1969).

15. G. G. Balint-Kurti and M. Karplus, Chem. Phys. Lett. **11,** 203 (1971).

16. G. G. Balaint-Kurti, Mol. Phys. **25,** 393 (1973).

17. J. L. Gole and R. N. Zare, Columbia Univ., New York, New York, private communication, 1972.

18. J. C. Polanyi, Appl. Opt. Suppl. 2 (1965).

19. J. V. V. Kasper and G. C. Pimentel, Phys. Rev. Lett. **14,** 352 (1965).

20. K. G. Anlauf, P. J. Kuntz, D. H. Maylotte, P. D. Pacey, and J. C. Polanyi, Disc. Faraday Soc. **44,** 183 (1967).

21. J. C. Polanyi and D. C. Tardy, J. Chem. Phys. **51,** 5717 (1969).

22. K. G. Anlauf, P. E. Charters, D. S. Harne, R. G. McDonald, D. H. Maylotte, J. C. Polanyi, W. J. Skrlac, D. C. Tardy, and K. B. Woodall, J. Chem. Phys. **53,** 4091 (1970).

23. W. W. Rice and R. J. Jensen, App. Phys. Lett. **22,** 67 (1973).

24. W. W. Rice and W. H. Beattie, Chem. Phys. Lett. **19,** 82 (1973).

25. D. R. Stull and H. Prophet, *JANAF Thermochemical Tables*, 2nd ed. (National Bureau of Standards, Washington, D.C., 1971).

26. R. L. Diebner and J. G. Kay, J. Chem. Phys. **51,** 3547 (1969).

27. M. Farber, O. M. Uy, and R. D. Srivastave, J. Chem. Phys. **56,** 5312 (1972).

28. R. C. Feber, "Heats of Dissociation of Gaseous Halides," Report LA-3164, (Los Alamos Scientific Laboratory, Los Alamos, New Mexico Nov. 1964).

29. A. G. Briggs and R. J. Kemp, J. Chem. Soc. D 1223 (1972).

30. V. G. Krisnamurty, Indian J. Phys. **27,** 354 (1953).

31. G. D. Rochester and E. Olsson, Z. Phys. **114,** 495 (1939).

32. R. F. Barrow and R. M. Clements, Proc. Roy. Soc. London Ser. A **322,** 243 (1971).

33. B. Rosen, *Tables Internationales de Constantes Sélectionées 17 Données Spectroscopiques Relatives aux Molécules Diatomiques*, (Pergamon, New York, 1970).

34. C. J. Cheetham and R. F. Barrow, Trans. Faraday Soc. **63,** 1835 (1967).

35. G. DeMaria, R. P. Burns, J. Drowart, and M. G. Inghram, J. Chem. Phys. **32,** 1373 (1960).

36. I. N. Godnev and A. S. Sverdlin, Izv. Vysshikh Uchebn. Zavedenii, Khim. Tekhnol. **9,** 40 (1966).

37. K. Ohwada, T. Soga, and M. Iwasaki, J. Inorg. Nucl. Chem. **34,** 363 (1972).

38. J. A. Hall and W. G. Richards, Mol. Phys. **23,** 331 (1972).

39. J. G. Phillips and S. P. Davis, Astrophys. J. **175,** 583 (1972).

40. B. Lindgren, J. Mol. Spectrosc. **43,** 474 (1972).

41. H. B. Palmer and C. J. Hsu, J. Mol. Spectrosc. **43,** 324 (1972).

42. J. G. Phillips and S. P. Davis, Astrophys. J. **167,** 209 (1971).

43. G. Herzberg, *Molecular Spectra and Molecular Structure I. Spectra of Diatomic Molecules,* 2nd ed. (Van Nostrand, New York, 1950).

44. H. H. Michels, "Theoretical Determination of Metal Oxide ƒ-Numbers," Report AFWL-TR-73-11 (Air Force Weapons Laboratory, 1973).

45. V. S. Zuev, S. B. Kormer, L. D. Mikheev, M. V. Sinitsyn, I. I. Sobel'man, and G. I. Startsev, JETP Lett. **16**(4), 157 (1972).

46. F. H. Webb, Jr., H. H. Hilton, P. H. Levine, and A. V. Tollestrup, *Exploding Wires, Vol. 2,* W. G. Chace and H. K. Moore, Eds. (Plenum, New York, 1962), p. 37.

47. H. D. Edelson and T. Korneff, *Exploding Wires, Vol 3,* W. G. Chace and H. K. Moore, Eds. (Plenum, New York, 1964), p. 267.

48. K. E. Moran, *Exploding Wires, Vol. 3,* W. G. Chace and H. K. Moore, Eds. (Plenum, New York, 1964), p. 285.

49. M. C. Coffman, *Exploding Wires, VOL. 4,* W. G. Chace and H. K. Moore, Eds. (Plenum, New York, 1968), p. 71.

50. M. J. Joncich and D. G. Reu, *Exploding Wires, Vol. 3,* W. G. Chace and H. K. Moore, Eds. (Plenum, New York, 1964), p. 353.

51. E. C. Cassidy and S. Abramowitz, *Exploding Wires,* Vol. 4, W. G. Chace and H. K. Moore, Eds. (Plenum, New York, 1968), p. 109.

52. W. G. Chace, *Exploding Wires,* W. G. Chace and H. M. Moore, Eds. (Plenum, New York, 1959), pp. 7–16.

53. B. Ya'akobi, *Exploding Wires, Vol. 4,* W. G. Chace and H. K. Moore, Eds. (Plenum, New York, 1968), p. 87.

54. H. Arnold and W. M. Conn, *Exploding Wires, Vol 2,* W. G. Chace and H. K. Moore, Eds. (Plenum, New York, 1962), p. 77.

55. M. S. Dzhidzhoev, V. T. Platonenko, and R. V. Khokhlov, Sov. Phys. Sp. **13,** 247 (1970).

56. G. C. Pimentel, Dan. Kemi. **50,** 1 (1969).

57. C. D. Jonah, R. N. Zare, and Ch. Ottinger, J. Chem. Phys. **56,** 263 (1972).

58. Photographs used with permission of W. Field, Los Alamos Scientific Laboratory.

59. E. M. Wescott, H. M. Peek, H. C. S. Nielsen, W. B. Murcray, R. J. Jensen, and T. N. Davis, J. Geophys. Res. **77,** 2982 (1972).

60. G. Haerendel, H. Foppl, R. Lüst, K. W. Michel, H. Neuss, R. Schöning, and J. Stöcker, "Symposium of Barium Releases as an Aeronomical Tool," Fifty-Second Annual Meeting of the American Geophysical Union, Washington, D.C., April 12–16, 1971; K. W. Michel, "Fluorescent Ion Jets for Studying the Ionosphere and Magnetosphere" (1974), to be published.

61.  J. M. Walsh, R. C. Shreffler, and F. J. Willig, J. Appl. Phys. **24,** 349 (1953).

62.  G. M. Grover, T. P. Cotter, and G. F. Erickson, J. Appl. Phys. **35,** 1990 (1964).

63.  D. B. Fradkin, A. W. Blackstock, D. J. Roehling, T. F. Stratton, M. Williams, and K. W. Liewar, AIAA J. **8**(5), 886 (1970).

64.  D. J. Roehling, T. F. Stratton, and D. B. Fradkin, J. Spacecraft and Rockets **9** (10), 725 (1972).

65.  W. H. Beattie, D. B. Fradkin, W. W. Rice, and R. J. Jensen (Los Alamos Scientific Laboratory, Los Alamos, New Mexico), unpublished data.

# INDEX